FOSSIL FUNGI

FOSSIL FUNGI

THOMAS N. TAYLOR

Department of Ecology and Evolutionary Biology and
Natural History Museum and Biodiversity Institute,
The University of Kansas,
Lawrence, Kansas

MICHAEL KRINGS

Department of Earth and Environmental Sciences, Palaeontology & Geobiology,
Ludwig-Maximilians-University, and
Bavarian State Collection for Palaeontology and Geology,
Munich, Germany

EDITH L. TAYLOR

Department of Ecology and Evolutionary Biology and
Natural History Museum and Biodiversity Institute,
The University of Kansas,
Lawrence, Kansas

AMSTERDAM • BOSTON • HEIDELBERG • LONDON
NEW YORK • OXFORD • PARIS • SAN DIEGO
SAN FRANCISCO • SINGAPORE • SYDNEY • TOKYO
Academic Press is an imprint of Elsevier

Academic Press is an imprint of Elsevier
32 Jamestown Road, London NW1 7BY, UK
525 B Street, Suite 1800, San Diego, CA 92101-4495, USA
225 Wyman Street, Waltham, MA 02451, USA
The Boulevard, Langford Lane, Kidlington, Oxford OX5 1GB, UK

ISBN: 978-0-12-387731-4

British Library Cataloguing-in-Publication Data
A catalogue record for this book is available from the British Library

Library of Congress Cataloging-in-Publication Data
A catalog record for this book is available from the Library of Congress

For Information on all Academic Press books
visit our website at http://store.elsevier.com/

CONTENTS

Preface *ix*
Acknowledgments *xiii*
About the Authors *xv*

CHAPTER 1
Introduction 1

Scope of this Volume 3
What Is Paleomycology? 4
History of Paleomycology 5
Naming Fossil Fungi 10

CHAPTER 2
How Fungal Fossils Are Formed and Studied 13

Preservation 13
Impressions 13
Compressions 14
Coal 15
Charcoal 17
Casts 18
Petrifactions and Permineralizations 18
Petrifactions 18
Permineralizations 21
Unaltered Material 23
Amber 24
Mummified Wood 24
Geochemical Evidence 25
Rock Surfaces 25
Nodules 25
Carbon Spherules 25
Unusual Preservation Types Containing Fungal Remains 26
Coprolites 26
Fungi and Crystals 26

CHAPTER 3
How Old Are the Fungi? 27

Phylogenetic Systematics 27
Molecular Clocks 28
Early Fossil Evidence 31
Fungi and the First Land Plants 33
Symbiosis: A Critical Component of Life 36

CHAPTER 4
Chytridiomycota 41

Fossil Chytrids 44
Precambrian Evidence 45
Rhynie Chert Chytrids 46
Carboniferous and Permian Chytrids 49
Problems in Naming Fossil Chytrids 51
Chytrids on Carboniferous Vegetative Tissues 52
Chytrids in and on Carboniferous Plant Spores 55
Chytrids in and on Paleozoic Pollen Grains 57
Chytrids and Seeds 58
Parasitic Chytrids and Host Responses 58
Epibiotic and Endobiotic Mycoparasites 58
Host Responses 59
Nothia aphylla – Multiple Host Responses 61
Hypertrophy 64
Mesozoic and Cenozoic Chytrids 65
Chytrid Summary 66

CHAPTER 5
Blastocladiomycota 69

Fossil Blastocladiomycota 69
Palaeoblastocladia milleri 70

CHAPTER 6
Zygomycetes 75

Precambrian Microfossils 76
Rhynie Chert 79
Zygosporangium-Gametangia Complexes and Sporocarps 80
Carboniferous 81
Triassic 85
"Sporocarps" and Other Fossils of Uncertain Affinities 88

"Sporocarps" 88
Other Enigmatic Fossils 97
Amber Fossils 99
Ichnotaxa 101
Conclusions: Zygomycetes 101

CHAPTER 7
Glomeromycota *103*

Biology of Glomeromycotan Fungi 104
Arum and *Paris* Morphological Types 105
Arbuscules 106
Ecology 107
Systematics and Biodiversity 107
Glomeromycotan Characters 108
Glomeromycotan Reproduction 110
Fossil Glomeromycota 111
Rhynie Chert Asexual Spores 112
Glomoid Spore Types 112
Gigasporoid Spore Types 114
Acaulosporoid Spore Types 115
Carboniferous and Permian Spores 116
Mesozoic, Cenozoic, and Subfossil Spores 117
Sporocarps 118
Root Nodules 118
Fossil Evidence of Arbuscular Mycorrhizae 119
Evolution of Arbuscular Mycorrhizae 126

CHAPTER 8
Ascomycota *129*

Geologic History of Ascomycota 131
Paleozoic Ascomycetes 133
Paleopyrenomycites devonicus 133
Palaeosclerotium pusillum 135
"Sporocarps" 136
Mesozoic and Cenozoic Ascomycetes 137
Taxa *Incertae Sedis* 137
Pezizomycotina 138
Eurotiomycetes 138
Sordariomycetes 141
Dothideomycetes 146
Leotiomycetes 162
Laboulbeniomycetes 163
Fungal Endophytes and Epiphylls 163
Endophytes 163
Fossil Fungal Endophytes 165
Epiphylls – Fungi on Leaf Surfaces 170
Conclusions: Endophytes and Epiphylls 171

CHAPTER 9
Basidiomycota *173*

Fossil Basidiomycetes 175
Agaricomycotina 177
Agaricales 178
Gasteroid Fungi 178
Gilled Mushrooms 179
Boletales 181
Hymenochaetales 182
Polyporales 183
Wood Rot 187
Fossil Wood Rot 188
Pucciniomycotina 193
Pucciniales 193
Fossil Rusts 194
Ustilaginomycotina 194
Fossil Smuts 194
Ectomycorrhizae 195
Fossil Ectomycorrhizae 197

CHAPTER 10
Lichens *201*

Thallus Morphology and Structure 202
Lichen Reproduction 204
Lichen Evolution 205
Precambrian Evidence of Lichens 205
Paleozoic Lichens 206
Mesozoic Lichens 210
Cenozoic Lichens 211
Fossils That Might Be Lichens 214

CHAPTER 11
Fungal Spores *221*

Naming Fungal Spores 222
Fungal Spores in Stratigraphy 223
Fungal Spores in Paleoecology 223
Fungal Spore Taxa 224
Amerospores 224
Basidiosporites 225
Diporisporites 225
Exesisporites 225
Hypoxylonites 225
Inapertisporites 226
Lacrimasporonites 226
Monoporisporites 227
Palaeoamphisphaerella 227
Didymospores 227

Dicellaesporites 227
Didymoporisporonites 228
Diploneurospora 228
Dyadosporites 228
Fusiformisporites 229
Phragmospores 229
Fractisporonites 229
Reduviasporonites 229
Brachysporisporites 229
Chaetosphaerites 230
Diporicellaesporites 230
Foveoletisporonites 231
Multicellaesporites 231
Ornasporonites 231
Pluricellaesporites 231
Polycellaesporonites 232
Dictyospores 232
Dictyosporites 232
Spinosporonites 232
Staphlosporonites 232
Tricellaesporonites 233
Scolecospores 233
Scolecosporites 233
Helicospores 233
Elsikisporonites 234
Involutisporonites 234
Staurospores 235
Tribolites 235
Frasnacritetrus 235
Other Fungal Spores and Structures 235
Hyphomycetes 236
Agonomycetes 237
Coelomycetes 237

CHAPTER 12
Fungal Interactions *239*

Fungus–Animal Interactions 239
Glomeromycota and Animals 240
Coprolites 241
Arthropod Coprolites 241
Large Animal Coprolites 242
Fungi and Arthropods 243
Fungus Gardens 245
Carnivorous Fungi 247
Fungi and Nematodes 247
Trichomycetes 248
Fungi in Eggs 250
Trace Fossils (Ichnofossils) 251
Fungus-Fungus Interactions 251
Fungus–Plant Interactions 251
Fungus–Geosphere Interactions 254
Bioerosion and Rock Weathering 254
Rhizosphere 255
Evidence of Biogenic Activity 256
Substrate Boring 256
Taxa 258
Paleoecology 259

CHAPTER 13
Bacteria and Fungus-Like Organisms *261*

Bacteria 261
Fossil Bacteria 262
Fossil Bacteria as Endosymbionts 264
Fossil Bacteria in the Rhynie Chert 264
Actinomycetes 265
Fossil Actinomycetes 265
Mycetozoa 267
Fossil Mycetozoa 267
Peronosporomycetes 268
Fossil Peronosporomycetes 269
Oogonium–Antheridium Complexes 273
Carboniferous–Permian 276
Peronosporomycetes: Conclusions 282

Glossary *283*
References *297*
Index *373*

PREFACE

After completing this volume, we now understand why there has been no prior attempt to fully document the fossil record of fungi. This statement, however, is not meant to lessen the impact of various papers, journal articles, and catalogs that have examined fossil fungi through time, which are discussed in this volume. What has made our attempt at chronicling fungi through geologic time such a challenge is the need to provide, on the one hand, a balance between fungi as they appear as fossils, including all the inherent limitations of preservation in the fossil record, and on the other hand, fungi as living organisms. There are many reasons that the discipline of paleomycology has not been at the forefront of paleobiology despite the importance of fungi and fungus-like organisms as driving forces in the evolution of other organisms and of ecosystems. One of these reasons certainly has been the idea that because fungi are relatively fragile organisms, the possibility that they are adequately preserved in the fossil record has been discounted. While this is certainly true with regard to certain modes of preservation, as we will present in the following pages, delicate fungal structures and entire organisms can, under certain circumstances, become preserved in extraordinary detail in the rock record.

Another significant factor in elucidating the fossil record of fungi concerns the interface between paleontologists, who study fossil organisms, and biologists, who study living fungi as well as other types of microorganisms. It comes as no surprise that most fossil fungi have been discovered by paleontologists by chance, while searching for and examining other organisms. Thus, the presence of fungal fossils may be less apparent when samples are collected and as a result may simply not be documented. The other component that we believe has adversely affected the growth of paleomycology concerns the background and wealth of knowledge among mycologists who study extant fungi and fungus-like organisms, knowledge that in most instances is lacking among paleobiologists. Thus, there is a major gulf between those who have the fungal fossils but lack knowledge about extant fungi, and those who have a wealth of knowledge about extant fungi, but lack the fossils. Added to this dilemma is the fact that those who might be in a position to collect certain types of fungal fossils probably discard the specimens that demonstrate what many fungi do best – degrade and decompose organic materials. How many times have you seen a slab of rock in a museum with scattered bits of organic matter that is labeled, "example of a Carboniferous fern that has been degraded by a saprotrophic fungus"? It is common practice in paleontology to collect the best-preserved specimens that provide the greatest amount of information about the particular fossil while at the same time discarding incomplete and poorly preserved specimens. The latter may be the result of fungal activity. Even in instances where the cells and tissue systems are preserved in great detail, such as in permineralized and petrified wood, symptoms of fungal activity, including poorly preserved cell walls, compacted wood, or even interruptions in the wood, may be discarded because complete and near perfect specimens are the basis of taxonomic or ecological descriptions. Occasionally an author may note that some of the cells contain fungal hyphae or symptoms of interactions with fungi, but this is not a common practice. Searching for "perfect" or the most diagnostic specimens is the operational mode in most paleobiological studies. For example, well-preserved leaf cuticles can provide information of various types based on the numbers, type, and patterns of epidermal cells that are faithfully reproduced in the cuticles. Many of the plants when they were alive, however, hosted various communities of microbes including fungi, some of which may still be preserved on the leaves, but they are often discarded for "cleaner" specimens.

Techniques used to study fossils, especially plants, have also hindered progress in paleomycology. While petrographic thin sections were once the primary source of information when examining permineralized plant material, this technique was largely replaced by acetate peels, which could be prepared far more quickly, required less instruction in the technique, and were of a more consistent thickness. The problem is that many fungi preserved in the rock matrix are dissolved away when making peels.

Although numerous fungi have been described in acetate peels prepared from permineralized peats, it is now understood that the old-fashioned thin section is superior when searching for fossil fungi in these materials. It takes time and patience, however, to produce thin sections of adequate quality. Some new techniques and imaging systems have been used to study a variety of paleontological specimens, including fungal fossils; where appropriate, we have noted these approaches throughout the text.

Despite our training primarily in paleobotany, we have tried to include enough factual information about extant fungi to make the discussion about the fossils more compelling, and to offer sufficient detail to our paleontological colleagues to expand their knowledge base about fungi, appreciate the potential in their research, and know what to look for. If an introductory course counts, the three of us have had some formal training in mycology, but we are certainly not professional mycologists. In trying to provide a volume that attempts to include both mycology and paleontology, and the various subdisciplines that intersect at multiple levels, we have provided what may sometimes be interpreted as overly basic information in either of these broad disciplines. In our writing we have tried to frame the discussion by asking whether a paleobiologist would understand this about living fungi, and conversely, whether a mycologist would understand that about paleobiology. In some instances readers might wonder, depending on their professional expertise, why we have provided so much information, especially about living fungi. The answer is not simply that we are fascinated by them but that we have tried to emphasize various features or hypotheses that might be identified in or applied to fossil fungi that have not yet been recorded. A simple example might be the consistent occurrence of a particular fungus in fossilized wood of a certain species that has an extensive record in both time and space. Is the co-occurrence a result of some type of coevolution, and do the host and the fungus show any independent evolutionary trends? We have also suggested questions that might be addressed if a larger data set of fossil fungal materials were available to be investigated, or incorporated into a broader picture of organism evolution, e.g., symbiosis. To that end we have provided an extensive and straightforward glossary of terms, with the intention that it will also assist those from other disciplines. We have included numerous illustrations of fossil fungi, many from our own papers and our previous volume – *Paleobotany: The Biology and Evolution of Fossil Plants,* which not only illustrate important characters preserved as fossils but also serve to demonstrate the kinds of features that can occur in the fossil record.

Another reason for our attempt to assemble information about fossil fungi within a single volume is that reports of fossil fungi in the literature have historically been published just about everywhere. In some instances articles on fossil fungi have occurred in journals that primarily cover living fungi, focus on applied mycology (e.g. brewing), or even in horticultural publications. When the focus is more aligned with the physical results of fungal activity in the rock record, research is presented in outlets that contain articles in geology, geobiology, and other closely related disciplines. Other reports about fossil fungi occur in journals that specialize in various subdisciplines in paleontology and paleobotany. Thus, tracking down the scattered information about fossil fungi is an issue for those interested in paleomycology. Another problem is the fact that, especially in older literature, references to fungi co-occurring with plant or animal fossils were often made in a cursory manner. Since key wording and extended indexing were not common practice at that time, such information is largely inaccessible to those who are not familiar with this style of report. Preparing this volume and examining the primary literature where possible has given us the opportunity to review some systematic interpretations and, where necessary, offer alternative or more up-to-date hypotheses.

Perhaps the most significant challenge we faced concerns the placement of various subject areas that we wanted to include in the volume. For example, does one talk about mycoparasitism within the context of when it first appears in the fossil record (i.e. in the discussion of Rhynie chert chytrids) or under a broader discussion of parasitism? Alternatively, is it better for the reader if the description of fossil basidiomycetes occurs within the phylogeny of the group today, or should they be discussed when the first fossils appear in geologic time? Even among ourselves we have had three different opinions – and they have constantly shifted regarding subject placement. This has resulted in a text that has attempted to present both perspectives while at the same time trying not to duplicate information.

We have also discussed a variety of fungus-like organisms, some of which were once included in the kingdom Fungi. However, we have limited this segment of the book (Chapter 13) to groups of organisms that are morphologically very similar to true fungi, and thus easily can be, and probably have been, confused with the latter. Examples of these might be the Peronosporomycetes and

Actinomycetes. The inclusion of fungus-like organisms in this book is also used to underscore a major problem faced by anyone attempting to identify fungal fossils – morphological similarity does not necessarily imply relatedness.

One of the major shortcomings of paleomycology is its inherent limitation to morphology as a tool in identification and systematic assignment. In addition to the above-mentioned factors that have resulted in the neglect of fungi in the fossil record, the inability to make confident systematic assignments has certainly rendered paleomycology unattractive to many paleobiologists and mycologists. We also appreciate that the phylogeny and classification of modern fungi, based primarily on molecular data, is a constantly changing target that can never be directly correlated with fossils; however, we see some movement in the appreciation of fossil forms relative to certain structural and morphological characters and their first appearance in time and space. In some cases our taxonomy differs from a phylogenetic classification, but we see the contents of this volume only as a starting point on a continuum of incorporating the geological history of fungal lineages with the diversity expressed by living fungi irrespective of the tools and techniques used.

ACKNOWLEDGMENTS

We owe an enormous debt of gratitude to many people who have influenced this work and contributed to the final product. This includes current and past colleagues who have freely contributed illustrations and unpublished data, and numerous professional organizations, publishers, and professional societies who have helped us in so many ways.

R. Agerer • L. Axe • F. Baron • D. Barr • K. Bauer • C. Beimforde • M.L. Berbee • J.A. Bergene • J.L. Black • J.E. Blair • R.A. Blanchette • J. Błaszkowski • B. Bomfleur • P. Bonfante • J.P. Braselton • G. Breton • A.W. Bronson • M.C. Brundrett • N. Butterfield • M.J. Cafaro • S.R.S. Cevallos-Ferriz • V. Chazottes • K. Clay • C.P. Daghlian • Y. Dalpé • J. Danek • R. Day • T. Demchuk • M.W. Dick • D.L. Dilcher • H. Dörfelt • N. Dotzler • D. Edwards • A. Galitz • J. Danek • J. Galtier • J.L. Garcia-Massini • J.F. Genise • T. German • P. Gerrienne • S. Golubic • A.K. Graham • C.J. Harper • S.T. Hasiotis • H. Hass • C. Haufler • D.L. Hawksworth • D.S. Hibbett • T.J. Hieger • H.-M. Ho • M. Holder • R. Honegger • V.B. Hosagoudar • J.M. Houts • J.J.C. Hower • A. Jumpponen • P. Kabanov • R.M. Kalgutkar • B. Kendrick • P. Kenrick • H. Kerp • S. Klavins • A.A. Klymuik • L. Krishtalka • C.C. Labandeira • J.D. Lawrey • B.A. LePage • R.W. Lichtwardt • J. Longcore • R. Lücking • P.C. Lyons • S.R. Manchester • K.K.S. Matsunaga • D.J. McLaughlin • S. McLoughlin • M.A. Millay • C.E. Miller • N.P. Money • S. Naugolnykh • A. Nel • G. Norris • J.M. O'Keefe • G.J. Ortiz • J.M. Osborn • M.G. Parsons • R.L. Peterson • C. J. Phipps • V. Podkovyrov • G. Poinar, Jr. • M.J. Powell • D. Redecker • M. Remshardt • R. Remy • W. Remy • G.J. Retallack • J. Rikkinen • G.W. Rothwell • G. Samoring • R. Saxena • A.R. Schmidt • J.W. Schopf • A. Schüßler • A.B. Schwendemann • M.-A. Selosse • R. Serbet • B.J. Slater • S.Y. Smith • D. Southworth • R.A. Stockey • P.K. Strother • C. Strullu-Derrien • S.P. Stubblefield • A.R. Sweet • S. Taliferro • J.V. Tassell • J.W. Taylor • A.M.F. Tomescu • J.M. Trappe • N.B. Trewin • N.H. Trewin • S.K.M. Tripathi • R.W.J.M. Van der Ham • K. Vogel • B.M. Waggoner • C.A. Wagner • C. Walker • H. Wang • T. Wappler • J.F. White, Jr. • If we have missed anyone, we sincerely apologize.

We are especially appreciative to Carla J. Harper and Rudolph Serbet for their extraordinary and invaluable help in the many tasks associated with assembling and photographing the specimens and illustrations used throughout the volume. Whether sizing illustrations, adding bar scales, searching for references or any of the myriad other tasks associated with completing this book, Rudy and Carla did so at the highest level of accomplishment and professionalism. Thank you both so very much. Also, a special thanks to Tim Hieger for his many contributions to this project. We are indebted to Sara Taliaferro who redrew the many illustrations of the fungal spores and fossil microthyriaceous ascocarps that appear in Chapters 8 and 11. Her dedication to detail and ability to view fossil specimens in a biological context is extraordinary. Of special acknowledgment is Jeannie Houts who has been associated with this project from the outset. Her dedication to the "book" and skillful assistance in so many ways has made the completion of this volume possible. Especially noteworthy has been her painstaking attention to the details associated with assembling the bibliography.

Professor Frank Baron and his colleagues at the Max Kade Center for German-American Studies and Department of Germanic Languages and Literature at the University of Kansas gave us assistance that is greatly appreciated. We acknowledge the wonderful hospitality provided by the Center when MK visited KU during phases of the writing of this book. Most importantly we value the friendship that has developed with Frank Baron.

We are grateful to the National Science Foundation and Alexander von Humboldt-Stiftung for providing financial assistance to our research and scholarship programs over the years that have contributed to studies of fossil fungi. Not only have these programs allowed us the opportunity to collect and study fossil fungi, but they have also provided the financial assistance to allow us to work together, to present research results at professional meetings, and to expand the network of biologists and geologists interested in fossil fungi and their activities.

Taking a book from the manuscript to the finished volume takes extraordinary talent that far exceeds that of the authors. We are especially appreciative for the help and guidance we have received from Ceinwen Sinclair, Freelance Copyeditor, and Roderick Crews, Freelance Proofreader, for their attention to detail during the editing stages. Melissa Read, Freelance Project Manager, kept all of the activities on time and skillfully organized. Working with Kristi Gomez, Acquisitions Editor, and Pat Gonzalez, Editorial Project Manager, both of Elsevier Inc., has been a genuine pleasure. These extraordinary individuals have given this project enormous assistance, and have done so with a level of professionalism and expertise that sets a very high standard. Kristi and Pat, thank you so very much – we simply could not have completed this project without your help.

We particularly wish to acknowledge the support from our families, friends, and colleagues. You have been patient, supportive, and loyal as we have trudged on with this project, sometimes never taking the time to say thank you. The institutions where we are employed – the University of Kansas, Lawrence, KS (USA), and both the Ludwig-Maximilians-University and the Bavarian State Collection for Palaeontology and Geology in Munich, Germany – have assisted in numerous ways for which we are grateful. Providing the opportunity for us to pursue our research and scholarship with fossil fungi has been a joyous opportunity. We extend a special acknowledgment to Hagen Hass, University of Münster, who has been the driving force in discovering many of the fossil fungi and fungal-like organisms in the Rhynie chert, to Renate and the late Winfried Remy who made some of our early studies possible, and to Hans Kerp for making recently discovered specimens available.

Numerous individuals, through their research interests and early studies, have influenced our pursuit of understanding the biology and evolution of fossil fungi and how they functioned in ancient ecosystems. Of special note are a few of the pioneers in paleomycology: David Dilcher, William Elsik, Jan Jansonius, Ramakant Kalgutkar, Robert Kidston, William Lang, Aloysius Meschinelli, Julius Pia, Kris Pirozynski, Sara Stubblefield, Bernard Renault, and Bruce Tiffney. During our careers we have been especially fortunate in having extraordinary teachers and colleagues who have always provided encouragement, loyalty, and unwavering support for us in our personal and professional activities. Thank you for always being there.

ABOUT THE AUTHORS

Thomas N. Taylor is a member of the National Academy of Sciences and holds the title of Distinguished Professor in the Department of Ecology and Evolutionary Biology, and Curator of Paleobotany in the Natural History Museum and Biodiversity Institute at the University of Kansas. He also has a courtesy appointment in the Department of Geology. He received his Ph.D. in botany from the University of Illinois and was a National Science Foundation postdoctoral fellow at Yale University. His research interests include Permian and Triassic biotas of Antarctica, early land plant–fungal interactions, the origin and evolution of reproductive systems in early land plants, symbiotic systems through time, and the biology and evolution of fossil microbes.

Michael Krings is Curator for Fossil Plants at the Bavarian State Collection for Palaeontology and Geology (BSPG) in Munich, Germany, and Professor of Plant Palaeobiology at the Ludwig-Maximilians-University, Munich. He also holds an affiliate faculty position in the Department of Ecology and Evolutionary Biology at the University of Kansas. He received his Ph.D. in botany from the University of Münster, Germany, and was an Alexander von Humboldt Foundation postdoctoral fellow at the University of Kansas. His research interests include Carboniferous, Permian, and Triassic seed plants from Europe and North America, and the biology and ecology of microorganisms in late Paleozoic terrestrial ecosystems.

Edith L. Taylor has held the title of Professor of Ecology and Evolutionary Biology and Senior Curator of Paleobotany in the Natural History Museum and Biodiversity Institute at the University of Kansas since 1995, and also serves as a Courtesy Professor of Geology. She received her Ph.D. in paleobotany from the Ohio State University, where she was an American Association of University Women Dissertation Fellow. She is the author of seven books or edited volumes, more than 180 publications, and is a fellow of the American Association for the Advancement of Science. Her research interests include fossil wood growth and paleoclimate, Permian and Triassic permineralized plants from Antarctica, distribution and diversity of Permian–Triassic Antarctic floras, the structure and evolution of fossil phloem, and fossil microorganisms.

1

INTRODUCTION

SCOPE OF THIS VOLUME 3
WHAT IS PALEOMYCOLOGY? 4
HISTORY OF PALEOMYCOLOGY 5
NAMING FOSSIL FUNGI 10

Fungi are extraordinary. Although fungi represent the largest life forms on Earth (Angier, 1992), and are instrumenta to the health and success of almost every ecosystem, they are perhaps the most underappreciated group of organisms. There are approximately 100,000 species of fungi and fungal-like organisms currently recognized, and as a result of molecular techniques it is estimated that there may be more than 5 million species of fungi inhabiting the Earth (e.g. Blackwell, 2011). Since there are estimated to be more than 400,000 species of flowering plants (angiosperms), the group we depend upon for food (e.g. Joppa et al., 2011), this means that there are more than 10 times more fungi than angiosperms. To realize and accurately document the biodiversity of fungi in the future it will be as important to develop and integrate data-based platforms (e.g. Halme et al., 2012) as it will be to document fungi in tropical forests and unexplored habitats (Hawksworth, 2004). Fungi impact just about every aspect of life and have played a major role in the evolution of life on land (e.g. Blackwell, 2000). In fact, the colonization of land by plants is hypothesized to have occurred as the result of symbiotic interaction(s) between a fungus and a photosynthesizing organism. Today this association continues in the form of mycorrhizae, which occur in and around the roots of vascular plants and increase nutrient uptake in an estimated 90% or more of all plant species.

Fungi range from single celled to complex multicellular organisms and can be found in all aerobic ecosystems where they colonize numerous substrates and perform multiple functions, some of which are poorly understood (e.g. Cantrell et al., 2011). For example, they occur in habitats ranging from Antarctica (e.g. Onofri et al., 2007; Kochkina et al., 2012) to salt flats to laminated organo-sedimentary ecosystems termed microbial mats. They have even been reported as epiphytes on unicellular or multicellular plant hairs or trichomes (Pereira-Carvalho et al., 2009), in deep-sea ecosystems (e.g. Schumann et al., 2004; Le Calvez et al., 2009; Nagano and Nagahama, 2012), and in 100 m of deep sea mud (Biddle et al., 2005). Fungi are the primary degraders of complex organic compounds such as lignin and cellulose and, through this recycling, are important in returning minerals to the soil. As a result of decomposition, carbon dioxide is returned to the atmosphere. They are especially important decomposers in acidic soils where bacteria often cannot exist. Fungi also bind soil particles to their mycelium, thus contributing to soil structure. In all of these functions they are crucial in the maintenance of the biosphere. Fungi, together with soil fauna, roots, and other microorganisms, are major drivers of ecosystems and in this capacity are involved in the critical processes of landscape sustainability, including nutrient cycling, in which they accumulate, store, and distribute minerals and carbon (e.g. Johnston et al., 2004). Most are filamentous, which allows them to readily explore and exploit new habitats including their own parental hyphae (e.g. Gadd, 2007).

Because fungi are carbon heterotrophs they have, by necessity, mastered various levels of cooperation with

T.N. Taylor, M. Krings & E.L. Taylor: Fossil Fungi.

DOI: http://dx.doi.org/10.1016/B978-0-12-387731-4.00001-3

other organisms: they partner with certain types of algae and cyanobacteria to form lichens, and their intimate relationships with land plants include mycorrhizal associations. They also occur as endophytes in some vascular plants and in such cases can be involved in contributing to pathogen resistance. What may be interpreted as the "dark side" of the fungal kingdom is that as plant pathogens they impact broad areas of agriculture by infecting all major crop plants and producing various mycotoxins involved in food contamination and in some cases human disease. Fungi are also known to negatively affect animals and humans as causative agents of diseases (e.g. Sternberg, 1994; Fisher et al., 2012). One of these diseases, chytridiomycosis, has made it into the news because it is believed to have caused the decline and even extinction of some amphibian populations on several continents (e.g. Longcore et al., 1999; Weldon et al., 2004).

In buildings, fungi and other microbes contribute to altering and erasing cultural structures by deteriorating historical stone in the form of statues, tombstones, historic buildings, and archaeological sites (McNamara and Mitchell, 2005; Mitchell and McNamara, 2010; Steiger and Charola, 2011). In addition, microbial activities and fungi are involved in the deterioration of paintings, wood, paper, textiles, metals, waxes, polymers, and various types of coatings (Koestler et al., 2003; Martin-Sanchez et al., 2012). Fungi also show no respect for some of the "pillars" of our modern society such as utility poles (Wang and Zabel, 1990).

Fungi also enter into mutualistic associations with animals; some even thrive within the animal, in anaerobic environments (Mountford and Orpin, 1994). Especially interesting are fungi that play a pivotal role in the rumen by physically and enzymatically attacking the fibrous plant material ingested by the ruminant animal (e.g. Kittelmann et al., 2012). By breaking down plant cell wall carbohydrates, such as cellulose and hemicellulose, these anaerobic fungi deliver readily accessible nutrients, mainly acetate, propionate, and butyrate, to their ruminant hosts (e.g. Gordon and Phillips, 1998).

Fungi affect our lives in innumerable ways; some are even believed to have shaped civilizations (e.g. Money, 2007; Dugan, 2008). For example, mushrooms are quite frequently depicted in ancient, medieval, and modern fine art. A group of ethnomycological rock paintings in the Sahara Desert (Tassili n' Ajjer National Park, Algeria) – the work of pre-Neolithic early hunter-gatherers some 7,000–9,000 years old – shows the offering of mushrooms, and large masked "gods" covered with mushrooms, suggestive of an ancient hallucinogenic mushroom cult (Figures 1.1, 1.2). These paintings may reflect the most ancient human culture yet documented in which the ritual use of hallucinogenic mushrooms is explicitly represented (Samorini, 1992). The Temptation panel in Mathias Grünewald's famous Isenheim Altarpiece, painted between 1510 and 1515, is believed to represent a depiction of ergot poisoning from rye flour contaminated by the fungus *Claviceps purpurea* (e.g. Battin, 2010). Ergot poisoning has also been suggested to be the reason for the strange behavior of women accused of witchcraft in Salem, Massachusetts Colony, in the late 1690s (Caporael, 1976), although some have refuted this hypothesis (Spanos and Gottlieb, 1976). Ergotism is known to cause jerky movements, which are characteristic of Sydenham's chorea or St. Anthony's fire, as it was called in the Middle Ages (Lapinskas, 2007). Today fungi are also used in

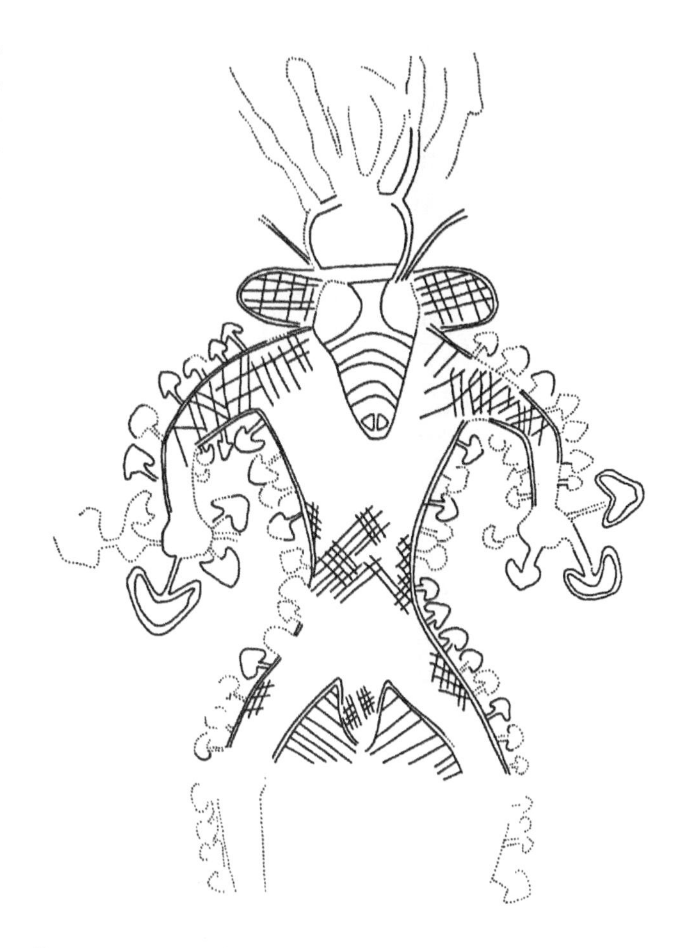

FIGURE 1.1 Outline of ethnomycological rock painting from the Sahara Desert Matalem-Amazar site dated at 9,000–7,000 BP showing man (a shaman?) with masked head and numerous mushrooms scattered over the body (figure ~0.8 m tall). (Courtesy G. Samorini.)

FIGURE 1.2 Original prehistoric painting from the site in Figure 1.1. (Courtesy G. Samorini.)

criminal investigations, i.e. forensics (Hawksworth and Wiltshire, 2011).

The extraordinary modern dance company *Pilobolus* is named after the phototropic, mucoralean dung fungus of the same name that has the capacity to propel a cluster of spores a great distance. Other fungi are simply inspirational and actually reach into the stratosphere (Wainwright et al., 2006)! Some fungi simply make you feel good by their production of non-hallucinogenic indole compounds and derivatives including serotonin (Muszyńska et al., 2013); others, such as some common mushrooms, sometimes called magic mushrooms, contain hallucinogenic compounds that can alter thought and mood. A number of people get great pleasure simply dressing up in homemade mushroom costumes at the annual Telluride Mushroom Festival in Telluride, Colorado, USA, an event that also includes a feast of mushrooms collected by the revelers and prepared by expert chefs (Lincoff, 1996).

Fungi are known to improve the acoustic quality of violins by degrading the cell walls of the wood of *Picea abies* (Norway spruce) and *Acer pseudoplatanus* (European sycamore maple), which is used to make the top plates of the instruments. As a result of treating the wood with wood-decay fungi that break down cell wall components, the decrease in cell wall thickness makes the violin not only louder but capable of producing more resonant tones (Schwarze et al., 2000). Infection by the ascomycete fungal parasite *Phaeoacremonium parasiticum* (formerly *Phialophora parasitica*; Crous et al., 1996) of the heartwood of the trunk and roots of evergreen trees (*Aquilaria, Gyrinops*) results in the formation of what is termed agarwood. This unique fungal-infected wood is known by multiple names (e.g. *ud, oud* in Arabic countries; black eaglewood in Tibet; *gaharu* in Indonesia; *agar* in Pakistan) depending on the culture. As a result of infection the tree produces an aromatic resin that becomes embedded in the heartwood making it more dense and black in color. The resin has been highly sought after because of its distinctive fragrance in perfume, incense, and aromatherapy products and can be very expensive (Hansen, E., 2000). One can only wonder whether some of the fungal-infected fossil woods held such olfactory properties.

While most fungi are involved in organic material decomposition, others are involved in formation of a variety of materials in an indirect way. For example, dried mycelia have been suggested as a substitute for Styrofoam and for making bricks that can be used in construction. Mycotecture, architecture based on the structure of the mycelium, has also been used in art and the manufacture of furniture. Others have employed living fungi as "templates" in the synthesis of highly ordered structures of nanoparticles (e.g. Bigall and Eychmüller, 2010; Sabah et al., 2012).

SCOPE OF THIS VOLUME

In order to document the number of living fungi, the mycological community has continued to enlist the help of an increasing number of biologists to record fungal diversity (e.g. Blackwell, 2011). Today this diversity is being measured in a number of ways, including molecular approaches that sample genome diversity within a phylogenetic framework. With increasing attention directed to fungal genomics there can be little doubt that these data will provide an unparalleled opportunity to better understand the biology and evolution of the fungal kingdom (e.g. Galagan et al., 2005). In addition, researchers are assembling these data into larger networks and using a coordinated infrastructure to address problems of fungal identification (Bruns et al., 2008). While these techniques are not applicable to examining the diversity of fossil fungi, we do believe that increasing the awareness of the fossil record of fungi will greatly expand the database of reports of these organisms in time and space. What makes the study of fossil fungi especially difficult is that the report of these organisms has generally been a secondary focus of paleontological studies until recently. Further compounding the dilemma of understanding the diversity of fossil fungi is that reports of their existence are not focused within specific disciplines but are found in broadly defined areas of biology and geology and the numerous subdisciplines within these areas.

What is inescapable, however, is the fact that fungi and fungal-like organisms were present in the earliest ecosystems on the Earth. As is true today, they were primary drivers of many crucial ecosystem functions and their record parallels, and sometimes predates, the diversity of other life forms in time. The challenge of recording the history of fungi through time is the primary focus of this book, but we will also be demonstrating some of the numerous interactions with other components of the ecosystems in which they lived and discussing abiotic aspects of the environment that fungi occupied. Nevertheless, there will continue to be new groups of extant eukaryotic organisms described that have some of the features of fungi but lack others and thus do not conform to the typical fungal realm. For example the Cryptomycota are an ecologically widespread and highly diverse group once thought to lack chitin in the cell wall and therefore not thought to be fungi (Jones et al., 2011a); now, however, they are considered to be members of the fungal kingdom (Jones et al., 2011b). Recording all of these evolutionarily successful organisms in the fossil record may be impossible using the techniques available today; however, having an appreciation of their existence in modern ecosystems will be significant as ideas about the fungal tree of life continue to evolve.

WHAT IS PALEOMYCOLOGY?

We define paleomycology as the study of the accumulated fossil evidence of fungi and their activities as measured in geologic time. The field includes study of fossilized fungi themselves as well as the full complement of their activities, which range from mineral weathering to both direct and indirect effects and interactions on and with other organisms. In this book we have also included the evidence of activities of a variety of fungus-like organisms, some of which at one time were grouped with the fungi but today are classified in other groups, as these organisms also have a fossil record. Thus, paleomycology is at the intersection of a number of disciplines, many of which overlap, some to a large degree.

One of these overlapping subject areas is paleomicrobiology, defined as the detection, identification, and characterization of microorganisms in ancient remains (Drancourt and Raoult, 2005). In general, these studies focus on the diagnosis of past infectious diseases in human populations caused by bacteria, viruses, and parasites, and principally involve the use of molecular techniques. Other people, however, would broadly define paleomicrobiology as the study of microbial communities and their activities through time. Paleomicrobiology overlaps with paleopathology, which is defined as the study of ancient diseases based primarily on skeletal remains (Aufderheide and Rodríguez-Martín, 1998). The discipline has also been expanded to include plant paleopathology, which is defined as the consequence of physical and chemical responses by plants to invasive microorganisms (e.g. viruses, bacteria, nematodes, and fungi) or to imbalances in nutritional or environmental conditions (Labandeira and Prevec, 2013).

Another area that includes the activities of fungi in geologic time is geobiology, a discipline initially defined by Teilhard de Chardin (Galleni, 1995) that builds on a broad range of interdisciplinary and disciplinary research focusing on the interactions between the biosphere and geosphere, and ranging from biomarkers to paleontology (e.g. Knoll et al., 2012). Some define geobiology as knowledge of the interface between life and the Earth's environment. Geomicrobiology is concerned with the roles of microorganisms in geological and geochemical processes (e.g. Ehrlich and Newman, 2008), while geomycology is the scientific study of fungi in the processes of fundamental importance to geology, including weathering, soil formation, decomposition, mineral deposition, nutrient cycling, the influence on proliferation of microbial communities in mineral substrates, and several other processes (e.g. Burford et al., 2003a, b; Rosling et al., 2009a, b; Gadd et al., 2012). Some view geomycology as a subdiscipline of geomicrobiology (e.g. Gadd, 2007; Rosling et al., 2009a) and as a smaller component of geobiology, which includes all of the biotic interactions on physical aspects of the

Earth system including climatic, atmospheric, oceanic, and geospheric (e.g. Beerling, 2009; Rosling et al., 2009a).

Not only is the documentation of fossil fungi the only method to gain some appreciation for the diversity of these organisms through geologic time (e.g. Sherwood-Pike, 1991), but paleomycology also is the only way to plot the occurrence of fungal apomorphies in certain groups and thus test the validity of molecular clock assumptions about divergence rates and evolutionary lineages. In addition, fungal occurrences that are associated with particular groups of plants, and certain species, provide a necessary framework to consider a wide range of outcomes ranging from the coevolution of host and parasite to insights about the climate and other environmental parameters in which the fungus lived. Fossil fungi also provide an indirect measure of interactions with other organisms and a time frame for when some of these interactions initially occurred in geologic time.

FIGURE 1.3 Leo Lesquereux.

HISTORY OF PALEOMYCOLOGY

It is impossible to know exactly who recorded the first fossil fungus. The majority of early reports on fossil fungi were from Carboniferous rocks, especially those from Great Britain. One of these is by Steinhauer (1818) who reported on some longitudinal configurations that he suggested might be tissues of a plant or evidence of decay. Sternberg (1820) described *Carpolithes umbonatus*, which has been interpreted as some type of ascomycete, and *Algacites intertextus*, also thought be fungal but now interpreted as of uncertain affinity. *Daedaleites volhynicus* is a name credited to Eichwald (1830) that has been used as an early example of a basidiomycete. While these names, and other species, are included in the Index Fungorum (http://www.indexfungorum.org/), mostly within the category Fossil Fungi, it is difficult to be certain what they represent.

FIGURE 1.4 Heinrich Robert Göppert.

Another early account of a probable fungus was by Lindley and Hutton (1831–1833), who described an oval structure, approximately 2.5 cm in diameter, with what were interpreted as perhaps growth increments of a fungal basidiocarp. They used the name *Polyporites bowmanni* and with some reluctance suggested it was perhaps the hymenium of some polypore. However, *P. bowmanni* and a similar specimen, reported by Lesquereux (1877) (Figure 1.3) from the Carboniferous of North America, were later determined to be fish scales (Carruthers, 1889). Another early report (Göppert, 1836) (Figure 1.4) identified small structures on Carboniferous foliage as leaf-inhabiting fungi, and it is reported that Robert Brown, then Keeper of Botany at the British Museum, showed Charles Darwin a piece of silicified wood containing fungal hyphae sometime during the 1840s (Smith, W.G., 1884). Studies of fossil wood, perhaps of roots, sometimes included a note that there were hyphae in some of the cells (e.g. Conwentz, 1880). Generally, these were not given names or described in any detail. In his benchmark review of paleomycology, Kris Pirozynski (1976a) (Figure 1.5) noted that most

FIGURE 1.5 Kris A. Pirozynski. (Courtesy Ottawa Geological Survey.)

of the early reports and descriptions of fossil fungi were based on observations using a hand lens or the naked eye rather than transmitted light.

Historically, perhaps Berkeley (1848) was one of the first to examine microfungi with a compound microscope. The specimens he described were included in preparations of Eocene Baltic amber. Although he provided descriptions and illustrations, and included them in the genera *Penicillium* and *Streptothrix*, and a new genus, *Brachycladium*, these assignments were only tentative and today are dismissed. Other early reports from the Carboniferous include structures thought to represent fungi in the form of variously sized ramifying tubes and occasional oval spores at the ends of hyphae and within the matrix (Hancock and Atthey, 1869, 1870). Described as several species of *Archagaricon*, it is difficult to determine whether these specimens are in fact fungi, a concept that was later challenged (Smith, W.G., 1877). More convincing examples of fungi, or perhaps Peronosporomycetes, were described and illustrated from a permineralized specimen of the Carboniferous lycopsid *Lepidodendron* (Smith, W.G., 1877, 1878). These fossils consist of septate hyphae within the tracheids together with spherical units approximately 50 μm in diameter

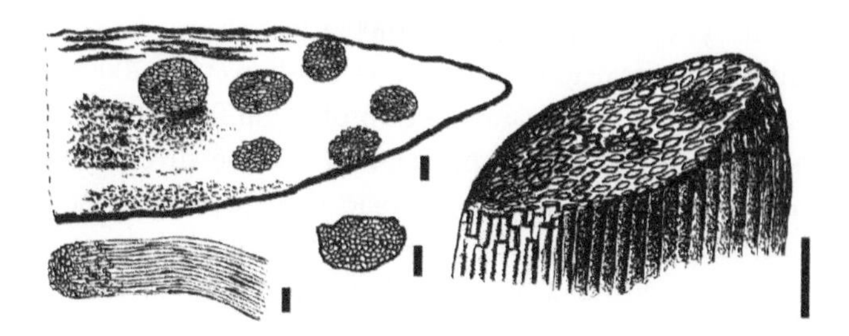

FIGURE 1.6 Representative figures of the putative Carboniferous polypore *Dactyloporus archaeus* showing distribution of the specimens and tube-like organization of the fungus. Bar = 1.0 cm. (From Herzer, 1893a.)

termed oogonia. Within the oogonia are seven or eight structures interpreted as zoospores, but the small number and size suggest that these are more likely oospores, if in fact the spore-like bodies are oogonia (James, 1893a, b). Another interpretation is that the internal contents represent some type of mycoparasite. Septate hyphae in some of the ground tissue cells of the fern *Zygopteris* from the Lower Coal Measures (Lower Pennsylvanian) of Halifax, West Yorkshire, have also been thought to represent some type of peronosporomycete (Cash and Hick, 1879).

There are a number of so-called fungi described in the late 1800s for which neither the description nor the illustration afford any suitable detail to suggest that the structure is a fungus. One of these, *Rhizomorpha 'sigillaria,'* a structure described from the shales of the Darlington Coal Bed in Beaver County, Pennsylvania (USA), and thought to represent a fungus (Lesquereux, 1877), was later interpreted as perhaps some insect burrow beneath the periderm of *Sigillaria* (James, 1885). This may also be the actual biological identity of *Incolaria securiformis* (Herzer, 1893b), the name applied to a putative fungus that comes from fissures in the extraxylary tissues of the Carboniferous lycophyte *Sigillaria.* Another Carboniferous form attached to a trunk and reported from the then No. 5 Coal from Tuscarawas County, Ohio (USA) consists of what is interpreted as a series of parallel tubes that form the pileus (Herzer, 1893a). No tissues or spores were recovered from *Dactyloporus archaeus* (Figure 1.6), and thus the affinities to some polypore were conjectural at best. Often these descriptions and sometimes generic designations are related to other supposedly fungal structures that were described earlier, but they are also dubious in terms of biological affinities.

In addition to fungi being reported in shales and coals, there are some early reports of fungi in amber discovered in certain coal seams. One of these is from the Ayrshire

FIGURE 1.7 Bernard Renault.

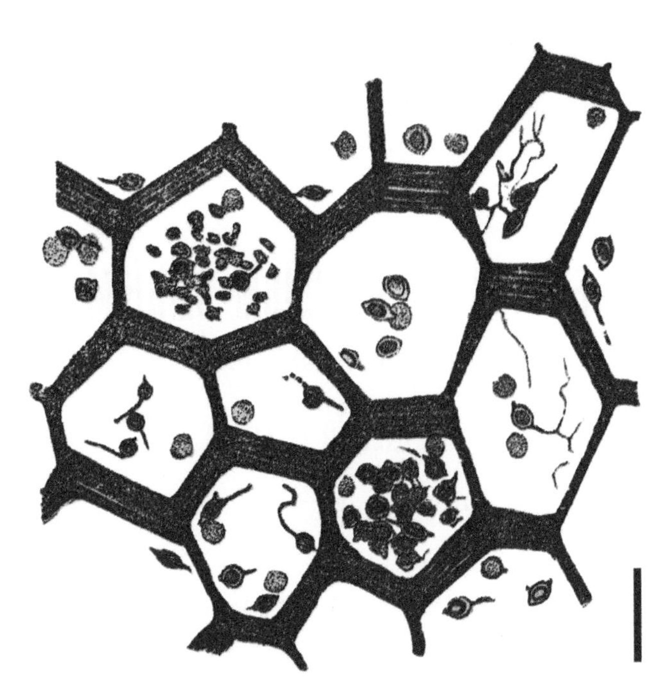

FIGURE 1.8 *Oochytrium lepidodendri.* Carboniferous, France. Bar = 20 μm. (From Renault, 1896.)

Coal Field (Great Britain) and includes rounded amber inclusions of various sizes that contain several types of plant parts and what are referred to as microfungi (Smith, J., 1896). Some of these consist of what may be hyphae with some type of attached reproductive unit; for these types the generic names *Peronosporoides* and *Leptonema* were established (Smith, J., 1896). Some of these drawings are extraordinary and leave little doubt as to the affinities of some forms (e.g. Smith, W.G., 1884). The fact that amber has also been found in the Carboniferous (Bray and Anderson, 2009) indicates that some of the plants at that time produced resins like those found today in some conifers and angiosperms, and that such deposits also represented a potential trap for fungi and other microorganisms.

With the increased use of the compound microscope there were a number of studies that looked at the detail of fungi preserved in various types of cherts, especially those from the Carboniferous of central France. Foremost among the researchers was the French paleobotanist Bernard Renault (Figure 1.7), who meticulously documented the morphology of a number of microorganisms including their distribution in plant tissues from permineralized cherts. Renault was not only a skilled technician who produced high quality thin sections; he was also interested in the correct systematic placement of the organisms he studied. Among his many studies are reports of algae (1903), Peronosporomycetes (1895b), and several types of fungi (1894a, 1895a,b, 1896b,c, 1899, 1900, 1903) (Figure 1.8). He also described a number of structures that he interpreted as bacteria (1894b, 1895c, 1896a–d, 1897, 1898, 1899, 1900, 1901); some of these are probably crystal patterns.

What is arguably the first compendium of fossil fungi was assembled by Meschinelli (1892, 1898) in which he recorded more than 60 genera and 350 species (Figure 1.9). This study emphasized descriptive taxonomy, but many of the identifications and interpretations were based on association of a particular type of symptom or characters that are no longer used (Pirozynski and Weresub, 1979). Meschinelli is also credited with adding "*-ites*" to the ending of modern generic names for fossil specimens he considered the same as living forms. Seward (1898) (Figure 1.10) was a little less charitable in noting that "certain species are of no botanical value, and should have no place in any list which claims to be authentic." In his four volume series, *Fossil Plants*, Seward (1898) provided a detailed analysis of many of the fossil fungi described to that date and challenged a number of the reports and their taxonomic placement. The nomenclatural status of several names allocated by Meschinelli, indicating they were believed to be fossil fungi (such as *Agaricites*), were also challenged (e.g. Holm, 1959). While many of these early reports may be suspect, perhaps what is most important is that rocks of

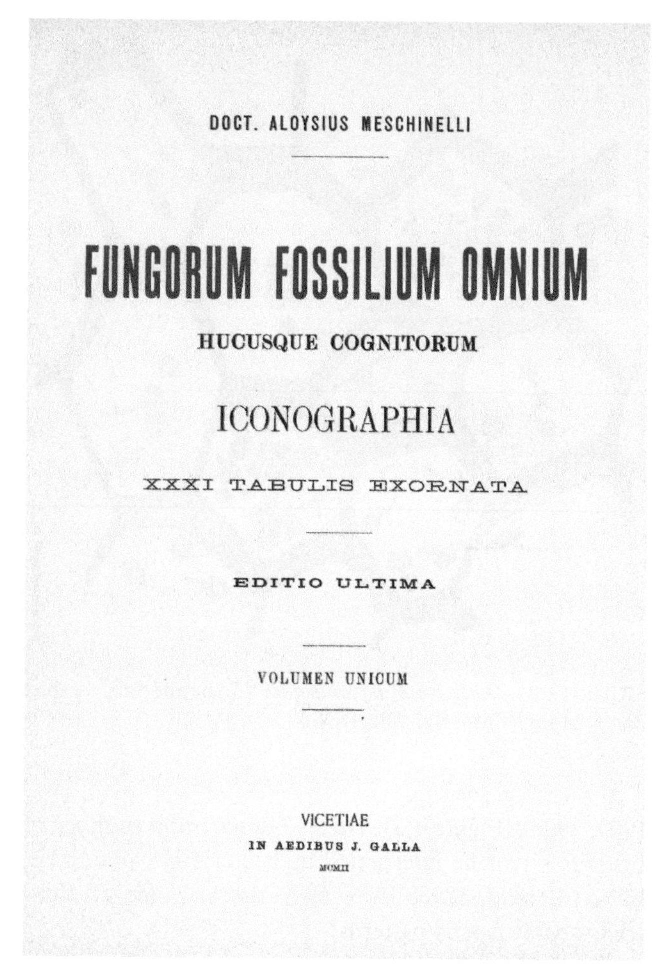
DOCT. ALOYSIUS MESCHINELLI

FUNGORUM FOSSILIUM OMNIUM

HUCUSQUE COGNITORUM

ICONOGRAPHIA

XXXI TABULIS EXORNATA

EDITIO ULTIMA

VOLUMEN UNICUM

VICETIAE

IN AEDIBUS J. GALLA

MCMII

FIGURE 1.9 Title page of Meschinelli's compendium on fossil fungi.

FIGURE 1.10 Albert C. Seward.

FIGURE 1.11 Julius von Pia.

FIGURE 1.12 Bruce H. Tiffney.

varying ages were being examined for evidence of microorganisms like fungi (e.g. Ellis, D., 1915).

Many of the early reports of fossil fungi are presented systematically by Julius Pia in Max Hirmer's textbook, *Handbuch der Paläobotanik* (von Pia, 1927) (Figure 1.11), including reproduced line drawings and, in several instances, photographs from the original papers. Of particular significance is the inclusion of the various forms within a higher taxonomic system at the family level. There are other scattered reports of fossil fungi that are catalogued within a taxonomic framework (e.g. Brabenec, 1909; Němejc, 1959).

The next major reference to fossil fungi was completed by the paleobotanists Tiffney and Barghoorn (1974) (Figures 1.12, 1.13). They provided a listing of more than 500 fossil fungal species that was annotated

FIGURE 1.13 Elso S. Barghoorn.

FIGURE 1.14 Constantine J. Alexopoulos.

in detail and, for the first time, the families were placed within a stratigraphic context. This contribution was later updated with additional fossil species, while others were modified or deleted (Taylor, T.N., 1993a). A slightly different approach was used by Stubblefield and Taylor (1988), who discussed fossil fungi relative both to their stratigraphic occurrence and to their possible roles as mutualists, saprotrophs, and parasites, hypothesizing a link between nutritional mode and environment. Also included was a brief discussion of techniques used to study fossil fungi and the nomenclatural approaches in naming them. Aspects of the nutritional modes of fossil fungi were also discussed in *The Biology and Evolution of Fossil Plants* (Taylor and Taylor, 1993) together with the inclusion of some of the new fossil representatives. Much of this information was further updated, and new fossil fungal reports included, in *Paleobotany: The Biology and Evolution of Fossil Plants* (Taylor et al., 2009). Many of the early accounts of fossil fungi occurred in volumes that focused on paleobotany; only a few books centering on living fungi have included fossils at all (e.g. Wolf and Wolf, 1947; Alexopoulos et al., 1996) (Figure 1.14).

In addition there have been multiple contributions that have considered the history of paleomycology (e.g. Pirozynski, 1976a), nomenclature (Wolf, 1969a; Pirozynski and Weresub, 1979; Ramanujam, 1982), specific groups of fungi (e.g. Pirozynski and Dalpé, 1989), fungi from particular regions (e.g. Teterevnikova-Babayan and Taslakhchian, 1973; Germeraad, 1979)

FIGURE 1.15 George O. Poinar, Jr.

or continents (Herbst and Lutz, 2001), specific fungi based on preservational mode (e.g. amber; see Poinar, 1992) (Figure 1.15), fungi as components of coal (e.g. Beneš, 1956, 1960; Lyons, 2000; Hower et al., 2010), and their stratigraphic position in multiple units within a narrow geographic setting (e.g. Singh et al., 1986). Publications have included discussions of basic terminology (e.g. Elsik, 1983) (Figure 1.16), collaborative approaches (Sherwood-Pike, 1990), sampling techniques (fungal palynomorphs) (e.g. Elsik, 1996; Kalgutkar and

FIGURE 1.16 William C. Elsik.

FIGURE 1.17 Ramakant M. Kalgutkar.

FIGURE 1.18 Jan Jansonius.

Jansonius, 2000 [Figures 1.17, 1.18]; Saxena and Tripathi, 2011), fungi associated with certain types of fossil plants (e.g. Rodríguez de Sarmiento et al., 1998), biodiversity (Dilcher, 1965), and fungi used as proxy indicators of

FIGURE 1.19 David L. Hawksworth.

certain events in Earth history, including continental drift (Nease and Wolf, 1975) and mass extinction (Visscher et al., 2011). There are also brief discussions of the evolution of fungi (e.g. Tripathi, 2012) and summaries of the major groups (Taylor et al., 2014). There are numerous other reports of fossil fungi in addition to these that deal with a variety of subject areas and these will be introduced in the appropriate chapters of this volume.

NAMING FOSSIL FUNGI

One discussion that directly concerns paleomycology relates to how fungi are named, an issue that will continue to impact our understanding of both living and fossil taxa (Hawksworth, 2011) (Figure 1.19). Historically, fungi were named based on features associated with sexual reproduction; however, some fungi only produce asexually, and still others reproduce both sexually and asexually. The majority of these forms are included in the Ascomycota, some in the Basidiomycota. This resulted in different names for the asexual stage (anamorph) and the sexual counterpart (teleomorph) if the stages could not be related to one another. More recently, the problem of fungal anamorphs has been exacerbated by the use of molecular tools linking the different stages (with different names) as parts of the same organism (holomorph) (e.g. Cannon and Kirk, 2000). For example, if a fungus has a name for both the anamorph and teleomorph, the holomorph assumes the name of the teleomorph, or sometimes the anamorph if certain conditions are met.

Adding difficulty to this problem in paleomycology is the fact that fragmentation and other effects brought about by the fossilization process often lead to incomplete preservation of the individual stages.

While fossil anamorphs and parts thereof are treated by the paleomycologist in a uniform manner, they in fact represent the result of various biological mechanisms within the life histories of ascomycete fungi (Seifert and Samuels, 2000). The fact that anamorphs are probably not homologous introduces another level of problems, especially when fungal palynomorphs are used to characterize taxa and thus identify species, and the presence of certain fungi (anamorphs) is used in defining stratigraphic and ecological conditions in deep time. The designation "synanamorph" is used for two or more morphologically distinct anamorphs produced by the same fungus (Hughes, 1979). Seifert and Samuels (2000) suggest the use of specific types of anamorphs to insure that those described are in fact homologous. For example, mycelial anamorphs lack stromatic elements but go through repeated cycles of asexual sporulation. A mononematous (i.e. production of conidia from superficial, exposed conidiogenous cells arising separately from vegetative hyphae or cells) anamorph, which is characteristic of many hyphomycetes included in the Moniliales, has well-differentiated conidiophores anchored in the substrate, but there are no stromatic elements. In conidial anamorphs there are fruiting bodies, while the designation "germinating anamorph" is used for conidiogenous cells emerging from ascospores or conidia. Thick-walled propagules termed sclerotia, aleuriospores, and chlamydospores are termed survival anamorphs. Yeast-like anamorphs are accumulations of blastically budding cells, while dispersive conidia that do not germinate are spermatial anamorphs. Where no propagules are produced, the term vegetative mycelium anamorphs is used. Teleomorph stages in fossil fungi are often far more characteristic and thus easier to name and classify systematically; however, fragmentation and the limits of optical resolution may hamper the definitive assignment. As a result morphology-based genera and species are used with fossil fungi that are defined based on morphological features of the teleomorph.

The problem of naming fungi, whether living or fossil, is far more complex than presented here, but since 2013 a single fungus has only one name, based on the nomenclatural principle of priority in which the oldest genus and species names must be combined irrespective of whether they were instituted as anamorph or teleomorph names (International Code of Nomenclature for Algae, Fungi and Plants [ICN], Melbourne Code; see McNeill et al., 2012). With regard to fungal fossils this means that isolated parts of the life cycle, anamorphs, and teleomorph stages must be merged into one biological species (a fossil fungus). This, however, is generally impossible based on the inherent limitations of the fossil record of fungi.

Throughout the book, the ages for geologic time periods have been taken from the most recent International Chronostratigraphic Chart (v. 2013/01, inside covers), prepared by the International Commission on Stratigraphy (IUGS) and available from www.stratigraphy.org. These dates are approximate, as they are defined by Global Boundary Stratotype Sections and Points (GSSPs) and not (in most cases) by absolute (radiometric) dating. They are also subject to change. See the website for updates and further information. See also Gradstein et al. (2012).

2

HOW FUNGAL FOSSILS ARE FORMED AND STUDIED

PRESERVATION 13
Impressions 13
Compressions 14
Coal 15
Charcoal 17
Casts 18
Petrifactions and Permineralizations 18
Unaltered Material 23
Rock Surfaces 25
Nodules 25
Carbon Spherules 25
Unusual Preservation Types Containing Fungal Remains 26

PRESERVATION

Evidence of the existence of biological systems at various points in geologic time can be recorded in the fossil record in many ways. Almost everyone has a visual image of dinosaur bones, or of the remains of mastodons that once roamed the Earth, which, having been carefully excavated were then reassembled to show the actual appearance of the organism. Some have collected fossils in the form of invertebrate shells or plant leaves, or have seen examples of these organisms as images or in museums or in private collections. But what about fossil fungi and other types of microorganisms? Where are these found, how are they preserved, and what methods can be used to study aspects of the paleomicrobial world? In this section we present the most common modes of preservation in which fungi might be found, and then discuss how some of these organisms are studied. Additional information about the preservation of plant fossils is available in Taylor, T.N., et al. (2009).

IMPRESSIONS

Impressions represent the negative relief of an organism that at one time was covered by sediment and thus fossilized. When the organism is "discovered" as a fossil millions of years later the actual organism is not present, having been decayed by microorganisms (in some instances by fungi) and affected by factors such as pressure, temperature, and sediment movement during fossilization. Unlike some other types of fossilization, impressions lack any organic remnants (e.g. carbon) of the original organism. Since fossils are defined as examples of the existence of life at a particular point in time and space, an impression may give some indication of the size, shape, and general morphology of the organism. In a modern example, leaves of trees sometimes fall and subsequently sink into the drying concrete of a sidewalk or building foundation. Over time the concrete hardens and the leaf disintegrates, but an imprint of the leaf remains in the concrete. The discovery of this imprint at some later point in time would represent an impression. In this instance nothing would be known about the tree that dropped the leaf in the wet concrete, but the fact that a living tree was involved would be recorded. It is obvious that the smaller the particle size of the sediment enclosing the organism, the more detailed the features of the organism that can be preserved.

T.N. Taylor, M. Krings & E.L. Taylor: Fossil Fungi.

DOI: http://dx.doi.org/10.1016/B978-0-12-387731-4.00002-5

The small size of most fungi makes them very difficult or even impossible to discover as impressions. Some of the early reports of leaves with various spots on them interpreted as fungi were no doubt impressions (Figure 2.1) (Unger, 1841–1847; Meschinelli, 1898). In addition, there have been reports of variously damaged leaves in which faint lines believed to be hyphae are present, or which otherwise show evidence of fungal decay (e.g. Iglesias et al., 2007). These fossils do not offer a great amount of detailed information but can be useful in documenting the occurrence of fungi of different ages. In general, the most useful way to view these fungi is via reflected light. Sometimes utilizing cross-polarized light is helpful in enhancing the contrast between the fossil and the matrix. In some instances it may be possible to use latex or other suitable material to prepare a mold of the negative relief of the surface, which could be examined with increased resolution such as via scanning electron microscopy (SEM).

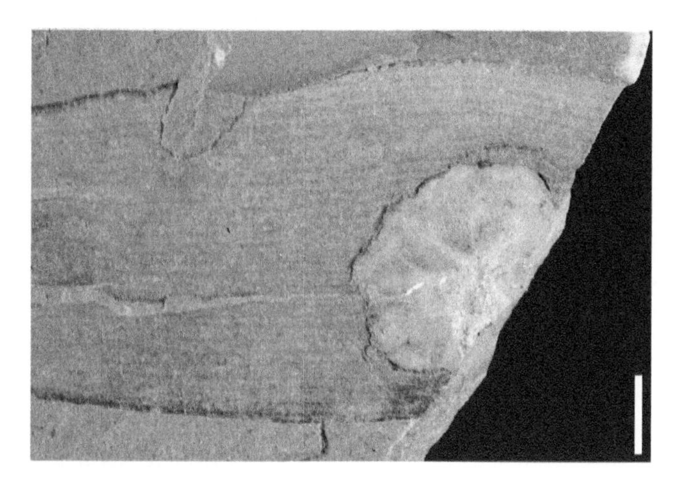

FIGURE 2.1 Impression of ascomycetous fungus *Xylomites asteriformis* on a *Podozamites* leaf. Jurassic, Germany. Bar = 0.5 cm. (Courtesy K. Bauer.)

COMPRESSIONS

Compressions are a type of plant preservation in which some of the original organic matter still remains, albeit usually altered in the form of coalification. Compression fossils may provide various insights into the fungi associated with plants. Small, opaque specks or black spots are relatively common on the surfaces of compression foliage throughout the late Paleozoic, as well as in geologically younger fossils. Based on size, shape, color, and spatial arrangement, numerous genera of such structures (e.g. *Excipulites*, *Hysterites*, *Sphaerites*; Figure 2.2) have over time been formally described (see Pirozynski and Weresub, 1979). The specks and dots have been typically interpreted as necrotic areas caused by phytopathogens or fruiting bodies of endophytic or epiphyllous fungi (e.g. Wagner and Castro, 1998). Studies based on cuticular analysis, however, indicate that not all of these structures are fungal in nature; some may represent the crystallized contents of secretory cavities (Krings, 2000). Nevertheless, it has been shown that during the chemical preparation of cuticles, fungi otherwise not seen become visible and, as a result, represent an expansive

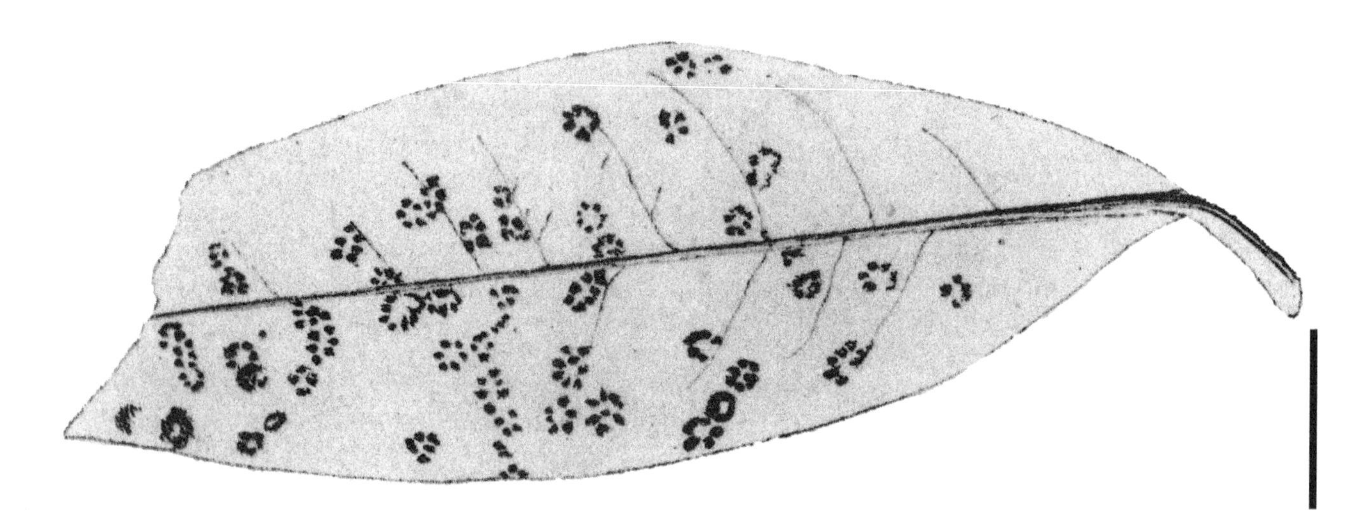

FIGURE 2.2 Leaf with *Sphaerites suessii* (perithecia?) scattered on the upper surface. Cenozoic, Slovenia. Bar = 10 mm. (From Meschinelli, 1902, redrawn from Ettingshausen, 1872.)

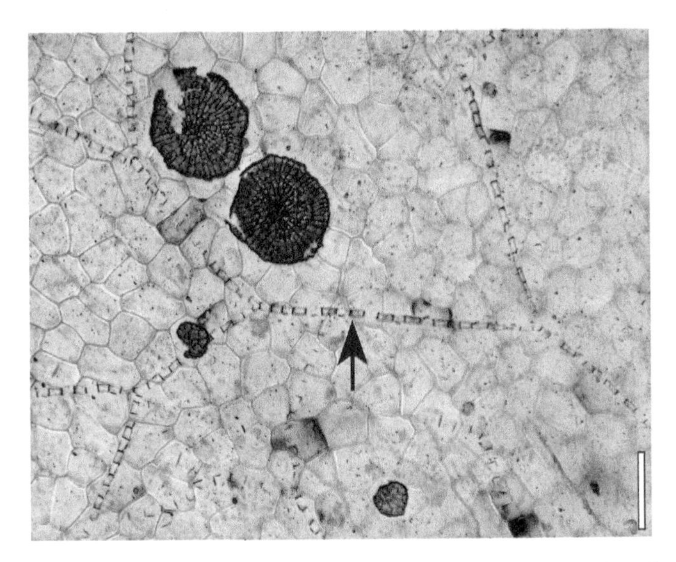

FIGURE 2.3 Leaf cuticle preparation with hyphae (*arrow*) and two epiphyllous thyriothecia on surface. Eocene, Tennessee, USA. Bar = 50 µm.

source of information about associations of endophytic and epiphyllous fungi (Figure 2.3) with land plants that were previously unknown (e.g. Dilcher, 1965). Cuticles are most commonly treated to remove opaque materials and then mounted on standard microscope slides for observation. In some instances such preparations can be studied via SEM (Alvin and Muir, 1970). In fact, our current knowledge of Cenozoic epiphyllous fungi associated with angiosperms is based primarily on chemically cleared leaf compressions. Moreover, leaf cuticles of many Cenozoic angiosperms can be carefully removed entirely from the rock matrix so that the cuticle comes away as an envelope that can be mounted and examined for fungi. Similar high-quality cuticle preservation has recently been observed in geologically older compression foliage fossils (Abu Hamad et al., 2008). Such cuticle envelopes are especially important as they may contain exact information on the spatial distribution of the fungi on (and perhaps within) the plant part as well as their spatial arrangement in relation to structural features of the leaf (i.e. types of veins, etc.). They can also provide information on mycelial morphology.

There is an increasing awareness of the presence of fungal remains on compressed plant material. For example, the epidermal anatomy of compressed plant parts from the lower Oligocene of Germany has been studied primarily to reassemble whole plant taxa from isolated parts, but fungi co-occurring with some of the remains were also described and interpreted as representing a community of saprotrophs (Kunzmann, 2012).

FIGURE 2.4 Surface of block of Holocene peat showing plant fragments, Canada. Bar = 1.0 cm.

COAL

Technically, coal represents a compression fossil, and historically has been a rich source of fungal fossils (e.g. Beneš and Kraussová, 1964). Coal is a heterogeneous mixture of macromolecular organic and inorganic compounds that have been compressed over extended periods of time (Scott, 1987) and today is found throughout the world (Belkin et al., 2009, 2010). In a general sense coal begins as peat (Figure 2.4), a soft organic material consisting of partly decayed plant material which forms in place in a swampy environment (bogs, mires, moors, muskegs, peat swamp forests, etc.) where it accumulates into thick beds under certain environmental conditions. In rare instances the plants that grew on the peat are preserved by some other means, such as in a tuff, and thus make it possible to reconstruct the vegetation based on the position of the individual plants (Wang et al., 2012). In other cases the peat mires are infiltrated by various minerals (e.g. calcium, pyrite, and siderite) that stop the decay process and entomb the plants into permineralizations, some of which are termed coal balls (Schweinfurth, 2009). In most instances, however, peat represents the initial stage in coal formation that then undergoes some degree of chemical alteration and physical compaction (Nadon, 1998). The organic materials within the coal are termed macerals, a term coined by the paleobotanist and social activist Marie Stopes (1935) (Figure 2.5), and the presence of these, together with the rank of the coal (i.e. the degree of coalification), is used to subdivide coal types, for example: vitrinite (Figure 2.6) – derived from coalified woody tissue; liptinite – derived from resinous and waxy parts of plants; and inertinite (including fusinite) – derived from charred and biochemically altered plant parts.

FIGURE 2.5 Marie C. Stopes.

FIGURE 2.7 Paul C. Lyons.

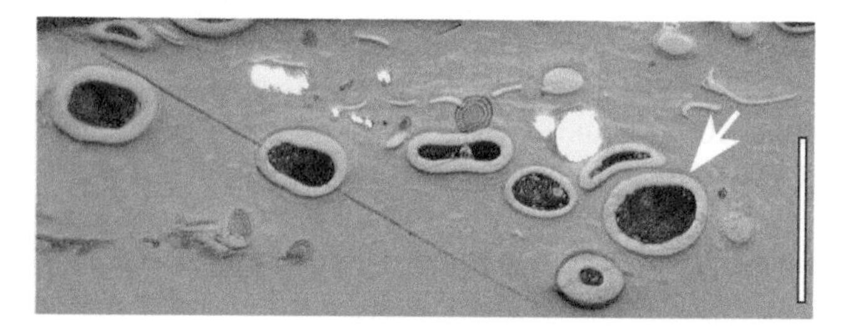

FIGURE 2.6 Reflectance photomicrograph of non-aperturate or monoporate fungal spores (*arrow*) in vitrinite. Bar = 50 μm. Oligocene, Romania. (Courtesy J.C. Hower.)

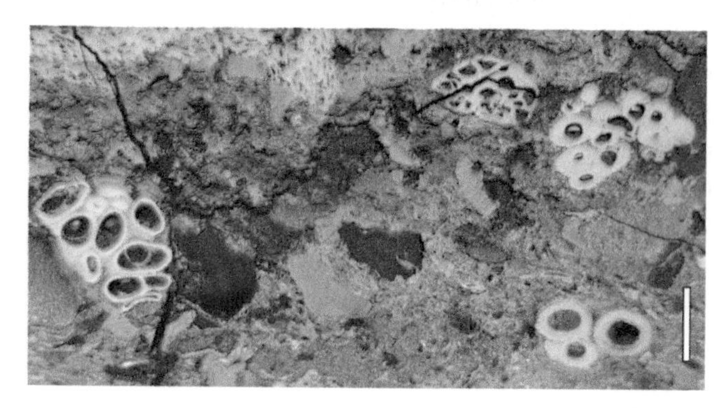

FIGURE 2.8 Petrographic thin section of coal showing fungal sclerotia in funginite. Cenozoic, India. Bar = 25 μm. (Courtesy J.C. Hower.)

The formation and analysis of coal macerals can be far more complex; however, there are techniques that can be used to tease apart the coal constituents and accurately determine the structure of the component plants (Kizil'stein, 2006). Bacteria, fungi, and insects can chemically and physically alter the macerals (Richardson et al., 2012). Low rank coals (e.g. lignites and sub-bituminous coals) have been a source of fungal palynomorphs because the plant material and microbes have not been highly metamorphosed. Today, even higher rank coals offer the opportunity to examine fungal remains using varying techniques and preparation procedures (O'Keefe et al., 2011a, b).

One of the most common and highly reflective inertinite macerals is funginite (Beneš, 1956; Lyons, 2000) (Figure 2.7); as the name implies, this is a coal maceral consisting mainly of fungal spores, hyphae, and sclerotia (Figure 2.8) (Thomas, K., 1848; Jeffrey and Chrysler, 1906; ICCP, 2001). The presence of funginite has been suggested as representing evidence of fossil or recent weathering (drainage) of peat; in association with other maceral types such as cutinite and suberinite, it is suggested as providing evidence of mycorrhizal symbioses (Hower et al., 2011a,b).

Some of the early evidence of fossil fungi was observed in petrographic thin sections of coal that were used, together with an analysis of the composition of the coal (based on the types of plants that were present in the ancient ecosystem), to infer how the coal was formed (Hacquebard and Donaldson, 1969; Figueiral et al., 1999; Moore and Shearer, 2003). In other studies the texture and morphology of certain peats have been used to measure fungal alteration in the formation of the coal (Moore et al., 1996). However, it is known that multiple types of vegetation may be represented in a thin band of coal and that the analysis of the ancient ecosystem based on fungal remains in coal is both biologically and chemically far more complex than previously thought (Scott, A.C., 2002). The paleomycologist must also keep in mind that the absence of fungal hyphae or other structures does not necessarily indicate the absence of fungi in a preparation, because there are some extant forms (e.g. brown rot fungi) that autolyze their hyphae during the decay process (Pujana et al., 2009). In addition, fungal hyphae might not survive the stages of coal formation.

Another maceral type in coal is resinite, which consists of ovoid and translucent accumulations of plant resins. Because some plants produce various types of protection against microbial degradation in the form of essential oils, alkaloids, lignin, tannins, and various resins, there are numerous instances in which resinite contains funginite (e.g. Hower et al., 2010). A number of these macerals can be identified based on chemical and physical mapping that can provide a more accurate picture of maceral type during peat formation and subsequent coalification stages (Bruening and Cohen, 2005; Chen et al., 2013). In some instances, what were thought to be fossil fungal remains have been reinterpreted as plant resins (Pareek, 1966).

Carboniferous and Cenozoic coals have been widely studied for their paleomycological content (Beneš and Kraussová, 1964, 1965; Mukherjee, 2012). In these studies some of the fungi were isolated from the coal and examined individually, while in other reports the analysis of the fungi was based on petrographic thin sections (Stach and Pickhardt, 1957, 1964). For example, several types of asci were reported from polished sections of Carboniferous coals and named *Clavascina orlovensis* and *Discoascina perforata* (Beneš, 1961). Based on the figures, it is difficult to determine whether either of these structures represents a fungus. In other studies of coals, the occurrence of fungi at particular stratigraphic levels was used as an indication that the increase in fungal elements was directly related to rapid fungal evolution and diversity (e.g. Beneš, 1978). In still other investigations the research focused on types of fungal spores present to determine whether the coals developed in place (autochthonous) or were allochthonous (moved to the site of deposition) in a deltaic system that may have had multiple marine incursions (e.g. Soomro et al., 2010).

The majority of the extensive coal deposits around the world formed during the Carboniferous–Permian (359–252 Ma) when there were large swamp-forest ecosystems that covered a great deal of Europe, North America, Asia, and Australia. The accumulation of these vast coal seams has historically been hypothesized as resulting from the inability of the saprotrophic organisms of the time to degrade plant material (including lignin and cellulose) in the anoxic conditions that existed in the swamp ecosystems. Using genomic tools and bioinformatics to examine wood-decaying fungi, especially white rot members of the Agaricomycotina (Robinson, 1990), an alternative hypothesis has been suggested (Floudas et al., 2012). These authors utilized molecular clock assumptions together with a comparative analysis of 31 fungal genomes to date the origin of lignin degradation near the end of the Carboniferous. The hypothesis infers that lignin degradation evolved only once and that within the Agaricomycotina group some members lost the ability to decay lignin and evolved other methods to obtain carbon (e.g. brown rot fungi, ectomycorrhizae). It should be noted, however, that most of the largest and most abundant Carboniferous coal-swamp plants utilized methods other than secondary xylem for support (Taylor, T.N., et al., 2009), and thus much of the plant biomass was not lignin derived; nevertheless, some in fact did produce wood (Krings et al., 2011a). There is also evidence that by the Late Devonian there were fungi associated with wood (Arnold, C.A., 1931) – and these had the capacity to degrade lignin (Stubblefield et al., 1985a) – at the same time as the first plants with secondary xylem appeared (Meyer-Berthaud et al., 1999, 2010). In addition, there are extensive coal deposits that formed during the Permian and Cretaceous, as well as in other time periods, and their presence seems to refute the hypothesis that Carboniferous coals formed due to a lack of wood-decaying organisms at that time. Irrespective of the validity of these competing ideas about carbon biomass accumulation, the utilization of genomic data with paleomycology represents a potentially exciting path forward in understanding multiple levels of global ecosystem evolution.

CHARCOAL

Another perhaps somewhat unlikely source of information on fossil and subfossil fungi is charcoal. For example, charcoalified pieces of *Prototaxites*, an enigmatic organism believed to represent a giant fungal fruiting body, have been reported from the Lower Devonian of the Welsh borderland and interpreted as products of wildfire activity (Glasspool et al., 2006). Moreover, the presence of fungal hyphae has been recorded for various occurrences of charcoalified wood (Scott, A.C., 2000; Uhl et al., 2007) and other charred plant parts (e.g. flowers; Hartkopf-Fröder et al., 2012). While some of these hyphae may represent fungi that colonized the coal, others are believed to represent fungi that colonized the plant in vivo. The characteristic biodeterioration symptoms of woods infected by various wood-rot fungi, as well as the fungi responsible for this decay, have also been documented in archaeological charcoal samples (Moskal-del Hoyo et al., 2010).

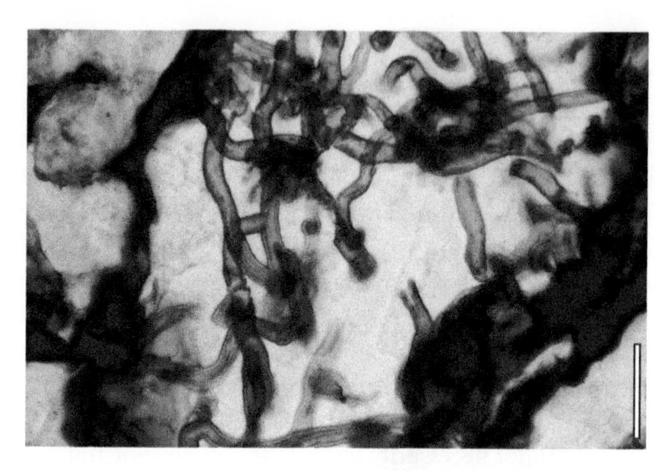

FIGURE 2.9 Cell filled with fungal hyphae. Pennsylvanian, Illinois, USA. Bar = 10 μm.

CASTS

Microborings, systems of galleries and tunnels in hard substrates such as rocks, calcareous shells, and ooids (Bundschuh, 2000), are sometimes caused by endolithic fungi, among other microbial endoliths (Vogel et al., 1987; Golubic et al., 2005, 2007). Microbial endoliths are known from terrestrial, freshwater, and marine environments today, where they are integral parts of various ecosystems (Glaub et al., 2007). In order to visualize the activities of these organisms in three dimensions, polymer resins are injected into the borings and then the rock or shell dissolved, revealing an artificial cast that then can be examined with SEM (Gatrall and Golubic, 1970; Golubic et al., 1970). The diameter of the borings, patterns of branching, and other features provide a set of morphological characters that can be compared with those from different fossil microbial endoliths, including fungi and cyanobacteria (Garcia-Pichel, 2006). Comparisons between associations of endolithic microorganisms today and those found as fossils (so-called ichnocoenoses) can be used to reconstruct a variety of paleoecosystem parameters such as paleobathymetry (Glaub, 1999; Chazottes et al., 2009), turbidity (Vogel et al., 1987), community succession (Gektidis, 1999), and organic erosion (Perkins and Halsey, 1971; Vogel et al., 2000).

PETRIFACTIONS AND PERMINERALIZATIONS

Petrifactions and permineralizations are two basic types of preservation that provide exceptional resolution of fossil organisms because the individual cells and tissue systems are preserved. In both types the plant or fungus was immersed in water with high concentrations of dissolved minerals; silica and calcium carbonate are two of the common ones in plant fossils, but pyrite and oxides may also be the preserving minerals. The details of the process of petrifaction and permineralization remain incompletely understood (Ballhaus et al., 2012; Läbe et al., 2012), but it is believed that the cell walls act as nucleation sites for mineral crystal growth (Dietrich et al., 2013). Once the mineral is precipitated and hardened, the plant material and anything associated with it, including fungi and other microbes, are entombed in the rock and very often faithfully preserved even down to the finest details (Figure 2.9). While fungal remains are usually found associated with plants (or animals) preserved by means of petrifaction or permineralization, there are also rare instances in which larger fungal structures (e.g. basidiocarps) are petrified or permineralized (e.g. Fleischmann et al., 2007)

Petrifactions

In petrifactions the original organic matter is replaced by minerals, and this secondary replacement can lead to the loss of delicate structures. Various types of woods are perhaps the most common petrifactions, including those from the Triassic Petrified Forest of Arizona (USA) and Jurassic Petrified Forest from Patagonia, Argentina, but bones can be preserved in this manner as well (Trueman and Tuross, 2002). The bright colors (Figure 2.10) of many petrified woods are the result of the presence of certain minerals. While there are some reports of fungi in the cells of wood preserved as petrifactions, there are far more fungi preserved in permineralizations. One of the best known deposits of petrifactions is the Lower Devonian Rhynie chert, which contains various types of silicified algae, fungi, plants, and animals.

FIGURE 2.10 Polished piece of petrified wood showing growth rings. Triassic, Utah, USA. Bar = 5.0 cm.

FIGURE 2.11 William Mackie.

RHYNIE CHERT. The Rhynie chert has figured prominently in discussions of fossil fungi for almost a century and represents an extraordinary view of fungus–plant interactions from approximately 410 million years ago (Berbee and Taylor, 2007). The deposit was initially discovered by William Mackie (Figure 2.11) in

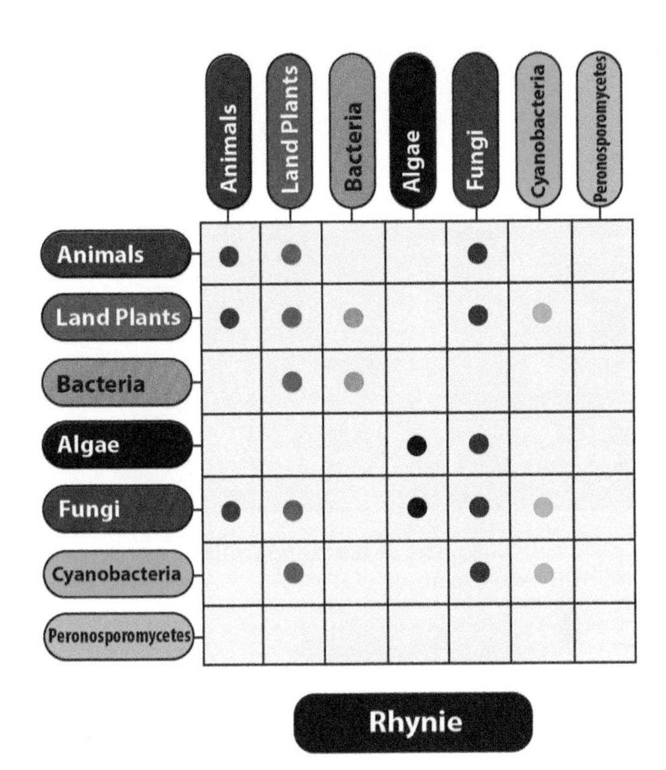

FIGURE 2.12 Distribution of interactions between major groups of organisms within the Rhynie chert ecosystem based on published results to date. (Courtesy G. J. Ortiz.)

1912 when he found some siliceous rocks in a stone wall near the town of Rhynie in Aberdeenshire, Scotland. These rocks turned out to contain various small plant axes with every cell preserved intact. In the intervening years numerous organisms have been described from the chert, including several types of early land plants (with and without true vascular elements), algae, various types of arthropods, cyanobacteria, lichens, fungi, and fungus-like organisms (Kidston and Lang, 1917, 1920a, b, 1921a, b). The Rhynie chert is today regarded as one of the most important Konservat Lagerstätten (deposits with exceptional preservation of fossils) because it is one of a few geological sites where preservation is so extraordinary that even certain organisms not typically seen in the fossil record are faithfully preserved (see Kerp and Hass, 2004). Moreover, there is simply no other early terrestrial ecosystem that offers the potential to examine interactions among organisms (Figure 2.12).

The Rhynie chert has been regarded as Lower Devonian (Pragian; Rice et al., 1995), but analysis of spores provides a slightly narrower age range of Early (but not earliest) Pragian–earliest Emsian (Wellman, 2004, 2006; Wellman

FIGURE 2.13 Suggested reconstruction of the Early Devonian Rhynie chert ecosystem.

FIGURE 2.14 Nigel H. Trewin.

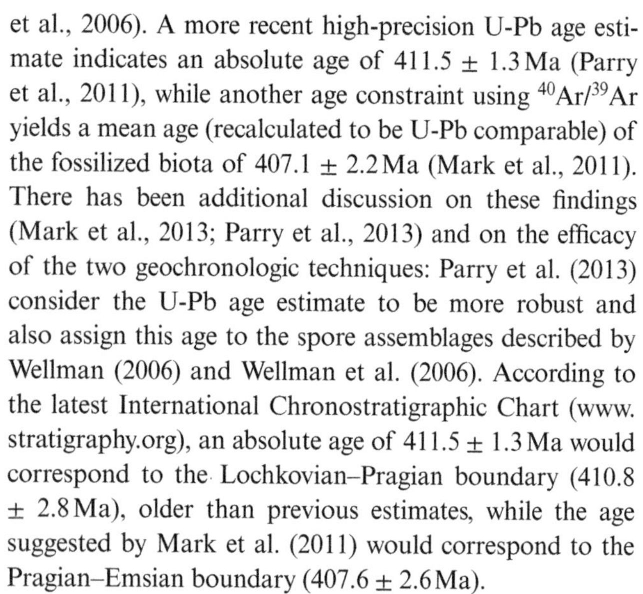

et al., 2006). A more recent high-precision U-Pb age estimate indicates an absolute age of 411.5 ± 1.3 Ma (Parry et al., 2011), while another age constraint using $^{40}Ar/^{39}Ar$ yields a mean age (recalculated to be U-Pb comparable) of the fossilized biota of 407.1 ± 2.2 Ma (Mark et al., 2011). There has been additional discussion on these findings (Mark et al., 2013; Parry et al., 2013) and on the efficacy of the two geochronologic techniques: Parry et al. (2013) consider the U-Pb age estimate to be more robust and also assign this age to the spore assemblages described by Wellman (2006) and Wellman et al. (2006). According to the latest International Chronostratigraphic Chart (www.stratigraphy.org), an absolute age of 411.5 ± 1.3 Ma would correspond to the Lochkovian–Pragian boundary (410.8 ± 2.8 Ma), older than previous estimates, while the age suggested by Mark et al. (2011) would correspond to the Pragian–Emsian boundary (407.6 ± 2.6 Ma).

The organisms preserved in the Rhynie chert lived in a subtropical setting in a portion of the large continent Laurussia (the Old Red Continent) that was located approximately 28° south of the equator (Rice et al., 2002). The rocks that contain the approximately 52 lenses of Rhynie chert are principally fine grained and accumulated on an alluvial plain with small ephemeral ponds and regional local lakes. The ecosystem (Figure 2.13) is now interpreted as a geothermal wetland (Channing et al., 2004), as there were associated hot springs that were part of a complex hydrothermal system (Rice et al., 2002). Both aquatic biota and terrestrial life became preserved as a result of temporary flooding of silica-rich water, or by silica-rich groundwater that percolated to the surface (Powell et al., 2000). It is hypothesized that water highly charged with dissolved silica rapidly entombed the organisms as the temperatures dropped and the mineralizing fluids invaded the cells and tissue systems, followed by deposition of opaline silica. The specific chemistry of the water and other physical parameters resulted in the extraordinary quality of the fossils from the areas around freshwater pools. Preservation of the organisms is interpreted as being nearly instantaneous (Trewin, 1996) (Figure 2.14), resulting in vascular plants and other organisms that are in growth position and, in the case of some of the fungi, closely associated with their host(s). The Rhynie chert contains both autochthonous (in growth position) and allochthonous preservation.

There is considerable interest today in precisely how the Rhynie chert plants were preserved and whether some or all of the biota were adapted to highly stressed environments (Channing, 2003; Channing and Edwards, 2009a). Research is underway examining a number of modern hot-springs ecosystems around the world (in places such as Yellowstone National Park, USA, and several sites in New Zealand and Patagonia, Argentina), including topics such as the role of microbes in silica precipitation (Jones et al., 1999, 2003, 2004; Phoenix et al., 2005; Jones and Renaut, 2007; Tobler et al., 2008; Channing and Edwards, 2009b). In addition, laboratory experiments are examining the silicification of cyanobacteria as a modern analogue to stages in biosilicification

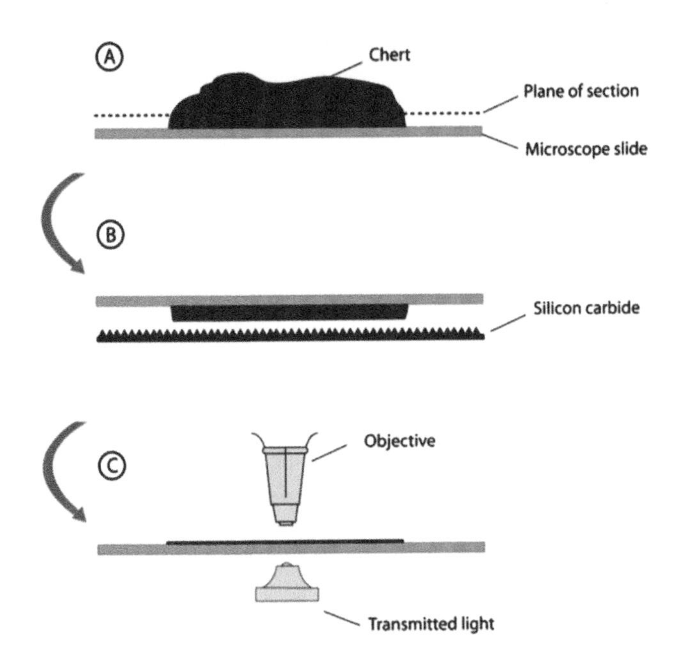

FIGURE 2.15 Stages in the preparation of a thin section. **A.** Cement fossil-bearing rock to glass microscope slide. **B.** Grind rock with abrasive. **C.** Examine section with transmitted light.

FIGURE 2.16 Surface of a Pennsylvanian coal ball. *Arrow* indicates partially exposed lycopsid cone on the surface. Illinois, USA. Bar = 1.0 cm.

seen in the fossil record (e.g. Benning et al., 2005). The geyser hot-spring system of Yellowstone National Park in Wyoming is perhaps the model system for studying the preservation process and for synthesizing information about the ecophysiology of the organisms that live in this specialized environment: Channing and Edwards (2009a) and Channing and Wujek (2010) hypothesized that silica accumulating around some of the Rhynie chert plants during life may have "predisposed" them to fossilization and that they were physiologically specialized so as to deal with osmotic and chemical stress in the landscape where they lived (Channing and Edwards, 2009a). An excellent study on the inferred structure/function relationships in the earliest unequivocal member of the genus *Equisetum* from the Jurassic San Agustín hot spring deposit in Argentina suggested that the adaptations and maintenance of ecophysiological innovation in a hot-spring environment may be less unique than previously thought (Channing et al., 2011).

While most of the information about the Rhynie chert organisms has been obtained from thin sections (Figure 2.15), there has been some interest in using the chert ecosystem as a model system to investigate the potential of biomarker reliability in deep time or as a potential in astrobiology. For example, Fourier transform infrared (FTIR) spectroscopy of the chert does show the presence of biomolecules; however, the overall resolution is poor because of the high percentage of silica (Preston and Genge, 2010).

PERMINERALIZATIONS

The principal difference between petrifactions and permineralizations is that the cell walls of the latter are still organic, although they have become altered to some degree. The process of permineralization is complex and may involve bacteria as the catalyst for certain chemical reactions (Daniel and Chin, 2010). In general, opaline silica (or another permineralizing mineral) is formed during permineralization and these films coat the surface of structures while colloidal silica permeates the cells. This preservation technique results in the extraordinary fidelity of various fossilized structures. One of the most common permineralization types is termed coal balls (Figure 2.16), plants preserved in calcium carbonate ($CaCO_3$) that are commonly found associated with Carboniferous coals in Euramerica (Europe and North America) and Permian coals in China.

COAL BALLS. Because coal balls are accumulations of (degrading) plant material (technically peat), they also are an excellent source of various forms of decaying organisms, including fungi. Numerous fungal remains

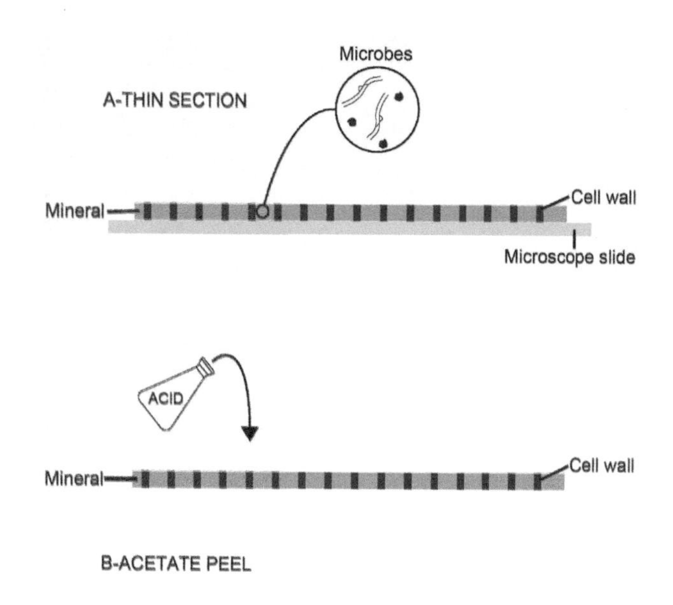

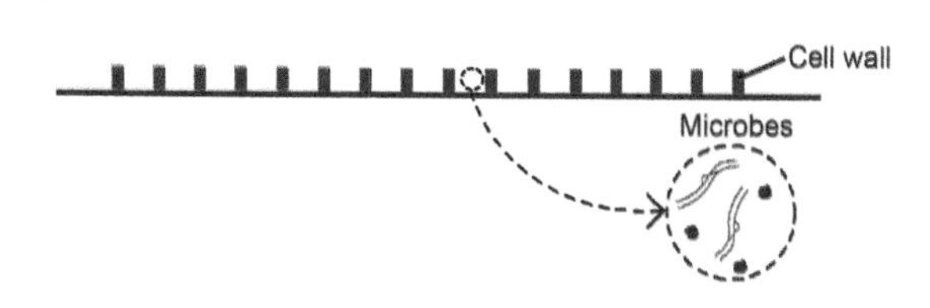

FIGURE 2.17 Comparison between methods for discovering potential microbes. **A.** Thin section, where the mineral matrix between cells can be examined, and **B.** Acetate peel, where the mineral matrix between the cell walls has been dissolved away.

have been found in coal balls, including hyphae, spores, and various types of reproductive structures. Initially fungi were discovered in thin sections along with the plant parts that were the primary focus of the research. Thin sections are prepared by cementing a piece of rock to a glass slide and then grinding the slice until it is thin enough (usually 50–150 μm) to be examined in transmitted light (Hass and Rowe, 1999). One of the primary advantages of thin sections is that they provide a slice of the entire matrix so that any microorganisms within the cell lumina or between plant parts can be viewed in sufficient resolution and magnification (Figure 2.17). However, this technique is quite time consuming and destroys a considerable amount of the fossil in the process. In instances where sufficient organic matter remains in the cell walls of the fossils (as in coal balls), the peel technique was developed, in which acid was used to dissolve the mineral matter, leaving the organic material standing in relief on the surface of the rock. Originally, a viscous mixture of nitrocellulose was added to the surface and allowed to dry, and then the peel pulled from the surface (Walton, 1928). Later, preformed sheets of cellulose acetate were used; with these it was possible to rapidly make numerous sections of the fossil, each about 50 μm thick, that did not waste much of the fossil (Joy et al., 1956; Galtier and Phillips, 1999). This process has been the standard where the fossil and matrix contained organic matter and has been the source of a great deal of information about Carboniferous fungi.

Despite the many advantages of using peels instead of thin sections, however, the process has one major drawback. During the etching process any small organisms preserved in the matrix, such as fungi and other microbes, can be lost (see Figure 2.17) (Taylor et al., 2011). It has generally been assumed that fungal diversity was low during the Carboniferous, due to low pH or other chemical aspects of the peat swamp ecosystem (Taylor, T.N., 1993b), but perhaps more likely is that our understanding of Carboniferous coal ball plants is biased due to the extensive use of the peel technique, which has eliminated many fungi during the etching process. Recent studies confirm that far more fungi are observed in Carboniferous permineralizations if thin sections rather than peels are prepared (Krings et al., 2007a, 2010a, 2011b; Dotzler et al., 2011), and the same is true of other permineralizations where peels have been the standard method of preparation (e.g. Permian and Triassic of Antarctica; Eocene Princeton chert of British Columbia). Many studies on permineralized plant fossils have concentrated on a particular organ, with the focus on determining other parts of the same plant, stages of development, phylogenetic information, or other anatomical and/or morphological directions. Fungi are sometimes noted as being present (e.g. Cevallos-Ferriz and Stockey, 1989) (Figure 2.18), but they are not described in detail because either there is insufficient material or they are interpreted as too general to be of interest. These fungi, however, may represent an important source of data because they occur in particular plant organs and thus may be useful in understanding various nutritional modes, types of plant–fungus interactions, and a variety of other research questions that can only be answered by studying fossil fungi.

Both thin-section preparations and acetate peels may provide the opportunity to utilize other types of imaging systems and data analysis platforms (e.g. spinning disc confocal microscopy, energy-dispersive X-ray microanalysis); however, these techniques are only now beginning to be used more routinely on permineralized fossils

FIGURE 2.18 Ruth A. Stockey. (Courtesy G. W. Rothwell.)

(Klymiuk et al., 2013a). Some of these techniques have provided an important source of new information about extant fungi (e.g. Wolinski, 2003), and it will be interesting to see how powerful they might be in helping us to tease out information about fossil fungi.

The development of Raman imaging, a non-destructive analytical technique, in multiple areas of paleobiology will be increasingly more important in the identification and characterization of the molecular structure of minerals and mineralized carbonaceous matter in petrographic thin sections (Schopf et al., 2005, 2011; Nasdala et al., 2012). Although many of these techniques have not been directed specifically at the identification of fungal fossils (e.g. Schopf et al., 2010a), they do hold great promise in correlating the morphology and geochemistry of ancient microorganisms, including fungi (Kudryavtsev et al., 2001; Schopf et al., 2006). The reader is referred to the volume *Raman Imaging – Techniques and Applications*, which provides a comprehensive review of the subject in multiple areas including those in the biological and geological sciences (Zoubir, 2012).

UNALTERED MATERIAL

Some fungal remains in the form of hyphae and spores could broadly be interpreted as evidence of unaltered material, especially specimens obtained from geologically young deposits such as the Pleistocene (Tehon, 1937; Wolf, 1969b; Chambers et al., 2011; Blaauw and

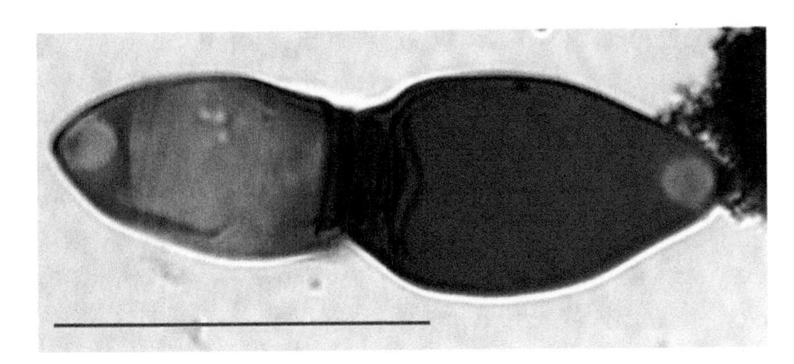

FIGURE 2.19 *Dicellaeporisporites poratus.* Eocene, Mackenzie Delta, Canada. Bar = 25 μm. (Courtesy M.G. Parsons.)

Mauquoy, 2012). Unfortunately, however, some of these fungal remains may also represent modern contaminants. From typical pre-Quaternary palynological preparations, fungal remains, including hyphae, sclerotia, and spores (Figure 2.19), have played an important role in interpreting the paleoecological changes and climate cycles that have taken place in broad geographic regions (van Geel, 1978). In other instances, determining the paleoenvironment based only on fungal spores has been difficult because many of the genera extend over a wide geographic region (e.g. Palacios-Chávez, 1983). Fungal spores have proven to be important in defining stratigraphic levels (biostratigraphy), and, in combination with other types of spores or pollen, have been useful as index fossils (Kar, 1990; Kalgutkar, 1993). Their presence or absence has also been used to suggest extinction and recovery events (Vajda and McLoughlin, 2004; Visscher et al., 2011).

Extant pollen and spores of various plants have been studied using SEM, and the technique has also been useful in examining modern fungal spores (Tsuneda, 1983; Read and Lord, 1991; Pilger and Thomasson, 2005). In addition, SEM imaging has been used to examine the ascomata of various fungi and to provide important details about fossil fungal interactions (van der Ham and Dortangs, 2005). The same technique with backscattered electron imaging (SEM-BSE) and CLSM (confocal laser scanning microscopy) has also been employed to better resolve fungi and other organisms in Lower Cretaceous amber (Ascaso et al., 2005; Schopf et al., 2011). Pyrolysis/gas chromatography-mass spectroscopy has also been used to identify the source of individual pieces of amber (Grimaldi et al., 2000a). As imaging techniques become more sophisticated it is worth contemplating whether subtle differences might be used to determine spore dimorphism in the fossil record of fungal spores (e.g. Young, T.W.K., 1969). Atomic force microscopy (AMF) has resolution that far exceeds the optical diffraction limit and may be useful at some stage in examining the surface of fossil fungal spores, either within a reproduction structure or in the dispersed state. The three-dimensional configuration of the fungal spore surface using AMF may offer character states that reflect developmental stages, allowing the paleomycologist to more accurately interpret what spore wall features represent, either developmentally or within the context of the life history of the fungus (e.g. Dufrêne et al., 2001; Dupres et al., 2010). The question will always be whether new techniques that have been developed for extant organisms can also be applied to fossils.

Amber

Perhaps the best evidence of unaltered material in the form of a matrix that preserves microorganisms is amber. Amber was extensively collected by many early scientists and collectors who understood that the substance was produced by different types of plants that occurred in many parts of the world (Conwentz, 1896). Fossil fungi preserved in amber can be easily observed and studied in detail by using various microscopy techniques (Speranza et al., 2010). It is therefore not surprising that fungi in amber were described as early as the nineteenth and early twentieth centuries (e.g. Göppert and Berendt, 1845a; Caspary and Klebs, 1907). Only recently, however, have these fossils received wider scholarly attention. Amber is an especially important source of paleomycological information because it contains a large diversity of organisms preserved from different terrestrial habitats (e.g. Grimaldi et al., 2000b; Girard et al., 2009; Schmidt et al., 2010a). Moreover, the translucent nature of the matrix makes it relatively easy not only to determine delicate features that are useful in systematics but also to ascertain how fungi interacted with other organisms. Today there are reports of representatives of many different groups of fungi in amber, including plasmodial slime molds (Waggoner and Poinar, 1992), a representative of the genus *Aspergillus* growing on a springtail (Dörfelt and Schmidt, 2005), and evidence of animal parasitism by fungi (Schmidt et al., 2007; Sung et al., 2008).

The extraordinary preservation potential that amber affords means it will continue to be an important avenue for research on fossil fungi. Amber pieces are usually quite small, however, and thus rarely provide information on the ecological configuration of the community in which the fungi (and their host organisms) lived, with some exceptions (Poinar et al., 1993; Perrichot, 2004; Schmidt et al., 2006, 2012; Schmidt and Dilcher, 2007). Moreover, almost all amber comes from Cretaceous and Cenozoic strata, which is too geologically recent to record the origin or early evolution of most major groups of fungi. Nevertheless, amber from the mid-Cretaceous of France (Charentes) has been used to reconstruct trophic levels of a 100 Ma forest soil (Adl et al., 2011). While amber has been reported from the Carboniferous (Bray and Anderson, 2009), it has not been found to contain direct evidence of fungi, with one possible exception (Smith, J., 1896). There are several comprehensive books available on amber, its origins, and the organisms found within it, including *Life in Amber* (Poinar, 1992), *Amber: Window to the Past* (Grimaldi, 1996), and *The Amber Forest: A Reconstruction of a Vanished World* (Poinar and Poinar, 1999). We have included examples of fungi preserved in amber in those sections of the book where a particular group is the primary focus. Fungi in amber can be studied by various techniques including transmitted and reflected light, together with image-enhancing software and X-ray synchrotron imaging for ambers and other materials that are opaque (Tafforeau et al., 2006; Lak et al., 2008). An interesting study looks at the carbon isotope signatures in resins and attempts to correlate these with the progression of insect infestation and the fungi that are included as a food source for the insects (McKellar et al., 2011).

Mummified Wood

This type of preservation can also be classified as unaltered because the wood is not mineralized and, in some cases, can still be cut and burned like extant wood. Preservation of this type occurs when woody trees are rapidly buried in hot or dry environments, maintaining the cells and tissue systems that were there when the plant was alive. Some sections of mummified conifer wood from the Miocene–late Pliocene at high latitudes in Alaska and Canada contain fungal hyphae (Wheeler and Arnette, 1994). An interesting study by Jurgens et al. (2009) reported modern biological degradation in the form of soft rot decay of mummified fossil woods of several dicots and conifers from the Paleocene–Eocene of the Canadian Arctic. This finding underscores the exceptional preservation under the mummifying conditions in which modern fungi can use fossil carbon. Several sites in the Canadian Arctic (e.g. Ellesmere Island and Axel

Heiberg Island) have yielded mummified wood which should be examined for fungi.

Geochemical Evidence

Moving forward, an area that will be increasingly important in paleomycology is geochemistry. There are numerous reports in the paleontological literature that have documented the presence of various biopolymers and unusual chemical signatures that have been used to support or refute the systematic affinities of various fossils. It is also well known, however, that when an organism enters the geological environment, multiple changes in the physical, chemical, and isotopic composition of the original biomolecules may begin, in addition to the potential of the original biochemicals being contaminated with exogenous organic matter (Fogel and Tuross, 1999). Biomolecules have not received a great amount of attention as a means of identifying fossil fungal activity to date; however, there is significant potential for using such biomarkers. One of these biomarkers is perylene, a polycyclic aromatic hydrocarbon that has been reported in various organisms and diverse ecosystems (Grice et al., 2009). Although there is no universal agreement, geochemical studies have been used to suggest that perylene has its origin with fungi, especially some wood decaying types (Jiang et al., 2000). Carbon isotopic studies of conifer wood samples from the Jurassic of Poland, as well as xylites, suggest the perylene values are consistent with an origin from wood-degrading fungi (Marynowski et al., 2013). One of the continuing problems with using geochemical markers to identify fungal activity in the fossil record is the degree of weathering and diagenesis of the specimens within the geological environment (Marynowski et al., 2011). As these parameters are better understood and new protocols developed, it is quite probable that a combination of geochemical, petrographical, and mineralogical studies will significantly increase our understanding of fungal evolution in time and space.

ROCK SURFACES

Fungi are agents of geochemical change and play an important role in mineral and rock habitats. The examination of bare rock surfaces has provided a wealth of information about microbial endoliths and interactions such as mineral weathering, but fungi from these sources in the fossil record have not been examined systematically. There may be some value in examining this potential source of information about ancient fungi, especially from rocks that might contain evidence of early terrestrial plant communities but which lack evidence of plant macrofossils (e.g. Ordovician, Silurian). For example, Rosling et al. (2009b) have proposed developing mycelia growth models that can be used to describe weathering by ectomycorrhizal fungi in mineral soils. If fungal endoliths are found in rock, would it be possible to develop similar models that might be used to extrapolate fungal growth at certain points in geologic time and to use this information to suggest affinities of the fungus?

NODULES

There are various types of concentric structures in modern soils that are termed nodules. They are usually characterized by shape, color, hardness, mineralogy, and elemental composition (e.g. Aide, 2005). These are sometimes termed concretions because they consist of concentric layers; many are produced by the activities of various microbes (Retallack, 2001). One type includes iron nodules from a modern marine terrace soil from Santa Cruz, California, that are up to 25 mm in diameter and consist of soil mineral grains cemented by iron oxides (Schulz et al., 2010); within the nodules are various fungal hyphae. Similar structures are known to be produced by microcolonial fungi from desert varnish ecosystems (Perry et al., 2007). It has been suggested that the iron nodules were precipitated by fungi as a method of sequestering primary mineral grains for nutrient extraction, perhaps influenced by seasonal wet and dry cycles (Schulz et al., 2010). Nodules of various morphologies and structures have also been reported from several sites in the Permian of Spain and Morocco (Aassoumi et al., 1992). They are interpreted as pedological concretions formed in hydromorphic soils that formed in tropical climatic conditions, perhaps associated with roots of the fossil gymnosperm *Cordaites*. Such structures are probably common in the fossil record.

CARBON SPHERULES

Carbon spherules and elongate particles are a common feature in modern soils and some of these structures are associated with wildfire events. Specimens may be up to 2 mm in diameter and some show internal structure. Specimens from Pleistocene–Holocene sedimentary sequences in the California Channel Islands indicate that many of these structures are structurally similar to modern fungal sclerotia (Scott et al., 2010). This interpretation is based on experimental studies using modern sclerotia that were charred and then examined using multiple imaging systems (reflected light microscopy,

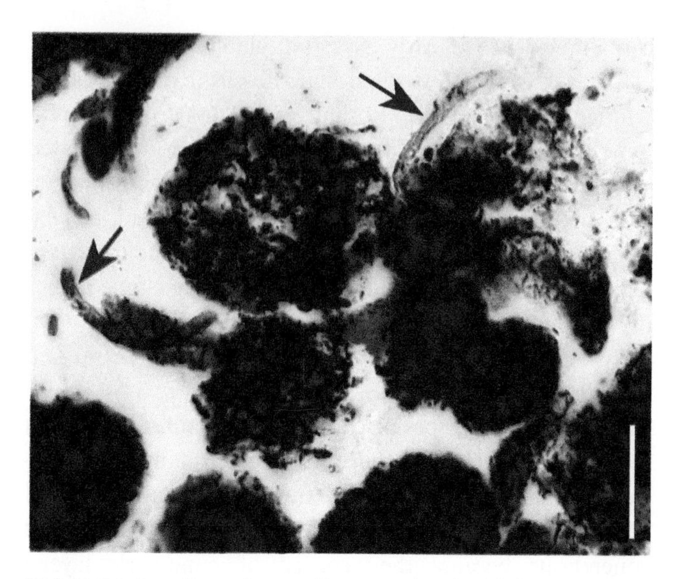

FIGURE 2.20 Several coprolites associated with fungal hyphae (*arrows*) in permineralized peat. Triassic, Antarctica. Bar = 25 μm. (Courtesy C.J. Harper.)

scanning and transmission electron microscopy, synchrotron radiation X-ray tomographic microscopy) and then compared to the fossil spherules.

UNUSUAL PRESERVATION TYPES CONTAINING FUNGAL REMAINS

Coprolites

Coprolites are fossilized feces or dung of animals and occur quite abundantly throughout the Phanerozoic. Fungal hyphae and spores are often well preserved in archaeological and fossil coprolites and may provide information on environmental conditions, animal diet, and sometimes even on plant–fungus interactions (Reinhard and Bryant, 1992). Some coprolites are permineralized or petrified (Figure 2.20) and in several instances have provided information about fungivory, for example in the Rhynie chert ecosystem (Habgood et al., 2004). Coprolites of herbivorous dinosaurs have been described that contain remains of phytoparasitic fungi and thus document the presence of this interaction in the fossil record (Kar et al., 2004a; Sharma et al., 2005). Eocene phosphatized coprolites from Kazakhstan contain various microorganisms involved in biological decomposition and transformation of feces (Zanin et al., 2002).

Fungi and Crystals

Another rather uncommon mode of preservation has been reported from the Rhenish Massif in Germany by Kretzschmar (1982) in the form of druses in ferruginous quartz that contain crystals preserving Oligocene to Miocene fungal mycelia encrusted with the mineral goethite. Similar types of preservation of microbial fabrics have also been reported from Scotland (Trewin and Knoll, 1999) and Belgium (Baele, 1999). Goethite-chalcedony fracture infills in Cenozoic volcanic rocks containing microbial filaments are also known from California (Hofmann and Farmer, 2000).

3

HOW OLD ARE THE FUNGI?

PHYLOGENETIC SYSTEMATICS 27
MOLECULAR CLOCKS 28
EARLY FOSSIL EVIDENCE 31
FUNGI AND THE FIRST LAND PLANTS 33
SYMBIOSIS: A CRITICAL COMPONENT OF LIFE 36

Fungi are no different from other lineages of organisms in that they are not represented in the rock record as a continuous and uninterrupted sequence of fossils. Nevertheless, as we learn more about them it becomes obvious that they are an ancient group that must have evolved relatively early in geologic time, perhaps more than one billion years ago (Parfrey et al., 2011). Despite the lack of fossil evidence for the earliest organisms that possessed a heterotrophic lifestyle, evolutionary biologists (including those practicing systematics, molecular ecology, genetics, and even paleobiology) have used other approaches and techniques to date certain events in geologic time and thus infer not only when organisms like fungi first appeared but also patterns in their evolutionary history (Figure 3.1). In addition to examining fossils at various points in geologic time, there are two principal methods that have become standard techniques in the study of evolutionary biology. One of these is the series of principles encompassed in phylogenetic systematics; the second uses this approach to date the divergence of organisms, and thus when groups first appeared on the Earth.

PHYLOGENETIC SYSTEMATICS

This area of biology uses living and fossil organisms to identify and understand the evolutionary relationships and patterns among organisms. Organisms are classified and then, using a well-defined set of cladistic principles, are grouped together into clades based on their shared derived characters (synapomorphies). Each of the characters, whether they be morphological or molecular, provides information that can then be used to generate an estimated relationship in the form of a cladogram or tree: a graphic representation of the best hypothesis of the evolutionary history of a species using multiple forms of data about the species. This approach views members of a group as sharing a common evolutionary history. Historically, morphological and other features of the organisms were used in phylogenetic analyses; however, today genetic sequencing data and computational phylogenetics are a focal point in most phylogenetic studies.

Because all organisms contain DNA, RNA, and proteins, closely related organisms possess a high degree of similarity in these molecules, while organisms more distantly related show a pattern of dissimilarity. These evolutionary changes result from mutations that may occur by chance or as a result of natural selection. There is a general belief that increased resolution of the relationships between organisms can be achieved by increasing the number of different organisms used in an analysis rather than the number of characters used in developing the phylogeny. In spite of the widespread usage of molecular tools in determining evolutionary relationships among groups of organisms, there are problems that continue to make some results questionable

T.N. Taylor, M. Krings & E.L. Taylor: Fossil Fungi.

DOI: http://dx.doi.org/10.1016/B978-0-12-387731-4.00003-7

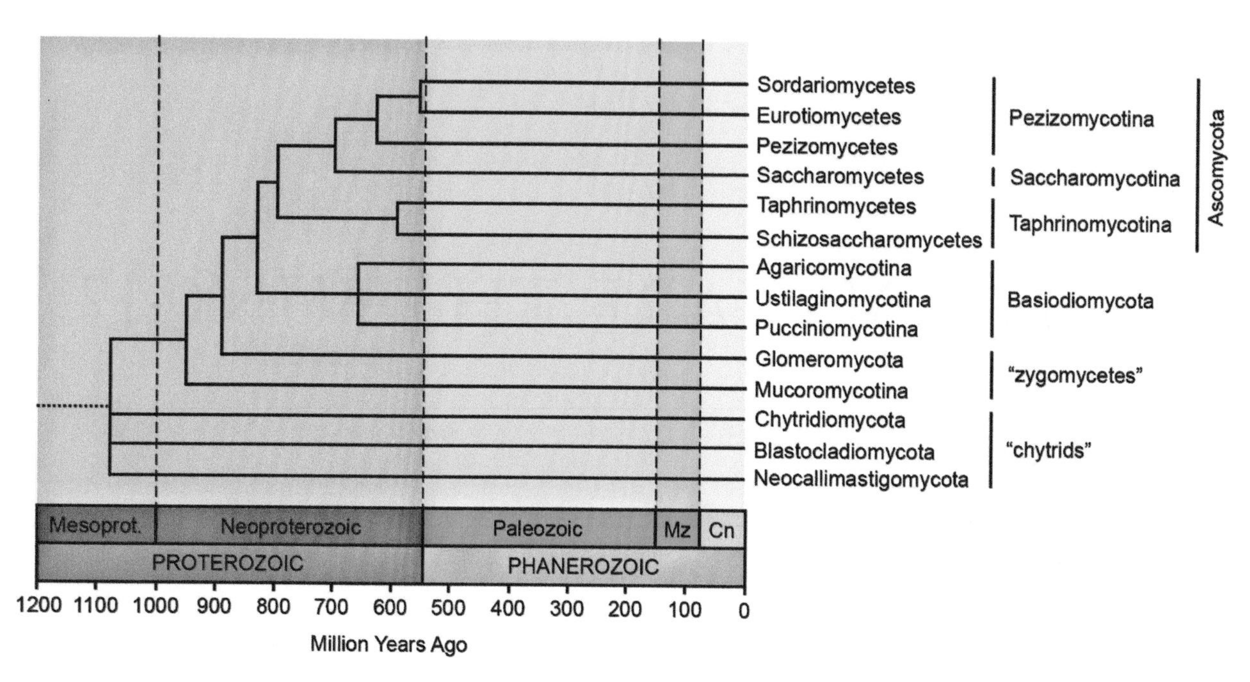

FIGURE 3.1 Timetree of fungi assembled for this volume by J.E. Blair. Divergence times are based on the following references: Berbee and Taylor, 2001; Heckman et al., 2001; Douzery et al., 2004; Hedges et al., 2004; Padovan et al., 2005; Berbee and Taylor, 2011; Parfrey et al., 2011; Floudas et al., 2012. Cn = Cenozoic, Mz = Mesozoic. (Unpublished data courtesy J.E. Blair.)

(e.g. Roger and Hug, 2006). One of these is the extensive horizontal gene transfer among organisms, which results in a different evolutionary history for some genes within the same organism (Rice et al., 2014). Nevertheless, the incorporation of genome-scale analyses will continue to increase the resolution of many phylogenetic approaches, including those of fungi (e.g. Leonard and Richards, 2012). One major step forward in understanding the relationships among fungi was the development of morphological and multilocus molecular datasets for the entire Kingdom Fungi (e.g. Blackwell et al., 2006; Hibbett et al., 2007).

MOLECULAR CLOCKS

Once a phylogeny for a clade has been assembled there are several approaches that can be used to infer the age of some of the organisms that are not necessarily represented as fossils. One of these is the molecular clock hypothesis, or evolutionary clock, which uses the constant accumulation of substitutions ("ticking of the clock") of DNA and protein sequence data among different organisms (e.g. Zuckerkandl and Pauling, 1962). The fundamental concept is that the genetic differences between any two species are proportional to the time since the species shared a common ancestor or when a new species diverged. Although molecular clocks have been used to assign absolute dates in geologic time, today it is understood that mutation rates are uneven both within and across genes, and among different lineages of organisms. This fact makes the comparisons among different organisms inaccurate and increases the importance of absolute dates based on fossils and the character states they demonstrate (e.g. Bromham and Penny, 2003; Welch and Bromham, 2005; Weir and Schluter, 2008; Andújar et al., 2012). Fossils are used, along with radiometric dating of the rocks containing the fossils, to calibrate the clock to absolute dates in geologic time. The concept of a constant rate molecular clock has been modified so that today most evolutionary biologists rely on a number of models that are more accurately termed "relaxed clocks" (e.g. Lepage et al., 2007). Most of these relaxed clocks are implemented in Bayesian statistical tools for inferring phylogenies, although penalized maximum likelihood implementations have also been used. Relaxed molecular clocks are formulated on the principle that while the rate is variable, the variation can be predictable. Either the rate varies around an average value or the evolutionary rate changes slowly over time because the variation is tied to other biological characteristics that undergo evolution (e.g. metabolism, certain aspects of physiology; Ho, 2008).

FIGURE 3.2 Mary L. Berbee.

FIGURE 3.3 John W. Taylor.

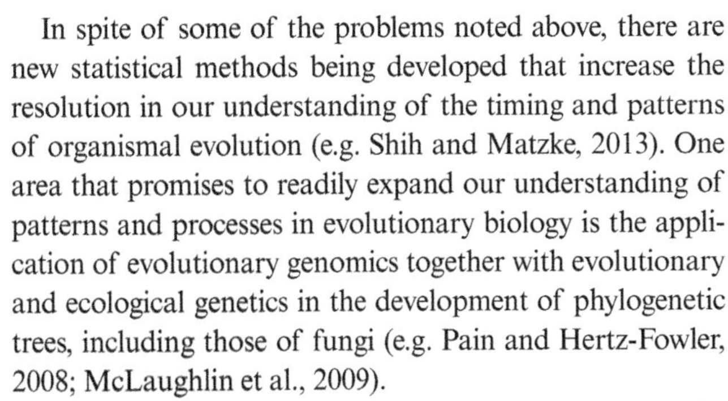

In spite of some of the problems noted above, there are new statistical methods being developed that increase the resolution in our understanding of the timing and patterns of organismal evolution (e.g. Shih and Matzke, 2013). One area that promises to readily expand our understanding of patterns and processes in evolutionary biology is the application of evolutionary genomics together with evolutionary and ecological genetics in the development of phylogenetic trees, including those of fungi (e.g. Pain and Hertz-Fowler, 2008; McLaughlin et al., 2009).

A critical step in correlating molecular phylogenies with geologic time is the use of fossils that provide minimum ages for divergences and represent the earliest known member of a lineage (Berbee and Taylor, 2010) (Figures 3.2, 3.3). There are, however, inherent problems with using fossils to calibrate molecular divergence-time estimates to those of an absolute age (e.g. Near and Sanderson, 2004). These include the incompleteness of the fossil record (Taylor et al., 2009), inaccurate taxonomic identification and the inclusion of these taxa in phylogenetic trees (e.g. Benton and Ayala, 2003), minimal age estimates applied to crown groups (e.g. Magallón and Sanderson, 2001), and inaccurate or poorly constrained dates for rock units containing the fossils (e.g. Conroy and van Tuinen, 2003).

The intense interest in developing new methods of relaxing the molecular clock has made it feasible to estimate the ages of internal nodes on a tree from information about the true evolutionary distance between taxa. The evolutionary distance is the number of evolutionary events (mutations or substitutions) that separates a pair of taxa. Unfortunately, these evolutionary distances are not directly observed; they must be estimated from data at the tips of a tree. The sophisticated methods currently available for relaxing the molecular clock do not solve the problems associated with estimating these evolutionary distances. In addition, molecular clocks have inherent errors that include problems with the assumed phylogenetic trees, which themselves may be inaccurate. These include: long, subtending phylogenetic branches that may be followed by multiple short ones; the actual models used in nucleotide or amino acid substitution; and how gene data are treated. These problems, and others, all contribute to difficulties in interpreting the distribution of geologic ages and absolute rates (e.g. Roger and Hug, 2006; Magallón et al., 2013).

A relaxed clock method can estimate relative ages of internal nodes on a tree but cannot, by itself, put the age estimates on an absolute timescale. Calibrating the tree using fossils is the most common method of arriving at absolute age estimates. Even if one ascertains the age of a fossil with very little error, the fossil must be attached to a phylogeny in order for it to provide calibration information for the rest of the tree. This process can be error prone, and it frequently entails identifying the taxonomic group for the fossil and using the fossil as a minimum age for the node that represents the most recent common ancestor of the clade to which the fossil is assigned (see Heath et al., 2013 for a new method that does not attach fossils to a specific internal node). Inferring a timescale for a tree when one has minimum age calibrations but no maximum age constraints usually results in estimates of node ages that are very broad. Typically, very old ages cannot be rejected by such analyses. To deal with some of these issues, and to increase the potential accuracy of incorporating fossils in age estimates, a number of technique have been employed to increase the reliability of the fossils (e.g. Douzery et al., 2004; Near et al., 2005; Warnock et al., 2012). Also necessary will be the use of increasingly sophisticated techniques such as bioinformatics to develop suitable well-vetted databases where fossil calibrations can be made available to molecular evolutionists (e.g. Brochu et al., 2004; Ksepka et al., 2011). This will require a coordinated effort by the paleobiological community who study fossil fungi from multiple perspectives, mycologists who understand the structure and development of fungal features, and geochronologists and stratigraphers who date the rocks, to increasingly develop and share data that will result in a greater degree of interdisciplinary research related to the divergence of fungal structures, morphologies, and lineages (Parham et al., 2012).

A number of molecular clocks using fossil fungi as calibration points have been considered (e.g. Redecker, 2002). An early example used 18S ribosomal RNA gene sequence data of 37 fungal species to calibrate a timescale that used glomeromycotan spores and other characters (e.g. clamp connections, septate hyphae) which appear in fossils (Berbee and Taylor, 1993). After normalizing results, these authors suggested a 1% nucleotide substitution rate per 10 myr, suggesting that the terrestrial fungi diverged from the Chytridiomycota approximately 550 Ma. This study was followed by others that used different techniques to deal with rate variation resulting in differing ages (e.g. Berbee and Taylor, 2006). In another approach that did not utilize fossils, Heckman et al. (2001) calibrated a phylogeny using 1600 Ma, the date often used to define the divergence of fungi, animals, and plants. The results obtained using this method placed the fungi as being much older, with the origin of the Glomeromycota between 1400 and 1200 Ma, and the separation of the Ascomycota and Basidiomycota around 1200 Ma. Like many concepts in paleobiology, identifying precisely when a group of organisms initially separated from its ancestor is difficult, and this is certainly the case when discussing fungi. In other sections of this volume, however, we will discuss some of the fossils that might be interpreted as early fungi, including some that extend back into the Proterozoic (2500–541 Ma).

As more fossils are discovered and new analytical techniques applied, the divergence dates for certain groups of fungi, as well as character states that define the groups, will obviously change (e.g. Berbee and Taylor, 2010). What is obvious in the case of fungi is that the major groups will be demonstrated to be far older than previously thought. Whether fossils are discovered that support these hypotheses, especially early divergences, is probably doubtful; however, the use of fossils will continue to play a critical role in understanding the evolution of fungi and the interactions they maintained in the paleoecosystems in which they functioned (e.g. Lücking et al., 2009). The following examples highlight the value of some these combined techniques in addressing important questions in the evolutionary biology of fungi using fossils and molecular clock assumptions.

The first of these is *Paleopyrenomycites devonicus,* a perithecial fungus preserved in the ~410 Ma Rhynie chert (Taylor et al., 1999, 2005a). This fossil is discussed extensively in Chapter 8 on Ascomycota and thus will not be described in detail here. While there is little doubt that the affinities of this fossil are within the Ascomycota, exactly where within the phylum the fossil belongs has been somewhat controversial (e.g. Berbee and Taylor, 2006). For example, the fossil has been used to calibrate the origin and diversification of the Pyrenomycetes (e.g. Redecker et al., 2000a; Heckman et al., 2001; Taylor and Berbee, 2006) and placed systematically in several classes. However, *P. devonicus* has also been suggested as a member of the Taphrinomycetes, a small group of plant parasites with both a yeast and filamentous state (e.g. Eriksson, O.E., 2005; Taylor and Berbee, 2006; Berbee and Taylor, 2007). Previously developed fungal phylogenies were rescaled using *P. devonicus* at three different calibration points. The results of this approach suggested: (1) the origin of the Ascomycota somewhere between 650–550 Ma; (2) the divergence of Chytridiomycota and zygomycetous fungi around 630 Ma, and (3) the divergence of the Glomeromycota around 600 Ma (Lücking et al., 2009). This latter study also suggested that *P. devonicus* is perhaps more accurately placed within the Pezizomycetes rather than the Sordariomycetes.

A second example of the use of molecular clock hypotheses in the study of fossil fungi focuses on the evolution of ectomycorrhizal (ECM) symbioses (Hibbett and Matheny, 2009). This symbiosis involves a number of extant woody plants in both temperate and tropical forests (e.g. pines and certain angiosperms) that are associated with a polyphyletic assemblage of mostly Agaricomycetes. The study utilized both the fossil record of ECM and relaxed molecular clock analyses of a large dataset based on protein sequences and large and small rRNA sequences of extant plants and fungi. While there remain many aspects to the evolution of ectomycorrhizae still to be determined, this study did indicate that there have been multiple ectomycorrhizal origins from autonomous saprotrophic fungi involving both angiosperms and gymnosperms, and that some of these associations may be quite old. In the genus *Amanita*, there appears to be a single origin of the ECM symbiosis (Wolfe et al., 2012). This conclusion is supported by the loss of critical cellulase genes, which suggests that these fungi can no longer decompose complex organic matter as saprotrophs and thus must rely on carbon from the host plant via the ECM mutualism.

The third example uses rock-inhabiting fungi that possess multiple adaptations, which allow them to live in extreme conditions (e.g. high UV radiation, high temperatures, low temperatures). Gueidan et al. (2011) examined two lineages of ascomycetous fungi because they are typically associated with early-diverging groups (Chaetothyriales and Dothideomycetes) and suggested that the ancestors were rock inhabitants. Based on relaxed molecular clock assumptions and some fossil specimens, it was suggested that the rock-inhabiting forms in the Dothideomyceta evolved during the Devonian, or perhaps slightly later in the Carboniferous, while the Chaetothyriales are hypothesized to have evolved somewhere in the Middle Triassic. While these hypotheses utilize phylogenetics and molecular clock assumptions – and frame these techniques against the backdrop of other biological activities, such as recovery following the Permian-Triassic mass extinction event and changing climatic parameters (Gueidan et al., 2011) – increased gene and taxon sampling together with the use of additional fossils will be necessary to substantiate or refute them. The examples noted above demonstrate the importance of using a combination of data sets and multiple approaches to investigate the processes and timing of major events in fungal biology and evolution.

EARLY FOSSIL EVIDENCE

Just when fungi first colonized the land surface remains an interesting enigma, and how this arrival corresponds with the evolution of land plants is an even more perplexing problem. While the land surface must have been available for colonization far earlier than there is adequate fossil evidence to prove, it has been suggested that the delay was related to oxygen levels (Berner et al., 2007) and the lack of an ozone shield to protect terrestrial life from UV radiation (Raven, 2000; Rozema et al., 2002). The initial terrestrial surface was no doubt rich in cyanobacteria and various heterotrophic bacteria, and, later, the fungi that preceded land plants into terrestrial ecosystems (e.g. Schopf and Klein, 1992; Stanevich et al., 2013; Wacey et al., 2013). This is based on limited fossil evidence and divergence times suggesting that fungi extend well back into the Proterozoic (Padovan et al., 2005; Hedges et al., 2006). Historically, there have been many reports of fungi and fungus-like organisms from the Precambrian that have subsequently been suggested to be non-fungal (e.g. Maithy, 1973). One of these is *Eomycetopsis*, which has been reported from numerous Neoproterozoic sites (e.g. Ragozina, 1993) including the Bitter Springs Formation (Love's Creek Member) in Australia, dated at approximately 830–800 Ma (Neoproterozoic) (Schopf, 1968; ~811 Ma according to Swanson-Hysell et al., 2012). Although *Eomycetopsis* has the appearance of fungal hyphae (2.5 μm in diameter), it was later reinterpreted as the tubular sheath of a filamentous cyanobacterium (Mendelson and Schopf, 1992a, b). Other examples of Precambrian microfossils that have been suggested as fungal in origin include *Caudosphaera* (Timoféev, 1970), *Huroniospora* (Darby, 1974), *Aimia, Eosaccharomyces, Majasphaeridium, Mucorites, Mycosphaeroides* (Hermann, 1979), *Shuiyousphaeridium* (Yan and Zhu, 1992), and a number of unnamed forms (e.g. Hofmann and Grotzinger, 1985; Allison, 1988; Weiss and Petrov, 1994). There are also a few fossils from the Precambrian that suggest a cyanobacterial/?fungus association in the form of a net-like structure that may represent some type of lichen (Yuan et al., 2005). By the Early Devonian the lichen symbiosis was well established (Taylor et al., 1997), indicating that fungi are even older. In addition, throughout the Proterozoic and into the Cambrian there are numerous reports of well-preserved microfossils that may represent some type of heterotrophic organization (e.g. Zang and Walter, 1992; Knoll et al., 2006; Porter, 2006). One of these is *Tappania*, an interesting fossil with branching filamentous processes that have been interpreted as possibly fungal-like (see Chapter 6, p. 78). Moreover, various types of filaments have also been reported from Ediacaran rocks of the Drook Formation in Newfoundland (Liu et al., 2012). They are evenly dispersed along the bedding plane as impressions and some of these show evidence of Y-branching. They range from 6–130 mm long and 0.5–1.0 mm in diameter. While many of the Ediacaran and younger microfossils represented as filaments are most probably cyanobacteria in various states of preservation, certain types of green algae, actinobacteria, and fungi are also potential candidates for these interesting fossils (e.g. Xiao et al., 2002; Sappenfield et al., 2011). It is unlikely that there are not multiple lineages of organisms included in the Ediacaran fossil assemblages.

It is interesting to note that even some of the much larger multicellular organisms of the Ediacaran biota are suggested to represent some type of fungus (Peterson et al., 2003). Specimens of *Aspidella*, *Charnia*, and *Charniodiscus* from the Avalon Peninsula of Newfoundland are interpreted as potential asexually reproducing saprotrophs that lived on sulfate-reducing bacteria (Peterson et al., 2003). These authors point out that just because it has been suggested that some of the Ediacaran taxa may represent fungi, this does not imply that the Neoproterozoic fossils represent a crown-group member of the Kingdom Fungi, but perhaps stem-group examples that are morphologically unrecognizable when compared to modern fungi. Although these Ediacaran fossils do not show obvious fungal features, the hypothesis that Peterson et al. (2003) put forward is especially important for several reasons. One is that rocks of this age show some evidence of an organism with a lichen-like organization (Yuan et al., 2005; see Chapter 10, p. 206), thus implying that some heterotrophic organism predated the lichen organization (e.g. Javaux, 2007). Second, and perhaps more important, is the reality that the earliest examples of fungi may have had little morphological resemblance to members of any modern groups. This latter concept sometimes gets lost in paleobiology, where morphological correspondence, especially with extant organisms, often by default represents the guiding principle.

Horodyskia is a problematic Mesoproterozoic fossil described as having a string-of-beads morphology (Fedonkin and Yochelson, 2002). The "beads" are circular to more elongate, sometimes of multiple shapes on the same string, and typically at uniform intervals. There have been a variety of interpretations as to what these structures represent (e.g. pseudofossil, dubiofossil, segmented sponge, prokaryotic colony, brown alga, foraminifera, slime mold sporangium, puffball, colonial eukaryote, extinct lineage). New specimens from the ~1.48 Ga Mesoproterozoic of Glacier National Park, Montana, USA, have been described as representing expanded cells (bladders) similar to those of modern *Geosiphon* (Retallack et al., 2013a), which is a glomeromycete and fungal symbiont of *Nostoc* (cyanobacteria). Grey et al., (2010), after examining *Horodyskia* specimens from 49 localities in Western Australia, concluded that their affinities are obscure, although their mode of preservation resembled other Ediacaran animal fossils. We are uncertain which of these ideas as to the affinities of *Horodyskia* has merit, if either, but as is the case with most fossil structures based on morphology alone, additional preservation modes will be needed to sharpen the focus of this unusual Precambrian structure.

Palaeorhiphidium is a Cambrian fossil that is described as fungiform and consists of anastomosing filaments, some with globose structures (Kolosov, 2013). It is suggested that perhaps this fossil is an early example of a Peronosporomycete; however, the identification of biflagellate zoospores is unconvincing. Stromatolites, and numerous other Precambrian organisms whose systematic affinities remain unknown, have been suggested as possible evidence of fungi and lichens. For example, fungal hyphae have been identified in stromatolites at various points in geologic time (e.g. Klappa, 1979a; Kretzschmar, 1982) and in rocks containing high concentrations of phosphate and calcium minerals (e.g. Dahanayake and Krumbein, 1985; Dahanayake et al., 1985). The report of Ediacaran paleosols containing various fossils interpreted as related to fungi in what is interpreted as growth position, while controversial (e.g. Knauth, 2013; Xiao, 2013), may lead to new interpretations of what these organisms actually represent and the environment in which they lived (Peterson et al., 2003; Retallack, 2013a, b).

Early terrestrial environments consisted of various crusts, biofilms, and microbial mats made up of cyanobacteria in addition to assorted microbial communities (e.g. Préat et al., 2000; Beraldi-Campesi, 2013). The diversity of this biomass was no doubt critical for the evolution of increasing complexity in fungal life histories. These early terrestrial ecosystems are also postulated to have had air-breathing arthropods that were present before the vascular plants, initially as carnivores and later as detritivores (e.g. Shear, 1991; Shear and Selden, 2001; Labandeira, 2005). In the early stages of these oligotrophic terrestrial ecosystems, the rock surface would have represented a relatively inhospitable environment for many forms of life. It is hypothesized that at least some of the initial fungi in these bare rock landscapes would have received carbon in the form of dust particles and as a component of biogenic sedimentary rocks that would have formed as a result of rock weathering and the release of mineral nutrients. Today fungi are components of all rock types where they interact with bacteria within biofilms and result in the formation of films, crusts, and varnishes on rock and mineral surfaces

(Gorbushina, 2007). There are also three-dimensional networks of filamentous and unicellular microorganisms, termed biodictyons, which are often embedded in sediment, soil, and rock (Gorbushina and Krumbein, 2000; Krumbein, 2003). Encrusted fungal hyphae have been reported to function as nucleation sites for various minerals (Latz and Kremer, 1996). Geochemical evidence in the form of biosignatures suggests that by mid-Cambrian time mycorrhizal fungi were also well established (Horodyskyj et al., 2012).

Fungi and lichens growing on bare rocks accelerate weathering (e.g. Stretch and Viles, 2002), often in association with endolithic fungal communities (Hoffland et al., 2004). Biophysical and biochemical weathering, together with the biogenic activities of microbial populations, also modify rock surfaces and begin the process of soil formation (e.g. Klappa, 1979b). Modern lichens produce organic acids that contribute to rock dissolution (Chen et al., 2000) and this must have helped to form early substrates for plant growth. Nevertheless, some lichen thalli can protect the rock surface from weathering for a period of time before chemical and physical processes of weathering begin (McIlroy de la Rosa et al., 2012, 2013). Lichens or lichen-like associations are also suggested to have played a major role in mineral recycling in the early Paleozoic (Shear, 1991; Gray and Shear, 1992).

FUNGI AND THE FIRST LAND PLANTS

The transition to life on land was a major evolutionary event in the history of photosynthetic organisms and involved various physiological modifications, including regulation of photosynthesis and photorespiration, as well as significant morphological and structural changes (Taylor et al., 2009). Most agree that the successful colonization of the terrestrial realm involved a symbiotic association with certain fungi (e.g. Pirozynski and Malloch, 1975). Whether this fungus–photoautotroph mutualism existed in aquatic ecosystems prior to the transition to land remains unknown.

Equally, no one knows precisely when the first land plants appeared on Earth or what they may have looked like. There are multiple opinions and lines of reasoning that have been used to mark when this transition may have taken place, with evidence suggesting that it may have occurred in the Cambrian (e.g. Chabasse, 1998) or even earlier during the Precambrian, approximately 900 Ma ago (Heckman et al., 2001), but certainly by Ordovician time (Taylor et al., 2009). It has also been suggested that mycorrhizal diversity subsequently evolved in response to climate change (Bellgard and Williams, 2011). Many of these hypotheses are not based on fossil evidence but, rather, are inferred from molecular divergence times among extant groups of organisms. The presence of various cryptospores in the early Paleozoic (Figure 3.4), plus details about the ultrastructure of the walls of these spores, suggests that by the Cambrian there were various photoautotrophic organisms, some believed to have been land inhabiting (e.g. Strother, 2000; Wellman and Gray, 2000; Strother et al., 2004; Yin et al., 2013). Although fossil spores predate evidence of the organisms that produced them, fragments of sporangia and spores from the Ordovician substantiate the existence of land plants at this time, some suggesting a liverwort grade of evolution (Wellman et al., 2003; Taylor and Strother, 2008, 2009; Rubinstein et al., 2010). Until spores are documented in sporangia that are attached to the organisms that produced them, there will be a continuing discussion as to whether these spores were produced by embryophytes or were mitotic and/or meiotic products produced by other organisms. The early Paleozoic is also the time when there is evidence of organisms, some constructed of variously aligned tubes and covered by a cuticle-like material (nematophytes), that existed in the desiccating environment of the early terrestrial ecosystem (e.g. Edwards and Axe, 2012). Regardless of what have

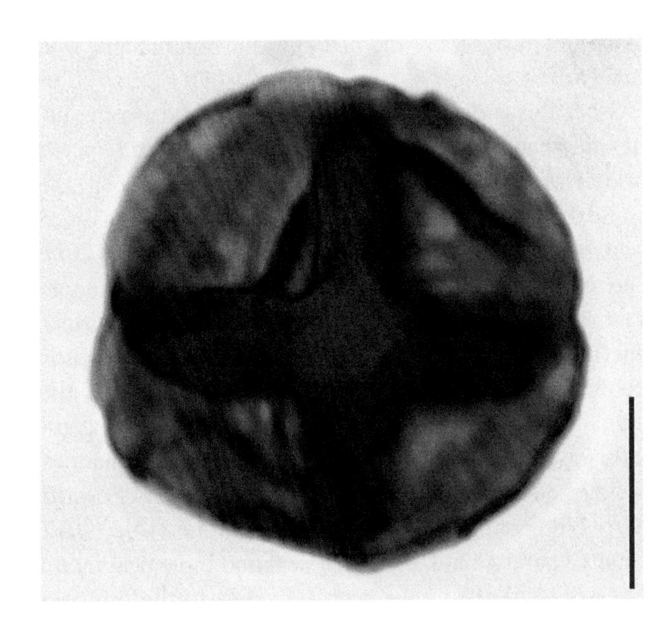

FIGURE 3.4 Quasiplanar cryptospore tetrad. Cambrian, USA. Bar = 5 μm. (Courtesy P. K. Strother.)

been interpreted as the earliest land plants, it is certain that heterotrophs initially played a major role in preparing and shaping the new niches that were available in the terrestrial environment (e.g. Selosse and Le Tacon, 1998; Hodkinson et al., 2002). The reader is referred to *Fungal Biology in the Origin and Emergence of Life* (Moore, D., 2013), a volume that includes a discussion of the possible role of fungi in shaping the early terrestrial ecosystem and some of the ideas that build on the hypotheses of Pirozynski and Malloch (1975).

An exciting avenue of research that directly relates to the timing of the terrestrialization of the Earth is the exhaustive analysis of Precambrian deposits and the fossil evidence they contain (e.g. Strother et al., 2011). As a result of such detailed studies it has been possible to plot not only the trajectory of eukaryotic diversity through time but also the first appearance of a number of biological innovations, such as multicellularity, cell and tissue differentiation, photosynthesis, biomineralization, the presence of certain biochemical products, sex, and embryogenesis, as well as heterotrophy (Javaux, 2007). These studies, which have focused on various cryptospore types, have also been used to hypothesize the types of habitat in which these organisms might have lived, for example freshwater and subaerially exposed (Strother et al., 2011). These benchmark innovations in eukaryote evolution will no doubt continue to change as new fossils are discovered and interpreted, and as techniques become more sophisticated and provide better resolution at multiple levels. Nevertheless, they represent a fundamental framework that is critical in deciphering which organisms were the first land inhabitants and when they evolved.

Other forms of evidence being used to address the timing of when plants became established on land include models of early landscapes that are used to see if they would have been able to support land plants based on available and consistent moisture (e.g. Morris et al., 2011). The role of vegetation and mycorrhizal fungi in enhancing weathering is also being considered, as is the question of whether information of this type can be used as a proxy record for analyzing the activities of fungi in deep time (e.g. Taylor, L.L., et al., 2012). Other approaches, together with fossils, will include developmental, biogeochemical, phylogenetic, and molecular data (e.g. Raven and Edwards, 2001; Niklas and Kutschera, 2010; Ligrone et al., 2012). Finally, there are numerous physical and biological factors that would need to have been overcome in the transition to land, including protection from the damaging effects of UV radiation, existence in a desiccating environment, and the ability to scavenge nutrients from the substrate (e.g. Taylor et al., 2009).

FIGURE 3.5 Several *Cooksonia* specimens on fractured surface of rock. Devonian, Brazil. Bar = 5 mm. (Courtesy P. Gerrienne.)

While there are earlier fossils of embryophytes, many point to the occurrence of the simple cooksonioids (Figure 3.5) from the late Silurian–Early Devonian, and perhaps even earlier in the mid-Silurian, as evidence for the existence of land plants containing vascular tissue (Gensel, 2008). These fossils are preserved as compressions, but detailed studies have provided information about some of their cell types, tissue systems, and even physiology (Edwards et al., 1992; Boyce, 2008). Simple Bauplans like that seen in the cooksonioids and the slightly younger land plant *Aglaophyton major* (Figure 3.6) from the Lower Devonian Rhynie chert (i.e. a system of rootless and leafless axes with terminal sporangia) have been used as prototypes and the starting point in discussions of land plant evolution, including the association of early plant life on land with fungi, especially arbuscular-mycorrhizal (AM) fungi. Kidston and Lang (1921b) were perhaps the first to suggest some biotrophic relationship between *A. major* and other Rhynie chert plants and some of the fungi they identified in the chert blocks. Later, Pirozynski and Malloch (1975) hypothesized that the colonization of land was the direct result of a symbiosis between a green alga and perhaps some aquatic fungus and that, as a result of this association, plants were able to successfully transition and ultimately exploit the terrestrial realm. The evolution of this relationship is further supported by various types of evidence, including its presence in all lineages of land

FIGURE 3.6 Reconstruction of *Aglaophyton major* from the Rhynie chert (Lower Devonian) showing prostrate axes attached to substrate at just a few sites. (Courtesy N.H. Trewin.)

plants, including bryophytes (e.g. Pressel et al., 2010). The hypothesis has also been supported by molecular studies that have demonstrated the presence of mycorrhizal genes in the common ancestor of land plants (e.g. Wang et al., 2010) and the demonstration of the presence of a diverse community of glomeromycotan fungi in the Rhynie ecosystem (Remy et al., 1994a; Dotzler et al., 2006, 2009). Since the initial concept, there have been other ideas offered regarding how this initial symbiosis may have evolved, including the transfer of fungal genes early in the evolution of embryophytes (e.g. Pirozynski, 1981; Atsatt, 1988, 1991; Jorgensen, 1993).

The extant green alga *Coleochaete* is considered by many as an excellent model for the type of green alga that adapted to a terrestrial environment (e.g. Graham, L.E., 1996; Graham et al., 2004). Together with the evolution of stomata, cuticle, conducting tissues, and matrotrophy, the plant–fungus symbiosis represents a major evolutionary innovation responsible for the exploitation of the surface of the Earth by land plants. The reader is referred to the point-by-point discussion of Graham et al. (2012), which combines geological, paleontological, and experimental evidence of the probable sequence of steps involved in the transition to a terrestrial habitat involving an organism like *Coleochaete*.

Although there are no fossil specimens showing fungus–plant associations earlier than those in the Rhynie chert, there is some indirect evidence that has a bearing on the initial relationship of the AM symbiosis and early land plants. For example, it has been suggested that late Silurian and early Devonian landscapes along the margins of Euramerica were unstable, and thus only plants at a rhyniophyte grade of evolution, and probably cryptospore-producing organisms, were able to successfully exist in these habitats along the margins of river channels (Edwards and Richardson, 2004). Multiple sources of evidence also indicate that the Rhynie chert plants lived in an extreme environment in which the water temperature was elevated. Moreover, the substrate contained high concentrations of heavy metals. The Rhynie chert Lagerstätte also contains other types of organisms, including algae, cyanobacteria, other types of fungi, and animals that apparently existed in this special ecosystem during the Early Devonian. What is perhaps unclear is whether the AM fungi in *Aglaophyton major* were necessary for the plants to survive in this habitat or simply represented a symbiosis that was a carryover from an association which evolved much earlier in some more stable ecosystem. It is interesting to speculate as to whether the high proportion of AM fungi to host tissue in *Aglaophyton* is a response to the harsh conditions of the Rhynie chert ecosystem, where increased fungal activity in scavenging nutrients would be an important adaptation. In this scenario the simple structure and morphology of *A. major* does not reflect a primitive condition in either the structure or morphology of the host but, rather, represents a steady state within this symbiosis for plants that were specialized physiologically (e.g. Edwards et al., 1998; Edwards, D., 2003). For some of the early land plants like *A. major* the only contact with the substrate occurred via rhizoids that developed in clusters along the horizontal axes. Does this intermittent contact with the substrate represent an adaptation that might have evolved to decrease infection from microorganisms, such as various fungi in the substrate, or an early stage in the land plant–fungus symbiosis?

As terrestrial plants and ecosystems evolved, they progressively grew in stature, biomass, nutrient demand, and rooting depth to extensively colonize uplifted continental land masses. During the middle Devonian, coevolution of the AM fungal symbiosis and rooting systems likely had a pivotal role in provisioning the expanding biomass, stature and biogeography of forested ecosystems with their rising demand for essential soil nutrients (Algeo and Scheckler, 1998; Field et al., 2012). The existence of diverse AM fungi in the Carboniferous clearly indicates that glomeromycotan fungi became adapted to other types of environments (e.g. peat swamps) and that symbiotic associations became established with other groups of plants.

A further complicating factor in tracing land plant symbiosis is the unknown role that climate may have played as more complex vegetation evolved that simply out-competed the rhyniophytes for niche space. The Early Devonian is also an interval in geologic time when atmospheric CO_2 concentrations were high (~1500 ppm), a condition hypothesized as increasing plant growth and reproduction and providing strong selective pressure for establishing mycorrhizal symbioses (Humphreys et al., 2010). In addition, it is possible that some of the fungal parasites in the Rhynie chert reflect a response to plants living in stressed conditions. Under this scenario the fungus–plant interactions, and perhaps even the fungi that are present, represent a community that has also evolved to live in this unusual ecosystem, that is, specialists rather than generalists. A modern analogue might include geothermal soils in Yellowstone National Park (USA) where molecular techniques demonstrate the presence of both generalist and unique phylotypes in the ecosystem (Appoloni et al., 2008). Moreover, it is hypothesized, based on modern fungus–plant interactions, that certain traits (e.g. asexual reproduction, synthesis of melanin-like pigments, flexible morphology) may function as adaptations to allow some fungi to exist in extreme habitats (Gostinčar et al., 2010). We wonder whether specific adaptations permitted certain glomeromycotan fungi to exist in stressful environments like the Rhynie ecosystem even before these fungi entered into symbioses with land plants.

SYMBIOSIS: A CRITICAL COMPONENT OF LIFE

Fungi are heterotrophic organisms and therefore require organic compounds as the source of energy for growth. Most fungi are saprotrophs that use dead organic matter as a carbon source, but others have established symbioses with other organisms, including microorganisms, plants, animals, and other fungi, from which they derive nutrients. Historically, symbiosis is sometimes defined as a permanent association between different species living together; others refer to the association as being prolonged or intimate. It was not until the nineteenth century (1867) that the Swiss botanist Simon Schwendener presented his "dual hypothesis," i.e. that lichens were formed by two organisms – an alga and a fungus (see Mägdefrau, 1992; Honegger, 2000). The term "symbiosis" is usually credited to the German surgeon, mycologist, and plant pathologist Heinrich Anton de Bary, although it was first coined by Albert Bernhard Frank (as "Symbiotismus") in 1877, a year before de Bary (who later credited Frank) used it publicly (Peacock, 2011). When symbiosis was initially defined (De Bary, 1879) as "the living together of unlike named organisms," it incorporated all nutritional modes of organismal interaction, that is, parasitism, mutualism, commensalism, and degrees or relationships that conveyed whether there was benefit or harm in the association. Today, however, the definition is rather diffuse; while some use the term exclusively for mutualistic associations, others believe that it can be used for any type of persistent biological interaction. De Bary (1879), however, recognized that there is a gradation from parasitism to mutualism throughout nature and realized that both extremes of the scale can be thought of as varieties of symbiosis (Sapp, 2004).

As a consequence, there are multiple views today of what constitutes a symbiotic relationship, and an equally diverse appreciation of how these interactions function in the biological world. Some interpret symbioses based on the ecological relationship of the partners, while others take a more mechanistic view that involves less subtle differences predicated on positive or negative effects during the association and precisely to what extent these benefits can be measured. These include whether an organism lives on, in, or is attached to another organism. From these seemingly simple organism–organism relationships, biologists are probing the highly complex associations between multiple types of organisms, and even how symbiosis may be involved in speciation (Brucker and Bordenstein, 2012). There can be no doubt, however, that symbiosis is a major force in evolution (e.g. Margulis, 1993) that relies on cooperation rather than competition, which is a fundamental in evolutionary theory. Some have argued that organisms function only in relation to other organisms and thus that symbiosis is of fundamental importance as a unifying theme in biology (Paracer and Ahmadjian, 2000). One example of this might be specific regions on flower petals where ultraviolet reflectance becomes a guide for bees to the source of nectar. In this symbiosis the bee obtains help in its search for food, while the plant increases the likelihood that pollination will occur. Nutrition is the end result in such associations, whereas in other symbioses shelter appears to be the primary benefit for one of the partners (e.g. galls). Another example that is perhaps more subtle involves certain rust and smut fungi that have evolved various strategies to attract pollinators to flowers they have infected so that the fungi have an additional method to disperse their spores (e.g. Roy, 1993, 1994). These fungal strategies include bright coloration, floral mimicry and scent (e.g. Raguso and Roy, 1998), as well as food reward (e.g. Kaiser, 2006), or the transfer of gametes via sugar solutions.

Another instance of a symbiotic association involving animals is fungus-growing ants in which a single fungal symbiont species is involved with multiple interactions among different ant lineages (Mikheyev et al., 2006). In other studies the ant species and fungus were found to be long-term partners that have remained stable over long periods of geologic time, a pattern that may drive the opportunity for coevolution (e.g. Mehdiabadi et al., 2012). A further example of this might be mycophagous animals that gain nutrition from the fungi they eat (especially fleshy fungi), while at the same time dispersing spores, which in turn may significantly alter and determine the structure of the plant community (e.g. Johnson, 1996). Some of the morphological features expressed in these associations certainly could be observed in fossils.

A number of fungal symbiotic associations involve more than two partners. An excellent example of this is the mutualism involving a fungal endophyte and a tropical grass that can both grow at elevated soil temperatures because the fungus is in turn infected by a virus (Márquez et al., 2007). Another example involves the entomopathogenic fungus *Metarhizium,* which is a common plant endophyte found in large numbers of ecosystems but is also known to be effective as a pathogen of numerous insects (e.g. Chouvenc et al., 2012). It has recently been demonstrated that in one form, *M. robertsii*, the endophytic capability and insect pathogenicity are coupled, because nitrogen derived from the insects is directly translocated to the plant hosts, thus providing an important source of this element in soils that are deficient in nitrogen (Behie et al., 2012). Mycophagous associations can also involve three partners, that is, a fungus, an animal feeding on the fungus, and a plant. For example, the ectomycorrhizal fungi *Laccaria trichodermophora* and *Suillus tomentosus* produce epigeous (above ground) fruiting bodies that are consumed by certain mice. Castillo-Guevara et al. (2011) and Pérez et al. (2012) have shown that passage through the digestive tract of one of the rodents increases the percentage of mycorrhizal formation by *S. tomentosus*. Moreover, the mice distribute the spores of the fungus to other areas where mycorrhizal associations can be established. Other studies have shown that animal mycophagy is important because it provides inoculum for diversifying the populations of mycorrhizal fungi for early successional plants in newly developing habitats (e.g. Cázares and Trappe, 1994; Schickmann et al., 2012). Another tripartite fungal symbiosis involves certain cool season grasses and clavicipitaceous fungi of the genus *Neotyphodium* (Pérez et al., 2013). In *Lolium perenne*, the presence of the endophyte *N. occultans* has been shown to significantly reduce infection by the pathogen *Claviceps purpurea* at population levels, suggesting that constitutive symbionts such as these systemic fungal endophytes may mediate the interaction between host grasses and pathogens. Similarly complex is the interaction between AM fungi and certain bacteria that are present in the hyphal cytoplasm, where they positively affect hyphal growth (Bonfante and Anca, 2009). The mycorrhizal fungi in turn are symbionts of plants. Moreover, some of the endophytic bacteria that have positive effects on the performance of the mycorrhizal fungi have antagonistic effects on plant pathogenic fungi (e.g. Sundram et al., 2011).

The ever-increasing application of molecular techniques indicates that maintenance of these symbioses, including the exchange of energy and nutrients between the partners, is a very complex process (e.g. Ghignone et al., 2012). While these techniques cannot be applied to fossils, at least not yet, the discipline of paleogenomics has the potential to be an important discipline (e.g. Abrouk et al., 2010). In both theory and practice, the sequencing of modern genomes can be used to interpret how genes functioned as seen in certain groups of fossils. The assumption is that the same genes that resulted in a particular character or feature today were in place in the fossil record of the group being examined, and/or historically there were complex gene networks that interacted in the fossils for which there are no modern analogues. While this approach sounds straightforward, there are multiple problems that are only beginning to be realized using this logic. Although fossil DNA has been reported from various plants, animals, and microorganisms, some dating back several hundred million years, many of these reports are not reliable (e.g. see discussion in Hebsgaard et al., 2005).

Mycorrhizal associations are perhaps the most common type of fungal symbiosis, with the most widespread being the arbuscular mycorrhizae in plant roots (Nicolson, 1967; Peterson et al., 2004). These associations are responsible for major changes and redistribution of inorganic nutrients (e.g. phosphorus) and carbon flow. Generally mycorrhizae are portrayed as a mutualism that benefits both the fungus, through a source of carbon, and the plant, which derives increased nutrient uptake, especially nitrogen and phosphorus, from the wide-ranging mycelium of the fungus, which can extend for 20,000 km in one cubic meter of soil (Pennisi, 2004). Others suggest that mycorrhizal function is better understood as a continuum between parasitism and mutualism and that the measure of this symbiosis is on plant fitness rather that nutrient gain (e.g. Johnson and Graham, 2013).

Precisely how the fungus is able to neutralize the host plant defenses in order to establish a symbiosis is still incompletely understood (e.g. Kloppholz et al., 2011), but the fact that AM fungi are the most common biotrophs of plant roots suggests that such mechanisms have been especially successful in these fungi, and for a long time. It is now known that some interactions between the fungus and plant roots require the presence of an effector protein in the fungus which provides a signaling mechanism that allows the mutualism to become established (Plett et al., 2011). This system also suggests a level of co-adaptation and codependence between the partners (e.g. Berch et al., 1984; Relman, 2008; Corradi and Bonfante, 2012). As might be expected, there has been considerable discussion about how the partners interact, whether resources are uniform in distribution, and how cooperation is maintained (e.g. Kiers et al., 2011). The life-history biology of the AM fungus begins with the germination of a spore and hyphae growing into certain cortical cells of the host to form specialized structures (arbuscules), and these are morphological structures that can be preserved in the fossil record and that indicate the presence of this type of mutualism in ancient plants. Not only do the structures produced by the fungus suggest the presence of mycorrhizal fungi but their position within the plant tissue may also provide evidence of this mutualism. Examples of fossil AM and ectomycorrhizae are described in other sections of this volume (see Chapters 7, 9). Mutualisms also occur at multiple levels of biological organization ranging from mitochondria in eukaryotic cells to various species, and even kingdom interactions, all increasing the complexity in the evolution of life. What drives the success of mutualism is and will continue to be a major field of research in evolutionary biology.

Another type of mycorrhiza, formed by an ascomycete and a few conifers (e.g. *Pinus* and *Larix*) and resembling ectomycorrhiza, is the ectendomycorrhizal association (Mikola, 1988). In this form a thin mantle that is embedded in a layer of mucilage and a Hartig net are formed in the cortical and epidermal cells, as well as intracellular hyphae (e.g. Yu et al., 2001; Peterson et al., 2004) (Figure 3.7). Because *Pinus* and *Larix* are known to be structurally preserved as fossils it will be interesting to see whether the ectendomycorrhizal pattern of development evolved early with these genera. Some extant plants such as the conifer *Wollemia nobilis* (Araucariaceae) have been described as containing both ectendomycorrhizal and AM fungi in the same plant (McGee et al., 1999). There has been no report from the fossil record of this type of mycorrhiza.

FIGURE 3.7 R. Larry Peterson.

There are other types of mycorrhizae that may be included under the terms endo- and ectomycorrhiza (Brundrett, 2004), some of which are quite specialized, including the ericoid, arbutoid, monotropoid, and orchid mycorrhizae (Peterson and Farquhar, 1994; Peterson et al., 2004; Imhof, 2009). These symbioses have been interpreted as being the result of various plant families that were evolving in new habitats and developed increasing nutrient efficiency. We are unaware of any of these having been reported in fossils, although indirect evidence has been used to suggest their existence in the Late Cretaceous (e.g. Cullings et al., 1996; Brundrett, 2002). The ericoid mycorrhiza appears to be confined to several angiosperm families in the Ericales (Read, 1983). A characteristic of the host roots in this mycorrhiza is that they are narrow and composed of a few epidermal, cortical, and conducting cells. Extraradical mycelia extend just a few millimeters from the surface of the root and the mantle is reduced to a few cells. The fungus colonizes the epidermal cells and then forms hyphal complexes in cells of the cortex. Some forms produce hyphae with capitate cystidia (e.g. Vohník et al., 2012). Arbutoid and monotropoid mycorrhizae also occur in the Ericales, but differ in structural features and in the fungal partners (e.g. Massicotte et al., 2008). Special basidiomycetous mycorrhizae are found in around 25,000 species of orchids (Rasmussen and Rasmussen, 2009), where they form intracellular hyphal coils termed pelotons that serve as the site of sugar and nutrient transfer. Moreover, since orchid seeds are often very small, lacking substantial endosperm- or seed-based nutrient reserves, they have an obligate requirement for mycorrhizal associations to provide nutritional support from germination to establishment (e.g. Arditti, 1967; Nurfadilah et al., 2013).

Finally under the category of symbiosis is the reverse situation in which the plant obtains carbon from the fungus. This system, termed mycoheterotrophy, is present in more than 800 species of plants, including many ancient lineages such as liverworts, lycophytes, ferns, gymnosperms, and angiosperms (Leake, 1994; Bidartondo, 2005; Merckx et al., 2013). These plants depend on mycorrhizal fungi but also obtain carbon from saprotrophic species (e.g. Hynson et al., 2013). Mycoheterotrophy is believed to have evolved from a mixotrophic ancestor, an organism that is both autotrophic and heterotrophic, rather than from autotrophy (Motomura et al., 2010). It is hypothesized that certain plants evolved mixotrophy to survive in low-light forest understory ecosystems (Selosse and Roy, 2009). This, is turn, is different from hemiparasitic plants such as mistletoe, which is capable of photosynthesizing but instead lives parasitically on other plants from which it obtains nutrients and water.

It would be misleading to suggest that the types of mycorrhizal associations noted above, and the groups of fungi involved, represent all of the major types of mutualisms in nature. Members of the rusts (Pucciniomycotina, Basidiomycota) are usually saprotrophs or parasites. However, certain members of the Atractiellomycetes, one class within the rust lineage, have recently been shown to enter into mycorrhizal associations with terrestrial and epiphytic orchids (Kottke et al., 2010). The occurrence of this mycobiont within a fungal group that otherwise contains only parasites and saprotrophs is interpreted as demonstrating a level of physiological flexibility that extends from saprophytism to the mutualism that is required for orchid mycobionts (Rasmussen and Rasmussen, 2009). Moreover, these studies also demonstrate that mycorrhizal interactions may be subtle and difficult to document.

Cyanobacteria are a major force in the evolution of hosts in providing photosynthesis, nitrogen fixation, UV protection, and toxic materials that can contribute to the defense of the host (Usher et al., 2007). Cyanobacterial symbioses with eukaryotes are ancient associations found in both terrestrial and aquatic ecosystems. The organisms enter into various symbioses with a large number of hosts including fungi, sponges, vascular plants, bryophytes, and protists (Adams, 2002; Adams and Duggan, 2008).

While the above examples of fungi in symbiotic interactions in the biological world are complex and many still remain poorly understood, especially at the molecular level, the concept termed "quorum sensing" renders all aforementioned symbioses appear simple by comparison (Miller and Bassler, 2001). This subdiscipline in microbiology focuses on how bacteria communicate with each other using chemical signal molecules to regulate numerous physiological activities including virulence, conjugation, production of antibodies, motility, sporulation, biofilm formation, and symbiosis. In this way they monitor the environment and alter behavior at the population level in response to changes in the number and species of other bacteria in the population (e.g. Waters and Bassler, 2005). While such interactions cannot be examined in the fossil record, the concept has far-reaching consequences to our understanding of the evolution of multicellularity and the role bacteria continue to play in a range of host–parasite interactions.

While symbioses are widespread in nature (e.g. humans contain only 10% human cells) and some are perhaps easily defined, the interactions that characterize these organisms as producers and consumers are very complex, especially at the microbial level, with some of the mechanisms involved only now beginning to be understood (Schmidt et al., 2013). Some of these examples of symbiosis are recorded in fossils because they are associated with particular morphological structures or tissues of the hosts, which can be studied if they are preserved in certain ways. For example, anatomical information has been used to classify extant types of mycorrhizae (e.g. Brundrett, 2004) and therefore can be inferred from the fossil record. In other instances the physical proximity of certain structures in the fossil record may be suggestive of an ancient symbiotic association. Many of these are discussed in other sections of this volume where the particular fungus is described in more detail.

4

CHYTRIDIOMYCOTA

FOSSIL CHYTRIDS 44
Precambrian Evidence 45
Rhynie Chert Chytrids 46
Carboniferous and Permian Chytrids 49
Parasitic Chytrids and Host Responses 58
Mesozoic and Cenozoic Chytrids 65
CHYTRID SUMMARY 66

Although primarily aquatic, modern Chytridiomycota (chytrids) are present in diverse habitats that extend from the tropics to the Arctic and occur in almost all types of terrestrial and aquatic ecosystems, including in high-elevation soils (Freeman et al., 2009). Thus it is not surprising that they have also been reported with some frequency from the fossil record. Some living forms occur in desert soils (e.g. Letcher et al., 2004; James et al., 2006a) and temporary ponds, while others are found in marine ecosystems as facultative or obligate parasites of algae and invertebrates (e.g. Amon, 1984; Gleason et al., 2011); still others have been reported from Antarctica (e.g. Ellis-Evans, 1985; Rao et al., 2012). Some can survive, but not grow, under anaerobic conditions (e.g. Gleason et al., 2007) and a variety of pH ranges (Gleason et al., 2010a). They are morphologically relatively simple and distinguished from other groups of fungi based on the possession of zoospores with a single flagellum on the posterior end; zoospores are produced in zoosporangia (Figure 4.1). The zoosporangium is the morphologically most prominent feature of chytrids. It is a sac-like structure in which internal divisions of the protoplasm result in production of the zoospores (James et al., 2006a). Chytrids consisting of a zoosporangium and filamentous rhizoids (Figures 4.2, 4.3) are termed eucarpic, while those in which the thalli are entirely converted to sporangia during reproduction are termed holocarpic. Monocentric thalli are characterized by a single zoosporangium, while polycentric forms consist of multiple sporangia produced on a network of rhizoids termed a rhizomycelium. The presence of zoospores with a single flagellum has been interpreted as evidence that the earliest fungi possessed flagellated zoospores that were subsequently lost in terrestrial fungi, which reproduce via nonmotile spores (Sekimoto et al., 2011). Other groups of so-called zoosporic fungi include the Blastocladiomycota, Cryptomycota, and Neocallimastigomycota.

Chytrids are important components of aquatic ecosystems today because they serve as a food source for certain zooplankton, decompose particulate matter, are parasites of aquatic plants and animals, and convert inorganic matter to organic compounds (e.g. Kagami et al., 2007; Gleason et al., 2008; Marano et al., 2011). In fact, most mycologists who teach about chytrids use fresh pollen grains as "bait" in water to collect living samples (Figure 4.4) of these organisms (e.g. Couch, 1939; Czeczuga and Muszyńska, 2001, 2004; Shearer et al., 2004). Most chytrids are parasites of vascular plants, rotifers, certain bryophytes, other fungi, and phytoplankton, and some anaerobic species also inhabit the digestive tracts of certain herbivorous mammals (Figure 4.5). The last, however, have more recently been excluded from the Chytridiomycota and placed in a separate phylum, the Neocallimastigomycota (Hibbett et al., 2007;

T.N. Taylor, M. Krings & E.L. Taylor: Fossil Fungi.

DOI: http://dx.doi.org/10.1016/B978-0-12-387731-4.00004-9

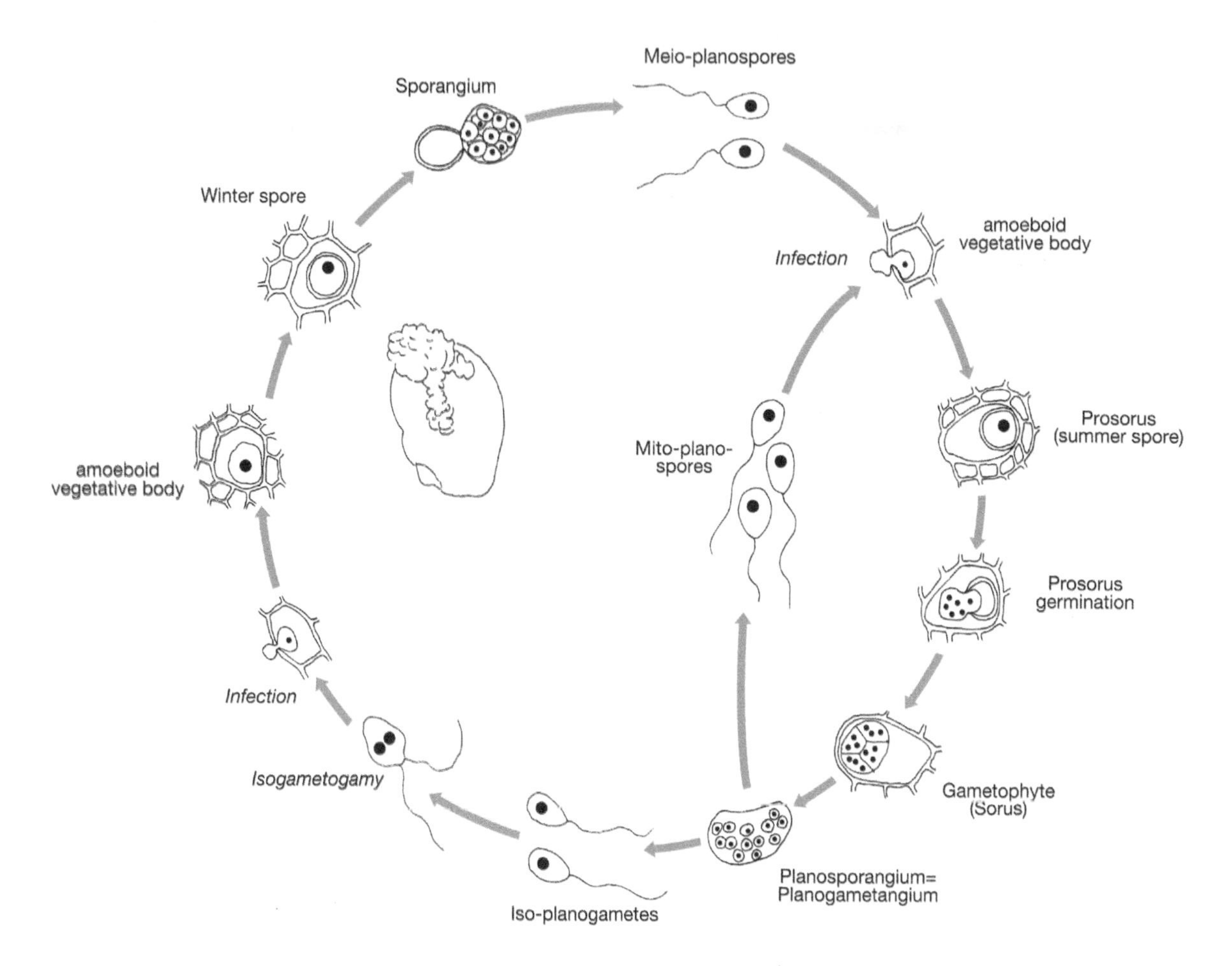

FIGURE 4.1 Life cycle of *Synchytrium endobioticum*. (After Esser, 1982.)

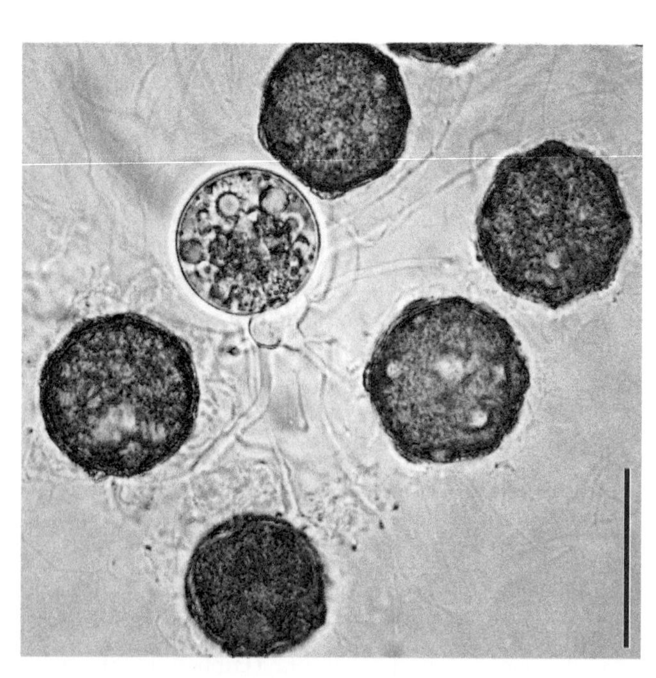

FIGURE 4.2 *Rhizoclosmatium aurantiacum* zoosporangium (*lighter colored sphere*) and rhizomycelium, growing between pollen grains. Extant. Bar = 25 μm. (Courtesy M.J. Powell.)

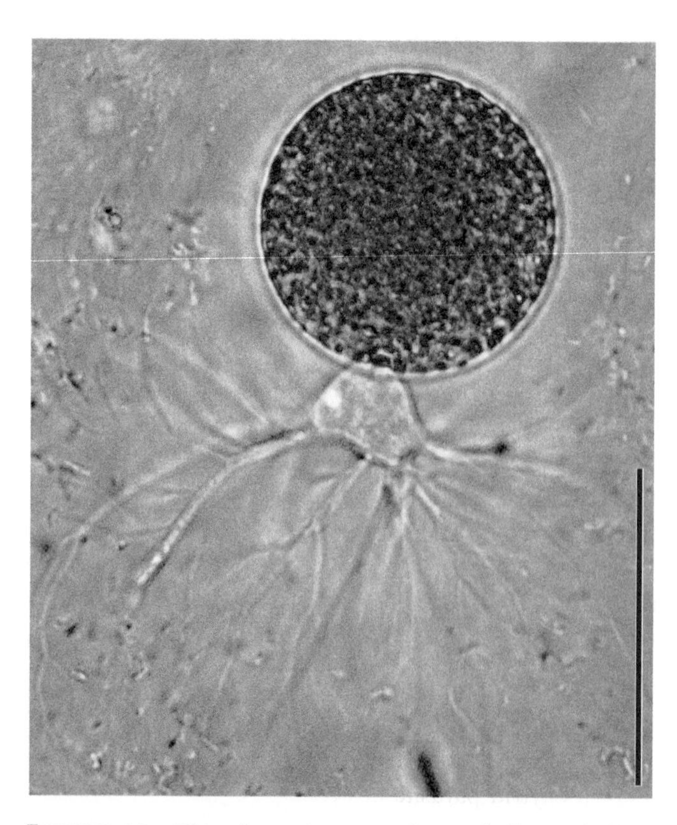

FIGURE 4.3 *Rhizoclosmatium aurantiacum* thallus and rhizomycelium. Extant. Bar = 25 μm. (Courtesy M.J. Powell.)

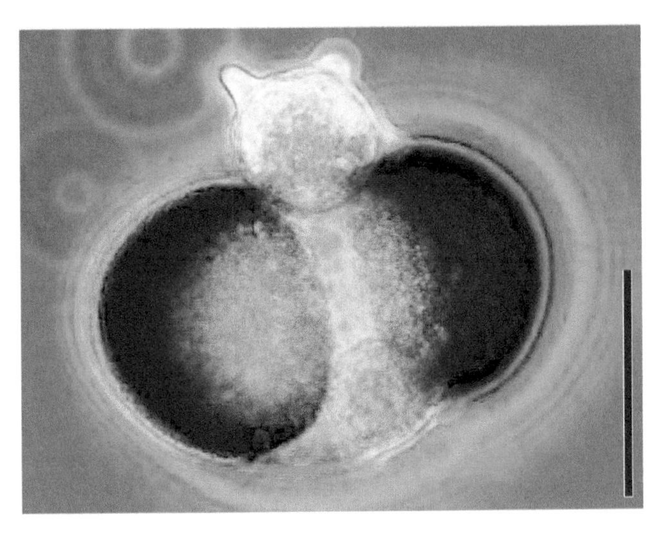

FIGURE 4.4 *Spizellomyces pseudodichotomus* zoosporangium with prominent discharge papillae attached to saccate pollen grain. Extant. Bar = 25 μm. (Courtesy D.J.S. Barr.)

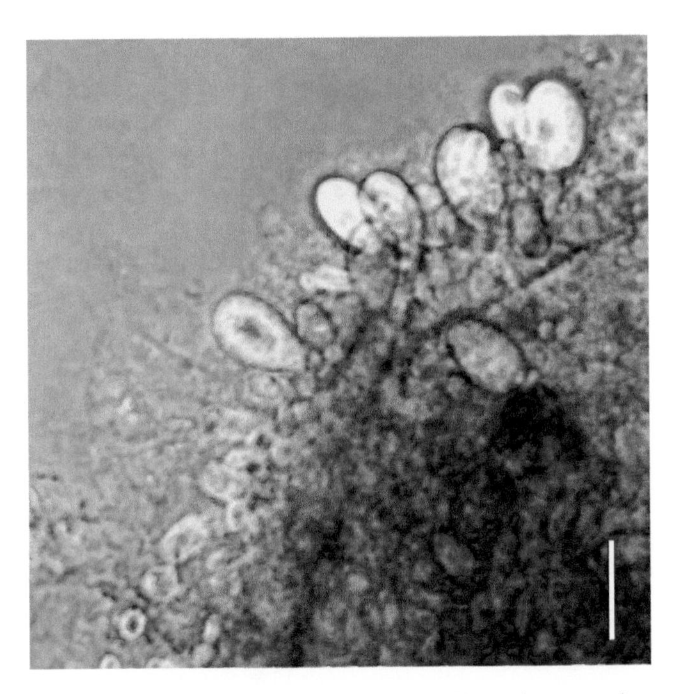

FIGURE 4.5 *Piromyces communis* (Neocallimastigomycota), a rumen fungus growing in culture on filter paper. Extant. Bar = 10 μm. (Courtesy D.J.S. Barr.)

FIGURE 4.6 Joyce E. Longcore.

FIGURE 4.7 John S. Karling (left) and James P. Braselton.

Nagpal et al., 2009). Chytrids are not major pathogens of economic importance, with the exception of those species of *Synchytrium* that cause potato diseases (Powell, 1993; Baayen et al., 2006). One chytrid, *Batrachochytrium dendrobatidis* (Longcore et al., 1999) (Figure 4.6), has gained international attention because it is the only known chytrid parasite of vertebrates and is responsible for a chytridiomycosis, a condition that causes thickening of the keratinized layer of the skin of certain amphibians, especially frogs (Berger et al., 1998, 2005). As a result of this disease, there have been major population declines and even extinctions of several amphibian species in many regions of the world (Retallick and Miera, 2007; Fisher et al., 2009; Martel et al., 2013), and there are worrisome indications that the disease is increasing in geographic range (e.g. James et al., 2009). Some studies have also suggested that there may be nonamphibian hosts for *B. dendrobatidis* (McMahon et al., 2013).

The two most authoritative volumes about the morphology, life history biology, systematics, and associations with other organisms of living chytrids are those by the eminent mycologist John Karling (1977, 1981) (Figure 4.7). All chytrids in which the life cycle is known

FIGURE 4.8 Donald J. S. Barr.

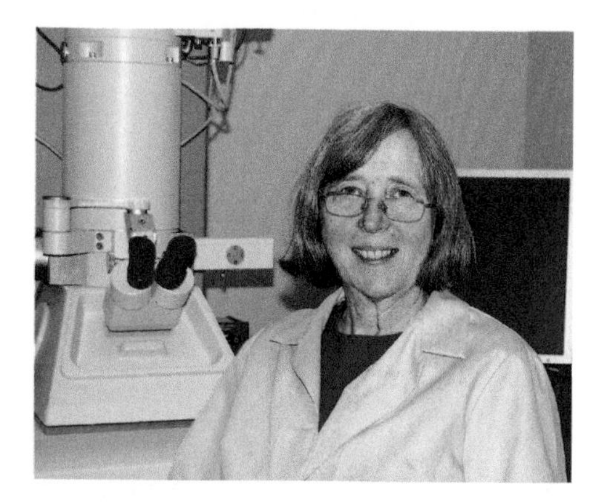

FIGURE 4.9 Martha J. Powell.

in detail are haploid, with the exception of the diploid resting spore (Barr, 2001) (Figure 4.8). Sexual reproduction has been described in many species (Sparrow, 1960); however, there are numerous modern chytrids in which the complete life history remains unknown. Barr (2001) notes, however, that not all resting spores in extant chytrids form as a result of sexual reproduction, with some representing thick-walled resistant sporangia. This has obvious implications in paleomycology when attempting to trace stages in the life history of a fossil chytrid. The widespread application of transmission electron microscopy (TEM) to the study of chytrids has further contributed to our knowledge of the biology and evolution of this group (e.g. Powell, 1976, 1978 [Figure 4.9]; Powell and Gillette, 1987; Barr, 2001), and the definition of major lineages has been supported by molecular data (Letcher et al., 2005).

Today there are more than 1250 named species of chytrids and the group is global in distribution (Shearer et al., 2007). The Chytridiomycota have long been regarded as the sister group to the remaining fungal phyla and were once thought of as monophyletic, based on features of the life cycle and characters of the zoospores. They have a long and interesting history regarding their placement on the tree of life. At one time they were considered true fungi, based on the chitin in their cell walls (Bartnicki-Garcia, 1970), but this interpretation was later modified and the group was thought to represent a transitional group between protists and Fungi because of the presence of motile zoospores (Barr, 1990). Subsequent phylogenetic studies based on ribosomal RNA confirmed the group as true fungi and they are now regarded as basal among the Fungi (James et al., 2000). Although earlier studies based on zoospore ultrastructure and certain morphologic features suggested the group to be monophyletic, some more recent research suggests that the chytrids are polyphyletic or paraphyletic (e.g. Tanabe et al., 2005; James et al., 2006a). In a comprehensive phylogenetic classification of fungi based on multilocus datasets, Hibbett et al. (2007) elevate the Blastocladiomycota and Neocallimastigomycota, two groups previously included within the Chytridiomycota, to the level of phyla (Figure 4.10). Based on these studies and those involving zoospore ultrastructure and morphology, some authorities now subdivide the remaining Chytridiomycota into six orders: Chytridiales, Cladochytriales, Lobulomycetales, Rhizophlyctidales, Rhizophydiales, Spizellomycetales; and two additional clades (Longcore, 2009). There will ultimately be additional revisions to fungal classifications as tools and analytical techniques continue to improve, especially in these basal fungal lineages (Hibbett et al., 2007), and the use of molecular tools indicates that chytrids are far more diverse than previously understood (Letcher et al., 2006, 2008a, b; Marano et al., 2012).

FOSSIL CHYTRIDS

Fossils do not normally display a suite of diagnostic features of sufficient clarity to allow a detailed comparison with extant chytrids and other groups of fungi and fungus-like microorganisms. Identifying chytrids based on fossil specimens is obviously a difficult task. This is due in part to the fact that there is extensive convergent evolution of morphological characters in this lineage of fungi, and thus certain ultrastructural and life history features, together with molecular and genetic data, are usually

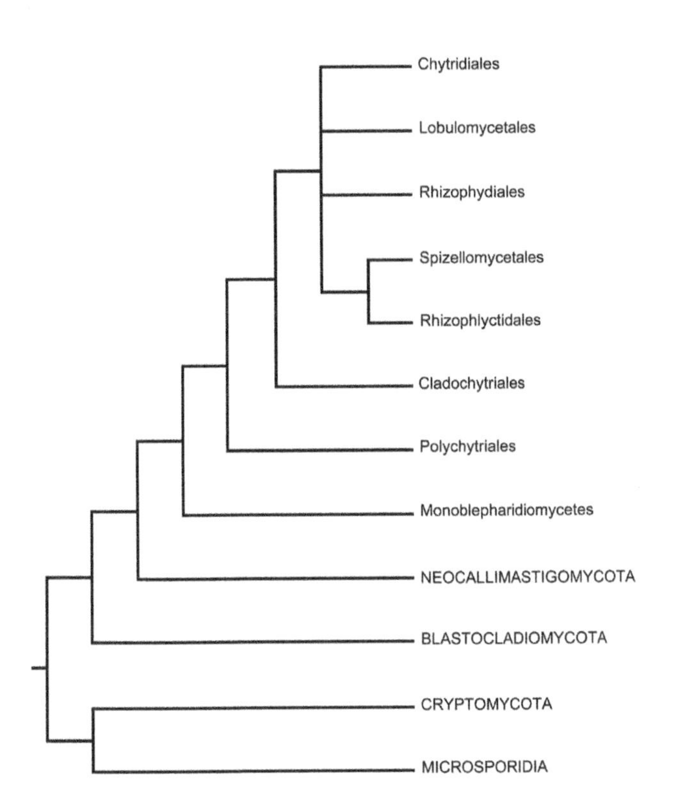

FIGURE 4.10 Suggested phylogenetic relationships within the chytrids.

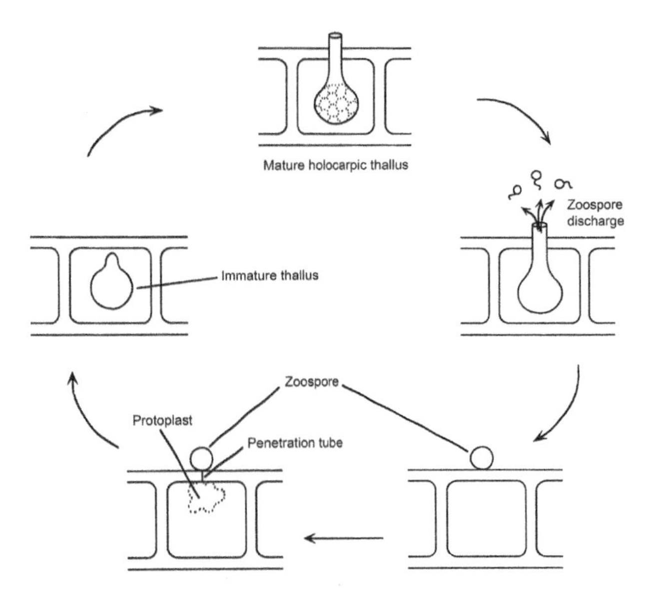

FIGURE 4.11 Suggested stages in the life history of a fossil holocarpic chytrid based on information collected from the Rhynie chert.

imperative in order to definitively ascribe a microorganism to the Chytridiomycota and determine its precise systematic position (e.g. James et al., 2000; Vélez et al., 2011). Such information can be obtained from living specimens but is virtually impossible to resolve with fossils. For example, rarely is there sufficient information about the life history of the organism, although in a few instances such details have been pieced together (Taylor et al., 1992a) (Figure 4.11), and to date at least, nothing about zoospore ultrastructure is known from the fossil record. The small size of the fossils certainly represents another factor that has contributed to a paucity of detailed information about ancient chytrids. Compounding the problem is the fact that structures produced by other groups of organisms, especially Hyphochytridiomycetes and certain Peronosporomycetes, are very similar, and thus often indistinguishable from structures seen in chytrids. Nevertheless, as is demonstrated in Chapter 5 on the Blastocladiomycota, there are features preserved in the fossils that do provide a level of resolution about the systematic position of at least a few fossils. Some scholars would suggest that it is perhaps more appropriate to refer to all fossil members as chytrid-like rather than chytrids. In a very real sense there is merit to this concept because paleomycologists generally cannot normally assign the fossils to a modern group, even at the level of order. Nevertheless, we will refer to the fossils described in the sections below as chytrids, and not chytrid-like, simply for reasons of convenience.

PRECAMBRIAN EVIDENCE

Molecular clock estimates hypothesize that the chytrids are an ancient group that inhabited the Earth at least 1.5 Ga (gigaannum = billion years) ago, or even earlier (Heckman et al., 2001; Hedges et al., 2004). These dates have continued to be pushed further back as new analytical techniques become available and more fungi are analyzed using molecular tools.

There are several reports on Precambrian microfossils (e.g. *Eomycetopsis*, *Vendomyces*, *Petsamomyces*) that have been interpreted as, or directly compared to, chytrids (e.g. Burzin, 1993; Belova and Akhmedov, 2006), but none of these is conclusive (Figure 4.12). Geologically slightly younger (Cambrian–Devonian), chitinozoans, which are now interpreted as representing a type of animal (e.g. Paris and Verniers, 2005), were at one time suggested to represent some type of fungal element (e.g. Locquin, 1981). Nevertheless, within the vast assemblages of Precambrian microfossils that have been described, it is highly probable that at least some of these organisms represent members

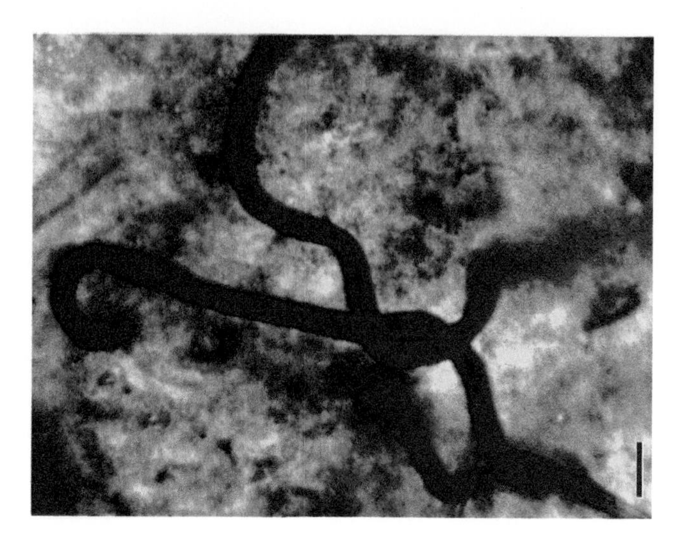

FIGURE 4.12 *Eomycetopsis robusta*. Proterozoic, Australia. Bar = 10 μm. (Courtesy J.W. Schopf.)

FIGURE 4.13 Robert Kidston.

of the Chytridiomycota, especially since there is fossil evidence for other clades of eukaryotes in the Proterozoic.

Microborings in different substrates that have been attributed to various groups of microorganisms (including fungi) have also been described in great detail from the Precambrian to the recent (e.g. Zebrowski, 1936; McLoughlin et al., 2007; Zhang and Pratt, 2008); however, the unusual mode of preservation of this evidence (trace fossils) presents a dilemma as to the precise attribution of the traces, especially in geologically older samples. The oldest unequivocal evidence of fossil chytrids comes from the Lower Devonian Rhynie chert and includes a variety of holocarpic and eucarpic forms that lived as saprotrophs or parasites of land plants, charophytes, and other fungi.

RHYNIE CHERT CHYTRIDS

The Rhynie chert Lagerstätte consists of several layers of fossiliferous beds containing shales and cherts that have been interpreted as a series of ephemeral freshwater pools within a hot-springs environment (Rice et al., 2002). See further information on the Rhynie chert and its biota in Chapter 2, pp. 19–21.

The earliest reports of chytrids preserved in the chert lenses are those of Kidston and Lang (1921b) in their extraordinary series of papers on the Rhynie chert (Kidston and Lang, 1917, 1920a, b, 1921a, b) (Figures 4.13, 4.14). They document a variety of micrometer-sized spherules and flask-like structures, some of which most certainly represent chytrid zoosporangia, including some that possess orifices suggestive of discharge pores. Especially interesting are forms that occur (clustered) on

FIGURE 4.14 William H. Lang.

larger fungal and land plant spores and probably represent (myco)parasites (Figures 4.15–4.17). Subsequent studies on the Rhynie chert noted the presence of several types of fungi in the interior tissues of several of the land plants, and some of these were also compared to modern chytrids (e.g. Harvey et al., 1969; Boullard and Lemoigne, 1971 [Figure 4.18]; Illman, 1984). More detailed analyses of the microfungal record from the Rhynie chert have identified several distinct chytrid morphotypes and life strategies, including holocarpic and

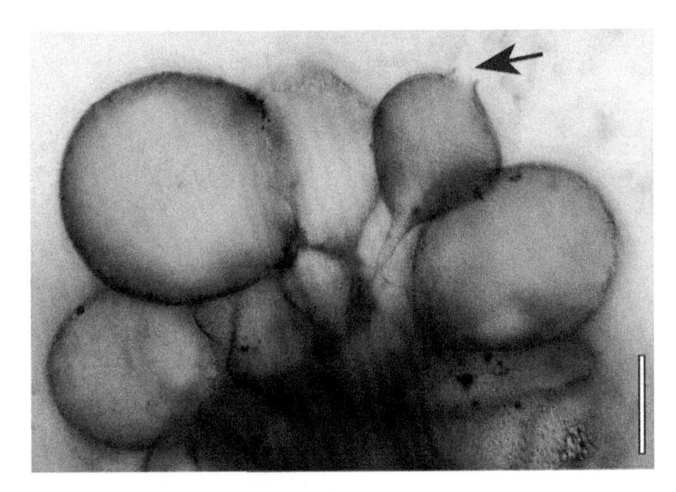

FIGURE 4.15 Several chytrid zoosporangia on the surface of an early land plant spore. Arrow indicates discharge pore. Lower Devonian Rhynie chert, Scotland. Bar = 10 μm.

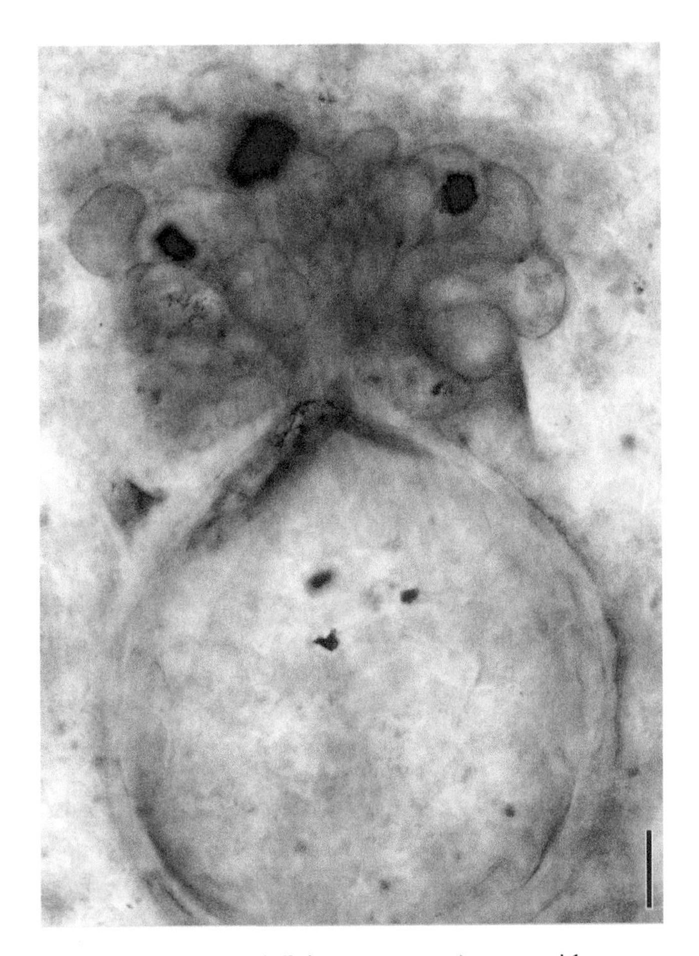

FIGURE 4.16 Fungal (?glomeromycotan) spore with numerous chytrids attached to distal surface. Lower Devonian Rhynie chert, Scotland. Bar = 20 μm.

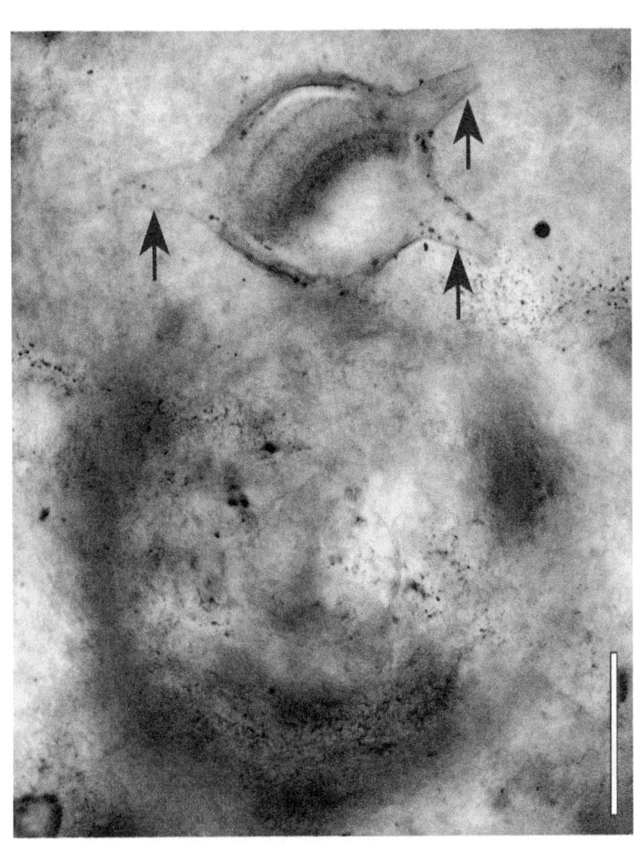

FIGURE 4.17 Chytrid zoosporangium with three prominent discharge papillae (*arrows*) on surface of large fungal spore. Lower Devonian Rhynie chert, Scotland. Bar = 15 μm.

FIGURE 4.18 Bernard Boullard. (Courtesy M.-A. Selosse.)

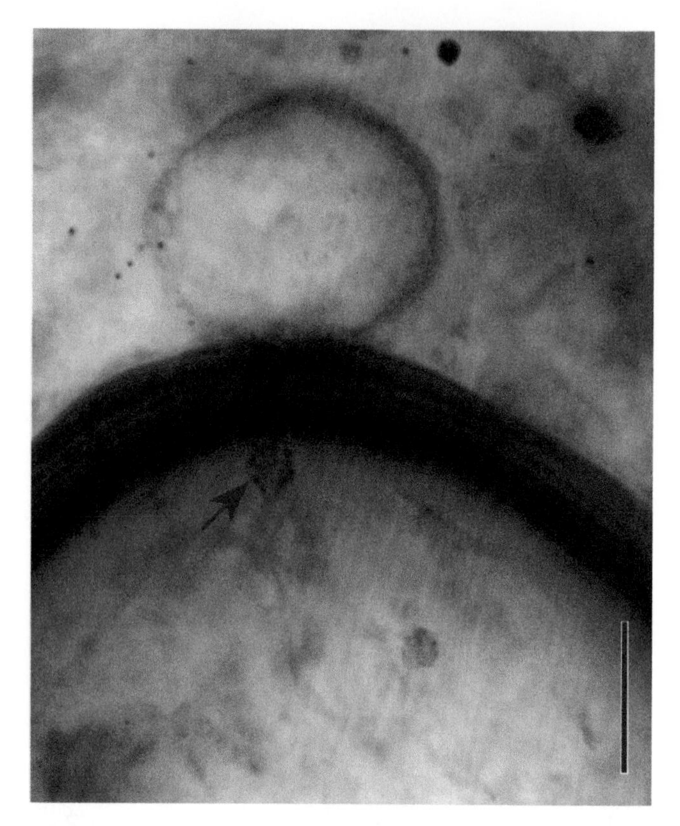

FIGURE 4.19 Chytrid attached to surface of thick-walled glomeromycotan spore. An apophysis (*arrow*) and a portion of the rhizomycelium is present extending into the spore lumen. Lower Devonian Rhynie chert, Scotland. Bar = 15 μm.

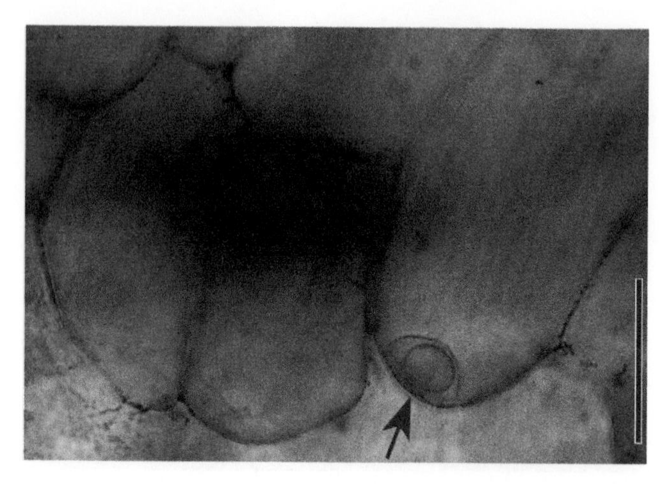

FIGURE 4.20 Holocarpic chytrid zoosporangium (*arrow*) attached to inner surface of plant cell. Lower Devonian Rhynie chert, Scotland. Bar = 15 μm.

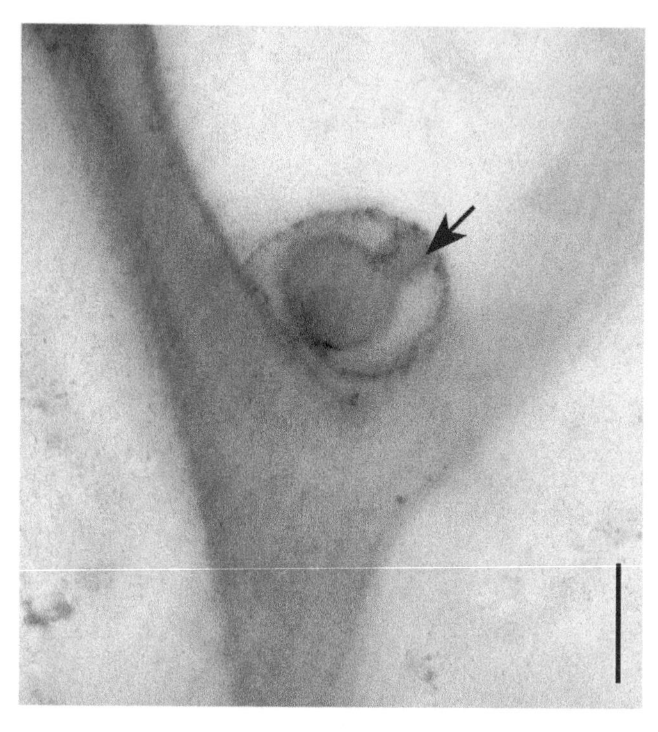

FIGURE 4.21 Chytrid zoosporangium attached to land plant axis. Arrow indicates discharge site. Lower Devonian Rhynie chert, Scotland. Bar = 5 μm.

eucarpic forms thriving on and in aquatic algae, land plants, and land plant and fungal spores (Figure 4.19) (Taylor et al., 1992a; Hass et al., 1994; Krings et al., 2009a, 2010b). Among the latter are both monocentric and polycentric forms.

For example, a distinctive, rather simple holocarpic form (Figures 4.20–4.25) is represented by a spheroidal to ovoid zoosporangium (up to 8 μm in diameter) that is surrounded by what appears to be a thick translucent wall (Taylor et al., 1992a). Zoospore discharge occurred through a single discharge tube with a slightly thickened collar (Figure 4.21). This organism frequently occurs attached to the walls of parenchyma cells of land plants, sometimes solitary (Figure 4.20) but more often in chains (Figure 4.22) or aggregates of >100 individuals (Figure 4.23); it has also been found on other fungi (Figure 4.24). A few of the specimens suggest that short rhizomycelial filaments extended into the host cell wall from the base of the zoosporangium (Figure 4.25).

In numerous fungal and land plant spores, the surface of the spore wall is densely covered in zoosporangia of monocentric chytrids, some of which suggest the presence of a rhizomycelium extending into the host lumen through the spore wall (Figures 4.15–4.17, 4.26) (Taylor et al., 1992a). In land plants this is especially true of the area around the distal suture, where the spore wall may be less ornamented and thinner as this is the germination surface and will be easier to penetrate. Some of

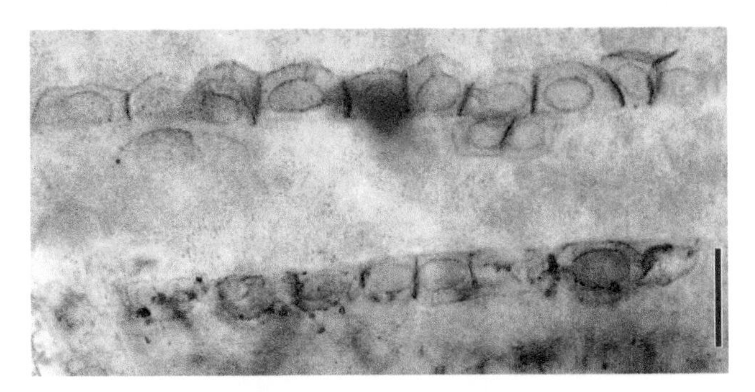

FIGURE 4.22 Zoosporangia of a holocarpic chytrid in narrow-diameter host cells (not preserved). Lower Devonian Rhynie chert, Scotland. Bar = 15 μm.

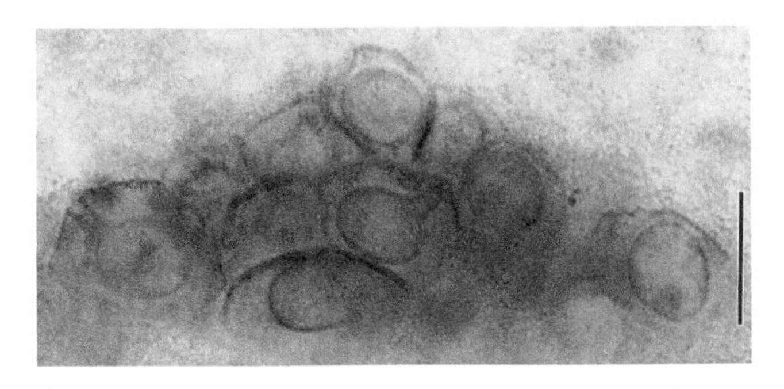

FIGURE 4.23 Cluster of holocarpic chytrid zoosporangia. Lower Devonian Rhynie chert, Scotland. Bar = 15 μm.

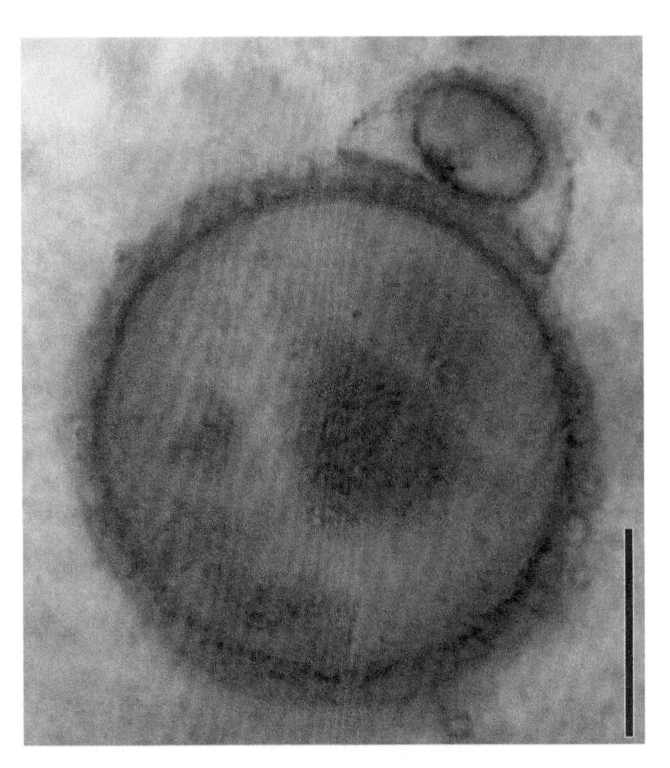

FIGURE 4.24 Holocarpic chytrid zoosporangium attached to surface of *Zwergimyces vestitus*. Lower Devonian Rhynie chert, Scotland. Bar = 10 μm.

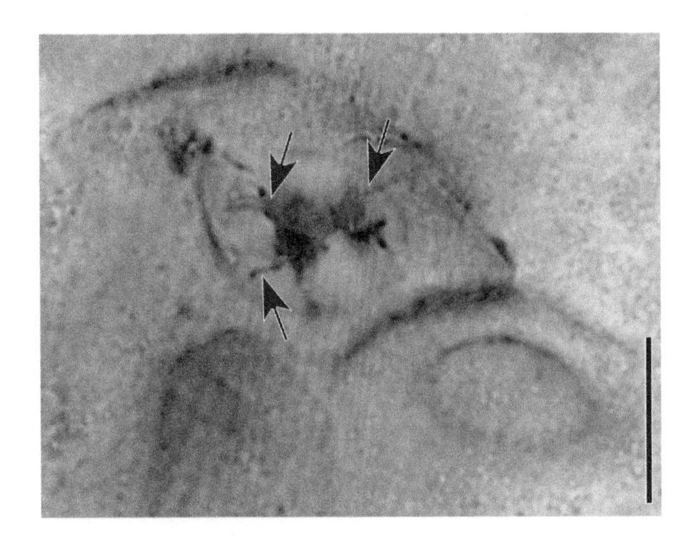

FIGURE 4.25 Holocarpic chytrid zoosporangium, underside showing rhizoids (*arrows*). Lower Devonian Rhynie chert, Scotland. Bar = 10 μm.

these epibiotic chytrids display a single discharge pore or papilla, whereas others show individual zoospores in the interior of the zoosporangium. In other zoosporangia refractive structures occur that have been interpreted as developing zoospores, and in at least one specimen, remains of a flagellum have been documented (Taylor et al., 1992a) (Figure 4.27). What appear to be encysted zoospores have also been reported in the spores of the land plant *Aglaophyton major* (Taylor et al., 1992a). On other land plant and fungal spores in the chert matrix there are irregularly shaped zoosporangia (Figure 4.28), some of which possess multiple discharge papillae. A number of Rhynie chert chytrids show some morphological similarity to modern members included in the Olpidiaceae and Spizellomycetaceae (Karling, 1977; Barr, 1980, 1984).

CARBONIFEROUS AND PERMIAN CHYTRIDS

The earliest microfossils interpreted as belonging to the chytrids are those reported from several Carboniferous plant parts preserved in Mississippian and Pennsylvanian cherts from France (Renault, 1895a, b). Some were given generic and specific names (e.g. *Grilletia sphaerospermi*; see below), even though the descriptions only included a few details about the size (Renault and Bertrand, 1885) (Figure 4.29; also see Figure 1.7). What is extraordinary

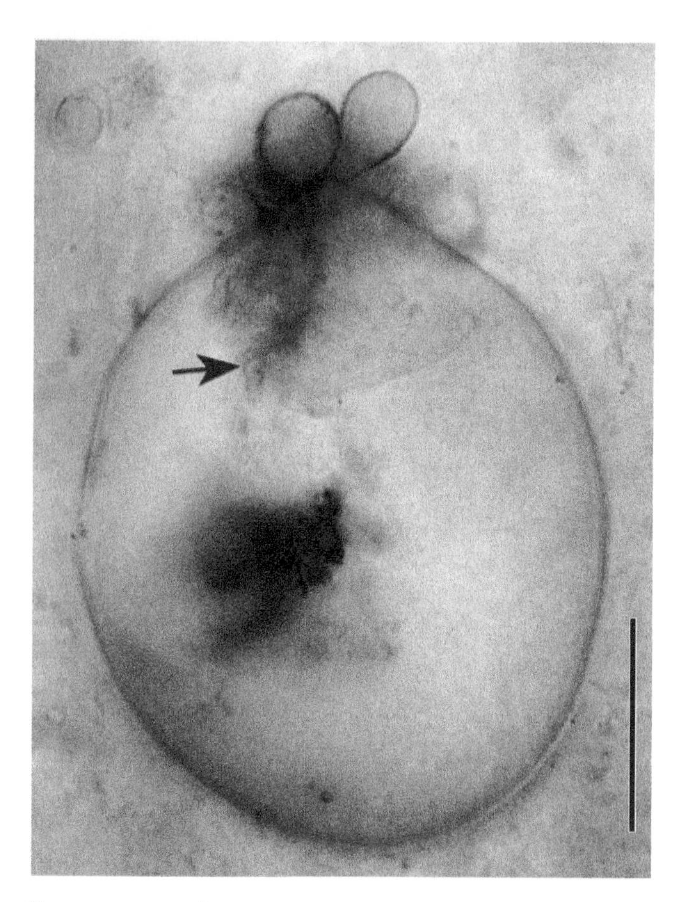

FIGURE 4.26 Spore with several epibiotic chytrid zoosporangia on the surface. Arrow indicates rhizoidal system in spore lumen. Lower Devonian Rhynie chert, Scotland. Bar = 25 μm.

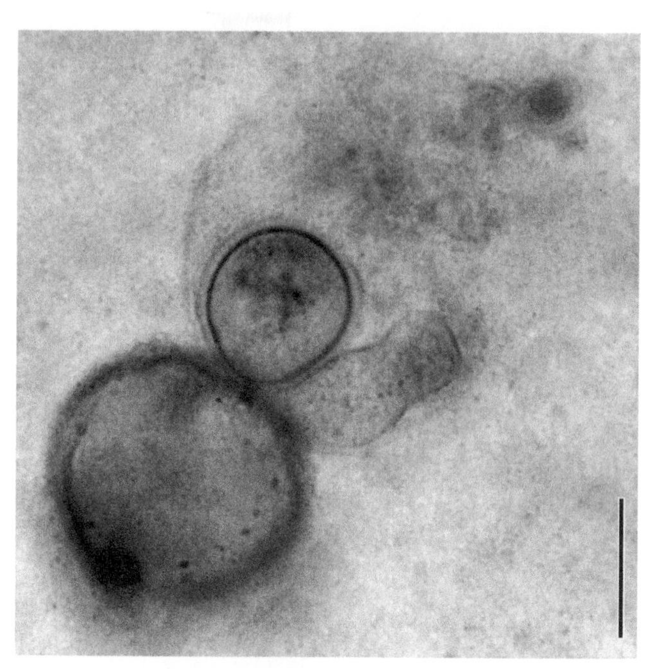

FIGURE 4.28 Large fungal spore with two different types of zoosporangia attached to the surface. Lower Devonian Rhynie chert, Scotland. Bar = 20 μm.

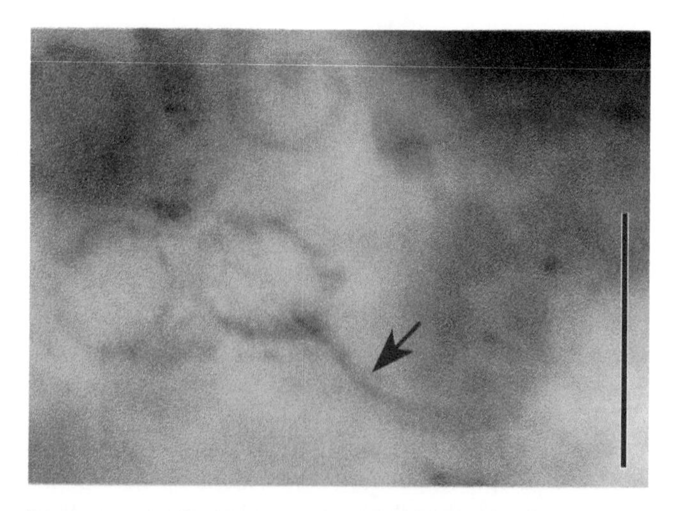

FIGURE 4.27 Chytrid zoospore showing slightly curved flagellum (*arrow*). Lower Devonian Rhynie chert, Scotland. Bar = 10 μm.

FIGURE 4.29 Paul Bertrand.

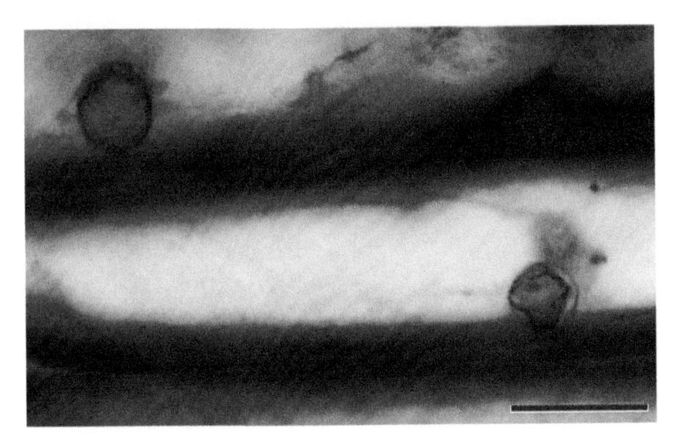

FIGURE 4.30 *Oochytrium lepidodendri* in the lumen of conducting elements of the arborescent lycopsid *Lepidodendron*. Mississippian, France. Bar = 20 μm.

about these early studies is the ability of the researchers to resolve fossil structures as small as chytrids with the transmitted light microscopy that was available at the time. In addition, these early studies placed the fossils within the context of extant organisms (e.g. chytrids) and even suggested their role in the ecosystem in which they occurred. In spite of resolution and magnification limits, several fossil types were described and illustrated from permineralized tissue of the arborescent lycophyte *Lepidodendron* as ovoid structures (12–15 μm in diameter) with a short papilla extending from one end (Renault, 1895a, b). Zoosporangia are slightly elliptical and up to 15 μm long. Renault suggested these structures were chytrids similar to modern forms of *Cladochytrium, Woronium*, and *Olpidium* and named them *Oochytrium lepidodendri* (Figure 4.30; also see Figure 1.8) (Renault, 1895a, b, 1896b). The significance of Bernard Renault's work on Carboniferous microorganisms undoubtedly parallels that of Kidston and Lang's (1921b) study of microscopic life in the Lower Devonian Rhynie chert. Unlike the Rhynie chert microorganisms, however, the existence of exquisitely preserved minute life forms in cherts from France has largely been forgotten since Renault's death and has only recently been revisited.

Problems in Naming Fossil Chytrids

Since the studies of Renault, almost none of the fossil chytrids published from the Carboniferous has been formally described and assigned scientific names, owing primarily to the various problems inherent in identifying fossil chytrids (see above). An excellent example of

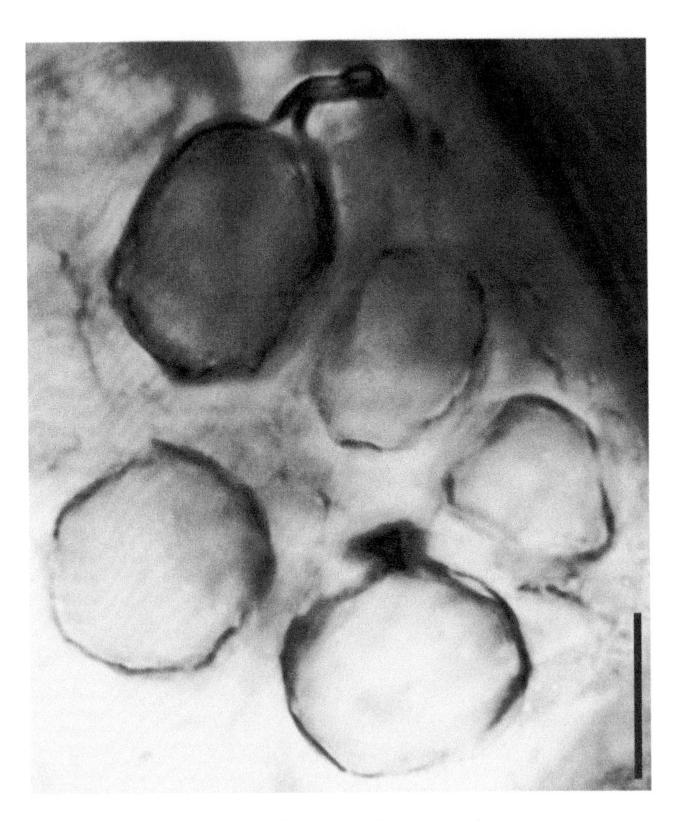

FIGURE 4.31 Cluster of chytrid-like microfungi in a sporangium of *Biscalitheca* cf *musata*. Upper Pennsylvanian, France. Bar = 10 μm.

this conundrum can be found in an assemblage of putative chytrids from the Upper Pennsylvanian Grand-Croix chert of central France that was discovered in the interior of a pteridophyte reproductive structure named *Biscalitheca* cf *musata* (Krings et al., 2009b). Within the tissues of *B.* cf *musata* occur four distinct types of microfossils resembling chytrid zoosporangia, distinguished by their overall morphology and occurrence in different tissue systems. The most common type consists of clusters of spherical to pear-shaped thalli, each up to 15 μm in diameter (Figure 4.31), some with a short stalk, that occur within individual host cells. In a second type the thalli occur more loosely scattered, have thicker walls, are slightly larger (18 μm in diameter), and are characterized by two prominent hollow projections, in one of which the distal end is slightly widened and open (Figure 4.32). It is suggested that the projections may represent some type of discharge tube like those in the modern chytrids *Diplophlyctis* (e.g. Sparrow, 1960; Pires-Zottarelli and Gomes, 2007), *Entophlyctis* (e.g. Barr, 1971; Chen and Chien, 1995), *Karlingomyces* (Blackwell et al., 2004) or

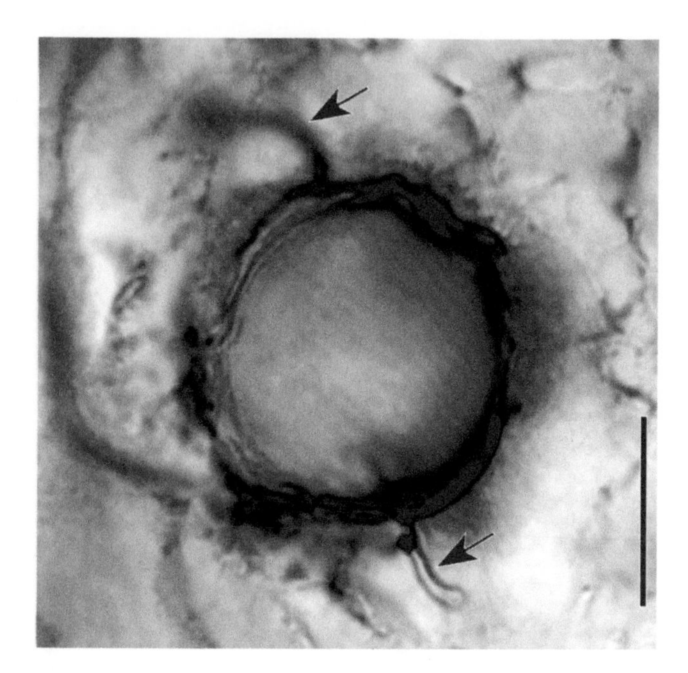

FIGURE 4.32 Chytrid-like microfungus in *Biscalitheca* cf *musata* sporangium showing oppositely positioned projections (*arrows*). Upper Pennsylvanian, France. Bar = 10 μm.

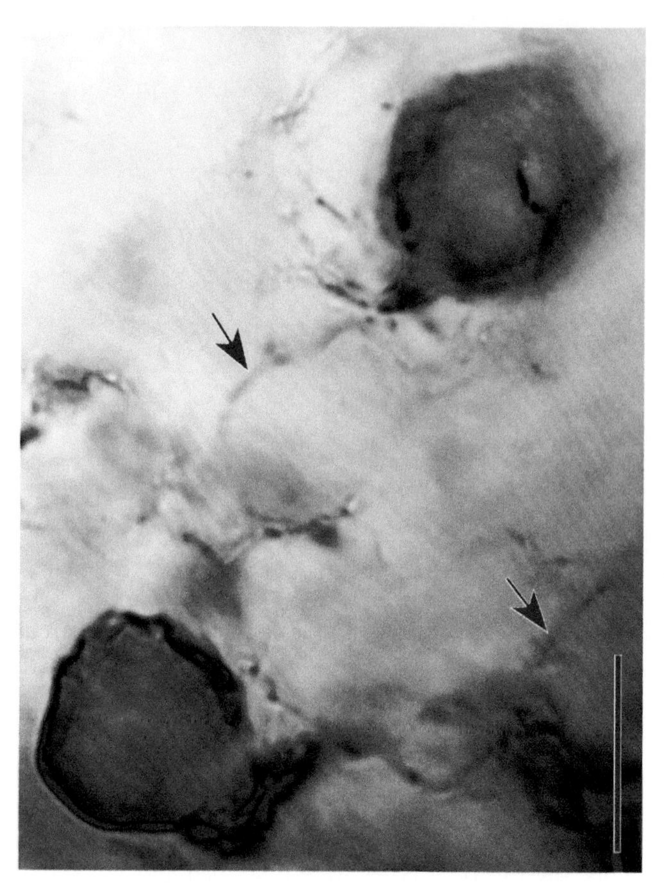

FIGURE 4.33 Two chytrid-like microfungi in *Biscalitheca* cf *musata* sporangium associated with multi-branched filaments (*arrows*). Upper Pennsylvanian, France. Bar = 10 μm.

Septochytrium (e.g. Pires-Zottarelli and Gomes, 2007). A third microfossil within this reproductive structure possesses multiple openings or pores, numerous minute pits in the wall, and delicate filaments that extend from the surface (Figures 4.33, 4.34). The fourth type is pear shaped, attached to the degrading cell walls of the host, and characterized by a distinct ornament of closely spaced setae. The absence of any distinct host response and generally very degraded nature of some of the host tissues suggest that the microorganisms in *B.* cf *musata* were saprotrophs. The putative chytrids in *B.* cf *musata* reveal another significant problem related to the precise interpretation of chytrid-like fossils that concerns the delineation of individual species among the fossils. Although the four morphotypes differ from one another in one to several characters, other features are shared. Since nothing is known about the range of morphological plasticity and life history of the putative chytrids associated with *B.* cf *musata*, it cannot be ruled out that several of the morphotypes actually belong to a single species. Conversely, it is also possible that the individual morphotypes each represent several morphologically similar species.

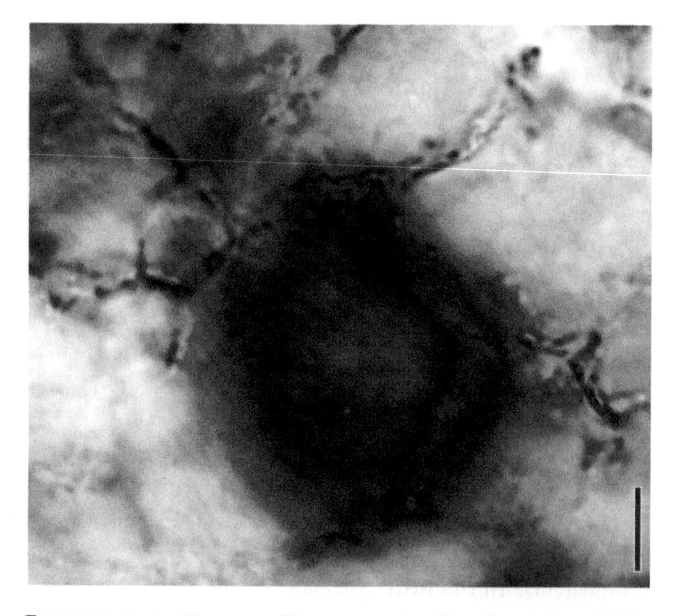

FIGURE 4.34 Tenuous filaments extending from chytrid-like zoosporangium. Upper Pennsylvanian, France. Bar = 5 μm.

Chytrids on Carboniferous Vegetative Tissues

There is an extensive diversity of chytrids preserved in a variety of plant parts from Carboniferous and

FIGURE 4.35 Francis W. Oliver.

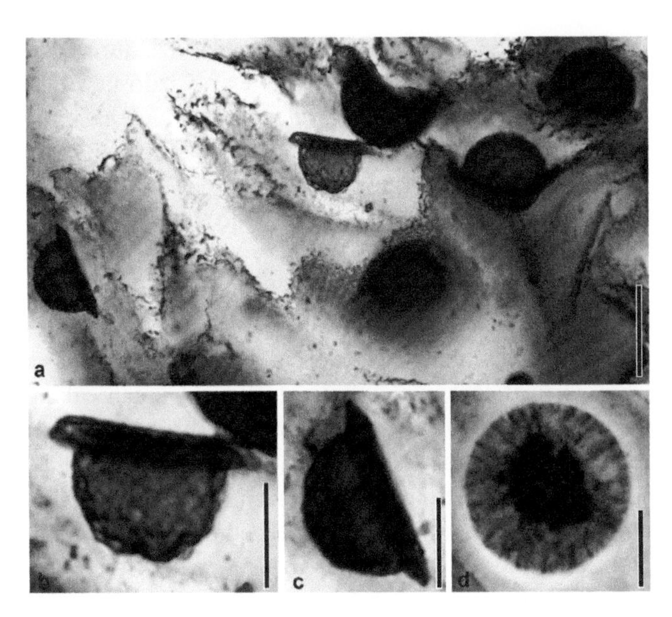

FIGURE 4.36 a) Hemispherical "bowler hat"-shaped chytrid zoosporangia in degrading plant tissue. Bar = 20 μm. b, c) Detail of zoosporangium in a). Bar = 10 μm. d) Saucer-shaped structure associated with cluster of zoosporangia. Mississippian, France. Bar = 10 μm.

Permian cherts and coal balls (e.g. Oliver, 1903) (Figure 4.35). Many of these occur as solitary resting spores and epibiotic and endobiotic zoosporangia. Some have been reported in plant megaspores (Krings et al., 2009c), while others occur in conducting elements (e.g. Williamson, 1888). Because of their widespread presence in Euramerican Carboniferous ecosystems (and by volume probably the largest source of biomass), tissues and organs of arborescent lycophytes such as *Lepidodendron* often contain numerous fossils with possible affinities within the Chytridiomycota (e.g. Krings et al., 2007a, 2009c, 2010a). Fossil chytrids have been described both from the cortical tissues, which make up the majority of the trunk, and in tracheids of the wood of arborescent lycophytes. For example, numerous microfossils with possible chytrid affinities occur in lycophyte xylem and periderm from the Mississippian cherts from Roannais and Esnost in central France in the form of stalked and sessile, spheroidal to pear-shaped (conical) structures that might represent zoosporangia. One form is spherical, 6–20 μm in diameter, attached to the cell wall by a short stalk (or apophysis), and usually has a single distal discharge pore. Another form is somewhat larger and attached to the host cell wall by a narrow stalk. Still other forms are unstalked, drop- or pear-shaped (conical), 10–20 μm high; some possess a pointed tip that is open distally or an apical cleft surrounding an opening. Some *Lepidodendron* tracheids also contain large numbers of small spore-like bodies and narrow filaments. The spore-like bodies, which appear suspended in the cell lumen or perhaps attached to the cell wall, range from 5 to 15 μm in diameter, and possess a thick translucent wall and two, usually oppositely positioned, circular openings. The spore-like bodies and filaments have previously been interpreted as belonging to a single organism, *Oochytrium lepidodendri*, which has been referred to the Chytridiomycota (Renault, 1895a, b, 1896b).

Other structures with possible affinities in the Chytridiomycota occur on largely degraded (?cortical) tissue of uncertain affinity from the same Mississippian chert deposits in France (Krings et al., 2009c). These structures are attached to the host cell wall and are up to 25 μm wide (including the rim) and 15 μm high; the wall is pustulose and appears to have been relatively robust. Some are spheroidal, whereas others range from hat- to bowl-shaped, constricted at the base, and have a wide distal opening surrounded by a slightly enrolled rim up to 3.5 μm wide (Figure 4.36a–c). Associated with these structures are circular, dorsiventrally compressed, saucer-shaped bodies, <20–23 μm in diameter (Figure 4.36d). These structures possess a prominent equatorial rim (up to 7 μm wide) that is ornamented by radially oriented irregular striae. There is a remarkable structural correspondence of these fossils with empty zoosporangia of the extant *Chytriomyces reticulatus* and *C. mammilifer* (Persiel, 1960). The broadly attached specimens may either represent immature zoosporangia or detached mature zoosporangia that adhered to the plant tissue with their distal ends. Moreover, *C. reticulatus* produces resting spores (Persiel, 1960) that resemble the saucer-shaped bodies associated with the fossil zoosporangia.

Fossils resembling chytrids have also been reported from cortical cells in the roots of Pennsylvanian calamites (Agashe and Tilak, 1970). They consist of mostly intercellular, nonseptate filaments and oval to spheroidal sporangium-like structures (~35 μm in diameter), the latter containing numerous globular bodies. The sporangium is thickened and appears to be a resting stage. The morphology of the resting sporangium has been used to suggest affinities of this fossil with members of the Synchytriaceae.

One of the most interesting root systems that existed during the Carboniferous and early Permian is represented by the root mantle of the marattialean tree fern *Psaronius* (see Taylor, T.N., et al., 2009). *Psaronius* root mantles represented a special habitat and were widely inhabited by fungi. For example, in an early Permian *Psaronius* root mantle from eastern Germany, more than 15 different morphotypes of intra- and intercellular fungal endophytes were discovered, including several types of spherical and pear-shaped chytrid zoosporangia (Figure 4.37), some with rhizomycelia extending into the host cell walls (Figure 4.38) (Barthel et al., 2010).

Spherical zoosporangia (5–8 μm in diameter), each with a single discharge pore surrounded by a faint collar, also occur in phloem mucilage cells of the Pennsylvanian fern *Botryopteris* (Smoot and Taylor, 1983; Klymiuk et al., 2013a). One of the zoosporangia shows what appears to be an operculum (Figure 4.39) and a tiny (~0.7 μm in diameter) structure interpreted as a zoospore (Klymiuk et al., 2013a). It is interesting to note that some of the fossil zoosporangia appear shrunken and in various shapes that mimic the morphologies encountered when extant chytrid zoosporangia are air dried, a condition suggested as representing a survival strategy (e.g. Gleason et al., 2004).

Various types of microfungal remains, some of which probably represent chytrids, have also been described in structurally preserved sphenophyte and cordaite leaves from the Upper Pennsylvanian of France (Dotzler et al., 2011; Krings et al., 2011b). For example, within the intercellular system of certain cordaite leaves occur intercalary

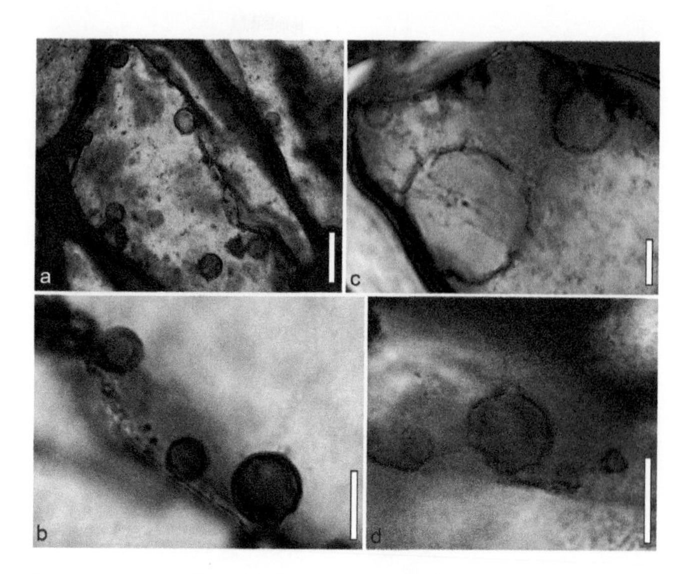

FIGURE 4.37 Chytrid-like zoosporangia associated with the Permian tree fern *Psaronius*. a) Several zoosporangia in cell lumen. Bar = 20 μm. b) Detail of zoosporangia attached to cell wall surface. Bar = 10 μm. c) Multi-porate zoosporangium on cell wall. Bar = 5 μm. d) Zoosporangium with discharge site and three small spheres. Bar = 10 μm. Permian, Germany.

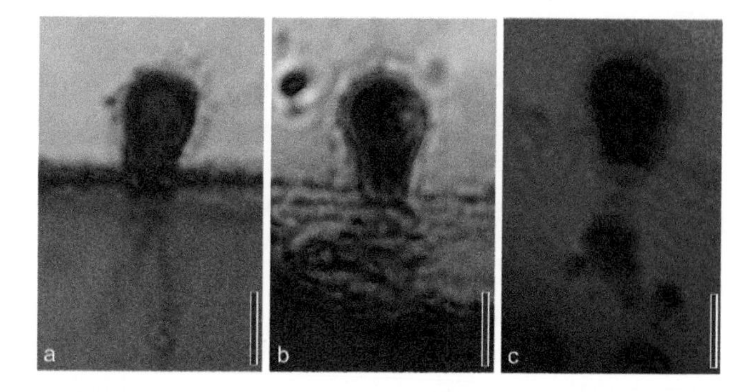

FIGURE 4.38 Epibiotic zoosporangia with rhizomycelium extending into cell wall of *Psaronius* root mantle cell. Permian, Germany. Bars = 5 μm.

and terminal swellings produced from an elaborate system of tenuous hyphae/filaments (Figure 4.40), as well as isolated putative (zoo)sporangia and clusters of small unicells. Moreover, spherical structures up to 18 μm in diameter are attached to outer host cell wall surfaces, preferentially to outer surfaces of bundle sheath cells. These structures are relatively thin walled and some distinctly ornamented. Both unornamented and ornamented spheres arise from, or give rise to, delicate mycelia extending out on the outer surface of the host cells to which the spheres are attached (Figure 4.41). Dense networks of tenuous, multibranched hyphae or filaments also occur in *Sphenophyllum* leaves.

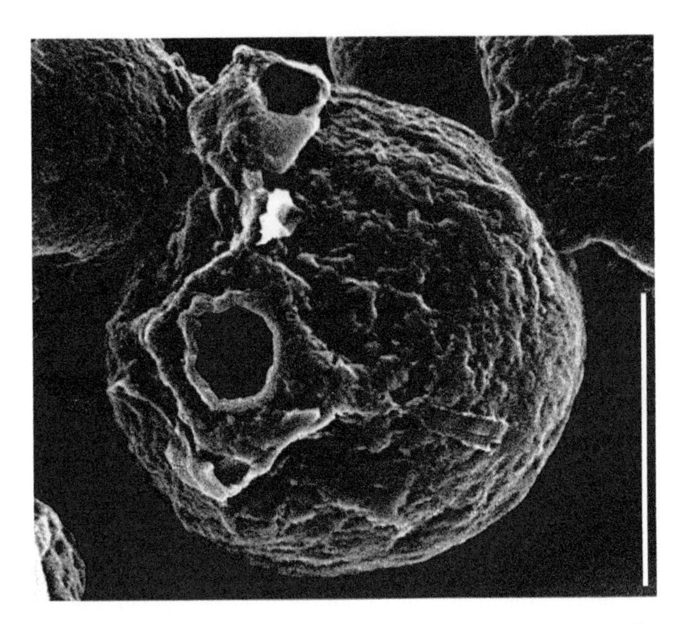

FIGURE 4.39 Zoosporangium with prominent discharge site within phloem cell of Carboniferous *Botryopteris* fern. Middle Pennsylvanian, USA. Bar = 3 μm.

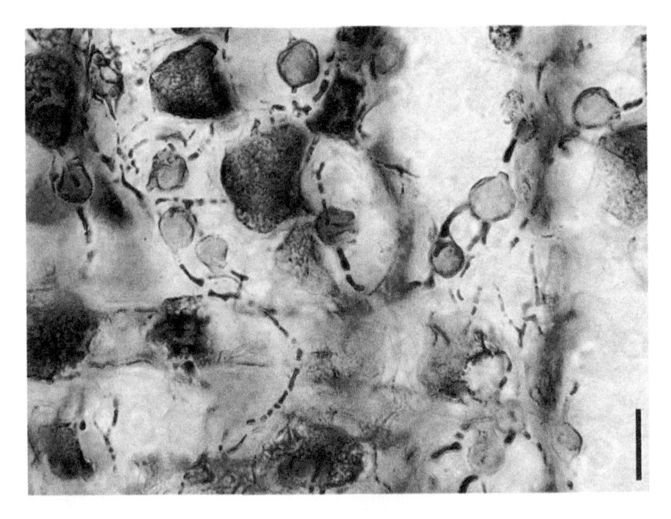

FIGURE 4.40 Polycentric microfungus in the mesophyll intercellular system of a cordaite leaf. Upper Pennsylvanian, France. Bar = 20 μm.

Unornamented, thin-walled, more or less spherical to pear-shaped structures, each up to 13 μm in diameter, occur in an intercalary position within the network of hyphae/filaments (Dotzler et al., 2011). Extending from one of these structures is what appears to be a discharge tube. Based on their overall morphology, some of these fossils may represent rhizomycelial systems and zoosporangia of monocentric and polycentric chytrids; however, affinities with other groups of Fungi or fungus-like microorganisms cannot be ruled out.

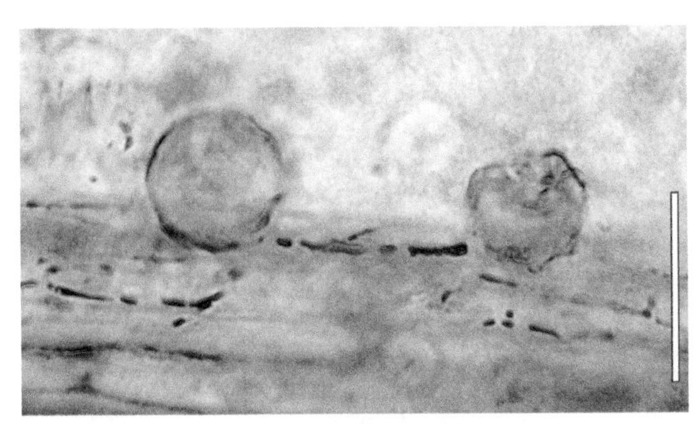

FIGURE 4.41 Chytrid zoosporangia (?) on the outer surface of bundle sheath cells within a cordaite leaf. Note basal mycelium extending out on host cell surface. Upper Pennsylvanian, France. Bar = 30 μm.

What are interpreted as parasitic *Synchytrium*-like chytrids have been recorded for several permineralized plant remains from the Permian of Antarctica (García-Massini, 2007a). The thallus of *Synchytrium permicus* is holocarpic and monocentric, and consists of thick-walled resting sporangia, thin-walled sporangia, and zoospores in different stages of development. The life cycle, proposed from the recognized developmental stages, begins when zoospores encyst on the host cell surface, subsequently giving rise to thin-walled sporangia with motile spores. Some zoospores (haploid) function as isogamous gametes that may fuse to produce resting sporangia (diploid). Roots, leaves, and stems of plants are among the tissues infected. Host response to infection includes hypertrophy.

Chytrids in and on Carboniferous Plant Spores

Present in many Carboniferous rocks are numerous lycophyte megaspores. Some of these spores contain one to numerous spheres up to 30 μm in diameter (Figure 4.42). Many spheres have a short fragment of the narrow subtending filament or hypha still attached, and some show what appears to be a circular apical discharge pore. The spherical structures have been described as *Palaeomyces majus* and referred to the "Mucorinées" by Renault (1896b). Krings et al. (2009c), however, suggest that they might well represent (immature) zoosporangia of a polycentric chytrid, perhaps a form that is comparable in basic structure to the extant *Cladochytrium* or *Nowakowskiella* (e.g. Marano et al., 2007). Spherical structures reminiscent of those in Carboniferous megaspores are also known to occur in large fungal (probably glomeromycotan; see Chapter 7, p. 113) spores from the

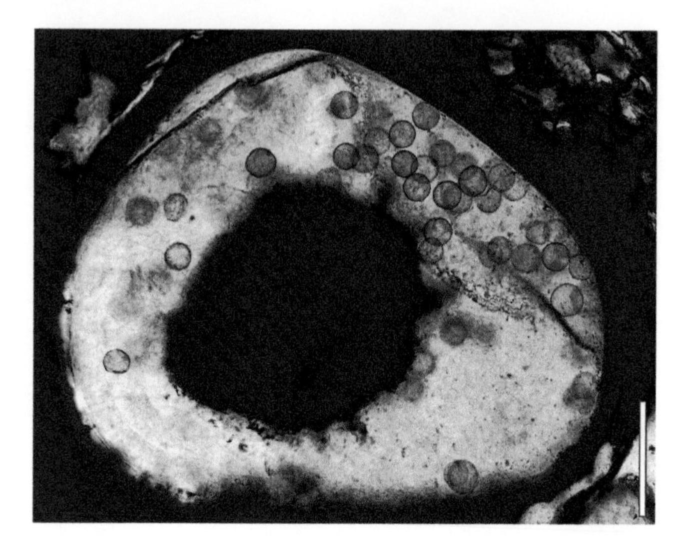

FIGURE 4.42 *Sublagenicula nuda*-type lycopsid megaspore containing numerous spheres and tenuous filaments of a putative polycentric chytrid. Mississippian, France. Bar = 150 μm.

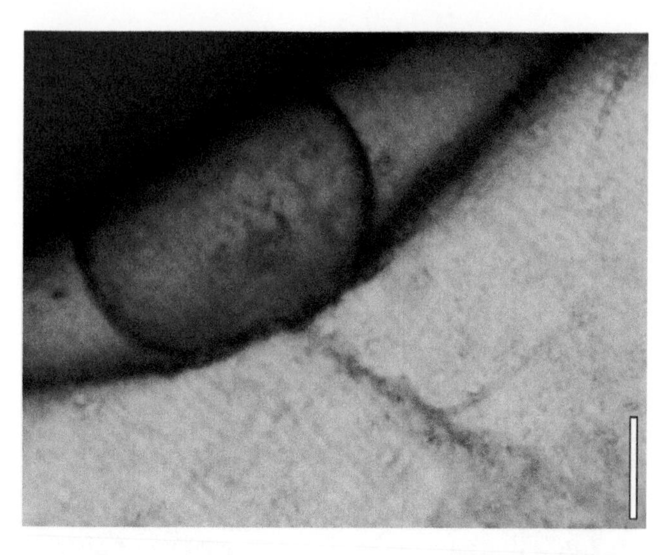

FIGURE 4.43 Chytrid zoosporangium with rhizomycelium extending through host megaspore wall. Mississippian, France. Bar = 20 μm.

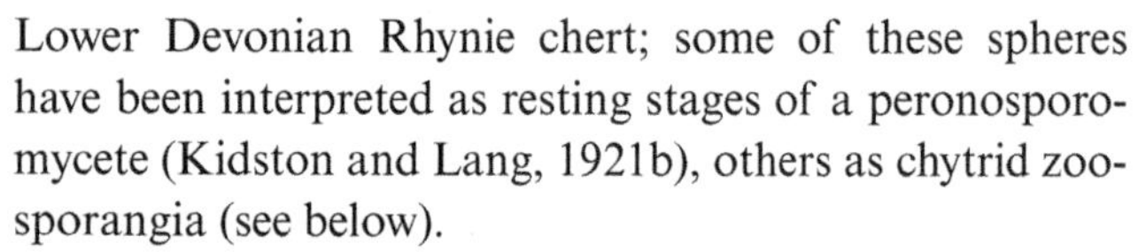

Lower Devonian Rhynie chert; some of these spheres have been interpreted as resting stages of a peronosporomycete (Kidston and Lang, 1921b), others as chytrid zoosporangia (see below).

One particularly interesting *Sublagenicula*-type lycophyte megaspore from the Mississippian of France has been reported to be colonized by a relatively large chytrid, which is one of very few Carboniferous examples where not only the generative portion (i.e. the zoosporangium) but also the rhizoidal system of the organism are preserved (Figure 4.43) (Krings et al., 2009c). The zoosporangium, which occurs between the separate inner wall layer and the remainder of the megaspore wall, is dorsiventrally compressed, smooth walled, and 62 μm wide by ~50 μm high. Extending from the zoosporangium is a rhizoidal system that penetrates the inner wall layer and extends into the spore lumen. This organism is one of only a few Carboniferous fossils that is assignable to the Chytridiomycota with confidence, based on a morphology that closely resembles that seen in many modern monocentric chytrids (e.g. species in the genera *Rhizophydium* and *Spizellomyces*; see Karling, 1977; Chen and Chien, 1998).

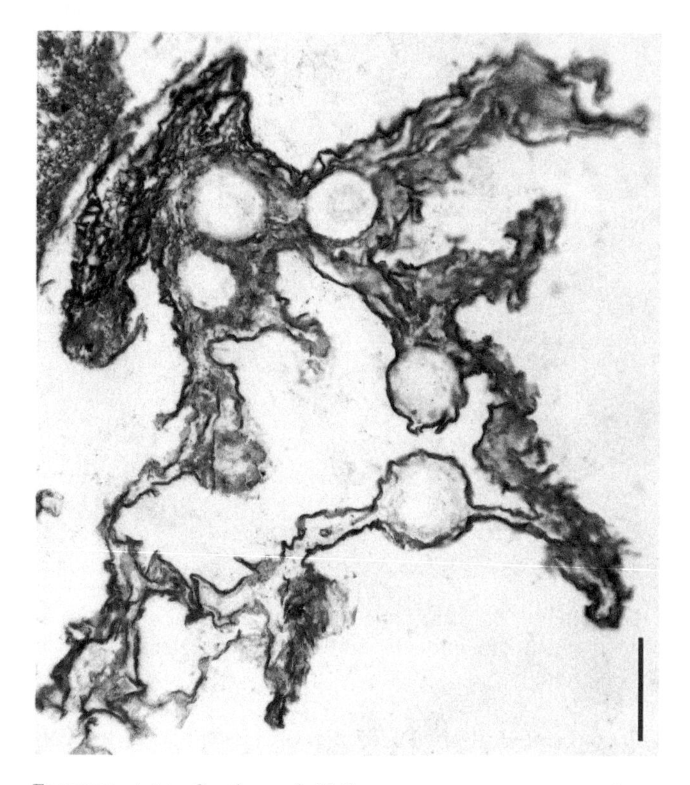

FIGURE 4.44 Section of *Triletes rugosus* megaspore showing several zoosporangia (?) or chlamydospores associated with stroma-like material in spore lumen. Middle Pennsylvanian Illinois, USA. Bar = 50 μm.

Another example of a possible chytrid associated with a spore comes from Pennsylvanian coal balls (Stubblefield and Taylor, 1984). Structures most certainly fungal in nature occur in the arborescent lycopsid megaspore *Triletes rugosus* (Figure 4.44). They consist of spherical units up to 70 μm in diameter that in some instances are associated with hyphae. Even using both transmission and scanning electron microscopy, together with transmitted light, however, the affinities of this fungus remain equivocal. Two possibilities include chlamydospores or chytrid zoosporangia.

FIGURE 4.45 Michael A. Millay.

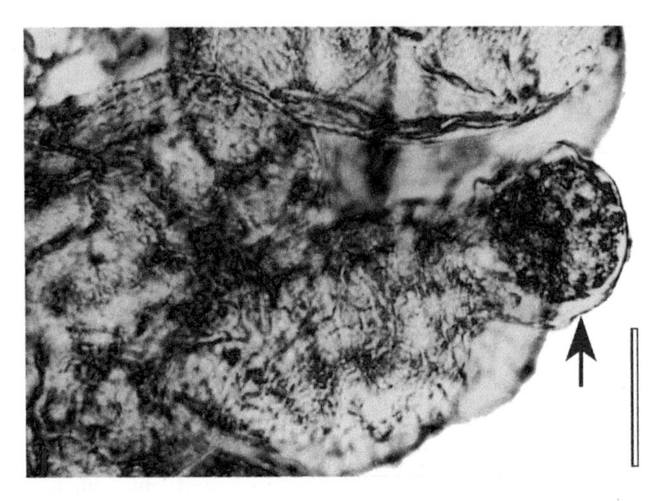

FIGURE 4.46 Epibiotic chytrid zoosporangium (*arrow*) attached to saccate pollen grain. Note two structures that may represent some type of discharge site. Pennsylvanian, Kentucky, USA. Bar = 10 µm.

Chytrids in and on Paleozoic Pollen Grains

Long used as bait for sampling of extant chytrids (Mueller et al., 2004), it is surprising that these microfungi have not been more commonly described from fossil pollen grains. The scarcity of documented evidence might be explained by the desire of palynologists to describe and illustrate pollen grains and spores that demonstrate a full complement of features that might be hidden by the presence of components of a fungal life cycle.

One well-preserved fossil pollen grain containing several morphological stages of both epibiotic and endobiotic thalli like those of modern chytrids comes from Carboniferous (Pennsylvanian) coal balls of eastern Kentucky (Millay and Taylor, 1978) (Figure 4.45). The pollen grains are saccate, and within the body are several immature zoosporangia, each up to 25 µm in diameter, some containing what are interpreted as zoospores (Figure 4.46). On the outer surface of the pollen are globose zoosporangia, each with multiple discharge papillae (Figures 4.46, 4.47); other zoosporangia have zoospores but no discharge papilla. The thallus in some of these chytrids is connected to the grain via a delicate rhizoidal system.

Degradation of pollen grains from the Permian of India showing spherical cells attached to the body, and skeletonization of the exine and saccus, was attributed to fungal degradation, probably by chytrids (Srivastava et al., 1999). Several morphologically distinct types of saccate pollen from the Permian of India have also been reported with opaque structures in the central body of the pollen grain (Vijaya and Meena, 1996). They range from 15–35 µm in diameter and are thought to represent stages in the development of the microgametophyte. Because of their size and occurrence in multiple genera of pollen, at least some may represent stages in the development of some type of chytrid zoosporangium. However, thalli of extant *Hyphochytrium catenoides*, a member of the Hyphochytridiomycetes, growing

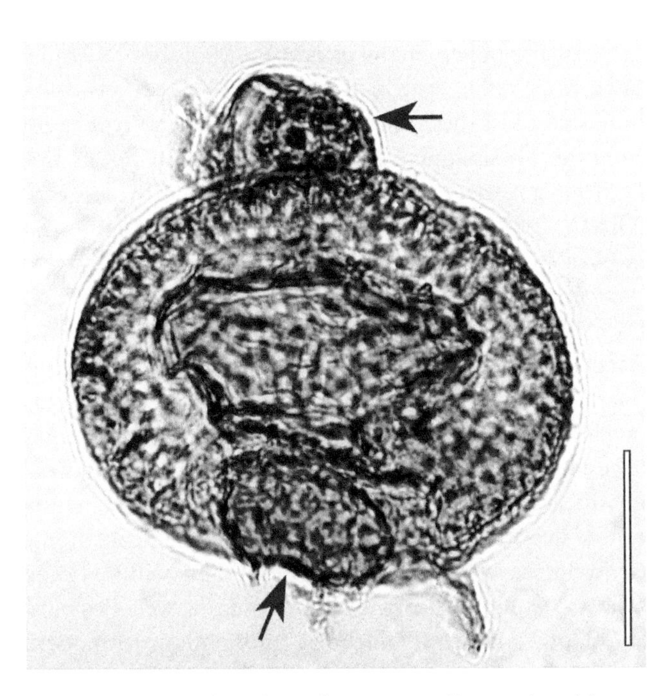

FIGURE 4.47 Polar view of saccate pollen grain with two chytrid zoosporangia (*arrows*) attached to surface. Pennsylvanian, Kentucky, USA. Bar = 25 µm.

saprophytically in pine pollen (e.g. Barr, 1970) are also remarkably similar to the spherical bodies found in the pollen grains from the Permian of India.

Chytrids and Seeds

Other fossil reproductive organs including seeds have been described as containing microfungi that probably represent chytrids. However, it is interesting to note that relatively few fungal remains associated with fossil ovules and seeds have been described to date (Krings et al., 2012a). Two fungi associated with seeds from the Upper Pennsylvanian of France are believed to represent members of the Chytridiomycota. One of these, *Grilletia sphaerospermii*, is represented by narrow hyphae/filaments and strings of sporangium-like vesicles that occur in the peripheral layers of the nucellus of a *Sphaerospermum* ovule (Renault and Bertrand, 1895), while the other consists of small ovoid bodies that have been noted on the surface of the nucellus, everywhere between the pollen chamber and the chalaza of another *Sphaerospermum* specimen (Oliver, 1903). Chytrid-like fungi have also been recorded for *Conostoma*-type seeds from the Mississippian of Burntisland, Scotland, as well as from a variety of other Carboniferous ovules/seeds (Oliver, 1903; Ellis, D.,1918). The precise affinities of most of these fungal fossils cannot be determined.

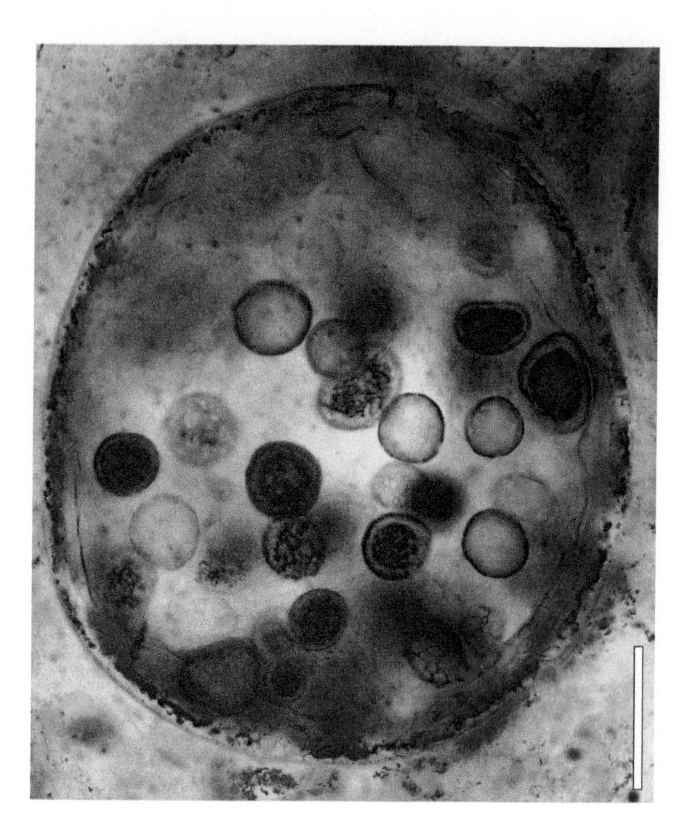

FIGURE 4.48 Numerous spores or zoosporangia within the lumen of a thick-walled spore. Lower Devonian Rhynie chert, Scotland. Bar = 50 μm.

PARASITIC CHYTRIDS AND HOST RESPONSES

Among the fossil fungi that have been described to date, one of the most complex problems relates to deciphering precisely the nutritional mode of fungi and how they may have interacted with other components of the ecosystem in which they existed. To some degree this information can be hypothesized from knowledge of the depositional environment where the fossils occur and within or on what particular host organism or tissue, as well as a variety of other indirect sources of information. In modern ecosystems, defense strategies of the host against harmful fungi may include the formation of some type of physical barrier, sometimes by programmed cell death, or by production of specialized substances with fungicidal properties that interfere with the biology of the pest; one fungus may suppress the activities of another, parasitic, fungus (e.g. Anderson et al., 2010).

Epibiotic and Endobiotic Mycoparasites

One of the interesting features of the Kidston and Lang (1921b) report on fungi from the Rhynie chert is the large number of fungal spores of various types that are associated with both epibiotic and endobiotic microfungi, some of which were probably saprotrophs but most of which were likely parasites. Several examples of these interfungal associations have been described (Kidston and Lang, 1921b; Taylor et al., 1992a; Hass et al., 1994), and one of these consists of large, apparently glomeromycotan, spores containing varying numbers of small (resting) spores or sporangia of other, intrusive, fungi (Figure 4.48). Glomeromycotan spores, which are among the largest spores known in the fungal kingdom, certainly represented particularly suitable habitats for parasitic microfungi because they contain abundant and easily accessible nutrients. The systematic affinities of most of the intrusive fungi remain elusive; however, differences in size and wall composition of the spores or sporangia, together with differences in the morphology of occasionally present subtending hyphae or filaments, suggest that a wide variety of microfungi in the Rhynie paleoecosystem lived, reproduced, and/or produced resting stages inside the spores of other fungi (Kidston and Lang, 1921b). Many of the intrusive fungi are quite similar to certain extant chytrids, although

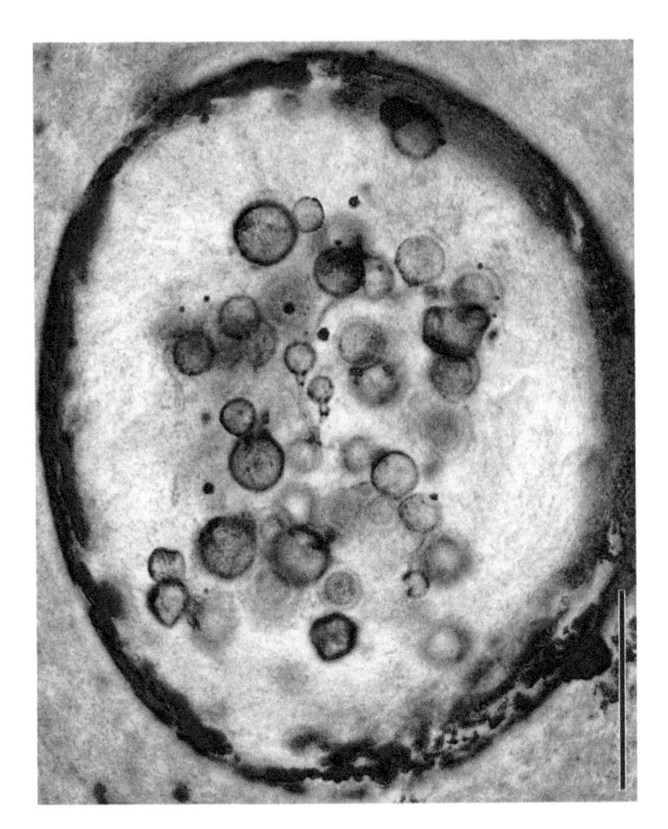

FIGURE 4.49 *Globicultrix nugax* inside thick-walled glomeromycotan spore. Lower Devonian Rhynie chert, Scotland. Bar = 50 µm.

some may represent other types of organisms, including other Glomeromycota. An interesting example is *Globicultrix nugax,* a polycentric thallus with a rhizomycelium of branched aseptate filaments and spherical sporangia that are exclusively terminal, usually positioned on a distinct subsporangial swelling of the subtending filament (Figure 4.49) (Krings et al., 2009a). Near the distal end of the sporangium is an apical or subapical discharge pore that may be slightly elevated and is papilla-like. The morphology and size of *G. nugax* has been compared to certain extant polycentric chytrids such as *Nowakowskiella* and *Cladochytrium*, both within the order Chytridiales (James et al., 2006a). Other types of intrusive microfungi in glomeromycotan spores from the Rhynie chert develop within the wall of the host spore (Figure 4.50) and show no evidence of a rhizomycelium extending into the spore lumen (Hass et al., 1994). In still others, only the distal end of the zoosporangia extends out from the cell wall.

Besides glomeromycotan spores, the vesicles of endomycorrhizal Glomeromycota from the Rhynie chert are frequently also colonized by a variety of intrusive

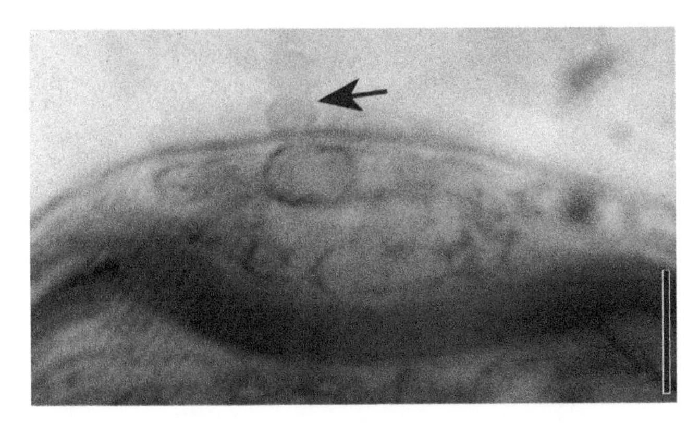

FIGURE 4.50 Mycoparasite thallus with protruding zoosporangium neck (*arrow*) developed between the wall layers of a glomeromycotan spore. Lower Devonian Rhynie chert, Scotland. Bar = 5 µm.

microfungi. One type produces tightly packed aseptate hyphae in the vesicle lumen; some specimens show evidence of a discharge papilla extending out from the vesicle surface (Hass et al., 1994). In others the filaments or hyphae are less tightly packed and irregularly constricted (Taylor et al., 1992a), and thus reminiscent of the strangulate rhizoids seen in members of the extant genus *Catenochytridium* (Berdan, 1941; Barr et al., 1987).

Host Responses

Perhaps the most direct source of information about the interactions between a microorganism and its fossil host is the presence of a host response. In the fossil record, such interactions indicate that the host was alive at the time of interaction with the fungus and responded in some way that can be interpreted from the fossil. There are of course some parasitic fungi that do not necessarily trigger a host response, and there are certainly host responses that cannot be measured because they may take place in other parts of the host organism or are not preserved in the fossil record (e.g. a stem parasite that causes leaf drop). For example, in extant fern roots, defensive cell wall appositions are constructed of a non-lignified polysaccharide matrix that serves as a template for the deposition of phenolic compounds which act as a barrier to microbial infections (e.g. Leroux et al., 2011). Moreover, it should be pointed out that wall appositions like those that are produced as a result of a biotic attack can also form as a result of mechanical wounding (e.g. bending of an axis) (Aist, 1977). Thus, when analyzing these structures in fossils, sources other than fungal parasites must also be considered.

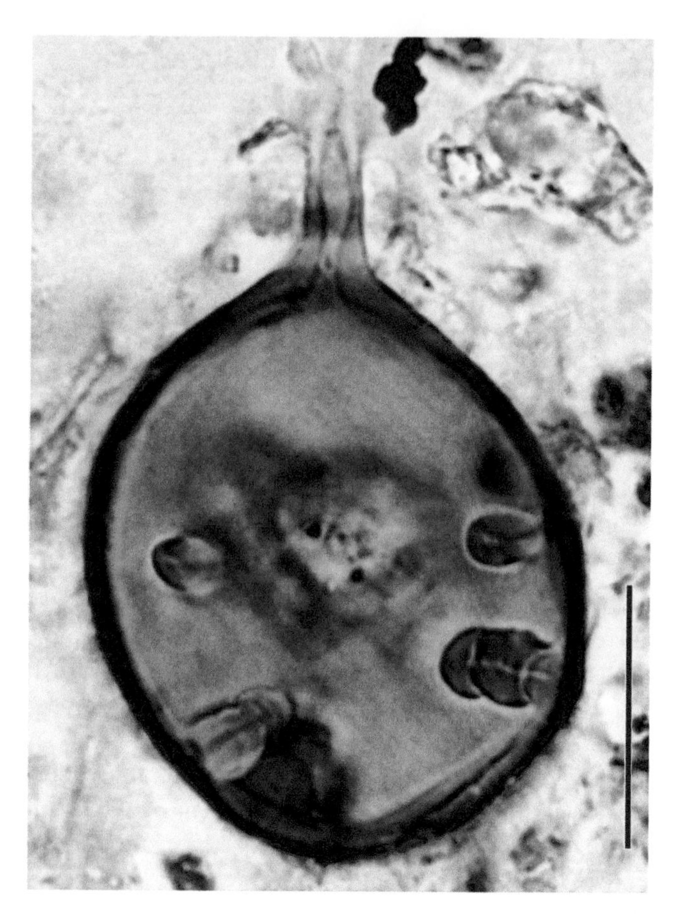

FIGURE 4.51 Chlamydospore with several papillae extending into the spore lumen. Note the longitudinal infection canal in the center of each papilla. Extant. Bar = 80 μm.

In the fossil record there are multiple examples of host responses caused by putative chytrids. One common type of host response recognizable from structurally preserved fossils is the development of structures termed callosities (Swart, 1975a), lignitubers (Mosse and Bowen, 1968), and various types of wall appositions (Collinge, 2009). Callosities assume a variety of shapes, from wart-like to elongate conical, and represent inwardly directed projections consisting of newly synthesized wall material that is formed by living plant cells (but also by certain fungal spores; see below) in response to invading fungi. Most are solitary but there are some that appear to branch. The fact that they are products of a living host is important for paleomycology since it indicates that the host was alive at the time of interaction with the fungus. Callosities encase the invading fungal hypha or filament, and it is widely interpreted that they are effective in preventing or retarding penetration by the parasite (e.g. Aist, 1977; Jeffries and Young, 1994). This type of host reaction has been observed in cells of numerous extant seed plants (e.g. Young, P.A., 1926; Rioux and Biggs 1994) but is also known to occur in ferns (e.g. Archer and Cole, 1986). In addition, it has been recorded for several fossil plants, including lepidodendralean lycophytes, *Astromyelon*-type calamitalean roots, a zygopterid fern from the Carboniferous of France (Krings et al., 2009b, c, 2010a; Taylor et al., 2012) and a microsporangiate gymnosperm strobilus from the Carboniferous of North America (Stubblefield et al., 1984).

In the fossil record, evidence of mycoparasitism is most commonly seen in large spores of members of the Glomeromycota that are abundant and widespread in the Early Devonian Rhynie chert ecosystem (Hass et al., 1994). It is interesting that none of the numerous glomeromycotan spores described and illustrated from the Rhynie chert by Kidston and Lang (1921b) shows evidence of callosities, although there are many spores that contain intrusive microfungi in the lumen (see p. 59). Many of the fossil callosities in glomeromycotan spores from the Rhynie chert are so well preserved that it is possible to follow a narrow, centrally located infection canal from the cell wall to the tapered end of the callosity (Figure 4.51). On the surface of the spore, the callosities appear as a circular scar about 10 μm in diameter, with a central translucent area that represents the infection canal. On either side of the canal is a series of chevrons representing the periodic synthesis of new wall material. One might view the formation of host callosities in these spores as being the result of a series of advances by the mycoparasite in the infection canal, each one prompting a response from the host in the form of the synthesis of new wall material.

The continued "arms race" between the parasite(s) and the host often results in the formation of multiple callosities in the same spore. Rarely does one find any evidence of the endoparasite in the spore lumen that also contains callosities. Although the precise affinities of the mycoparasites attacking the glomeromycotan spores in the Rhynie chert remain unknown, on many of the spore surfaces are chytrid zoosporangia, some directly over the infection canal (Figure 4.52). This provides strong evidence that at least some of the callosities formed as a result of parasitism by chytrids. Similar host responses in the form of callosities have been documented in both extant glomeromycotan spores (e.g. Swart, 1975a; Boyetchko and Tewari, 1991) and fossil glomeromycotan spores from the Carboniferous (Krings et al., 2009c).

While this example of a host response can be interpreted as indicative of a distinctly parasitic association, there are various types of microfungi residing in the

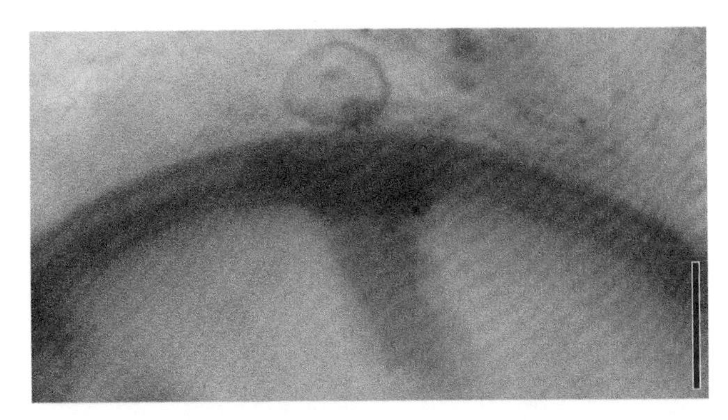

FIGURE 4.52 Chytrid attached to the outer surface of a glomeromycotan spore wall with host response papilla beneath. Lower Devonian Rhynie chert, Scotland. Bar = 5 μm.

walls and/or the interior of glomeromycotan (and other fungal) spores in the Rhynie chert that cannot be identified as to nutritional mode because there is no observable host response (Taylor et al., 1992a; Hass et al., 1994). Resolving the nutritional mode(s) of these associations would be particularly interesting with regard to better understanding the dynamics within the Rhynie paleoecosystems because, if the intrusive microfungi were parasites, they most likely impacted the number of viable glomeromycotan spores (see Purin and Rillig, 2008), reducing the number of mycorrhizal inoculations and thereby altering the structure of this early land plant community. In extant ecosystems the presence of mycoparasitism on glomeromycotan spores is known to result in a decrease in fitness. If certain species of mycorrhizal Glomeromycota in the ecosystem are less heavily infected because their spores are more resistant than others, parasitism could contribute to changing the structure of the plant community (Maherali and Klironomos, 2007). As true roots evolved with their absorptive capacity, perhaps there was less selectivity for a more efficient arbuscular mycorrhizal fungus than in early terrestrial ecosystems. While fitness cannot be measured using this parameter in a fossil ecosystem like the Rhynie chert, knowing what types (species) of fossil glomeromycotan spores were parasitized could provide some evidence relating to the resistance of some types of mycorrhizal fungi. For example, it would be interesting to know if there is any correlation between the construction of the asexual spore wall (layers and organization) and fungal parasites. Are some glomeromycotan spores structurally more resistant to chytrid attack because of their complex wall construction? As we learn more about the early diversity of glomeromycotan fungi in the fossil record, it will be interesting

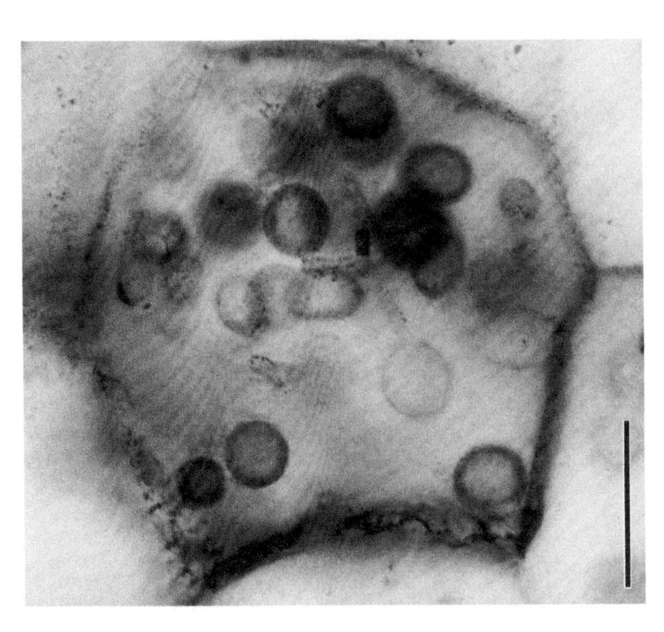

FIGURE 4.53 Cluster of microfungal spores in hypodermal cell from the rhizoidal ridge of *Nothia aphylla*. Lower Devonian Rhynie chert, Scotland. Bar = 30 μm.

to see if parasite resistance of glomeromycotan spores is in any way correlated with their structural organization and phylogenetic position.

Nothia Aphylla – *Multiple Host Responses*

Another example of a host response in the fossil record occurs in *Nothia aphylla*, a small Rhynie chert plant in the Zosterophyllophyta, an early group of vascular plants with affinities to various extant club mosses and the extinct Carboniferous arborescent lycophytes (El-Saadawy and Lacey, 1979). *Nothia aphylla* was 15–20 cm tall and characterized by dichotomously branched aerial axes arising from a system of non-stomatiferous, subterranean rhizomatous axes characterized by a prominent ventral rhizoidal ridge. The rhizoidal ridge, which is unique among Rhynie chert land plants, consists of a rhizoid-bearing epidermis, a multi-layered hypodermis, and files of parenchyma cells that connect to the stele; intercellular spaces are virtually absent (Kerp et al., 2001; Daviero-Gomez et al., 2005).

Three morphologically distinct fungal endophytes can be found in the rhizome and rhizoids of *Nothia*, including one type that is believed to be endomycorrhizal (Krings et al., 2007b, c) and one that may be chytrid-like. One is an intracellular endophyte. The hyphae are smooth-walled, aseptate, and occasionally branched. In hypodermal cells, the fungus produces clusters of spherical spores borne singly on short branches (Figure 4.53). This fungus

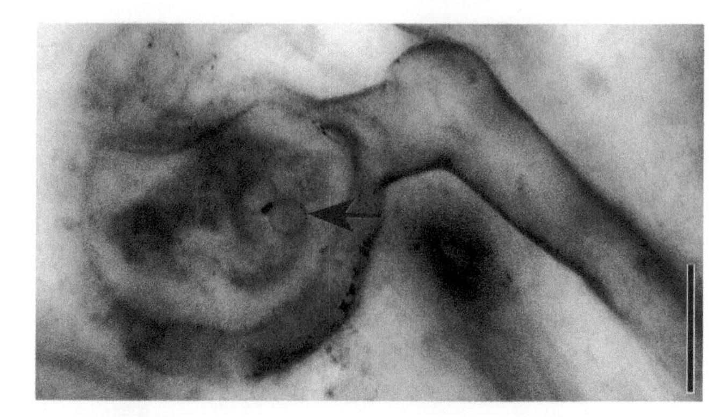

FIGURE 4.54 Inflated tip of *Nothia aphylla* rhizoid showing endophytic fungi (*arrow*). Lower Devonian Rhynie chert, Scotland. Bar = 60 μm.

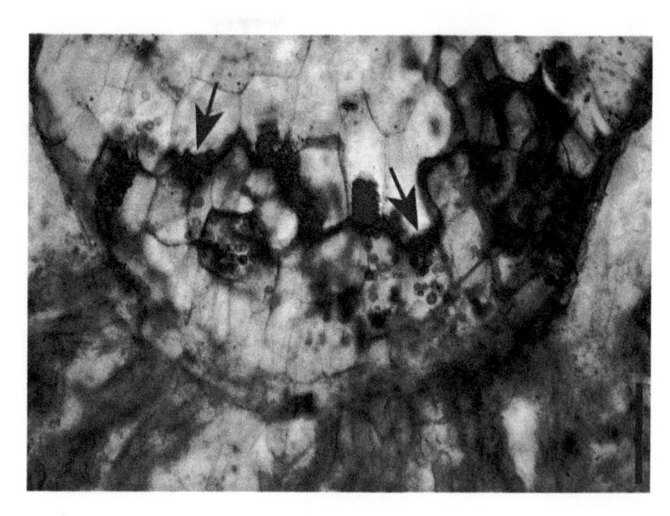

FIGURE 4.55 Section through *Nothia aphylla* rhizoidal ridge tissues showing zig-zag boundary line (*arrows*) of secondarily thickened cell walls that extends across hypodermis. Lower Devonian Rhynie chert, Scotland. Bar = 100 μm.

is abundant in all tissues of the rhizoidal ridge, including rhizoids, hypodermis, and conducting cells of the ridge. The fungus enters the axes through either the tip of the rhizoid or the rhizoid base. Typically it is the rhizoid tip that is infected and, apparently as a result of the infection, forms a characteristic bulge (Figure 4.54). As the fungus spreads, the lumina of hypodermal and connective cells become filled with hyphae and spores; in the most heavily infected *N. aphylla* specimen, approximately 95% of the cells of the rhizoidal ridge contain hyphae and/or spores of this endophyte. In cross sections of the most heavily infected axes, a distinct zigzag pattern of secondarily thickened anticlinal and outer periclinal cell walls is visible; it extends across almost the entire hypodermis of the rhizoidal ridge (Figure 4.55) and separates infected from uninfected tissues. Moreover, in two of the infected rhizoidal ridges the peripheral layers of the hypodermis (and the ventral epidermis in one instance) are disintegrated (Figures 4.56, 4.57).

The second fungal endophyte in *N. aphylla* is represented by large spherical to ellipsoid intracellular structures that are 50–120 μm in diameter, or 80–150 μm long and 40–60 μm wide, and may represent resting spores (chlamydospores) or zoosporangia. The fungus occurs in the rhizoids and all tissues of the rhizoidal ridge but is most abundant in the hypodermis of the ridge. Often the infected rhizoids and host cells are distinctly inflated. Also present in the prostrate axes are large, repeatedly branching aseptate hyphae of a third endophyte that occur intracellularly in rhizoids and the rhizoidal ridge, and which belong to a glomeromycotan fungus that might have been endomycorrhizal. Moreover, they are abundant in intercellular spaces of the cortex of both the prostrate and proximal portions of aerial axes. In the

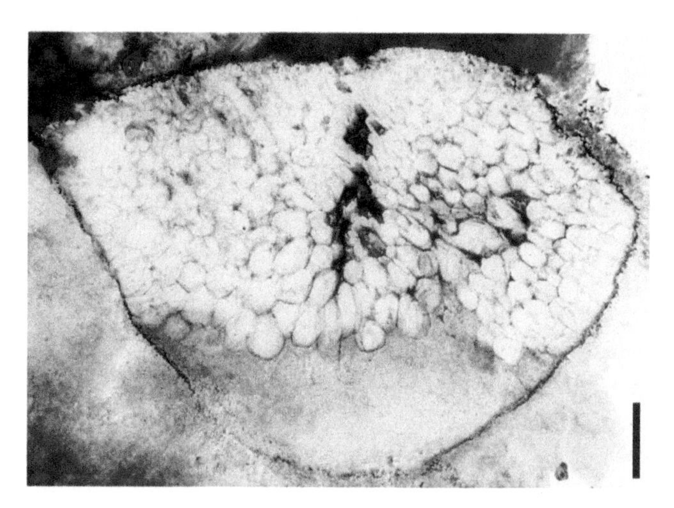

FIGURE 4.56 Section through *Nothia aphylla* rhizoidal ridge showing tissue void. A few cells along the margin are slightly larger in diameter. Lower Devonian Rhynie chert, Scotland. Bar = 250 μm.

intercellular system of the cortex, hyphae produce globose to irregularly shaped vesicles and scattered thick-walled spores up to 180 μm long and 150 μm wide. This fungus enters the axes via the rhizoid tip or base; infected rhizoids are not altered morphologically. In hypodermal cells, hyphae become sheathed with an encasement layer composed of cell wall material, and thus separated from the host cell protoplast (Figure 4.58).

The three fungal endophytes in *Nothia aphylla* are associated with characteristic cell and tissue alterations and host responses that are either specific (i.e. associated

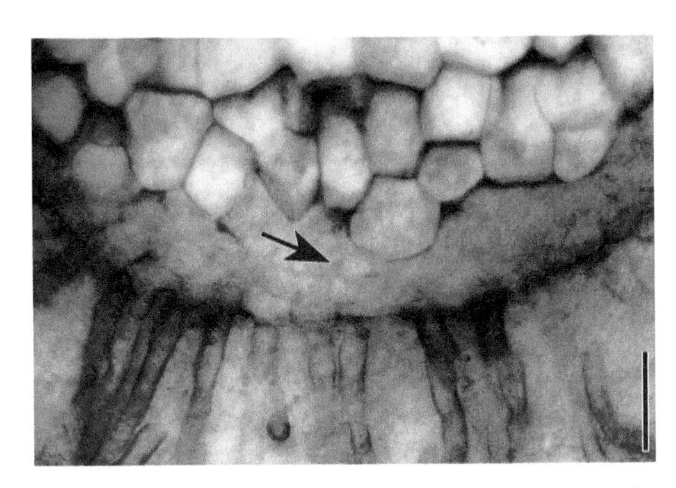

FIGURE 4.57 Section through *Nothia aphylla* rhizoidal ridge showing a void where hypodermal cells are degraded (*arrow*). Lower Devonian Rhynie chert, Scotland. Bar = 60 μm.

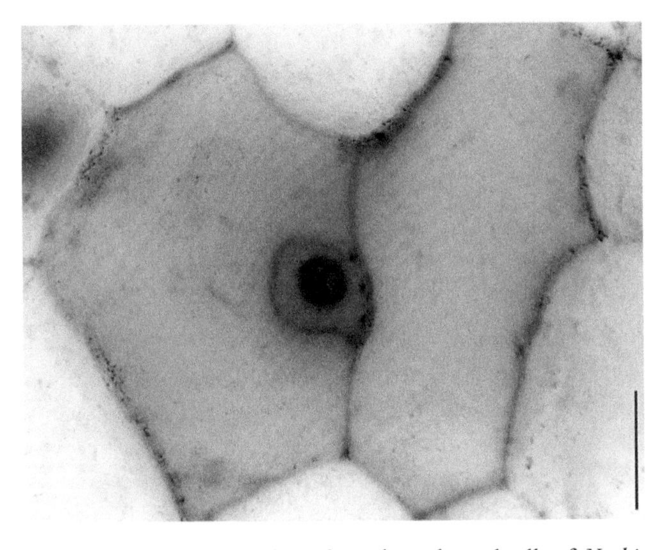

FIGURE 4.58 Cross section of two hypodermal cells of *Nothia aphylla*, one with encasement surrounding intracellular hypha. Lower Devonian Rhynie chert, Scotland. Bar = 20 μm.

with only one of the fungi) or non-specific (associated with two or all three fungi). Especially interesting is that the most heavily infected rhizoidal ridges are characterized by a hypodermal zigzag line composed of secondarily thickened cell walls (see Figure 4.55). This line marks the outer boundary of cells containing one of the fungi, and hence probably represents a specific host response that was effective in separating infected from uninfected tissues. Moreover, two of the infected rhizoidal ridges contain peripheral regions that are devoid of cells (see Figure 4.57). Krings et al. (2007b) suggest that this tissue degradation may have been effective as a defense mechanism based on the fact that, in some extant plants, phytopathogenic microorganisms may be deterred by programmed cell death around the infected areas that inhibit the microbe from spreading (Hammond-Kosack and Jones, 1996; Veronese et al., 2003; Glazebrook, 2005; Anderson et al., 2010). It is interesting that several of the Rhynie chert plants (e.g. *Aglaophyton major*) studied in detail possessed prostrate axes that are in contact with the substrate only via patches of rhizoids. In *N. aphylla*, however, the bilaterally symmetrical axis is totally subterranean with numerous rhizoids extending from the median ridge. One might speculate that this type of anchoring and conducting system would be advantageous to water uptake, especially in what has been interpreted as a sandy substrate (Kerp et al., 2001). This soft substrate may also have facilitated asexual reproduction, which was present in this plant. On the other hand, this type of anatomy may have resulted in an increased level of susceptibly to various types of endoparasites.

There is one additional aspect of the fungal interactions associated with *Nothia aphylla* involving the morphology and anatomy of the host that deserves mention. In a number of the Rhynie chert plants that have endomycorrhizae, the mycorrhizal fungus (see Chapter 7, p. 121) enters the plant via stomata on the aerial prostrate axes and then spreads out into the cortex to form a clearly defined zone of arbuscules (Remy et al., 1994a; Taylor et al., 1995a). In *N. aphylla*, however, the rhizomatous axes were subterranean and non-stomatiferous. This may have provided the selective pressure for an alternative mode of colonization by endomycorrhizal fungi (Krings et al., 2007a). The fungus entered the plant through rhizoids, probably because the axes were non-stomatiferous. Moreover, the morphology and radial arrangement of cells in the rhizoidal ridge, along with the virtual absence of intercellular spaces, perhaps did not provide an intercellular infection pathway into the cortex. Krings et al. (2007a) assumed that *N. aphylla* tolerated intracellular penetration in the rhizoids and within the tissues of the rhizoidal ridge in order to become inoculated. Tolerating (or even facilitating) intracellular penetration within a limited area of the axis may simultaneously have provided the plant with a means of recognizing and subsequently distinguishing the endomycorrhizal fungus from potentially harmful parasites. The parasites, once recognized, are confined in the tissues of the rhizoidal ridge by the host responses described above. This interpretation provides an interesting avenue of further research – to see, perhaps, not only how many

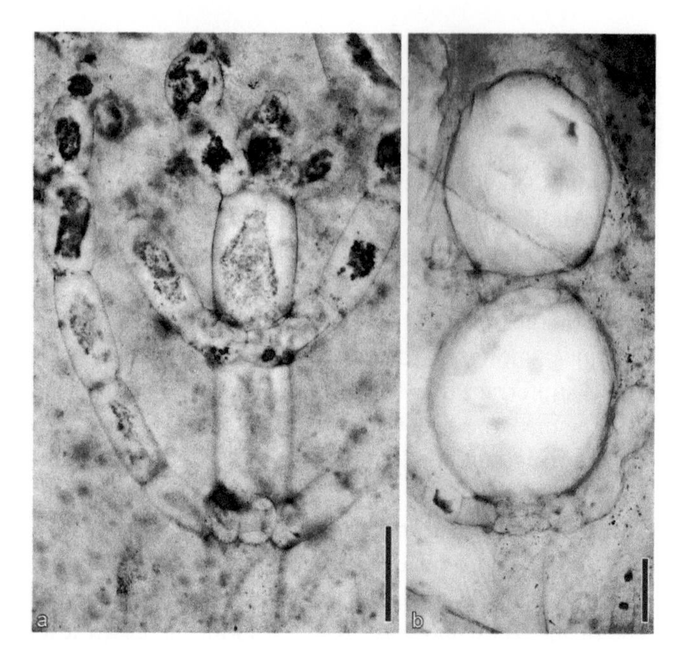

FIGURE 4.59 *Palaeonitella cranii.* a) Longitudinal section of normal axis with whorls of appendages arising at nodes. b) Two enlarged cells of infected axis. Compare cell size with internodal region cells in (a). Lower Devonian Rhynie chert, Scotland. Bars = 100 μm.

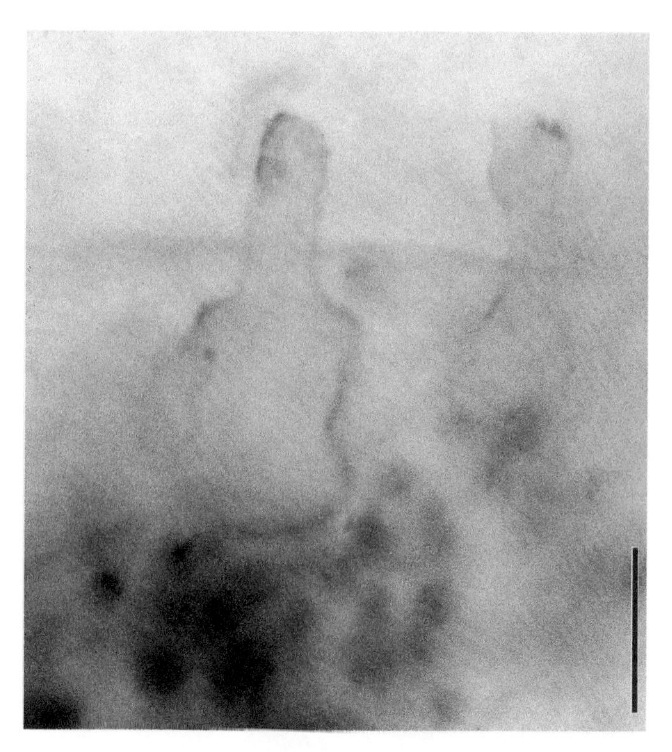

FIGURE 4.60 Two partially collapsed zoosporangia of *Milleromyces rhyniensis* with papillae extending through cell wall of *Palaeonitella cranii.* Note granular bodies beneath zoosporangia. Lower Devonian Rhynie chert, Scotland. Bar = 10 μm.

different patterns of mycorrhizal infection may have existed by Early Devonian time but also how these patterns might relate to structure–function relationships in early evolving land plants. Since there are a variety of cell types and tissue systems in land plants represented in the Rhynie chert ecosystem, it will be interesting to see if the microfungal endophyte signaling systems were equally diverse and whether the fungi adapted to the plants, or if the plants evolved in response to certain types of endophytes (Krings et al., 2012b).

Hypertrophy

The Rhynie chert provides the earliest evidence of hypertrophy (an increase in cell size as a result of an external stimulus) in the fossil record. It occurs in the charophycean green alga *Palaeonitella cranii*, which consists of an upright axis bearing regularly spaced whorls of lateral branches (Figure 4.59), some of which may branch further (Taylor et al., 1992b, c). On the surface of some cells of the alga are several thalli of different morphological types of chytrids that appear to be embedded in a thick surface layer; when the alga was living, this may have been some type of gelatinous coating on the cell surfaces. One of these chytrids was described as *Milleromyces rhyniensis* (Figure 4.60), a spherical to globose zoosporangium approximately 60 μm in diameter, which in some regions of the alga has a discharge tube that extends out from the cell wall. At the base of the thallus is a rhizoidal system (Taylor et al., 1992b). There are sufficient specimens preserved at various stages of development to enable the reconstruction of the life history of this chytrid (see Figure 4.11), including the presence of encysted zoospores, distinct discharge papillae, and the nature of the thallus. Also associated with *P. cranii* are two additional forms – *Lyonomyces pyriformis* (Figure 4.61) and *Krispiromyces discoides* (Figure 4.62); these differ in thallus morphology but are comparable with several extant chytrids that are parasites of freshwater algae, including species of *Entophlyctis* and *Phlyctochytrium* (Sparrow, 1960; Karling, 1977). The host response in *P. cranii* consists of several enlarged cells in the main axis (see Figure 4.59) that are approximately five times the diameter of normal ones (Taylor et al., 1992c). Hypertrophy is not the result of cell division but rather of the increased production of macromolecules that result in an increase in cell size. This host response also occurs in various extant algae that are parasitized by chytrids or other microorganisms, and results in a large

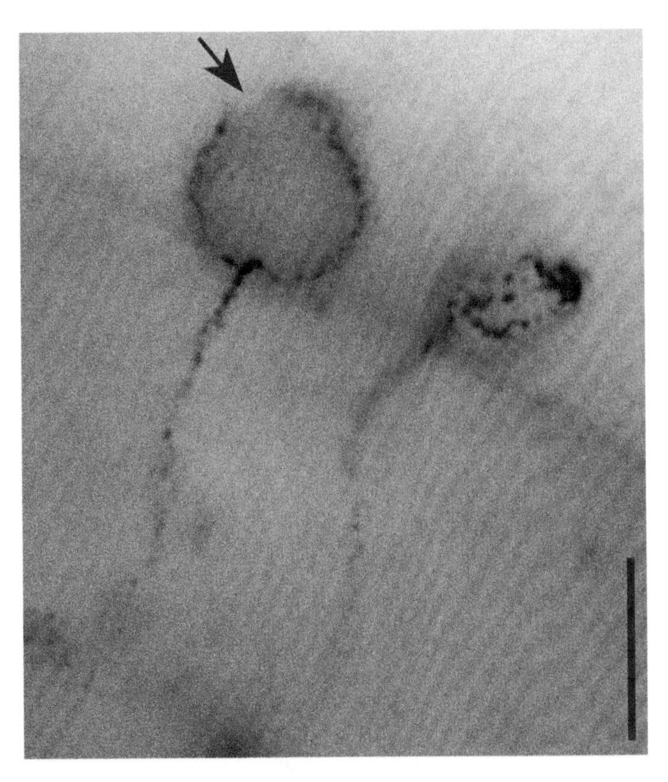

FIGURE 4.61 Two thalli of *Lyonomyces pyriformis* on cell of *Palaeonitella cranii.* Note rhizomycelium extending through host cell wall. Arrow indicates possible discharge pore. Lower Devonian Rhynie chert, Scotland. Bar = 15 µm.

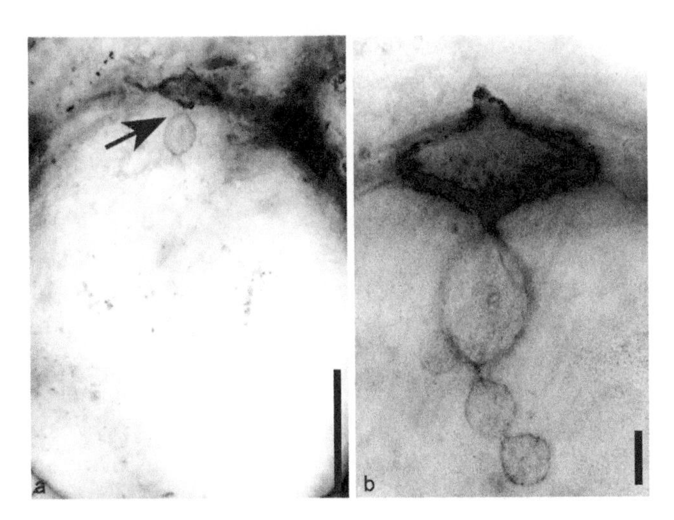

FIGURE 4.62 Thallus of *Krispiromyces discoides* (*arrow*) extending through cell wall of *Palaeonitella cranii.* Bar = 80 µm. b) Detail of a) showing well-defined zoosporangium and rhizomycelium. Lower Devonian Rhynie chert, Scotland. Bar = 10 µm.

volumetric increase in cell size (Moore-Landecker, 1982). This same pattern in cell increase has been reported in modern *Chara*, a close relative of *Palaeonitella* (Karling, 1928).

In the Carboniferous, hypertrophy has been reported in the underground parts (stigmaria) of the arborescent lycophyte *Lepidodendron* (Weiss, 1904a). The specimen contains the normal complement of tissues as observed in transverse section: a central vascular bundle surrounded by various cell types that make up the cortex. The cortex shows some indication of successive rows of cells that appear callus-like in organization. Within the outer cortex is a large elliptical cell that is much larger than the surrounding cells and contains a thin-walled spore. It is suggested that the parasite is morphologically similar to the extant chytrid genus *Urophlyctis*, which causes extensive hypertrophy and hyperplasia in roots, perhaps similar to gall formation (Weiss, 1904a).

MESOZOIC AND CENOZOIC CHYTRIDS

Chytrid zoosporangia have also been reported from a Jurassic hot-spring deposit (San Agustín) in Patagonia, Argentina (García-Massini et al., 2012). In this silicified deposit, which has some similarities to the much older Rhynie chert (Guido and Campbell, 2011), are various fragmented fungal remains (e.g. pycnidia of coelomycetes, thyriothecia of ascomycetes), including structures that resemble chytrid life history stages. They consist of spherical to pyriform structures that are similar to zoosporangia of several modern genera (e.g. *Chytridium, Nowakowskiella, Olpidium, Rhizophydium, Synchytrium*). Some thalli display what appears to represent the remains of a rhizomycelium. Within the fossil assemblage are both holocarpic and eucarpic forms, and in some are structures believed to be zoospores (García-Massini et al., 2012). This new discovery and the demonstration of a potentially high diversity of fungi in the Jurassic will be an important source of new information about host responses and fungus–plant interactions. The fact that the organisms from the Jurassic San Agustín hot spring ecosystem are silicified promises an important window into the fossil fungi of the Mesozoic.

There are a number of fossil chytrids that have been described from geologically younger sediments. Some of these occur in Eocene rocks of the famous Green River Formation (50 Ma), a deposit covering approximately 25,000 square miles in the western United States. The formation represents a series of lakes that were the repositories of an extraordinary diversity of organisms living in a climate interpreted as moist temperate to subtropical. There are thousands of interbedded layers of lake deposits and within some of these are fossils interpreted as chytrids. They are preserved in oil-containing shales, which are lithified and compacted organic oozes that accumulated in shallow lake bottoms and can be studied in thin sections of the rocks (Bradley, 1931). Within some of these sediments are structures that were initially interpreted as some type of alga (Davis, f1916; Köck, 1939; Kirchheimer, 1942), but subsequently understood to be zoosporangia of holocarpic chytrids (Bradley, 1967). Some range up to 45 μm in diameter and have discharge tubes of various lengths. Others have thicker walls in some regions and are lobed; these are suggested to represent dehisced zoosporangia. These fossil forms were included in the extant genera *Entophlyctis* and *Pleotrachelus* as new species, an assumption that is probably unwarranted (see Sherwood-Pike, 1988). The specimens of *Entophlyctis* reported from the Cenozoic (probably Miocene) of India are probably some epiphyllous fungal germlings (Jain and Gupta, 1970). Various microfossils have also been described and illustrated from thin section preparations of Jurassic and Cretaceous rocks of Great Britain (Ellis, D., 1915). Some of these are reported as being similar to filaments that were initially described by Renault (1895a, b, 1896b) as *Oochytrium* and found in conducting elements of *Lepidodendron.*

Pollen grains and spores sometimes contain structures and bodies that have uncertain affinities. No doubt some of these may represent chytrids that were preserved along with the pollen grain but have not been reported because the focus of the research was on the pollen rather than the contents. For example, *Rhizophidites triassicus* was described as a parasite on some Triassic spores; these chytrids resemble the extant *Rhizophydium pollinis-pini,* which is a parasite of pine pollen (Daugherty, 1941). An example that may represent such an occurrence was reported from the Deccan Intertrappean cherts (Chitaley and Sheikh, 1971) (Figure 4.63). Other suggested evidence of chytrids is reported from the Lower Rhine Valley Brown Coal (Cenozoic) of Germany in a few spores and pollen grains (e.g. Neuy-Stolz, 1958).

It is interesting to note that no persuasive evidence of chytrids has to date been found in amber (e.g. Girard and Adl, 2011), despite the fact that many other groups of microorganisms are well represented in this mode of preservation (e.g. Martín-González et al., 2009; Saint Martin et al., 2012, 2013) and the translucent nature of the matrix makes it relatively easy to discern even the delicate features that are useful in systematics, as well as those that can assist in determining interactions with other organisms.

FIGURE 4.63 Shya D. Chitaley.

CHYTRID SUMMARY

In modern ecosystems, chytrids are a critical component of the microbial world since many function as both saprotrophs and plant parasites. Using the Rhynie chert as a model ancient ecosystem, it is apparent that chytrids were a major component of aquatic communities by Early Devonian time and, at least from structural and morphological features that can be observed in the fossils, were already highly diversified (Taylor and Taylor, 1997; Taylor et al., 2004). This is consistent with molecular phylogenies that regard the group as very ancient and the basal group for all fungi (e.g. James et al., 2006b). To date the Rhynie chert has been the oldest ecosystem in which chytrids have been positively identified and characterized by multiple structural and morphological features. Although these Lower Devonian chert deposits have been studied for almost 100 years, there are many types of chytrids and chytrid-like fossils that still need to be recorded. In addition to expanding our knowledge about various thallus and reproductive features, there are several additional paths of future research that will be especially important moving forward in the study of fossil chytrids. One of these will be the utilization of multiple specimens in order to reconstruct the life history of particular types. This will be perhaps easiest for holocarpic forms that are small and have heavily infected the same tissue systems so that multiple specimens can be analyzed with some degree of confidence that the thalli represent the same organism. A second major thrust of studies dealing with fossil chytrids, regardless of the age, needs to be focused on interactions with other organisms. To date there are relatively few of these, but it is obvious that the occurrence of parasitic forms at approximately 411 Ma indicates that the necessary signaling mechanisms involved in parasitism were in place. We know relatively little about how these organisms interacted with other potential hosts, especially vascular plant lineages that evolved since the Early Devonian and that changed the dimensions and interactions of the ecosystem. For example, will it be possible to examine the effects of chytrid parasites on the germination of glomeromycotan spores that are a biological component of endomycorrhizae in relation to effects on the population of early land plants? And are there any other relationships between these zoosporic fungi and mycorrhizal fungal associations (e.g. Letcher and Powell, 2002)?

5

BLASTOCLADIOMYCOTA

FOSSIL BLASTOCLADIOMYCOTA 69
Palaeoblastocladia milleri 70

This clade of organisms was included as an order of the Chytridiomycota until it was elevated to the phylum level based on a life cycle with sporic meiosis and several ultrastructural and molecular features (James et al., 2006a, b; Porter et al., 2011). There are approximately 140 species in the group, which are characterized by eucarpic thalli that may be monocentric, polycentric, or mycelial. They include forms that are both saprotrophs and obligate parasites of both plants and animals, and there is increasing data that these zoosporic fungi are very important in food web dynamics (e.g. Gleason et al., 2010b). Zoospores are characterized by a nuclear cap consisting of ribosomes and anteriorly located lipid globules, which have been interpreted as ancestral characters in sporic fungi (Sparrow, 1960). Several ultrastructural features (e.g. absence of an electron-opaque plug in the transitional region of the flagellum, centriole arranged at right angles to the basal body) have been used to support the phylogenetic position of the Blastocladiomycota.

Unlike all other fungi, members of the Blastocladiomycota exhibit a complete alternation of generations – the life cycle consists of a haploid gametothallus that alternates with a diploid sporothallus (Figure 5.1). The gametothallus is typically characterized by two gametangia, one distal to the other. The upper one is smaller, often represents the male structure, and produces smaller gametes than the larger, proximal gametangium. Zoospores are produced in a single terminal zoosporangium on the sporothallus and they in turn give rise to new gametothalli. The sporothallus can also produce thick-walled resting sporangia in response to harsh environmental conditions. The use of both ultrastructural and molecular techniques has been especially valuable in more finely resolving groups within the phylum (e.g. Letcher et al., 2006, 2008a, b). Currently, the phylum contains the single order: Blastocladiales (Peterson, H.E., 1910). Included are the following families: Blastocladiaceae (exclusively saprotrophs), Catenariaceae (saprotrophs and pathogens), Coelomomycetaceae (pathogens of invertebrates), Physodermataceae (obligate parasites of plants), and Sorochytriaceae (pathogens of tardigrades), with only the Coelomomycetaceae and Physodermataceae interpreted as being monophyletic (Porter et al., 2011).

FOSSIL BLASTOCLADIOMYCOTA

Although there are some reports of blastocladialean fossils, these lack a sufficient suite of features to allow definitive placement in the group (e.g. Gangwar and Shamshery, 1982). To date there is only a single fossil that has been placed within the Blastocladiomycota with a high degree of confidence (Remy et al., 1994b; Taylor et al., 1994a). Although it has generally been difficult, some might say impossible, to place fossil taxa into existing groups (especially families or orders) based on morphological features, this is not the case with *Palaeoblastocladia milleri* (Remy et al., 1994b).

T.N. Taylor, M. Krings & E.L. Taylor: Fossil Fungi.

DOI: http://dx.doi.org/10.1016/B978-0-12-387731-4.00005-0

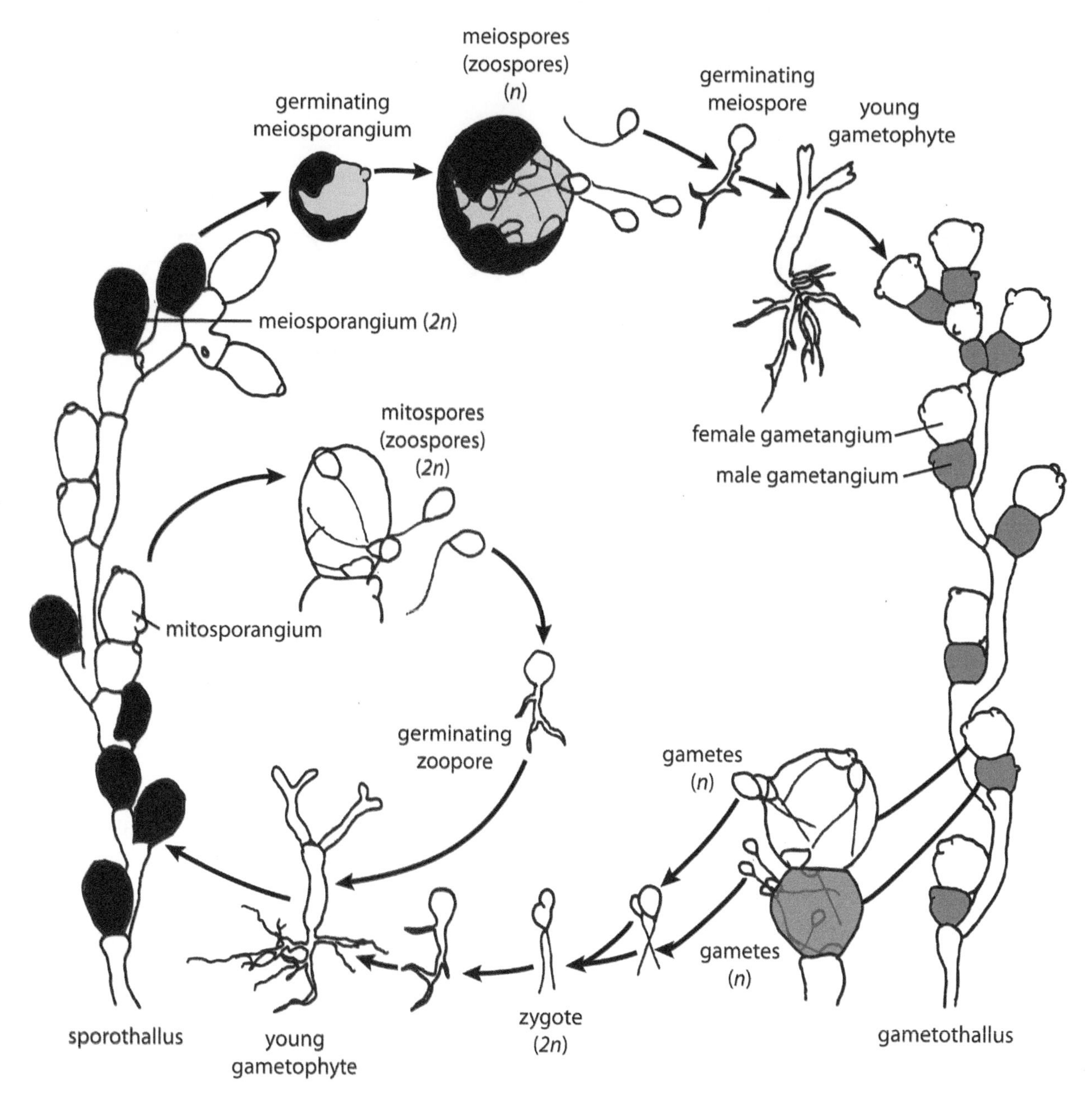

FIGURE 5.1 Life cycle of *Allomyces arbuscula*. Extant. (Modified from Raven et al., 1999.)

PALAEOBLASTOCLADIA MILLERI

This fungus occurs in the Rhynie chert in the form of tufts that arise from stomata and between the epidermis and cuticle of *Aglaophyton major*. The tufts extend out from axes and are approximately 200 μm in length (Figure 5.2). Thalli are of two nearly identical morphological types that consist of aseptate hyphae bearing branched rhizoids, each up to 16 μm in diameter (Remy et al., 1994b). On one type (sporothallus) are terminal bulb-shaped (Figure 5.3), thin-walled zoosporangia (= mitosporangia), each approximately 30 μm in diameter and attached to the hypha by a distinct flattened septum.

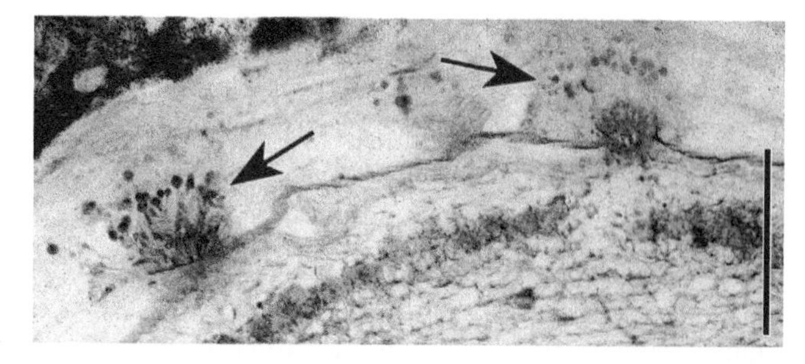

FIGURE 5.2 Portion of an axis of *Aglaophyton major* in transverse section showing two tufts of *Palaeoblastocladia milleri* (*arrows*) extending from the epidermis of the plant. Lower Devonian Rhynie chert, Scotland. Bar = 400 μm.

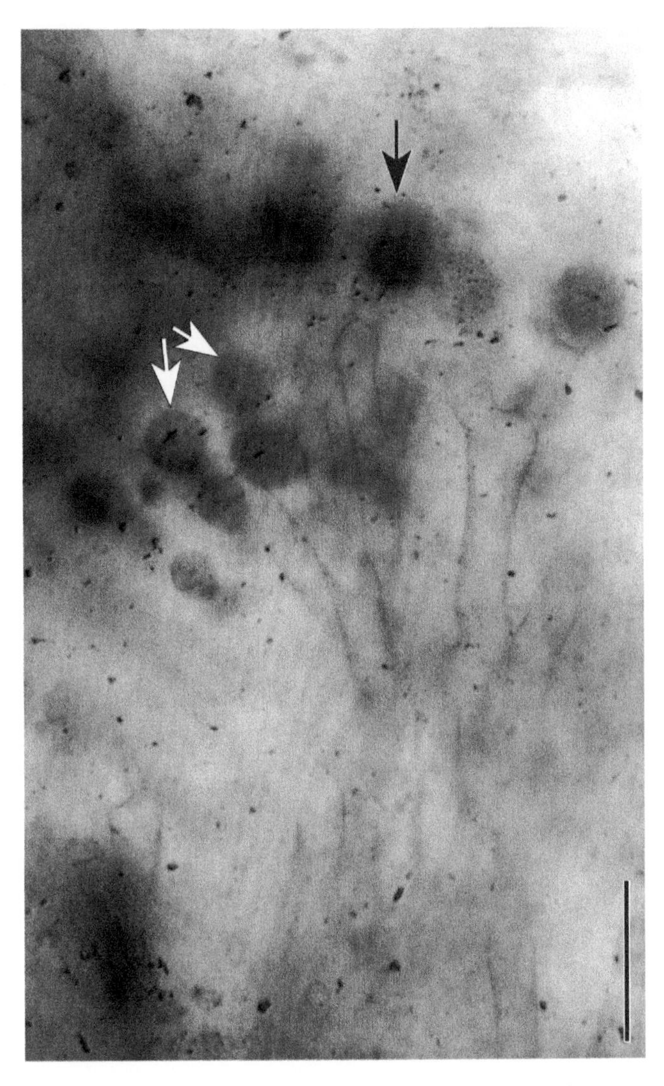

FIGURE 5.3 Several sporothalli (*black arrow*) and gametothalli (*white arrows*) of *Palaeoblastocladia milleri*. Lower Devonian Rhynie chert, Scotland. Bar = 50 μm.

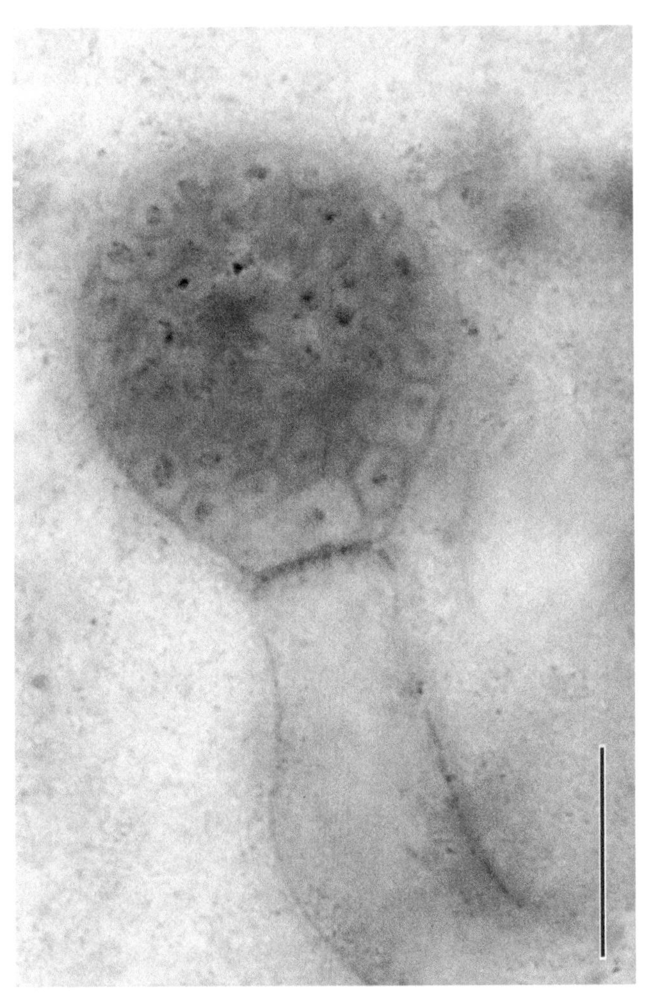

FIGURE 5.4 Zoosporangium of *Palaeoblastocladia milleri* showing dark inclusion in each zoospore. Lower Devonian Rhynie chert, Scotland. Bar = 10 μm.

Some of the zoosporangia contain contents that range from early stages in zoospore cleavage to zoospores with two dark inclusions (Figure 5.4). These opaque structures are present in the majority of zoosporangia and are morphologically similar to the nuclei and/or nucellar cap characteristic of blastocladialean zoospores (e.g. Ritchie, 1947; Sparrow, 1960; Barr, 2001). None of the zoosporangia shows evidence of discharge papillae, a condition that is apparently uncommon in extant members of the group; however, there are clumps of up to 25 zoospores in the matrix suggesting that perhaps they were released by the dissociation of the zoosporangium wall. Also associated with the sporothalli are meiosporangia, or resting sporangia. They also lack discharge papillae, are up to 44 μm in diameter, and possess a patterned surface ornament of delicate depressions or punctae.

The other reproductive structures present on the gametothallus of *P. milleri* are barrel-shaped gametangia that are smaller than the zoosporangia and are distinguished by their organization in pairs (Figures 5.3 and 5.5); sometimes a stack of three is present. None shows evidence of discharge papillae. The contents of the gametangia consist of circular to lenticular-shaped bodies up to 6.5 μm in diameter. In the matrix are several larger (up to 12 μm) structures that may represent zygotes.

The zoospores within the zoosporangia of *P. milleri* need to be critically examined in imaging systems other than transmitted light in order to see if some information might be obtained about the structures suggested to be possible nuclear caps or nuclei. This may be an especially

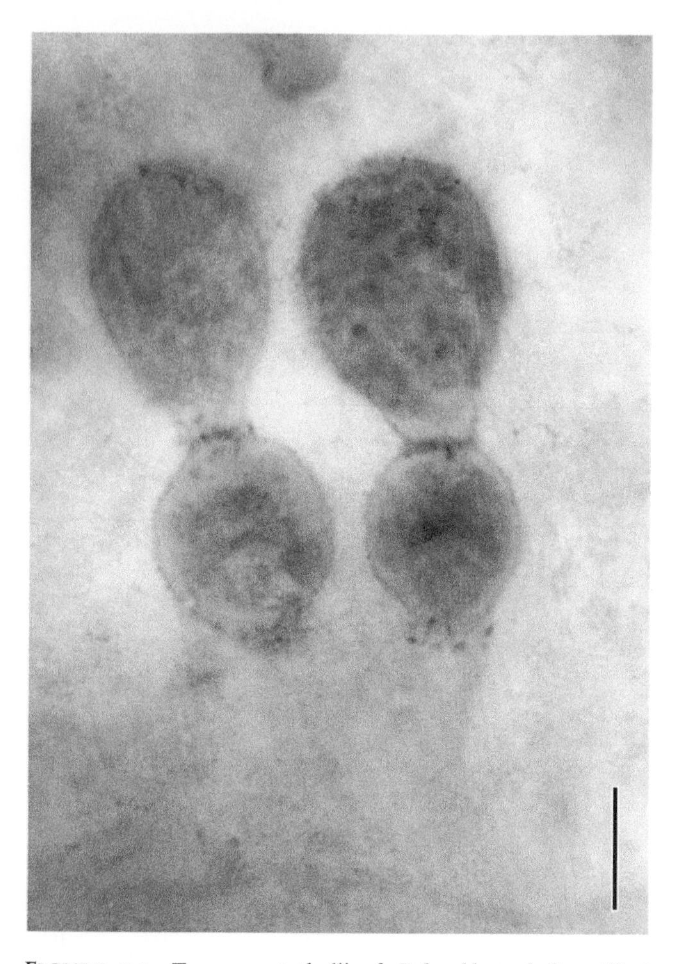

FIGURE 5.5 Two gametothalli of *Palaeoblastocladia milleri*, each consisting of a pair of gametangia. Note the difference in size between terminal and basal gametangia. Lower Devonian Rhynie chert, Scotland. Bar = 10 μm.

rewarding endeavor since zoospore ultrastructure has been interpreted as a highly conserved character (Barr, 2001). If there are any useful character states preserved, it will be interesting to see if the fossil evidence correlates in any way with phylogenies based on molecular data (Powell and Roychoudhury, 1992; James et al., 2006a, b).

On some of the host axes of *Aglaophyton major* are structures that may represent early stages in thallus formation. Based on morphological studies and the presence of two distinct types of thalli (Figure 5.6), *P. milleri* shares a number of features with some members of modern *Allomyces* and closely related species (Emerson, 1941; Emerson and Robertson, 1974; Prabhuji et al., 2012). The collective suite of characters of *P. milleri* is most similar to that of the *Eu-Allomyces*-type, including *A. arbuscula* (see Figure 5.1) and *A. macrogynus*. Especially important is the fact that the *Eu-Allomyces* group is characterized by an isomorphic alternation of generations, which is also present in the fossil *P. milleri*. Among the differences between the fossil and the extant taxa is the absence of any observable discharge mechanism for either the zoospores or gametes in the fossils. It is speculated that perhaps this mechanism may have evolved later in the group. Regardless of the direct comparisons that might be made between this fossil and modern members of the Blastocladiales, the complement of morphological and structural features seen in *P. milleri,* together with the evidence of an isomorphic alternation of generations, underscores the fact that, within the ancient "aquatic fungi," considerable diversity was already present by Early Devonian time, even in patterns of life history biology.

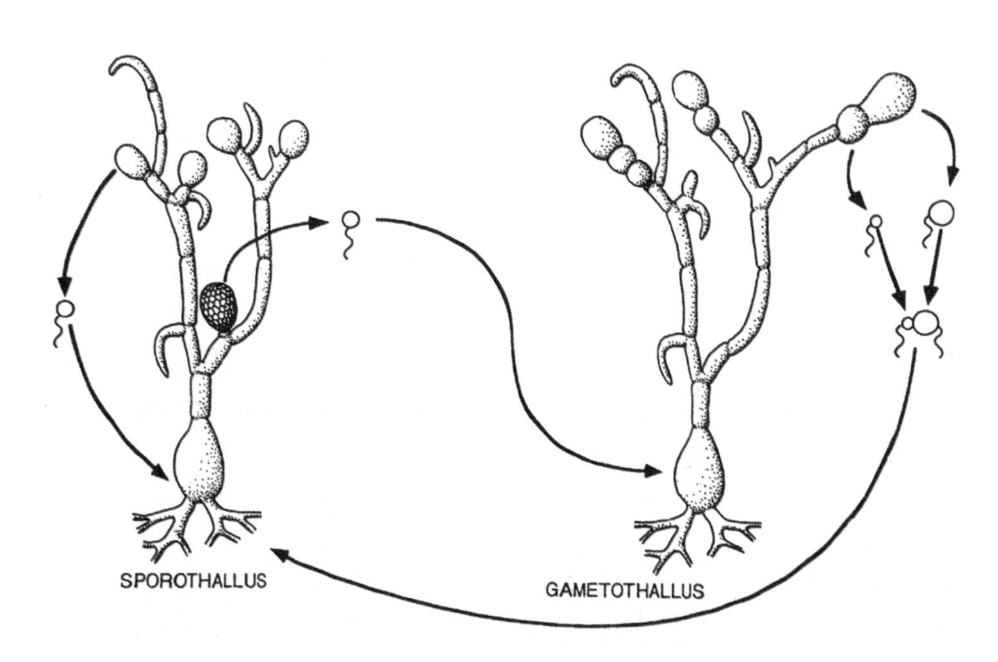

FIGURE 5.6 Suggested reconstruction of the life cycle of *Palaeoblastocladia milleri.*

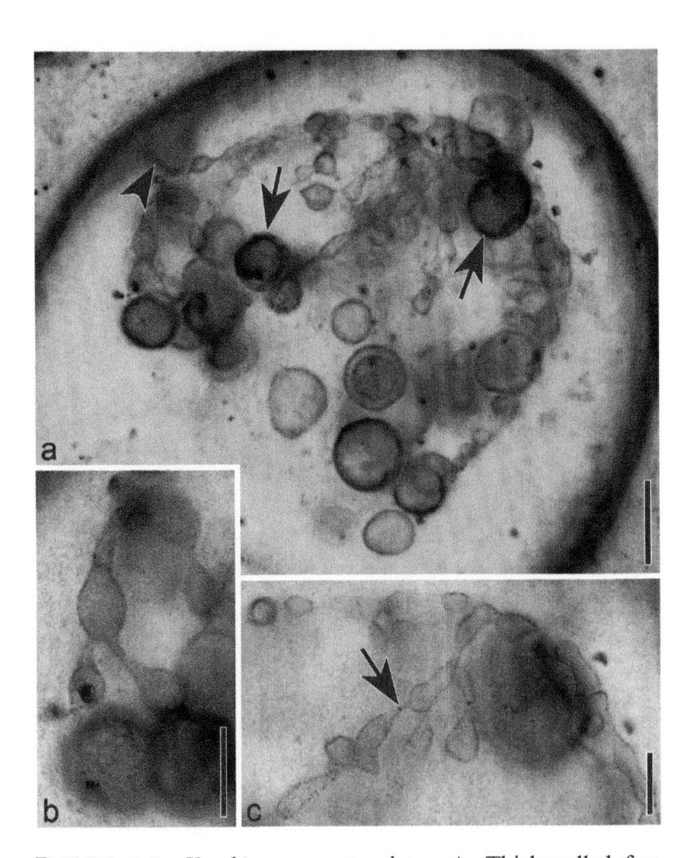

FIGURE 5.7 *Kryphiomyces catenulatus*. **A.** Thick-walled fungal (?glomeromycotan) spore containing thallus with reproductive structures (*arrows*). Arrowhead indicates position of papilla extending from the inner surface of the host spore wall. **B.** Hyphal branch with terminal spheroidal structures. **C.** Hyphae with (?pseudo-)septum (*arrow*) and catenulate swellings. Lower Devonian Rhynie chert, Scotland. Bars = 20 μm (A), 10 μm (B, C).

Another interesting microfungus from the Rhynie chert is *Kryphiomyces catenulatus*, an endobiotic mycelial thallus in a glomeromycotan spore (Figure 5.7A), which is constructed of pseudoseptate hyphae (Krings et al., 2010b). The unusual features of this microfungus include variously shaped crenulate swellings (Figure 5.7C) along the hyphae and the formation of what are interpreted as reproductive structures of some type at the tip ends of some hyphae (Figure 5.7B). These range up to 17 μm in diameter and are formed by a distinct constriction of the parental hypha. What type of organism *K. catenulatus* represents is difficult to interpret. Hyphal morphology is reminiscent of that in certain extant Hyphochytridiomycota, Chytridiomycota, Blastocladiomycota (i.e. *Gonapodya*), and even Ascomycota, but specific features that can be used to assign the fossil to any modern group are absent. Whether this interesting fungus was a mycoparasite of some type or a saprotroph that developed inside the spore after death is currently unknown.

6

ZYGOMYCETES

PRECAMBRIAN MICROFOSSILS ... 76
RHYNIE CHERT ... 79
ZYGOSPORANGIUM-GAMETANGIA COMPLEXES AND SPOROCARPS ... 80
Carboniferous ... 81
Triassic ... 85
"Sporocarps" and Other Fossils of Uncertain Affinities ... 88
AMBER FOSSILS ... 99
ICHNOTAXA ... 101
CONCLUSIONS: ZYGOMYCETES ... 101

The zygomycetous fungi (formerly Zygomycota) comprise about 1% of the true fungi; approximately 900 living species have been described (Kirk et al., 2001). They are an ecologically heterogeneous, paraphyletic, or polyphyletic assemblage of predominantly terrestrial organisms, which are generally placed near the base of the fungal tree of life (e.g. White et al., 2006; Liu et al., 2009; Liu and Voigt, 2010). The vegetative mycelium lacks septa except where reproductive units are produced and in regions of old or injured hyphae in Mucorales and Zoopagales, but plugged septa are regularly formed in Dimargaritales and Kickxellales (Benny et al., 2001). Zygomycetous fungi reproduce asexually via nonmotile endospores formed in sporangia, sporangiola (small sporangia), or merosporangia, or by the formation of chlamydospores, arthrospores, and yeast cells, and sexually (where documented) by the formation of zygospores in zygosporangia (in some forms clustered in sporocarps, e.g. Figure 6.1) following gametangial fusion, or azygospores without prior gametangial conjugation (Benjamin, 1979; Benny et al., 2001). Sexual reproduction in heterothallic species requires two mating types, (+) and (−), to be crossed in order for zygospores to be produced, whereas in homothallic species zygospores are formed from a single mycelium (Benny, 2012). A germinating zygospore can produce a germ sporangium, germ sporangiolum, or germ merosporangium in which spores are produced; in water a germ tube is formed (Benny et al., 2001; Benny, 2012). Most zygomycetous fungi thrive as saprotrophs, others as parasites and disease causative agents in other fungi, plants, animals, and even humans (White et al., 2006; Rogers, 2008; Richardson, 2009; Benny, 2012);

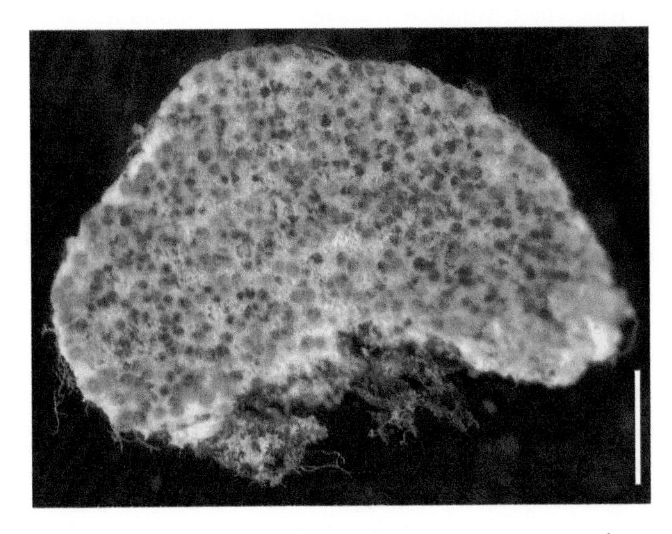

FIGURE 6.1 Section of *Endogone* sp. sporocarp showing numerous zygosporangia embedded in a yellowish gleba. Extant. Bar = 1.0 mm. (Courtesy D. Southworth.)

T.N. Taylor, M. Krings & E.L. Taylor: Fossil Fungi.

DOI: http://dx.doi.org/10.1016/B978-0-12-387731-4.00006-2

still others enter into mutualistic associations (mycorrhizae) with plants (Fassi et al., 1969; Walker, C., 1985).

Molecular phylogenetic studies indicate that the group informally termed the zygomycetes consists of several molecularly distinct groups (James et al., 2006b; Benny, 2012). As a result, the phylum Zygomycota was rejected as non-monophyletic and five apparently monophyletic lineages were recognized, including the Glomeromycota (Schüßler et al., 2001) and four subphyla *incertae sedis* (i.e. Entomophthoromycotina, Kickxellomycotina, Mucoromycotina, and Zoopagomycotina). For the latter taxa, the informal umbrella term "zygomycetous fungi" or "zygomycotan fungi" is frequently used (Hibbett et al., 2007; Liu and Voigt, 2010). A fifth subphylum, the Mortierellomycotina, has more recently been separated from the Mucoromycotina (Hoffmann et al., 2011). More extensive gene-based analyses indicate that the Entomophthoromycotina are quite distinct from all other fungi and thus deserve to be recognized at the level of a new fungal phylum, Entomophthoromycota (Humber, 2012; Gryganskyi et al., 2012).

Molecular studies have been used to suggest that the first zygomycetous fungi appeared during the Precambrian, approximately 1.4–1.2 Ga (Heckman et al., 2001; Blair, 2009); more conservative estimates place the divergence at about 800 Ma (Berbee and Taylor, 2001). If these estimates are accurate, zygomycetous fungi were certainly important elements in ancient terrestrial ecosystems and must have existed before many other terrestrial organisms. Nevertheless, well-documented evidence of fossil zygomycetes continues to be rare. Not even the famous Rhynie chert, which is the single most important source of information on fossil fungi to date relative to paleoecosystem functioning (see Taylor et al., 2004), has produced conclusive evidence of zygomycetous fungi. As a result, efforts to reconstruct the evolutionary history and phylogeny of the zygomycetous fungi or of lineages within this group are to date based exclusively on the analysis of extant members (e.g. Nagahama et al., 1995; White et al., 2006; Liu et al., 2009; Petkovits et al., 2011).

The scarcity of fossil evidence of zygomycetous fungi remains perplexing, especially since habitats conducive to the growth of these organisms, together with depositional environments conducive to their preservation, were available at least by the Paleozoic (e.g. Krings et al., 2012b). The scarcity of reports appears to be related in part to the nature of the fossil record of fungi in general, which typically results in the preservation of isolated parts or stages of the life cycle. Most structures formed during the zygomycetous life cycle are nondiagnostic at the level of resolution available with transmitted light. Only mature zygosporangia or zygospores with attached gametangia and/or suspensors (plus perhaps certain asexual reproductive structures such as the columellate sporangia of the Mucoraceae) represent components of the life cycle that lend themselves to preservation in a recognizable form and thus can be used to positively identify a fossil zygomycetes. As with the identification and documentation of other fossil fungi, however, the recognition of these structures as fossils relies heavily on the mode of preservation and technique(s) used to prepare samples.

To date calcium carbonate coal balls and silicifications represent the only sources of compelling evidence of fossil zygomycetes (Krings et al., 2013c). Other types of preservation, including amber, have not yet yielded unequivocal evidence of zygomycetous fungi, with a few possible exceptions. Another reason for the scarcity of reports on fossil zygomycetes may be that most of these fungi are saprotrophs. While biotrophic fungi often trigger some type of host responses, and/or possess specific infection and penetration structures (e.g. appressoria), along with special features facilitating nutrient extraction from the host (e.g. arbuscules, haustoria), saprotrophs do not normally possess special structures that allow for positive recognition as fossils. As a result, far more definitive evidence of biotrophic rather than saprotrophic fossil fungi has accumulated. Below, we provide some examples of fossils that have been interpreted as, or compared to, zygomycetous fungi. Since the systematic affinities remain more or less inconclusive for most of these fossils, they are presented largely according to their geological appearance rather than their systematic affinities.

PRECAMBRIAN MICROFOSSILS

Although molecular clock estimates indicate that the zygomycetous fungi originated in the Precambrian, compelling fossil evidence to corroborate this hypothesis has not been produced to date. Nevertheless, there are a few enigmatic Precambrian microfossils that have been compared with zygomycetous fungi; all of these are from the Proterozoic.

The geologically oldest fossils that have been suggested as displaying certain morphological similarities with extant zygomycetous fungi are organic-walled microscopic structures (termed amoeboid thalli) that have been discovered in ~2 Ga-old black shales from the Paleoproterozoic Pechenga Complex of the Kola Peninsula, Russia, and have been named *Petsamomyces polymorphus* (Belova and Akhmedov, 2006). Specimens occur singly or attached to what appear to be subtending hyphae or filaments. On average they are 20 × 48 μm in diameter and vary from irregularly ovoid and sac-shaped to complex; all have a solid, somewhat transparent outer wall. So-called hyphae co-occurring with the amoeboid thalli are 3–6 μm wide and multi-branched. Irregular thickenings are common and supposedly represent incipient thalli. Reproductive structures resembling cysts and sporangia have also been described. The biological affinities of *Petsamomyces* remain equivocal. The fossils show a mixture of structural features that occur in members of different groups of modern fungi and fungus-like organisms, including Myxomycetes, Peronosporomycetes, Chytridiomycota, and zygomycetous fungi. Other Precambrian fossils have been assigned to the now obsolete class Phycomycetes (e.g. Timoféev, 1970), a group that today may include members of the Chromista as well as some fungi traditionally placed in the Chytridiomycota and zygomycetes (Kirk et al., 2008).

Hermann (1979), Hermann and Podkovyrov (2006), and German and Podkovyrov (2011; Figure 6.2) described and gave the name *Mucorites ripheicus* to some carbonaceous compressions of irregularly aggregated filaments, globules, and what appeared to be fused filaments from the Lakhanda microbiota (late Mesoproterozoic/Riphean; ~1030–1020 Ma) of the Uchur-Maya region of southeastern Siberia (Figures 6.3–6.5). The fossils are believed to represent different life cycle stages of a mucoralean zygomycete in which gametangial fusion and azygo- or zygospore formation are virtually identical to those observed in the modern *Mucor tenuis*. Numerous spheroidal fossils interpreted as zygospores co-occur with pairs of distally swollen hyphae suggestive of gametangial fusion. Other specimens are morphologically similar to zygospore germination and development of an incipient sporangium (see Figure 6.4). Structures interpreted as sporangiophores were also found (see Figure 6.5). Hermann and Podkovyrov (2006) noted that the morphology of the fossils could be used to establish their systematic affinities with the zygomycetes. Slightly older than the Lakhanda fossils are dispersed remains from the middle Riphean Debengdinskaya

FIGURE 6.2 Tamara German (left) and Victor N. Podkovyrov (right).

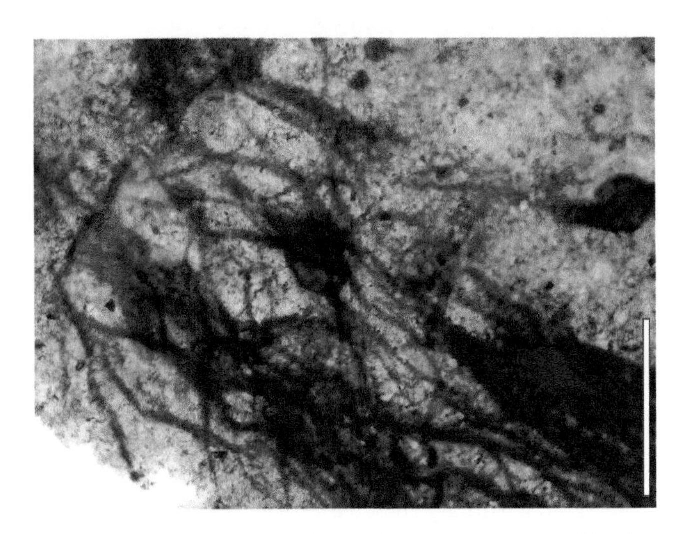

FIGURE 6.3 *Mucorites ripheicus* showing aggregate of hyphae and putative zygospores. Proterozoic, Siberia. Bar = 200 μm. (Courtesy T. German and V. Podkovyrov.)

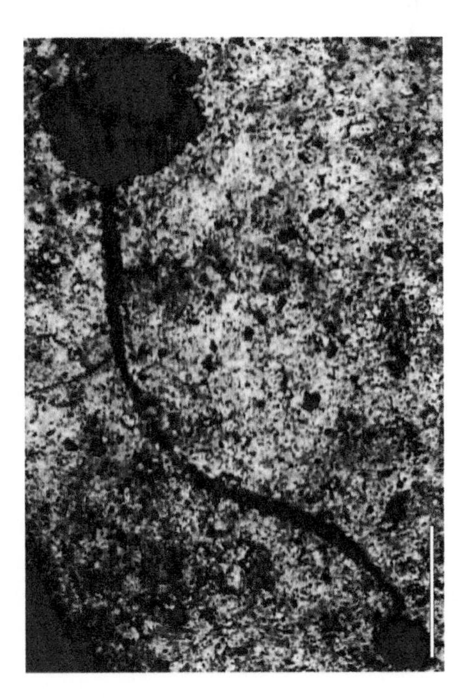

FIGURE 6.4 *Mucorites ripheicus.* Zygospore germination and development of an incipient sporangium. Proterozoic, Siberia. Bar = 100 μm. (Courtesy T. German and V. Podkovyrov.)

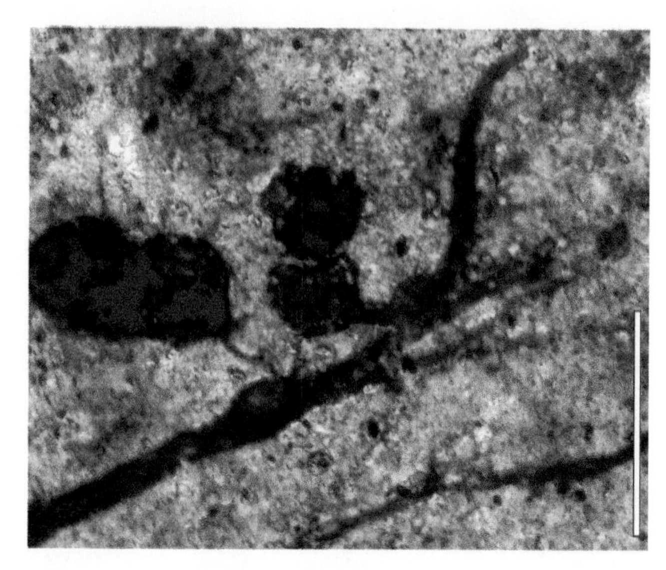

FIGURE 6.5 *Mucorites ripheicus.* Putative sporangiophores. Proterozoic, Siberia. Bar = 200 µm. (Courtesy T. German and V. Podkovyrov.)

Formation (~1300–1200 Ma; middle Mesoproterozoic) of the Olenekskiy uplift in Siberia described by Stanevich et al. (2007). Among these latter fossils are several thick-walled spherical structures (named *Lophosphaeridium* sp. 1) that have been compared to zygosporangia seen in modern members of the Mucorales.

Another Proterozoic fossil that has variously been interpreted as representing some level of fungal organization is the Mesoproterozoic–Neoproterozoic (~1600–542 Ma) *Tappania* (Figure 6.6), an organism initially described as an acritarch (e.g. Yin, 1997; Javaux et al., 2001; Butterfield, 2005a, b; Nagovitsin, 2009). *Tappania* ranges among the oldest ornamented microfossils known (Javaux et al., 2004; Huntley et al., 2006; Moczydłowska et al., 2011) and has been reported from sites in Canada (Butterfield, 2005a, b; Figure 6.7), China (Xiao et al., 1997; Yin, 1997), Australia (Javaux et al., 2001), India (Prasad and Asher, 2001), and Russia (Nagovitsin, 2009). The most prominent species, *T. plana*, is characterized by irregularly septate filamentous processes, each up to 60 µm long, that form a series of anastomoses surrounding a central, spheroidal, ovoid or slightly elongate vesicle, which may reach up to 160 µm in diameter. The processes, which may be branched or unbranched, appear to communicate freely with the vesicle (Knoll et al., 2006).

Javaux et al. (2001) interpreted *Tappania* as an actively growing vegetative cell or germinating structure rather than a metabolically inert resting stage (cyst), what many acritarchs are believed to represent, based on the irregular morphology and asymmetric distribution of processes (Knoll

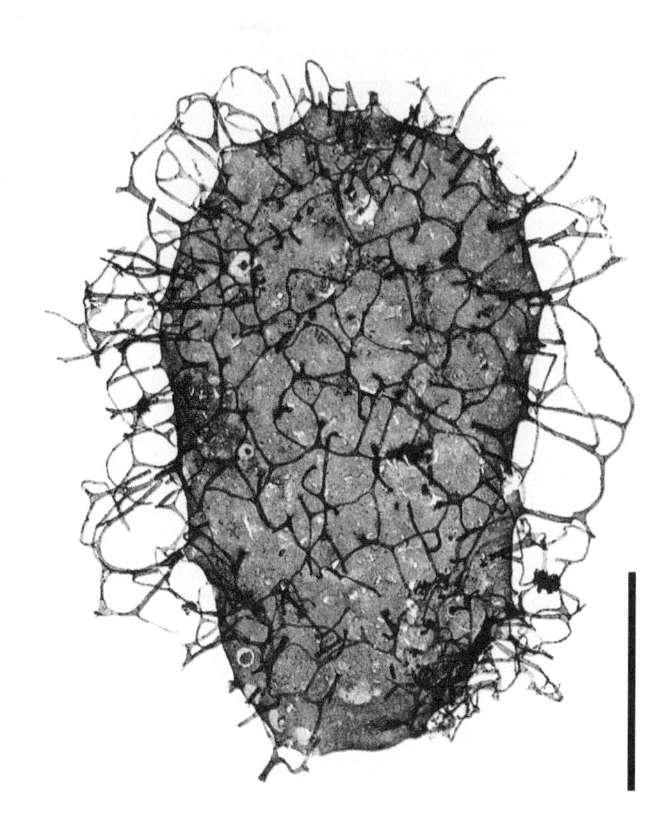

FIGURE 6.6 *Tappania* sp. showing anastomosing external ornament. Neoproterozoic, Canada. Bar = 100 µm. (Courtesy N.J. Butterfield.)

FIGURE 6.7 Nicholas J. Butterfield.

et al., 2006). Butterfield (2005a, b) used the high level of organization to suggest that *Tappania* represents an early fungus that occupies a position somewhere between the Ascomycota and zygomycetous fungi. Berbee and Taylor (2010) view *Tappania* as an interesting puzzle, but have challenged Butterfield's interpretation of the fossil as being fungal. Knoll et al. (2006) regard affinities of *Tappania* with the fungi as possible but prefer to view the fossil as problematic because of the limited number of systematically informative characters. These last authors, however, have agreed with Butterfield (2005a, b) that *Tappania* was a eukaryote with a complex cytoskeleton, probably preserved in an actively growing phase of its life cycle, and at least plausibly heterotrophic. On the other hand, Moczydłowska et al. (2011) dismiss the fungal affinities of *Tappania* because of its resistant cell walls and certain structural features that do not conform to those of fungi. Rather, these authors favor affinities of *Tappania* with some algal group. Moreover, they interpreted collar-like extensions occurring in some specimens of *Tappania* as algal excystment structures, rather than attachment sites of parental hyphae. Retallack et al. (2013a), however, suggested that *Tappania* may have been a free-living fungus associated with cyanobacterial mats, perhaps belonging to the Archaeosporales (Glomeromycota). Establishing the biological affinities of *Tappania*, including whether it is, in fact, fungal, will require more definitive evidence, for example on the parental structure, if any, in/on which it was produced.

RHYNIE CHERT

Within the Early Devonian Rhynie chert are abundant, exquisitely preserved fossil fungi (Taylor et al., 2004), including members of the Chytridiomycota (e.g. Taylor et al., 1992a; Krings et al., 2009a), Blastocladiomycota (Remy et al., 1994a), Glomeromycota (e.g. Taylor et al., 1995a; Dotzler et al., 2009), and Ascomycota (Taylor et al., 2005a), as well as representatives of the fungus-like Peronosporomycetes (Taylor et al., 2006; Krings et al., 2012c). To date, however, the Rhynie chert has not produced conclusive evidence of the presence of zygomycetous fungi; those fungi that had been assigned to that group historically (e.g. the arbuscular mycorrhizal fungus *Glomites rhyniensis*; see Taylor et al., 1995a) are today accommodated in the Glomeromycota (see Chapter 7). Although no definitive zygomycetes are known from the Rhynie chert, there are several fossils that have been noted as bearing some similarities. For example, it was recently suggested that certain mycorrhizal fungi in the Rhynie chert land plant *Horneophyton lignieri* might represent zygomycetous fungi with affinities to the Endogonales (Strullu-Derrien et al., 2014).

It has been suggested that the mycobiont of the lichen-like organism *Winfrenatia reticulata* from the Rhynie chert

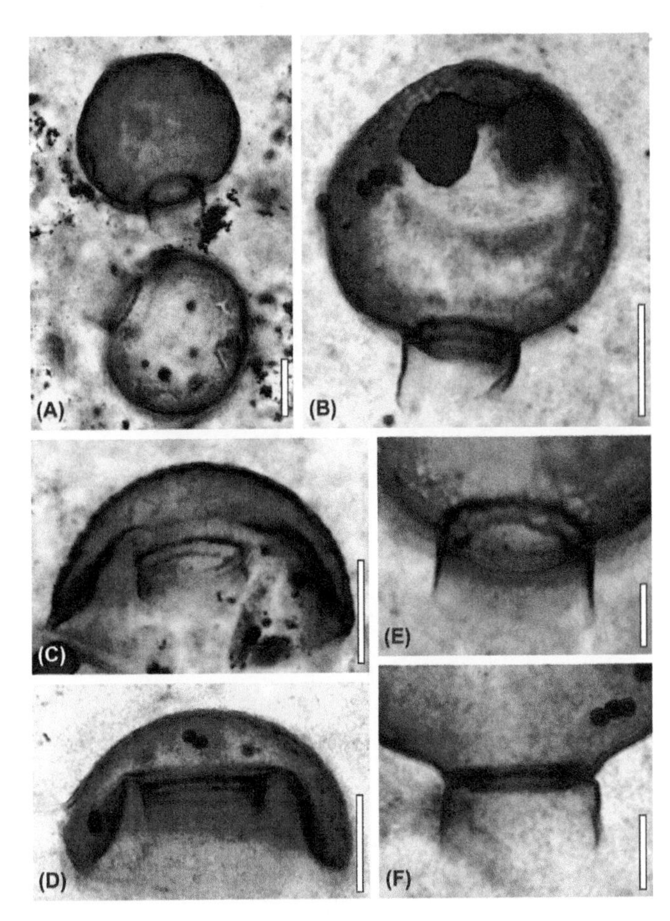

FIGURE 6.8 Microfossils from the Rhynie chert resembling columellae of the extant mucoralean zygomycete *Rhizopus*. See explanation of figures in text. Lower Devonian Rhynie chert, Scotland. Bar (**A–D**) = 20 μm; (**E–F**) = 10 μm.

(Taylor et al., 1997; Karatygin et al., 2009) was a member of the zygomycetous fungi based on the presence of aseptate hyphae and thick-walled, sculptured spores (Taylor et al., 1997). *Winfrenatia reticulata* (see Chapter 10, p. 208 for detailed description) consists of a mycelial mat constructed of interwoven hyphae. Along the upper surface of the mat are numerous shallow depressions, within which are coccoid unicells that are morphologically similar to certain extant cyanobacteria. Hyphae of the fungus extend into the depressions and become intertwined with the cyanobacteria.

Another Rhynie chert fossil that might represent part of a zygomycetous fungus occurs in the form of tiny, globose to subglobose structures, uniform in size and shape and 50–60 μm in diameter. These structures (Figure 6.8A, B) occur singly or in groups dispersed in the chert matrix, close to, but never in, degraded land plant axes and sporangia. Some of the specimens are "collapsed," i.e. the proximal half appears deflated with the distal half depressed on

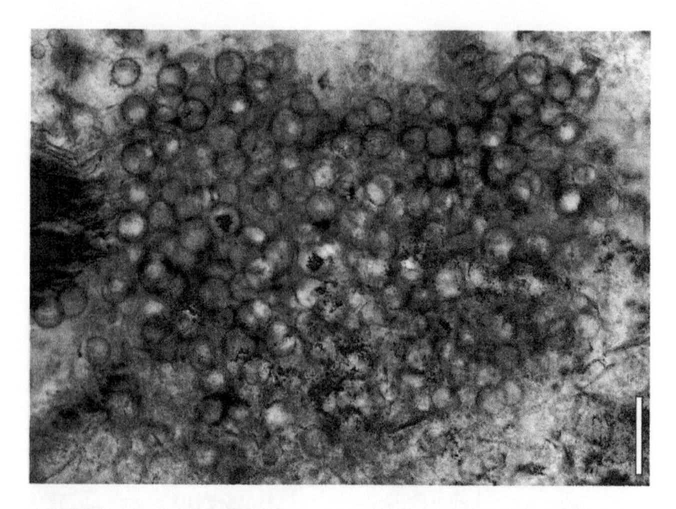

FIGURE 6.9 Aggregation of *Zwergimyces vestitus* from the Rhynie chert. Lower Devonian Rhynie chert, Scotland. Bar = 100 µm.

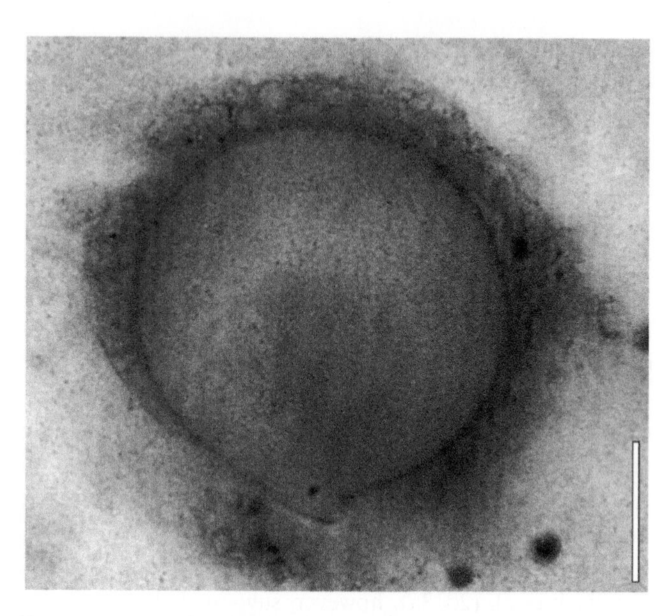

FIGURE 6.10 Specimen of *Zwergimyces vestitus* with thin-walled inner investment element. Lower Devonian Rhynie chert, Scotland. Bar = 10 µm.

top, giving the whole structure an umbrella-like configuration when viewed laterally (Figure 6.8C, D). On the proximal side of each of the spheres is a circular orifice, which is surrounded by a conspicuous collar-like structure (Figure 6.8E, F). The proximal end of the collar appears irregular, suggesting that it may have been mechanically separated. The fossils are morphologically similar to columellae seen in members of the extant genus *Rhizopus* (Mucorales). If this assignment is accurate then the orifice would represent the attachment site of a sporangiophore, and the collar-like appendage, the proximal portion of a peridium that was repositioned downwards.

Zwergimyces vestitus (initially described as *Palaeomyces vestitus*; Kidston and Lang, 1921b) is a small, spheroidal fungal reproductive unit, 40–50 µm in diameter, which occurs in degraded plant axes from the Rhynie chert (Krings and Taylor, 2013). Specimens occur singly or in aggregates, sometimes embedded in a confluent meshwork of hyphae (Figure 6.9), and are composed of a central cavity sheathed by an outer hyphal and inner non-hyphal investment (or mantle). The outer investment is typically prominent, whereas the thickness of the inner investment element varies considerably between specimens (from <1 to >5 µm) (Figures 6.10 and 6.11). *Zwergimyces vestitus* represents the oldest evidence of a hyphal mantle in fungal reproductive units. The fossil is morphologically similar to the mantled zygosporangia of certain extant zygomycetous fungi, but also the spores of some Glomeromycota. Evidence of gametangial fusion, which could be used to positively identify *Zwergimyces* as a zygomycete, however, is lacking.

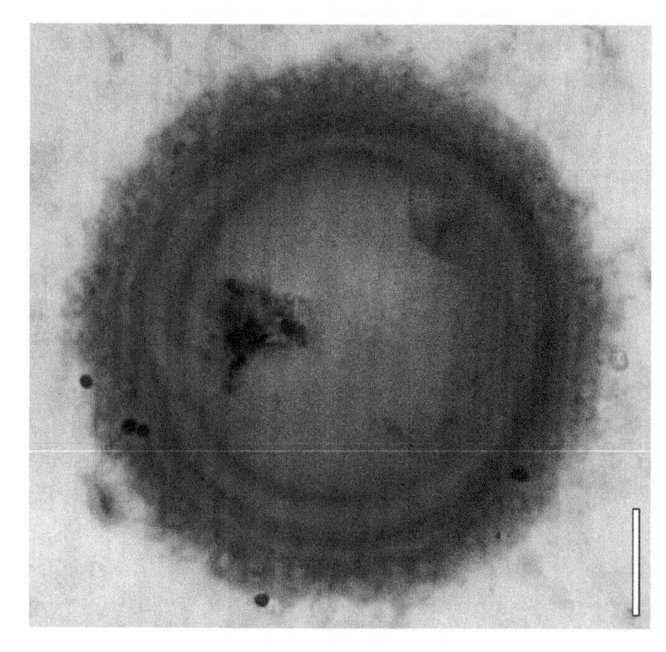

FIGURE 6.11 Specimen of *Zwergimyces vestitus* with massive non-hyphal inner investment element. Lower Devonian Rhynie chert, Scotland. Bar = 20 µm.

ZYGOSPORANGIUM-GAMETANGIA COMPLEXES AND SPOROCARPS

Additional records of fossil mantled fungal reproductive units are exclusively Carboniferous and Triassic, no doubt due to the presence of well-preserved coal balls from Euramerica (Carboniferous) and silicified peat from Antarctica (Triassic). Most of these fossils have been referred to the zygomycetous fungi based on the presence of associated structures interpreted as apposed gametangia. Other spherical structures morphologically similar to *Z. vestitus* include the Carboniferous and Triassic microfossils that have collectively been termed fungal "sporocarps."

CARBONIFEROUS

There are several reports of Carboniferous spores (chlamydospores) that historically have been assigned to the zygomycetous fungi (e.g. Wagner and Taylor, 1982; Stubblefield and Taylor, 1988 and references therein; Taylor et al., 1994b), but that today are accommodated in the Glomeromycota (see Chapter 7). Compelling fossil evidence of true zygomycetous fungi from the Carboniferous occurs in the form of three different types of structurally preserved reproductive units interpreted as (mantled) zygosporangia with apposed gametangia that are preserved in coal balls from the Lower Pennsylvanian (~323–315 Ma, Carboniferous) of Great Britain.

The first of these occurs in a gymnosperm ovule (Krings and Taylor, 2012b). Ten specimens of this reproductive unit have been detected in the space that the nucellus and megaspore once occupied in the seed. The fungus consists of a smooth-walled, near-perfect sphere (55 μm in diameter) attached to a hollow, dome-shaped structure that is open at its wide end. At the tip of the dome-shaped structure is a smaller element, which may be more or less spherical, drop- or dome-shaped (Figure 6.12). This structure, which is 5–8 μm in diameter, also appears to be open at one end. The lumen of the larger sphere and dome-shaped structure, as well as the lumen of the dome-shaped structure and smaller element, are interconnected.

The second fossil consists of an assemblage of around 40 reproductive units that occur in the tracheids of a fragment of degraded wood (Krings and Taylor, 2012a). The reproductive units occur singly or in aggregations. Single specimens are spherical to oval in outline and up to 90 μm in diameter, while aggregated individuals are more variable in shape. All reproductive units are composed of a central cavity sheathed in a prominent investment (Figure 6.13). The investment is constructed of two different types of elements, with the outer, prominent one composed of hyphae, and the inner element non-hyphal. The outer investment is formed of tightly interlaced hyphae (Figure 6.14); septa are present but appear to be relatively rare. The inner layer is recognizable as a

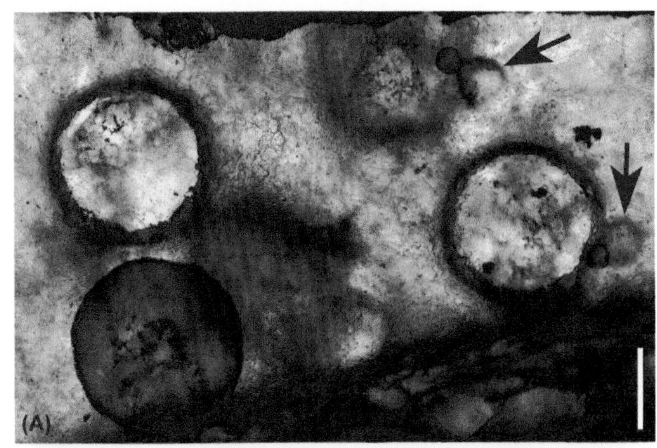

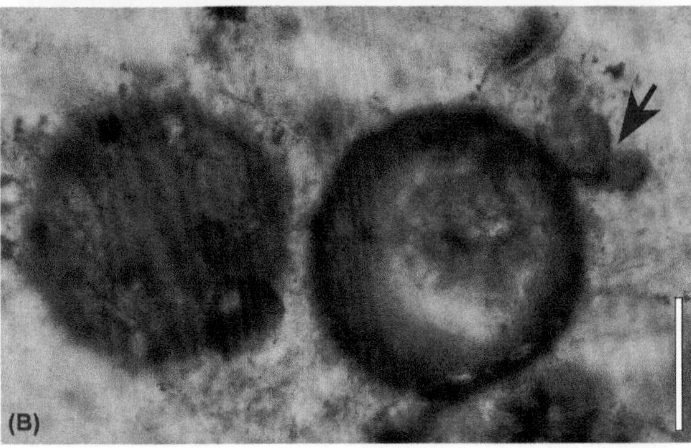

FIGURE 6.12 Putative zygosporangia in a cordaite (gymnosperm) ovule. **A.** Note remains of apposed gametangia still attached (*arrows*) in some zygosporangia. Bar = 25 μm. **B.** Detail of gametangia. Bar = 20 μm. Pennsylvanian, Great Britain.

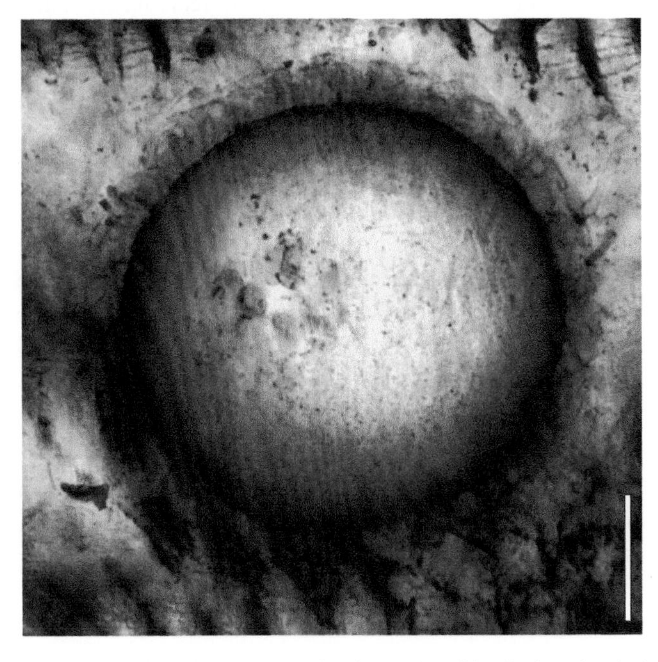

FIGURE 6.13 Fungal reproductive unit with distinct hyphal mantle. Pennsylvanian, Great Britain. Bar = 20 μm.

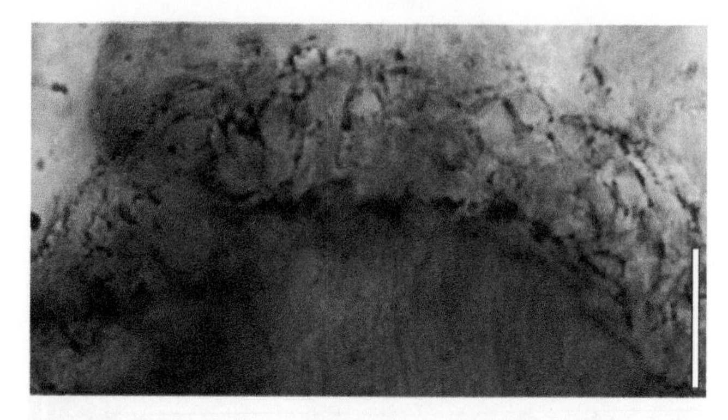

FIGURE 6.14 Detail of Figure 6.13 showing hyphal investment. Pennsylvanian, Great Britain. Bar = 5 μm.

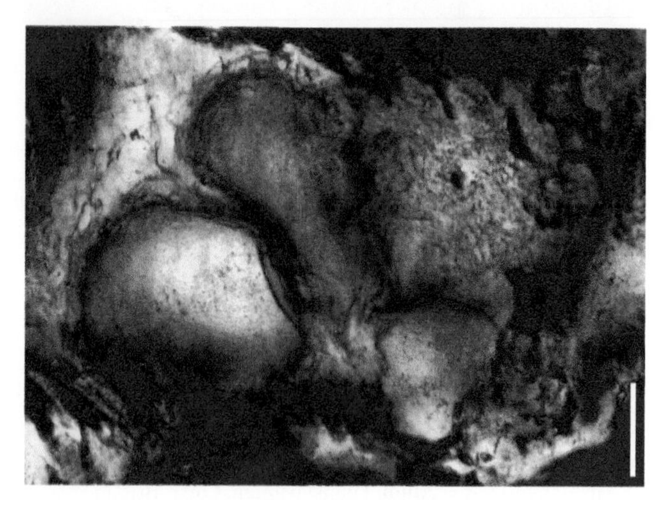

FIGURE 6.15 Aggregation of mantled reproductive units. Pennsylvanian, Great Britain. Bar = 20 μm.

dark line extending along the inner surface of the hyphal mantle. Aggregates of specimens may additionally be surrounded by a confluent meshwork of wide aseptate hyphae (Figure 6.15). Closely associated with many of the reproductive units are smaller spherical to elongate structures. In most specimens, one associated structure is recognizable, but in some they occur in pairs (*arrows* in Figure 6.16). The two associated structures forming a pair appear to be organically connected. Narrow subtending hyphae indicate that the associated structures are not formed as outgrowths of the reproductive units.

The third fossil, which was formally described as *Halifaxia taylorii* (Krings et al., 2013a), occurs in the xylem of a fern axis. The reproductive units (Figure 6.17) occur singly and consist of a sphere subtended by an inflated structure that is termed in the original description informally as a "subtending structure." An irregularly shaped element, termed the "smaller element," is found attached to the proximal portion of the subtending structure in some

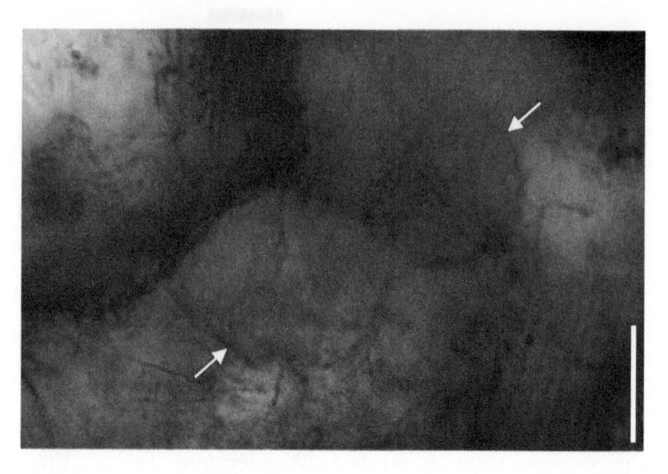

FIGURE 6.16 Paired associated structures (?gametangia) (*arrows*) of one of the mantled reproductive units. Pennsylvanian, Great Britain. Bar = 10 μm.

of the specimens. The sphere (Z in Figure 6.18) is 85–90 μm in diameter and composed of a central cavity surrounded by a hyphal mantle. The subtending structure (see MG in Figure 6.18) is sac-like or primarily conical, and in most specimens is ensheathed by loosely interwoven hyphae. The smaller element (mG in Figure 6.18), which lacks a hyphal investment, clasps the proximal portion of the subtending structure and then produces one stout branch that extends further up along the outer surface of the subtending structure. The tip of this branch appears to fuse laterally with the subtending structure. A transverse septum separates the distal portion of the branch from the rest.

All three reproductive units have been interpreted as zygosporangium-apposed gametangial complexes. The spherical component is believed to represent the zygosporangium, which, in two of the fossils, is covered by a hyphal mantle. The associated structures accordingly represent the two gametangia, each subtended by a suspensor. In two of the fossils, the gametangia differ from each other in size, and thus are termed macro- and microgametangia. The condition seen in the fossils closely corresponds to that in certain modern representatives of the Endogonaceae (Bucholtz, 1912; Thaxter, 1922; Yao et al., 1996). Moreover, it has been observed that in certain Endogonaceae the gametangium walls increase in thickness after gametangial fusion and thus may remain intact even until zygosporangium maturation (e.g. Bucholtz, 1912). This observation may explain why both the large and small associated structures in the fossil found within a gymnosperm ovule are open at one end (see Figure 6.12). The open ends would correspond to the attachment sites of the gametangia to the subtending suspensors, which do not have secondarily thickened walls and thus rapidly

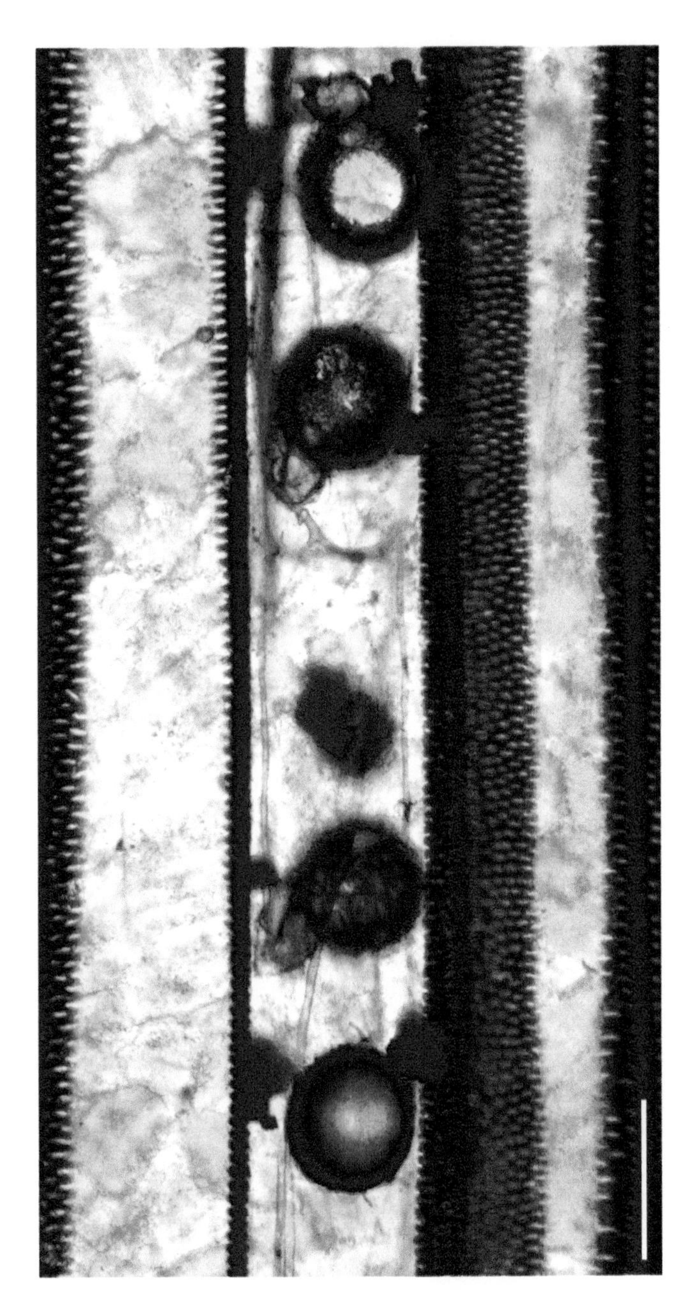

FIGURE 6.17 Hyphae and several reproductive units of *Halifaxia taylorii* in the xylem of a fern axis. Pennsylvanian, Great Britain. Bar = 100 μm.

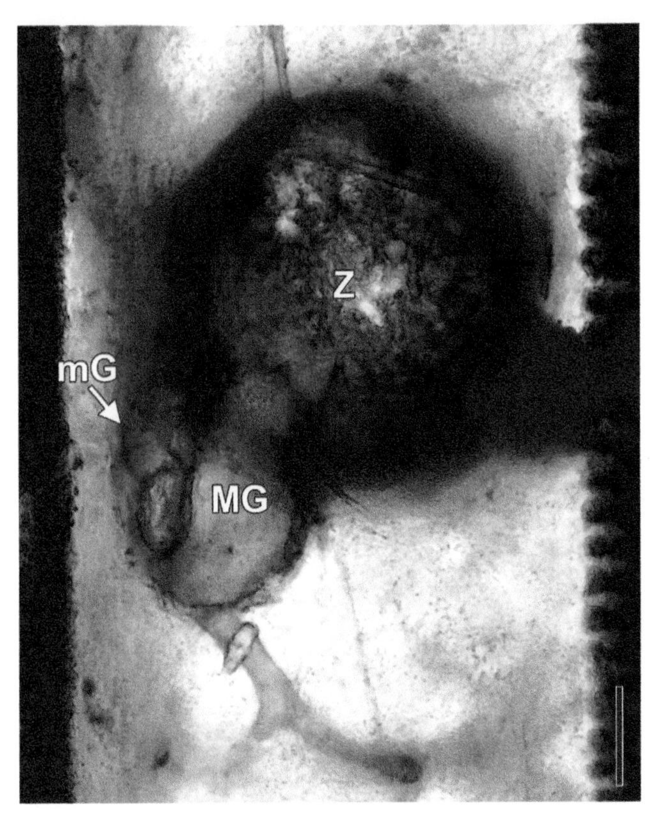

FIGURE 6.18 Mantled zygosporangium (Z) of *Halifaxia taylorii* with macrogametangium (MG), and smaller microgametangial branch (mG). Pennsylvanian, Great Britain. Bar = 20 μm.

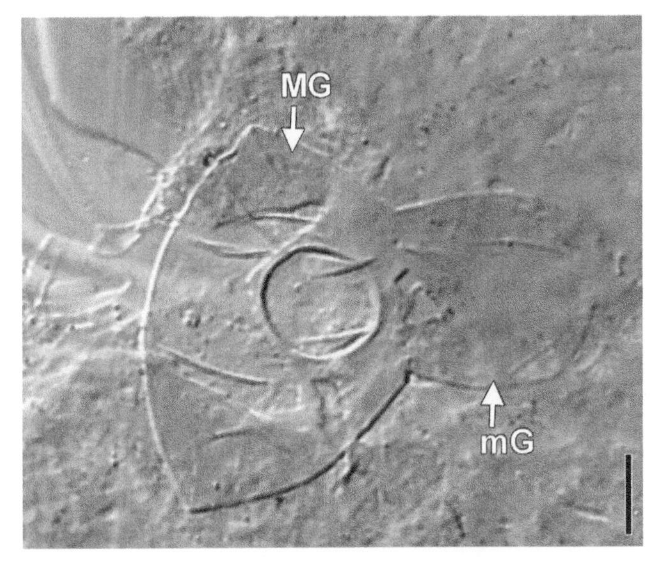

FIGURE 6.19 *Endogone lactiflua* showing fused gametangia (MG, megagametangium, mG, microgametangium) and distinct pore in the megagametangium through which the developing zygosporangium emerges. Extant. Bar = 10 μm. (Courtesy J. Błaszkowski.)

disintegrate following maturation of the zygosporangium and zygospore. Adding support to this interpretation is the fact that the configuration exhibited by these fossils is virtually identical to that seen in several of the zygosporangia with attached paired gametangia of extant *Endogone* species (e.g. Figure 6.19). A structural feature of *H. taylorii* that does not occur in Endogonaceae is the smaller element subtending the microgametangial branch and clasping around the proximal portion of the subtending structure

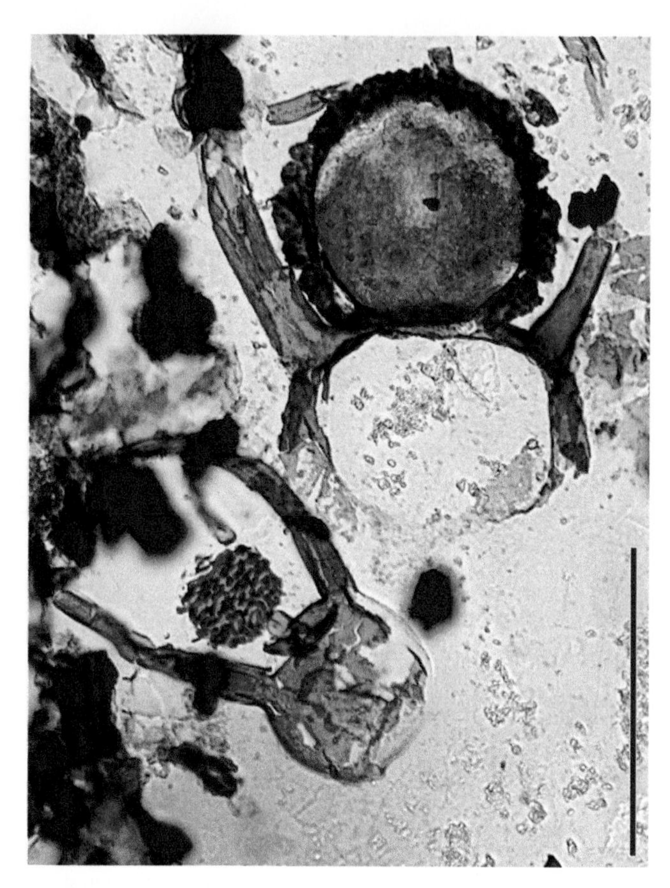

FIGURE 6.20 *Protoascon missouriensis.* Pairs of conjoined spheroids with filamentous appendages arising from the proximal spheroid enveloping the distal spherical structure. Middle Pennsylvanian, Missouri, USA. Bar = 50 μm.

(see Figure 6.18). However, a somewhat similar feature was reported as occurring during sexual reproduction in *Mortierella capitata*, in which the microprogametangium initially develops a branched structure that entwines densely around the elongating, club-shaped macroprogametangium (Degawa and Tokumasu, 1997).

A geologically slightly younger fossil interpreted as an azygosporangium-gametangium or zygosporangium-gametangia complex is *Protoascon missouriensis*, an assemblage of some 50 fungal reproductive units that occur clustered in a seed preserved in a coal ball from the Middle Pennsylvanian (~315–307 Ma) of Missouri, North America (Batra et al., 1964; Baxter, 1975; Taylor et al., 2005b). Each of the reproductive units consists of a pair of conjoined spheroids, 50–150 μm in diameter, in which the distal spheroid is thick walled and ornamented, while the proximal spheroid is relatively thin walled (Figure 6.20). Up to twelve filamentous appendages arise from near the apex of the proximal sphere and

FIGURE 6.21 Filamentous appendages of *Protoascon missouriensis* arising from the proximal spheroid. Middle Pennsylvanian, Missouri, USA. Bar = 50 μm.

envelop the distal unit (Figure 6.21). Each pair of spheroids measures approximately 250 μm from the base of the proximal spheroid to the tip of the enclosing appendages. It appears that, in one of the specimens, a second, smaller sphere (*arrows* in Figure 6.22) is attached to the ornamented sphere in the opposite position to the proximal one.

As the name suggests, *P. missouriensis* was initially thought to be a member of the Ascomycota (Batra et al., 1964) but later reinterpreted as belonging to the Chytridiomycota (Baxter, 1975). Subsequent studies (Pirozynski, 1976a; Taylor et al., 2005b), however, have reinterpreted *P. missouriensis* as belonging to some zygomycetous lineage. According to Taylor et al. (2005b), the basic structure of *P. missouriensis* is comparable to azygosporangia produced by *Mucor azygosporus* (Benjamin and Mehrotra, 1963; O'Donnell et al., 1977), although the suspensor in the latter fungus does not produce appendages. The appendages in *P. missouriensis*, however, are remarkably similar to those seen in *Absidia spinosa* (Figure 6.23) (see Lendner, 1907; Boedijn, 1958), of which there is also one variety, *A. spinosa* var. *azygospora* (Figure 6.24) – that produces abundant azygospores (Boedijn, 1958). In accordance with the interpretation of the ornamented spheroids containing single spheres of *P. missouriensis* as azygo- or zygosporangia, the proximal spheroid would be homologous with a prominent suspensor, and the enclosing aseptate extensions that envelop the distal structure would be homologous with suspensor appendages. A distal opening or thinned area in the

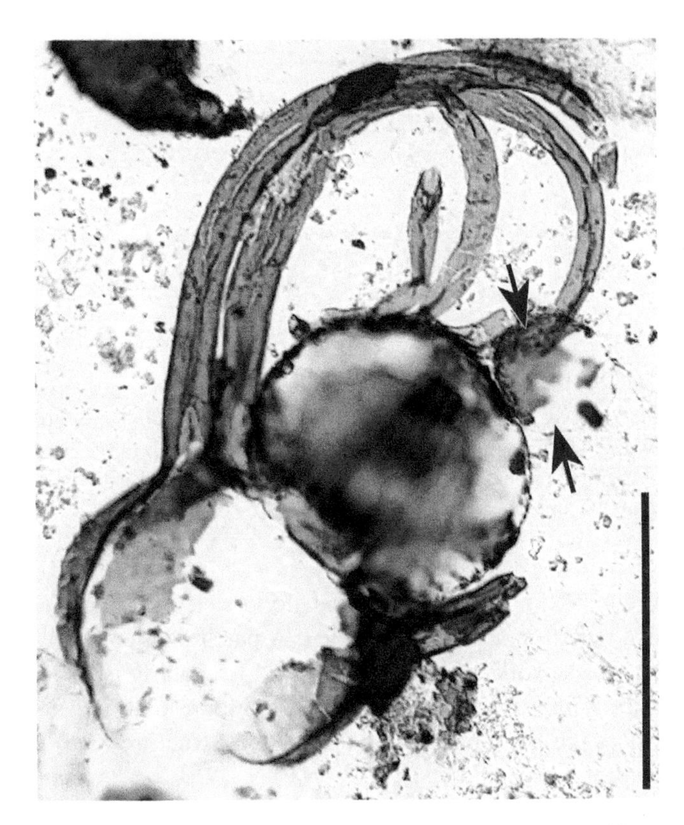

FIGURE 6.22 *Protoascon missouriensis* specimen with a smaller spherical structure (*arrows*) attached to the distal unit. Middle Pennsylvanian, Missouri, USA. Bar = 50 μm.

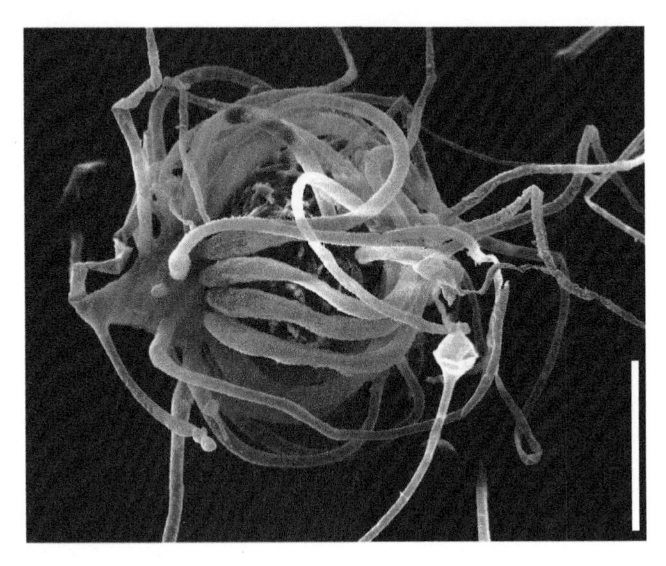

FIGURE 6.23 Zygosporangium of *Absidia spinosa* enveloped by prominent suspensor appendages. Extant. Bar = 35 μm. (Courtesy H.-M. Ho.)

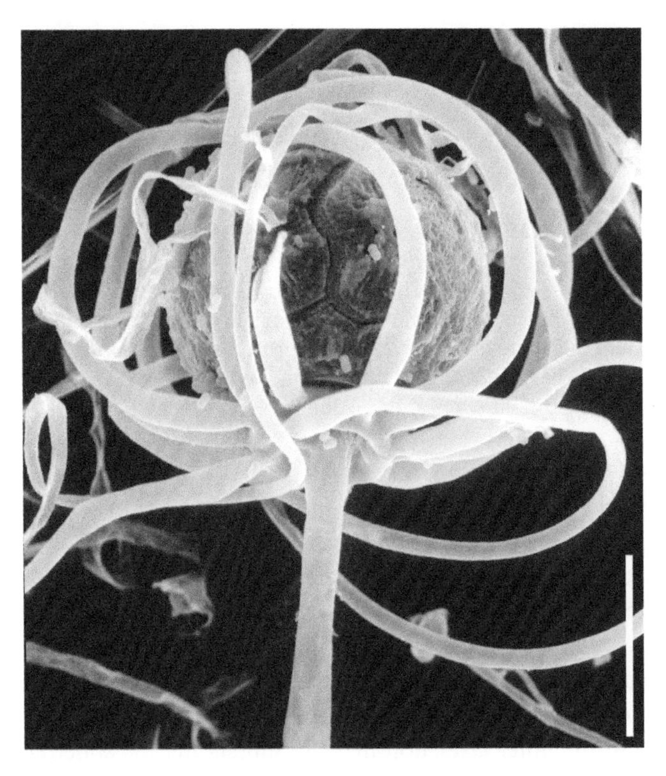

FIGURE 6.24 Azygosporangium *of Absidia spinosa* var. *azygospora*. Extant. Bar = 35 μm. (Courtesy H.-M. Ho.)

sporangium wall of several *P. missouriensis* specimens was suggested to represent the attachment site of a second, perhaps diminutive and short-lived suspensor. In at least one specimen there is a smaller distal structure attached to the thick-walled sporangium (see *arrows* in Figure 6.22) that might represent a second suspensor.

Although the occurrence of abundant zygomycetous fungi from the Carboniferous was postulated more than 100 years ago by the British paleontologist R.C. McLean (1912), only four putative Carboniferous representatives of this group have been documented. It is interesting to note that all Carboniferous zygomycetes described to date occur within the confines of plant parts. This is unusual since most modern zygomycetes produce zygospores aerially, on or in the soil, or on organic debris (Benny et al., 2001). As to whether the occurrence of the Carboniferous zygosporangium-gametangia complexes within plant parts represents a preservation bias in which only those specimens protected by plant tissue are preserved in a recognizable form, or reflects some life history strategy of zygomycetous fungi in the Carboniferous, remains to be determined.

TRIASSIC

Probably the most persuasive fossil representative of the Endogonaceae was discovered in permineralized peat from the Middle Triassic (~247–237 Ma) of Antarctica, and formally described as *Jimwhitea circumtecta* (Krings

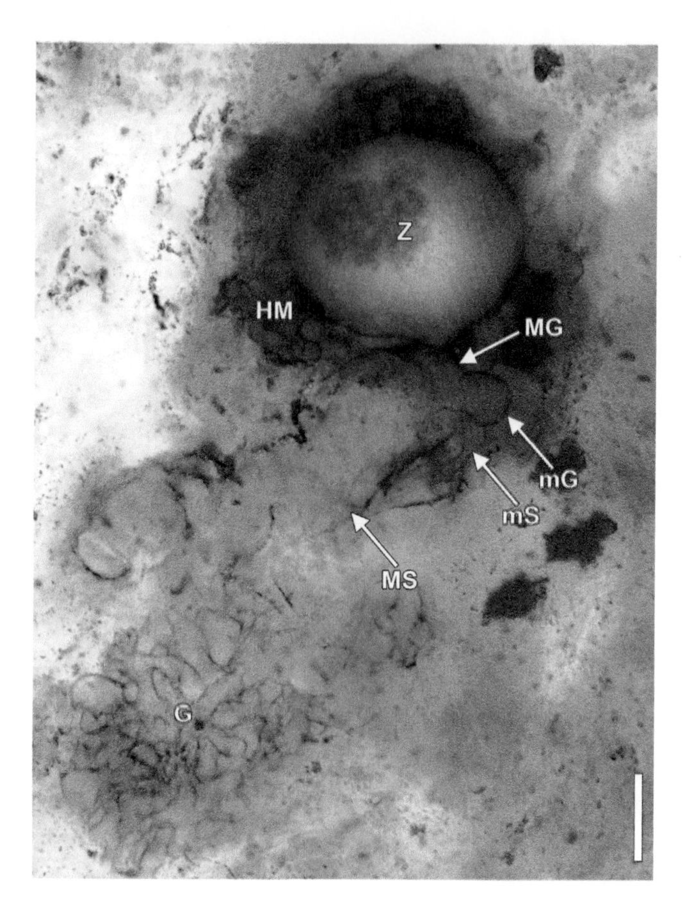

FIGURE 6.25 *Jimwhitea circumtecta* zygosporangium (Z) surrounded by hyphal mantle (HM). The sporangium arises from a megagametangium (MG) subtended by a megasuspensor (MS). Laterally attached to the megagametangium is the microgametangium (mG) subtended by a microsuspensor (mS). The entire complex appears to have developed from a loose network of hyphae (G). Triassic, Antarctica. Bar = 20 μm.

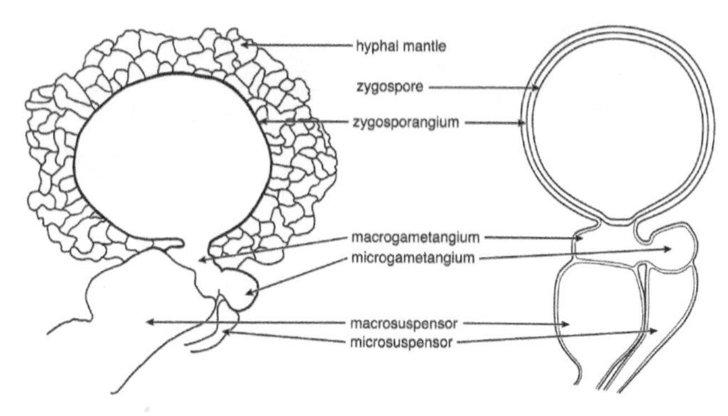

FIGURE 6.26 Comparison of *Jimwhitea circumtecta* (left) with Thaxter's illustration of a zygosporangium-apposed gametangia complex of *Endogone lactiflua* (right, redrawn from Thaxter, 1922.)

et al., 2012d). This fossil (Figure 6.25) consists of a spheroid borne on an inflated, sac-like structure to which is attached a smaller globose element subtended by a distally widened hypha. The spheroid (see Z in Figure 6.25) is 85 μm in diameter and composed of a central cavity surrounded by a prominent, two-layered mantle, with the outer layer composed of hyphae (see HM in Figure 6.25), and the inner layer non-hyphal. Subtending the spheroid is a smooth-walled sac-like structure; a direct connection exists between the central cavity of the spheroid and the lumen of the sac-like structure. The distal portion of the latter (see MG in Figure 6.25) is separated from the rest (see MS in Figure 6.25) by a septum. Physically connected to the tip region of the sac-like structure is a much smaller globose element (see mG in Figure 6.25), which is subtended by a hypha-like structure (see mS in Figure 6.25). The lumina of the globose element and sac-like structure are interconnected. A patch of a conspicuous meshwork of multi-branched, irregularly shaped, tightly interlaced hyphae (see G in Figure 6.25) occurs where the proximal end of the sac-like structure used to be (not preserved).

The configuration seen in *J. circumtecta* (Figure 6.26) virtually parallels that depicted by Thaxter (1922) for the zygosporangium-apposed gametangia complex of the extant *Endogone lactiflua*. This figure (see Figure 6.26, right side) shows a zygosporangium containing a zygospore budding from a large gametangium (macrogametangium) which is subtended by a macrosuspensor. The macrogametangium, separated from the suspensor by a septum, is fused laterally with a smaller gametangium (microgametangium) that is subtended by a microsuspensor. Because of the structural correspondences between *E. lactiflua* and *J. circumtecta* (see Figure 6.26, left side), the spheroidal component of the fossil was interpreted as a zygosporangium, with the hyphal investment representing the mantle and the inner, non-hyphal layer the sporangiothecium (Krings et al., 2012d). The sac-like structure accordingly represents the macrogametangium, which is subtended by a macrosuspensor (with a septum between the two structures), while the small globose element attached to the tip of the sac-like structure is interpreted as a microgametangium subtended by a microsuspensor. The meshwork of tightly interlaced hyphae at the proximal end of the macrosuspensor likely represents the gleba (spore-bearing inner mass) that gives rise to the gametangia.

Another fossil from the Middle Triassic of Antarctica that is quite similar to *J. circumtecta* was described and informally named Fungus No. 4 by White and Taylor

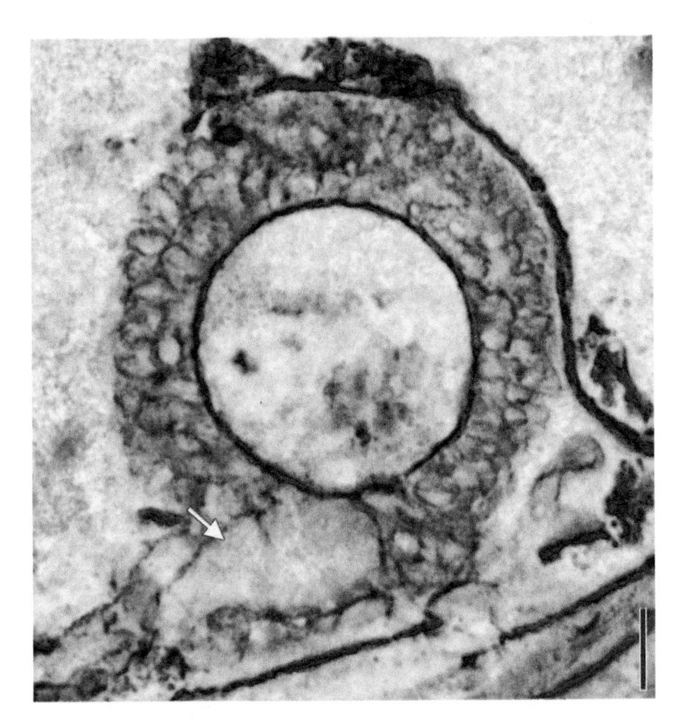

FIGURE 6.27 Putative zygosporangium subtended by inflated megagametangium/suspensor (*arrow*). Triassic, Antarctica. Bar = 10 μm.

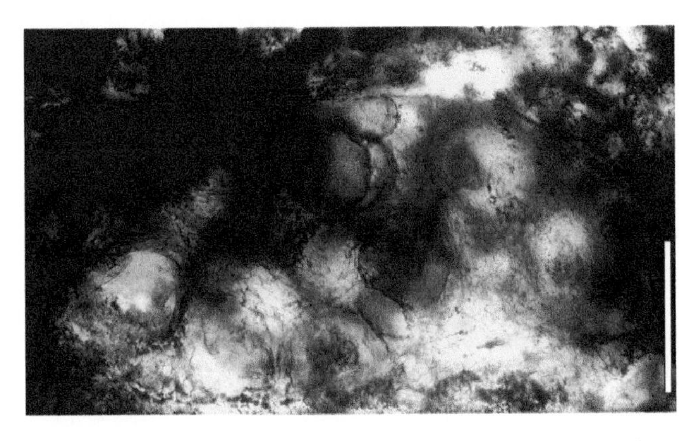

FIGURE 6.28 Sporocarp portion in same slab as *Jimwhitea circumtecta* showing sporangia embedded in gleba. Triassic, Antarctica. Bar = 100 μm.

(1989b). This fossil (Figure 6.27) consists of a mantled sphere (~60 μm in diameter) subtended by an inflated, sac-like structure (see *arrow* in Figure 6.27), which likely represents the macrogametangium and macrosuspensor. Evidence of gametangial fusion, however, is lacking.

Co-occurring with *J. circumtecta* in the same chert block is a sporocarp portion (Figure 6.28) that is bounded on the outside by a narrow peridium or pseudoperidium

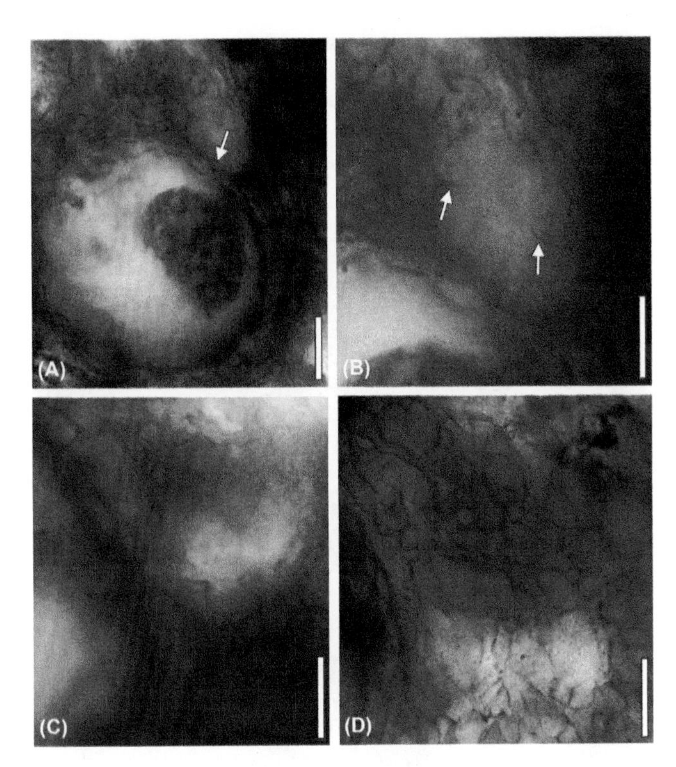

FIGURE 6.29 Detail of the sporocarp in Figure 6.28. **A.** Sporangium showing putative (mega)gametangium and suspensor (*arrow*). Bar = 20 μm. **B.** Septum (*arrows*) between gametangium and suspensor. Bar = 10 μm. **C.** Area of attachment between (mega)gametangium and sporangium. Bar = 10 μm. **D.** Gleba. Bar = 10 μm. Triassic, Antarctica.

(Krings et al., 2012a). The sporocarp contains 12 sporangia or spores, which are embedded in a gleba of irregularly swollen, thin-walled hyphae (Figure 6.29D). The individual sporangia/spores are (sub)globose or ovoid and up to 60 μm in diameter. Some sporangia/spores are surrounded by what appears to be a developing hyphal mantle that is incomplete (i.e. not traceable around the entire sporangium). Several of the sporangia/spores are physically connected to sac-like structures (Figure 6.29A, C); those contained in the sporocarp are approximately the size of the zygosporangium of *J. circumtecta*. Moreover, several of the sporangia in the sporocarp are borne on sac-like structures, which are interpreted as gametangia or suspensors; in one specimen, remains of what appears to be the septum separating the gametangium from its subtending suspensor are present (*arrows* in Figure 6.29B). Gametangial fusion, however, has not been observed. In addition, the patch of interlaced hyphae interpreted as a gleba closely associated with *J. circumtecta* (see G in Figure 6.25) is structurally similar to the sporocarp gleba.

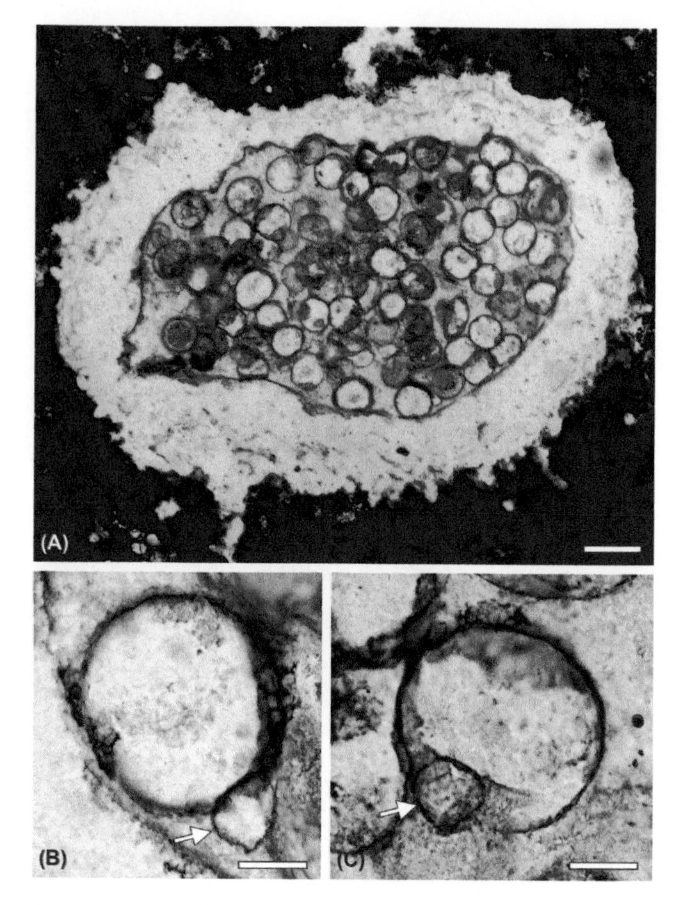

FIGURE 6.30 Sporocarp with possible affinities to the Endogonales. **A.** Massive peridium enclosing numerous, tightly packed spores. Bar = 100 μm. **B, C.** Individual spores with attached suspensor cells (*arrows*). Bar = 20 μm. Triassic, Antarctica.

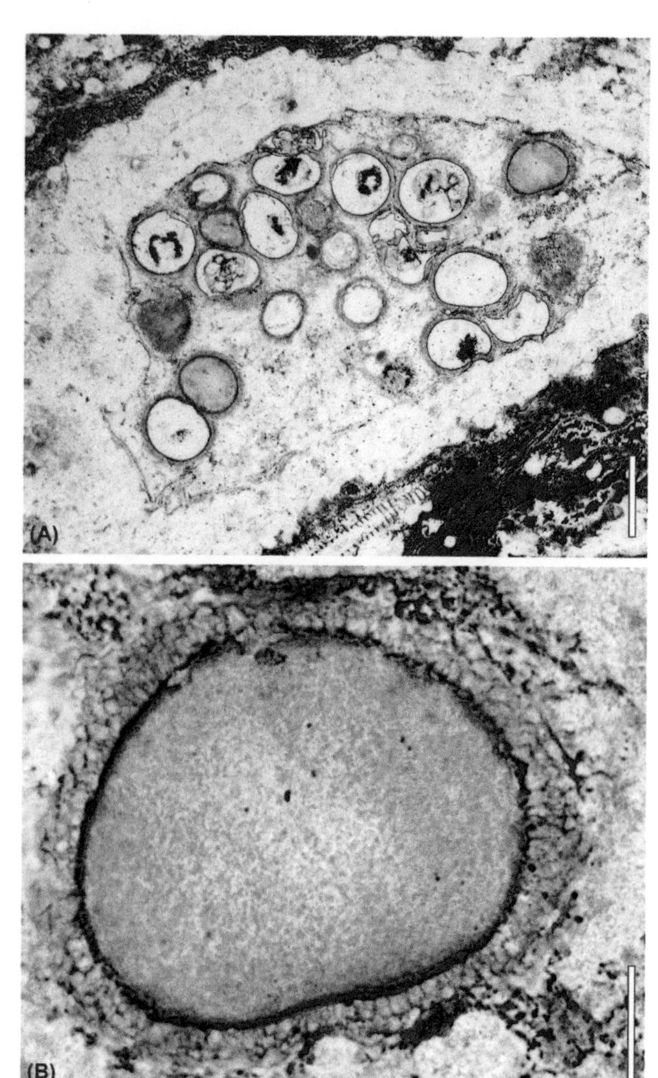

FIGURE 6.31 **A.** Sporocarp with thick-walled peridium containing numerous sporangia/spores. **B.** Single spore from sporocarp in (A) showing prominent mantle. Triassic, Antarctica. Bars = 100 μm (A), 20 μm (B).

All these correspondences strongly suggest that this sporocarp also belongs to *J. circumtecta*.

Two additional sporocarps containing sporangia or spores with suggested affinities to the Endogonales have been described from the Middle Triassic of Antarctica and informally named Fungus No. 2 and Fungus No. 3 (White and Taylor, 1989b). Fungus No. 2 (Figure 6.30A) is 600 × 1000 μm in diameter and composed of numerous spores surrounded by a mycelial peridium composed of interwoven hyphae. Individual spores are globose and 60–67 μm in diameter; some possess a spherical or drop-shaped associated structure 18–20 μm in diameter (*arrows* in Figure 6.30B, C), which has been interpreted by these authors as a suspensor cell. Fungus No. 3 (Figure 6.31A) is similar to Fungus No. 2; however, individual spores are characterized by a prominent mantle composed of tightly interlaced hyphae (Figure 6.31B). The sporocarp portion co-occurring with *J. circumtecta* differs from both sporocarps described by White and Taylor (1989b) with regard to peridium thickness, which is up to 180 μm in Fungus No. 2 and up to 90 μm in Fungus No. 3. Moreover, a gleba was not reported in either Fungus No. 2 or Fungus No. 3.

"SPOROCARPS" AND OTHER FOSSILS OF UNCERTAIN AFFINITIES

"SPOROCARPS"

Within Carboniferous coal balls and chert from Europe and North America are a variety of small (usually <1 mm in diameter) spherical structures, including

FIGURE 6.32 Robert W. Baxter.

FIGURE 6.33 Gilbert A. Leisman.

some that are ornamented, which have collectively been termed sporocarps (e.g. Spencer, 1893; Hutchinson, 1955; Baxter, 1960 [Figure 6.32]; Davis and Leisman, 1962 [Figure 6.33]; Stubblefield et al., 1983 [Figure 6.34]). Krings et al. (2011e), however, have suggested that the collective use of the term sporocarp for these fossils may be inaccurate, and thus, if used, the name should be put in quotation marks. "Sporocarps" may be solitary, but there are many specimens in which several individuals are clustered together (e.g. Williamson, 1880 [Figure 6.35]; McLean, 1922; Hutchinson, 1955; Stubblefield et al., 1983). All are composed of a central cavity surrounded by an investment or mantle of loosely arranged interlacing and/or tightly compacted hyphae, which may be septate or aseptate. In all types, there is evidence to suggest that the investment is bounded on the inside by a narrow non-hyphal layer.

Based primarily on investment composition and surface ornamentation, several fossil genera have been introduced for these fossil structures: *Mycocarpon* (Figure 6.36) is up to 600 μm in diameter and characterized by an investment of interlaced hyphae up to four layers thick (Hutchinson,

FIGURE 6.34 Sara P. Stubblefield. (Courtesy G.W. Rothwell.)

FIGURE 6.35 William C. Williamson.

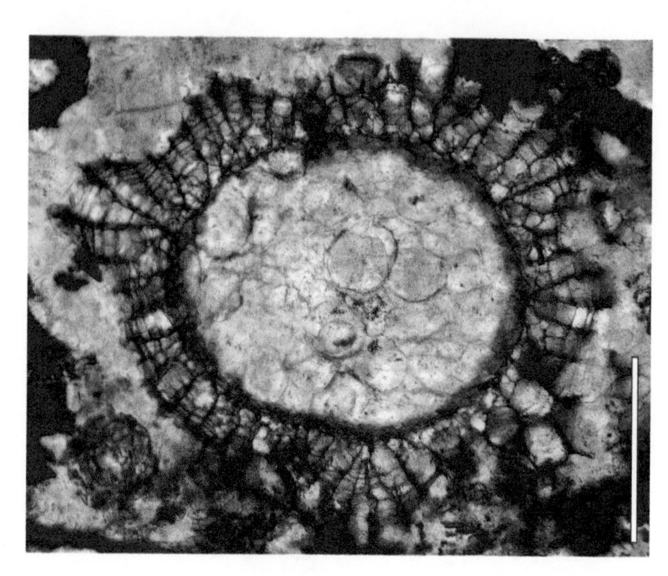

FIGURE 6.37 *Sporocarpon cellulosum* showing pseudoparenchymatous investment. Pennsylvanian, Great Britain. Bar = 150 μm.

FIGURE 6.38 *Dubiocarpon* sp. showing investment and ornamentation of prominent spines. Pennsylvanian, Great Britain. Bar = 100 μm.

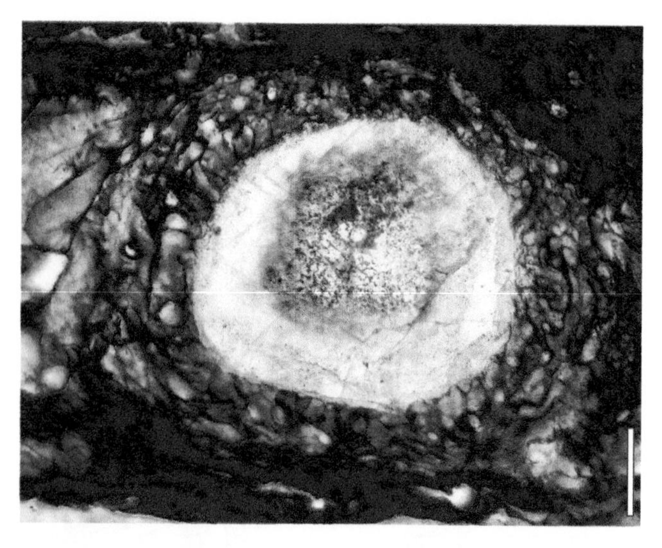

FIGURE 6.36 *Mycocarpon* sp. showing investment of interlaced hyphae. Pennsylvanian, Great Britain. Bar = 50 μm.

1955; White and Taylor, 1991). Specimens of *Sporocarpon* (Figure 6.37) are similar in size but possess a pseudoparenchymatous investment that extends outward into narrow, conical processes (Baxter, 1960; Stubblefield et al., 1983). A third type, *Dubiocarpon* (Figures 6.38–6.40), is up to 700 μm in diameter and distinguished by an investment constructed of radially elongated segments and spines (Figure 6.39A) extending out from the surface (Stubblefield et al., 1983; Gerrienne et al., 1999).

The most prominently ornamented type is *Traquairia* (Figure 6.41), initially described as a radiolarian rhizopod (Carruthers, 1873). Specimens may occur singly or in groups of four or five. Individual specimens are nearly spherical and range up to 1.0 mm in diameter. Because the outer portion of

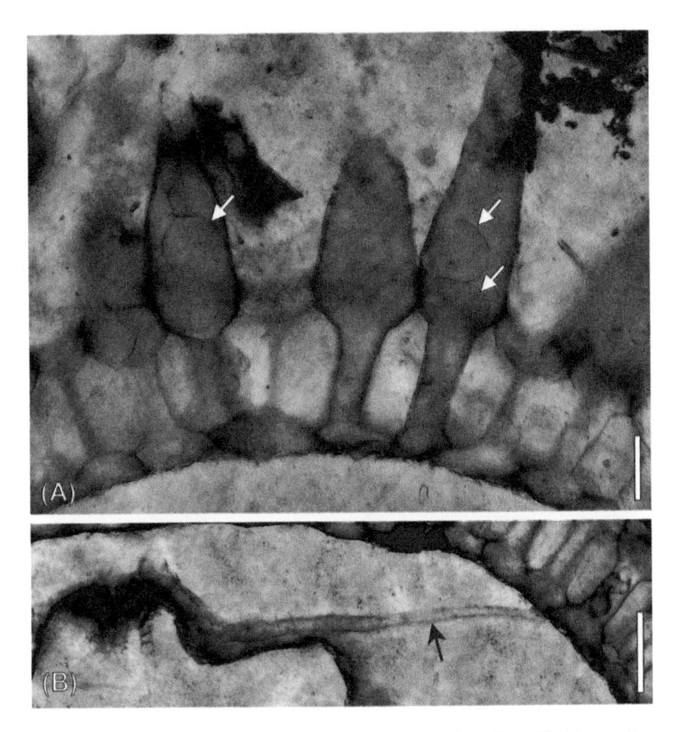

FIGURE 6.39 Detail of *Dubiocarpon* sp. showing (**A**) intrusive fungi (*arrows*) in several of the spines, and (**B**) hypha (*arrow*) of an intrusive fungus extending into cavity. Pennsylvanian, Great Britain. Bars = 20 μm.

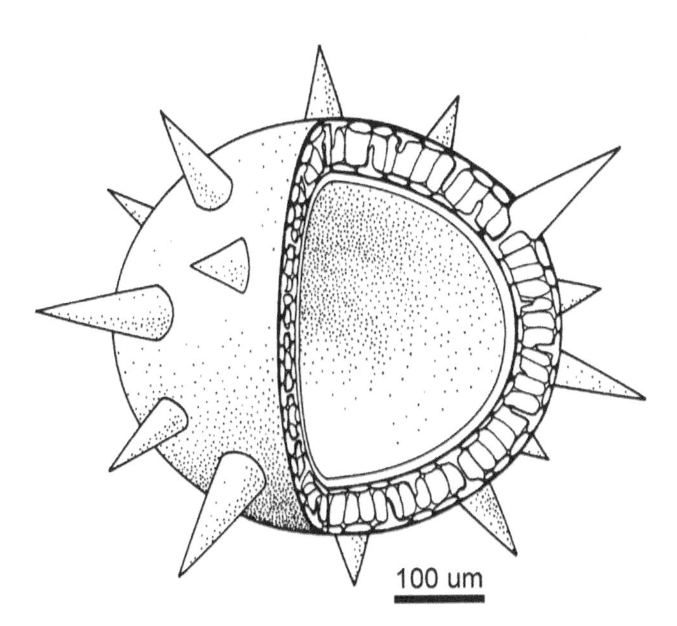

FIGURE 6.40 Suggested reconstruction of *Dubiocarpon elegans* (from Gerrienne et al., 1999).

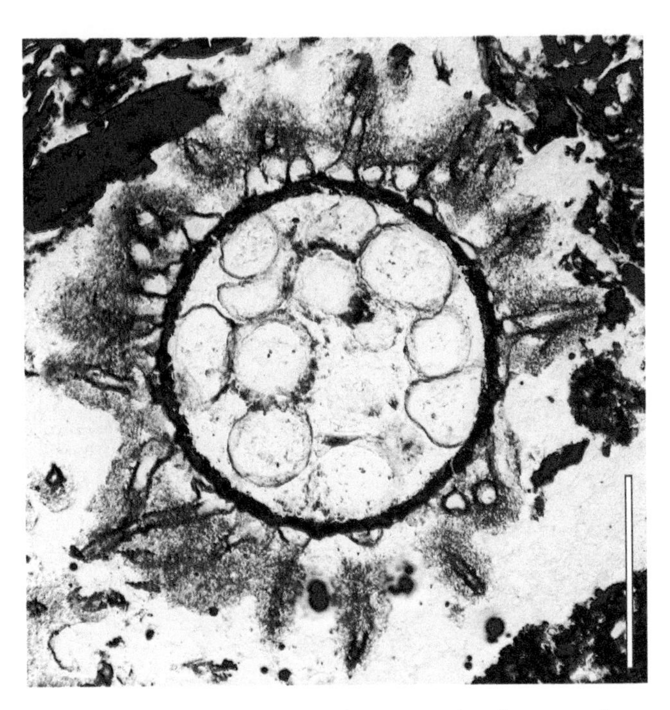

FIGURE 6.41 *Traquairia williamsonii* showing prominent ornament and multiple spherical structures in the central cavity. Middle Pennsylvanian, Lewis Creek, Kentucky, USA. Bar = 200 μm.

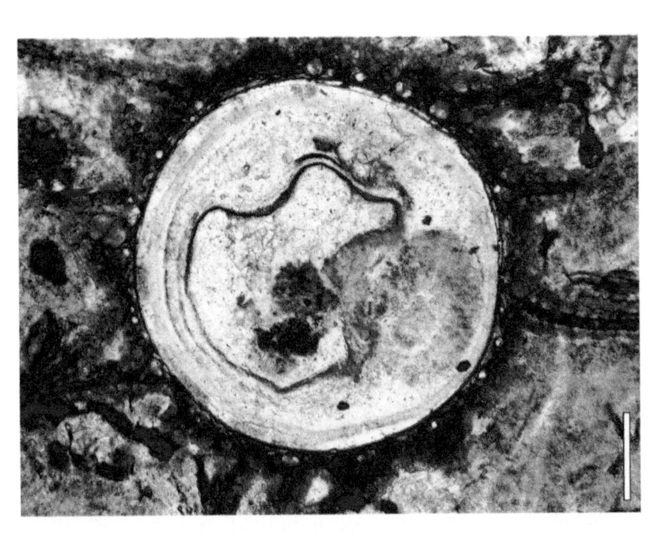

FIGURE 6.42 *Traquairia* sp. showing prominent ornament and single large sphere in the cavity. Pennsylvanian, Great Britain. Bar = 100 μm.

the investment is often not complete, measurement of specimens sometimes includes the central cavity, which may be 150–550 μm in diameter. The investment of *Traquairia* is complex, with the outer layer constructed of branching hyphae (Figures 6.42 and 6.43), some of which are organized into hollow spines or echinations (Scott, R., 1911; Stubblefield and Taylor, 1983). Proximally, spines branch extensively to form a compact, interlacing network. The wall of each spine projects at many points, forming tubular secondary processes that become increasingly narrow as they branch. The secondary

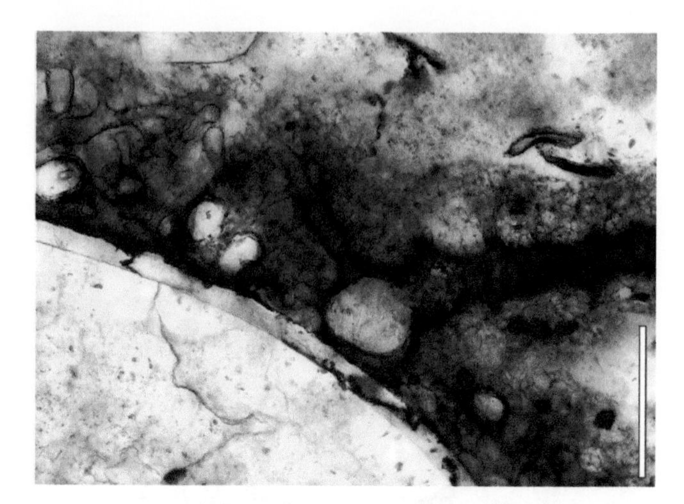

FIGURE 6.43 Detail of Figure 6.42 showing complexity of investment composed of branching hyphae, some of which form prominent spines. Pennsylvanian, Great Britain. Bar = 50 μm.

processes are continuous, with a dense weft of branching hyphae ensheathing the whole "sporocarp" (Figures 6.41 and 6.44), and specimens often appear to be embedded in an amorphous layer, which earlier workers referred to as a "spongy substance" (Carruthers, 1873), "plastic substance" (Williamson, 1880), or "gelatinous envelope" (Scott, R., 1911), and suggested it was secreted by the spines. Stubblefield and Taylor (1983), however, demonstrated by using scanning electron microscopy (SEM) that this layer is consistently composed of small, branching hyphae, too fine in most cases to be resolved in transmitted light. Several species have been delimited based on the organization of the wall and spines, and on the contents of the central region.

Specimens of *Traquairia carruthersii* from the Mississippian of Burntisland, Scotland are up to 250 μm in diameter and have unbranched spines evenly spaced over the surface (Scott, R., 1911). Slightly smaller forms with evenly spaced, but longer, spines have been included in *T. williamsonii* from the Pennsylvanian Lewis Creek (Kentucky) coal ball site (Stubblefield and Taylor, 1983). Within the central cavity are small spheres ranging from 33 to 55 μm in diameter. Even longer (250 μm) spines have been reported in *T. ornatus*; the tips of these spines typically bifurcate and are sometimes curved. Specimens of *T. spenceri* have perhaps the longest spines (337 μm) with a slightly inflated base and unbranched distal ends. The central cavity also contains smaller spheres, although this species is known from a single specimen.

While "sporocarps" are relatively common in Pennsylvanian (~323–299 Ma) deposits, they have rarely

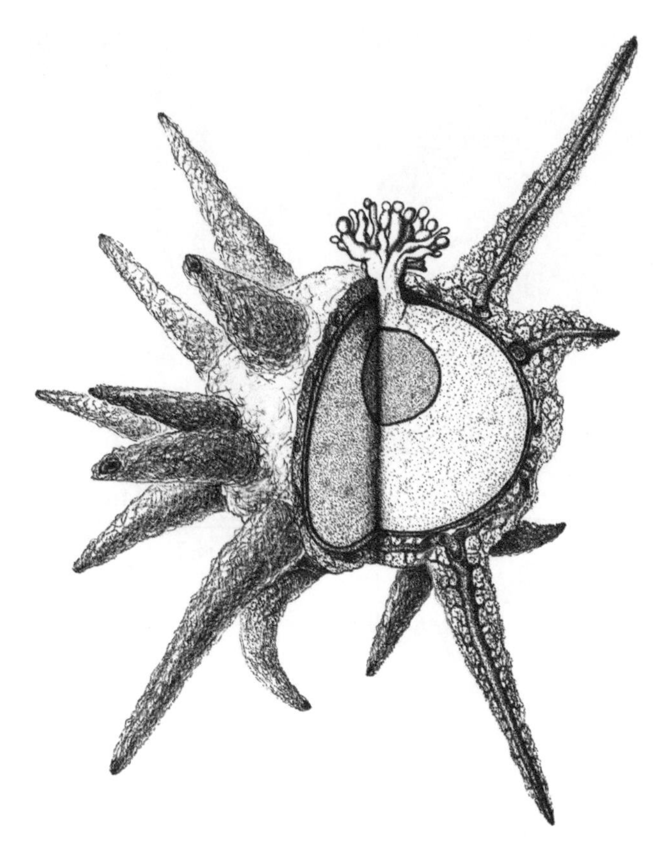

FIGURE 6.44 Cutaway view of suggested reconstruction of *Traquairia* sp. showing organization of investment and fascicle of radially oriented structures extending from preformed aperture (based on specimen in Figure 6.52).

been reported from geologically older or younger strata. The oldest evidence, *Mycocarpon rhyniense*, comes from the Lower Devonian Rhynie chert (Krings et al., 2014). Several forms are known from the Mississippian (~359–323 Ma). One of these, *Roannaisia bivitilis*, occurs in a Visean (~331 Ma) chert from the Roanne area in France (Taylor et al., 1994b), while a second, *Mycocarpon cinctum*, comes from Esnost (Rex, 1986; Krings et al., 2010c), another French locality yielding Visean cherts (Galtier, 1971; Scott et al., 1984). *Roannaisia bivitilis* (Figure 6.45) is large, between 500 and 600 μm in diameter, and characterized by a two-parted hyphal investment (Figure 6.46), the outer component (up to 100 μm thick) of which is constructed of interlaced hyphae with irregular swellings and bounded to the outside by a delicate, epidermis-like covering. The inner component (up to 50 μm thick) is constructed of narrower hyphae. The "sporocarp" contains a single spore, which has a multilayered wall. On the other hand, *M. cinctum* (Figure 6.47A) is the smallest "sporocarp" described to date, with a diameter of <100 to 200 μm. The outer component of the

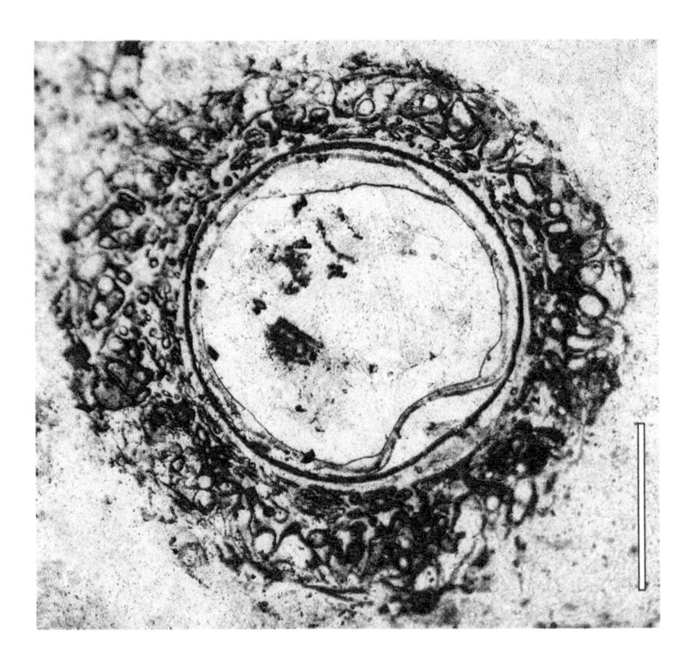

FIGURE 6.45 Section of *Roannaisia bivitilis.* Mississippian, central France. Bar = 175 μm. (Courtesy J. Galtier.)

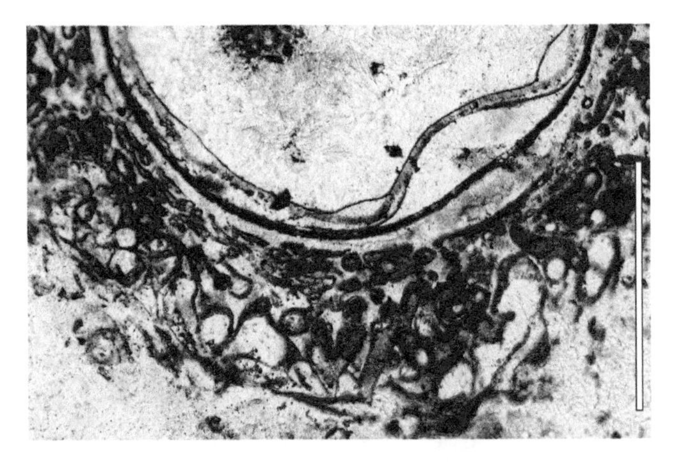

FIGURE 6.46 Detail of two-layer hyphal investment of *Roannaisia bivitilis*. Mississippian, central France. Bar = 175 μm. (Courtesy J. Galtier.)

two-parted hyphal investment consists of interlaced, thin-walled, septate hyphae extending around the circumference of the structure (Figure 6.47B); the inner component is constructed of densely spaced and interwoven, thick-walled hyphal branches produced by the hyphae of the outer wall. Intermixed with the thick-walled elements are thinner-walled hyphal branches, suggesting that the inner wall expanded by additional branches intruding between the pre-existing ones (Figure 6.47C). Moreover, hyphal lysis along the interior wall surface appears to have been an integral process in sporocarp development.

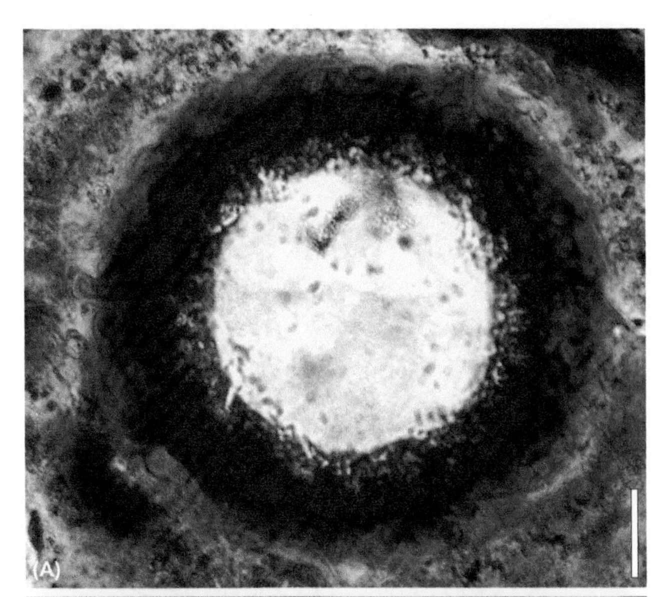

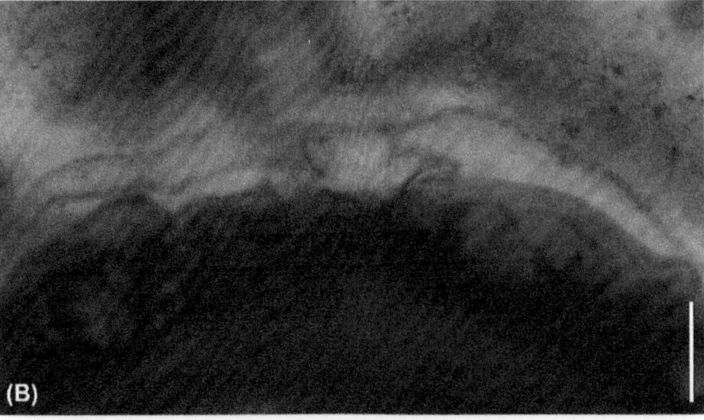

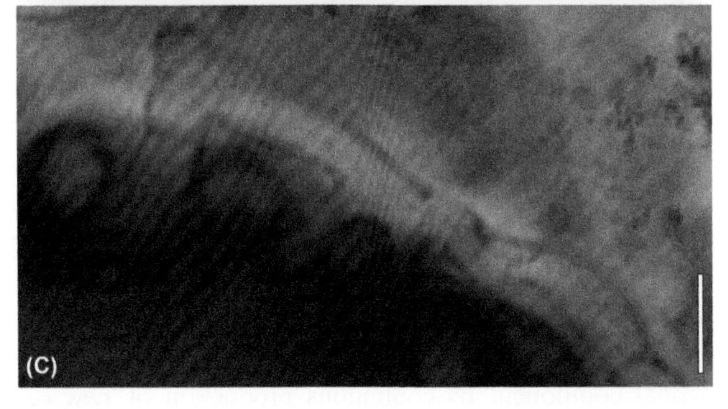

FIGURE 6.47 *Mycocarpon cinctum*. **A.** Section showing massive hyphal investment. **B.** Detail of thin-walled, peripheral hyphae. **C.** Detail of hyphal branch extending into the inner investment layer. Bar = 10 μm. Mississippian, France. Bar (A) = 20 μm; (B, C) 5 μm.

An interesting "sporocarp" similar to forms known from the Carboniferous is *Mycocarpon asterineum* from the Middle Triassic of Antarctica (Taylor and White, 1989). Mature specimens of this fossil (Figure 6.48) are between 200 and 450 μm in diameter and characterized by

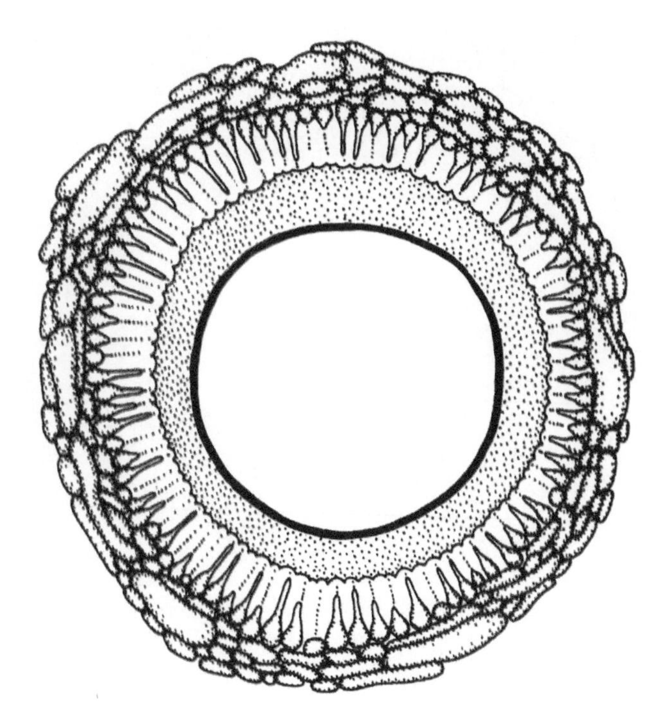

FIGURE 6.48 Suggested reconstruction of *Mycocarpon asterineum* showing two-layered outer hyphal investment, non-hyphal inner investment layer, and central spore. (From Taylor and White, 1989.)

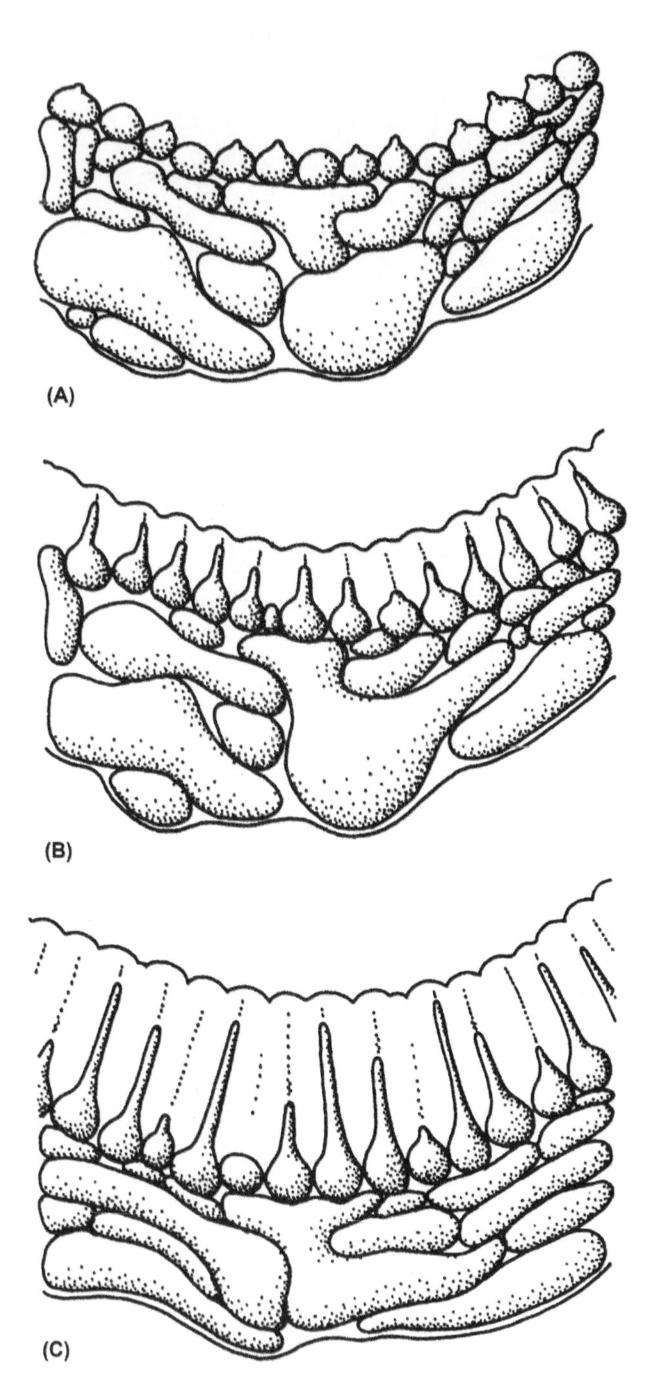

FIGURE 6.49 Sequence of developmental stages in the investment of *Mycocarpon asterineum*. **A.** Immature wall and compact mycelia. **B.** Increasing inner layer with cells of the inner margin becoming acuminate. **C.** Mature investment with outer hyphal layer less compact. (From Taylor and White, 1989.)

an investment constructed of an outer hyphal and inner non-hyphal component. The outer investment is 30–40 μm wide and composed of several layers of interwoven hyphae, whereas the inner, non-hyphal component is up to 35 μm thick and consists of numerous radially oriented channels that give the structure a striate appearance. Hyphae from the outer component extend into the channels along the outer margin. Immature individuals possess investments with a crenate inner margin surrounded by densely interwoven hyphae (Figure 6.49A). Taylor and White (1989) interpreted this margin as having produced the inner non-hyphal component by continuous production of new cell wall material; each cell apparently became acuminate as new wall material was produced (Figure 6.49B, C). Krings and Taylor (2013) speculated that a similar process might also have led to the increase in thickness of the inner, non-hyphal investment layer during maturation of the Early Devonian mantled reproductive unit *Zwergimyces vestitus* (see Figures 6.10 and 6.11). However, what relationship, if any, exists between the "sporocarps" and *Z. vestitus* remains unknown. Although the "sporocarps" and *Z. vestitus* are similar in overall morphology, the latter is only half the size of the smallest "sporocarp" (i.e. *M. cinctum*) described to date. Moreover, the hyphal investment of *Z. vestitus* is less complex than that seen in most "sporocarps."

In *Endochaetophora antarctica* (Figure 6.50), a spherical fossil from the Middle Triassic of Antarctica that is

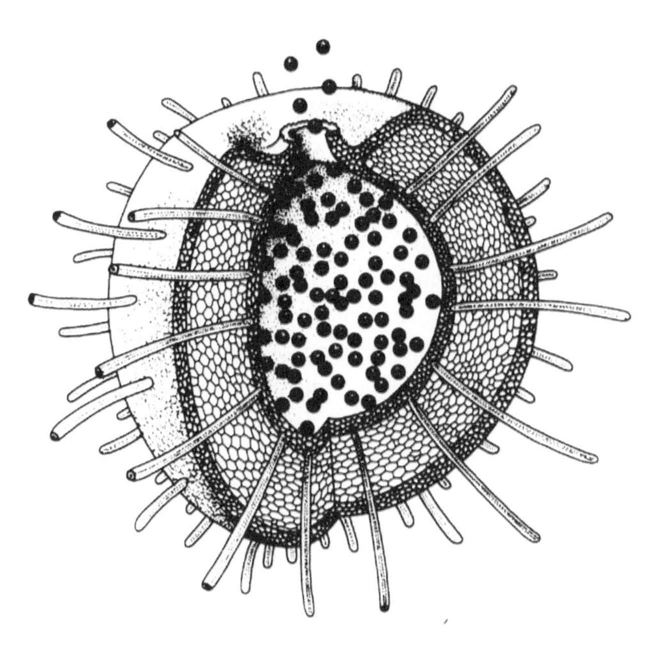

FIGURE 6.50 Suggested reconstruction of *Endochaetophora antarctica*. (From White and Taylor, 1988.)

FIGURE 6.51 James F. White, Jr.

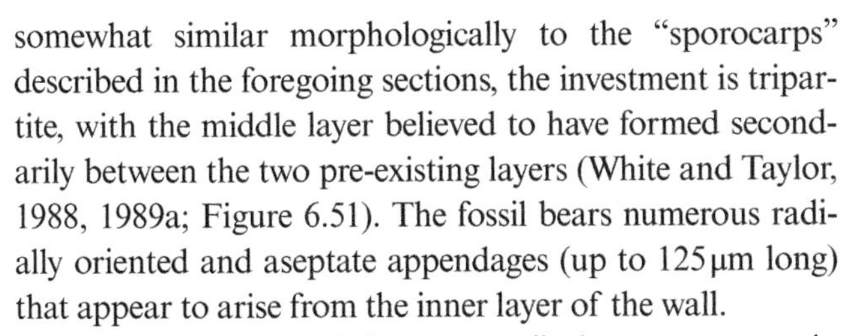

somewhat similar morphologically to the "sporocarps" described in the foregoing sections, the investment is tripartite, with the middle layer believed to have formed secondarily between the two pre-existing layers (White and Taylor, 1988, 1989a; Figure 6.51). The fossil bears numerous radially oriented and aseptate appendages (up to 125 µm long) that appear to arise from the inner layer of the wall.

A small number of "sporocarps" show structures in the investment that may correspond to some type of aperture. In only two of these, however, is there any evidence of a structure extending from the aperture. One of these is a *Traquairia* from the Lower Coal Measures (Lower Pennsylvanian) of Great Britain (Figure 6.52; also see Figure 6.44) which shows a preformed aperture from which emerges a fascicle of elongate, radially oriented structures (Krings et al., 2011e). These structures (Figure 6.53) are up to 115 µm long and appear to arise from a common, elevated region; some appear to branch at the base. The distal end of each is rounded and characterized by a subdistal constriction, which suggests that individual units are being formed terminally. This interpretation is further supported by the formation of a transverse septum in the region of the constriction in some of the structures. A second example is a specimen of *Sporocarpon* illustrated by McLean (1922) and consisting of a cluster of spiny, spore-like bodies united by delicate filaments and present just above an orifice in the investment. Another interesting specimen is *T. carruthersii* from the British Pennine Lower Coal Measures Formation (Scott, R., 1911). This fossil shows a separation in the inner non-hyphal layer of the investment that is covered by a small plug.

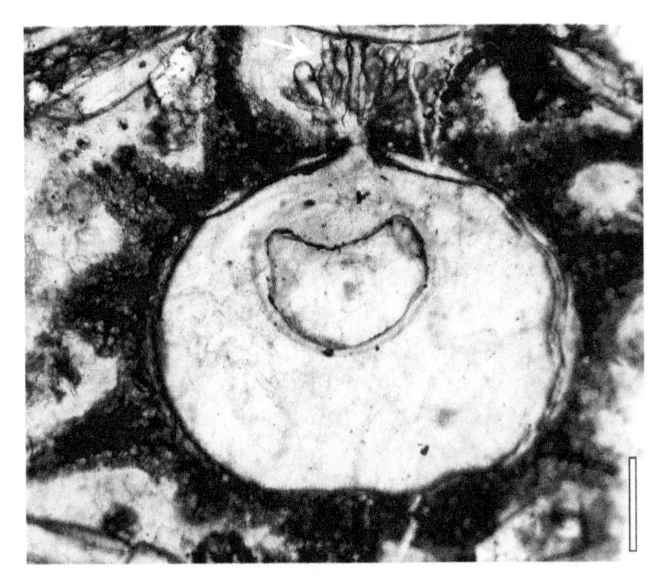

FIGURE 6.52 *Traquairia* sp. with fascicle of radially elongate structures (top) extending from distal aperture. Pennsylvanian, Great Britain. Bar = 100 µm.

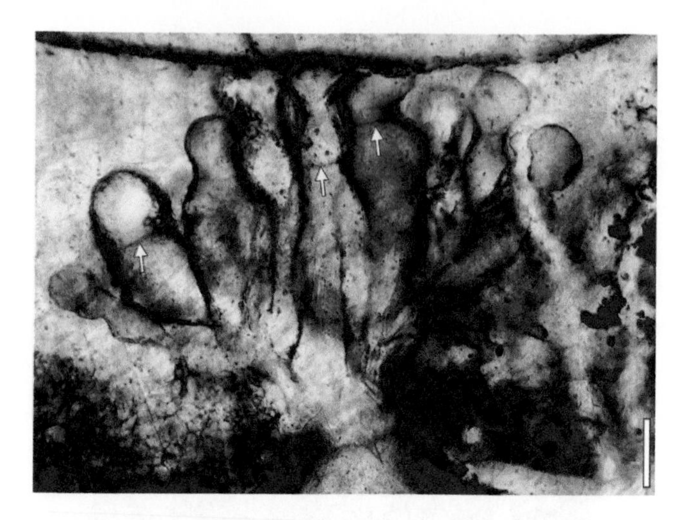

FIGURE 6.53 Detail of elongate structures in Figure 6.52. Note distal constriction of some elements and transverse septa (*arrows*). Pennsylvanian, Great Britain. Bar = 20 μm.

A few other Carboniferous "sporocarps" have been described with an interruption in the investment making it difficult to determine whether the space is natural and represents a preformed aperture or is simply the result of some biotic or abiotic force (e.g. Davis and Leisman, 1962; Taylor et al., 1994b). A structural feature in the investment suggestive of a dispersal pore is also present in *E. antarctica* (see Figure 6.50). In this fossil, there is also an apparent breakdown of the investment associated with the formation of the aperture (White and Taylor, 1988). Although there are no obvious structures that may be involved in spore formation, the central lumen of at least one *Endochaetophora* specimen contains thin-walled spores.

"Sporocarps" have been generally interpreted as fungal in origin based on the investment constructed of hyphae; the precise systematic affinities of these structures, however, remain elusive. The most controversial aspect concerns the nature of spherical structures present within the central cavity. In some specimens the cavity is empty, but more often it contains one to several spheres (e.g. Figure 6.41) that have been the basis for several hypotheses regarding the affinities of all "sporocarps" in general. One idea suggests affinities with the Ascomycota based on specimens containing one large sphere believed to represent an ascus, which in turn contains several smaller spheres interpreted as ascospores (Stubblefield et al., 1983). According to this idea, the "sporocarp" would represent a cleistothecium. An alternative interpretation views the large sphere as a zygospore, and the entire structure is interpreted as a reproductive structure (i.e. a mantled zygosporangium) of a member of some zygomycetous group (Pirozynski, 1976a; Taylor and White, 1989; Krings et al., 2010c). If this latter interpretation is accurate then the smaller spheres found within the large sphere in some specimens would represent some type of intrusive microfungus.

There is an increasing body of circumstantial evidence to corroborate the latter hypothesis. For example, evidence of mycoparasitism occurs in a specimen of *Dubiocarpon* from the Lower Pennsylvanian of Great Britain (Krings et al., 2011a). The parasitic fungi are represented by spherical structures (see Figure 6.39A) as well as by hyphae forming appressorium-like swellings at the contact region with host walls (see Figure 6.39B). This discovery supports the suggestion that many of the small spheres present in other Carboniferous "sporocarps" may in fact represent stages of mycoparasites. Moreover, the radially oriented structures emerging through the preformed aperture in a specimen of *Traquairia* from Great Britain (see Figures 6.44, 6.52, 6.53) are morphologically similar to conidiophores bearing terminal conidia of certain extant fungi in the order Entomophthorales (Figure 6.54) and thus might suggest affinities of *Traquairia* with the zygomycetous fungi. The presence of structural features resembling those of modern entomophthoralean fungi in Carboniferous fossils concurs with recent molecular analyses indicating that the Entomophthoromycota originated 405 ± 90 million years ago (Gryganskyi et al., 2012). Moreover, these authors have suggested that entomopathogenic lineages in Entomophthoraceae probably evolved from saprobic or facultatively pathogenic ancestors during or shortly after the evolutionary radiation of the arthropods.

Additional circumstantial evidence corroborating the hypothesis that at least some "sporocarps" represent mantled zygosporangia include the presence of a confluent hyphal meshwork extending around and between clustered specimens in several representatives of *Mycocarpon* (e.g. McLean, 1922; Stubblefield et al., 1983). This suggests that these structures were produced in groups of two to several, possibly within sporocarps. Finally, Taylor and White (1989) suggest that the inner, non-hyphal investment component of the Triassic "sporocarp" *M. asterineum* was produced by a layer of special hyphae along the inner surface of the outer wall layer through continuous secretion of wall material. As the structure expanded, the outer wall layer became successively compacted. It is interesting that a somewhat similar developmental sequence has also been reported in the zoosporangium mantle of certain extant

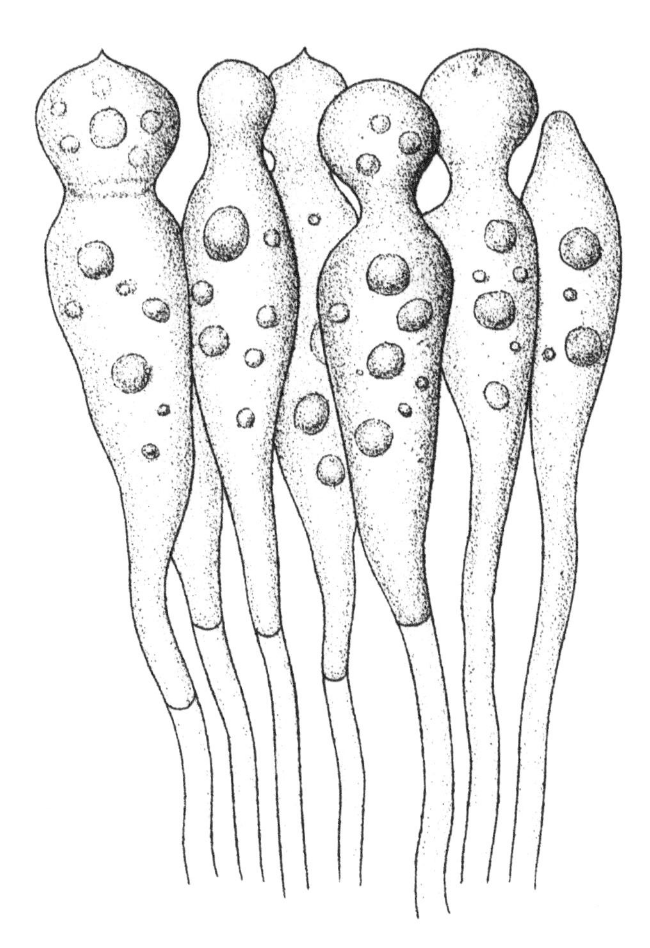

FIGURE 6.54 *Entomophthora* sp. conidiophores bearing terminal conidia. Extant. (From Thaxter, 1888.)

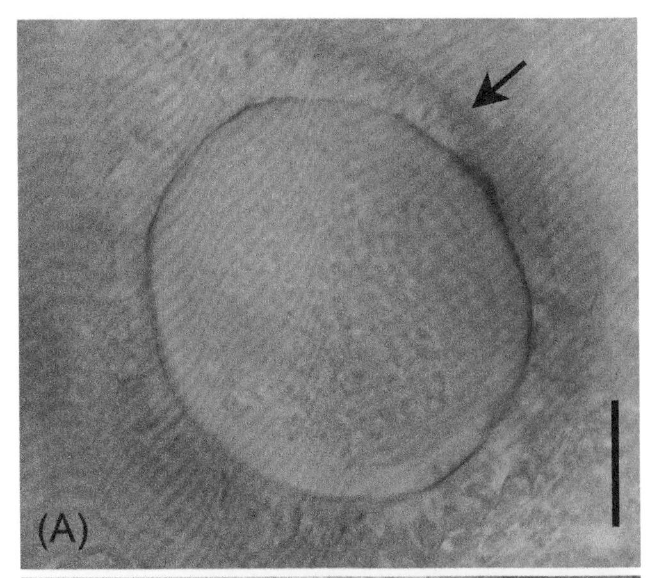

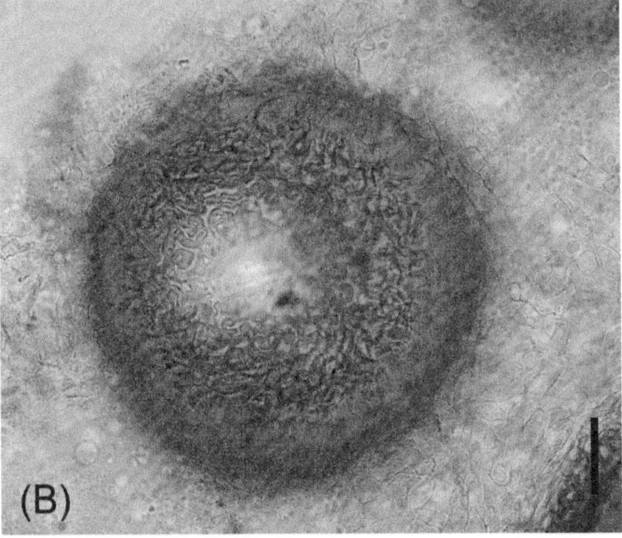

FIGURE 6.55 Zygosporangium of *Endogone* sp. showing mantled surface (*arrow*) at midlevel (**A**) and external surface (**B**). Extant. Bars = 25 μm. (Courtesy D. Southworth.)

Endogone species (Figure 6.55) (Bonfante-Fasolo and Scannerini, 1976).

Structural features confirming the zygomycetous affinity of the "sporocarps" have not yet been conclusively documented. Determining the precise affinities of these fossils has been hampered by the fact that virtually all of the specimens discovered to date appear to be at approximately the same stage of development, that is, fully developed structures. Immature structures would certainly be influenced by preservational bias, and we also believe that they may be rather difficult to accurately identify. A second problem relates to the fact that "sporocarps" always occur isolated or in relatively small clusters and thus cannot be related to the system on/in which they were produced.

Other Enigmatic Fossils

There are several other (micro)fossils in the rock record for which the biological affinities remain unresolved but that have variously been referred to or compared with zygomycetous fungi. Several types of small, ornamented structures, which occur in abundance in coal balls from the Lower Pennsylvanian of Great Britain (Williamson, 1878, 1880, 1883), have also been discovered in coal balls and chert deposits from elsewhere (Krings et al., 2011d) and have been named *Zygosporites* (Williamson, 1878, 1880). Specimens of *Zygosporites* are either spherical (Figure 6.56) or ovoid to elongate and characterized by prominent, antler-like extensions on the exterior surface and a truncated, collar-like extension. *Zygosporites* was

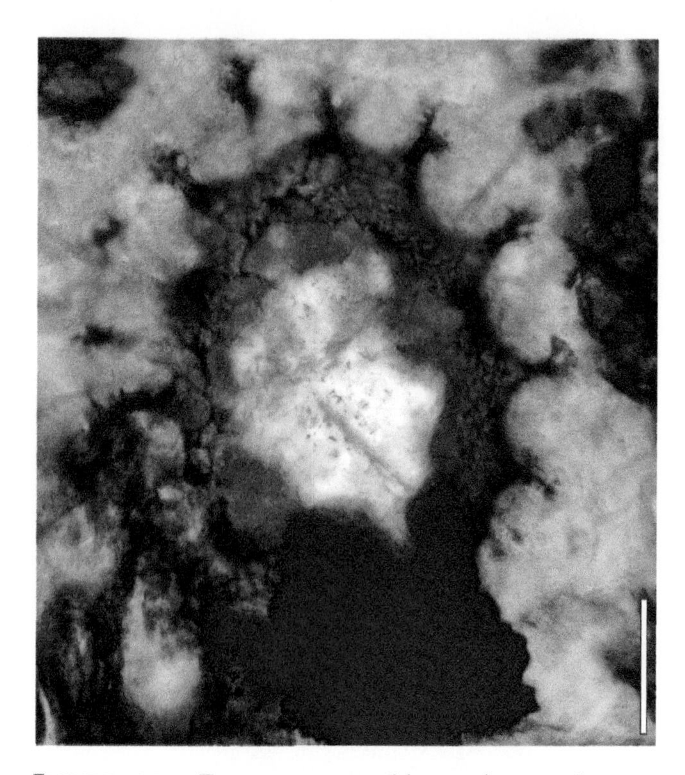

FIGURE 6.56 *Zygosporites* sp. with prominent wall extensions. Pennsylvanian, Great Britain. Bar = 50 μm. (Courtesy N. Dotzler.)

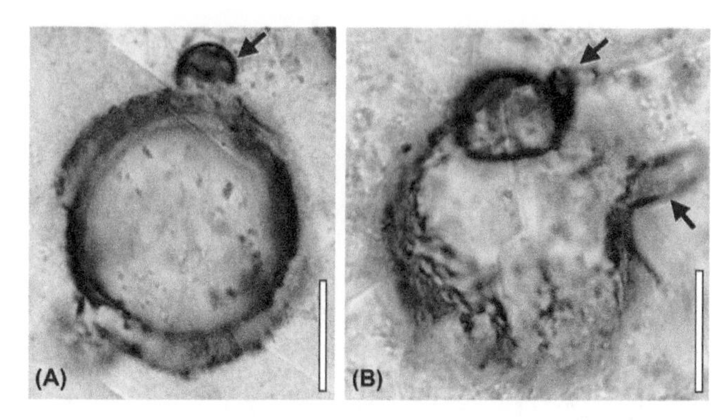

FIGURE 6.57 Two distinct spherical structures associated with smaller round units (*arrow* in **A**) and subtending hyphae (*arrows* in **B**). Pennsylvanian, Great Britain. Bars = 10 μm.

initially believed to represent some type of land-plant spore (Williamson, 1880) or the zygote of some type of zygnematophycean green alga (e.g. Spencer, 1893). McLean (1912) noted similarities between *Zygosporites* and the zygospores of *Phycomyces nitens* (Mucorales). A recent hypothesis (Krings et al., 2011d), however, suggests that *Zygosporites* may represent oogonia of peronosporomycetes (see Chapter 13), based on the remarkable correspondence of the surface ornamentation patterns to those of the peronosporomycete *Combresomyces*, which is known from the Carboniferous (Dotzler et al., 2008; Strullu-Derrien et al., 2011) and the Triassic (Schwendemann et al., 2010).

Krings et al. (2010a) describe two different types of spherical structures (up to 30 μm in diameter) that occur terminally on narrow hyphae from the Lower Pennsylvanian of Great Britain. Many of these spheres are closely associated with, and most likely also physically attached to, a second, slightly or distinctly smaller (up to 15 μm in diameter) spherical structure (Figure 6.57A, B). In one specimen there appears to be a pore in the small sphere through which an organic connection with the large sphere is maintained. The systematic affinities of these fossils remain unknown. One hypothesis offered by the authors regards the spheres as peronosporomycete oogonia to which are adpressed single paragynous antheridia (see Chapter 13). Alternatively, they might represent members of the Zygomycetes, in which the large sphere would be the zygo- or azygosporangium, while the smaller ones represent a suspensor.

Another enigmatic fossil that has been referred to the zygomycetous fungi is *Mucor combrensis* (Renault, 1896b), later renamed *Mucorites combrensis* (Meschinelli, 1898), from the upper Visean of France. This fossil occurs in the form of a net-like structure within a lycophyte megaspore (Figure 6.58). However, the structure, interpreted as a mycelium by Renault (1896b), may well be a preservational artifact. *Palaeomyces gracilis* and *P. majus* (genus later renamed *Palaeomycites* by Meschinelli, 1898), are two additional fungal fossils from the Visean of France that were attributed to the zygomycetous fungi by Renault (1896b). Both taxa occur in the form of spheroidal to ovoid vesicles (up to 38 μm in diameter in *P. gracilis* and 96 μm in *P. majus*) interpreted as (chlamydo)spores that are borne on largely aseptate hyphae in a terminal or intercalary position (e.g. Figure 6.59). Most of the fossils described under the name *Palaeomyces* or *Palaeomycites*, however, including *P. gracilis* and *P. majus*, cannot be assigned systematically with confidence because diagnostic features are absent (see also Chapter 7, p. 113). Another Carboniferous fossil believed to represent a coenocytic, zygomycetous mycelium with affinities in the Mucoraceae has been described from sapromyxite coal from Russia as *Mucorodium palaeomycoides* (Zalessky, 1915). Spheroidal structures interpreted as azygotes, as well as pairs of distally swollen filaments suggestive of gametangial fusion, occur in some of the specimens (*arrows* in Figure 6.60).

FIGURE 6.58 *Mucorites combrensis* (*arrow*). Mississippian, France. (From Renault, 1896b.)

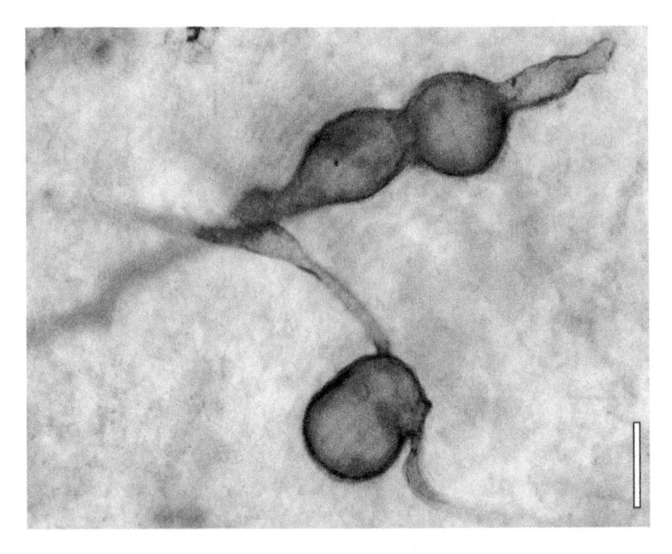

FIGURE 6.59 *Palaeomyces gracilis* showing aseptate hyphae and intercalary swellings. Mississippian, France. Bar = 20 µm. (Originally figured by Renault, 1896b.)

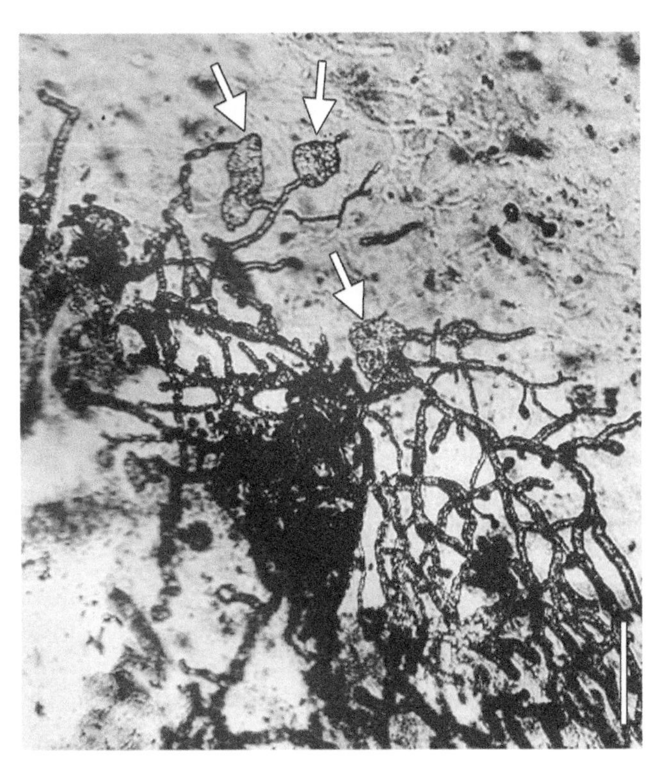

FIGURE 6.60 *Mucorodium palaeomycoides* filaments with spheroidal structures and swellings suggestive of gametangial fusion (*arrows*). Carboniferous, Russia. Bar = 30 µm. (From Zalessky, 1915.)

Lithomucorites miocenicus is an interesting dispersed microfossil from the Miocene (~23–5 Ma) of India (Kar et al., 2010; Saxena and Tripathi, 2011). It represents what appears to be a fungal sporangium that is apophysate, flask shaped, or (sub)globose, and measures 25–52 × 22–49 µm; some of the specimens occur on the tip of a sporangiophore. The wall is closely ornamented with baculae, pilae, and verrucae. The fossil has been interpreted as a zygomycetous fungus because of coenocytic hyphae, asexual reproduction by means of sporangiophores, and the absence of flagellate cells. *Mycozygosporites laevigatus* is another dispersed microfossil from the Miocene of India interpreted as zygomycetous (Kar et al., 2010). The fossil consists of a thick-walled sphere (believed to represent a zygosporangium) with two tubular hyphae attached opposite to each other.

AMBER FOSSILS

Although there are numerous reports of representatives of many different groups of fungi and fungus-like organisms in amber, accounts of zygomycetous fungi are comparatively rare, perhaps due in part to the fact that most Zygomycetes are saprotrophic organisms living in habitats not normally preserved in amber. However, there are several vegetative hyphae and mycelia in amber that have been referred to the Zygomycetes (e.g. Grüss, 1931; Peñalver et al., 2007; Speranza et al., 2010; Girard and Adl, 2011). One well-preserved mycelium occurs in amber from the El Mamey Formation (Miocene) in the Dominican Republic (Rikkinen and Poinar, 2001). Hyphae are smooth walled and largely coenocytic; septa are mainly present to delimit reproductive structures. Several hyphae at the periphery of the mycelium bear apical or subapical swellings, 5–20 μm in diameter, that apparently represent gametangia and/or sporangia, while others develop prominent spherical structures containing single, spore-like objects. Still other hyphae are disarticulated into catenulate (chainlike) hyphal elements or arthroconidia. Certain structural features of this mycelium, including the coenocytic hyphae and the putative sporangia, may suggest affinities with the zygomycetous fungi.

Particularly interesting are the thick, aseptate hyphae that occur within nematodes preserved in upper Oligocene-Miocene amber from Mexico (Jansson and Poinar, 1986). These hyphae resemble the assimilative hyphae of certain modern nematophagous Zygomycetes; however, the fossils are too incomplete to be assigned systematically. Another interaction between fungi and animals is preserved in Early Cretaceous (~120–115 Ma) Álava amber from Spain (Speranza et al., 2010). This interaction occurs in the form of a thrips containing a vegetative mycelium and reproductive structures believed to be similar to those seen in certain extant Zygomycetes. From the same amber, Speranza et al. (2010) also reported on structures that they believed to represent zygospores.

Poinar and Thomas (1982) described an entomophthoralean fungus living on the abdomen and thorax of a winged termite preserved in amber from the Dominican Republic (Miocene). The body of the animal is covered with a white mat composed of closely appressed, apparently coenocytic hyphae. A layer of conidia lines the surface of the mycelial mat. Some of these conidia are budding, and a number of smaller ones (secondary conidia) were present in the amber just adjacent to the mycelial mat. Based on spore size and shape, these authors concluded that the conidia fall between the "*fresenii*" and "*lampyridarum*" groups of *Entomophthora*, as defined by Hutchison (1963), and the "*culicis*" group, as recognized by Waterhouse (1975).

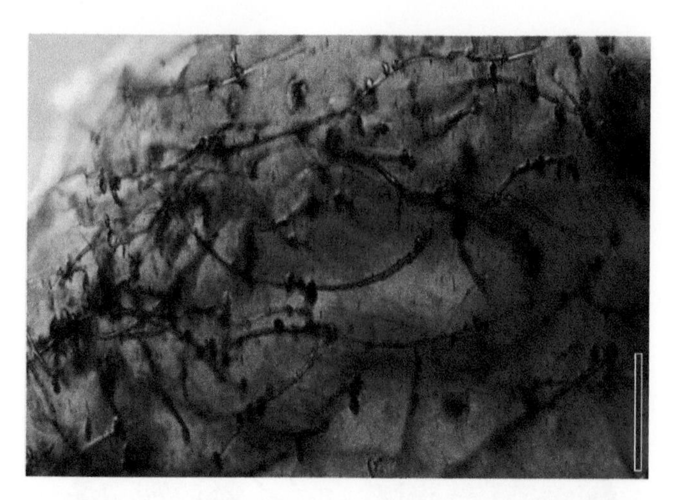

FIGURE 6.61 *Palaeodikaryomyces baueri* showing aseptate hyphae and vesiculi. Cretaceous, Germany. Bar = 100 μm. (Courtesy H. Dörfelt and A. Schmidt.)

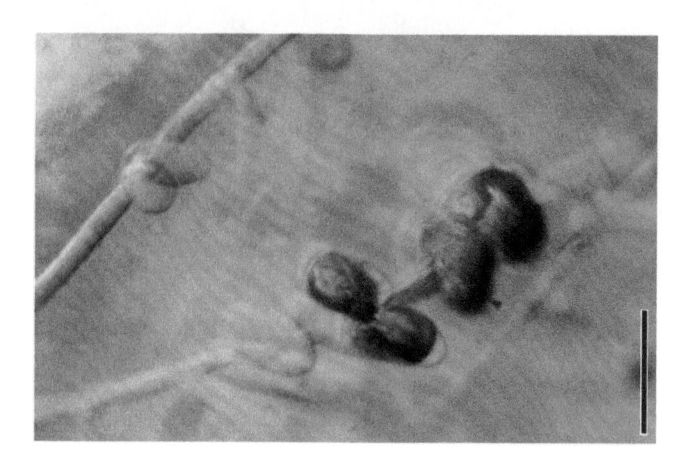

FIGURE 6.62 Detail of *Palaeodikaryomyces baueri* in Figure 6.61 showing single hypha. Cretaceous, Germany. Bar = 20 μm. (Courtesy H. Dörfelt and A. Schmidt.)

Palaeodikaryomyces baueri is a fossil fungus preserved in Cenomanian (Upper Cretaceous; ~101–94 Ma) Schliersee amber (Dörfelt and Schäfer, 1998) that is believed to represent a saprotrophic organism occupying a basal position between the Ascomycota and Basidiomycota on the one hand and the zygomycetous fungi on the other. The fossil (Figures 6.61 and 6.62) is characterized by aseptate hyphae and vesiculi, developing septa, branches at the vesiculi, clamps or loops, and cysts at the loops (Schönborn et al., 1999). *Palaeodikaryomyces baueri* is believed to have preserved the essential characters of the primary Dikaryomycetidae, not differentiated into Ascomycota and Basidiomycota. Schmidt et al. (2001) hypothesized that *P. baueri* was an archaic fungus that persisted into the late Mesozoic.

ICHNOTAXA

Ichnotaxa (trace fossils) are fossil taxa that are based on the fossilized activities of organisms rather than the actual organisms (body fossils). One ichnotaxon, *Stolophorites lineatus* from the Upper Triassic (~237–201 Ma) of North America, has been attributed to the zygomycetous fungi (Bock, 1969). It consists of several groups of small casts, about 5 mm long, resembling pear or club-shaped forms; that is, they are composed of a cone-shaped stalk interpreted as a sporangiophore and terminating in an oval or obtuse head thought to represent a sporangium (Figure 6.63). The individual structures are evenly spaced and arranged in straight rows, and appear to be attached to a stolon-like base. Bock (1969) compared the fossils with the extant *Rhizopus nigricans*. Subsequent workers, however, have regarded the casts as indeterminable (e.g. Olsen and Baird, 1990).

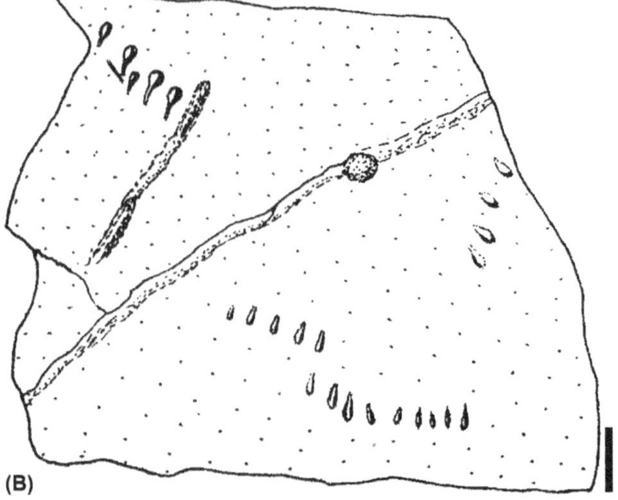

FIGURE 6.63 **A.** Several elongate casts of *Stolophorites lineatus*. **B.** Distribution of *Stolophorites lineatus* on specimen. Triassic, USA. Bars = 5 μm (A), 10 μm (B). (From Bock, 1969.)

CONCLUSIONS: ZYGOMYCETES

Some of the fossils of zygomycetous fungi described in this chapter demonstrate that, with suitable preservation, these fungi can be documented in great detail. Such fossils are also of great importance as a source of information that can be used to accurately calibrate molecular clocks and define minimum ages for various fungal lineages. Moreover, it is becoming quite apparent that the fossil record of various lineages of fungi is not only ancient but also demonstrates a high diversity of forms, some of which closely parallel extant counterparts, even to details relating to micromorphological (cytological) features associated with reproduction. Such comparisons can be valuable in the discussion of the evolution of developmental stages of putatively sexual structures in ancient fungi that heretofore have not been recognized. This will not only increase our understanding about various groups of fungi in time and space but also give us some idea about when various features evolved.

On the other hand, enigmatic fossils such as the columella-like structures from the Lower Devonian Rhynie chert and the "sporocarps" from the Carboniferous and Triassic represent interesting components of ancient ecosystems that continue to result in speculation as to their systematic affinities and biological significance. Within these structures there are basic similarities in size and organization that suggest at least some may belong to the same higher taxonomic category, perhaps a lineage of the zygomycetous fungi. Like so many aspects of paleomycology, one specimen often is the single necessary segment of information that helps to elucidate the affinities that have remained elusive. We are certain that this will be the same trajectory regarding all enigmatic fossils detailed in this chapter as they continue to be reported and studied.

7

GLOMEROMYCOTA

BIOLOGY OF GLOMEROMYCOTAN FUNGI 104
Arum and *Paris* Morphological Types 105
Ecology 107
Systematics and Biodiversity 107
GLOMEROMYCOTAN CHARACTERS 108
GLOMEROMYCOTAN REPRODUCTION 110
FOSSIL GLOMEROMYCOTA 111
Rhynie Chert Asexual Spores 112
Carboniferous and Permian Spores 116
Mesozoic, Cenozoic, and Subfossil Spores 117
Sporocarps 118
Root Nodules 118
Fossil Evidence of Arbuscular Mycorrhizae 119
Evolution of Arbuscular Mycorrhizae 126

The Glomeromycota is a monophyletic group (Figure 7.1) of soil-borne fungi that are among the most important microorganisms on the Earth, not only because they form intimate associations with plants but also because they are believed to have been crucial in the initial colonization of the terrestrial realm by land plants (e.g. Pirozynski and Malloch, 1975; Humphreys et al., 2010; Sanders and Croll, 2010; Wang, B., et al., 2010; Field et al., 2012). The fungi in this group were originally included in the Glomales as an order in the former Zygomycetes (Gerdemann and Trappe, 1974; Figure 7.2). However, as a result of molecular phylogenetic studies, a new phylum, the Glomeromycota, was established (Schüßler et al., 2001; Figure 7.3). The history of the taxonomy and systematics of the Glomeromycota have been summarized by Stürmer (2012).

Glomeromycota occur in almost all terrestrial ecosystems, including temperate and tropical forests, deserts, grasslands, sand dunes, and a variety of agroecosystems, where they have a major impact (e.g. Gianinazzi et al., 2010; Muthukumar and Tamilselvi, 2010; Schüßler and Walker, 2011); at least one form has been isolated from soil from Antarctica (Cabello et al., 1994). Approximately 250 species have been formally described to date (e.g. Schüßler and Walker, 2010; Oehl et al., 2012; INVAM, 2013). They form obligately biotrophic associations with a vast number of terrestrial plants (embryophytes), perhaps as many as 90% (Parniske, 2008; Smith and Read, 2008), and thus have the potential to significantly influence plant biodiversity and fitness (e.g. Veresoglou et al., 2012); for example, some are effective in controlling fungal and other microbial pathogens (e.g. Newsham et al., 1995; Schüßler et al., 2001; Vannette and Hunter, 2009; Li et al., 2012; Ji et al., 2013). Although the focus of glomeromycotan research has been principally on their occurrence in seed plants, including some aquatic and marshy forms (e.g. Søndergaard and Laegaard, 1977; Radhika and Rodrigues, 2007; Baar et al., 2011), from a paleomycological perspective it is interesting to note that they also form associations with various vascular cryptogams, including lycopsids and ferns (Turnau et al., 2005; Winther and Friedman, 2007, 2008, 2009; Fernández et al., 2008; Horn et al., 2013), as well as bryophytes (e.g. Schüßler, 2000; Pressel et al., 2010). There have also been several reports of arbuscular mycorrhizae in the roots of *Equisetum* (e.g. LaFerrière and Koske, 1981; Koske et al., 1985; Dhillion, 1993 Fernández et al., 2008), but these have been viewed

T.N. Taylor, M. Krings & E.L. Taylor: Fossil Fungi.

DOI: http://dx.doi.org/10.1016/B978-0-12-387731-4.00007-4

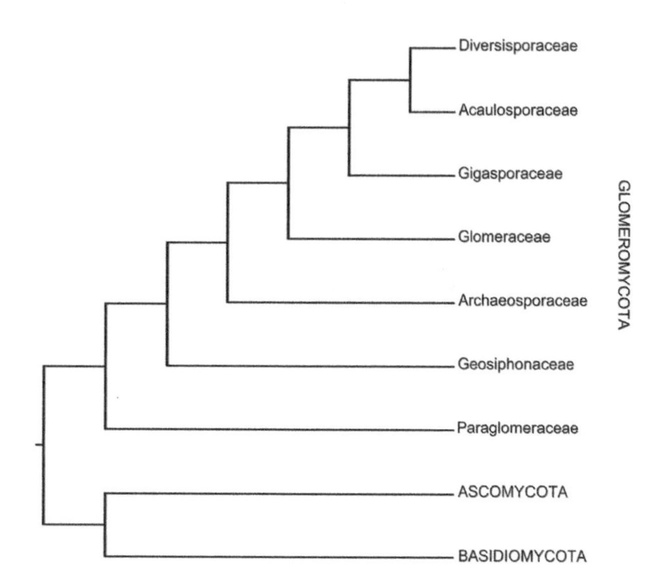

FIGURE 7.1 Suggested phylogenetic relationships within the Glomeromycota.

FIGURE 7.2 James W. Gerdemann (left) (Courtesy Janice Gerdemann) and James M. Trappe (right).

by others as facultative associations at best (e.g. Brundrett, 2002). Berch and Kendrick (1982) suggested that if extant horsetails are in fact non-mycorrhizal, their diverse ancestors may have shared this peculiarity, and their reduction in diversity was perhaps a result of competition from plants equipped with root-inhabiting fungi. More recently, however, a study by Hodson et al. (2009) that examined multiple *Equisetum* specimens found about 30% of the plants contained hyphae, arbuscules, and vesicles characteristic of arbuscular mycorrhizae.

FIGURE 7.3 Arthur Schüßler.

BIOLOGY OF GLOMEROMYCOTAN FUNGI

Glomeromycotan fungi enter into arbuscular mycorrhizal (AM) symbioses (sometimes also referred to as vesicular-arbuscular mycorrhizal or VAM symbioses) with their plant hosts where they serve as links for nutrient exchange between the above- and below ground parts of the plant. All of these fungi are obligate biotrophs that grow in plant roots (or thalli in bryophytes) without causing the plant any adverse symptoms (e.g. Hildebrandt et al., 2002). Through their symbiotic association, the fungi develop an extensive hyphal network to expand the absorptive function of the plant roots, which is apparently superior to that of coarser root hairs. There is also evidence that AM diversity is significant at the functional level, in which one kind of arbuscular mycorrhiza is complementary to the host root system because of root hair length and density, its capacity to solubilize certain inorganic phosphorus (P) sources, and its ability to proliferate in nutrient-rich patches of soil (Koide, 2000). AM fungi (AMF) are not, however, host specific. Rather, they are to some extent host-preferential (e.g. Yang, H. et al., 2012). Moreover, there is a high selectivity from host plants for AMF. This may indicate that certain types of AMF–host plant combinations are better adapted to certain types of ecosystem variability (e.g. Sanders and Fitter, 1992; Bever et al., 2001; Torrecillas et al., 2012). A study by Klironomos (2000) indicates that AM fungi differ in their intra- and extra-radical morphologies as well as in their effects on plant nutrition, growth, and susceptibility to infection by other soil-borne fungi. The complex interactions between mycorrhizal fungi and their hosts are governed by positive and negative feedback mechanisms (e.g. Bever, 2002; Zhang et al., 2010). AMF can also inhabit soil types that may be less stable than those of other types of mycorrhizae (Kottke, 2002; Lambers and Teste, 2013) and AM community composition and structure is further correlated with soil pH or pH-driven changes in soil chemistry and disturbance (Lekberg et al., 2011). For example, in one study of the conifer *Sequoiadendron giganteum* it was found that AM fungal colonization was more influenced by the availability of plant-assimilated carbon to the fungus than by the fungal supply of mineral nutrients to the roots (Fahey et al., 2012). A significant number of AMF are associated with achlorophyllous, mycoheterotrophic plants (e.g. Merckx et al., 2009, 2012); these plants obtain all their carbon from the mycorrhizal networks via indirect exploitation of nearby autotrophic plants (e.g. Courty et al., 2011).

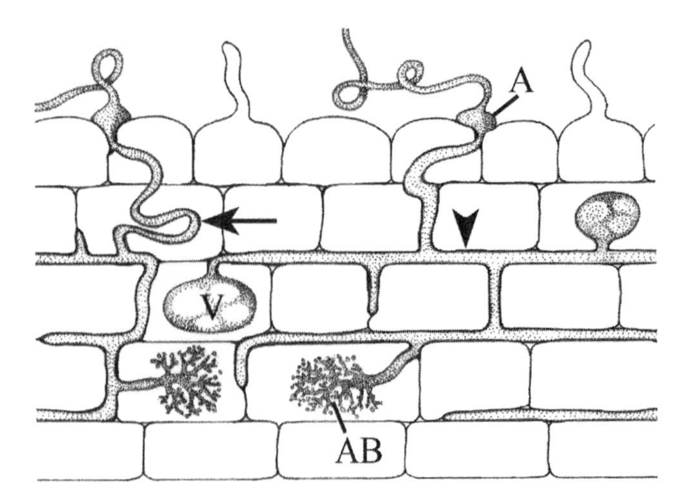

FIGURE 7.4 Diagrammatic representation of root containing *Arum*-type arbuscular association. Infection hypha penetrating root epidermal cell (A) and forming a coil (*arrow*) before entering intercellular system (*arrowhead*). AB = arbuscule; V = vesicle. (Modified from Peterson et al., 2004.)

ARUM AND *PARIS* MORPHOLOGICAL TYPES

Historically there have been two basic morphological types of AM colonization, dependent on intracellular contents (Gallaud, 1905). While these are broad categories that lack a systematic framework, they appear to have some significance with regard to metabolic activity (e.g. see Matekwor Ahulu et al., 2005; van Aarle et al., 2005; Harikumar and Potty, 2009), and their recognition does include morphological features that can be preserved in the fossil record. Both the structure of the root cortex and type of arbuscular mycorrhizae help determine whether the *Arum* or the *Paris* types develop in the host. In the *Arum* type (Figure 7.4), hyphal growth occurs intercellularly in a longitudinal pattern in the root, with short side branches giving rise to arbuscules at nearly right angles (e.g. Armstrong and Peterson, 2002). On the other hand, the *Paris* type (Figure 7.5) is generally defined by intracellular hyphae that form coils with arbuscules forming along the coil (Kubota et al., 2005). To further define arbuscular mycorrhizae in all plants, including those from the fossil record, Strullu-Derrien and Strullu (2007) proposed using the term paramycorrhizas for those colonizing thalli and shoot systems that are generally above the substrate level, and eumycorrhizas for those found in root systems beneath the soil.

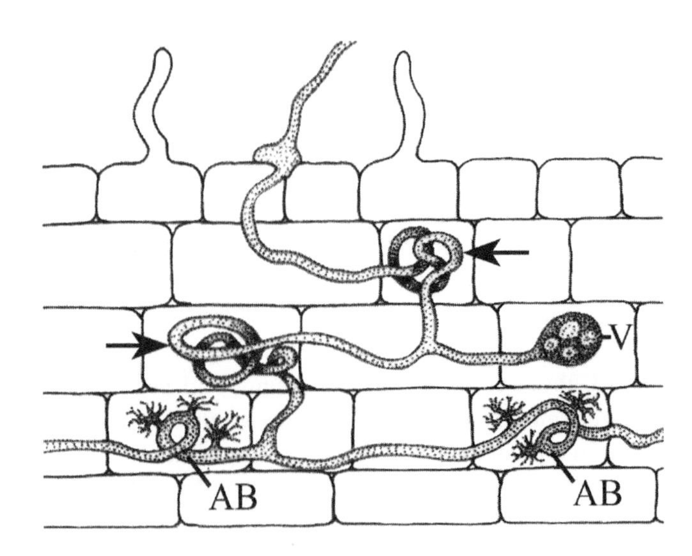

FIGURE 7.5 Diagrammatic representation of root containing *Paris*-type arbuscular association. Arrows indicate extensive coiling in cortical cells and arbusculate coils (AB) and vesicles (V). (Modified from Peterson et al., 2004.)

There are additional features of each type at the cellular level that probably cannot be resolved within fossils. It

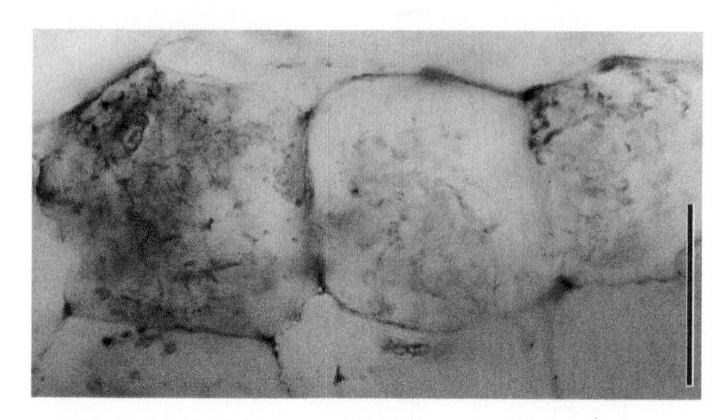

FIGURE 7.6 Arbuscules in the cells of the early land plant gametophyte *Lyonophyton rhyniensis*. Lower Devonian Rhynie chert, Scotland. Bar = 25 µm.

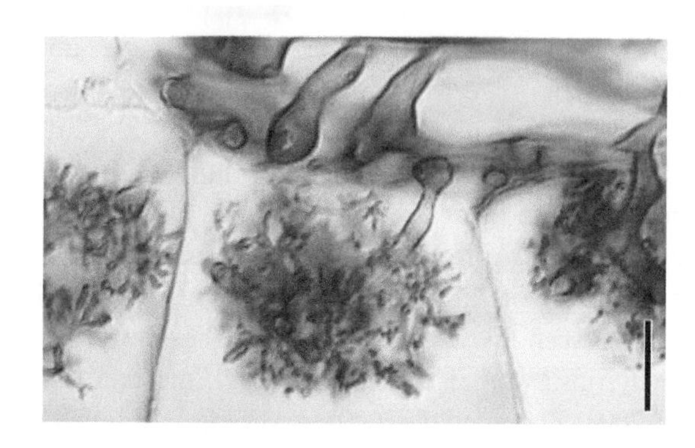

FIGURE 7.7 Highly branched arbuscules and hyphae in the inner cortex of an *Asarum canadense* root. Extant. Bar = 10 µm. (Courtesy M.C. Brundrett.)

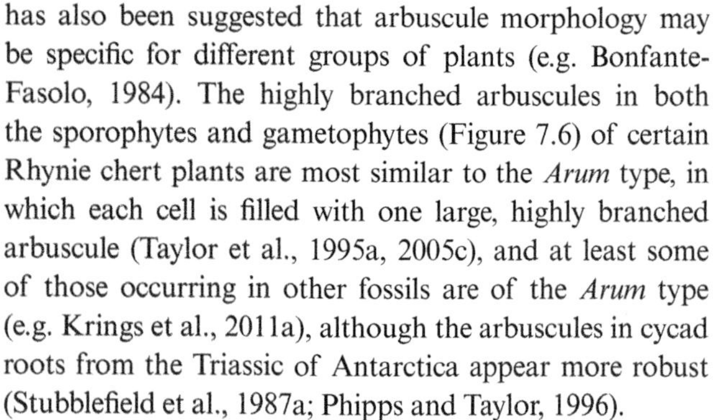

has also been suggested that arbuscule morphology may be specific for different groups of plants (e.g. Bonfante-Fasolo, 1984). The highly branched arbuscules in both the sporophytes and gametophytes (Figure 7.6) of certain Rhynie chert plants are most similar to the *Arum* type, in which each cell is filled with one large, highly branched arbuscule (Taylor et al., 1995a, 2005c), and at least some of those occurring in other fossils are of the *Arum* type (e.g. Krings et al., 2011a), although the arbuscules in cycad roots from the Triassic of Antarctica appear more robust (Stubblefield et al., 1987a; Phipps and Taylor, 1996).

An inherent issue in defining mycorrhizal structures as being either of the *Arum* or *Paris* morphological types is the fact that some arbuscular mycorrhizae can produce either type depending upon the host (e.g. Bedini et al., 2000; Kubota et al., 2005), the fungus, and perhaps the environment (Dickson, 2004; Feddermann et al., 2010). These studies suggest that characterizing either the *Arum* or *Paris* type as the primitive condition has little phylogenetic importance. It is especially difficult in paleomycology where there are typically a limited number of specimens, and thus extant and fossil arbuscule morphology may be difficult to relate to particular groups of plants or families. Nevertheless, because the environment in which certain fossil roots grew is increasingly well known, based on sedimentology, there may be opportunities to correlate arbuscule morphology with both biological (e.g. types of plants) and environmental parameters (e.g. certain types of soils, light regime; see, for example, Harper et al., 2013). There is evidence to suggest that some extant AMF have the ability to switch between root endophytism and mycorrhizae (e.g. Demchenko et al., 2004), but almost nothing is known about how the arbuscule evolved morphologically.

ARBUSCULES

The most important feature of arbuscular mycorrhizae, together with the molecular signaling that takes place between the partners, is the arbuscule (Figure 7.7) (e.g. Cox and Sanders, 1974; Franken and Requena, 2001; Harrison, 2005; Ercolin and Reinhardt, 2011). Arbuscule formation is a coordinated process of subcellular development of the host cell and the AM fungus. Arbuscules are developed upon penetration of cortical cells via invagination of the plasma membrane. The "trunk" hyphae are formed first, and these are bound by the cell wall of the host. Subsequently, the trunk hypha branches repeatedly to develop the arbuscule, which as a whole remains enveloped by a special extension of the host plasma membrane, the periarbuscular membrane (Ivanov et al., 2012). This membrane, which is continuous with the plasma membrane of the cortical cell but differs from the latter in the presence of certain proteins, greatly increases the surface area of the plasma membrane. The periarbuscular membrane is composed of two domains (Pumplin and Harrison, 2009): the arbuscule branch where the specific phosphate transporters are located (e.g. Harrison et al., 2002; Bonfante and Anca, 2009) and the arbuscule trunk, which contains a different protein (Pumplin and Harrison, 2009). As arbuscules develop there is an increase in the number of organelles in the cortical cells (Hause and Fester, 2005). Research has shown that strigolactones, a small group of compounds that occur in the rhizosphere, are not only the signaling molecules that regulate the establishment of symbiosis between the host and AMF but also the growth regulators in the plant (e.g. Akiyama and Hayashi, 2006; Parniske, 2008; Ruyter-Spira et al., 2013). It has also been suggested that gibberellins are involved in determining arbuscule formation (Foo et al., 2013). It is now known that plant cell activity of the host is responsible for preparing the intracellular environment at the time of infection (Genre et al., 2005) and that a penetration system also occurs in the cortical cells (Genre et al., 2008).

Once formed, arbuscules persist for 4–12 days before senescing (e.g. Alexander et al., 1989), perhaps as a result of their inability to continue to deliver phosphate and other nutrients (Parniske, 2008). This provides the opportunity for the same cortical cell to later become re-infected, perhaps by a different AM fungus that is more efficient, a strategy that is controlled by the host (Alexander et al., 1989). Because nutrient availability will change in the soil over time, this strategy presumably provides the host with the opportunity to rapidly capitalize on those fungal networks that are the most productive.

ECOLOGY

Understanding the biology and ecology of the Glomeromycota will continue to be important, not only because AMF are associated with the roots of most economically important crop plants (e.g. Ruiz-Lozano et al., 1995; Smith and Smith, 2011) but also because genetically modified plants will interact with AMF in the future, perhaps with unforeseeable effects (e.g. Cheeke et al., 2012). AMF are also critical in global phosphorus, nitrogen, and carbon cycling (Fitter, 2005; Krüger et al., 2009), and in terrestrial ecosystem functioning at the soil–plant interface (Augé, 2004; Khalvati et al., 2005; Heimann and Reichstein, 2008; van der Heijden et al., 2008; Lambers et al., 2009). It has generally been hypothesized that the increase in atmospheric carbon dioxide results in increasing soil carbon sequestration and protection of organic carbon from decomposition (e.g. Drigo et al., 2010); however, there is now evidence that arbuscular mycorrhizae may diminish carbon pools in the top soil levels (Cheng et al., 2012; Kowalchuk, 2012).

Arbuscular mycorrhizae increase plant productivity and therefore increase litter accumulation (e.g. Verbruggen et al., 2013). Some nitrogen is also transferred from the soil to the plant via the mycorrhizal system; however, much remains unknown about this process (e.g. Smith and Smith, 2013). Protozoa grazing on bacteria also provide nitrogen within the mycorrhizal system (Koller et al., 2013). Inorganic nitrogen taken up by the fungus outside of the roots can also be incorporated into amino acids in the form of arginine and transferred to the plant without carbon (Govindarajulu et al., 2005).

There is compelling evidence that the presence of arbuscular mycorrhizae strongly influences many plant community factors because the fungi can colonize multiple plants forming a common mycorrhizal network (e.g. Selosse et al., 2006 [Figure 7.8]; Thompson, 2006; Taylor, 2006; Brundrett, 2009). They can also form hyphal anastomoses between different species (e.g. de la Providencia et al., 2005; Voets et al., 2006). In addition, AMF can have negative interactions with non-host plants (Veiga et al., 2013). In some instances the fungi demonstrate strong evidence for host preference within the AM communities (e.g. Husband, et al., 2002), and invasive plants may affect the abundance and diversity of AM species (Shah et al., 2010). AMF can operate at multiple phylogenetic levels (Maherali and Klironomos, 2007), as well as influencing various biotic and abiotic factors, including resource distribution, host-plant stress tolerance, soil structure, microbial equilibrium, and fungal competition (Schüßler and Walker, 2011 and literature therein). Although there are other types of mycorrhizal associations, the arbuscular mycorrhiza is the most widespread, and without these fungal partners terrestrial plants would perhaps not be able to compete.

FIGURE 7.8 Marc-André Selosse.

SYSTEMATICS AND BIODIVERSITY

It is generally believed that the biodiversity of arbuscular mycorrhizae is far greater than is presently understood (e.g. Öpik et al., 2008) and that some of the problems of identification relate to the absence of a sufficiently large suite of morphological characters, the potential for dimorphic spores, and the difficulty in deciphering the life history biology (Krüger et al., 2009). For example, the number of species based on spore morphology is probably much larger than the approximately 250 species that have been delimited (Redecker and Raab, 2006). New species are continually being identified (e.g. Błaszkowski et al., 2012; INVAM, 2013), and it is apparent that even when using spore morphology and mode of spore formation as characters, the results may be at odds with results obtained using molecular tools (Schüßler and Walker, 2010; Figures 7.3 and 7.9). Schüßler and Walker (2010) provide a continually updated, online species list of Glomeromycota (http://schuessler.userweb.mwn.de/amphylo/). There are many research questions that remain to be answered about the Glomeromycota, including refining the systematic position of many of the species (e.g. Oehl et al., 2011b) and examining the biogeography, impact on community structure, genome organization, DNA ploidy levels, host preferences, and the evolution of these various structural features (e.g. Brundrett et al., 1999; Walker et al., 2007; Krüger et al., 2012; Schüßler et al., 2011). The Glomeromycota currently include four orders (Archaeosporales, Diversisporales, Glomerales, and Paraglomerales), which may include ten families (Błaszkowski, 2012; Redecker et al., 2013).

GLOMEROMYCOTAN CHARACTERS

Glomeromycotan fungi are characterized by coenocytic (aseptate) to sparsely septate hyphae, large, multinucleated spores, and highly branched structures termed arbuscules that fill certain cells in the cortex of the host plant root and appreciably increase the surface area at the host–fungus interface. Glomeromycota in AM symbiosis display two distinctive complements of structural features: one outside the host root, termed extraradical (Figure 7.10), and the other one, termed intraradical, inside the host (e.g. Brundrett, 2004). Outside the host, the fungus forms an extensive network of distributive and absorptive hyphae (in some forms with clustered swellings, i.e. auxiliary cells; Figure 7.11), and may also produce spores. Inside the host root, the fungus produces intercellular hyphae, vesicles (Figure 7.12), and in many instances spores, as well as intracellular arbuscules. The spores are produced for the most part ectocarpically in the substrate or in the

FIGURE 7.9 Christopher Walker.

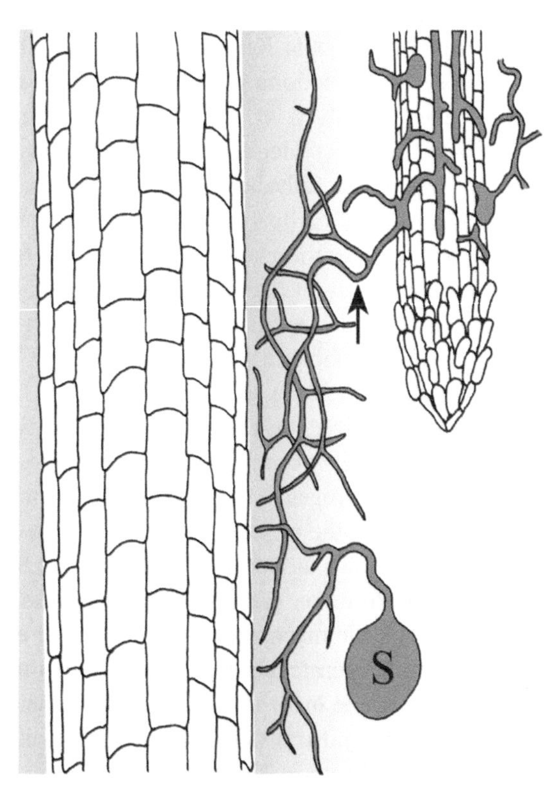

FIGURE 7.10 Colonization of root surface from germinating spore (below, S) or other root (*arrow*). (From Peterson et al., 2004.)

roots of the host plants. Some species produce spores in unstructured dense masses or in structured sporocarps (Figure 7.13) at or near the surface of the soil (Schüßler and Walker, 2010). Spores are spheroidal and variable in size (40–800 μm) and characterized by various wall layers and components that have been used to define species (e.g. Walker, C., 1983; Morton, 1985, 1988, 1990a, b; Morton and Benny, 1990; Maia and Kimbrough, 1994; Morton and Bentivenga, 1994; Walker and Schüßler, 2004; Walker et al., 2007; Oehl et al., 2011a; INVAM, 2013). There are three basic kinds of spores among the Glomeromycota (Walker, 1987). The majority of species form spores by blastic inflation and thickening of a subtending hypha (glomoid) (Figure 7.14). Another group produces large spores by initial production of a small bulbous base followed by blastic expansion, with or without the production of flexible inner wall components (gigasporoid; Figure 7.15). A third type of spore (acaulosporoid) is produced within an initial, relatively thin-walled blastic saccule, either laterally or centrally in the narrowed saccule neck, or rarely completely filling the expanded saccule lumen. Moreover, gigasporoid and acaulosporoid spores may possess a distinct mode of spore germination in which germ tube formation is preceded by the development of a germination shield (Figure 7.16). There is considerable interspecific and intergeneric variation with regard to size and shape of the germination shield, which may range from small, simple coils to prominent, profoundly infolded or lobed structures. The reader is referred to the beautifully illustrated volume on the Glomeromycota by Błaszkowski

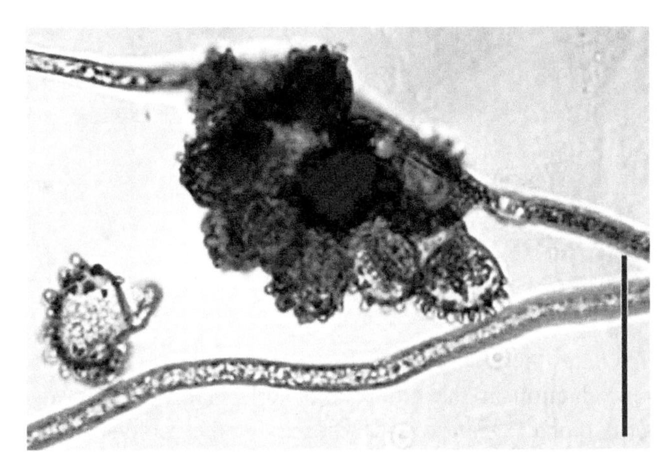

FIGURE 7.11 Auxiliary cells of *Gigaspora gigantea*. Extant. Bar = 50 μm. (Courtesy C. Walker.)

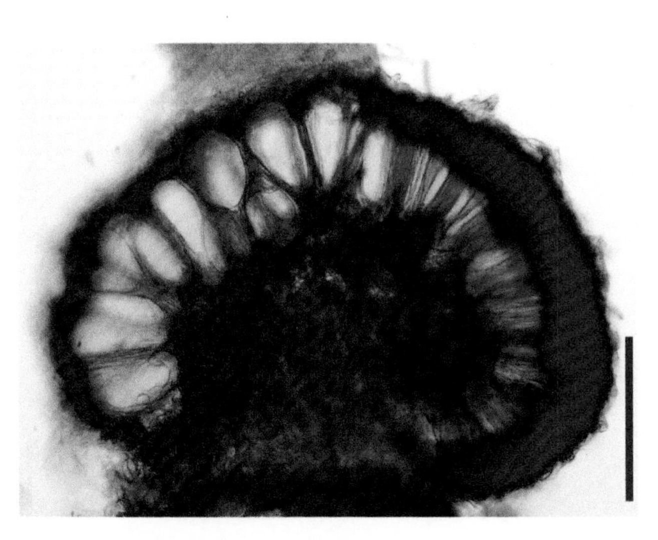

FIGURE 7.13 Sporocarp of *Sclerocystis*. Extant. Bar = 50 μm. (Courtesy C. Walker.)

FIGURE 7.12 *Glomus* vesicles. Extant. Bar = 50 μm. (Courtesy M.C. Brundrett.)

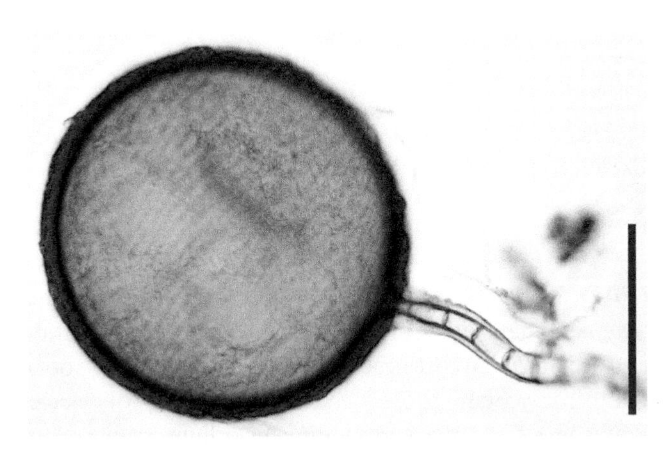

FIGURE 7.14 Glomoid spore. Extant. Bar = 50 μm. (Courtesy C. Walker.)

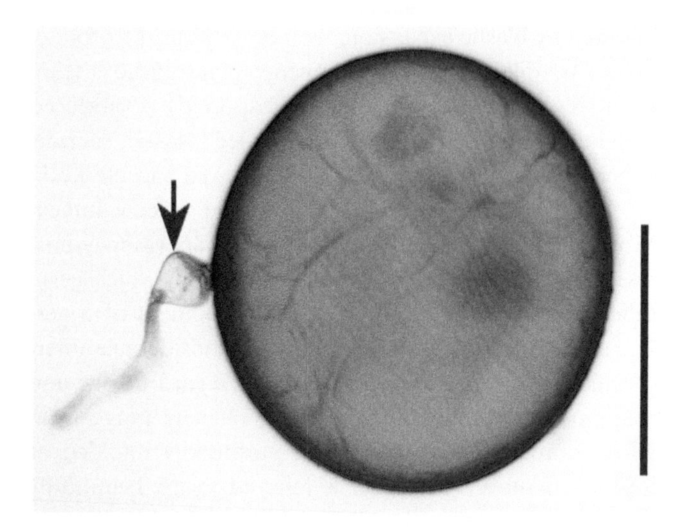

FIGURE 7.15 *Gigaspora ramisporophora* with bulbous base (*arrow*). Extant. Bar = 150 μm. (Courtesy C. Walker.)

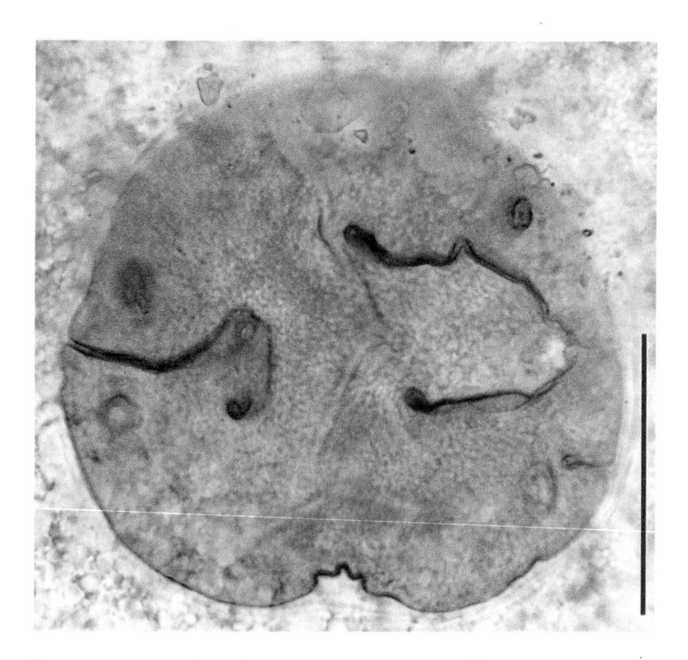

FIGURE 7.16 Germination shield of *Scutellospora cerradensis*. Extant. Bar = 25 μm. (Courtesy C. Walker.)

(2012) for detailed descriptions of these features. Spores may germinate either inside root tissue or in the substrate to form new mycorrhizae by inoculation of new roots. Glomeromycotan spores contain hundreds of nuclei that come from the surrounding mycelium, rather than being the result of divisions of a single nucleus (Jany and Pawlowska, 2010; Gianinazzi-Pearson et al., 2012).

FIGURE 7.17 Paula Bonfante.

GLOMEROMYCOTAN REPRODUCTION

To date there has been no direct evidence of sexual reproduction in the group, and molecular variation suggests clonal expansion is the basis for the ecological success of these fungi (Pawlowska, 2005; den Bakker et al., 2010). Using large sequence datasets, however, Riley and Corradi (2013) and Corradi and Bonfante (2012) (Figure 7.17) showed that many genes commonly linked with sex in a number of groups of fungi across the genome, occur in species of the AM fungus *Glomus*. This result has also more recently been reported by Tisserant et al. (2013) based on 28,232 genes in *Rhizophagus irregularis* (= *Glomus intraradices*). Moreover, Halary et al. (2011) found that 85% of the core meiotic genes (essential genes found in animals, plants, and fungi that code for proteins that jointly make up the conserved meiotic recombination machinery of eukaryotic cells) known in the yeast *Saccharomyces cerevisiae* were also present in AMF. This indicates that Glomeromycota possess all the machinery necessary for successful meiosis (see also Sanders, I.R., 2011). Ehinger et al. (2012) demonstrated that surprising amounts of genetic and phenotypic variation occur successively from a single AM fungus spore through clonal growth. This is an unexpected amount of variation to be derived from a clonal organism, and for which these authors see no other parallels in nature. Another study (Marleau et al., 2011) concluded that the spores of AMF contain nuclei with two origins: those that migrate into the spore and those that arise by mitosis in the spore. Therefore, these spores do not represent a stage in the life cycle with a single nucleus, raising the possibility that AMF, unlike all other known eukaryotic organisms, lack the genetic bottleneck of a single-nucleus stage. It is hypothesized that the developmental patterns of sporogenesis have adaptive significance for dealing with the accumulation of deleterious mutations in this system (Jany and Pawlowska, 2010) and that selection is further characterized by intra-individual polymorphism of the nuclear ribosomal RNA coding genes and various protein sequences, which reflect the historically ancient genome of the group (e.g. Simon et al., 1993; Heckman et al., 2001; Pawlowska, 2007). Moreover, it has also been demonstrated that AMF are promiscuous in that hyphal connections in the form of hyphal fusion can develop with other members of the population, which may enable further genetic exchange (e.g. Giovannetti et al., 2004; Croll et al., 2009; Young, J.P.W., 2009; Sanders and Croll, 2010) as well as nutrient foraging and nutrient recycling in the plant communities (Jakobsen, 2004). Polyploidy has also been demonstrated in at least one species of *Glomus*, suggesting another explanation for the long-term evolutionary success of this clade in the absence of sexual reproduction (Pawlowska and Taylor, 2004).

FIGURE 7.18 Yolande Dalpé.

FOSSIL GLOMEROMYCOTA

There are several reports of enigmatic microfossils that have been interpreted as possible Glomeromycota based on overall appearance (Pirozynski and Dalpé, 1989; Figure 7.18). Some of these appear similar to vesicles in size and general shape but were interpreted as sporangia because they contained spores (Locquin, 1983); in others (*Paleocatenaria disjoncta*) it is impossible to tell what they represent. Specimens of *Paleobasidiospora taugourdeauii* were collected from a suite of rocks stratigraphically interpreted as Cambrian–Ordovician, as well as from some subsoils in the Sahara; the latter probably represent contamination (Pirozynski and Dalpé, 1989). Another form, "*Archechytridium operculatum*" (Locquin, 1982), appears with a hyphal attachment; this structure may represent a chytrid with a rhizoid. Pirozynski and Dalpé (1989) further questioned whether *P. taugourdeauii* and *A. operculatum* are even fungal, suggesting that these fossils may represent some type of acritarch.

An interesting fossil that certainly represents an asexual glomeromycotan spore was reported from the Ordovician of Wisconsin, USA (Redecker et al., 2000a; Figure 7.19). Because the specimens were discovered in a rock maceration, there is no information about the potential plant that was associated with this fungus. *Palaeoglomus grayi* consists of aseptate hyphae up to 5 μm in diameter that may variously branch and produce terminal globose-subglobose spores, each up to 95 μm in diameter (Redecker et al., 2002). Transmitted light observations suggest that in most of the spores the wall is a single layer thick, but in some specimens another inner layer is present. Some spores are attached to a short hyphal segment that is slightly widened where attachment occurs. In general most of the spores are associated with hyphae, a condition which is not common in fossil preparations. While there can be little doubt that *P. grayi* is morphologically similar to a number of asexual spores of glomeromycetes, including *Glomus*, whether the specimen is Ordovician in age may be more problematic. The specimens are certainly well within the age range of the group based on both molecular clock assumptions (e.g. Hedges and Kumar, 2004; Berbee and Taylor, 2006) and fossil evidence suggesting that the group was already

FIGURE 7.19 Dirk Redecker.

well diversified by the Early Devonian (Schüßler and Walker, 2011). However, the fossils were extracted by hydrochloric acid (HCl) maceration from a dolomite, a carbonate sedimentary rock that often represents a post-depositional replacement of limestone (Vasconcelos et al., 1995). Despite the fact that the spores were macerated from the center of a block to avoid contamination, the possibility still exists that in a porous rock like dolomite, contamination may have already taken place from extant Glomeromycota (e.g. Tang and Lian, 2012) or from fungi that were redeposited from younger overlying sediments. Several lines of inquiry point to the importance of bacteria and biofilms in the formation of dolomite, which could cause further microbial contamination (e.g. Roberts et al., 2004; Krause et al., 2012). For many reasons it will be especially important to detail the sedimentological setting and depositional history, especially the geochemistry, of this sequence of rocks.

There are a number of reports of fossils that might be interpreted as Glomeromycota for which the age is questionable. For example, fungal filaments were described from a Devonian limestone in Alberta, Canada (Fry and McLaren, 1959). The filaments were observed after chemical maceration of the rock and consist of hyphae, some up to 3.3 μm in diameter, that branch at right angles. Whether these fungi represent some type of contamination is difficult to determine; however, later in this volume (Chapter 12) we will discuss endolithic organisms, including fungi, that have the capacity to bore into calcium carbonate.

Some type of potential contamination is also a question with regard to *Rhizophagites acinus*, a spore that shows some resemblance to extant *Glomus* from the Cretaceous of Alberta, Canada (Srivastava, 1968; Berch, 1985). *Rhizophagites* has also been reported from the Pliocene (e.g. Norris, 1986) and Pleistocene (e.g. Butler, 1939; Rosendahl, 1943; Wilson, L.R., 1965). Some authorities use the name *Palaeomycites* (Kalgutkar and Jansonius, 2000). The fungus has also been recorded from Upper Jurassic rocks from Tanzania (Schrank, 1999), but is a common contaminant in palynological surface samples from carbonates and limey soils (Wood et al., 1996). This does not mean that all spores circumscribed by *Rhizophagites* are necessarily contaminants, but that caution must be exercised because many of the specimens have the appearance of extant *Glomus* species and may represent an extant species or slightly older fungus that has intruded into the roots (e.g. Koske, 1985). There are numerous reports of *Glomus*-like spores from Cenozoic sediments around the world that were reviewed by Pirozynski and Dalpé (1989). Some of these occur in thin-section preparations (e.g. Smith, W.G., 1878), while others have been described from palynological preparations (e.g. Jen, 1958).

RHYNIE CHERT ASEXUAL SPORES

Although only a few of the early land plants from the Lower Devonian Rhynie chert have been examined in sufficient detail to document the presence of mycorrhizal symbioses (see below), there is a strong indication that Glomeromycota were in some way associated with all land plants in this paleoecosystem. One type of evidence occurs in the form of glomeromycotan spores (Figure 7.20) and structures interpreted as sporocarps (Figures 7.21 and 7.22) within the cortical tissues of these plants; often associated with the spores are hyphae that terminate in thin-walled vesicles. Several basically different types of glomeromycotan spores have been described from the Rhynie chert:

Glomoid Spore Types

Spheroidal or globose spores, some with a segment of a subtending hypha attached, are placed in the genus *Palaeomyces* (see Figure 7.20). The taxon was initially

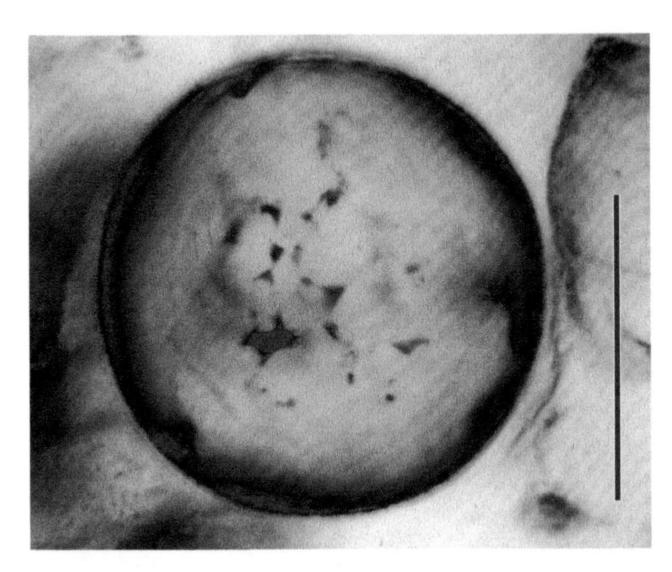

FIGURE 7.20 Thick-walled spore (*Palaeomyces* sp.). Lower Devonian Rhynie chert, Scotland. Bar = 10 µm.

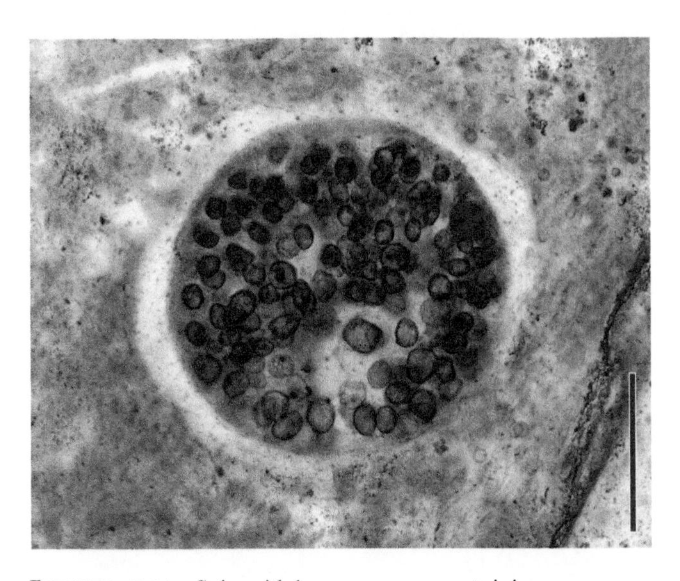

FIGURE 7.21 Spheroidal sporocarp containing numerous spores. Lower Devonian Rhynie chert, Scotland. Bar = 50 µm.

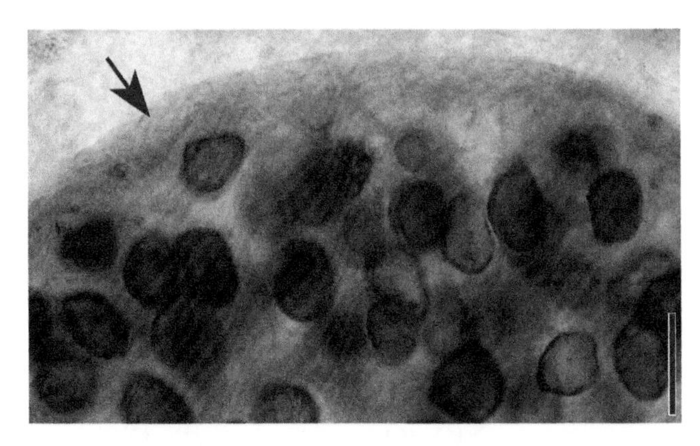

FIGURE 7.22 Detail of Figure 7.21 showing the hyphal peridium (*arrow*). Lower Devonian Rhynie chert, Scotland. Bar = 10 µm.

introduced by Renault (1896b) for spores occurring in the cortex of *Lepidodendron* from the Carboniferous of Esnost, France. *Palaeomycites* is a substitute name for *Palaeomyces* that was introduced by Meschinelli (1898). While both names are used, only *Palaeomyces* is validly published. The original description of the genus also included a discussion of branched hyphae 3–4 µm wide, and the presence of a slight hyphal swelling between the spore and hypha (which would be indicative of gigasporoid spore formation, but see below). In addition to the spores, there were elongate, thinner walled vesicle-like structures in the preparations. In *P. majus* there is a structure extending from the spore that is interpreted as a germination tube (Renault, 1896b).

Palaeomyces is widespread in the Rhynie chert and associated with several land plants, including *Aglaophyton major* and *Asteroxylon mackiei*, as well as degraded plant material, but also occurs as solitary spores, vesicles, and hyphae in the matrix (Kidston and Lang, 1921b; Sharma et al., 1993). Many use the genus as an appropriate repository (morphogenus or fossil genus) for the fungal remains in the chert until the systematic affinities can be more accurately determined. Of the 15 fungal types Kidston and Lang (1921b) described from thin sections of the chert, five were given species names (i.e. *Palaeomyces gordoni, P. asteroxyli, P. horneae, P. vestita, P. simpsoni*) and one was termed a variety (*P. gordoni* var. *major*). Spore size and morphology are variable, and in some specimens they described multiple layers of the wall. Since most of the spores assigned to *Palaeomyces* are borne terminally on a subtending hypha that lacks a bulbous base, we regard these spores as being of the glomoid type. Consequently, some of the spores have been transferred to the fossil taxon *Glomites* (see below). The original descriptions by Kidston and Lang (1921b) were accompanied by extraordinary photographs of the fungi, including some structures that are still not well understood. To some degree their descriptions were also based on the types of plants with which the spores were associated, where they occurred in the tissues, size of spores and hyphae, and other features. While a number of the fungi in the Rhynie chert were understood to be saprotrophs by Kidston and Lang (1921b), they did

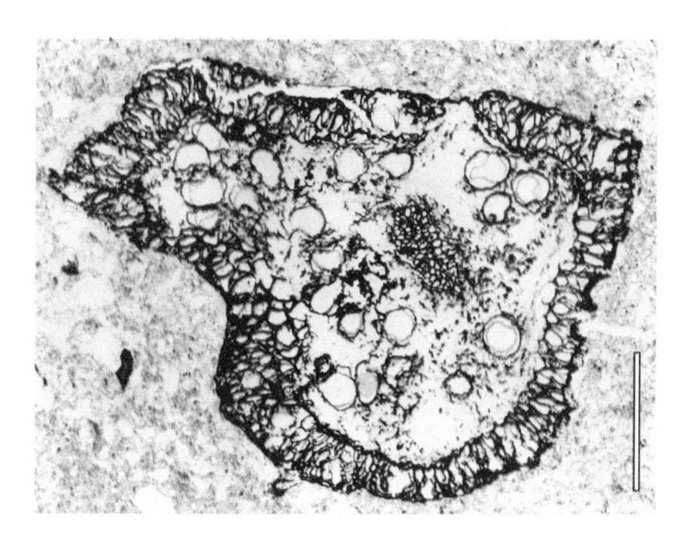

FIGURE 7.23 Transverse section of early land plant *Psilophyton dawsonii* showing numerous chlamydospores within the cortical tissues. Lower Devonian, Gaspé, Canada. Bar = 500 µm. (From Stubblefield and Banks, 1983.)

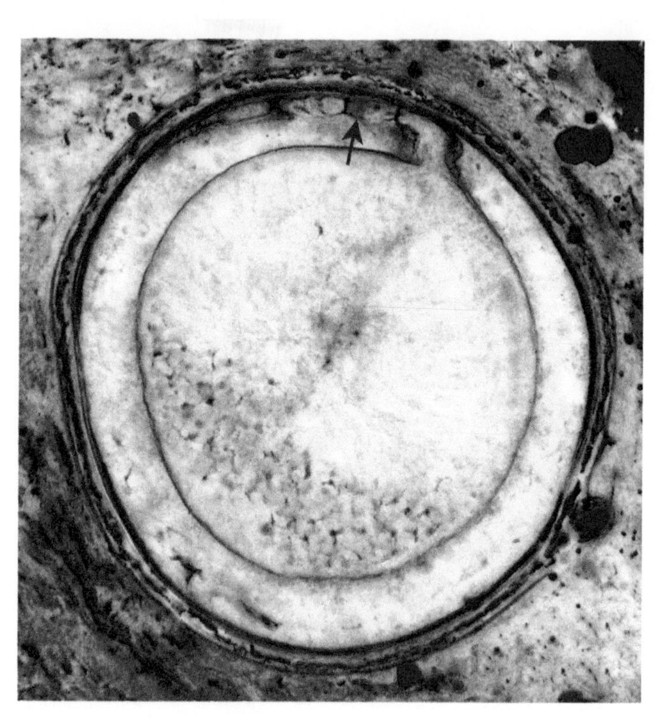

FIGURE 7.24 *Scutellosporites devonicus* with germination shield (*arrow*) in section view. Lower Devonian Rhynie chert, Scotland. Bar = 50 µm.

not discount some being mycorrhizal, an especially significant speculation for the time. These authors also recognized that some of the spores (= chlamydospores) contained mycoparasites.

Other Devonian glomoid spores in plant tissue (Figure 7.23) have been reported from the Emsian of the Gaspé, Canada (Stubblefield and Banks, 1983; Banks and Colthart, 1993). They range from 60 to 175 µm in diameter and are characterized by a double wall and evidence of hyphal attachment at a single site. They are interpreted as possible oogonia, chlamydospores, or some type of resting spore, but appear most similar to the spores of glomeromycotan fungi.

Gigasporoid Spore Types

In the cortical tissues of *Asteroxylon mackiei* are a number of globose to subglobose spores, formally described as *Scutellosporites devonicus* (Figure 7.24), that range from 260 to 350 µm in diameter and lack any distinctive ornament on the surface (Dotzler et al., 2006; Figure 7.25). The wall is subdivided into two wall groups, with the outer wall group well preserved, up to 18 µm thick and two- or three-layered, and the original thickness of the inner group difficult to estimate. A translucent region, up to 30 µm thick, occurs between the dark layer and outer surface of the spore lumen. The most interesting feature of these spores is the presence of a germination shield that extends along the inner surface of the dark layer. The germination shield is round or oval in outline, ~140 µm

FIGURE 7.25 Nora Dotzler.

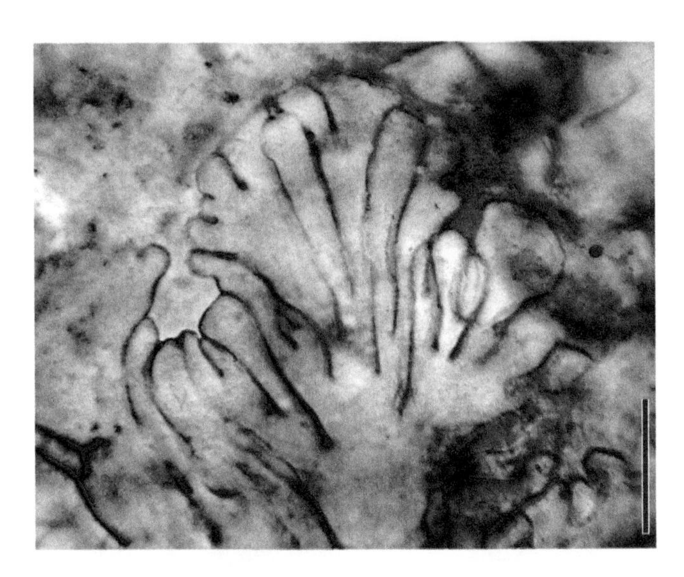

FIGURE 7.26 Palmately lobed germination shield in surface view. Lower Devonian Rhynie chert, Scotland. Bar = 20 µm.

in diameter, up to 15 µm high, and distinctly lobed, with each of the lobes 25–33 µm wide, or displaying a complex infolding along the margins. In one of the axes, a structurally similar but somewhat smaller spore without a germination shield is attached to a slightly bulbous subtending hypha. Dotzler et al. (2006) regard these Early Devonian spores as similar to the spores produced by members of the modern genus *Scutellospora*, especially with regard to the presence of the prominent, circular germination shield with a lobed margin.

Acaulosporoid Spore Types

The detailed study of glomeromycotan spores from the Rhynie chert has also produced another spore type with a complex germination shield (Figure 7.26) (Dotzler et al., 2009). These spores are present in the cortical tissues of *Aglaophyton major* and are quite different from *Scutellosporites devonicus* in that they develop laterally within the neck of a thin-walled sporiferous saccule like those in modern acaulosporoid AMF (Figure 7.27). The saccules appear as long-necked, balloon-shaped structures up to 700 µm long. The saccule wall is double, with a persistent inner component (wc2) and, on about 60% of the specimens, a thick (5 µm) outer component (wc1), lacking or present only as an amorphous coating on the remaining specimens and therefore interpreted as evanescent in the sense of Walker, C. (1983). The saccule wall, mainly seen as wc2 after wc1 has disintegrated, continues to become a component of the outer spore wall group (Figure 7.28). In some specimens, the intact saccule wall components

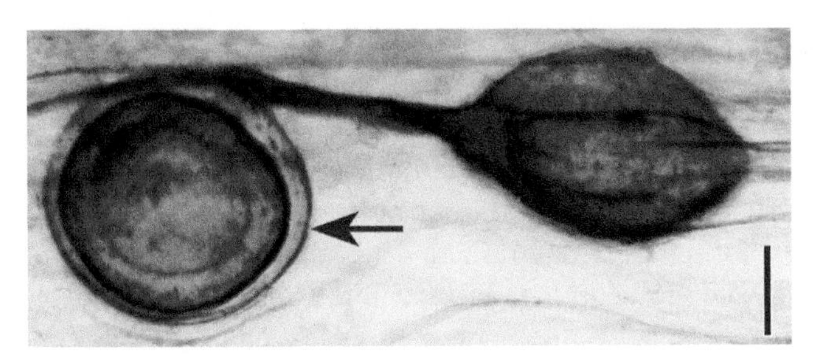

FIGURE 7.27 Spore (*arrow*) formed in the neck of a saccule. Lower Devonian Rhynie chert, Scotland. Bar = 100 µm.

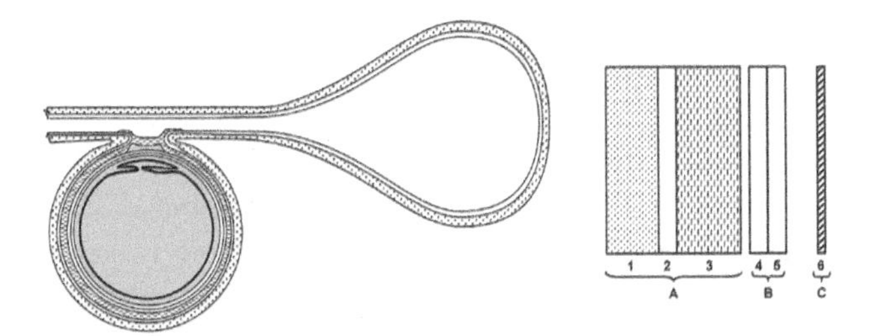

FIGURE 7.28 Diagram of the saccule and spore wall structure (left), and murograph (right) of the spore wall structure interpretation from an acaulospore and sporiferous saccule from the Lower Devonian Rhynie chert. Shading in the illustration is intended only to clarify the separate components. The murograph shading follows the patterns established by Walker (1983). The nature of the inner two components could not be related to present-day descriptions and therefore has not been shaded. (Modified from Dotzler et al., 2009.)

extend around the spore to form the outer component of the outer wall group. In other specimens only wc2 can be seen beyond the point of spore development, and in yet others, little or no evidence of the saccule neck remains. The wall structure of the acaulospore is complex and has been suggested to comprise three major parts, including an outer group consisting of the lateral expansion of the saccule neck and an adherent laminated component, a middle group of two adherent components, and an innermost wall group of a single, thin, apparently flexible component. The germination shield is formed by extrusion of wc6 and extends along the inner surface of the wc5. The size and shape of the germination shield, as well as the extension of the structure along the inner surface of wc5, vary. In most specimens it represents a profoundly lobed and infolded structure (see Figure 7.26), with one of the lobes sometimes enlarged to form an elongate, tongue-shaped extension, which is sparsely lobed along the lateral margins and deeply fringed or palmately lobed distally. The largest germination shields occupy most of the inner surface of the middle wall group in plan view.

The studies of Rhynie chert glomeromycotan spores detailed above have provided important new information at several levels of inquiry. First, they demonstrate the wealth of structural and developmental data about the asexual spores that can be used to compare them with extant forms. Second, these data can be used to pinpoint the oldest occurrence of certain features useful in phylogenetic analyses of modern groups (such as germination shields), and to test whether such structures are plesiomorphic (primitive) or apomorphic (advanced). Finally, the spore studies provide a template to examine the spores of geologically younger AMF to see if there are any changes that can be plotted over time, and whether any relationship exists between spore type and specific plant lineages (lycophytes, certain seed plant clades, etc.). The report by Oehl et al. (2011a) that combined spore morphology/wall structure and germination, together with existing genetic information, identified two new extant glomoid genera (*Septoglomus* and *Viscospora*) and underscored the importance of how structural spore features might be used with fossils.

FIGURE 7.29 Frederick E. Weiss.

CARBONIFEROUS AND PERMIAN SPORES

Our understanding of Carboniferous fossil plants has historically been enhanced by examining thin sections of European permineralizations or coal balls. Within many of these preparations are large globose spores that were interpreted as evidence of some type of mycorrhizal associations, many associated with tissues of the arborescent lycophytes that dominated the Carboniferous landscape in Euramerican coal swamps. Because many of the fossil spores have the same morphology as those of modern members of the Glomeromycota, especially *Glomus*, their presence has been used historically as a proxy record for the presence of arbuscular mycorrhizae in fossils (e.g. Weiss, 1904b [Figure 7.29]; Osborn, 1909; Wagner and Taylor, 1981). Wagner and Taylor (1982) examined approximately 2000 spores in acetate peels from coal balls associated with three Carboniferous coals of the Midcontinent Basin in the United States using both transmitted light and scanning electron microscopy. The spores showed size variations among localities and stratigraphic levels (Middle Pennsylvanian Lewis Creek, Kentucky: 100–252 µm; Middle Pennsylvanian West Mineral, Kansas: 169–304 µm; and Upper Pennsylvanian Berryville, Illinois: 150–405 µm), but there was also some overlap. Most of the spores (83%) occurred in cortical tissues of roots and other underground organs (e.g. *Stigmaria* appendages, which are homologous with leaves). Some spores possessed a hyphal attachment and the presence of a septum near the spore base. Despite the presence of hyphae in the sections, none of the spores was attached to an extensive mycelium; rather they occurred in small clusters or, more commonly, solitarily. In many of the spores it was possible to resolve multiple layers of the wall (e.g. Miller and Jeffries, 1994). Fine details of the walls could be seen when acetate peels (Figure 7.30) were prepared using long-term etching in HCl and then the rock surface adhering to the peel was examined with the scanning electron microscope (Stubblefield et al., 1985b). To date the ultrastructure of fossil fungal spores has provided relatively little information of a systematic nature. Although the structural features revealed by these techniques underscore the diversity in spore wall organization, additional work is necessary to separate any development features and preservation artifacts from diagnostically useful structural features.

Glomorphites intercalaris is a Permian spore type discovered in silicified peat from Antarctica and compared to some species of *Glomus* (García Massini, 2007b). The spores occur in clusters of up to 150, perhaps within the epidermis of a root, and may be terminal or intercalary

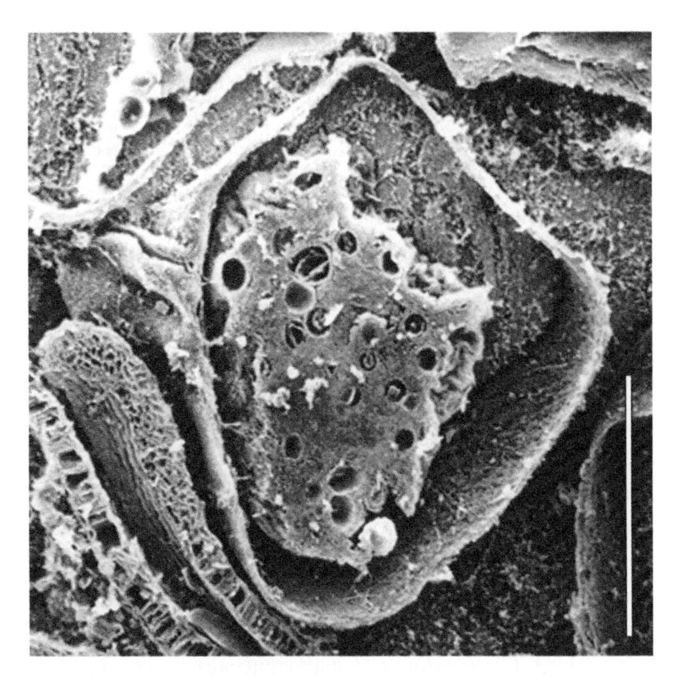

FIGURE 7.30 Over etched spore showing multiple wall layers. Middle Pennsylvanian, Kentucky, USA. Bar = 25 μm.

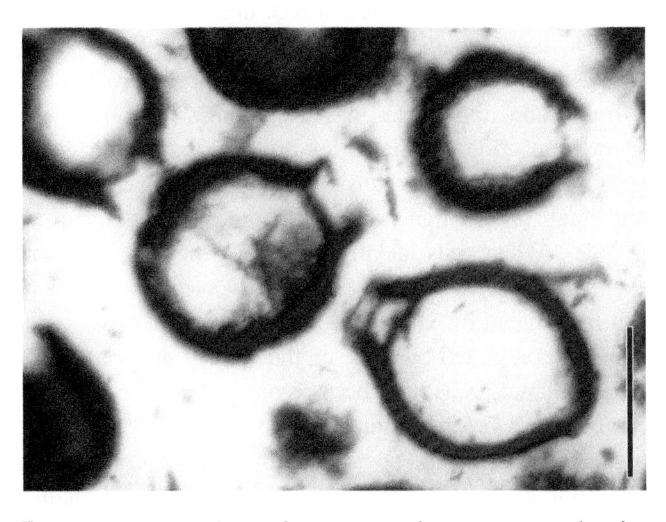

FIGURE 7.31 *Glomorphites intercalaris* spores showing intercalary (lower right) and terminal attachment. Permian, Antarctica. Bar = 25 μm.

(Figure 7.31). Individual spores are ellipsoidal, pyriform, or globose and approximately 50 μm in diameter. Spores are attached to a fragment of a subtending hypha in various patterns and the spore wall may be constructed of three layers. One of the interesting aspects of *G. intercalaris* is the fact that within the clusters there is a large diversity of spore types, perhaps suggesting that they represent different levels of development or that some may represent some type of sporocarp.

FIGURE 7.32 Juan L. García-Massini.

MESOZOIC, CENOZOIC, AND SUBFOSSIL SPORES

We are not aware of any reports of isolated glomeromycotan spores in the roots of Mesozoic and Cenozoic plants because there are relatively few localities of this age where plant parts (other than isolated wood) are preserved as either permineralizations or petrifactions. There is also the issue of the relatively small number of paleobotanists who study roots. However, there are a few accounts of isolated, putatively glomeromycotan spores from these periods of time. For example, García Massini et al. (2012) (Figure 7.32) reported on permineralized pyriform chlamydospores bearing fragments of subtending hyphae reminiscent of the genus *Glomus* from the Middle–Upper Jurassic San Agustín hot spring deposit in Argentina. Moreover, clusters of subcircular glomoid spores (up to 150 μm in diameter) borne on angular hyphae have been macerated from dinosaur coprolites from the Upper Cretaceous of India (Kar et al., 2004a, b).

Subfossil evidence of AMF occurs in the form of late Quaternary *Glomus*-like spores from Ontario, Canada (Berch and Warner, 1985), and specimens belonging to *Glomus fasciculatum* have been identified in lake sediment cores from Gould Pond (central Maine) and Upper South Branch Pond (north central Maine; Anderson et al., 1984), as well as in numerous palynological samples (e.g. Kholeif and Mudie, 2009; Kiage and Liu, 2009). Some subfossil remains of AMF are believed to represent valuable proxy indicators for paleoenvironment reconstructions (e.g. van Geel, 2001; Musotto et al., 2012). For example, the occurrence of *Glomus* sp. has been shown to be

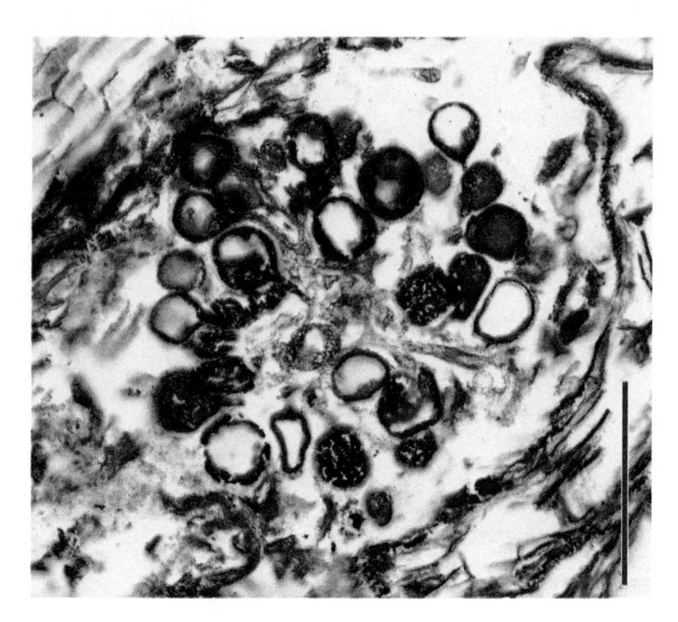

FIGURE 7.33 Radially arranged spores of a *Sclerocystis*-like fungus. Triassic, Antarctica. Bar = 15 μm.

indicative of active soil erosion processes in the catchment area (Medeanic and Silva, 2010) and correlates well with decreases in soil moisture levels (Chmura et al., 2006). Its presence has also been used as an indicator of soil conditions associated with aridity and stressed environments (Limaye et al., 2007).

SPOROCARPS

While the majority of spores that are thought to belong to members of the Glomeromycota occur isolated or in small clusters, there is one silicified specimen from the Triassic of Antarctica that appears to be most similar to some type of sporocarp (Stubblefield et al., 1987b). The single specimen is represented by an incomplete cluster of aseptate hyphae that radiate out from a common region (hyphal plexus); each hyphal segment terminates in an ovate-spherical chlamydospore up to 54 μm in diameter (Figure 7.33). Surrounding the fossil is a delicate layer that may represent the remains of a peridium. At its widest point the cluster is approximately 350 μm in diameter. This putative sporocarp is not associated with any plant debris in the matrix. The fossil is morphologically similar to *Sclerocystis*, a modern genus in the Glomeraceae (Glomerales) that is characterized by producing sporocarps (e.g. Almeida and Schenck, 1990); however, some species of *Glomus* are also known to produce similar structures (Morton, 1988). Some have suggested that the sporocarpic habit represents an advanced clade in *Glomus* and that morphological features of the sporocarp show a continuum of character states (Almeida and Schenk, 1990; Wu, C.-G., 1993). These data imply that most *Sclerocystis* species belong in *Glomus*, a suggestion that is also supported by molecular data (Redecker et al., 2000b). Other descriptions of *Sclerocystis* fossils are perhaps more likely examples of mycoparasitism (Sharma and Tripathi, 1999).

ROOT NODULES

Small non-N_2 fixing protuberances (modified lateral roots), often termed root nodules or mycorrhizal nodules, are known to occur in several families of conifers (Phyllocladaceae, Podocarpaceae, Araucariaceae; Khan and Valder, 1972), as well as in some angiosperms (e.g. Khan, 1972; Duhoux et al., 2001; Scheublin et al., 2004; Imaizumi-Anraku, et al., 2005; Scheublin and van der Heijden, 2006; Dickie and Holdaway, 2010). In the flowering plant *Gymnostoma* (Casuarinaceae) there are both mycorrhizal nodules and N_2-fixing actinorhizal root nodules (Duhoux et al., 2001).

AMF have been reported in root nodules of the conifer *Notophytum krauselii* from the Lower Triassic of Antarctica (Schwendemann et al., 2011). Nodules are approximately 1.0 mm long and occur in pairs (Figure 7.34); they are slightly embedded in the cortex of small (0.5 mm in diameter) roots. The fungi consist of hyphae with arbuscules and small vesicles that occur in the outer cortex of mature nodules. When cells lacking direct evidence of fungi are examined with energy dispersive X-ray spectroscopy there is a high signal of calcium recorded, suggesting that fungi may be represented by some of the opaque material in the cells. Calcium levels have been recorded from root hairs in extant plants before contact and infection of AM hyphae (e.g. Parniske, 2008), suggesting the potential for using this tool in certain preservation types of fossil plants. Earlier reports of fossil root nodules based on impression/compression specimens and plants preserved as molds dated the oldest persuasive evidence of these structures as early as the Cretaceous (e.g. Cantrill and Douglas, 1988; Cantrill and Falcon-Lang, 2001; Banerji and Ghosh, 2002). Not only does the Triassic report indicate that this type of mycorrhizal association is at least 100 Ma older than previously thought; it also shows the evolution of an early structural relationship between the fungus and the host, and provides evidence that the nodule-root symbiosis evolved from AM functions (Parniske, 2008). Arbuscules have also been found in non-nodular immature roots of *N. krauselii* (Figure 7.35).

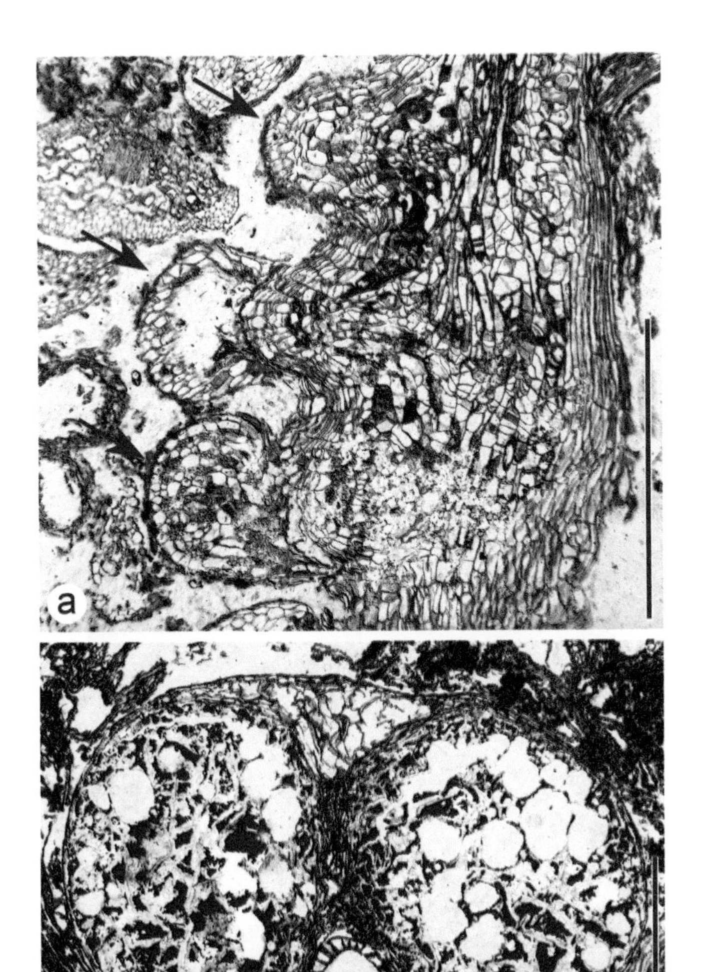

FIGURE 7.34 Root nodules (*arrows*) on fossil conifer *Notophytum krauselii* (**A**). Detail of a pair of nodules (**B**). Triassic, Antarctica. Bars = 30 μm (A), 500 μm (B). (Courtesy A.B. Schwendemann.)

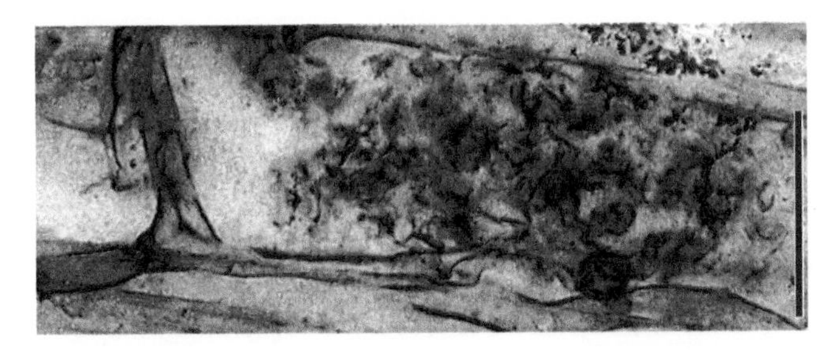

FIGURE 7.35 Detail of *Notophyton krauselii* non-nodule root cell containing arbuscule. Triassic, Antarctica. Bar = 25 μm. (Courtesy C.J. Harper.)

What is the function of the mycorrhizal root nodule? One hypothesis suggests that the nodules simply permit the plant to increase the extent of arbuscular mycorrhizal infection while minimizing the cost that would be associated with developing an extensive root system (McGee et al., 1999). Adding support to this hypothesis is a study by Dickie and Holdaway (2010), which found that nodules effectively double the root volume in podocarps and thus permit a much greater extent of AM infection than would be possible without nodules. The form of nodules is ideally suited for this purpose – a sphere creates the greatest possible increase in cortical volume with the lowest possible investment in vascular or epidermal tissue. It has also been suggested that mycorrhizal nodules stimulate phosphate uptake (e.g. Morrison and English, 1967). For example, podocarps in New Zealand live in rainforests that are often waterlogged and extremely low in available phosphate (Russell et al., 2002); the ability of these plants to thrive in these conditions is considered to be due to their symbiosis with mycorrhizal fungi, which inhabit the nodules on their roots (Baylis et al., 1963). Other studies have postulated, however, that the nodules are able to fix small amounts of atmospheric nitrogen in the presence of a prokaryotic symbiont (e.g. Furman, 1970), while still others are uncertain of the function (Duhoux et al., 2001). One apparent difference between the cells containing AMF in the form of arbuscules and those cells containing them in nodules is that the cells of the latter die and the nodules are abscised, while the cortical cells invaded by typical AMF remain alive after the arbuscules have collapsed, and they can subsequently host new arbuscules.

FOSSIL EVIDENCE OF ARBUSCULAR MYCORRHIZAE

Historically there have been several reports of *in situ*-preserved arbuscular mycorrhizae, especially in Carboniferous plants, that have documented the presence of the physiological exchange structure (arbuscule) in the host plant cells. For example, several early studies of cordaitalean rootlets from the Carboniferous of Great Britain reported the presence of aseptate hyphae in cells of the cortex that were interpreted as evidence of mycorrhization (e.g. Weiss, 1904b; Lignier, 1906; Osborn, 1909; Halket, 1930). Osborn (1909) suggested that some of the hyphae he described might represent arbuscules, and, at least in the line drawings that are presented, some look in fact like *Paris*-type mycorrhizae. The line drawings presented by Halket (1930) of the apices of *Amyelon* roots containing mycorrhizae may also represent *Paris*-type arbuscules; in others, the

FIGURE 7.36 Arthur A. Cridland.

FIGURE 7.38 Henry N. Andrews, Jr.

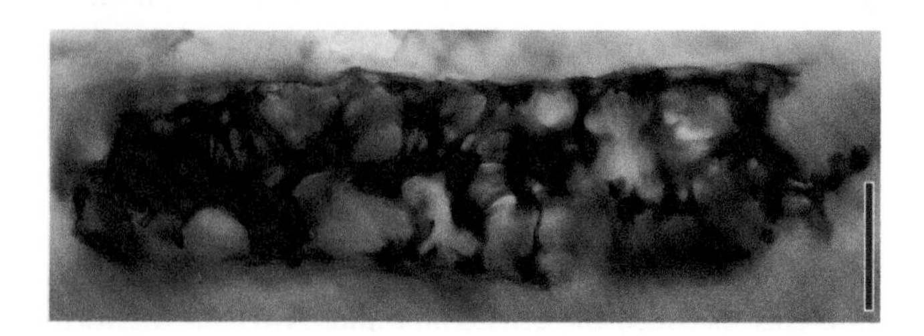

FIGURE 7.37 Cell containing degraded cell contents or biomimetic structure. Middle Pennsylvanian, USA. Bar = 10 μm. (Courtesy A.A. Klymiuk.)

FIGURE 7.39 Hagen Hass.

structures interpreted as arbuscules represent degraded cell wall material. Cridland (1962) (Figure 7.36) suggested that in many of these early reports what were interpreted as arbuscules were in fact degraded cell contents (Figure 7.37), secretory substances, parasitic or saprophytic fungi, or some type of unusual cell, and thus could not be regarded as persuasive evidence of the presence of arbuscular mycorrhizae in these plants. Fungi thought to be mycorrhizal were also reported from the cortical cells and tracheids of a Carboniferous marattialean fern root preserved in coal balls from North America (Andrews and Lenz, 1943; Figure 7.38). In some of the cells were spores and what might be interpreted as vesicles, while in others the cells were found to be packed with hyphae.

The oldest unequivocal evidence of arbuscules comes from the Rhynie chert (Remy et al, 1994a). Perhaps the best documented example of arbuscular mycorrhizae in a fossil land plant occurs in the sporophyte generation of *Aglaophyton major* (Taylor et al. 1995a; Remy and Hass, 1996 [Figures 7.39 and 7.40]). One of the interesting aspects of the organization of *A. major* is the degree to which the prostrate axes are in contact with the substrate. Typically in most land plants, a substantially large portion of the plant body grows through the substrate in the form of roots or rhizomes. In *A. major*, however, the contact between the plant body and substrate is restricted to small areas of the sinuous prostrate axes that produce rhizoids

FIGURE 7.40 Renate Remy (left) and Winfried Remy. (Courtesy D. Remy.)

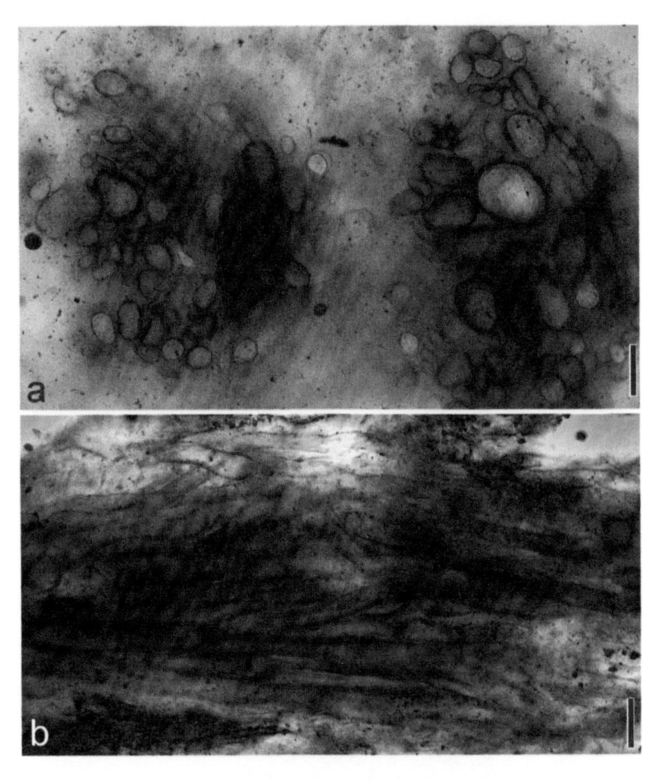

FIGURE 7.41 Section of glomeromycotan rhizomorph showing individual hyphae in transverse (A) and longitudinal sections (B). Lower Devonian Rhynie chert, Scotland. Bars = 15 μm (A) and 30 μm (B).

(Remy and Hass, 1996); no part of the plant body grows within the substrate and all axes are stomatiferous. The endomycorrhizal fungus found in this early land plant, *Glomites rhyniensis*, includes both extraradical and intraradical aseptate to sparsely septate hyphae; extraradical hyphae may form massive rhizomorphs (Figure 7.41) in which individual hyphae occasionally anastomose (Figure 7.42). The stomata on the prostrate axes of *A. major* provide the entrance sites for the fungus. From the substomatal chamber the fungus spreads throughout the intercellular system of the outer cortex (Figure 7.43) (Taylor et al., 1995a). Some hyphae terminate in elongate-globose multilayered spores, while others penetrate cell walls to form arbuscules (Figure 7.44), exclusively within a well-defined narrow zone of tissue, one or two cells thick, between the outer and middle cortex. These ancient arbuscules are highly branched and some possess inflated tips, precisely as in modern ones. One of the interesting features of *G. rhyniensis* and its host is the fact that the arbuscule zone (Figures 7.45 and 7.46) occurs in both prostrate and upright axes throughout the entire length to the apex. The presence of what are interpreted as collapsed arbuscules within the arbuscule zone suggests that these structures were ephemeral, as in modern arbuscular mycorrhizae (e.g. Bonfante-Fasolo, 1984; Bonfante and Genre, 2010). It has been reported that as arbuscule branches collapse they are coated with a fibrillar material

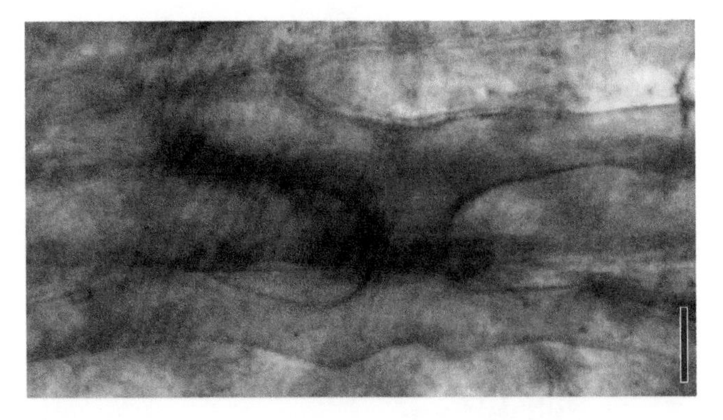

FIGURE 7.42 Anastomosing glomeromycotan hyphae. Lower Devonian Rhynie chert, Scotland. Bar = 15 μm.

composed of polysaccharides produced by the host cell (Dexheimer et al., 1979). While difficult to determine, it is also possible that some of the fossil cell contents referred to as arbuscules in Carboniferous rootlets and interpreted as degraded cell contents (see above), might in fact represent some type of host response to the breakdown of the arbuscules.

FIGURE 7.43 Hyphae (*arrows*) in intercellular spaces of *Aglaophyton major* cortex. Lower Devonian Rhynie chert, Scotland. Bar = 30 μm.

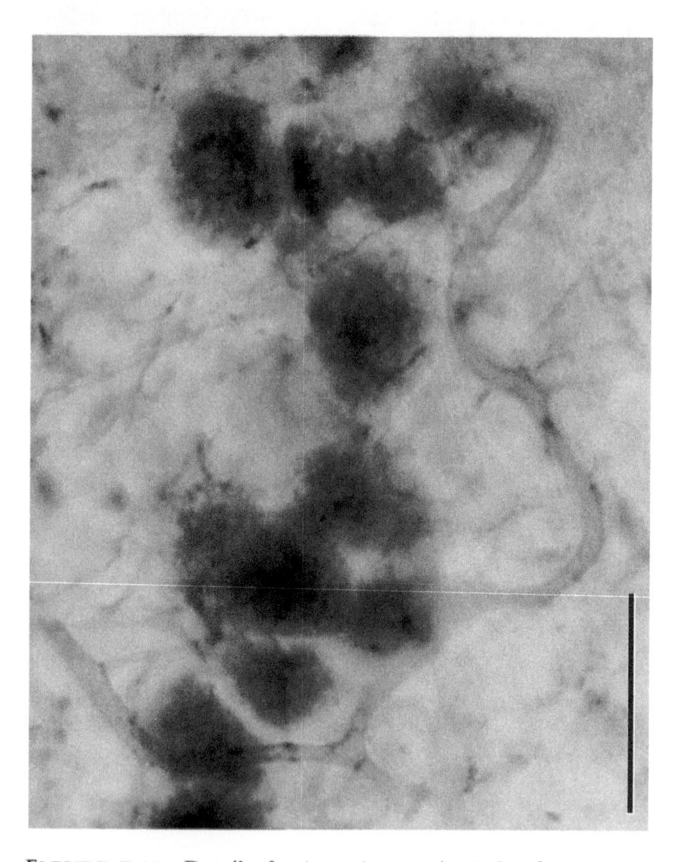

FIGURE 7.44 Detail of arbuscule zone in *Aglaophyton major* showing trunk hyphae and arbuscules. Lower Devonian Rhynie chert, Scotland. Bar = 100 μm.

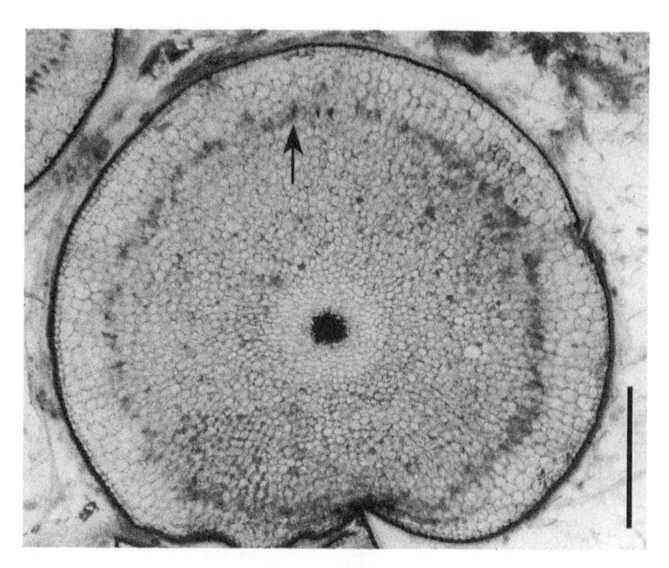

FIGURE 7.45 Transverse section of *Aglaophyton major* upright axis showing extensive cortical tissues and small central conducting strand. Arrow indicates position of arbuscule zone. Lower Devonian Rhynie chert, Scotland. Bar = 0.5 mm. (Courtesy H. Kerp.)

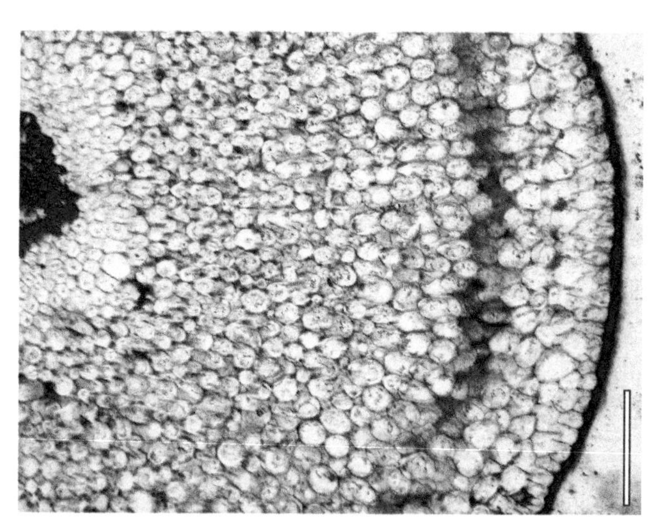

FIGURE 7.46 Detail of *Aglaophyton major* arbuscule zone in Figure 7.45. Lower Devonian Rhynie chert, Scotland. Bar = 250 μm. (Courtesy H. Kerp.)

Nothia aphylla, another early land plant from the Rhynie chert, is characterized by prostrate axes that are bilaterally symmetrical in cross section and extend through the substrate much like the rhizomes of extant plants (Kerp et al., 2001; Daviero-Gomez et al., 2005). Moreover, stomata do not occur on these axes, nor does the multilayered hypodermis contain an extensive intercellular system. Nevertheless, the subterranean axes of *N. aphylla* are colonized by a glomeromycotan fungus that is similar to *G. rhyniensis* morphologically and also believed to be mycorrhizal (Krings et al. 2007b, c). In contrast to *A. major*, however, the mycorrhizal fungus enters the *N. aphylla* axes through the rhizoids that occur on the ventral side along what is called the rhizoidal ridge. Apparently because intercellular spaces are virtually absent in the hypodermis, the fungus extends through this tissue as an intracellular endophyte until it reaches the cortex where intercellular spaces are present. In the cortex, the fungus forms an extensive intercellular network of hyphae and produces vesicles and thick-walled spores. The most interesting aspect of this putative endomycorrhizal association is that the intracellular growth of the fungus in the hypodermis is somehow controlled by the host through the production of cell wall sheaths around the fungal hyphae. As a result the fungus appears to be "guided" through the hypodermis (without being able to extract nutrients from the host) and into the cortex where intracellular penetration is no longer possible. To date arbuscules have not been identified in *N. aphylla*.

Several of the Rhynie chert plants have been demonstrated to have free-living, centimeter-sized gametophytes anchored to the substrate by delicate rhizoids, as well as sexual reproductive structures (antheridia and archegonia) borne in a slightly elevated position (e.g. Remy and Remy, 1980a, b [Figure 7.46]; Remy and Hass, 1996; Kerp et al., 2004). These minute gametophytes are also mycorrhizal and contain arbuscules (see Figure 7.6) arising from trunk hyphae in a confined zone in the upright axis (Taylor et al., 2005c). While it has been known for some time that the free-living gametophytes of a number of extant so-called lower vascular plants (e.g. ferns, lycopsids) are mycorrhizal (e.g. Lang, 1899; Duckett and Ligrone, 2005), only recently has it been demonstrated that the same fungal symbionts are shared by both the sporophyte and gametophyte (Winther and Friedman, 2007, 2008, 2009; Leake et al., 2008a). This suggests that the molecular and cellular basis involved in the mutual recognition of plant host and fungal partner in early land plants, as well as the subsequent accommodation of the AMF, were already well established 410 Ma ago (Bonfante and Genre, 2008).

Another species of *Glomites*, *G. sporocarpoides*, also occurs in the Rhynie chert ecosystem in moribund and dispersed tissues of *Rhynia gwynne-vaughanii* (Karatygin et al., 2006). In this taxon, *G. sporocarpoides* arbuscules were not identified, but there were large, glomoid sporocarps (each up to 350 μm in diameter) containing numerous spores (20–24 μm in diameter) in the tissues. In some sporocarps there are spores that appear to be germinating. It is suggested that *G. sporocarpoides* was a pathogen and that the sporocarps developed in necrotic zones of tissue (Karatygin et al., 2006).

FIGURE 7.47 Longitudinal section of *Stigmaria ficoides* appendage showing central vascular strand (tracheids with scalariform wall thickenings) and vesicles (*arrows*). Bracketed area (A) shows cortical cells with arbuscules. Pennsylvanian, Great Britain. Bar = 100 μm.

Arbuscules have also been reported in structurally preserved *Radiculites*-type narrow cordaitalean rootlets of >1 mm in diameter from the Carboniferous Grand'Croix chert in France (Strullu-Derrien et al., 2009). The fungus colonizes a discontinuous fungal zone in the inner portion of the cortex, where it produces coiled hyphae within host cells that extend from cell to cell intracellularly. Small arbuscules occur in some of the cortical cells; where the arbuscule-forming hypha enters the host cell, a slight thickening (apposition) is recognizable. Based on a comparison with the arbuscules of extant plants, those in *Radiculites* are most similar to the *Paris*-type.

Arbuscules are also known to occur in the filiform appendages of the underground organs (*Stigmaria*) of arborescent lycophytes preserved in coal balls from the Carboniferous of Great Britain and Germany (Krings et al., 2011a; Figure 7.47). This AMF occurs near the tip of the appendages, where the middle cortex is still partly intact, and occupies a well-defined zone around the vascular strand. This zone is located within the inner portion of the middle cortex, which is composed of elongate, relatively narrow parenchyma cells. The fungus includes trunk hyphae that grow along the long axis of the appendage. Extending from these hyphae are narrower branches that may produce vesicles or spores (Figure 7.48). Other branches penetrate individual cells of the cortex to form

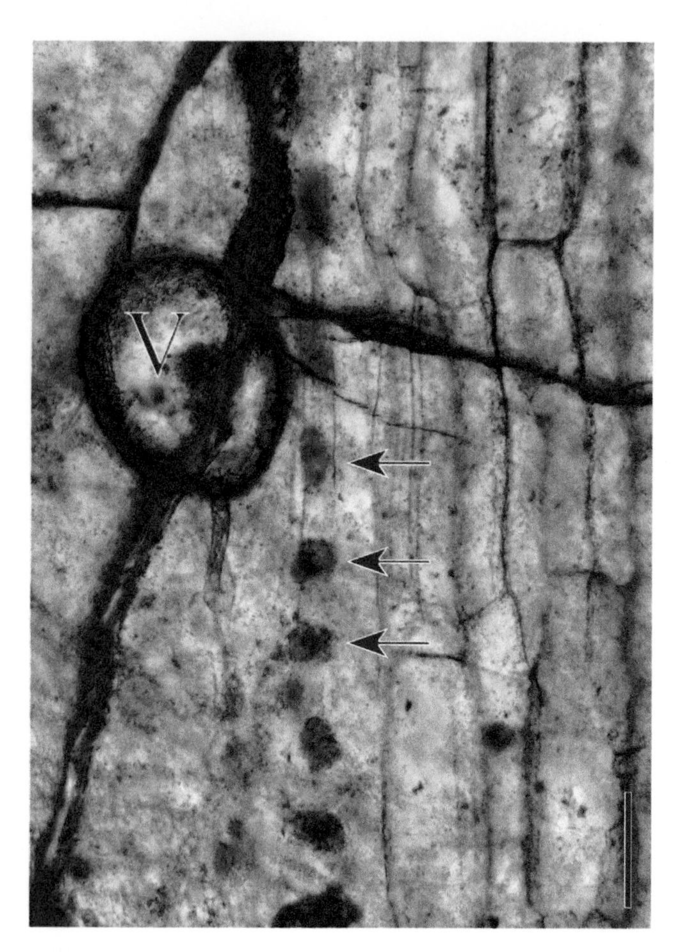

FIGURE 7.48 Detail of Figure 7.47 showing vesicle (V) and several arbuscules in narrow-diameter cells of the cortex (*arrows*). Pennsylvanian, Great Britain. Bar = 50 μm.

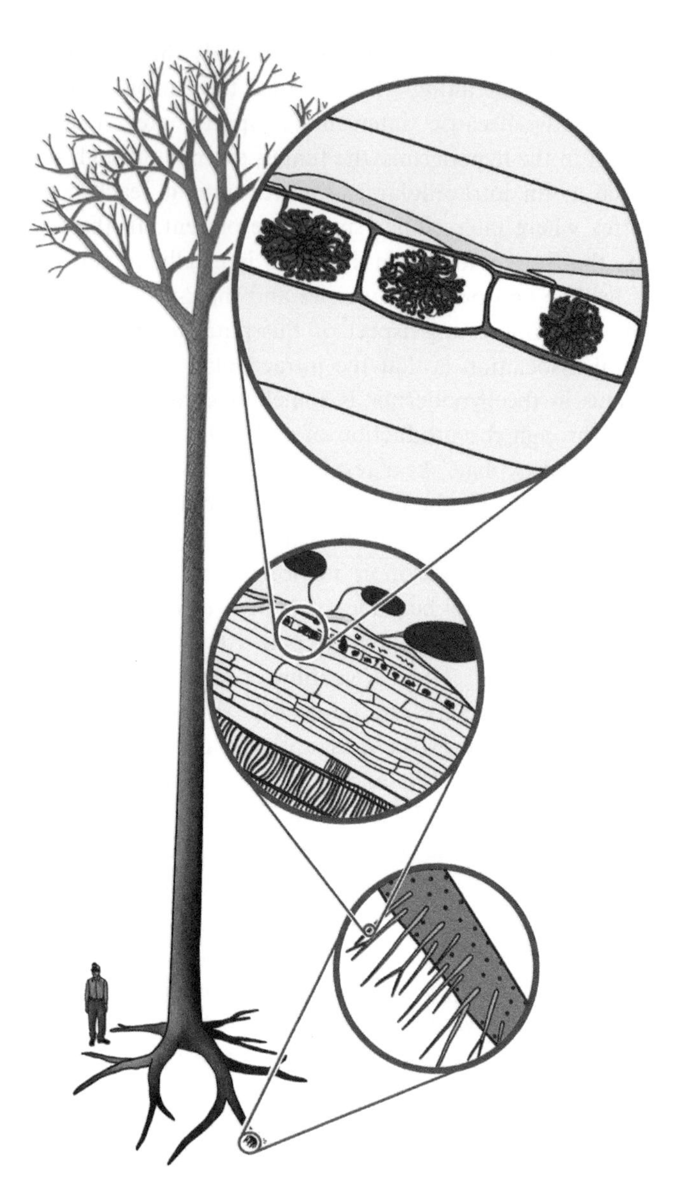

FIGURE 7.49 Reconstruction of arborescent lycopsid showing helically arranged appendages on *Stigmaria ficoides* (lower circle), position of arbuscule zone (middle circle), and detail of arbuscules (upper circle).

multibranched structures that are interpreted as arbuscules. The arbuscule-like structures appear in longitudinal section in one or two rows (Figure 7.49). As development of the stigmarian appendage continues and the middle cortex disintegrates (mature appendages show a characteristic void in this region), the fungus is represented only by vesicles and spores, and by occasional trunk hyphae that remain in the void. This developmental sequence appears to be consistent with one of the key points in Brundrett's (2004) (Figure 7.50) definition of mycorrhizae, namely synchronized plant–fungus development. As is the case with modern arbuscular mycorrhizae, the active association is confined to the growing tips of the appendages, while the more proximal (mature) regions only contain trunk hyphae, spores, and vesicles. What is especially interesting about the arbuscular mycorrhizae in the stigmarian appendages is that, although these host structures are functioning as roots, structurally and developmentally they are in fact homologous with leaves. This fossil evidence supports the hypothesis that it is not the type of plant organ that dictates the establishment of mycorrhizal associations but rather the functional environment provided by the organ (Brundrett, 2002; Bonfante and Genre, 2008; Krings et al., 2011a). It further strengthens the idea that the fungi involved in the formation of arbuscular mycorrhizae were highly adapted and able to successfully colonize different types of plant organs irrespective of their evolutionary stage of development.

FIGURE 7.50 Mark C. Brundrett.

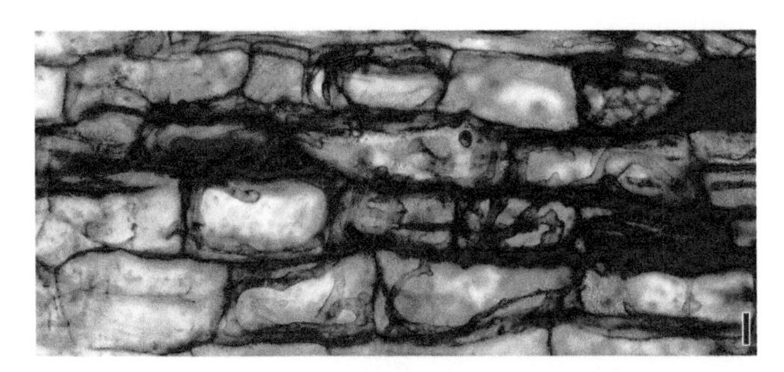

FIGURE 7.51 Section of glossopterid root (*Vertebraria* sp.) cortex showing robust intracellular hyphae of *Glomites vertebrariae*. Permian, Antarctica. Bar = 25 μm.

The seed ferns are a polyphyletic grouping of extinct late Paleozoic and Mesozoic fossil plants (Doyle, 2006; Hilton and Bateman, 2006). A great deal is known about their morphology and anatomy (e.g. Taylor, T.N. et al., 2009) but almost nothing about their mycorrhizal associations. These plants were characterized by gymnospermous secondary xylem (in some cases) and reproduction via seeds, but they had foliage that was mostly fern-like in organization (at least in the Carboniferous forms). The Glossopteridales are a group of seed ferns recognized by entire leaves with a unique type of anastomosing venation and distinctive reproductive organs; the group dominated most ecosystems in Gondwana (the southern part of the supercontinent Pangea) during the Permian. Although much is known about the glossopterids, only recently has a mycorrhizal association (Figure 7.51) been identified in plant remains from Antarctica (Harper et al., 2013). Small (~1.0 mm in diameter) silicified roots of *Vertebraria*, the underground organ of the glossopterids, contain a zone that is two or three cell layers thick in which the cells contain septate hyphae. Intracellular hyphae are approximately 5 μm in diameter and may possess swellings; in the root cells the hyphae form large coils and occasionally vesicles. This fossil fungus, named *Glomites vertebrariae*, is of the *Paris*-type based on hyphal diameter, degree of coiling, and the presence of knobs or swellings. While arbuscule morphology can be highly variable, this is the first persuasive demonstration that the largest group of Permian seed plants was mycorrhizal. Because these Antarctic glossopterids grew at high paleolatitudes under a unique light regime (4 months of light, 4 months of darkness, 4 months of diffuse light), it is hypothesized that perhaps during the periods of darkness, the plants dissociated from their fungal partners because of the decrease in their ability to produce carbon (Harper et al., 2013).

AMF have been reported in the roots of the cycad *Antarcticycas* from the Triassic of Antarctica (Stubblefield et al., 1987a; Phipps and Taylor, 1996). Two types of AMF are reported from this plant, both evenly distributed in the cortex with neither occupying a distinct zone. One of these is *Gigasporites myriamyces* (Figure 7.52); it is characterized by intercellular and intracellular hyphae, approximately 6 μm in diameter, that form loops and coils in the cells. Septa are present in hyphae and arbuscules possess a thick trunk and delicate ultimate branches. In the second type, *Glomites cycestris* (Figure 7.53), the hyphae have slightly narrower diameters and are associated with terminal, thin-walled, elongate vesicles up to 56 μm long. Interestingly, living cycads have been reported with multiple types of AMF in their roots (Iqbal and Shahbaz, 1990; Muthukumar and Udaiyan, 2002) and these have been compared to the *Arum* type.

The presence of AMF has also been reported in the conifer *Metasequoia milleri* (Stockey et al., 2001) from the Eocene Princeton chert site (British Columbia, Canada). The terminal mycorrhizal roots range from 2 to 5 mm in diameter. They are diarch with a prominent inner cortical zone surrounded by a parenchymatous

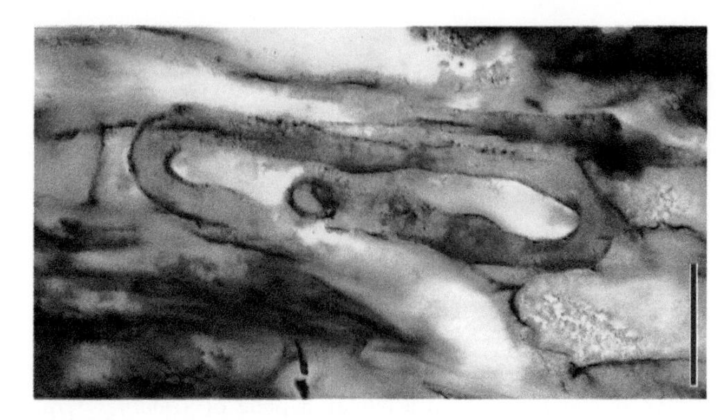

FIGURE 7.52 Detail of *Gigasporites myriamyces* hyphal coil in fossil cycad *Antarcticycas schopfii* root cortical cell. Triassic, Antarctica. Bar = 25 μm.

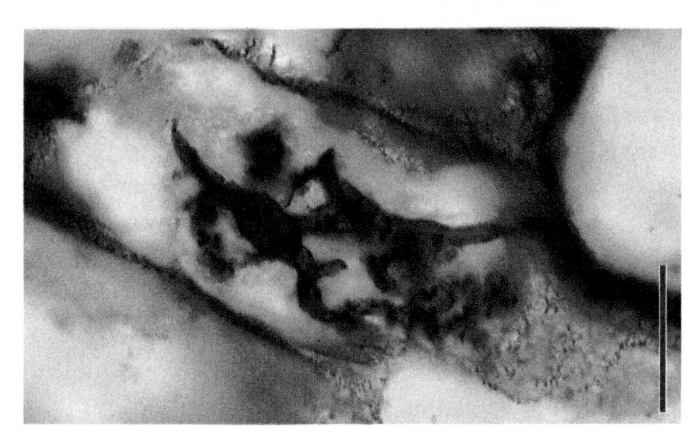

FIGURE 7.53 Detail of *Glomites cycestris* arbuscule in fossil cycad *Antarcticycas schopfii* root cortical cell. Triassic, Antarctica. Bar = 25 μm.

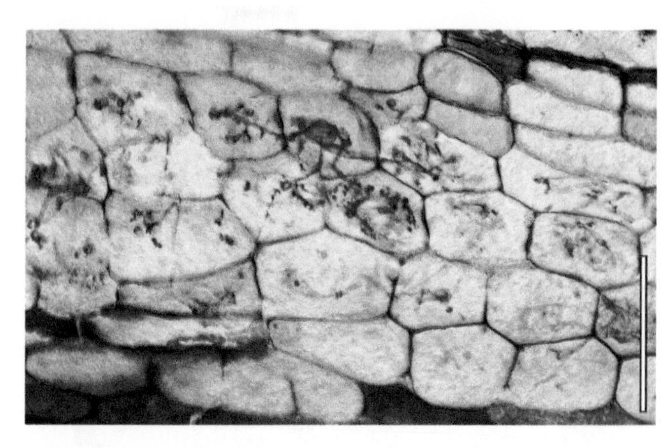

FIGURE 7.54 Section of fossil conifer *Metasequoia milleri* root cortex showing branching hyphae and arbuscules. Eocene, Canada. Bar = 100 μm. (Courtesy R.A. Stockey.)

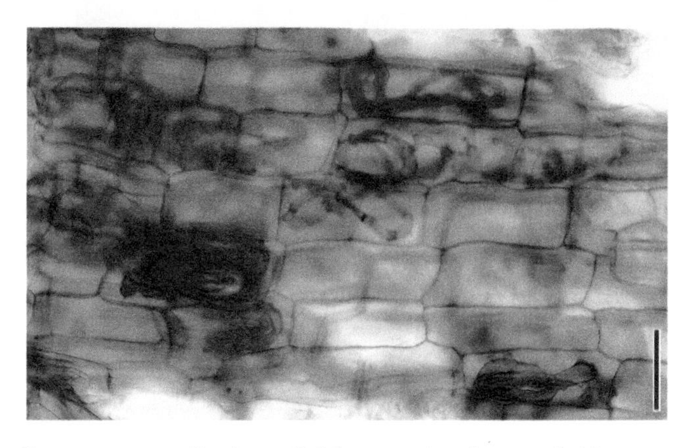

FIGURE 7.55 Section of *Metasequoia glyptostroboides* root cortex showing branching hyphae and intracellular coils. Extant. Bar = 50 μm. (Courtesy R.A. Stockey.)

cortex and epidermis; root hairs are absent. The outer two or three layers of cortical cells are usually devoid of contents, whereas the next layers inward are more or less filled with aseptate fungal hyphae. Hyphae are irregularly branched and extend intracellularly from cell to cell; most form coils within individual cells. These coils, from which arbuscules arise, characterize *Paris*-type mycorrhizae. The mycorrhizal hyphae produce abundant multi-branched arbuscules that terminate in dark, apparently swollen tips (Figure 7.54). The mycorrhiza in the fossil *M. milleri* roots shows some morphological similarity to that seen in the extant *M. glyptostroboides* (Figure 7.55) (e.g. see Böcher, 1964). In both the extant and fossil forms the hyphae are intracellular and produce arbuscules with slightly inflated tips. While the *Paris*-type of mycorrhiza occurs in numerous groups of plants, including conifers (e.g. Kough et al., 1985; Smith and Smith, 1997), the presence of this symbiont in *M. milleri* suggests that this association has persisted relatively unchanged since the Eocene, at least in the taxodiaceous conifer *Metasequoia*.

EVOLUTION OF ARBUSCULAR MYCORRHIZAE

The origin and evolution of AMF remain controversial. The AM symbiosis is distinctive in that it is obligate for the fungus but in most cases facultative for the plant host (Helgason and Fitter, 2009). It is widely held that the apparent lack of, or at least poor development of, roots in the earliest land plants, in tandem with a scarcity of available mineral nutrients (particularly phosphorus) in the rudimentary soils, necessitated the evolution of AM symbiosis as a means of colonization of land by plants (Cairney, 2000). Some authors have suggested that a saprophytic or parasitic fungus that had already acquired the enzyme system to penetrate cell walls may have

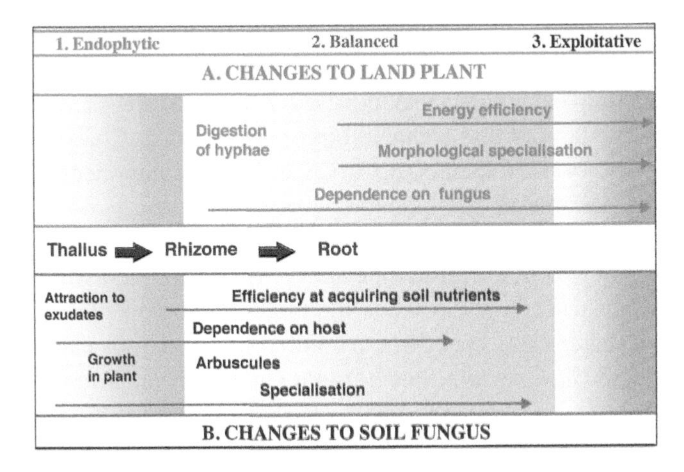

FIGURE 7.56 Hypothetical order of (**A**) changes to land plants (upper half) and (**B**) fungi (lower half) during three stages in the evolution of vesicular-arbuscular mycorrhizal associations. (From Brundrett, 2002.)

served as the first fungal partner (Morton, 1990b; Taylor and Osborn, 1996). This could have occurred initially by increasing access to plant exudates and organic substances after the death of the host (Brundrett, 2002). The presence of various fungal parasites associated with some of the earliest land plants and associated AMF may have also served to necessitate growth of the latter within the host tissues (Krings et al., 2007b, c), perhaps to control growth of the parasites. Once within the plant tissue of the rhizome and/or rhizoids, the absorption of carbohydrates by the AMF would become an accessible source of energy, and a concomitant increase in absorptive capacity for the host from the widespread fungal mycelium. These steps, as well as others – including fungus–host recognition, plants more efficiently hosting mycorrhizae, increased nutrient uptake, and cell fungus–host interface in the evolution of arbuscules – are detailed by Brundrett (2002) and are graphically portrayed in Figure 7.56; they suggest that the initial stages in the evolution of this fungus–plant symbiosis required a wide range of biological and ecological modifications within the fungus, and later to the plants (Brundrett, 2002). Some have suggested that the delicate balance achieved by the fungus and plant may have evolved so as to reward both partners and not tip the balance in either direction. For this tightly controlled mutualism to continue there must be a simultaneous exchange across a common interface (Brundrett, 2002; Delaux et al., 2013).

Another hypothesis uses the cyanobacterial–glomeromycotan mutualism as seen in *Geosiphon* as a model or early precursor to mycorrhizal evolution, based on the presence of arbuscule-like structures (e.g. Schüßler and Kluge, 2001; Brundrett, 2009). Other authors support the hypothesis that the AM symbiosis evolved from some endophyte already associated with a green alga, the group most closely associated with land plants (Pirozynski and Malloch, 1975; Lewis, 1991; Lewis and McCourt, 2004). The recent discovery of components involved in the pre-symbiotic dialog in the closest green algal relative to land plants (Delaux et al., 2013) has the potential to open up a new window in the study of the evolution of AM symbiosis. Delaux et al. (2013) hypothesized that advanced charophytes may be capable of forming a type of symbiosis with microbes that has been overlooked until now or that these components are playing other, nonsymbiotic, roles in the algae. Still another suggestion is that the ancestral fungus was a pathogen that invaded the plant tissues, but because the host phosphorus content was so low the fungus also had a phosphorus deficiency and thus imported the element via its mycelial network. Accordingly, the symbiosis increasingly evolved as a result of some phosphorus leaking into the apoplasts of the host (Helgason and Fitter, 2005).

The AM symbiosis is regarded as the ancestral type since it occurs in most plant groups, especially in the early diverging major clades of land plants (Wang and Qiu, 2006), including liverworts and hornworts (e.g. Read et al., 2000; Russel and Bulman, 2005). There are two hypotheses that have been suggested regarding the timing of these events in land plant evolution. One is that glomeromycotan fungi co-evolved with liverworts prior to the evolution of interactions with vascular plants, perhaps sometime in the Ordovician (Kottke and Nebel, 2005). The second model suggests that arbuscular mycorrhizae evolved with early vascular plants and subsequently enlarged the fungal evolutionary niches by developing symbiotic associations with other plants like liverworts (e.g. Ligrone et al., 2007). Both ideas have merit, and the exploitative nature of the Glomeromycota suggests that both models may be equally parsimonious. It has also been hypothesized that arbuscular mycorrhizae have evolved into other types of mycorrhizal associations (Brundrett, 2002; Wang and Qiu, 2006). For example, a study by Krause et al. (2011) provides evidence of a basidiomycete fungus associated with several liverwort genera. This has stimulated ideas about the pattern of sequential fungal colonization stages that could lead to the evolution of a mycorrhizal thallus.

Further underscoring the complexities of the evolution of arbuscular mycorrhizae is the suggestion that perhaps Glomeromycota were not the first fungal mycorrhizal partners. This idea is not based on fossils but rather on the occurrence of members of the Mucoromycotina, for example *Endogone*, in the most primitive lineages of extant land plants, that is, liverworts (Bidartondo et al., 2011). Although there is currently little conclusive fossil evidence (see Strullu-Derrien et al., 2014) to test this hypothesis, molecular ecology and electron microscopy of a limited number of extant liverworts and hornworts suggest that perhaps some members of the Mucoromycotina were the first group to establish some type of endomycorrhizal system in the earliest land plants. Although molecular clock assumptions to date show no distinct age difference between *Endogone* and the Glomeromycota, both are interpreted as predating the origin of land plants (Bidartondo et al., 2011). Is it possible that some members of the Mucoromycotina were early mutualists of emerging terrestrial land plants and that they were simply outcompeted by the Glomeromycota because zygomycetous fungi like *Endogone* relied on sexual reproduction? Or, perhaps both groups initially relied on sexual reproduction, but the loss in the Glomeromycota in some way made their asexual reproduction more evolutionarily stable. Another question that may ultimately be answered by the fossil record is whether the habitats of the earliest land plants in some way influenced their interactions with early mutualists. All of these hypotheses relating to the evolution of arbuscular mycorrhizae will be discussed and tested in the future using new tools and, hopefully, older fossils. Whether the evolution of arbuscular mycorrhizae occurred once or multiple times, or multiple times in different biological systems, the AM symbiosis was certainly a complex process that involved numerous abiotic and biotic factors that included both the plant (e.g. Chabaud et al., 2011; Maillet et al., 2011) and some component of the genome of the glomeromycotan fungi (e.g. Corradi and Bonfante, 2012).

There is no direct fossil evidence that could be used to reconstruct how the arbuscule might have evolved. One idea, however, uses the Carboniferous fossil root endophyte *Cashhickia acuminata* (see Chapter 8) as a model system (Taylor et al., 2012). Perhaps a fungus similar to *C. acuminata* initially extended through the apoplast and intercellular spaces of the host cortex and later gained the capacity to penetrate the cell wall but was unable to breach the cell membrane. With an increasing number of endophyte hyphae passing through the cell wall, but not the cell membrane, there would be an increased surface area where the membrane and hyphae are in contact. As a result, the membrane–fungus interface increased to a level where greater exchange was facilitated. Of the several steps postulated in the evolution from endophytic to mycorrhizal association, one involves the ability of the fungus to penetrate cells without causing damage to the host (Brundrett, 2002; Genre and Bonfante, 2005). Morphologically, *C. acuminata* satisfies these requirements. Finally, as host cells and the fungus continued to evolve, multiple sites of hyphal penetration became restricted to a single site (arbuscule trunk domain of Pumplin and Harrison, 2009) and the increased surface area afforded by the multiple cell penetrations became replaced by the highly dichotomous arbuscule.

8

ASCOMYCOTA

GEOLOGIC HISTORY OF ASCOMYCOTA 131
PALEOZOIC ASCOMYCETES 133
Paleopyrenomycites devonicus 133
Palaeosclerotium pusillum 135
"Sporocarps" 136
MESOZOIC AND CENOZOIC ASCOMYCETES 137
Taxa *incertae sedis* 137
Pezizomycotina 138
FUNGAL ENDOPHYTES AND EPIPHYLLS 163
Endophytes 163
Epiphylls – Fungi on Leaf Surfaces 170
Conclusions: Endophytes and Epiphylls 171

The Ascomycota is a monophyletic phylum that includes the subphyla Saccharomycotina (budding yeasts) (Suh et al., 2006), the Pezizomycotina (generally filamentous fungi), and the monophyletic or paraphyletic Taphrinomycotina (Archiascomycetes *sensu* Nishida and Sugiyama 1993, 1994) (Spatafora et al., 2006; Hibbett et al, 2007; Liu et al., 2009); together with the Basidiomycota they form the subkingdom Dikarya (Figure 8.1).

The Ascomycota are the largest and most diverse group of extant fungi with perhaps 65,000 formally described species that may thrive as saprotrophs, or parasites of plants, animals, and other fungi, or form symbiotic associations in lichens; they also form mycorrhizal relationships with vascular plants (Egger, 2006; Tedersoo et al., 2006). Ascomycetes are worldwide in distribution and exist in a variety of habitats; some highly reduced forms even grow obligately on the gametophytes of various bryophytes, including the rhizoids (e.g. Duckett et al., 1991; Döbbeler, 1997). Some have harmful effects, such as the well-known Dutch elm and chestnut blight diseases in North America, powdery mildews that affect grapevines and ornamental lilacs, ergot of rye and other cereals that can cause the production of carcinogenic alkaloids, and athlete's foot. There are also those ascomycetes that have a more "positive" interaction with humans, including *Penicillium*, baker's yeast, and truffles (not the chocolate ones!).

The principal character of the group is the production of an ascus (Figure 8.2), a sac-like structure (technically a meiosporangium) in which fusion of nuclei takes place, followed by meiosis and the formation of non-motile meiospores, also termed ascospores (Pöggeler et al., 2006). Some ascomycetes are polyploid (e.g. Albertin and Marullo, 2012). Numerous members of this group are asexual (anamorphic), that is, they lack a sexual phase (perfect stage); at least, a sexual stage has not been described or observed to date in these forms. Such forms, which have also been termed "Fungi Imperfecti" or "imperfect fungi," reproduce by means of conidia formation and were once lumped in the Deuteromycota (see Taylor, J.W., 1995), together with anamorphic species from other fungal taxa (e.g. Basidiomycota). Today, however, they are classified based on morphological and/or physiological similarities to ascus-bearing taxa and molecular data. For example, the presence of a lamellate hyphal wall of two layers has been a useful character to unite both teleomorphic and anamorphic states in the absence of asci (Kirk et al., 2008). Ascomycotan hyphae are typically septate, with septal pores that provide continuity throughout the hypha. On either side of the septum are spherical to hexagonally shaped membrane-bound

T.N. Taylor, M. Krings & E.L. Taylor: Fossil Fungi.

DOI: http://dx.doi.org/10.1016/B978-0-12-387731-4.00008-6

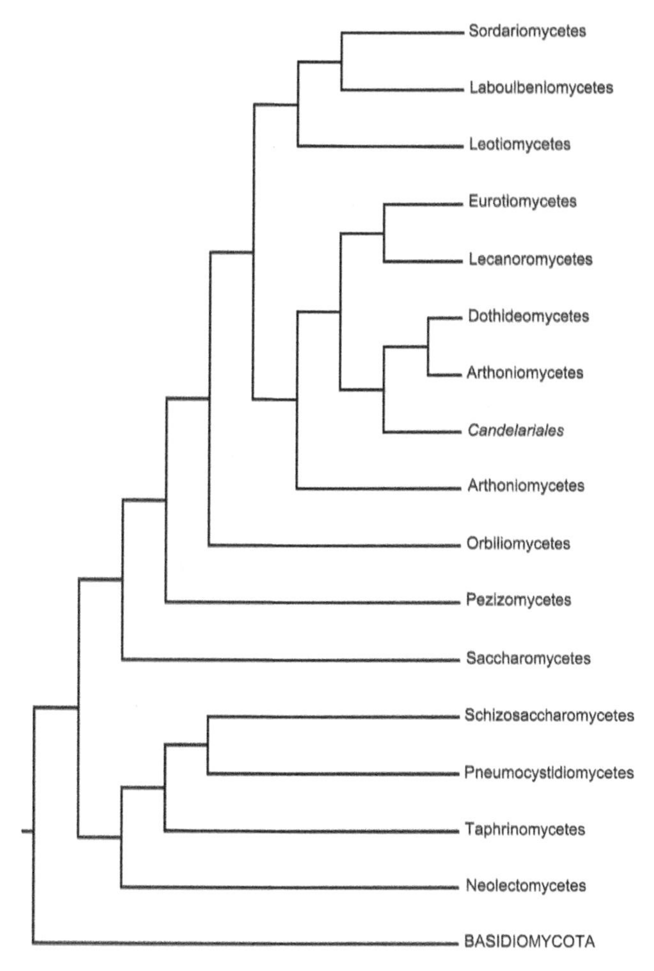

FIGURE 8.1 Suggested phylogenetic relationships within the Ascomycota. (After Schoch et al., 2009a.)

structures termed Woronin bodies that can block pores when hyphae are ruptured (e.g. Trinci and Collinge, 1974; Markham and Collinge 1987).

The Saccharomycotina (ascomycete yeasts) have played a significant role in human activities for millennia; for example, there are records from the Middle East and China that depict brewing and bread-making 8,000–10,000 years ago (Suh et al., 2006; Kurtzman et al., 2011). They usually grow as single cells, with division by budding or (less frequently) by fission. No members of the Saccharomycotina produce fruiting bodies (ascocarps or ascomata); rather, they form naked asci (Eriksson and Winka, 1997; Kurtzman, 2011).

Members of the Taphrinomycotina range from simple and yeast-like to filamentous (e.g. Liu et al., 2006). Members of the plant pathogenic Taphrinales, the largest order within the Taphrinomycotina, are dimorphic, having a saprotrophic yeast phase and a filamentous pathogenic phase (Blackwell and Spatafora, 2004). *Neolecta* is the only member known to produce an enclosing ascocarp (Landvik et al., 2001, 2003; Liu and Hall, 2004). Sexual reproduction in *Neolecta* is ascogenous but does not involve the formation of ascogenous hyphae. Included in the Taphrinomycotina is also a relatively new group of filamentous, cryptically reproducing ascomycetes, the Archaeorhizomycetes, which are soil fungi often found around tree roots. They do not form ectomycorrhizal structures, however, but rather appear to be saprotrophs (Rosling et al., 2011). Like the discovery of fossil fungi, this group of extant forms further underscores the extraordinary level of unknown biodiversity in the fungal kingdom, and the levels of biological interaction that remain to be understood.

Development of the ascomata (or ascocarps), including the structure and dispersal nature of asci, together with molecular sequence data, have been used in recognizing phylogenetic relationships and higher taxonomic categories within the Ascomycota (e.g. Lutzoni, et al., 2004; Blackwell et al., 2006; James et al., 2006b; Zhuang and Liu, 2012). The most common types of ascomata include: the apothecium, an open, cup-shaped ascocarp; the cleistothecium, a globose, completely closed structure; and the perithecium, a flask-shaped ascocarp with an open pore or ostiole. In early classifications the various groups were delimited based primarily on ascoma types – for example, the Discomycetes with disc-like apothecia, the Pyrenomycetes with ostiolate perithecia, and the Plectomycetes with closed, spherical cleistothecia (see Schmitt et al., 2005). The problem with this classification system, however, was that it became increasingly clear that fruiting-body characters can converge (e.g. Berbee and Taylor 1992). As a result, other features such as ascoma development, ascus structure, and ascospore morphology became increasingly important in classification. For example, the concept of the Loculoascomycetes was introduced for those pyrenomycetes with ascolocular development and bitunicate asci (e.g. Luttrell, 1951, 1955, 1973; Barr, 1987). In ascolocular development, the asci develop in pockets or locules formed by undifferentiated somatic hyphae; in ascohymenial development, the asci occur within a true hymenium developed from generative ascogenous hyphae. Nevertheless, with the advent of molecular tools these characters and features were found to be polyphyletic, and the Loculoascomycetes are now included by many researchers in two taxa: the Chaetothyriomycetidae and Dothideomycetes (= Loculoascomycetes, bitunicate ascomycetes) (e.g.

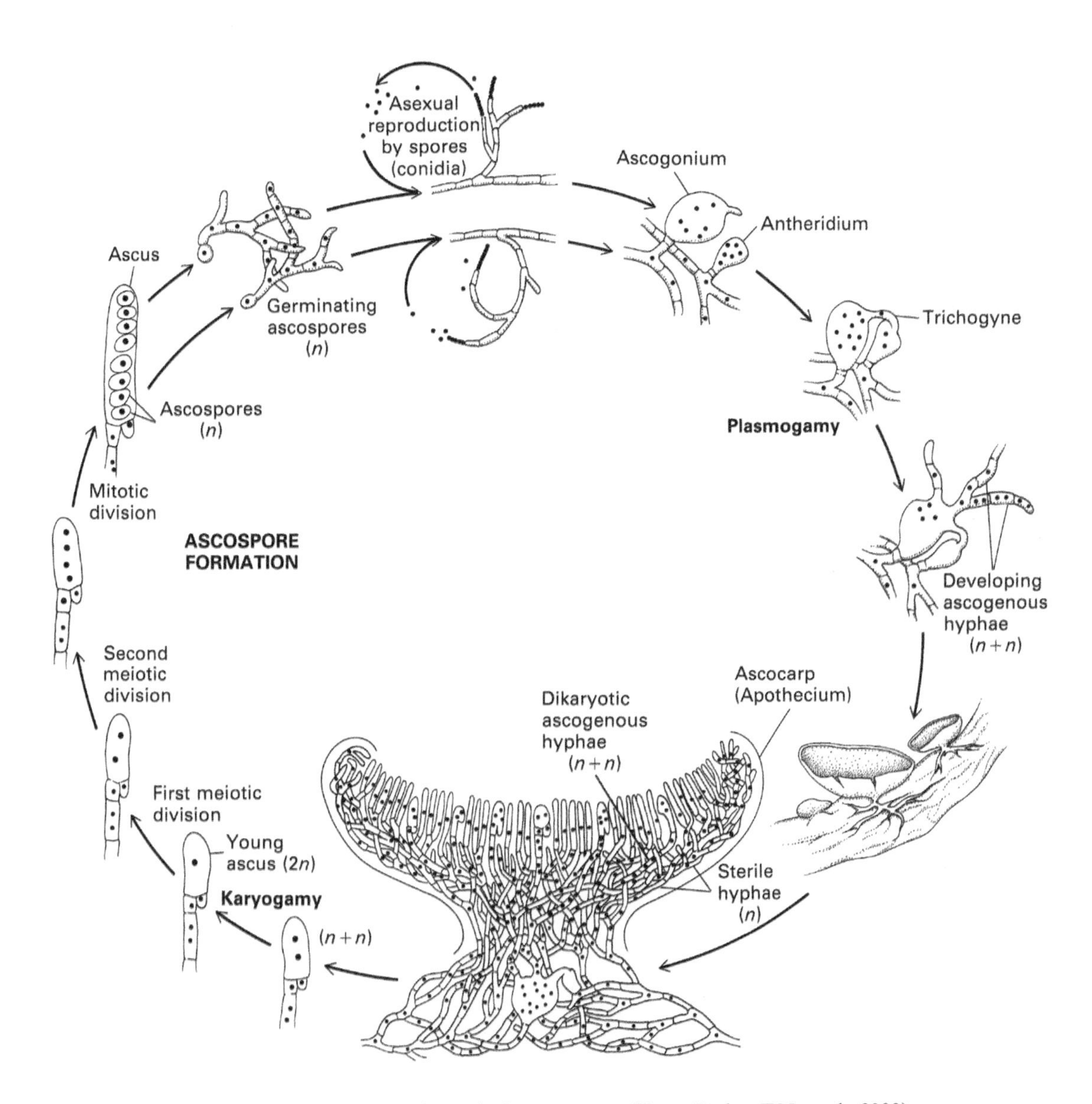

FIGURE 8.2 Life cycle of a typical ascomycete. (From Taylor, T.N. et al., 2009).

Lumbsch and Huhndorf, 2007). While there is general agreement based on molecular characters that perithecial ascomycetes evolved several times from apothecial ancestors, the pattern of ascus dehiscence appears to be a relatively conservative feature (Schoch et al., 2009a).

GEOLOGIC HISTORY OF ASCOMYCOTA

Although molecular clock evidence suggests that Ascomycota and Basidiomycota split ~1000–900 Ma ago, which would place their origin in the mid-Proterozoic (Hedges et al., 2006; Prieto and Wedin, 2013), like other fungal groups, the geologic history of the ascomycetes based on credible fossil evidence only extends back into the Paleozoic (e.g. Taylor, T.N., 1994). There are a few records of Precambrian macrofossils that have variously been interpreted as lichens based on circumstantial evidence (e.g. Retallack, 1994; Retallack et al., 2013b). If some of these fossils were lichens then it might be possible that they also contain ascomycetes as mycobionts (see Chapter 10 for details on some of these fossils). To date the oldest persuasive fossils that are thought to be ancient members of anamorphic Ascomycota have been recovered from mid-late Silurian rocks from Sweden (Sherwood-Pike

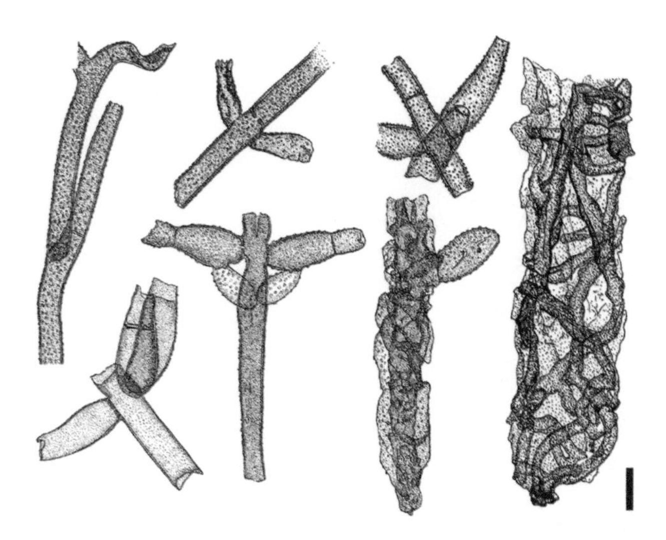

FIGURE 8.3 Various forms of hyphae recovered from bulk macerations. Silurian, Sweden. Bar = 10 μm. (Modified from Sherwood-Pike and Gray, 1985.)

and Gray, 1985). The fossils consist of hyphae (Figure 8.3) and spores that were recovered from drill core samples; as a result, nothing is known about other organisms in the ecosystem, or the nutritional mode of these fossils. The most conspicuous fungal element in the assemblage consists of ovoid, multiseptate spores up to 55 μm long and 25 μm in diameter. The presence of distinctive scars on some of the spores are like those of certain conidial fungi. Other fungal remains include perforate septate hyphae with short branches; these share some morphological resemblance to conidiophores that produce bottle-shaped conidiogenous cells termed phialides. The samples did not include ascocarps or ascospores. Interestingly phialides are the most common type of conidiogenous cell among extant conidial fungi.

Ornatifilum is the artificial generic name assigned to Devonian specimens from the Anglo-Welsh basin that consist of small branched tubes approximately 10 μm in diameter and morphologically resemble some of the fossils from Sweden (Burgess and Edwards, 1991). In *O. granulatum*, the septate hyphae branch at obtuse angles whereas, in *O. lornensis,* the surface is covered by various forms of ornament (e.g. grana, coni, spines). Similar specimens are known extending from the Silurian into the Devonian (Wellman, 1995; Lakova, 2000; Lakova and Göncüoglu, 2005). Whether these tubes represent fungi continues to remain an unresolved issue.

Fungal remains have also been described from lower Silurian (Llandoverian) rocks recovered from bulk macerations in Virginia, USA (Pratt et al., 1978). This assemblage is diverse and includes various types of banded tubes, membranous cellular sheets, cuticle, various spore types including those in tetrads, and septate hyphae. There is a superficial resemblance of these Silurian fossils to specimens of *Ornatifilum* (see above). Some hyphae have phialides up to 20 μm long that may be appressed to the wall of the parent hypha. It is suggested that these hyphae are perhaps most closely related to a hyphomycete. Irrespective of the precise affinities of these Silurian anamorphic fungi, their presence demonstrates that fungi were present in ecosystems at a time when terrestrial plants were becoming established.

There are numerous enigmatic fossils ranging from the Silurian into the Devonian that are characterized by various tube-like structures in which the individual tubes may contain multiple types of thickenings. Associated with some are sheets of cuticle-like material, and in some instances, spores that may or may not be related to the tubes. While the systematic position of most of these fossils remains a mater of debate, at least some are morphologically and structurally similar to certain types of lichens (Honegger et al., 2013). One of these is *Orestovia*, an Early–Middle Devonian organism isolated from paper coals of western Siberia. It consists of naked, unbranched, distally tapering axes that are up to 20 cm long x 2 cm wide (Ergolskaya, 1936; Ishchenko and Ishchenko, 1980; Krassilov, 1981). Most specimens are preserved as hollow envelopes, often displaying an epidermis-like, (?pseudo-) cellular pattern. Extending the length of each axis is a delicate strand of elongate tubes with annular to reticulate thickenings on the internal wall. In a few specimens, spores occur in the cortex of the axes and range from 150 to 190 μm in diameter. Arising from the surface are structures interpreted as thyriothecia, each about 1.0 mm in diameter and including clavate hyphopodia (Krassilov, 1981). From the specimens figured it is difficult to determine whether these structures were produced by *Orestovia* (and thus that *Orestovia* was a fungus) or by some organism colonizing *Orestovia*, or represent some type of ascostromata, or are simply artifacts of preservation.

Despite the hypotheses (based on phylogenetic analyses) that plants with a bryophytic grade of organization should have been a common component of early terrestrial ecosystems prior to the advent of vascular land plants, the fossil record of these plants continues to remain meager (e.g. Taylor, T.N., et al., 2009). None of the fossils are sufficiently preserved in a manner to indicate whether they harbor endophytic fungi of any type;

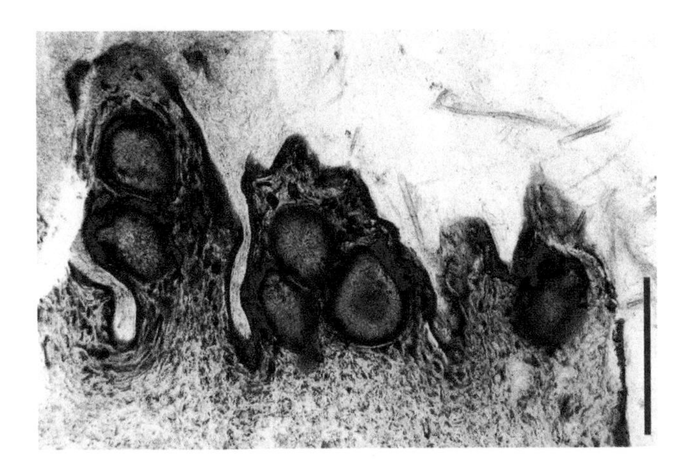

FIGURE 8.4 Section of a portion of the stem of *Asteroxylon mackiei* showing several perithecia of *Paleopyrenomycites devonicus* within enations. Lower Devonian Rhynie chert, Scotland. Bar = 500 µm.

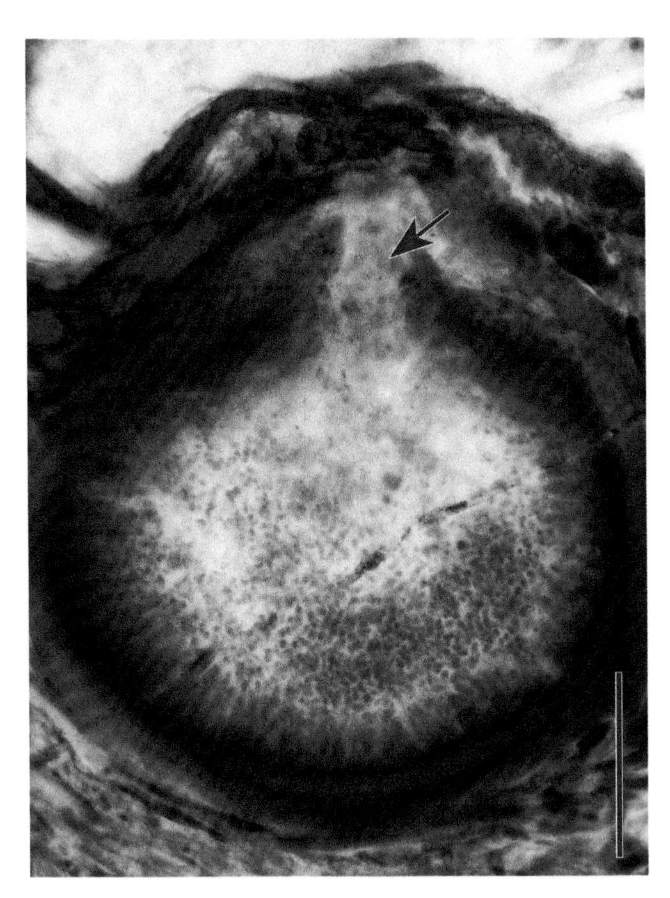

FIGURE 8.5 Longitudinal section of *Paleopyrenomycites devonicus* perithecium showing ostiole (*arrow*) and multilayered wall. Lower Devonian Rhynie chert, Scotland. Bar = 100 µm.

however, the fact that delicate structures such as rhizoids are preserved in the Rhynie chert and that extant leafy liverwort rhizoids contain relatively easily documented ascomycetous fungi (e.g. Pressel et al., 2008a) holds promise that fungal interactions with early thalloid organisms could be demonstrated.

PALEOZOIC ASCOMYCETES

PALEOPYRENOMYCITES DEVONICUS

To date, the oldest structurally preserved ascomycete is *Paleopyrenomycites devonicus* from the Lower Devonian Rhynie chert (Taylor et al., 1999, 2005a). Specimens may be solitary or in small clusters of both the anamorphic and teleomorphic stages. The fungus is preserved in the cortex of enations, stems, and rhizomes of *Asteroxylon mackiei*, a vascular plant with affinities to the early lycophytes (Taylor et al., 1999). Globose perithecia are scattered just beneath the epidermis and appear as circular dark areas that range up to 500 µm in diameter (Figure 8.4). It is especially interesting that the perithecia are often positioned within the substomatal chambers with the ostiole positioned directly beneath the stomatal pore. Each perithecium is constructed of a multilayered wall of hyphae. Arising from the upper surface of the perithecium is a short neck, about 50 µm long, through which the ascospores are released (Figure 8.5). Lining the inner surface of the perithecium are paraphyses that are scattered among the asci, both projecting toward the center of the structure. Paraphyses are aseptate, up to 50 µm long, and characterized by a tapered tip. Asci are about the same length and widest at the mid level; at the distal end the wall is single and there is a central pore. The number of ascospores per ascus is difficult to determine in the fossils, but the upper limit appears to be 16 (Figure 8.6). Each spore is circular in outline, about 10 µm long, and lacks ornament. What are interpreted as mature spores are 1–5 septate (Figure 8.7), and some show evidence of germination. There is no indication that ascospores were forcibly ejected, a condition in other ascomycetes that is considered a derived character. Some of the fossil perithecia contain mycoparasites in the form of coiled hyphae.

An especially interesting feature of *P. devonicus* is the presence of tufts of conidiophores arising from acervuli that are scattered along the axes intermingled with the immature perithecia. These are interpreted as the anamorphic or asexual phase of the fungus (Taylor et al., 2005a). Stages in the development of the anamorph indicate they are formed beneath the cuticle and consist of clusters of

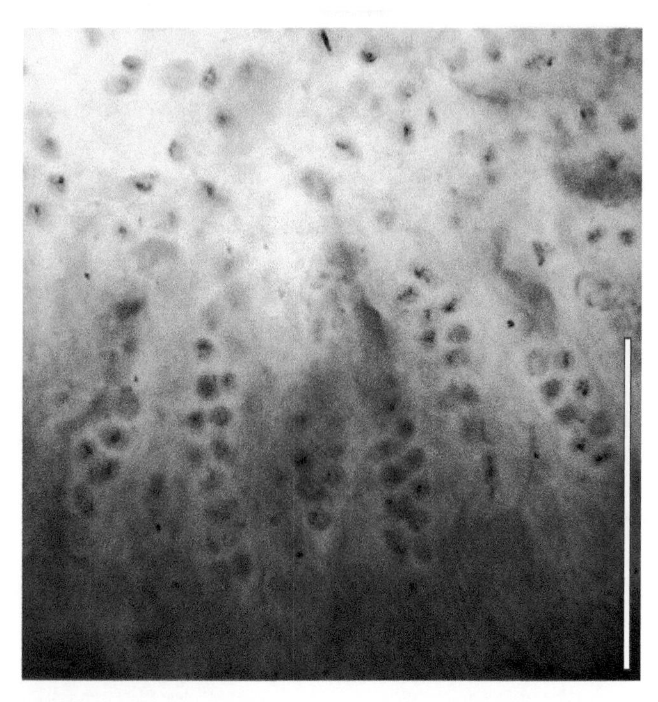

FIGURE 8.6 Longitudinal section through perithecium of *Paleopyrenomycites devonicus* showing several asci containing ascospores. Lower Devonian Rhynie chert, Scotland. Bar = 30 μm.

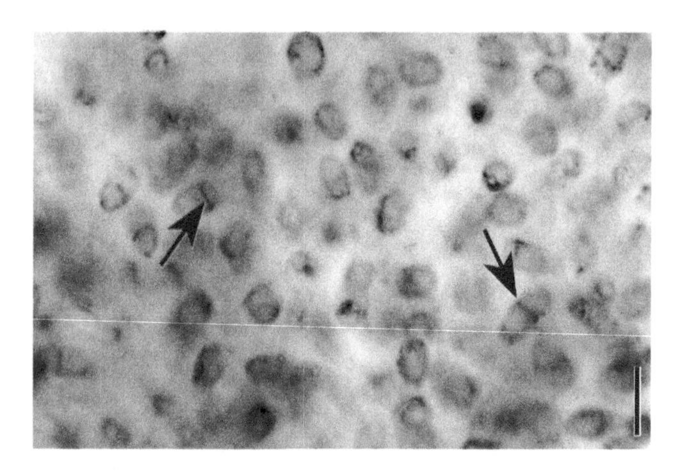

FIGURE 8.7 Cluster of *Paleopyrenomycites devonicus* ascospores in perithecium showing transverse septa (*arrows*). Lower Devonian Rhynie chert, Scotland. Bar = 5 μm.

up to 20 conidiophores that may extend 600 μm out from the epidermis. Conidiogenesis is thallic, and conidia are holoarthric and basipetal with the most mature conidium at the tip (Figure 8.8). Mature arthrospores are round to cuboidal and 4–5 μm in diameter.

Despite the excellent preservation of *P. devonicus* and a large suite of characters that can be used to more clearly define the systematic position of this Devonian fungus, there have been a number of conflicting views regarding its taxonomic placement and how it can be used in evolutionary relationships of filamentous ascomycetes (e.g. Padovan et al., 2005). Taylor et al. (2005a) suggested that *P. devonicus* may be most closely related to a pyrenomycete, perhaps a member within the Sordariomycetes (e.g. Eriksson, 2006). This group of mostly terrestrial fungi in the Pezizomycotina includes 16 orders comprising more than 1,119 genera and 10,500 species of non-lichenized ascomycetes, and is one of the largest monophyletic clades within the ascomycetes (Zhang et al., 2006). All have unitunicate asci produced in perithecia (Kirk et al., 2008). The three subclasses (Sordariomycetidae, Hypocreomycetidae, Xylariomycetidae) are generally distinguished by the color of the perithecia, details of the ascus tip, the color of the ascospores, and the role of these fungi with the host – all features that are impossible to discern with current paleobotanical techniques. Living members of the group function as pathogens, endophytes, and saprotrophs in most terrestrial ecosystems; life in aquatic (freshwater and marine) ecosystems is considered derived (Samuels

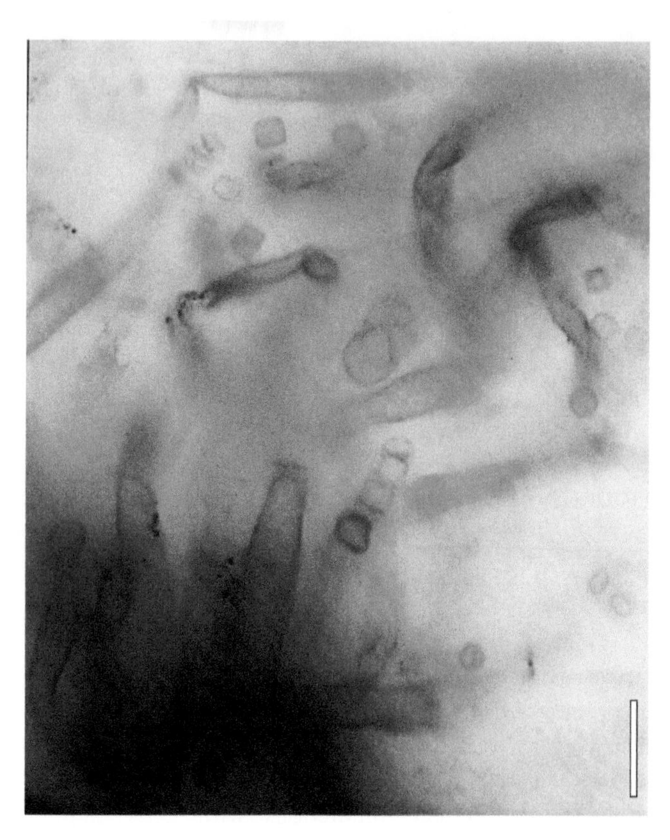

FIGURE 8.8 Conidiophores of *Paleopyrenomycites devonicus* showing holothallic development. Lower Devonian Rhynie chert, Scotland. Bar = 20 μm.

and Blackwell, 2001). A phylogeny based on 42 genomes suggests that the Sordariomycetes is a sister group to the Leotiomycetes, a group that produces asci in apothecia (Fitzpatrick et al., 2006).

Another interpretation is that *P. devonicus* is a member of the Taphrinomycotina (Taylor and Berbee, 2006), a group that lacks ascomata and is considered to include the earliest diverging taxa in the group (Sugiyama et al., 2006). In general, molecular studies suggest that the earliest ascomycetes were yeast-like and lacked complex ascospore-producing structures (e.g. James et al., 2006b; Zhuang and Liu, 2012); however, clavate ascomata up to 7 cm tall are formed from the tips of conifer roots in one of the three species of *Neolecta* (Neolectales), where they may function as weak parasites (Redhead, 1979; Landvik et al., 2001). In *N. vitellina* the apex of the ascus has an annular sub-apical thickening, and ascospores are forcibly ejected (Landvik et al., 2003). In placing *Neolecta* with the early-diverging lineages of ascomycetes, molecular studies may indicate that filamentous growth and complex fruiting structures may be the ancestral stage (Landvik et al., 2001). Sugiyama et al. (2006), however, note that the Taprhinomycotina are diverse and cannot be delimited based on phenotypic characters. While it might be tempting to suggest that *P. devonicus* is an early member of the Taphrinomycotina, paraphyses are not produced in *Neolecta.*

Yet another hypothesis views *P. devonicus* as being related to the lichens. Termed the Protolichen Hypothesis (Eriksson, O.E., 2005), *P. devonicus* is thought to have evolved from lichen ancestors that were symbionts with microalgae or cyanobacteria. While it is highly probable that ancient saprotrophs lived on dead algal material and some association between a fungus and an alga may have existed, suggesting that *P. devonicus* is a pre-lichen is probably not accurate because lichens are now believed to have evolved from nonlichenized, ascomycetous ancestors after the evolution of land plants and not from "pre-lichens" (e.g. Lücking et al., 2009; Schoch et al., 2009a).

Other groups of ascomycetes that form perithecia similar in some respects to those in *P. devonicus* include the Dothideomycetidae, Chaetothyriomycetidae, and certain other members of the Pezizomycotina (e.g. Lücking et al., 2009). We considered *P. devonicus* in the section dealing with the use of molecular clocks (Chapter 3, p. 30) as it represents an excellent example of the conflict that sometimes arises between molecular data sets and the fossil record.

One of the inherent problems in most areas of paleobiology is the relatively small number of specimens that are available to study. Sometimes only a single specimen is known, which limits not only the number of techniques that can be used to examine it, but also the ultimate interpretation, since there is no opportunity to examine stages of development or discern what the specimen may represent in the life history and biology of the organism. Sometimes there are multiple specimens that can be studied using various techniques, but we still cannot accurately determine the affinities of the fungus (e.g. Stubblefield and Taylor, 1984).

PALAEOSCLEROTIUM PUSILLUM

An excellent example of the dilemma mentioned above is the fossil *Palaeosclerotium pusillum*, described from the matrix of Middle Pennsylvanian coal balls (Rothwell, 1972). The spherical fossils are interpreted as sclerotia that are up to 1.20 mm in diameter and constructed of three distinct zones of tissue. The central region consists of branched and interlaced hyphae, with some hyphae characterized by short, bud-like branches (Figure 8.9). Hyphae are up to 4 μm in diameter and possess septa at branch points. In some specimens the central area is devoid of hyphae, a condition interpreted to be a result of preservation. Others contain small cavities (40 μm) in the central region that contain amorphous material. The

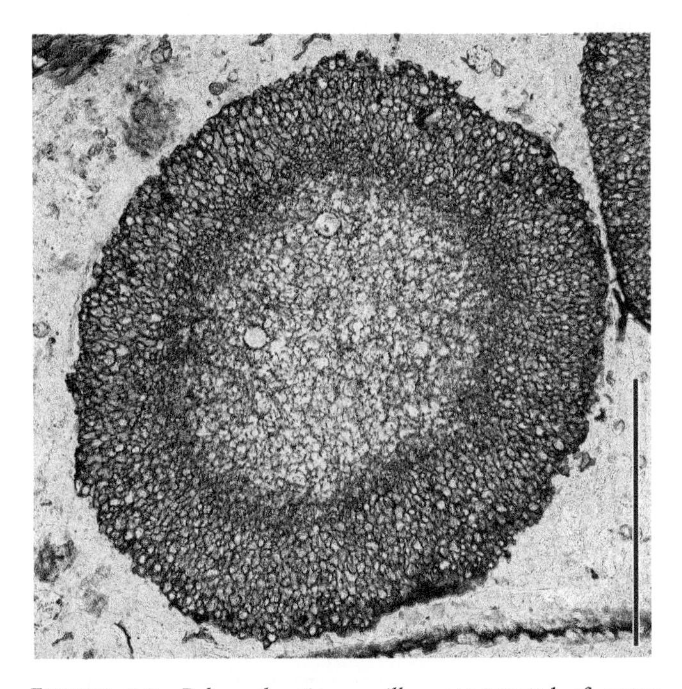

FIGURE 8.9 *Palaeosclerotium pusillum* constructed of outer pseudoparenchymatous zone and central region containing branched, septate hyphae. Middle Pennsylvanian, Illinois, USA. Bar = 400 μm.

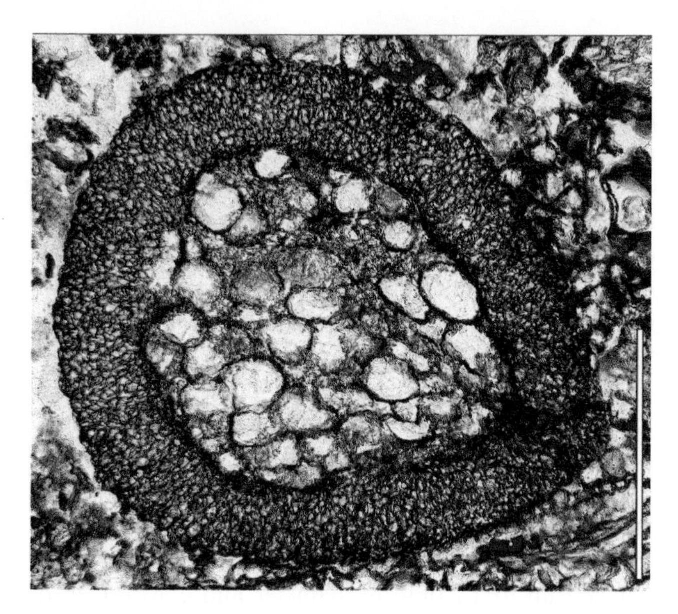

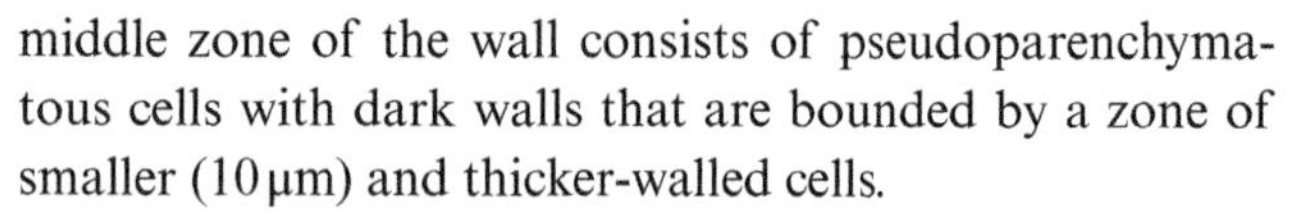

FIGURE 8.10 Central region of *Palaeosclerotium pusillum* containing spore-like bodies. Middle Pennsylvanian, Illinois, USA. Bar = 300µm.

FIGURE 8.11 David J. McLaughlin.

middle zone of the wall consists of pseudoparenchymatous cells with dark walls that are bounded by a zone of smaller (10µm) and thicker-walled cells.

The discovery of additional specimens of *P. pusillum* provided more information about the organization and structure of this interesting fossil (Dennis, 1976). It was suggested that the individual sclerotia each represented a cleistothecium that contained asci with between four and eight ascospores (Figure 8.10). This fossil is especially interesting as some of the hyphae have simple clamp connections attached to other hyphae that possess complex septal pores (dolipores). This interesting combination of features resulted in *P. pusillum* being interpreted as a fungus that was perhaps intermediate between an ascomycete (asci and ascospores) and a basidiomycete (hyphae with clamp connections) (Dennis, 1976). An additional interpretation regarding the biological affinities of *P. pusillum* was provided by McLaughlin (1976) (Figure 8.11), who suggested that perhaps the fossil represented more than a single organism. He based his interpretation partly on the nature of the septal pores but also considered whether the hyphae of the fruiting body and those with clamp connections were reliable. The suggestion that the affinities of *P. pusillum* were close to the Eurotiales was supported by Singer (1977), but discounted by Pirozynski and Weresub (1979), who suggested that the fossil represents some type of early dikaryotic fungus that is neither ascomycetous nor basidiomycetous, but rather an intermediate that links basidiomycetes with some type of lichen-like symbiosis. They suggest that the asci described by Dennis (1976) in some specimens may in fact represent cavities in the stroma, and that the spores are like those produced after meiosis. We are unaware of any additional studies or interpretation of this interesting fossil; however, like those before us (e.g. Stubblefield and Taylor, 1988) we believe that new techniques might yield important results as to the biological affinities of these relatively common Pennsylvanian fossils.

"SPOROCARPS"

Historically there have been a number of fossil microscopic structures (between <100µm and 1.5mm in diameter), predominantly of Carboniferous age, that have been termed "sporocarps." They are generally spherical, may possess some degree of ornamentation, and appear isolated or in clusters, especially in permineralized peat. Although their affinities have been variously construed, the presence of internal spheroidal bodies resulted in many of these structures being interpreted as ascomycete cleistothecia (e.g. Hutchinson and Walton, 1953; Stubblefield and Taylor, 1988). Today most of these structures have been reinterpreted as zygomycetes, and a full description of them may be found in that section (Chapter 6, p. 88–89).

MESOZOIC AND CENOZOIC ASCOMYCETES

TAXA *INCERTAE SEDIS*

A number of reports of Mesozoic and Cenozoic fungi have suggested affinities with some type of ascomycete but the precise systematic placement could not be determined based on the specimens at hand. One of these is *Onakawananus varitas*, a form used for loosely arranged branching hyphae that form mycelial masses in lignites from what are interpreted as Cretaceous sediments of Canada (Radforth, 1958). The fungi were suggested to be some type of ascomycete associated with conifer wood, but this assignment remains equivocal.

Ascodesmisites malayensis is the name of an Eocene ascomycete that is believed to have affinities within the Pezizomycotina (Trivedi et al., 1973). In *A. malayensis* the specimens consist of septate hyphae bearing conidia on short conidiophores. The authors suggested that within the fossil are both anamorph and teleomorph stages, the former closely associated with the extant genus *Desmidiospora* (e.g. Thaxter, 1891), with *Ascodesmis* as the sexual stage (e.g. Seaver 1916). The absence of asci, however, led Korf (1977) to note that the specimens are not convincing and should not have been assigned to the Ascomycota.

Rossellinia congregata is the name used for small, millimeter-sized opaque spots or pustules arising from specimens of Oligocene wood (Engelhardt, 1887), but the taxon has also been reported from the Miocene (e.g. Skirgiełło, 1961). The structures are flat at the point of attachment with the host and then slightly tapering until they become truncated with a flat surface. The distal surface appears partially opened and is regarded as a perithecium with slightly elongate spores.

Cribrites aureus is a flattened shield-like fungus of pentagonal- to hexagonal-shaped cells of similar size, with each cell containing a single pore (Lange, 1978a). The pores appear to be arranged in a radial pattern. Specimens were recovered from middle–upper Eocene sediments from Australia. What have been termed manginuloid hyphae are also present in this assemblage.

Another fossil fungus that has been assigned to the Pezizomycotina is *Diplodites sweetii* from the Upper Cretaceous Deccan Intertrappean Beds, India (Kalgutkar et al., 1993). The specimens occur in permineralized fruits of *Viracarpon* (angiosperm) in the form of mycelia and fruiting bodies. Hyphae are septate, smooth, and variously branched. Pycnidia are globose, up to 108 µm in diameter, and ostiolate; they may appear singly or in small clusters and are often surrounded by stromatic tissue. Two types of conidia are present that appear to correlate with the size and shape of the pycnidia; the two-celled conidia are up to 18 µm long, while the single cell forms may be 7 µm long. Other fossils assigned to *Diplodites* are known based on pycnidia (e.g. *D. sahnii*; Singhai, 1974) or two-celled spores arising from branched conidiophores, such as *D. mohgaoensis* (Kalgutkar et al., 1993; originally *Botryodiplodia mohgaoensis*, Barlinge and Paradkar, 1982), while in others conidia are unbranched (*D. rodei*). *Diplodites* specimens are typically compared to modern forms of *Sphaeropsis* (= *Diplodia*), a pathogen that is common in coniferous forests, especially pines, around the world (e.g. Stanosz et al., 1997). Infection typically occurs at the tip of needles, or from hyphal aggregates that enter the stem through the anticlinal walls (e.g. Chou, 1978).

A similar fossil form, *Palaeodiplodites yezoensis*, was described associated with a bisexual cone of *Cycadeoidella* (a bennettitalean) from the Upper Cretaceous (middle Turonian) of Japan (Watanabe et al., 1999). This species is characterized by acervuli that are up to 500 µm in diameter and have ellipsoidal conidia with a broad base and thick septa. Two additional ascomycetes that have been reported from the Cretaceous of Japan are *Pleosporites shiraianus* (Suzuki, 1910) and *Petrosphaeria japonica* (Stopes and Fujii, 1911). Other fossils attributed to *Diplodia* (*Sphaeropsis*) have been reported from permineralized chert from the Deccan Intertrappean beds of India (Mahabalé, 1968), and from a monocot axis from the same strata (Varadpande and Sampath, 1981). The permineralized specimens of *D. intertrappea* have two-celled spores up to 13 µm long, and thick-walled, oval pycnidia. Dispersed fossils resembling *Sphaeropsis* perithecia have been reported from a number of Cenozoic sediments, including the Eocene of Jamaica (Germeraad, 1979) and Miocene of India (Singh and Chauhan, 2008),

The name *Pezizites* was used for cup-shaped apothecia from the Miocene (Göppert and Berendt, 1845) and Oligocene of Germany (Ludwig, 1859) that superficially resemble the modern genus *Peziza*, a polyphyletic genus of saprotrophic cup fungi (Hansen et al., 2001, 2002). The fossils are associated with decayed plant material and are up to 1.0 mm in diameter. *Pezizasporites taiwanensis* is used for elliptical ascospores from the Miocene of Taiwan that are up to 20 µm long (Huang, 1981).

While the majority of ascomycotan fungi that have been reported from the fossil record are represented as compression specimens on leaves, or are recovered in palynological samples (e.g. Ambwani, 1982), there are others preserved in woody tissues. One of these occurs in silicified dicot wood from the Cretaceous Deccan Intertrappean beds of India (Patil and Singh, 1974). What are termed ostiolate pycnidia occur in subepidermal tissue and have chains of two to four conidiophores. Spores are characterized by swollen ends. The specimens are interpreted as being similar to *Sirophoma* (anamorphic member in the Pezizomycotina; Höhnel, 1917). In other wood types from cherts of the Deccan Intertrappean beds, fungal spores occur in aggregations like those in some extant hyphomycetes (e.g. Shete and Kulkarni, 1977).

Another interesting fungus-like structure that occurs in the Late Cretaceous of the Netherlands is *Geleenites fascinus*, originally attributed to a megaspore with a series of spines radiating from the top of the structure (Dijkstra, 1949). Subsequent examination (Jansonius et al., 1981) indicates that *Geleenites* is a perithecium with perithecial hairs extending from the distal end much like those in the extant fungus *Ascotricha xylina* (Xylariaceae) (Roberts et al., 1984).

Birsiomyces pterophylli is the name used for slightly elongated asci from the Upper Triassic of Switzerland (Neuewelt near Basel), some with ascospores and microconidia, associated with the cuticle of *Pterophyllum longifolium* leaves (Schaarschmidt, 1966). Especially interesting in this study is the fact that sections were prepared of the cuticle so that some details of the asci could be resolved.

Felix (1894) described *Perisporiacites larundae* as a fungus that occurred in the tracheids of *Taenioxylon* (?Fabaceae) wood from the Eocene. The fossils consist of spheroidal to elliptical bodies interpreted as perithecia and illustrated with an irregular surface pattern. Other fossils assigned to *Perisporiacites* have been reported from the Miocene of Argentina (García-Massini et al., 2004), the Deccan Intertrappean beds of India as *P. varians* (Sahni and Rao, 1943), the Cretaceous of China as *P. shaheziensis* (Zheng and Zhang, 1986), and from the Cretaceous of the Russian Far East as *P. zamiophylli* (Krassilov, 1967). Together with *P. varians*, which is characterized by more or less spherical opaque bodies of a dark color, Sahni and Rao (1943) also described *Palaeosordaria lagena* for flask-shaped fruiting bodies. *Epicoccum deccanensis* is the name used for spore/conidia-like bodies that occur in masses in the vessels of fossil wood of *Barringtonioxylon* from the Cretaceous–Paleocene Deccan Intertrappean beds of India (Srivastava et al., 2009). Associated hyphae are septate, frequently branched, and 3–5 μm in diameter. Conidia (spores) are circular and 10–30 μm in diameter; young conidia are without septation, but mature ones are multicellular (dictyoconidia) and have a funnel-shaped base and attachment scar.

Xylomites cycadeoideae is suggested to represent stromatic masses in a Cretaceous bennettitalean cone (Chrysler and Haenseler, 1936), but all that is preserved are septate hyphae approximately 3 μm in diameter. In the absence of any spores or other diagnostic features, the systematic placement of this fungus remains equivocal.

There are some reports of fossil ascomycetes for which there is sufficient information to place the fossils with a phylogenetic context based on recent phylogenetic classifications. In the following sections we attempt to include some of the fossils within classes and orders based on suggestions of the original authors or our interpretations of the fossils. Some of these judgments will ultimately change as ideas about the phylogenetic position of living fungi are modified and better resolved, and when new information becomes available about the fossils. Nevertheless, some fossils contain morphological features that can be useful in phylogenetics because they can be used in addressing character evolution.

PEZIZOMYCOTINA

The subphylum Pezizomycotina is sometimes termed the Euascomycetes or filamentous ascomycetes and is characterized by septate hyphae, each with a specialized pore and Woronin bodies that can seal the pore when the hypha is damaged. They are widespread ecologically and are characterized by morphologically distinct and complex fruiting bodies. Members of this paraphyletic group produce ascogenous hyphae. Among the ten classes currently recognized, the Pezizomycetes (operculate asci) and Orbiliomycetes (non-poricidal asci) are thought to be early diverging groups (Spatafora et al., 2006; Schoch et al., 2009a; Kumar et al., 2012).

Eurotiomycetes

These fungi are sometimes referred to as plectomycetes or cleistothecial ascomycetes, and the class conservatively contains more than 3000 species. Two subclasses are recognized: the Eurotiomycetidae, which produce enclosed ascomata with prototunicate asci, and the Chaetothyriomycetidae, which form ascomata with an opening like that in the Dothideomycetes or

FIGURE 8.12 Alexander R. Schmidt.

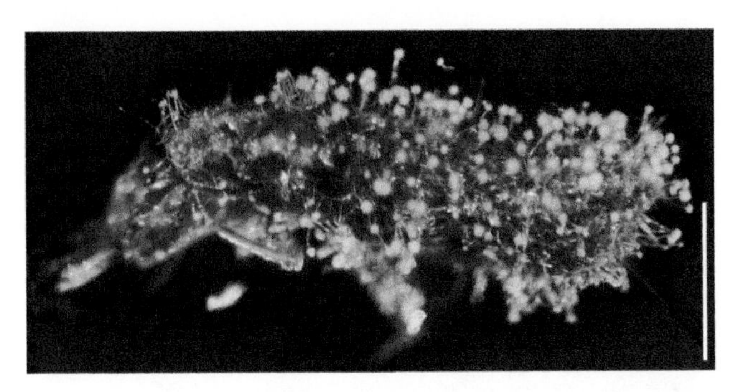

FIGURE 8.13 Numerous conidial heads of *Aspergillus collembolorum* attached to the surface of a springtail arthropod. Eocene amber, Russia. Bar = 0.5 mm. (Courtesy A.R. Schmidt.)

Sordariomycetes (Geiser et al., 2006; Schoch et al., 2006). Included in the Eurotiomycetidae is the genus *Aspergillus* (Trichocomaceae), common molds that can be found in a variety of habitats.

FOSSIL EUROTIOMYCETES. Interestingly, *Aspergillus* has been reported as a fossil encased in amber from the Palo Alto mine, Dominican Republic (Thomas and Poinar, 1988), which is probably lower Miocene (Burdigalian), or 20–15 Ma according to Iturralde-Vinent and MacPhee (1996) and Poinar (2010). The fossil specimens consist of smooth conidiophores up to 2.0 mm long that terminate in a head; conidia are approximately 8 μm in diameter. The fossil is included in an extant taxon, *A. janus*, because all of the morphological features that are preserved in the fossil are identical to modern forms. This systematic placement suggests that stasis, at least for this species, has lasted for approximately 40 myr. Another report of *Aspergillus* comes from Eocene Baltic amber, this fungus, attached to a springtail (Collembola), consists of hyphae and clusters of conidiophores (Dörfelt and Schmidt, 2005; Figure 8.12). Conidial heads of *A. collembolorum* produce radial chains of conidia 35–70 μm in diameter (Figure 8.13). The nature of the resin inclusion suggests that the fungus may have penetrated and parasitized the living organism and that sporulation by the fungus may have taken place after being covered by liquid resin (Dörfelt and Schmidt, 2005). *Aspergillites torulosus* is a name used for chains of up to five conidial spores, each circular and ranging from 13 to 17 μm in diameter, from the Cenozoic of Malaysia (Trivedi and Verma, 1970). Kalgutkar and Jansonius (2000), however, regard the spore type *Cercosporites* (Salmon, 1903) as being a more appropriate designation than *Aspergillites*. Species of this spore range from the Paleogene to the Neogene. Soomro et al. (2010) described an interesting microfossil from a Paleocene brown coal from Pakistan as *Palaeo-Aspergillus multiseriate*. The specimen consists of a fragment of a hypha that terminates in a dispersed head approximately 15 μm in diameter. Extending from the head are what appear to be chains of conidia between 4.5 and 6 μm in diameter.

Another fossil that has tentatively been aligned with the Eurotiomycetes is *Cryptocolax clarnensis* from the late Eocene Clarno Formation of north central Oregon, USA (Scott, R.A., 1956). This fossil consists of spherical cleistothecia, up to 90 μm in diameter, and was found in dicot wood (?*Castanopsis*); some of the cleistothecia are flattened. The multilayered wall is compact and formed of pseudoparenchymatous to parenchymatous polygonal cells, and within the cleistothecium are several asci, each with eight elongate ascospores that are 10 μm long. Associated with these cleistothecia are chains of up to 18 conidia, some of which are free in the matrix. These spores are up to 12 μm long and associated with septate hyphae 2–5 μm in diameter. In the vessels of another dicot wood specimen (Magnoliaceae) are larger cleistothecia (120 μm) that are included in another species, *C. parvula*, with asci that contain more globose ascospores 7.5 × 6.0 μm (Scott, R.A., 1956). The morphology of the cleistothecium is used to suggest affinities with the extant genus *Cephalotheca*, a saprotroph that

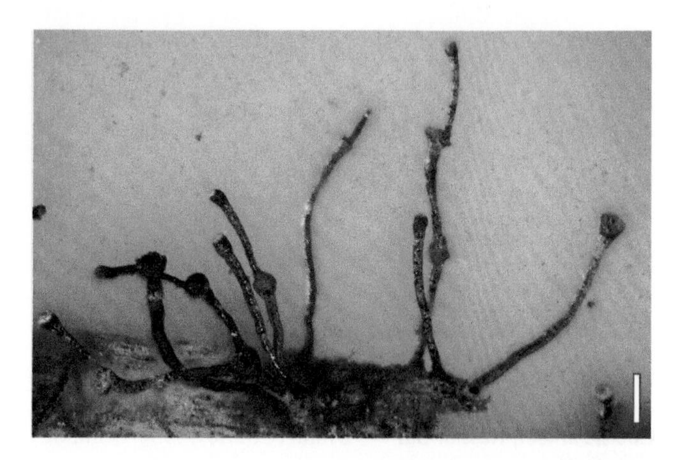

FIGURE 8.14 Fossil *Chaenothecopsis* sp. showing erect ascomata at varying developmental stages. Eocene Baltic amber, Germany. Bar = 250 μm. (Courtesy A.R. Schmidt.)

grows on wood and is included in the Sordariales (e.g. Suh and Blackwell, 1999; Huhndorf et al., 2004).

A new extant species of *Chaenothecopsis* (Mycocaliciaceae), found growing on exudates from the conifer *Cunninghamia,* has been compared to similar forms from Eocene Baltic amber (Tuovila et al., 2013). The fossil specimens consist of stipitate ascomata, approximately 1.0 mm long, that are in various stages of development (Rikkinen and Poinar, 2000) (Figure 8.14). Ascospores are dicellate and elliptical. Other specimens from amber have ascomata that do not branch. In *C. bitterfeldensis* (Oligocene–Miocene boundary) the fruiting body is morphologically similar to modern species (Figure 8.15), including the germination of the ascospores (Rikkinen and Poinar, 2000) (Figure 8.16). One difference is that the elliptical spores in the fossil are much larger (up to 16 μm long) than those of resinicolous extant species of *Chaenothecopsis*. Modern species include those that are parasites of algae and lichens, saprotrophs on dung, wood, and exudates of certain plants such as resins of conifers (e.g. Titov and Tibell, 1993; Tuovila et al., 2011; Messuti et al., 2012). The sequence of resin deposition in the fossil suggests that in this instance some of the fungi actually were growing on the resin prior to additional resin accumulation that entrapped them. Another important contribution of this study is that it provides an opportunity to consider the early diversification of species of *Chaenothecopsis* in relationship to the production of plant resins.

Petrified wood from the Eocene La Meseta Formation of Seymour Island, Antarctica, provides information about a fungal association in the trunk of an araucarian tree, *Araucaria marenssii* (Cantrill and Poole, 2005). Extending throughout the pith region and inner surface of the wood cylinder are numerous branched, septate hyphae and flask-shaped, ostiolate perithecia, each up to 220 μm in diameter. The majority of the perithecia are empty; however, a few contain elliptical, bicelled spores, some approximately 20 μm long. The distribution of the fungus in and on the trunk suggests that fungal degradation was initiated soon after the tree had fallen in the forest ecosystem. While the primary focus of this study was the description of an Eocene species of *Araucaria*, the morphology and distribution of the fungus was carefully described and illustrated, and this represents an important source of information about saprotrophs from high polar paleolatitudes.

FIGURE 8.15 *Chaenothecopsis bitterfeldensis* fruiting body. Eocene amber, Germany. Bar = 100 μm. (Courtesy A. R. Schmidt.)

FIGURE 8.16 Jouko Rikkinen.

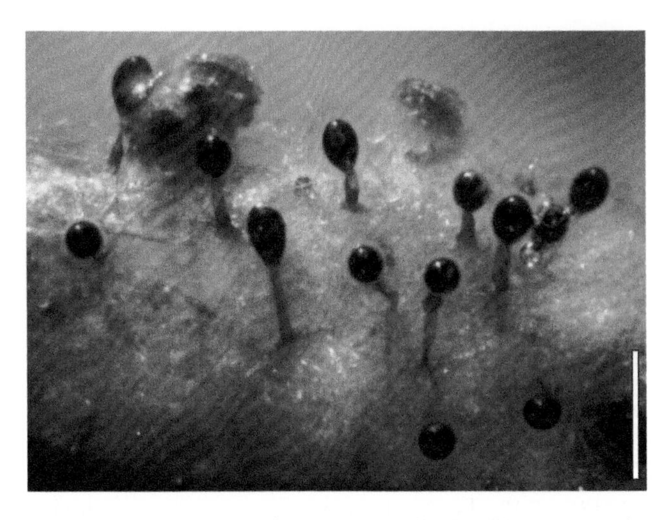

FIGURE 8.17 Colony of *Monotosporella doerfeltii* anamorphs. Eocene amber, India. Bar = 50 µm. (Courtesy A.R. Schmidt.)

SORDARIOMYCETES

Included in this group are more than 600 genera of fungi, including most of the non-lichenized ascomycetes. Most members are characterized by perithecial (more rarely cleistothecial) ascomata and inoperculate unitunicate or prototunicate asci containing eight ascospores (Zhang et al., 2006). Historically this group has been referred to as the Pyrenomycetes. They are present in most ecosystems as endophytes and pathogens of both plants and animals, and as saprotrophs involved in decomposition and nutrient cycling. Various members in one genus, *Sordaria*, have become important model organisms in genetics and developmental biology (e.g. Engh et al., 2010).

FOSSIL SORDARIOMYCETES. There have been numerous reports of fossil fungi in amber from around the world. One form that is believed related to the Sordariomycetes is included in Eocene amber from a lignite mine in India (Sadowski et al., 2012). *Monotosporella doerfeltii* is the name of an anamorph of scattered colonies made up of hyphae up to 3.5 µm wide (Figure 8.17). The specimens occur on a longitudinally ribbed structure of dense mycelia that may represent the teleomorph in the form of a decayed fruticose lichen. Conidiophores may be single or in clusters, occasionally with a single septum. Conidia are developmentally terminal and monoblastic, and mature spores are 1–3 septate and 15 × 28 µm in maximum size. The fossils are compared with extant members of *Monotosporella* that grow on degraded wood or resin flows. Extant species of *Monotosporella* have septation of the conidia and other features that parallel those in the fossil *M. doerfeltii*, including the manner in which conidiophores proliferate. Sadowski et al. (2012) provided an excellent comparison of several morphological features of a number of extant species that are useful in relating the fossil to modern forms.

An interesting fungus, formally described as *Spataporthe taylorii* and included in the Sordariomycetes, was discovered in Lower Cretaceous deposits from Vancouver Island, British Columbia, Canada (Bronson et al., 2013). The specimens are permineralized in marine calcium carbonate and consist of numerous spheroidal perithecia up to 470 µm wide, embedded within the tissues of a conifer leaf (Figure 8.18). Three-dimensional imaging indicates that more than 70 ascocarps were preserved in the incomplete leaf. The wall of each perithecium is two-layered and includes paraphyses at the base that are associated with clavate to tear-drop shaped, inoperculate asci. Well-defined ascospores could not be unequivocally identified in *S. taylorii*, although there are small subunits within the matrix and one ascus that is up to 10 µm in diameter. This fossil is especially important because it demonstrates that there are sufficient characters preserved to make it possible to place the specimen within an order (Diaporthales) and perhaps family (Gnomoniaceae) with a high degree of confidence.

Another fossil fungus assigned to the Sordariomycetes provides the oldest fossil evidence of animal parasitism

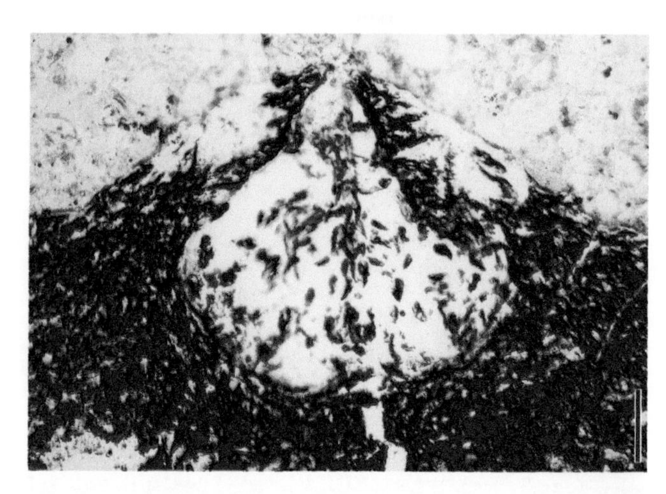

FIGURE 8.18 Longitudinal section through perithecium of *Spataporthe taylorii* with detached asci inside. Lower Cretaceous, Vancouver Island, Canada. Bar = 100 μm. (Courtesy A.W. Bronson and A.M.F. Tomescu.)

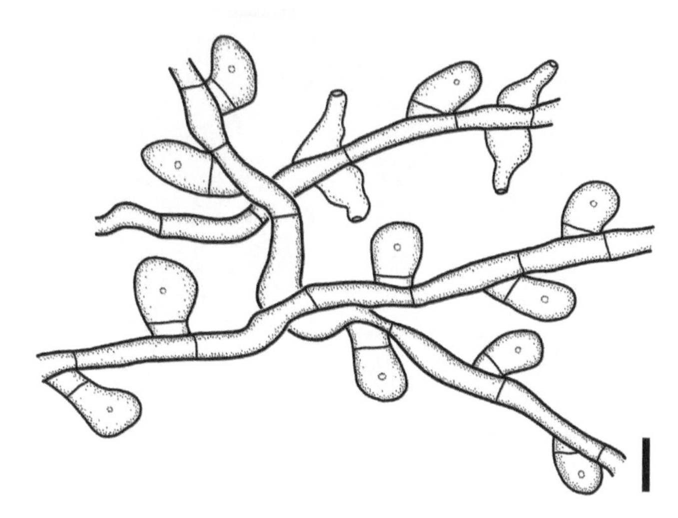

FIGURE 8.19 *Meliola cyclobalanopsina*. Extant. Bar = 10 μm. (Redrawn from Hosagoudar, 2008.)

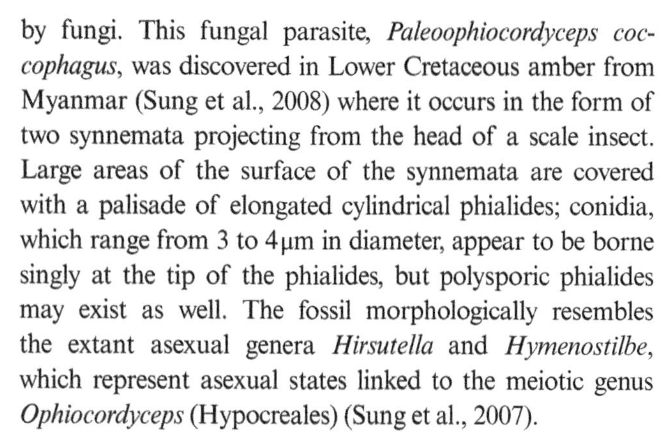

by fungi. This fungal parasite, *Paleoophiocordyceps coccophagus*, was discovered in Lower Cretaceous amber from Myanmar (Sung et al., 2008) where it occurs in the form of two synnemata projecting from the head of a scale insect. Large areas of the surface of the synnemata are covered with a palisade of elongated cylindrical phialides; conidia, which range from 3 to 4 μm in diameter, appear to be borne singly at the tip of the phialides, but polysporic phialides may exist as well. The fossil morphologically resembles the extant asexual genera *Hirsutella* and *Hymenostilbe*, which represent asexual states linked to the meiotic genus *Ophiocordyceps* (Hypocreales) (Sung et al., 2007).

MELIOLALES. This group of Sordariomycetes *incertae sedis* is generally regarded as tropical and subtropical in distribution and is generally absent from warm arid regions. Extant members of the group are obligate biotrophic parasites and are found on mature leaves and young branches of angiosperms; they have also been reported on some gymnosperms and a few ferns (Parbery and Brown, 1986). Termed black mildew fungi, they either occur in discrete colonies ranging up to 1.0 mm in diameter or, in some instances, form multiple, overlapping colonies that cover the leaf surface. These fungi form conidiogenous cells or phialides that are unicellular and bottle shaped. The ascomata or perithecia are globose and may be ostiolate with asci that are unitunicate. Some meliolaceous fruiting bodies are thought to be specific for the site of colonization either on the adaxial or abaxial surface of the leaf. Ascospores are 1–4 septate, some constricted at the point of the septum, and brown at maturity. Some systematists recognize eight genera and more than 1900 species, members of which are known to parasitize 155 host plant families (Saenz and Taylor, 1999); others include nine genera (Hosagoudar and Agarwal, 2008). At least 230 species have been described from China (Song et al., 2003), and new forms are being described from all over the world (e.g. Brazil; Hosagoudar and Archana, 2010; Pinho et al., 2012). Individual colonies of *Meliola*, the largest genus, consist of dark, hyphopodiate hyphae that radiate out from a central group of globose ascomata; setae may be present on some hyphae. Appressoria typically consist of two cells and phialides are unicellular (Figure 8.19).

The Meliolaceae is the largest family of the order Meliolales, and most species occur on tropical trees. Comprehensive reviews of the Meliolales can be found in Hansford (1961, 1963), Hosagoudar (1996, 2003a,b, 2006, 2008), and Hosagoudar et al. (1997) (Figure 8.20). Hosagoudar and Agarwal (2008) have published an identification manual of the order including families of hosts. Two families are recognized: the Meliolaceae has phialidic hyphae, clavate hyphae, and 3–4 septate ascospores, while in the Armatellaceae the hyphae are nonphialidic (Figure 8.21), asci are typically round, and the ascospores have one or two septa (Gäumann, in Eriksson and Hawksworth, 1986). These authors also assembled a listing of 56 extant angiosperm families and one gymnosperm group (Gnetaceae) on which Meliolaceae have been identified, including descriptions of the fungi. Data

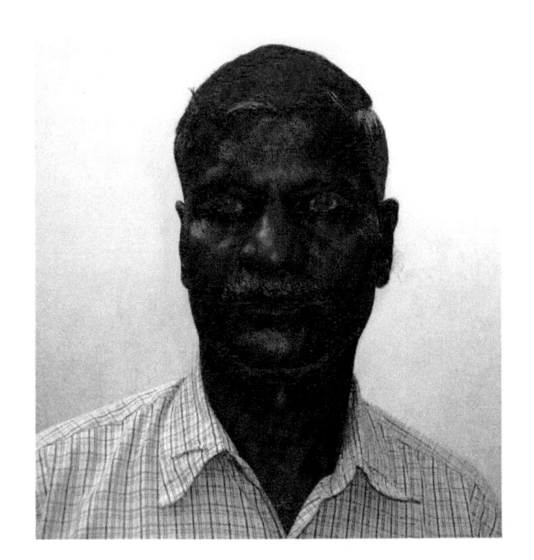

FIGURE 8.20 Virupakshagouda B. Hosagoudar.

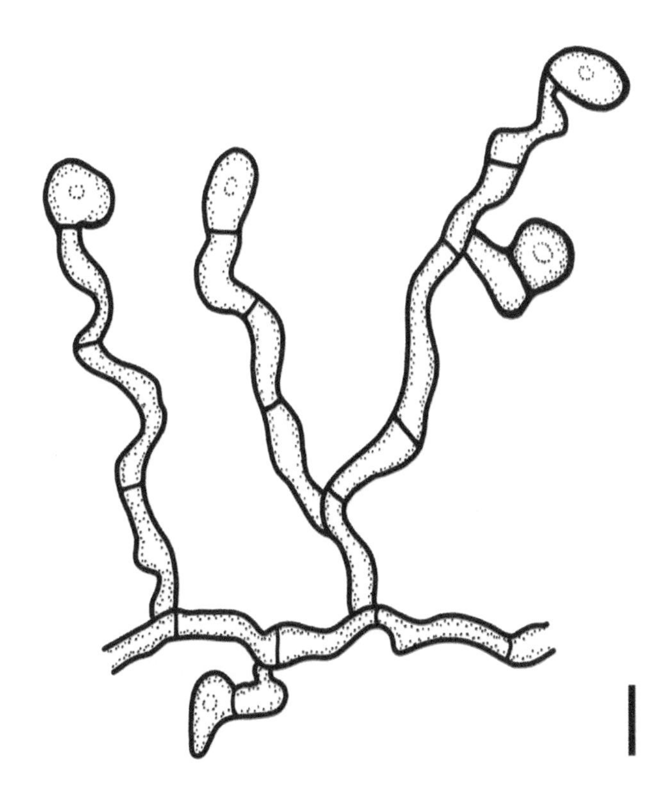

FIGURE 8.21 *Armatella caulicola*. Extant. Bar = 10 μm. (Redrawn from Hosagoudar, 2008.)

of these types will be especially useful for the paleomycologist to see if the fossil forms show any similar pattern of host specificity, but they also provide paleobiogeographic and range information for the fungi.

The number of wall layers in the ascus has been used as a morphological feature in the classification of the Meliolaceae. In some accounts the family was placed with the Pyrenomycetes on the basis of the unitunicate ascus (e.g. Yarwood, 1973), while in others they were associated with the bitunicate Dothideales (e.g. Arx and Müller, 1975; Eriksson, O.E., 1981). Phylogenetic analysis based on small subunit rRNA sequences suggests that the Meliolaceae are perhaps most closely related to the unitunicate Sordariomycetes (Saenz and Taylor, 1999).

One of the phenetic techniques used to describe extant fungi in the Meliolaceae is a numerical code devised by Beeli (1920). This formula involves eight digits that are separated by a decimal point. Those to the left of the decimal point define the features used in delimiting the genus and species (septation of ascospores, color of mature spores, ascocarp shape, hyphopodia position, hyphopodia cell number), whereas to the right of the period are measurements of the asci and ascospores (Farr, 1971). Other characters have subsequently been suggested (e.g. colony shape, position of certain cells) and the number of digits has been expanded to twelve or fifteen (Thomas, J., 2009).

Fossil Meliolales. Fossils of the melioloid type have been reported exclusively from the Cenozoic. One of the earliest reports is that of Colani (1920), who described some epiphyllous "thallophytes" on Cenozoic leaves of *Taxus* from Indochina that appeared to be morphologically similar to *Meliola* (for additional information on epiphytes, see the end of this chapter). Other early specimens were reported by Köck (1939) from Eocene brown coals of the Geisel valley, Germany and the Cenozoic of Poland (e.g. Skirgiełło, 1961). Some of these include septate spores and hyphae with hyphopodia. In other sediments there are extensive colonies on fragments of cuticles in which the leaves are not identified. These specimens are similar to *Meliola* and consist of colonies up to 1.5 mm in diameter together with capitate hyphopodia and spores. This material also included some five-celled spores, each up to 34 μm long (Köck, 1939). Dilcher (1965) suggested that at least superficially some of the features are comparable to forms described from the Eocene of Tennessee (USA); however, since Köck's specimens are not adequately described, a critical comparison cannot be made.

One of the major contributions to megafossil paleobotany that directly impacted the study of fossil fungi is a series of studies on exceptionally preserved angiosperm

FIGURE 8.22 David L. Dilcher.

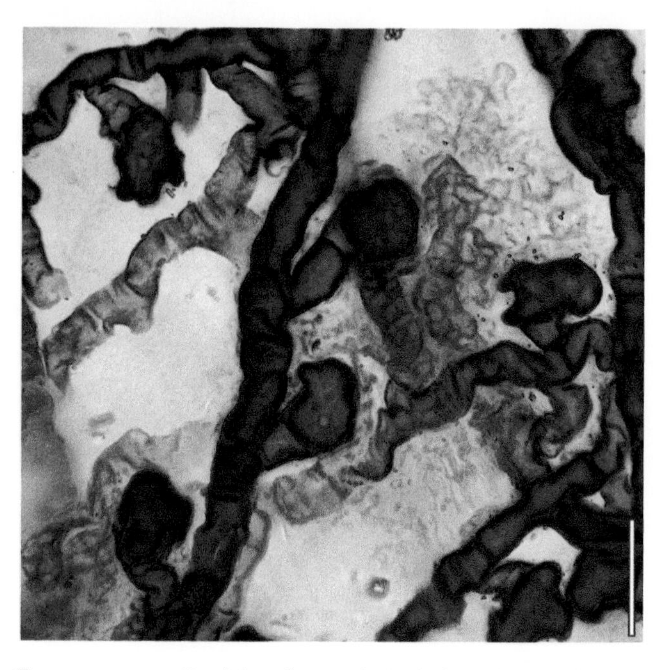

FIGURE 8.23 Partial colony of *Meliola anfracta* showing sinuous nature of hyphal cells and irregularly shaped hyphopodia. Eocene, Tennessee, USA. Bar = 10 μm. (Courtesy Florida Museum of Natural History.)

leaf compressions initiated by David Dilcher (Figure 8.22) in 1965. Although fossil leaves were initially reported from a number of Eocene clay deposits in western Tennessee by Berry (1916), Dilcher "opened up" the clay pits to the scientific community with remarkable studies detailing these early Eocene sites and their floras. The leaves are preserved as both impressions and compressions, with the latter having in some instances complete cuticle envelopes that can be peeled from the rock surface. Such cuticles provide detailed information about the type and distribution of epidermal cells, the nature and spatial arrangement of veins, and the occurrence of specialized epidermal structures such as glands and various types of hairs. Dilcher also recognized and described numerous types of epiphyllous fungi from more than 500 leaves (Dilcher, 1963, 1965). These were placed in five families (Meliolaceae, Microthyriaceae, Micropeltaceae, Tuberculariaceae, Dematiaceae) and included several species nearly identical to modern forms. He also provided a listing of other potential epiphyllous fungi reported from the fossil record. Interestingly, some of the megafossils from the Eocene sites were mummified (e.g. Dilcher, 1969), and thus provided a potential source of new information about possible endophytes in some of the plants. Moreover, Dilcher was able to trace the development of fungal colonies by examining the stages of the fungus on different leaves and then piecing together the sequence to reflect the nearly complete life history of the fossil. In instances where only hyphae were present, and were not associated with reproductive structures, the generic name *Sporidesmium* was used (Dilcher, 1965).

One of the fungal fossils described by Dilcher is *Meliolinites* (*Meliola*) *anfracta*, which consists of colonies up to 3 mm in diameter that are often densely branched. Hyphal cells are sinuous along the lateral walls and up to 9 μm wide; stalk cells are variously angled and hyphopodia are distal (Figure 8.23). The smooth-walled spores are four celled and up to 50 μm long. In another form, *M. spinksii*, capitate hyphopodia arise from the hyphae in a forward and upward position; spores are four septate. Also present in the fossil are mucronate hyphopodia that occur oppositely and are approximately 18 μm long. Interestingly, germinating spores and colonies occur only on the lower surface of the leaf. Some have suggested that caution needs to be considered in assigning those colonies that lack mycelial setae to the Meliolaceae and have used the genus *Meliolinites* instead (e.g. Selkirk, 1975). Selkirk (1975) added several new types from Miocene sediments in New South Wales, Australia. The genus is known from the Eocene to the Miocene.

Meliolinites dilcheri is a well-preserved fossil from the lower Eocene Rockdale Formation of Texas, USA (Daghlian, 1978; Figure 8.24). The site is regarded as an interdistributary backswamp or open lake deposition

FIGURE 8.24 Charles P. Daghlian.

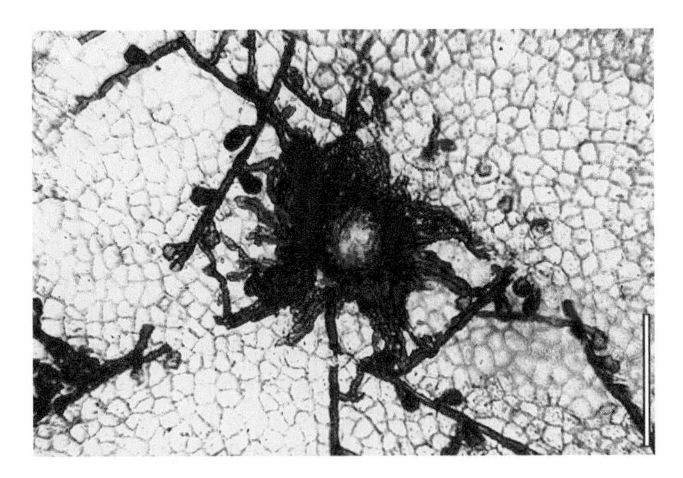

FIGURE 8.25 Colony of *Meliolinites dilcheri* showing hyphal branching and capitate hyphopodia. Eocene, Texas, USA. Bar = 60 μm. (Courtesy D.L. Dilcher.)

in which fragments were deposited in a thin lens of fine-grained clay. The fossils are organized in colonies (up to 2.0 mm in diameter) on the upper surface of leaves and consist of hyphal branches that sometimes alternate with capitate hyphopodia (Figure 8.25). Head cells are about 15 μm in diameter and possess a single haustorial pore. Perithecia are approximately 150 μm in diameter, and like the mycelium, lack setae; spores are five celled. Developmental stages of *M. dilcheri* are similar to those that have been reported from extant *Meliola* (e.g. Ward, 1883; Graff, 1932; Goos and Palm, 1979). The precise affinities of the leaf fragment remain equivocal, although some type of lauraceous plant has been suggested. The host of a Neogene species, *M. siwalika*, is also believed to be lauraceous (Mandal et al., 2011).

An interesting report by Van Geel et al. (2006), which demonstrated the potential of examining multiple host–parasite interactions among organisms in the fossil record, noted the subfossil occurrence of *Isthmospora spinosa*, a hyperparasite of various genera within the Meliolaceae (e.g. see Dubey and Moonnambeth, 2013). Specimens were recovered from a Holocene bog and demonstrated the tripartite relationship between the host *Calluna vulgaris* (heather) and the parasitic fungus *Meliola ellisii*, which is in turn parasitized by *I. spinosa*.

PHYLLACHORALES. This is a small order of tropical ascomycetes that produce perithecia embedded in the tissue of the host, generally within a stroma and protected by a shield-like structure. Most members are obligate parasites that are of economic importance (Silva-Hanlin and Hanlin, 1998). Ascospores are generally aseptate and the ascus has an apical ring (Hawksworth et al., 1995). Molecular data indicate that the order is polyphyletic with some of the genera having associations with the Xylariales, Hypocreales, and Sordariales (Wanderlei-Silva et al., 2003).

Paleoserenomyces allenbyensis is a structurally preserved ascomycete from the Eocene Princeton chert found in leaf tissue of the palm *Uhlia allenbyensis* (Currah et al., 1998). The fungus occurs on both the adaxial and abaxial surface and consists of a stroma approximately 1.0 mm thick which is composed of multiple locules (Figure 8.26) lined with tissue of thin-walled hyphae; an ostiole arises from a rounded papilla. None of the locules contain asci or ascospores. Fungal symptoms of this type are sometimes termed tar spots. The fossil shares similar features with the extant fungus *Serenomyces*, which also forms spots on palm leaves (Barr et al., 1989; Hyde and Cannon, 1999). Present in some of the locules of *P. allenbyensis* are globose ascomata, which are up to 120 μm in diameter and contain asci (some up to 50 μm long) that are interpreted as hyperparasites. Within some locules, asci contain eight uniseriate ascospores. The hyperparasite, formally described as *Cryptodidymosphaerites princetonensis*

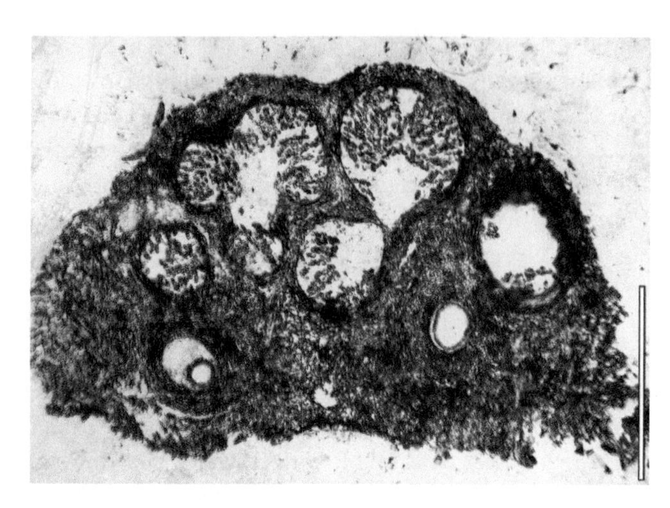

FIGURE 8.26 Section of stroma of *Paleoserenomyces allenbyensis* with endoparasite in locules. Eocene, Canada. Bar = 0.5 mm. (Courtesy R.A. Stockey.)

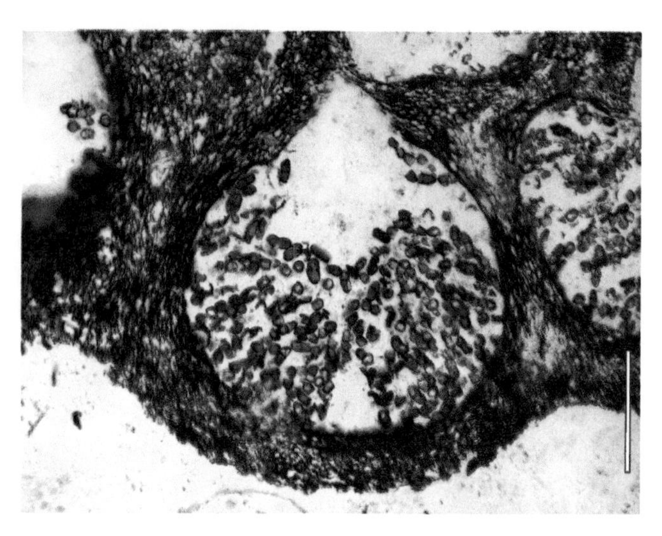

FIGURE 8.27 Section of *Paleoserenomyces allenbyensis* locule with asci of endoparasite *Cryptodidymosphaerites princetonensis* inside. Eocene, Canada. Bar = 100 µm. (Courtesy R.A. Stockey.)

(Figure 8.27) shares a number of morphological characters with *Didymosphaeria*, a plant pathogen of multiple hosts that is included in the Pleosporales (Aptroot, 1995). Pyriform perithecia up to 150 µm in diameter were also reported from the Eocene palm *Latanites* (Fiore, 1932). Asci up to 18 µm long are described as containing once-septate ascospores.

DIAPORTHALES. Included in this order of extant ascomycetes are saprotrophs and a number of plant pathogens, with perhaps chestnut blight (*Cryphonectria parasitica*) being the most recognized, as least in North America (Castlebury et al., 2002). Within the group are more than 90 genera and perhaps 500 species characterized by dark perithecia immersed in a stroma, lack of paraphyses at maturity, and unitunicate asci that are free within the centrum at maturity (e.g. Rossman et al., 2007). The order includes nine families (Rossman et al., 2007). Where known, anamorphs are coelomycetes that produce spores with either one septum or none.

There are numerous reports of leaf spots on fossil leaves that may represent at least some coelomycetes that have modern counterparts within the Diaporthales (e.g. Crous et al., 2012). One of these is *Protocolletotrichum* reported on leaves from the Cretaceous of India (Kar et al., 2004a). The fossil specimens of *P. deccanensis* are preserved on various fragments of cuticle and consist of acervuli, each surrounded by 20–35 unbranched, pointed setae. Morphologically, the fossil shares a number of features with the extant genus *Colletotrichum*, a common pathogen of a wide variety of woody and herbaceous plants, including many that are grown as crops (e.g. Cannon et al., 2012). A rather unusual source of fossil evidence of Diaporthales is dinosaur dung from the Cretaceous Lameta Formation in India (Kar et al., 2004b; Sharma et al., 2005). Macerations of these coprolites have yielded abundant, isolated subcircular *Protocolletotrichum* acervuli bearing setae along the margin. The setae are septate, slightly swollen at the base, and have a pointed tip.

Dothideomycetes

This is the largest and most diverse group of ascomycetes with more than 19,000 species of parasites, saprotrophs, and occasional lichen-forming types (Kirk et al., 2008; Hyde et al., 2013). Ascocarp morphology and structure are variable; they arise from cavities in the stromatic tissue of the host. Fungi in this class have ascolocular development, in which asci develop in already formed pockets or locules, as compared to other ascomycetes that have ascohymenial development, where an ascocarp is formed from coiled hyphae. Asci are bitunicate and fissitunicate (i.e. the ascospores are released by the inner wall turning outward, while the outer wall ruptures); conidia and ascospores are typically septate (Schoch et al., 2006). The group, previously included in the Loculoascomycetes, is currently divided into two subclasses, the Dothideomycetidae and Pleosporomycetidae (see Boehm et al., 2009). Phylogenetic analyses indicate 10–12 orders in which there have been multiple transitions from saprotrophic to plant-associated and lichenized lifestyles, and from terrestrial to aquatic habitats (Schoch et al., 2009b; Suetrong et al., 2009; Ohm et al., 2012). Some of these organisms are parasites of other fungi and animals. Also within the group are rock-inhabiting fungi that are believed to lack sexual reproductive structures and that form slow-growing, compact melanized colonies on nutrient-poor rock surfaces (Ruibal et al., 2009). These fungi appear to have also evolved a strategy of altering their metabolism under changing temperatures (e.g. Tesei et al., 2012).

PLEOSPORALES. The Pleosporales is perhaps the largest order in the Dothideomycetes, and a group characterized by flask-shaped pseudothecia in most instances (e.g. Mugambi and Huhndorf, 2009). The numbers of families in the Pleosporales are variable (e.g. Wang et al., 2007; Kirk et al., 2008) and some authorities include more than 300 genera and 4700 species in the order. Zhang et al. (2009) suggest that morphological characters, including the size, shape, and immersion degree of ascomata, as well as characters of the ostiole, can sometimes be of phylogenetic value. Nutritionally the order includes saprotrophs, parasites, pathogens, epiphytes, and endophytes.

Pteropus brachyphylli is an ascomycete found on leaves of the fossil conifer *Brachyphyllum patens* from the Upper Cretaceous (Maastricht Formation, Valkenburg Member) of Belgium (Van der Ham and Dortangs, 2005). Hypostromata (Figure 8.28) are found in the stomata, but spore-producing fruiting bodies are absent. Structurally, the hypostroma consists of a mass of hyphae that fills the stomatal pit and other cells of the stomatal complex. Other stromata not associated with stomatal complexes consist of thyriothecium-like structures that are up to 400 μm in diameter and may contain two-celled spores, but no asci. Opaque inclusions in cells of the leaf are interpreted as some type of host response. The hypostromata found in the stomata are like those of the extant fungal teleomorph *Phaeocryptopus* (Venturiaceae) (e.g. Stone and Carroll, 1985; Stone, J.K. et al., 2008), a weak parasite described on the leaves of several conifer taxa (Van der Ham and Dortangs, 2005). Structurally preserved perithecia have also been reported on leaves of *Cryptomeriopsis mesozoica*, a conifer from the Upper Cretaceous of Hokkaido, Japan (Suzuki, 1910). This fungus, *Pleosporites shirainus*, is up to 150 μm in diameter and occurs in the hypodermis. Septate hyphae range from 2–5 μm in diameter and within the fruiting body are asci and perhaps paraphyses. No spores were preserved.

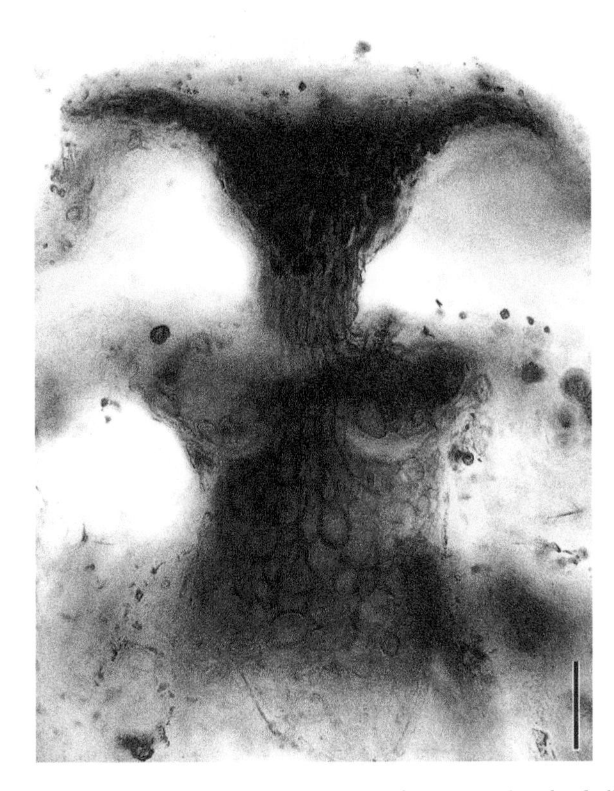

FIGURE 8.28 Section of hypostroma of *Pteropus brachyphylli*. Upper Cretaceous, Belgium. Bar = 20 μm. (Courtesy R.W.J.M. Van der Ham.)

A fossil from British Columbia that shows some features of the Pleosporales is *Margaretbarromyces dictyosporus* (Figure 8.29), a permineralized pseudothecium that consists of uniloculate, pyriform ascomata 390 μm in diameter and 420 μm tall (Mindell et al., 2007). The wall is two layered and pseudoparenchymatous, with a short neck arising from the distal surface; there is a suggestion of periphyses along the inner wall of the ostiolar canal. Hyphae are septate and highly branched. Asci are clavate and arranged basally. Ascospores in *M. dictyosporus* may be uni-biseriate, multiseptate, and up to 90 μm long (Figure 8.30). Chert from the Upper Cretaceous–Paleocene Deccan Intertrappean beds of India containing wood of a dicot, similar to members of the Ebenaceae, contains fungi that are morphologically similar to modern *Phoma* (Chitaley and Patil, 1972). Various types of fungal spores have also been reported from some of these sites (e.g. Chitaley, 1957).

Fossil conidia of the pleosporalean anamorph *Tetraploa* (Tetraplospheriaceae; Hatakeyama et al.,

FIGURE 8.29 Longitudinal section through the ascoma of *Margaretbarromyces dictyosporus* showing the ostiole and ascospores. Eocene, Canada. Bar = 100 µm. (Courtesy R.A. Stockey.)

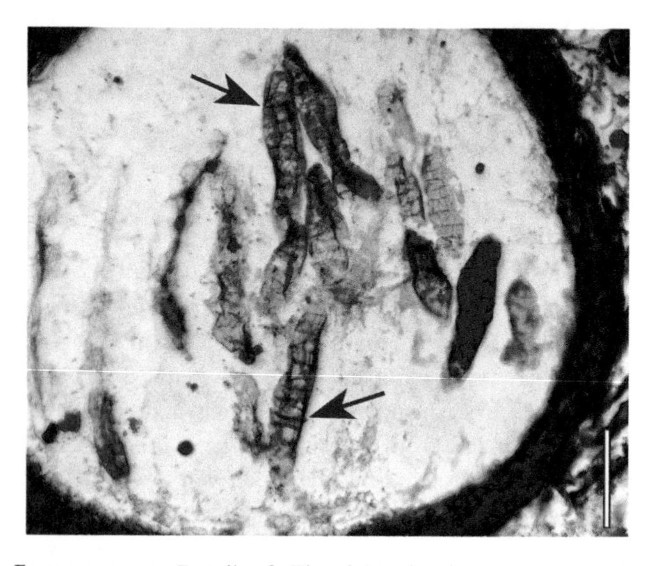

FIGURE 8.30 Detail of Fig. 8.29 showing several septate ascospores (*arrows*) of *Margaretbarromyces dictyosporus.* Eocene, Canada. Bar = 50 µm. (Courtesy R.A. Stockey.)

2005; Tanaka et al., 2009) are definitely known from the Upper Cretaceous and are regularly encountered in Miocene and younger palynological samples worldwide (e.g. Germeraad 1979; Mudie et al., 2002; Samant and Mohabey, 2009; Saxena and Ranhotra, 2009; Worobiec et al., 2009; Karpińska-Kołaczek et al., 2010); in Cenozoic samples, they are associated with abundant pollen of Poaceae. *Tetraploa*-like spores from the upper Miocene of Poland are associated with bamboo pollen (Worobiec et al., 2009). There is also one report from the Permian of India (Jha and Aggarwal, 2011). While most palynologists readily use the name *Tetraploa* for fossil forms, some prefer to assign them to a different genus, *Frasnacritetrus* (e.g. Taugourdeau, 1968; Saxena and Sarkar, 1986). Worobiec et al. (2009) suggested that the occurrence of *T. aristata* in fossil pollen samples indicates a warm temperate climate. Wolf and Nease (1970) reported *Tetraploa* from late Pleistocene sediments in North Carolina (USA), together with another fossil fungus they identified as *Cordana pauciseptata*. This ascomycete, which was discovered on leaves of a member of the Urticaceae, is compared to the extant *C. musae*, a pathogen on banana foliage widely distributed throughout the tropics (Höhnel, 1924). Fossils morphologically similar to *Tetraploa* from the Miocene are known, each with four branches arising from one end and they are interpreted as aquatic forms of certain hyphomycetes like *Ceratosporella* (Kar et al., 2006).

MICROTHYRIACEOUS FUNGI (MICROTHYRIALES). It is impossible to know exactly who recorded the first fungus on a fossil leaf, but the chances are great that this fungus was an ascomycete, and probably some microthyriaceous type. The group has historically represented an order of epiphyllous fungi (i.e. the Microthyriales) that has included various numbers of genera and species of ascomycetes with flattened fruiting bodies (thyriothecia) in which the upper wall cells of most members radiate in parallel arrangement from a central ostiole, and the bitunicate asci contain two-celled ascospores. Today the majority of these fungi live on broad-leaved angiosperms. They are considered to be the best known of all fossil fungi, with the earliest forms recorded from the Jurassic (e.g. García Massini et al., 2012) and Cretaceous (e.g. Martínez, 1968; Alvin and Muir, 1970), and many genera and species reported from the Cenozoic (e.g. Gupta, 1994); some specimens have also been reported from the Permian (Bajpai and Maheshwari, 1987). They are easily recognized because of their peltate morphology and thus are routinely identified in palynological samples (Elsik, 1978). Fungi in this group are often referred to as foliar epiphytes because it is not certain whether they derive nutrients from the host or receive all nutrients from rain runoff or airfall. Today these fungi are found in rainforest areas on flowering plants, but some have been reported from conifer needles (Pirozynski and Shoemaker, 1970). The growth of microthyriaceous fungi is specifically related to moisture rather than temperature (e.g. Tripathi and Saxena, 2005). Kirk et al. (2008) suggested that perhaps the group is polyphyletic and this hypothesis is supported by a detailed molecular study of the Microthyriaceae in which 11 of the 50 genera were moved into other families (Wu et al., 2011). Historically, most fossil microthyriaceous fungi have been included within the Ascomycetes (e.g. Ferreira et al., 2005), and some have been related to extant forms.

Hyphae are sometimes the only record of epiphyllous fungi on the surface of a fossil leaf. These sometimes ramify over the leaf surface sparsely and demonstrate dichotomous branching; secondary branching is lateral. The hyphae of these forms rarely tangle and the septa are thick and dark. In some preparations the lateral walls of the hyphae are often hyaline so that only the septa are visible. These remains of fossil epiphyllous fungi are often termed manginuloid hyphae and have been organized in groups that can be compared with modern equivalents (Lange, 1978a). Such studies provide another dimension to determining the identity of fossil fungi and may be useful in helping to trace environmental changes in the past based on shifting patterns of fungal colonization of host taxa.

For example, *Manginula* is a genus that includes extant and fossil forms (e.g. Arnaud, 1918). The genus was later emended by Lange (1969) to include *Shortensis* (Dilcher, 1965). Subsequently, Selkirk (1972) further revised the taxonomy by placing *Manginula* (*Shortensis*) *memorabilis* into *Vizella* (Figure 8.31) and some Eocene species of *Manginula* into *Entopeltacites*, a form genus in which no reproductive spores were known (Phipps, 2007). In some reports similar fungi were placed in the Vizellaceae (Lange, 1978a). Currently *Entopeltacites* includes five species from the Southern Hemisphere: *E. osbornii* and *E. maegdefravii* from the Eocene of Australia (Lange, 1969) and *E. attenuatus, E. irregularis*, and *E. cooksoniae* from the Miocene of the same continent (Selkirk, 1972; Phipps, 2007 [Figure 8.32]). From North America, specimens of *E. remberi* come from the famous early Miocene Clarkia leaf locality (Priest River Basalt) in Idaho, USA and consist of extensive webs of hyphae (Figure 8.33) spread across the leaf surface of *Persea pseudocarolinensis* (Phipps, 2007). The septate hyphae are approximately 5 μm in diameter, with single-celled, pyriform hyphopodia approximately 7 μm in diameter. The reproductive structures are scattered on the surface and consist of circular, ostiolate empty units. No specimens showed evidence of ascomata, pycnidia, or spores. The fossils are similar to several extant genera in the Vizellaceae (e.g. the plant parasite *Vizella*), a group within the extant Dothideomycetes that is difficult to classify (e.g. Swart, 1975b).

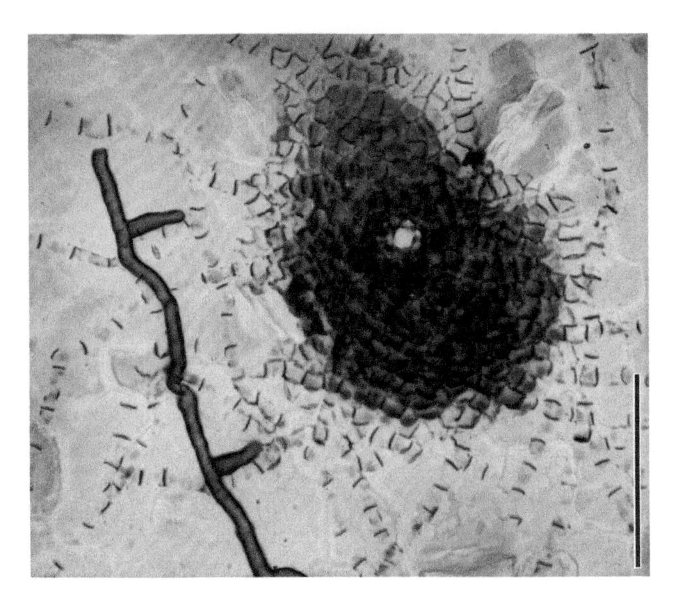

FIGURE 8.31 Ascoma of *Vizella memorabilis* on angiosperm leaf showing position of ostiole. Eocene, Tennessee, USA. Bar = 50 μm. (Courtesy Florida Museum of Natural History.)

FIGURE 8.32 Carlie J. Phipps.

It is interesting to note that 80% of the leaves from the Miocene Clarkia site have epiphyllous fungi present (Phipps and Rember, 2004; Figure 8.34). The presence of these epiphyllous fungi has been interpreted as indicating a warm temperate climate with annual precipitation greater than 100 cm per year (Lange, 1978a; Phipps and Rember, 2004; Phipps, 2007). This is confirmed by both geologic evidence and environmental variables based on the

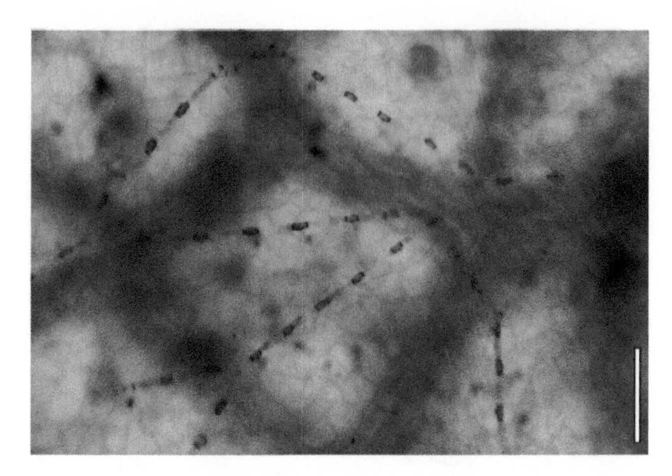

FIGURE 8.33 Epiphyllous hyphae of *Entopeltacites remberi*. Miocene, Idaho, USA. Bar = 50 µm. (Courtesy C.J. Phipps.)

FIGURE 8.34 William C. Rember.

existence of extant members of the Vizellaceae. Specimens of an extant species of *Vizella* (*V. metrosideri*) share features with other genera, making the assignment of fossil fungi in this group especially difficult (Johnston, 2000).

There are numerous reports of microthyriaceous fungi for which details of the fungus and the host are known. One of these is *Mariusia andegavensis*, described on the cuticle of *Frenelopsis* (conifer; Cheirolepidiaceae) from the Upper Cretaceous of France (Pons and Boureau, 1977). The ascoma is ostiolate, up to 100 µm in diameter, and with a fimbriate margin; ascospores are up to 10 µm long and three-septate. Infected (with *M. andegavensis*) and uninfected shoots of *F. alata* collected from the Cenomanian (Cretaceous) of the Czech Republic were analyzed by gas chromatography–mass spectrometry to see if there were biomarkers that could be used to identify activities of fossil fungi (Nguyen Tu et al., 2000). In addition to the presence of common compounds in both the uninfected and infected fossil leaves, hydroxysuccinic acid and benzoic compounds in the infected fossil leaves were interpreted as degradation products of lignin, suggesting that these fungi were able to degrade lignin. Since the fossils were compression specimens it is also possible that contamination from other rock layers may have altered the chemical profiles.

Specimens such as *Asterinites colombiensis* from the Paleocene of Colombia are defined principally on the basis of hyphal morphology, including hyphopodia (e.g. Doubinger and Pons, 1973). Another form that was found on an Early Cretaceous leaf of *Frenelopsis* is the micropeltaceous *Stomiopeltites cretacea* (Alvin and Muir, 1970). The specimens come from the Isle of Wight, UK. This species is characterized by a radiating weft of epicuticular anastomosing hyphae that are up to 3 µm in diameter. Thyriothecia are scattered on the surface, each about 250 µm in diameter and with a central ostiole. What are interpreted as pycnidia are 25–40 µm in diameter and constructed of a plectenchymatous wall. Spores are absent. Another younger species that comes from the Eocene of Tennessee, USA is *S. plectilis* (Dilcher, 1965). The specimens occur on the lower (abaxial) epidermis of *Sapindus* leaves and include fruiting bodies up to 210 µm in diameter. The plectenchymatous hyphae are sinuous and produce large cells of the fruiting body wall; asci and spores were not identified. A geologically younger form has been reported on *Betula*, *Smilax*, and *Zizyphoides* leaves from the Miocene of the Clarkia site in Idaho, USA (Phipps and Rember, 2004). Specimens of *Stomiopeltites amorphos* have pseudoparenchymatous perithecia that are up to 180 µm in diameter (Figure 8.35). The ostiole is slightly elevated and may be large (9–21 µm in diameter). A number of authors have suggested that *Stomiopeltites* is closely related to the extant

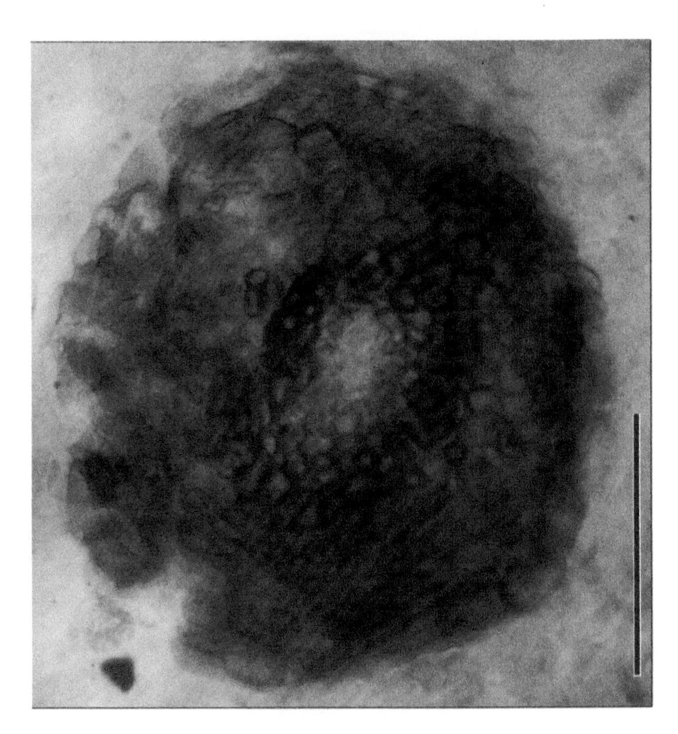

FIGURE 8.35 Surface view of *Stomiopeltites* sp. perithecium. Miocene, Idaho, USA. Bar = 50 μm. (Courtesy C. J. Phipps.)

FIGURE 8.36 Oswald Heer.

genus *Stomiopeltis* (e.g. Dilcher, 1965; Luttrell, 1973; Phipps and Rember, 2004); however, it should be noted that most of the diagnostic characters of the extant species are based on features of the ascus, a structure in the fossil that may be either difficult to resolve or lost during preparations (Phipps and Rember, 2004). The fact that *S. plectilis* (Eocene) is reported on the abaxial epidermis of *Sapindus* leaves (Dilcher, 1965; Kalgutkar and Jansonius, 2000) parallels at least one extant member (*S. aspersa*) and no doubt others that are limited to the abaxial surface of leaves of Lauraceae from India (Luttrell, 1946). *Muyocopron* is another extant ascomycete that has some similarities to *Stomiopeltites*. Conspicuous shield-like and nearly round structures reminiscent of *Stomiopeltites*-like thyriothecia have also been described from the Jurassic San Agustín hot spring ecosystem in Argentina (García Massini et al., 2012).

Microthyriaceae. This family includes foliar epiphytes, biotrophs, and saprotrophs of leaves of numerous plants; the fungi may or may not have a superficial mycelium, and some of them form hyphopodia. Thyriothecia are small (<200 μm in diameter), flat, and cells on the upper surface radiate out from a central ostiole. The manner in which the thyriothecia split or dissolve to release spores is also an important character of some forms. Pseudoparaphyses are arranged around the edge of the ascomata. Asci are bitunicate, elongate, and produce eight spores. Ascospores are one-septate and may have some type of appendages extending from the surface. Estimates of the number of extant genera range from seven (Wu et al., 2011) to 54 (Kirk et al., 2008), and from 220 to 278 species; all are worldwide in distribution.

Fossil Microthyriaceae. Some of the earliest forms of microthyriaceous fungi were reported on angiosperm leaves and, based on overall morphology, were compared with modern epiphyllous types. Many were reported as small pustules, termed perithecia or sporangia, and given names such as *Xylomites* (e.g. Unger, 1841; Ettingshausen, 1851; Heer, 1876 [Figure 8.36]). This taxon is probably the most species-rich genus used for fossil putative fungal remains. It was introduced by Unger (1841) for flat and more or less circular spots or specks on impressed/compressed leaves (e.g. see Figure 2.2) that, typically, are lighter in the center and darker towards the margin. Multiple forms have been added to the genus since Unger's initial description; most come from the Cretaceous and Cenozoic, but a few have also been described from the Triassic and Carboniferous (see Pia, 1927). Due to the vague circumscription of *Xylomites*, the genus has over time become an excellent example of a "waste-basket or trash taxon." One interesting form, *X. intermedius* from the Triassic of Sweden (Nathorst, 1879; Figure 8.37), has

FIGURE 8.37 Alfred G. Nathorst.

FIGURE 8.38 Isabel C. Cookson.

been suggested to have affinities with the Phacidiales (discomycetes). An equally uncertain genus used in the past for specks and spots on fossil wood or bark that were believed to be fungal in nature is *Sphaerites* (Figure 2.2). With regard to the usability of these taxa, Knowlton (1923, p. 143) noted that "all these leaf spot fungi are so similar that without the essential organs it is almost impossible to be certain of identification."

The number of fossil microthyriaceous ascocarps is large and, because of certain characteristic features, many of the fossils have been persuasively related to modern families and even genera (e.g. Doi and Uemura, 1985). They range from the Early Cretaceous to the Holocene from around the world. Many are well defined and described in detail; others are reported as being present on fossil leaves but are not the primary focus of the publication.

Other possible microthyriaceous fossils have been reported from Oligocene brown coals of Germany (Potonié and Venitz, 1934; Köck, 1939; Kirchheimer, 1942; Frantz, 1959), the Pliocene of central Europe (Altehenger, 1959), and the Miocene of North America (e.g. Sherwood-Pike, 1988), Hungary (Kedves, 1959; Simoncsics, 1959), and lignite from India (e.g. Ramanujam, 1982). Interestingly, some of these fossils are morphologically similar to macroconidia of *Desmidiospora,* an anamorph in the Pezizomycotina (Clark and Prusso, 1986).

Classification of Fossil Microthyriaceae. There have been numerous classification systems proposed for fossil microthyriaceous ascocarps (e.g. Rosendahl, 1943; Cookson, 1947 [Figure 8.38]; Rao, A.R., 1958; Dilcher, 1965; Venkatachala and Kar, 1969; Jain and Gupta, 1970; Elsik; 1978; Pirozynski, 1978; Saxena and Tripathi, 2011; Tripathi, 2012). The classification modified by Elsik (1978) is used by many workers and involves morphological and structural features such as: the presence (porate) or absence (aporate) of pores; the colony being radiate or non-radiate, and, if non-radiate, whether the margin is with or without spines. Radiate forms are further classified as to the degree of elaboration of the margin. Among the radiate forms a further subdivision is based on the presence or absence of an ostiole, and in some genera, on how the pore is organized (Saxena and Tripathi, 2011; Figures. 8.39 and 8.40). We will present a number of the genera listed by Tripathi (2012) as examples of the classification noted above (e.g. Elsik, 1978) and rely heavily on the extraordinary data set (approximately 300 genera and 950 validly published species including new genera and combinations) presented in Kalgutkar and Jansonius (2000) as well as on forms restricted to India (Saxena and Tripathi, 2011). Elsik (1978), Kalgutkar and Jansonius (2000), and Phipps and Rember (2004) provided an excellent discussion of some of the issues relating to the classification of fossil microthyriaceous fungi, both extant and fossil. In the discussion that follows we will use the generic designations in dealing with these fossil fungi where possible.

FIGURE 8.39 Rajendra K. Saxena.

FIGURE 8.40 Suryakant M. Tripathi.

Asterothyrites. In these epiphyllous fungi the mycelium is superficial and persistent. Ascomata are circular, flat, and constructed of radiating hyphae that are interlaced to form a pseudoparenchymatous tissue (Figure 8.41). The central cells are modified, and the ascocarp is provided with a star-shaped opening, which may be formed as a result of the dissolution of loosely arranged central cells. The genus has sporadically been reported from the Cretaceous (e.g. Krassilov, 1967; Kalgutkar and Braman 2008) and regularly from Cenozoic sediments (e.g. Dilcher, 1965; Gupta, 1994). It was originally proposed by Cookson (1947) for several specimens described from

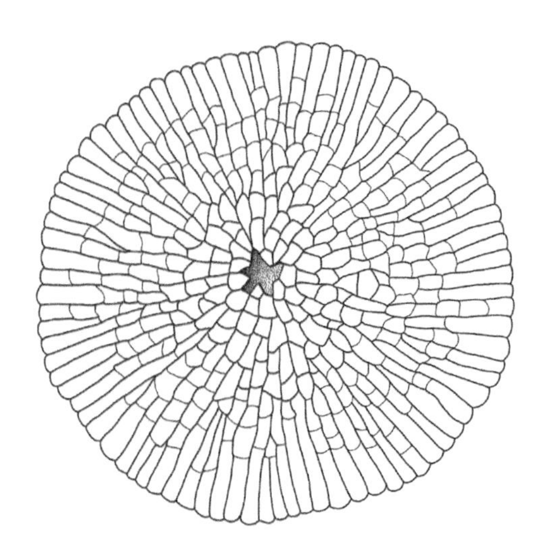

FIGURE 8.41 Microthyriaceous ascocarp of *Asterothyrites tennesseensis*. Eocene, USA. (Redrawn from Kalgutkar and Jansonius, 2000.)

Cenozoic sediments from the Kerguelen Archipelago, New Zealand, and Australia. For example, ascomata in the non-hyphopodiate *Asterothyrites minutus* are scattered and characterized by a single central cubical or hexagonal cell that may represent a star-shaped opening. In *A. canadensis* from the upper Paleocene–lower Eocene of the Yukon Territory, Canada, the ascomata range from 55 to 115 μm in diameter and have a distinct opening. The frequent presence of *Asterothyrites ostiolatus* on *Agathis* and *Araucaria* (Araucariaceae) leaves from the Cenozoic of Australia and Tasmania was viewed by Cookson and Duigan (1951) as reflecting a climate that was both warmer and more humid than today.

Brefeldiellites. Epiphyllous fungi of this type possess a large, sometimes fan-shaped, membranous, hyaline stroma composed of radiating cells that divide near the periphery. Fertile areas or ascomata occur at the margin (Figure 8.42). The central ascoma cells break away as a dehiscence mechanism. Fossil specimens are regarded as most similar to the extant genus *Brefeldiella* (Spegazzini, 1889). In *Brefeldiellites fructiflabella* from the lower Eocene of Tennessee, USA, the stroma is up to 675 μm in diameter and formed by cuboidal cells that produce a fimbriate margin to the colony (Dilcher, 1965). Ascomata develop near the margin of the stroma and are 150 μm wide × 130 μm long; a ring of opaque cells forms the ostiole. In *B. argentina* from the Lower Cretaceous Baqueró Formation of Santa Cruz Province, Argentina, the stroma may be up to 1.5 mm in diameter, with a lobed or

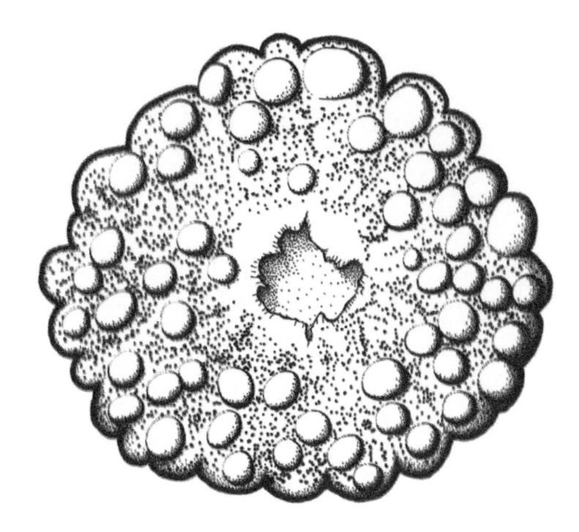

FIGURE 8.42 Microthyriaceous ascocarp of *Brefeldiellites argentina*. Cretaceous, Argentina. (Redrawn from Kalgutkar and Jansonius, 2000.)

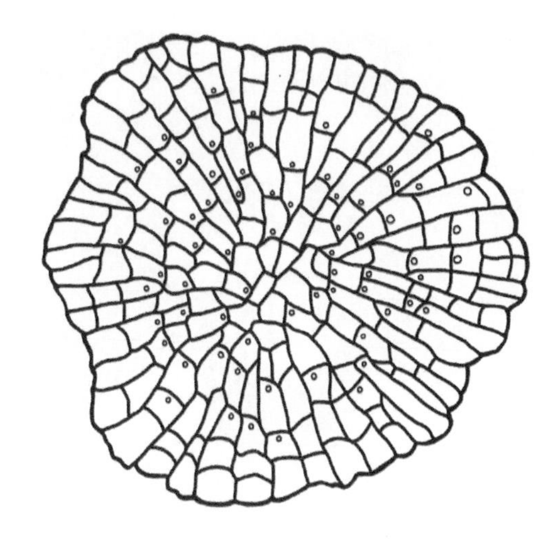

FIGURE 8.43 *Callimothallus pertusus*. Eocene, USA. (Redrawn from Kalgutkar and Jansonius, 2000.)

crenate margin (Martínez, 1968). Ascomata are produced near the stroma margin where growth is active. In both species details of the spores are lacking. Specimens of *B. fructifabella* are reported on leaves of *Chrysobalanus* (Rosales, Chrysobalanaceae), whereas those of *B. argentina* are reported as preserved on *Podocarpus* (but see Kalgutkar and Jansonius, 2000), the cycadophyte foliage type *Dictyozamites*, the seed fern *Ruflorinia*, and on an undetermined conifer, all from the Baqueró Formation (Martínez, 1968).

Callimothallus. If all fossil fungi names ended in *-ites*, it would make them easy to distinguish from extant organisms, but this is not the case. Multiple species of the genus *Callimothallus* have been reported that range from the Cretaceous (e.g. Doubinger and Pons, 1975) to the Cenozoic (e.g. Ramanujam and Rao, 1973; Tripathi, 1988; Kalgutkar and Jansonius, 2000). In *C. pertusus* (Figures 8.43 and 8.44) from the lower Eocene of Tennessee, the stroma is circular to subcircular, up to 250 μm in diameter, and there are no free hyphae. Individual cells each possess a single pore (Dilcher, 1965); in fact *Callimothallus* is the only microthyriaceous fungus that is multiporate. Cells of the peripheral region of the stroma are often slightly thickened. In *C. australis* (late–middle Eocene) the distinguishing feature is an aggregation of free hyphae that are the attachment point to the leaf surface (Lange, 1978a). A flattened oval ascoma, which is up to 90 μm in diameter and has a crenulate margin and occasional lobes, is characteristic of *C. corralesensis* from the Cretaceous of Colombia (Doubinger and Pons, 1975). Central cells are opaque to translucent and isodiametric.

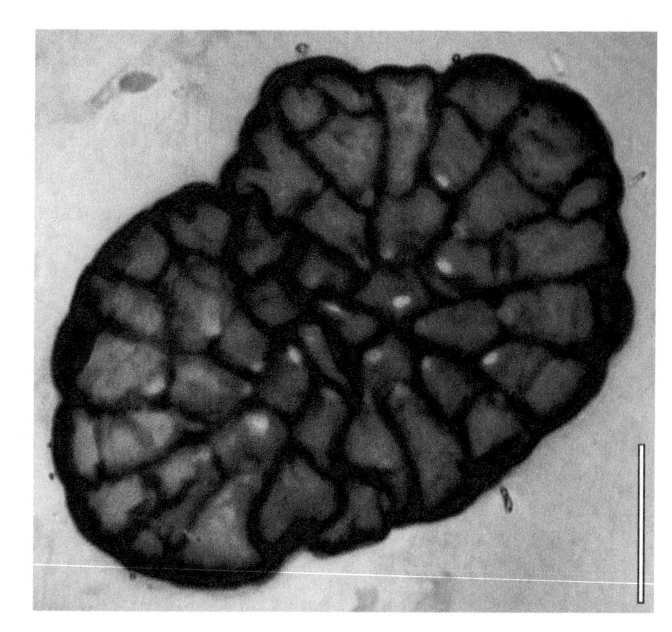

FIGURE 8.44 Stroma of *Callimothallus pertusus*. Eocene, Tennessee, USA. Bar = 10 μm. (Courtesy Florida Museum of Natural History.)

A Miocene form from the Quilon Formation (India), *C. dilcheri*, is characterized by a proximally located pore in each cell. Specimens range from 75 to 120 μm in diameter and possess bottle-shaped cells near the margin of the ascomata (Rao and Ramanujam, 1976). A species from the Late Cretaceous of the State of Coahuila, Mexico, initially described as *Paramicrothallites irregularis* (Martínez-Hernández and Tomasini-Ortiz, 1989), is now known as *C. irregularis*.

Not all of the species descriptions of *Callimothallus* include information about the host plant. Exceptions are the report of *C. pertusus* on the abaxial surface of fossil leaves of the monocot *Smilax* (Smilacaceae) from the upper Miocene of Zhejiang Province, China (Ding et al., 2011). Spores in this species are described as being 7–9 µm long and unicellate. Fossil leaves of *Cunninghamia* (Cupressaceae) from the upper Miocene of southeastern China have also served as the host for *C. pertusus* (Du et al., 2012), as have possible leaves of *Diospyros* (Ebenaceae) from the middle Miocene along the Himalayan foothills of India (Phadtare, 1989). Both of the latter authors also noted that *C. pertusus* on *Diospyros* is directly or closely associated with costal areas of the leaf epidermis, which is suggested to denote nutritional dependability. Clusters of ascomata have also been reported from both leaf surfaces of the conifer *Cephalotaxus* (Cephalotaxaceae) from the Oligocene of Guangxi in South China (Shi et al., 2010). Characters of the ascomata are identical with those of *C. pertusus*. It should be noted that some of the fossil epiphyllous fungi are morphologically and structurally similar to fossil representatives of the chlorophycean green alga *Ulvella*, which can also be found in Cretaceous and Cenozoic rocks (Hansen, J.M., 1980). Thus, it is important to provide sufficient characters when noting the occurrence of fossil epiphyllous fungi, especially when they are not attached to leaves.

Euthythyrites. Members of this genus of epiphyllous fungi are characterized by linear, sometimes fan-shaped ascomata (some up to 540 µm long), in which the lateral margins are uneven and dehiscence occurs by a longitudinal slit (Figure 8.45) (Cookson, 1947). Information about the spores is lacking. Specimens were originally described from Cenozoic angiosperm leaves of *Oleinites* (Oleaceae) from Australia. Specimens of *E. keralensis* from the upper Miocene of India have small peg-like hyphopodia (Ramanujam and Rao, 1973). Several of the fossils placed in *Euthythyrites* are reported as being similar to the modern forms *Aulographum* and *Lembrosiopsis*, which may also occur on oleaceous leaves.

Microthallites. Tripathi (2012) included this genus in his review of fossil fungi. The artificial genus was initially used for forms that were too poorly preserved to place them more exactly within the Microthyriaceae (Cookson, 1947) and for forms that could not precisely be related to modern or fossil genera (Dilcher, 1965). Since then, however, two of the species – *Microthallites lutosus* (Dilcher,

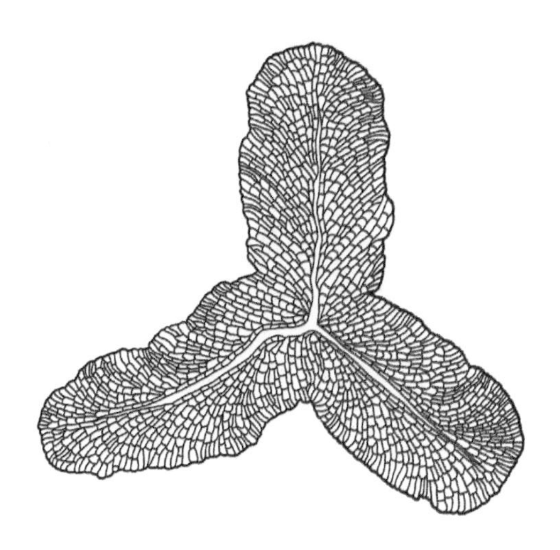

FIGURE 8.45 Ascoma of *Euthythyrites oleinitis*. Oligocene–Miocene, Australia. (Redrawn from Kalgutkar and Jansonius, 2000.)

1965) and *M. cooksoniae* (Rao and Ramanujam, 1976) – have been transferred to *Phragmothyrites* (Kalgutkar and Jansonius, 2000), while a third, *M. spinulatus* (Dilcher, 1965), has been moved to *Asterothyrites*, based on features of the ostiole (Kalgutkar and Jansonius, 2000). Kalgutkar and Jansonius (2000) noted that Jain and Gupta (1970) did not provide an emended diagnosis, but removed the forms with an ostiole, leaving only forms without an ostiole still in *Microthallites*. *Paramicrothallites* is a genus introduced for *M. spinulatus* and similar forms from the Cenozoic of India by Jain and Gupta (1970). Nevertheless, the name *Microthallites* is retained for poorly preserved specimens lacking an ostiole that are found in palynological preparations.

Microthyriacites. These epiphyllous fungi consist of radiate ascomata each composed of two regions: a central area, which consists of uniformly shaped, squarish cells that lack a radial pattern, surrounded by a zone of hyphae forming a pseudoparenchymatous thallus organized in a radial pattern. The genus was proposed by Cookson (1947) and later emended by Kalgutkar and Jansonius (2000); it is mainly Cenozoic in geological range (e.g. Straus, 1961; Givulescu, 1971; Gupta, 1994). *Microthyriacites baqueroensis* (Figure 8.46) is an Early Cretaceous form from Santa Cruz Province, Argentina, that is found on foliage of *Dictyozamites* (Martínez, 1968). The specimens are generally large (1.0–1.2 mm) and consist of two well-defined zones. Two forms have also been described on Early Cretaceous cycad leaves from China (Zheng and Zhang, 1986) and dispersed

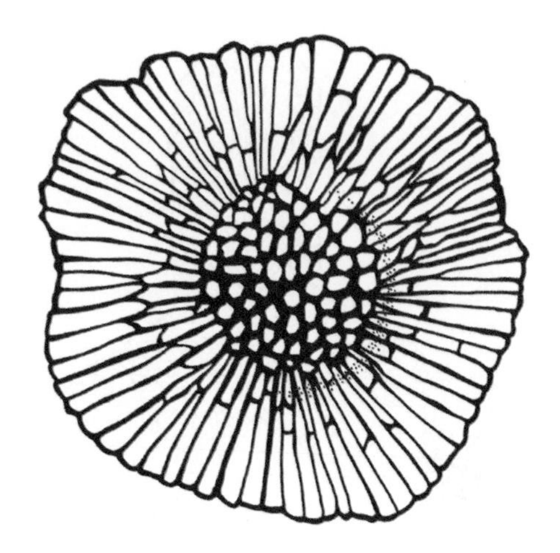

FIGURE 8.46 *Microthyriacites baqueroensis*. Cretaceous, Argentina. (Redrawn from Kalgutkar and Jansonius, 2000.)

FIGURE 8.47 Microthyriaceous ascocarp of *Parmathyrites turaensis*. Cenozoic, India. (Redrawn from Kalgutkar and Jansonius, 2000.)

in Early Cretaceous palynological samples from Japan (Legrand et al., 2011). Yet another form from the Lower Cretaceous of Southern Primorsk, Russia, occurs on the abaxial surface of *Cephalotaxus* (conifer) and was named *Notothyrites cephalotaxi* (Krassilov, 1967). This species has been moved to *Microthyriacites* because of the distinct difference between the central cells and those toward the periphery, and the dubious nature of the ostiole (Kalgutkar and Jansonius, 2000). *Notothyrites* has also been reported from Cenozoic sediments in Finland (Tynni, 1977; Eriksson, B., 1978) and India (Jain and Gupta, 1970), and from cycad cuticles from the Lower Cretaceous of China (Zheng and Zhang, 1986). The oldest evidence of *Notothyrites* comes from the Permian of India (Jha and Aggarwal, 2011).

Parmathyrites. In this form the central cells are conspicuous and isodiametric, and the hyphae are radially arranged to form a pseudoparenchymatous tissue (Jain and Gupta, 1970). The outer radial cells are prominent and possess thickened walls, and the flattened ascomata may or may not possess an ostiole. One of the distinguishing features of this genus is the presence of a peripheral sheath of spines on the fruiting body. In *Parmathyrites indicus* from the Miocene of Western Ghats, India, the spines around the periphery number up to 70, each 20–50 μm long (Jain and Gupta, 1970). The peripheral spines in *P. ramanujamii* (Miocene of Assam, India) are up to 15 μm long and possess a broad base (Singh et al., 1986). Specimens of *P. turaensis* (Figure 8.47) have spines described as having a bulbous base (Kar et al., 1972). A relatively small form with an inconspicuous ostiole has been described from the Neogene of India as *P. tonakkalensis* (Patil and Ramanujam, 1988).

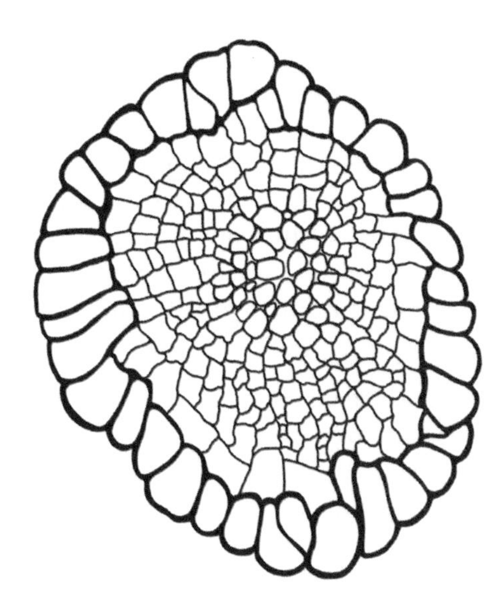

FIGURE 8.48 Microthyriaceous ascocarp of *Phragmothyrites kangukensis*. Paleocene–Eocene, Canada. (Redrawn from Kalgutkar and Jansonius, 2000.)

Phragmothyrites. The ascoma in this form is circular to subcircular and astomate (without an ostiole) (Edwards, W.N., 1922). Hyphae are radially arranged, organized into a pseudoparenchymatous tissue, and sometimes show various developmental stages in certain regions of the ascoma (Figure 8.48). The central cell or cells

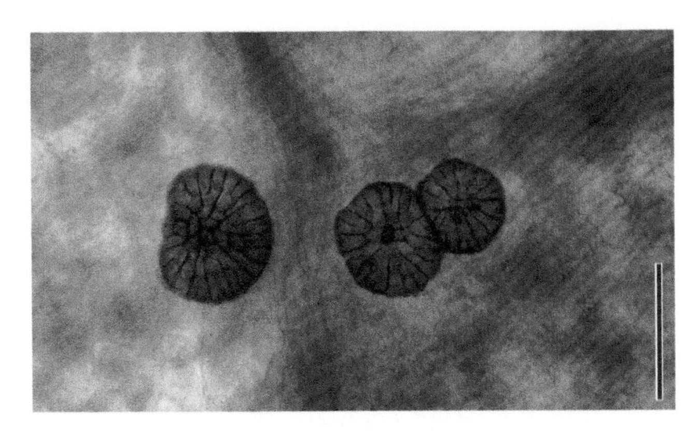

FIGURE 8.49 Three ascocarps of *Phragmothyrites concentricus* associated with leaf veins. Miocene, Idaho, USA. Bar = 25 μm. (Courtesy C.J. Phipps.)

FIGURE 8.50 Detail of *Phragmothyrites concentricus* ascocarp. Miocene, Idaho, USA. Bar = 50 μm. (Courtesy C.J. Phipps.)

are isodiametric (Kar and Saxena, 1976). Hyphae are not free nor are there any pores in the cells (Saxena and Tripathi, 2011). The absence of a central region of thick-walled cells in the ascomata has resulted in the transfer of several species of *Microthyriacites* to *Phragmothyrites* (e.g. Dilcher, 1965; Kar et al., 1972). *Phragmothyrites cooksoniae* from the Eocene and Miocene of India is up to 245 μm in diameter and has the two-part organization; there is an inner region of uniform cells and an outer zone of more radial cells (Rao, A.R., 1958). *Phragmothyrites ramanujamii* from the early Eocene of India is up to 120 μm in diameter, with the individual cells thick-walled and the cells along the margin hyaline and smooth (Samant, 2000). An interesting species, *P. concentricus* (Figure 8.49), has been reported from the lower Miocene of the Clarkia beds of Idaho, USA (Phipps and Rember, 2004). In this species the ascoma is flattened, circular (up to 150 μm in diameter), and has a smooth, entire margin. Cells form concentric rings at 6–15 μm intervals as a result of regular and synchronous tangential cell divisions (Figure 8.50); free hyphae are lacking. An ostiole and pores in the cells are absent. The Miocene specimens were located on angiosperm cuticles of *Betula* (beech), *Smilax* (willow) (Williams, J.L., 1985), and *Zizyphoides* (Trochodendraceae, a vesselless dicotyledon) (Phipps and Rember, 2004), while Sherwood-Pike and Gray (1988) reported similar fungi from the same site on *Quercus* (oak), *Smilax*, and *Cephalotaxus* (conifer). W.N. Edwards (1922) reported specimens on Eocene leaves of *Pityophyllum* (conifer). An excellent reference to species of *Phragmothyrites* from India, including provenance and age, has been compiled by Saxena and Tripathi (2011). An interesting *Phragmothyrites*-type fungus, *Polyhyphaethyrites giganticus*, from the Deccan Intertrappean beds (Mohgaon Kalan; probably latest Cretaceous) of India is an important discovery because it is permineralized (Srivastava and Kar, 2004). The specimen is associated with plant tissues in chert and has been described based on thin sections. The fungus is up to 4.0 mm in diameter but has no hole or ostiole in the middle. Hyphae (8–25 μm) are intertwined and form a net-like structure. Fungal fossils resembling *Phragmothyrites* have also been reported from Miocene amber from northeastern Peru (Antoine et al., 2006) and Jurassic chert from Argentina (García Massini et al., 2012). A cleistothecium somewhat reminiscent of *Phragmothyrites* comes from the Miocene of India and has been named *Kalviwadithyrites saxenae* (Rao, M.R., 2003). The cleistothecia are circular to subcircular, ranging from 105 to 115 × 95–110 μm, dimidiate (an ascomatal wall that only covers the top half), and non-ostiolate; free hyphae are absent. The marginal cells are rectangular to polygonal, whereas the central cells are squarish and isodiametric.

Plochmopeltinites. This form is characterized by an ostiole with a conspicuous rim and a membrane made up of irregularly branched, sinuous, and intertwining hyphae aligned in a radial pattern (Cookson, 1947). The original specimens were recovered from the Oligocene–late Miocene of Australia, but the taxon has since also been recorded for Cenozoic sediments elsewhere, including the Miocene of Argentina (García Massini et al., 2004) and India (Ramanujam and Rao, 1973). The

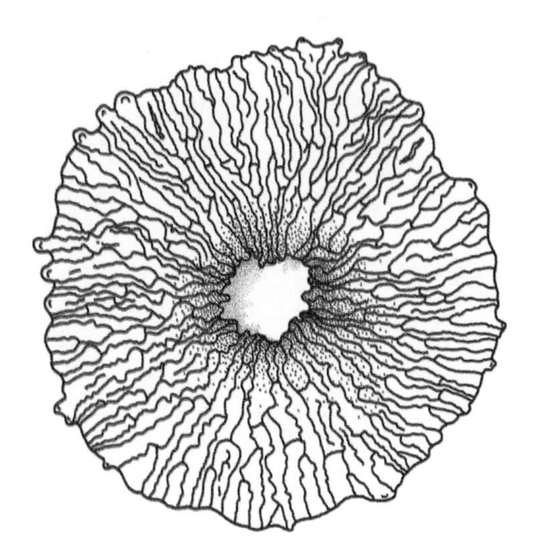

FIGURE 8.51 Microthyriaceous ascocarp of *Plochmopeltinites cooksoniae*. Miocene, India. (Redrawn from Kalgutkar and Jansonius, 2000.)

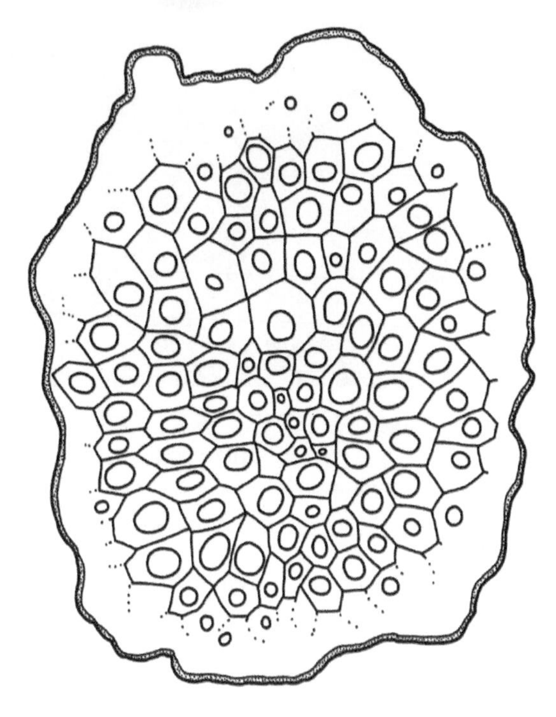

FIGURE 8.52 Microthyriaceous ascocarp of *Ratnagiriathyrites hexagonalis*. Neogene, India. (Redrawn from Kalgutkar and Jansonius, 2000.)

discoid ascomata in *P. cooksoniae* (Figure 8.51) from the Miocene of Tamil Nadu, India, are up to 166 μm, irregular on the periphery, and have occasional free hyphae extending from the cells of the margin (Ramanujam and Rao, 1973). Hyphal branches may terminate against other hyphae, with branching more conspicuous near the margin of the ascomata (Reddy et al., 1982). In *P. masonii* the ascomata are up to 200 μm in diameter and possess a sinuous margin, and the membranes are described as being prosenchymatous (Selkirk, 1975). The specimens were recovered from Oligocene–Miocene rocks from New South Wales, Australia.

Ratnagiriathyrites. Specimens of this type are characterized by ascomata that are subcircular to irregular in outline and by the absence of an ostiole (Figure 8.52) (Saxena and Misra, 1990). Unlike many of the other forms, this genus is distinguished by the absence of radially arranged cells containing pores (Kalgutkar and Jansonius, 2000). The single species, *R. hexagonalis*, is 114 × 90 μm and has hexagonal cells that increase in size toward the periphery. Specimens have been reported almost exclusively from the Miocene of India (Saxena and Tripathi, 2011).

Trichopeltinites. This fungus is characterized by ascomata which develop as thickened areas of the thallus that are highly variable in size and shape (Cookson, 1947). The forms described by Köck (1939) and Kirchheimer (1942) as *Phycopeltis* (*Phykopeltis*) may represent examples of this form. The type specimen, *Trichopeltinites pulcher*, dehisces by an irregular ostiole. In *T. fusilis* (Figure 8.53) from the Eocene of Tennessee, USA, some specimens may exceed 500 μm in size and are organized as rows of hyphae extending from a central area (Dilcher, 1965). Stromata possess ascomata that appear as thickened areas up to 50 μm in diameter, located in the lobes of the thallus. Thallus variability is a characteristic of *T. kiandrensis* from the lower Miocene of New South Wales, Australia (Selkirk, 1975). Some thalli are ribbon like, whereas others are circular with extending lobes. The fruiting bodies are characterized by a central pore that is thought to mimic the lysigenous pseudo-ostiole seen in extant forms (Selkirk, 1975). The genus is also known from the Lower Cretaceous of Russia (Krassilov, 1967). Specimens of *T. nilssonioptericola* are characterized by a fattened, fan-shaped stroma up to 3.0 mm long. The fruiting bodies can be identified as thickened areas approximately 58 μm in diameter. The fungus is reported on leaves of *Nilssoniopteris* and *Cladophlebis* (Krassilov, 1967). Other interesting specimens come from the Cretaceous–Cenozoic transition on Seymour Island, Antarctic Peninsula (Upchurch

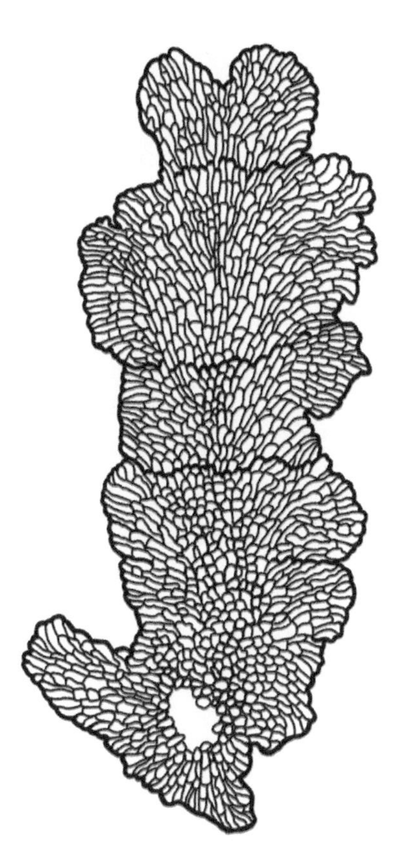

FIGURE 8.53 Microthyriaceous ascocarp of *Trichopeltinites fusilis*. Eocene, Tennessee, USA. (Redrawn from Kalgutkar and Jansonius, 2000.)

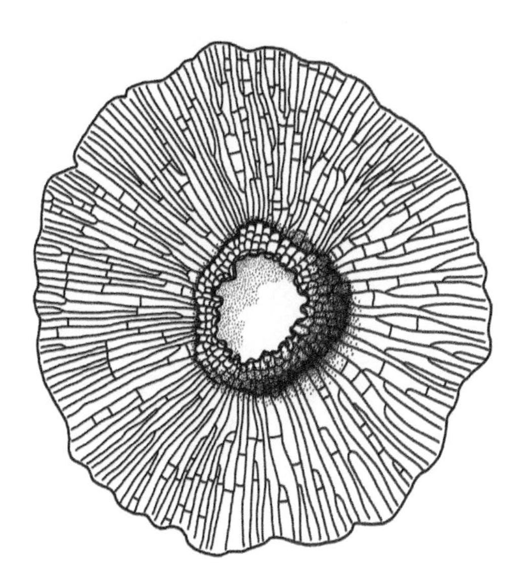

FIGURE 8.54 Microthyriaceous ascocarp of *Trichothyrites padappakarensis*. Miocene, India. (Redrawn from Kalgutkar and Jansonius, 2000.)

and Askin, 1989). A further interesting aspect of *Trichopeltinites* is that in certain successions in North America, it disappears at the Cretaceous–Paleogene boundary, presumably along with the megafloral species that served as hosts (references in Nichols and Johnson, 2008).

Trichothyrites. Defining features of this form are the presence of square, radiating cells and an elevated collar of two to six cells, some with minute projections (Figure 8.54). The thyriothecia are disc shaped as a result of compression. In *T. pleistocaenicus* (Pleistocene of Minnesota, USA) the specimens are circular and characterized by an adaxial and abaxial membrane (Rosendahl, 1943). Pleistocene *Trichothyrites* from Hungary are particularly abundant on leaves of *Buxus pliocenica* (Erdei and Lesiak, 1999). Thyriothecia may be up to 200 μm in diameter and have an elevated collar of cells around the ostiole (Smith, P.H., 1980). Some of the cells that make up the collar have delicate setae. Specimens of *T. airensis* from the Oligocene–Miocene of several Southern Hemisphere sites are up to 160 μm in diameter and have a well-defined ostiole (Cookson, 1947; Kalgutkar and Jansonius, 2000). *Trichothyrites ostiolatus* was isolated from a fragment of an *Olenites* leaf, but it was initially described as a species of the epiphyllous fungus *Asterothyrites* (Cookson, 1947; Kalgutkar and Jansonius, 2000). As the generic name suggests, these fossils have been interpreted as being morphologically similar to the extant forms in *Trichothyrina* (Petrak, 1950), a group of tropical fungi that are often hyperparasites (Ellis, J.P., 1977). P.H. Smith (1980) described isolated thyriothecia from the upper Eocene of southern England as *Trichothyrites hordlensis*, and he noted that it is difficult to assign fossil taxa to living species as the mode of nutrition is often unknown. J.P. Ellis (1977) noted that both the host substrate and shape of the ascospores are necessary characters needed for assignment to extant genera. Additional species from India can be found in Saxena and Tripathi (2011).

Developmental Stages. Dilcher (1965) provides a detailed historical record of microthyriaceous fungi, listing some of the fossil genera and species that might belong in the Microthyriaceae and describing the immature developmental or germling stage of several forms, including germinating spores (Figures 8.55 and 8.56) (Elsik and Dilcher, 1974). Similar early developmental stages have also been reported from leaves obtained from lower

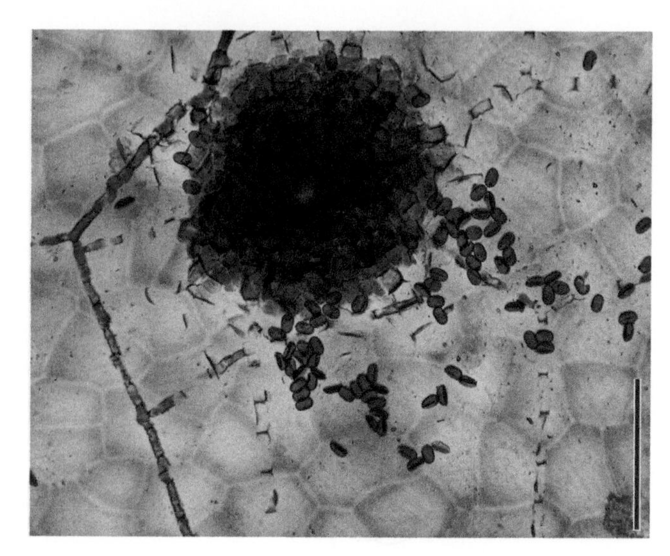

FIGURE 8.55 Pycnidium with numerous pycnidiospores on cuticle surface. Eocene, Tennessee, USA. Bar = 50 μm. (Courtesy Florida Museum of Natural History.)

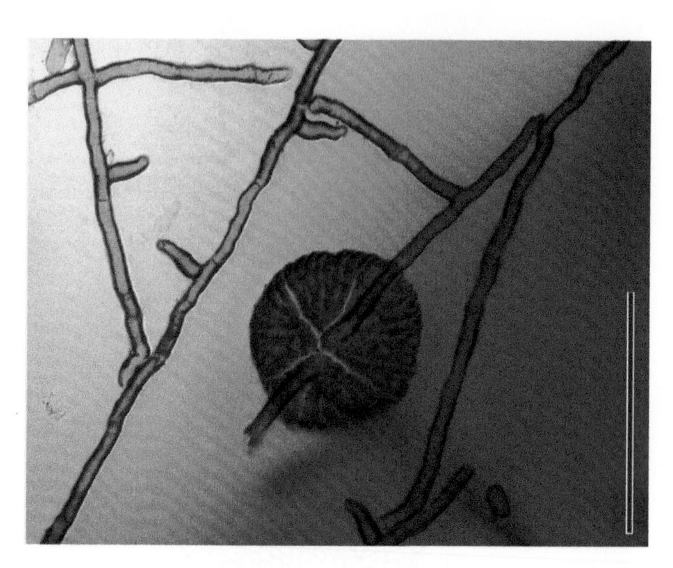

FIGURE 8.57 Immature fruiting body of *Asterina eocenica*. Eocene, Tennessee, USA. Bar = 50 μm. (Courtesy Florida Museum of Natural History.)

FIGURE 8.56 Germinating spores of *Meliola anfracta* showing development of initial hyphopodium (*arrow*). Eocene, Tennessee, USA. Bar = 25 μm. (Courtesy Florida Museum of Natural History.)

Miocene lignites of New South Wales, Australia, and demonstrate spore germination and perithecium development in other ascomycetes such as *Asterina kosciuskensis* (Selkirk, 1975).

One of the interesting aspects of the ascomycete fruiting bodies that have been described and illustrated from fossil leaf cuticle fragments and palynological samples is the fact that only a few lack the diagnostic features that allow them to be aligned with modern taxa. While paleomycologists dealing with these fossils have proposed generic names ending with *-ites* to signify that the taxon is a fossil, the absence of the actual reproductive structures (ascogenous hyphae and asci) and spores in most of these fossils has hindered a more complete understanding of the total complement of features about the fossil fungus. The fossils have been described based on morphological and structural features that relate to the size, shape, and general organization of the often shield-shaped fungal structures. Hyphal diameter and the level and type of branching are often used as generic features, together with the presence or absence of an ostiole (Figure 8.57) and whether cells possess pores. For many of the fossils the degree of opacity is also used as a character in defining particular forms. Rarely is there any information provided about the reproductive structures (e.g. asci and pycnidia) and almost never details about the types of spores that were produced by a particular fossil fungus.

The fact that these reproductive features are absent from the majority of the fossils presents an interesting dilemma, especially in cases where the other fossils in the assemblage are well preserved, notably the plant cuticles on which most are described. There has been a variety of suggestions as to why these structures are not preserved. One of these is that the life history of at least some of the forms may have been completed prior to fossilization. Another may be related to the ephemeral nature of the asci in some groups, which may have deliquesced immediately after spore formation and thus were not preserved (Phipps and Rember, 2004). Perhaps the most plausible explanation may be related to the manner in which the fossil specimens are recovered and treated prior to examination. Most palynological preparations use various types and concentrations of acids and bases to prepare the specimens so that hyphae and other structures are more translucent when examined in transmitted light. This treatment many simply dissociate spores from the reproductive structures that once contained them. This rather simple explanation was also found to be the case when examining cellulose acetate peels for fossil fungi and comparing them to petrographic thin sections. Fungi had been etched or washed away in the preparation of the peel but were still present in thin section preparations (Taylor et al., 2011). Supporting the suggestion that treatment of the fossils is the primary contributor to losing information about asci and spores is the fact that when these epiphyllous fungi are examined with scanning electron microscopy, more structural features of the fungus can be seen (e.g. Alvin and Muir, 1970; Daghlian, 1978).

CAPNODIALES. Sooty molds are a large group of fungi, most of which are nonparasitic, nonpathogenic colonizers of leaves of a single growing season or may extend for several seasons on wood; others, however, are pathogens or saprotrophs (Crous et al., 2009). Members of one family, the Piedraiaceae, colonize hair shafts of mammals (de Hoog et al., 2000). With regard to paleomycology, the Capnodiaceae represent a particularly interesting group because they are epiphyllous saprotrophs that are often associated with sap-sucking insects. The honeydew exuded by these insects serves as the nutritive substrate for the molds (Hughes, 1976; Chomnunti et al., 2011). They are characterized by superficial ascomata with fasciculate asci, and hyaline to dark, septate ascospores (Crous et al., 2009). Anamorphs are dematiaceous, and include mycelial (phragmo- to dictyoconidia), spermatial and pycnidial synanamorphs (Hughes, 1976; Cheewangkoon et al., 2009). Hyphae are constricted at the septa and taper at the ends (moniliform).

A few early reports described sooty molds on conifer twigs preserved in Eocene Baltic amber (e.g. Menge, 1858; Caspary and Klebs, 1906, 1907). An updated distribution of the group extending from the Cretaceous into the Cenozoic has been provided by Schmidt et al. (2014). Several fossils that morphologically resemble sooty molds have been reported from amber that ranges from 45 to 22 Ma (Rikkinen et al., 2003). The fossils consist of dark hyphae that are highly branched and form cylindrical to subglobose cells that are delineated by the constriction at each septum. Some cells may be up to 15 μm in diameter (Figure 8.58). The fossils are assigned to the extant species *Metacapnodium succinum*. Co-occurring with some of the hyphae are fossils assignable to the form genus *Capnophialophora*, conidiogenous cells that form two to four phialides (Hughes, 1966). The presence of conidial stages in the fossil sooty mold *M. succinum* further confirms the taxonomic placement of this fungus.

FIGURE 8.58 *Metacapnodium succinum* showing tapering moniliform hyphae. Eocene amber, Germany. Bar = 20 μm. (Courtesy A.R. Schmidt.)

There are several reports of fossil fungi that are similar to the extant fungus *Cladosporium* (Davidiellaceae), one of the largest genera of dematiaceous hyphomycetes and among the most common molds that are pathogens, parasites, and allergens. Members of the genus are characterized by a coronate scar structure, conidia in acropetal chains, and *Davidiella* teleomorphs (e.g. Braun et al., 2003; Bensch et al., 2010). *Cladosporites bipartitus* from the Eocene has elliptical to pyriform conidia, approximately 10–12 μm long, that have a single septum (Felix, 1894). Fossil fungi forming small tufts have also been described from the lumen of vessels of the angiosperm wood *Laurinoxylon* from the middle Eocene of Texas, USA (Berry, 1916). The hyphae of *C. fasciculatus* are thin (1.3 μm in diameter) and have tufts of simple conidia that are up to 12 μm in diameter. *Cladosporites oligocaenicum* is reported from *Palmoxylon* wood from the early Oligocene of Mississippi, USA. The specimens are highly branched, with spherical conidia that often appear in pairs but sometimes consist of up to four linear spores (Berry, 1916). Both single and paired spores have been reported in *C. ligni-perditor*, a fossil from the Pliocene of Nebraska, USA (Whitford, 1914).

Leotiomycetes

The Leotiomycetes, colloquially referred to as discomycetes, are an ecologically diverse group of non-lichenized fungi that produce small apothecia (rarely cleistothecia) characterized by inoperculate, unitunicate asci (Wang et al., 2006; Hustad and Miller, 2011). The group contains five orders and approximately 500 genera based on morphological and molecular characters (Hibbett et al., 2007).

ERYSIPHALES. This economically significant order of fungi are plant pathogens that form a white, powdery dust on the vegetative and reproductive parts of flowering plants (e.g. Amano, 1986; Glawe, 2006). The powdery mildews are obligate biotrophs that are worldwide in distribution and presently include 16 genera and approximately 900 species (Braun, 2011; Braun and Cook, 2012). It is reported that there are more than 9,838 hosts for the powdery mildews, all on angiosperms, and of these, the majority (9,176) on dicotyledons. Of the remaining 662 monocotyledonous hosts, 634 species are associated with grasses (Takamatsu, 2013). None has been reported on ferns or gymnosperms. All members have closed ascocarps (cleistothecia) with persistent asci and a basal hymenium.

Fossil Erysiphales. While there are numerous fossil spore-like structures that possess variously shaped appendages like those in some Erysiphales, there are only a few reports of fossils that can be regarded with any confidence as members of this group. Fossil structures like *Protoascon missouriensis* and *Traquairia* spp., both from the Carboniferous, and once suggested as belonging to the Erysiphales, are now believed to represent other groups of fungi (e.g. Taylor et al., 2005b; Krings et al., 2013a). If the conidial specimens described by Barthel (1961) represent fossil Erysiphales, then the order does in fact extend back to the Carboniferous. The name *Erysiphites protogaeus* has been used for spherical to dorsiventrally somewhat compressed structures that are characterized by a reticulate surface ornament and four or five distal openings that occur densely spaced on *Ficus kiewiensis* leaves from the Paleocene of Russia (Schmalhausen, 1883; Meschinelli, 1892). These fossils were originally placed in the extant genus *Erysiphe*. Specimens thought to belong to the group have also been reported from the Miocene of Italy as perithecia (Pampaloni, 1902); however, it remains difficult to determine, even after examining the type material, whether this fossil (named *E. melilli*) represents a powdery mildew or even a fungus (Salmon, 1903). *Uncinulites baccarinii* is another fossil from the Miocene of Italy that was originally proposed as a powdery mildew (Pampaloni, 1902). The specimens are up to 22 µm in diameter and interpreted as perithecia that are ornamented by straight to curved appendages. Others have described a number of these fossils simply as fungal spores that probably have little to do with powdery mildews (e.g. Jansonius and Hills, 1977; Salard-Cheboldaeff and Locquin, 1980; Kalgutkar and Jansonius, 2000). One form, *U. artuziae* from the Cenozoic of Turkey, is up to 40 µm in diameter, irregularly circular, non-septate, and inaperturate, and bears numerous spines that are 5–7 µm long (Ediger and Alişan, 1989). Other appendage-bearing spores from the Cretaceous have been described as *Erysiphe*- and *Uncinula*-like (Kar et al., 2004b), or under the names *Lithouncinula* and *Protoerysiphe* (Sharma et al., 2005). Septate hyphae, some bearing globose conidia 7–9 µm in diameter, have been reported and named *Ovularites barbouri* from a Cretaceous leaf fossil from Nebraska (Whitford, 1916). Although the fossil fungus is interpreted as being similar to the arthroconidial hyphomycete *Ovularia*, this assignment is not warranted based on the fossils.

Another approach that has been employed to better understand the geologic history of the Erysiphales relates to the hosts (Takamatsu et al., 2010). Because extant Erysiphales are only associated with angiosperms that are thought to have first appeared close to the Early Cretaceous, some of the appendage-bearing fossils reported earlier are probably not members of this group. The presence of certain angiosperm hosts during the Late Cretaceous, however, has been used as indirect evidence of the initial radiation of the group. Nevertheless, it should be possible to identify members of this group on the surface of fossil leaf cuticles since many of the ascocarps are highly ornamented. When reported, the fossils will be especially important in calibrating molecular phylogenies of the group and may also be useful in tracking host relationships and geographic distributions that have been well documented in extant members (e.g. Amano, 1986; Takamatsu, 2013). The fact that the group is confined only to angiosperms has been used to infer that the Erysiphales have evolved from a single ancestral taxon that developed the ability to parasitize plants only

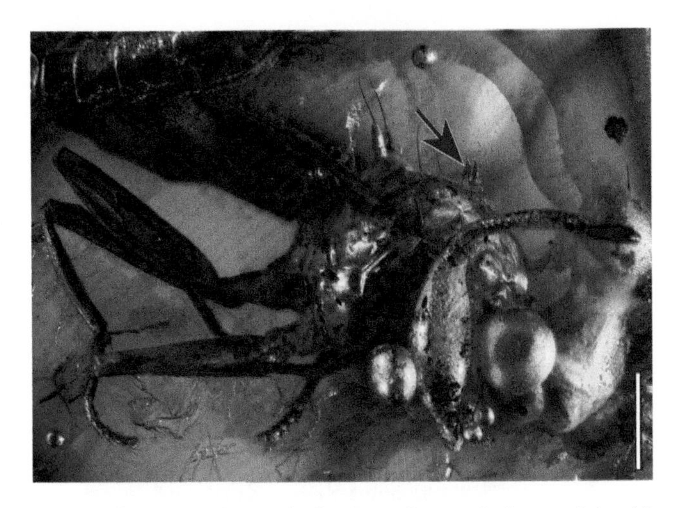

FIGURE 8.59 Stalk-eyed fly *Prosphyracephala succini* with a tuft of the fungus *Stigmatomyces succini* (*arrow*) attached to thorax. Eocene amber, Germany. Bar = 500 um. (Courtesy A.R. Schmidt.)

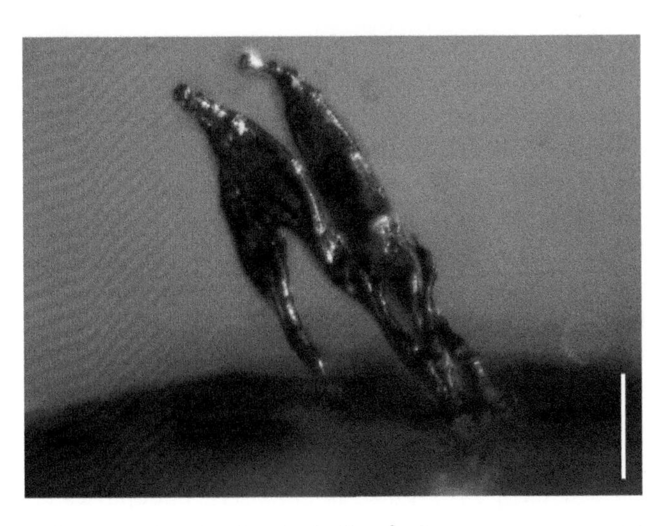

FIGURE 8.60 Detail of thalli of *Stigmatomyces succini*. Eocene amber, Germany. Bar = 100 μm. (Courtesy of A.R. Schmidt.)

once (Niinomi et al., 2008). Interestingly, although studies of the powdery mildews indicate the group contains a very large genome (around 120 million base pairs), it is a parasite with few infection strategies of the host, suggesting to some that it is at an evolutionary dead end (e.g. Spanu et al., 2010).

Laboulbeniomycetes

These fungi are obligate and mostly host-specific ectoparasites of aquatic and terrestrial arthropods, including insects, mites, and millipedes (Santamaria, 2001; Weir and Blackwell, 2005). They do not form a mycelium, but instead produce thalli directly from sticky ascospores that glue themselves to the host integument. Most forms are <1 mm long and are found on antennae, mouthparts, or other body regions of the host; some forms produce haustoria that extend into the animal (Haelewaters et al., 2012). The entire life cycle takes place on the outside of the host, but they become visible to the naked eye only when their fruiting bodies emerge. The Laboulbeniales were initially interpreted as abnormal cuticular outgrowths of arthropods, later as belonging to the acanthocephalans (parasitic worms), Basidiomycota, zygomycetous fungi, and Ascomycota (Weir and Blackwell, 2001). A fossil representative of *Stigmatomyces* (genus in the Laboulbeniaceae; members are parasitic on Diptera) was formally described as *S. succini* (Figures 8.59 and 8.60). It was discovered on the thorax of a stalk-eyed fly (*Prosphyracephala succini*, Diopsidae) enshrined in Neogene amber from Bitterfeld, Germany (Rossi et al., 2005) and represents the only compelling fossil evidence of the Laboulbeniomycetes to date.

FUNGAL ENDOPHYTES AND EPIPHYLLS

ENDOPHYTES

Endophytes are those fungi, mostly ascomycetes, but also some basidiomycetes, that produce their vegetative parts entirely within plant tissues. They may grow within roots, stems, and leaves, emerging to produce spores only at plant or host-tissue senescence (e.g. Stone et al., 2004; Singh et al., 2011; Unterseher, 2011; Weiss et al., 2011). Some may become effective as saprotrophs after plant senescence (e.g. Zhou and Hyde, 2001; Osono, 2006; Hyde and Soytong, 2008). Fungal endophytes are defined functionally by their occurrence within asymptomatic tissues (i.e. lacking visible disease symptoms at the moment of detection; see Schulz and Boyle, 2005) of plants (but see Hyde and Soytong, 2008, for alternative definitions). They occur in all major lineages of land plants, including bryophytes and the free-living gametophytes of pteridophytes (e.g. Boullard, 1979, 1988; Swatzell et al., 1996; Pressel et al., 2008b; Muthukumar and Prabha, 2012; Zhang et al., 2013), and in communities ranging from the Arctic to the tropics (Arnold et al., 2000; Arnold, A.E., 2007), but they have also been reported from marine macroalgae (e.g. Flewelling et al., 2013). Fungal endophytes may enter into varied and variable interactions with their hosts, including mutualism, cryptic commensalism, and latent and virulent pathogenicity (Schulz and Boyle, 2005 and references therein), in which they may significantly affect various degrees of host performance (e.g. Omacini et al., 2001, 2006; Arnold et al., 2003;

Arnold and Engelbrecht, 2007; Sieber, 2007; Rodriguez et al., 2009; Davitt et al., 2010).

All plants are thought to contain at least one fungal endophyte. Moreover, there appears to be a relationship between endophyte community composition and phylogenetic relatedness of the hosts (Saunders et al., 2010). In other words, closely related plants have similar endophyte communities and similar relationships with their fungal inhabitants. For example, closely related hosts may share a chemical or physical defense trait that is not shared with a more distantly related plant species. The diversity of endophytic fungi is thought to be enormous, especially when based on molecular analyses (e.g. Gazis and Chaverri, 2010; Gazis et al., 2011). Endophytes are nonpathogenic by definition (see above) and often share nutrients with the host. Despite being hidden in plant tissues, endophytes can have major consequences for the host plants and the communities in which they live (e.g. Carroll, 1988), and they are therefore used in agriculture and biotechnology, where their presence often conveys a selective advantage to the host in the form of increased resistance to fungal pathogens and herbivorous insects (e.g. Bayman, 2007; Saunders et al., 2010; Martin et al., 2012; Estrada et al., 2013). Research also indicates that certain endophytes not only enhance the disease resistance of their host in some systematic manner, but can also interact directly with certain fungal pathogens (e.g. Dingle and McGee, 2003; Raghavendra and Newcombe, 2013). In addition, interactions between extant plants and arbuscular mycorrhizal fungal communities can be variously modified by endophytes (Omacini et al., 2006). It is also known that phylogenetically diverse bacterial endosymbionts occur within the hyphae (endohyphal) of endophytic fungi (Hoffman and Arnold, 2010). These bacteria can alter the morphology and physiological traits of the fungus, thus affecting the association between the fungus and its plant host in diverse ways.

Extant endophytic ascomycetes have been divided into two major groups: (1) those termed C-endophytes (clavicipitaceous endophytes; i.e. members of the ascomycete family Clavicipitaceae; Class 1 of Rodriguez et al., 2009), which colonize cool- and warm-season grasses (Poaceae) and a few sedges (Cyperaceae), and (2) the non-clavicipitaceous endophytes (NC-endophytes), which are found in most other plant groups (e.g. bryophytes, ferns, conifers, and most flowering plants) (Rodriguez et al., 2009).

Clavicipitaceous endophytes are characterized by the systemic fungal colonization of the host plant by only one

FIGURE 8.61 Keith Clay.

fungal species and the establishment of a long-term symbiosis (Tadych et al., 2009). Despite the small number of C-endophytes, the group has received considerable attention because the association results in the accumulation of fungal metabolites that are effective in increasing the host plant's drought tolerance, resistance to vertebrate and invertebrate pests, resistance to fungal diseases, and tolerance to poor soil conditions, among other biotic and abiotic stresses (e.g. Clay, 1992 [Figure 8.61]; Clay and Schardl, 2002; Bischoff and White, 2005; Kuldau and Bacon, 2008). This is especially important because the grasses lack the biosynthetic capacity for the production of such secondary metabolites (Kuldau and Bacon, 2008).

NC-endophytes include a polyphyletic assemblage of ascomycetes that cause discrete and symptomless infections in the aerial tissues of plants. They are difficult to characterize and it is also difficult to understand how they interact within terrestrial ecosystems. Unlike the C-endophytes, which are restricted to grasses and are often transmitted vertically via host seeds, the NC-endophytes infect a broad range of plant hosts belonging to disparate groups and families, and they are transmitted horizontally by spore dispersal (Devarajan and Suryanarayanan, 2006).

Some have been reported as having the capacity to switch from an endophyte to a free-living lifestyle (Vasiliauskas et al., 2007; Maciá-Vicente et al., 2008) and this has made it even harder to understand the roles of these fungi. Three general functional groups have been suggested for extant NC-endophytes based on life history, ecological interactions, transmission patterns between host generations, where they occur in the plant, and what types of plants they infect (Classes 2–4 of Rodriguez et al., 2009). Although all three classes have broad host ranges, Class 3 endophytes colonize only shoots and form highly localized infections (Higgins et al., 2011), and therefore they may be the easiest to identify in the fossil record. All of the fossil endophytes thus far documented are apparently of the NC-type. This may reflect the fact that the grasses are relatively young geologically (see Kellogg, 2001). Based on phytoliths, it has been suggested that the grass family evolved in the Late Cretaceous (e.g. Piperno and Sues, 2005; Prasad et al., 2005), although Crepet and Feldman (1991) proposed the early Eocene, based on megafossils.

Fossil Fungal Endophytes

The term fungal endophyte is usually used today to refer to fungi that exist in living plant tissues without causing observable disease symptoms at the time they are detected. With fossil material, however, this definition is problematic because the identification of fungal endophytes based on fossils is challenged by the inherent difficulties of determining the condition of the host plant at the time of colonization – was it alive and fully functional or in the process of senescence or decay? For example, intact fossil plant tissue systems containing fungi may be suggestive of an endophytic association (Figure 8.62), whereas fragmented and partially degraded tissue systems containing fungi may signal saprotrophism. However, it is particularly difficult to determine whether the fragmentation and decay was initiated prior to or after fungal colonization, or whether the decay is a preservational artifact. Moreover, some fungal endophytes may become effective as saprotrophs after plant senescence. Since fungi present in fossil plants may have utilized similar nutritional strategies, it is almost impossible to distinguish fungal endophytes (as defined above) from saprotrophs in the fossil record. To deal with some of these issues, Krings et al. (2009d) adopted the initial suggestion of Brown et al. (1998), who designated a fungal endophyte in a strictly descriptive sense, and have suggested that, with fossils, the term fungal endophyte be used for all fungi that occur within intact plant cells or tissues in which there are no visible disease symptoms. In fossil specimens where there is an obvious host response or disease symptom, or other structural evidence that may signal an interaction, the endophyte may be more specifically defined as a parasite, mutualist, or pathogen.

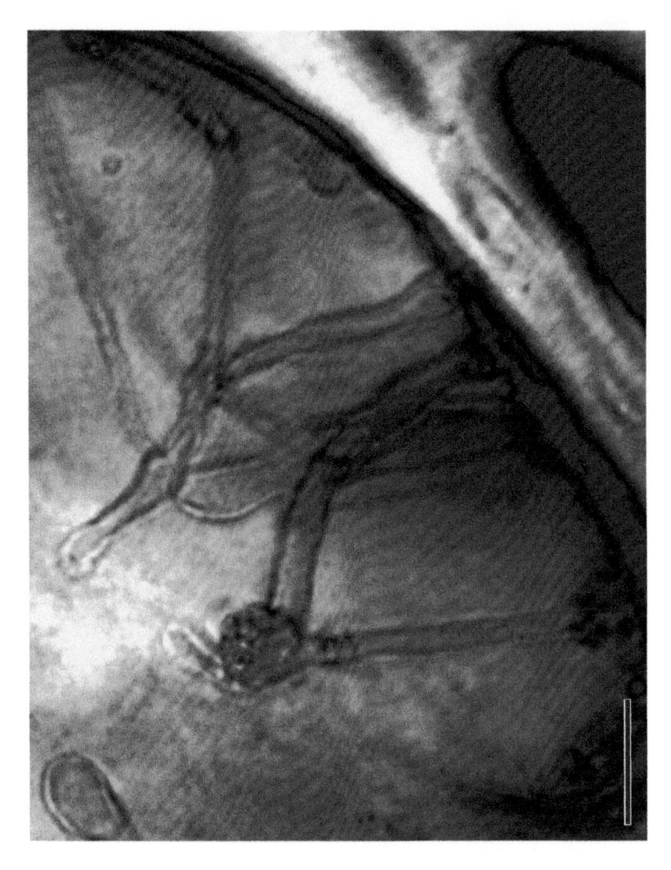

FIGURE 8.62 Hypodermal cell of fern pinnule (*Pecopteris* sp.) containing hyphae of endophytic fungus. Upper Pennsylvanian, France. Bar = 10 μm.

Molecular evidence suggests that all major groups of fungi were already well diversified by the time the first land plants with conducting elements appeared on Earth. Most of the early land plants were constructed of small prostrate or upright axes that were generally rootless and leafless (Taylor, T.N., et al., 2009); a few forms, however, demonstrate a more complex organization in which there was a tendency toward organ differentiation. Land plants with true leaves and roots first appeared later during the Early Devonian (Raven and Edwards, 2001). Fungal endophytes can be documented from the oldest structurally preserved land plants that lacked a differentiation of the plant body into shoot axes, leaves, and roots. We have included most fossil examples of fungal endophytes in the systematic treatment of the major groups of fungi.

FIGURE 8.63 Ari Jumpponen.

Below are given a few examples of fungal root and leaf endophytes that have been described based on exquisitely preserved fossils but which cannot be assigned to any of the major groups of fungi.

ROOT ENDOPHYTES. Root endophytes internally infect host tissues, are relatively inconspicuous, and are transiently symptomless (Stone et al., 2000). They interact with host root cells in various ways depending on plant species and the environment in which colonization occurs (e.g. Addy et al., 2005). Usually, the fungi are intracellular and invade root cells without triggering obvious structural host reactions (Peterson et al., 2008). Some are thought to represent a continuum from parasitism to mutualism, and perhaps in some conditions they are mycorrhizal (e.g. Jumpponen, 2001 [Figure 8.63]; Yuan et al., 2010). The ecological role of these fungi is now beginning to be explored in detail (Jumpponen and Trappe, 1998b; Schulz et al., 2006). For example, some root endophytes are believed to significantly enhance nitrogen uptake (Zijlstra et al., 2005; but see Mayerhofer et al., 2013). Since these endophytes colonize root tissue of apparently healthy plants (Schulz and Boyle, 2005) concomitantly with mycorrhizal fungi, their precise identities and host preferences often remain largely unknown. It has been suggested that ecological interactions between root endophytes and mycorrhizal fungi might exist and that these interactions could be important in understanding the complex assembly processes of belowground fungal communities (Toju et al., 2013). For example, one study suggests that specific characteristics of different *Neotyphodium* root endophytes of *Schedonorus phoenix* (tall fescue, a cool-season perennial bunchgrass native to Europe) may be responsible for negative effects on the arbuscular mycorrhizal fungal colonization of roots of other neighboring plant species, even upon the death of the host, through long-term leaching into soil (Antunes et al., 2008).

One group of root endophytic fungi that has received considerable scholarly attention is termed dark septate endophytes (DSEs), a group that is polyphyletic, with most forms representing conidial or sterile ascomycetous fungi. They colonize plant roots both intracellularly and intercellularly, forming melanized septate hyphae (hence the name) and microsclerotia (Jumpponen, 2001; Mandyam and Jumpponen, 2005). Although there is considerable interest in DSE fungi, there are no clearly defined criteria to determine whether a fungus belongs to this group (Knapp et al., 2012). It has been suggested that DSE fungi be termed septate endophytes since some have hyaline hyphae (O'Dell et al., 1993); others suggest that certain types of DSEs be further characterized (Barrow, 2003). To date DSEs have been reported in approximately 600 extant plant species in 114 families and are worldwide in distribution (Jumpponen and Trappe, 1998a), where they often occur in high-stress environments (Knapp et al., 2012). DSEs have been thought to play an important role as mutualists in the ecophysiology of plants; however, little about this potential role is known (Usuki and Narisawa, 2007; Rodriguez et al., 2009; Alberton et al., 2010). For example, certain DSEs can reduce disease intensity caused by phytopathogenic peronosporomycetes in Norway spruce (*Picea abies*) seedlings (Tellenbach and Sieber, 2012). Another study has shown that DSEs have adverse effects on the growth of Norway spruce seedlings, but also that mycorrhization can compensate for the adverse effects of DSEs, probably through the reduction of DSE density (Reininger and Sieber, 2012). It has been suggested that the presence of melanin in the hyphae of DSE fungi confers some selective advantage for living in a diverse environment under extreme conditions. One interesting report indicated that, under ionizing radiation, living melanized fungal cells increased their growth, which may give this fungus an advantage in survival (Dadachova et al., 2007).

Fossil Root Endophytes. Identifying root endophytes in the fossil record is (at best) a difficult problem for several reasons. While there are many hyphae in fossil plant organs, including roots (e.g. Halket, 1930; Cridland, 1962), establishing their life strategies is equivocal. Structures like spores and arbuscules have been useful in identifying certain types of mycorrhizae and the hyphae associated with them, while the presence of a host response has been useful in identifying examples of parasitism. Since host responses are rarely encountered when endophytic fungi are present in modern plants (see pp. 165–167), it is probable that the same situation existed in the fossil record. The melanin in the hyphae of living DSE fungi, and their presence in roots, which helps in identification, is also a problem with fossils since there are various preservational and diagenetic processes that may result in changing the color of the hyphae. In spite of these limitations there are a few fossils that may in fact represent root endophytes.

One of these fungi occurs in the root mantle of the Carboniferous–Early Permian marattialean tree fern *Psaronius*. Although other ferns, both fossil and extant, possess a root mantle, that of *Psaronius* is unusual because it consists of several layers of intertwining aerial roots that at some levels fuse by proliferation of the cortex. As a result of this developmental pattern, the root mantle forms an ensheathing structure around the actual stem that may become extensive over time. In an Early Permian *Psaronius* stem from Germany, more than 15 different types of intra- and intercellular fungi have been discovered, ranging from chytrid-like zoosporangia to basidiomycetous hyphae with exceptionally well-preserved clamp connections (Barthel et al., 2010). One of the most interesting fungi in these roots is an intracellular mycelial system of uncertain affinity that extends through large portions of the proliferating root cortex. It consists of hyphae that form prominent appressoria on host cell walls and arbuscule-like structures in the cell lumen (Figure 8.64). The inability to place this fungus systematically precludes details about its nutritional mode and relationship with the host.

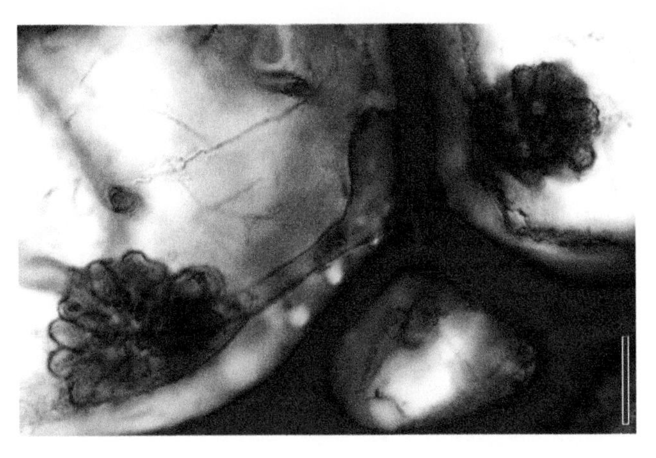

FIGURE 8.64 Prominent arbuscule-like structures in root mantle cells of the permineralized tree fern *Psaronius*. Lower Permian, Germany. Bar = 10 μm.

While the *Psaronius* example documents endophytes in aerial roots, the general organization and functioning of substrate roots is quite different. One of the inherent difficulties in studying substrate roots in the fossil record, irrespective of the quality of preservation, is their generally disarticulated occurrence, which makes it difficult to determine the source plant, especially if the roots contain only primary tissues. It is precisely these roots that one would expect to contain evidence of the colonization by endophytic fungi. One of these is a fungus from the Upper Pennsylvanian of France that consists of a meshwork of mycelia in the cortex of the calamite root *Astromyelon* (Taylor, T.N., et al., 2012). This fungus, *Cashhickia acuminata*, consists of aseptate, branched hyphae, each up to 6 μm in diameter, with acuminate tips that appear to enter the host from the outside and initially extend through the apoplast of the cortical cells. At some point, however, the hyphae invade individual cortical cells but are confined to only the outer periclinal and anticlinal cell walls (Figure 8.65). As a result, the hyphae are always directed toward the center of the root, perhaps somehow related to the position of the central conducting strand. The extent of intracellular penetration is variable among the various root specimens. While some display an extensive intracellular presence of the fungus with almost all of the cortical cells affected, others show more localized infections in which only one to a few adjacent cells contain hyphae. There is no consistent host response present when *C. acuminata* penetrates the cortical cell wall except for the presence of a delicate sheath of new wall (?) material around some hyphae (Figure 8.66). It is difficult to determine what the nutritional mode of *C. acuminata* may be. For example, this asymptomatic root endophyte may have conferred some level of nutrient exchange as a result of the increased surface area of the cell-wall-penetrating hyphae (Taylor, T.N., et al., 2012). Another interpretation is that the fungus represents a parasite or pathogen. To date *C. acuminata* has only been found in the roots of one group of vascular cryptogams (Sphenophyta); the group includes a single modern genus, *Equisetum*, which is known to contain a variety of fungal endophytes (e.g. Hodson et al., 2009).

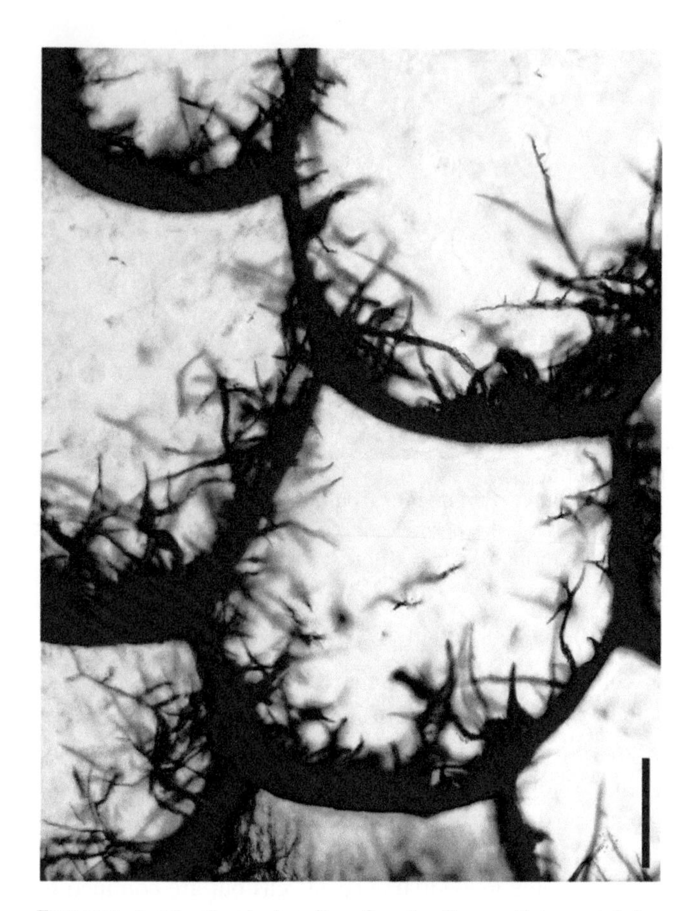

FIGURE 8.65 Cortical cells of calamite rootlet containing aseptate, branched intracellular hyphae of *Cashhickia acuminata* that arise from outer periclinal and all anticlinal walls, but not from inner periclinal cell walls. Upper Pennsylvanian, France. Bar = 20 μm.

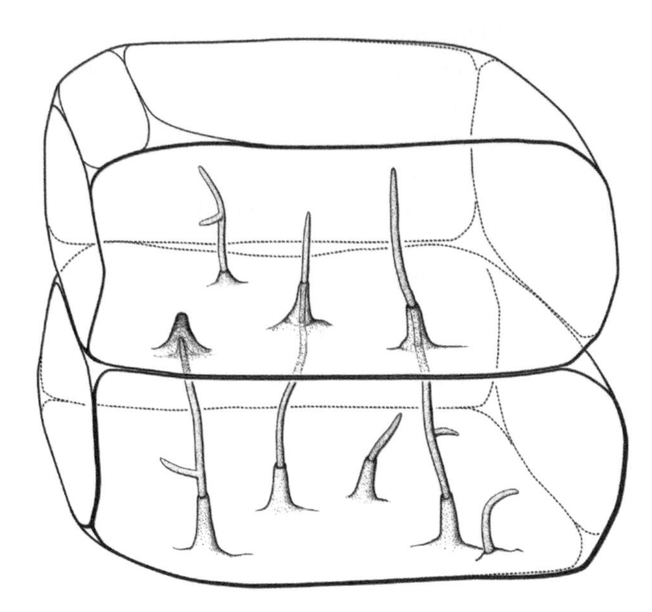

FIGURE 8.66 Diagrammatic reconstruction of two partially infected cortical cells, showing *Cashhickia acuminata* hyphae penetrating cell above and callosity formation around the basal portions of the hyphae. Center of root at top of diagram. (From Taylor, T.N., et al., 2012.)

FIGURE 8.67 Two microsclerotia in cells of the aquatic angiosperm *Eorhiza arnoldii* showing site of attachment to septate hypha. Eocene Princeton chert, British Columbia, Canada. Bar = 10 μm. (Courtesy A.A. Klymiuk.)

Klymiuk et al. (2013b) described a fossil DSE from the lower Eocene Princeton Chert of British Columbia, Canada. Microsclerotia (Figure 8.67) and monilioid hyphae (Figure 8.68) were found within roots of the vascular plant *Eorhiza arnoldii* and compare favorably to those of DSE fungi in extant plants; in addition, the plant host was asymptomatic for the infection.

LEAF ENDOPHYTES. Although leaves constitute a harsh habitat for fungi because nutrient availability is limited and leaves undergo extreme fluctuations in humidity, temperature, gas exchange gradients, and ultraviolet radiation (Goodman and Weisz 2002; Zimmerman and Vitousek, 2012), leaf endophytes are believed to represent a major component of today's fungal associations with plants (e.g. Arnold, A.E., 2007; Rodriguez et al., 2009; Wang, Z. et al., 2009). For example, more than one hundred species, mostly filamentous ascomycetes, have been identified in the leaves of the conifer *Pseudotsuga menziesii* (Carroll and Carroll, 1978), and an equally large number of fungi are associated with the leaves of *Pinus taeda* (Arnold et al., 2007). Fungal diversity in leaves includes both pathogenic and nonpathogenic forms (Dighton, 2003), and the relationships between the types may be based on the temporal and spatial diversity of resources (Bélanger and Avis, 2002). Irrespective of the tools (molecular) or data type analyzed (morphological), it is obvious that foliar fungal endophytes are ubiquitous among all major lineages of terrestrial plants and

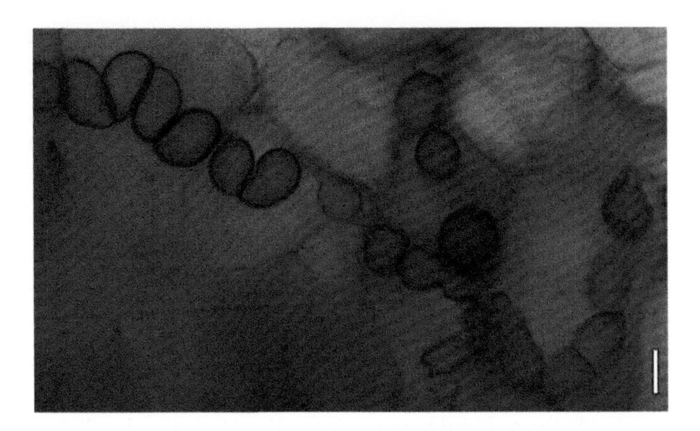

FIGURE 8.68 Monilioid hyphae in host cells of the aquatic angiosperm *Eorhiza arnoldii.* Eocene Princeton chert, British Columbia, Canada. Bar = 10 µm. (Courtesy A.A. Klymiuk.)

exist in a wide range of communities. It is estimated that the greatest diversity of fungal endophytes today occurs in tropical forest trees (e.g. Arnold and Lutzoni, 2007; Arnold, A.E., 2008; Gazis et al., 2011) and that in many, transmission occurs horizontally via raindrops.

Like roots, true leaves (microphylls and megaphylls) had evolved by the end of the Early Devonian (Taylor, T.N., et al., 2009), and it is reasonable to assume that endophytic fungi rapidly exploited these new niches. Fungal remains in leaves, however, are very rarely reported in late Paleozoic and early Mesozoic fossils (i.e. prior to the Cretaceous; e.g. Van der Ham and Dortangs, 2005), despite the fact that foliage fossils are perhaps the most intensively studied plant organs. Krings et al. (2012b) speculated that the absence of a well-defined record may be due in part to the fact that the majority of foliage fossils are preserved as impressions and compressions, which do not normally lend themselves to the presence of microscopic remains on the surface or in the interior. Another explanation may be related to the construction and tissue system organization in certain types of leaves that might influence preservation, and the impact on the potential habitat of the endophyte.

Fossil Leaf Endophytes. Perhaps more is known about the leaves of Carboniferous plants than any other group in the fossil record, including angiosperms. This is because these leaves have been collected in association with coal mining activities for several centuries, were relatively easy to obtain, and in general could be found as large specimens. Moreover, Carboniferous plants have been extensively studied anatomically because of their occurrence in the form of permineralizations (coal balls) and petrifactions (chert), and many of these studies focused exclusively on the leaves (e.g. Laveine, 1986; Laveine and Belhis, 2007; Reihman and Schabilion, 1976a,b).

One of the first reports of fungi in tissues of Carboniferous foliage was in pinnules of *Alethopteris* (Oliver, 1903), a leaf type produced by the Paleozoic seed fern *Medullosa.* The endophytes occur in small locules within the mesophyll cells close to the abaxial surface of the leaf, between veins. The structures were initially interpreted as sporangia (Renault, 1883) but later identified as fungal and named *Urophlyctites oliveranus* (Magnus, 1903). The fungus consists of delicate hyphae that are attached to solitary thin-walled spores, each with a delicate surface ornament. The fungus has been suggested to be some type of parasite, but there is no mention of any host response.

Another leaf-inhabiting fungus has been discovered in a structurally preserved fern pinnule fragment from the Upper Pennsylvanian of central France that is assignable to *Pecopteris*, a taxon used for pinnules that were produced by marattialean ferns, some true ferns, and at least one seed fern (Taylor, T.N., et al., 2009). The fungus found in this fragment is an intracellular endophyte of uncertain affinity, and it inhabits the cells of the hypodermis (Krings et al. 2009d). The fungus consists of branched, septate hyphae that produce long-necked hyphal swellings and structures that probably represent conidia (see Figure 8.62). Also present are irregular thick-walled structures, possibly representing some type of microsclerotia. Despite the excellent preservation of this fungus numerous details are still lacking, which makes it impossible to define the systematic affinities of this specimen. It may represent some type of ascomycete since most modern fungal leaf endophytes, including those that share features with this fossil specimen, belong to this group.

A few other leaf endophytes have been discovered in compression fossils through cuticular analysis; the evidence for the presence of fungi in these leaves consists of hyphae and spores in the parenchymatous mesophyll and conducting tissues (e.g. Barthel 1961; Krings 2001). For example, macerations of the seed-fern foliage type *Mariopteris nervosa* show anastomosing hyphae 6–15 µm wide and up to 18 µm long; no reproductive structures or spores are reported with the specimens (Barthel, 1961). The inability to place these fungi within the context of life-history biology precludes a determination of their affinities or their nutritional mode.

EPIPHYLLS – FUNGI ON LEAF SURFACES

Microbiologists often use the term phyllosphere to describe the leaf surface or total above-ground surfaces as a site for the establishment of various microorganisms including bacteria, yeasts, protozoa, and fungi (e.g. Leveau, 2006; Whipps et al., 2008). Our knowledge of the microbiology of the phyllosphere has historically trailed behind our knowledge of the microbiology of the rhizosphere, particularly with respect to fundamental questions regarding which microorganisms are present and what is their function (Vorholt, 2012). In general, the phyllosphere is interpreted as the surface of vascular plants; however, there is considerable microbial biomass also associated with bryophytes (Davey et al., 2009), and understanding the latter associations will be especially important because of the geological age of certain bryophyte lineages and the early terrestrialization of the Earth. Examining and testing microecological theories and concepts about the phyllosphere is a rapidly expanding area of study that uses proteogenomics to discover and evaluate the most abundantly expressed genes in the phyllosphere environment (Meyer and Leveau, 2012). Some researchers, however, prefer to use the term phyllosphere for the immediate area of the leaf and refer to the actual leaf surface as the phylloplane. Floroplane is a term used for the microbial community associated with flowers (e.g. Shade et al., 2013). We will use phyllosphere for the leaf surface as well as its interior when considering microbes and fungi associated with fossil leaves. Studying microorganisms on leaf surfaces must also include the tissue systems of the leaf, including the various (functionally different) types of cells (e.g. mechanical or resin secreting) as well as epidermal derivatives (e.g. trichomes) and other surface features (e.g. Shepherd and Wagner, 2012).

The outer surface of the leaf is covered by a waxy layer of insoluble polymers and lipids termed the cuticle, which also includes the cuticular membrane. It is estimated that the interface between the cuticle and the atmosphere is greater than the surface of all the continents if they were combined (Riederer and Schreiber, 1995) or, stated differently, there are approximately one billion square kilometers of worldwide leaf surface. Because surface topography directly affects the potential microhabitats that are available, techniques have been used to map habitats of extant leaf surfaces (e.g. Mechaber et al., 1996). So far as we are aware this has not been tried with fossil leaves or cuticle envelopes. These approaches underscore the extraordinary diversity of potential microbial and fungal niches in the ecosystems of the world today (Lindow and Brandl, 2003; Leveau, 2006), and presumably in the past. For example, the distribution of fossil fungi on *Fagus* (beech) leaves from the Neogene of Japan proved useful for making comparisons with extant epiphyllous fungi on modern beech leaves (Doi, 1983).

The physical environment of the leaf surface is constantly changing (e.g. changes in temperature and humidity, nutrient availability, UV radiation) and the difference between air and leaf temperature affects the development of fungal pathogens (e.g. Bernard et al., 2013). This results in a hostile environment for pathogenic fungal spores (e.g. Braun and Howard, 1994). One of the principal reasons for the success of these organisms in spite of the hostile conditions on host leaf surfaces is their ability to locate appropriate surfaces and then to elaborate specialized infection structures (Tucker and Talbot, 2001; Ikeda et al., 2012). Current research suggests that glycoproteins and proteoglycans are the main adhesives not only in microscopic fungi but also in the Peronosporomycetes (Epstein and Nicholson, 2006). Adding to the difficulty of fungi gaining a foothold on leaf surfaces is the fact that plants may secrete antimicrobial chemicals, including proteins, that represent a defensive strategy designed to inhibit fungal spore germination (Shepherd and Wagner, 2007). Plants also produce specialized epidermal derivatives in the form of oil and salt glands, laticifers, resin ducts, hydathodes, and trichomes, all structural barriers (but at the same time also potential entryways) in the leaf/fungus "arms race"; many of these structures also exude or are linked to the production of certain antimicrobial chemicals (Calo et al., 2006). In addition fungi and fungus-like organisms have effective means of penetrating the cuticular surface, including enzymatic destruction and the release of extracellular materials to the leaf surface, as well as other methods to penetrate the plant epidermis (Wang, Z. et al., 2009).

Most of the fossil epiphyllous fungi described to date belong to the Ascomycota and are therefore covered above. A fossil leaf-colonizing fungus obtained through cuticular analysis is represented by rosette-like thalli recovered from the Mississippian of Germany (Hübers et al., 2011). One of the thalli has dimensions up to 170 × 135 μm and is composed of a single layer of radially oriented, probably aseptate, hyphae. Hyphae either terminate within the thallus or dichotomize repeatedly to form fan-shaped structures composed of multiple (elongate) lobes at the thallus margin. Hyphal tips are rounded and usually slightly or distinctly wider than the subtending portion; some are more-or-less club shaped and characterized by two or three minute distal lobes. The presence of more pronounced cutinization around the margin of the thallus is interpreted as evidence to suggest that the fungus was a parasite, with the thickened cuticle representing a host response. In the center of the thallus are pores that may represent sites of penetration, and there is in fact some resemblance of the whole structure to hyphopodia. It is hypothesized that, as in modern leaf-inhabiting fungi, the fungus migrates through the apoplast and then through the anticlinal cell wall of neighboring epidermal cells (e.g. Koch and Slusarenko, 1990). Fungi have also been reported on the surfaces of carbonaceous *Glossopteris* leaves in the form of spores and hyphae, based on examination with a scanning electron microscope (Srivastava, 1993). Disruption of the tissues is the basis for the suggestion that these leaves were parasitized.

CONCLUSIONS: ENDOPHYTES AND EPIPHYLLS

Reports like some of those noted above suggest that if the focus of research were to be on fungi rather than leaves there might be far more information available about Paleozoic and Mesozoic leaf-colonizing fungi than is currently available. This in turn would expand information about potential host preference and would constitute an important source of information about fungal paleoecology, especially in the Carboniferous. On the other hand, if it can be documented that many genera of Paleozoic leaves actually lacked fungi, then a series of hypotheses might be advanced, including that perhaps some Paleozoic plants produced secondary metabolites that had fungicidal properties (Taylor and Osborn, 1996). Such a hypothesis would have to be tested within the constraints of a relatively consistent depositional system and with taphonomy and certain diagenetic parameters well delimited. Once such an analysis had been completed for a particular site, companion localities of the same or different age could be compared, as could depositional environments. As paleomycology moves forward there will no doubt be increasing use of chemical profiles for both the sediments and fossils, thus perhaps providing another method that might be useful in determining the existence of plants that were able to compete against certain types of fungal parasites.

Many of the structural and morphological features of modern fungi that colonize leaves today should also be in evidence in the past. Fossil leaves preserved in a variety of modes are no doubt the most common of all fossil plant parts. As a result, one would expect a far greater number of reports of both direct and indirect evidence of fossil leaf-inhabiting fungi, perhaps just as conspicuous a component of the paleoecosystem as they are today (Lindow and Brandl, 2003). As such, these fossils may provide some insight into a variety of ecological and evolutionary points of reference or questions. For example, among some groups it is known that certain leaf endophytes produce small, reduced fruiting bodies immersed in fungal or dead host tissue, while saprotrophic species have large intricate fruiting bodies (Wang et al., 2009). Other avenues of inquiry would focus on special structural mechanisms present in fossils, including life history strategies or patterns of development, which are similar in modern forms or morphologically closely related types. Are there any differences in host preference? Does a particular leaf type always have the same epiphylls or endophytes? Some modern leaf fungi colonize the midrib region of the leaf to a greater extent than the lamina, and closer to the tip than the leaf base (Cannon and Simmons, 2002). Is there any evidence in fossils for this type of tissue preference? These and many other areas will also be important sources of new information as paleomycology increases its contribution to ecosystem dynamics in the fossil record. Finally, at some localities anatomically preserved plants have been extensively studied and many of the plants reconstructed and now understood in great detail (e.g. Eocene Princeton chert locality; see Cevallos-Ferriz et al., 1991; Klymiuk et al., 2013b,c). For sites such as these, it will be especially productive in the future to seriously look for fungi and compare them with modern forms that occur on or in certain types of plant organs (e.g. seeds) or in association with certain types of plants (e.g. aquatic angiosperms, conifers).

9

BASIDIOMYCOTA

FOSSIL BASIDIOMYCETES 175

AGARICOMYCOTINA 177

Agaricales 178

Boletales 181

Hymenochaetales 182

Polyporales 183

Wood Rot 187

PUCCINIOMYCOTINA 193

Pucciniales 193

USTILAGINOMYCOTINA 194

Fossil Smuts 194

ECTOMYCORRHIZAE 195

Fossil Ectomycorrhizae 197

The Basidiomycota is a monophyletic group belonging to the crown group of fungi, with more than 31,000 species described to date (Kirk et al. 2008; Blair 2009); however, molecular and genetic studies indicate that there is an extraordinary diversity within this group yet to be discovered (e.g. Arnold et al. 2007; Kemler et al. 2009). Together with the Ascomycota, they comprise the subkingdom Dikarya (=higher fungi) within the kingdom Fungi (Figure 9.1). It is estimated that approximately 37% of the described fungi belong to the Basidiomycota. The diversity of cellular constructions in hyphal systems and basidiocarps is the expression of a long evolutionary history for the group. Basidiomycota today are important contributors to multiple levels of ecosystem functioning and have developed substrate dependencies of enormous ecological importance, especially in forest ecosystems (Oberwinkler, 2012). For example, they are effective in litter degradation and as degraders of different components in wood, including lignin (e.g. Tanesaka et al., 1993; Adl, 2003). The ability to remove lignin from plant materials, a key process for carbon recycling in forest ecosystems, is in fact mainly found in Basidiomycota (e.g. Peláez et al., 1995). Moreover, basidiomycetes represent an important food source for certain animals (fungivory), especially insects (e.g. Schigel, 2012). The macroscopic basidiocarps are particular rich in carbohydrates (structural polysaccharides) and proteins, including all the amino acids necessary for insect development and growth (Lundgren, 2009). Many basidiomycetes are parasites and causative agents of diseases in plants, while others produce a number of biochemical compounds that can be either beneficial or toxic to animals and humans. Some forms enter into mutualistic associations with a variety of other organisms; distinctive among these are ectomycorrhizae (e.g. Bonfante and Genre, 2008; Rinaldi et al., 2008), and symbioses with liverworts (e.g. Kottke et al., 2003), lichen symbioses (e.g. Lawrey et al., 2007), and fungus-farming (fungiculture) by ants and termites (e.g. Chapela et al., 1994; Mueller et al., 2005). A spectacular fossil example of fungus farming is seen in the galleries, nests, and fungus combs of fungus-growing termites discovered in upper Miocene–lower Pliocene sandstones of the northern Chad basin of Africa (Duringer et al., 2006, 2007). The fossil comb, termed *Microfavichnus alveolatus*, is a small-scale alveolar mass (3–8 ×2–4 cm). The base of the construction is flat to concave. It is entirely composed

DOI: http://dx.doi.org/10.1016/B978-0-12-387731-4.00009-8

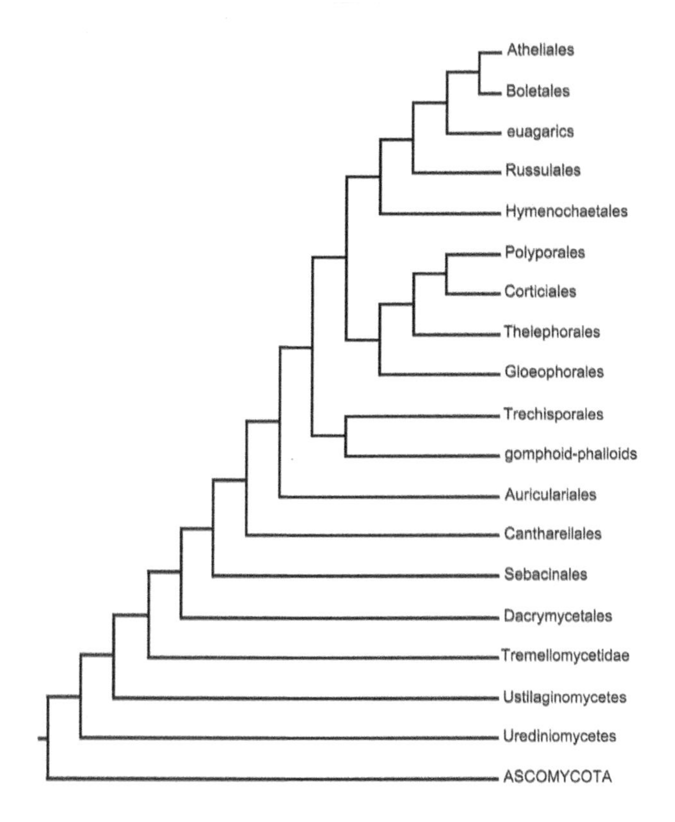

FIGURE 9.1 Suggested phylogenetic relationships within the Basidiomycota.

of a sub-horizontally layered, alveolar-like structure composed of a regular juxtaposition of millimeter-scale round pellets or balls (mylospheres).

Members of the Basidiomycota, sometimes termed the club fungi, are highly variable and include both terrestrial and aquatic species. This variability has evolved to the point where there is no single feature that is constant or unique to the group (Swann et al., 2001). The most diagnostic feature of the group is sexual reproduction via the production of meiospores on delicate projections (termed sterigmata) of a club-shaped structure termed the basidium (hence the name) (Figure 9.2). When basidiospores are forcibly ejected they are termed ballistospores (Money, 1998 [Figure 9.3]; Pringle et al., 2005). Two other features generally attributed to the Basidiomycota are clamp connections on some hyphae and the dikaryon as the extended vegetative phase (in contrast to the Ascomycota in which the dikaryon condition is usually restricted to the ascogenous system). The terms dikaryon and dikaryophase circumscribe a condition in which two compatible nuclei are brought together by cytoplasmic fusion (plasmogamy) without nuclear fusion (karyogamy; Anderson and Kohn, 2007). The two nuclei subsequently divide synchronously and, as a result, the binucleate condition is maintained. The clamp connection is a hyphal protrusion that develops at cell division to facilitate maintenance of the dikaryon condition by providing a bypass for one of the nuclei. Karyogamy only occurs within the basidium, leading to the formation of a zygote that immediately undergoes meiotic division.

The mycelium in the Basidiomycota is composed of septate hyphae; however, septum morphology is not uniform within the group (Van Driel, 2007). While some, such as the Pucciniomycotina, have septa with a pore morphology similar to that seen in the filamentous Ascomycota (see Chapter 8), but without Woronin bodies, others (i.e. Ustilaginomycotina) possess a septum with a septal pore that may have a slightly swollen rim around the pore and may also be associated with membrane caps or membrane bands. In addition, the Agaricomycotina have a barrel-shaped swelling around the pore, the dolipore (also called a *Verschlussband* or parenthesome), which is generally associated with a septal pore cap. Morphological features that have historically been used to characterize major lineages include the morphology of the basidium, certain ultrastructural features, and biochemistry (e.g. see references in Yang, 2011).

Currently, the phylum includes three major clades – the Agaricomycotina (mushrooms, jelly fungi, bracket fungi, and others), Pucciniomycotina (rusts), and Ustilaginomycotina (smuts and others) (Hibbett et al., 2007) – as well as the isolated Wallemiomycetes (Matheny et al., 2006; Padamsee et al., 2012). Until recently, a subdivision into Homo- or Holobasidiomycetes and Heterobasidiomycetes was also used. The former group is estimated to contain approximately 13,500 species and includes the gilled mushrooms and puffballs that produce meiotic spores on nonseptate basidial cells; the homobasidiomycetes are believed to have evolved multiple times (Hibbett et al., 1997a). The fruiting body of these fungi includes the most complex and highly evolved forms; however, approximately 15% of the described species of homobasidiomycetes are represented as simple crust-like forms (Hibbett and Binder, 2002). Interestingly, these latter resupinate forms could easily be overlooked in the fossil record. Other petrified fossil specimens thought to represent some type of gilled mushroom have been confused with burls, irregular woody structures produced by some trees in response to injury caused by various agents (e.g. M'Alpine, 1903). The Heterobasidiomycetes is a

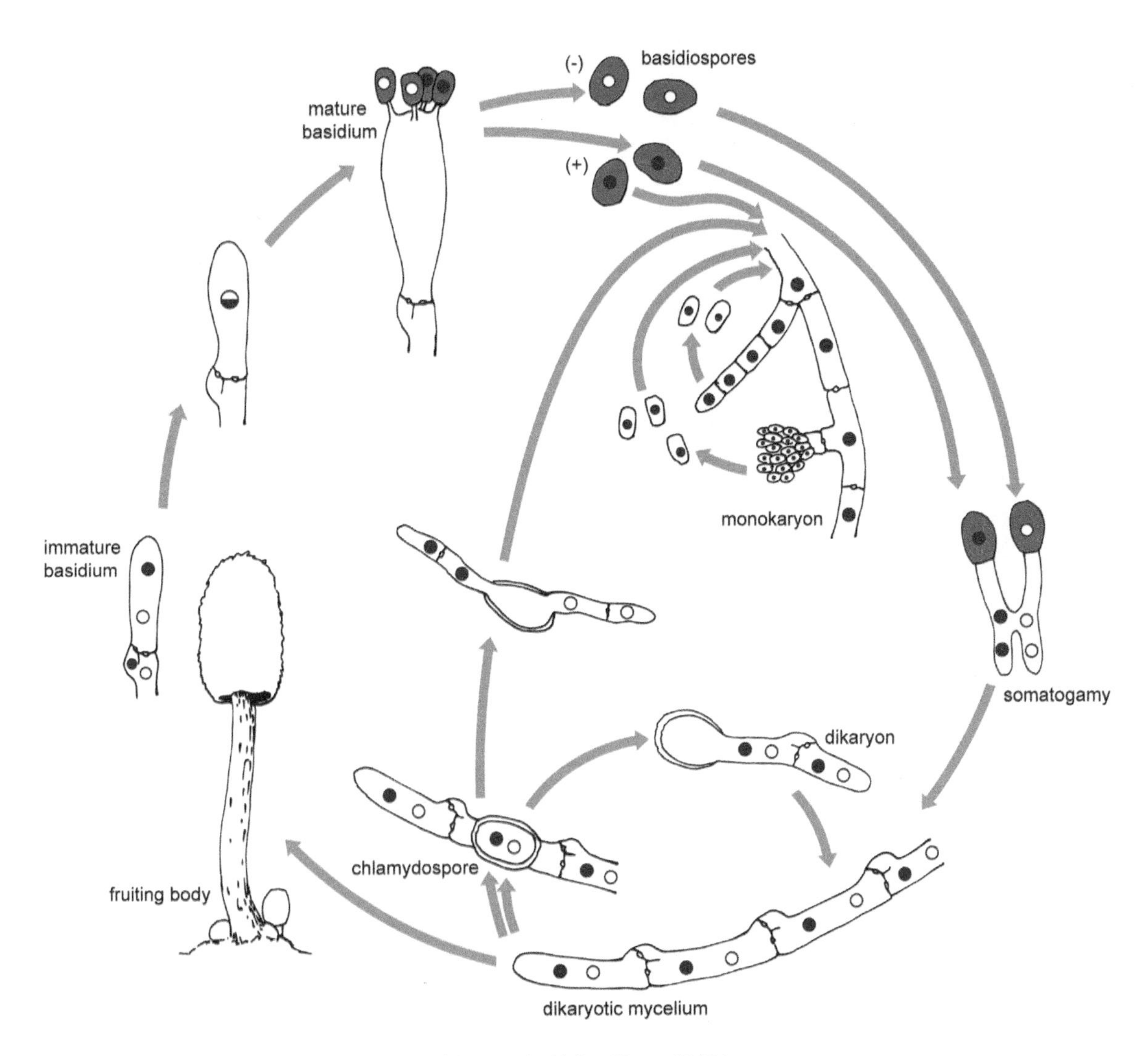

FIGURE 9.2 Life cycle of a typical basidiomycete (*Coprinus*). (After Esser, 1982.)

highly paraphyletic group characterized by septate basidia with large, irregularly shaped sterigmata and spores. In some classifications this group contains the jelly fungi, rusts, and smuts; however, phylogenetic studies indicate that this basidial character is best used only in a descriptive context and not to refer to formal taxa (Hibbett, 2007).

In the following sections, we survey the relatively small fossil record of the Basidiomycota that to date includes several fossils *incertae sedis*, and representatives of the Agaricomycotina and Pucciniomycotina. Persuasive fossil evidence of the Ustilaginomycotina has not been produced to date, with the exception of a few Cenozoic and subfossil spores (e.g. Kalgutkar, 1993; McAndrews and Turton, 2010).

FOSSIL BASIDIOMYCETES

Molecular clock estimates have been used to suggest that the first Basidiomycota arose during the Paleozoic, some 500 million years ago (Berbee and Taylor, 2001; Oberwinkler, 2012); other hypotheses suggest an even earlier origin (Heckman et al., 2001) depending on the calibration points used (see Taylor and Berbee, 2006). Based on this estimate of 500 Ma as a minimum age for Basidiomycota as a distinct lineage, it is reasonable to assume that these fungi should have been important elements in ancient continental ecosystems at least from the Late Silurian–Early Devonian onward. However, the extant diversity of Basidiomycota is in marked contrast

FIGURE 9.3 Nicholas P. Money.

FIGURE 9.4 Robert L. Dennis. (Courtesy G.W. Rothwell.)

to our current understanding of their paleobiodiversity. The preservation bias of fungi in general has certainly been a major influence in any attempts to document the paleodiversity and evolution of these organisms based on the fossil record. For such a large group of fungi, with several relatively distinctive morphological features (e.g. clamp connections, production of basidiocarps), it is interesting that the fossil record of the Basidiomycota has remained poor and the identity of the few documented fossils (especially those from the Paleozoic) largely equivocal. For example, the enigmatic nematophyte *Prototaxites* (Silurian–Devonian) has been suggested to represent a basidiomycete (Hueber, 2001), but there is no compelling evidence to date to demonstrate typical basidiomycete features (see Chapter 10, p. 214). There have also been a number of reports from the nineteenth and early twentieth centuries of Carboniferous fossils that were initially interpreted as basidiomycetes based on superficial resemblance to basidiocarps of modern polypores (e.g. Lindley and Hutton, 3 vols., 1831–1837; Lesquereux, 1877; Herzer, 1895; Hollick, 1910). Almost all of these reports were later questioned, however, and the specimens reinterpreted as non-fungal.

One structural feature that has been used to identify fossil basidiomycetes in the absence of sexual reproductive structures is the clamp connection. The first generally accepted basidiomycete body fossils with this feature occur within a structurally preserved fern rachis (*Botryopteris antiqua*) from the upper Visean (Mississippian) of France, a fossil approximately 330 Ma old (Krings et al., 2011f). Hyphae with clamp connections are common in the cells of the inner portion of the outer cortex but also occur in the inner cortex, xylem, and peripheral cortical layers. Clamp-bearing hyphae may be straight or slightly bent, or appear as sinuous tubular running hyphae (up to 2 μm wide) that pass from cell to cell largely unbranched. Clamp connections occur at irregular intervals. If branching occurs, it is usually (but not always) by hyphal proliferation through clamp formation. Other hyphae with clamp connections branch repeatedly and form more or less dense clusters that are confined to individual host cells or small groups of adjacent cells. Morphologically similar hyphae with clamp connections have been reported in a fern (*Zygopteris illinoiensis*) rhizome from the Middle Pennsylvanian (~305 Ma) of North America and formally described as *Palaeancistrus martinii* (Dennis, 1969, 1970; Figure 9.4). The straight septate hyphae of this fossil are up to 4.8 μm in diameter and branch at right angles. Individual cells are conspicuously elongate and bear well-defined clamp connections. Some hyphae terminate in swollen cells thought to be chlamydospores. Dennis (1970) compared *Palaeancistrus martinii* with hyphae of the extant *Panus tigrinus* (= *Lentinus tigrinus*), a wood-decaying polypore (e.g. Hibbett et al., 1994), and suggested that the fossil was a saprotroph. Other reports of hyphae

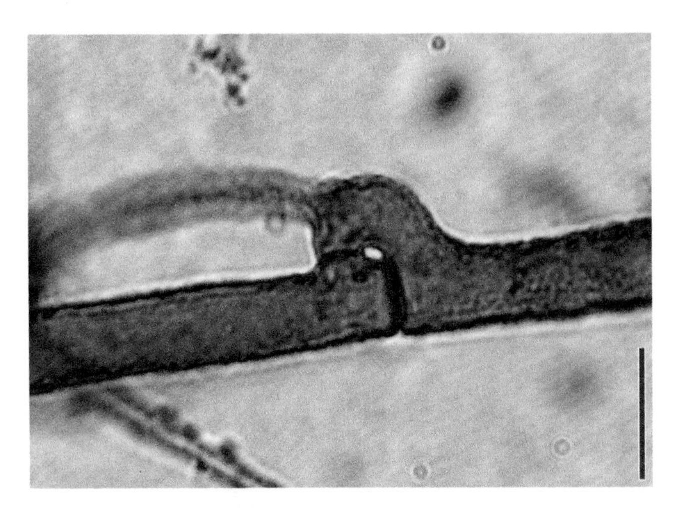

FIGURE 9.5 Hyphae with clamp connection and narrower hyphal branch extending from the clamp in the root mantle of a fossil fern (*Psaronius* sp.). Permian, Germany. Bar = 5 µm.

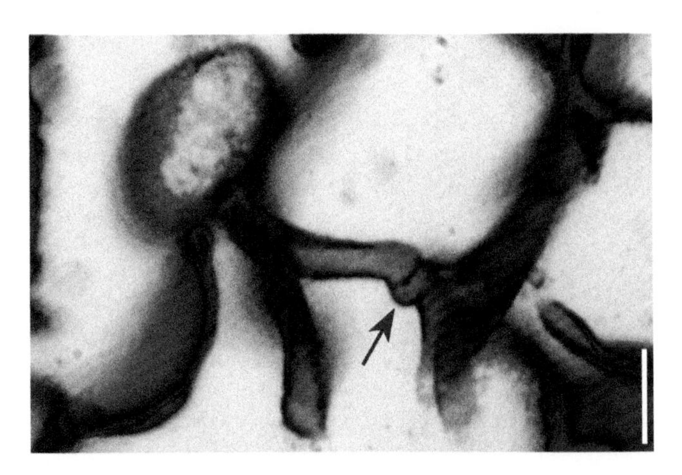

FIGURE 9.6 Clamp connection (*arrow*) of *Palaeofibulus antarctica*. Triassic Antarctica. Bar = 20 µm.

with clamp connections from the Paleozoic include hyphae growing within the root mantle of a *Psaronius* tree fern from the Lower Permian of Germany (Figure 9.5 Barthel et al., 2010).

Palaeofibulus is the name of a clamp-bearing fungus (Figure 9.6) from permineralized peat from the Triassic Fremouw Formation of Antarctica (Osborn et al., 1989; Figure 9.7). Septate hyphae range up to 7.5 µm in diameter, and extending from some of the hyphae are chains of ellipsoidal to globose spores up to 31 µm in diameter (Figure 9.8). While *P. antarctica* occurs in the same peat deposit as white rot fungi, the affinities and nutritional mode cannot be conclusively resolved.

FIGURE 9.7 Jeffrey M. Osborn.

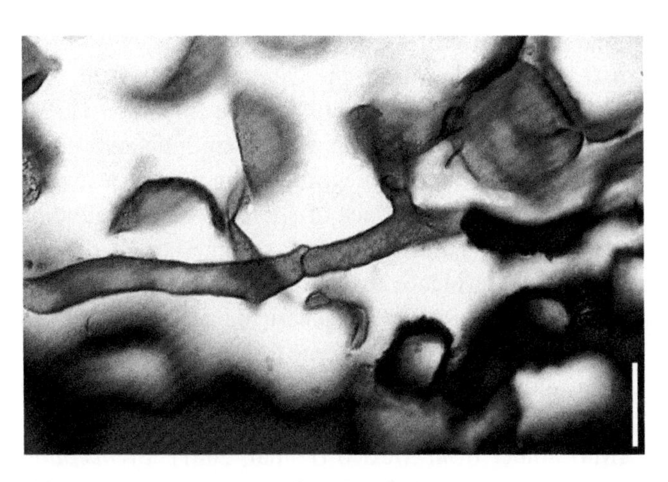

FIGURE 9.8 Overview of *Palaeofibulus antarctica*. Triassic Antarctica. Bar = 20 µm.

AGARICOMYCOTINA

This subphylum is one of the three major clades of the Basidiomycota and contains approximately 70% of the known basidiomycetes, with most of these including mushrooms, bracket fungi, puffballs and a few others. Included in this group are fungi that are parasites or saprotrophs, and some that form mycorrhizal associations.

AGARICALES

Gasteroid Fungi

The gasteroid fungi ("stomach fungi") is an artificial and broadly defined group of taxa not closely related to each other but historically placed in the obsolete class Gasteromycetes. They are, however, all characterized by basidia and basidiospores produced inside the basidiocarps, which consist of a central gleba (fleshy spore-bearing mass) surrounded by a peridium (e.g. Pegler et al., 1995). Most gasteroid fungi lack the ability to discharge spores ballistically, and therefore spores are released by the breakdown of the sporocarp. Included in this group are several orders including earthstars, stinkhorns, and puffballs. Molecular-based phylogenies suggest that these fungi and other forms with enclosed spore-bearing structures evolved at least four times (Hibbett et al., 1997a).

Several reports of fossils that have been included in the gasteroid fungi are viewed today as problematic or have since been discounted. For example, *Gasteromyces farinosus* was described as a Carboniferous gasteromycete associated with coals from the Urals (Ludwig, 1861). This interpretation was later questioned, however, and the fossil reinterpreted as representing "nothing more than an aggregate of spores and spore tetrads of some archegoniate plant" (Solms-Laubach, 1891). *Geaster florissantensis* was described from the Oligocene Florissant beds of Colorado by Cockerell (1908). The fossil consists of flattened ray-like structures up to 3.0 cm long and 1.1 cm wide. Although the rays are arranged in a pattern suggestive of an earthstar, the lack of spores in the surrounding matrix and the nature of the rays suggests that there is not sufficient evidence to consider this fossil a gasteromycete (Tiffney, 1981). The generic name *Geasterites* has also been used for this Oligocene fossil (Pia, 1927).

The oldest persuasive fossil evidence of gasteroid fungi comes from upper Oligocene–lower Miocene (~26–22 Ma) amber from Mexico (Poinar, 2001). *Lycoperdites tertiarius* is represented by a set of 23 specimens at different stages of maturity within a single piece of amber. The fruiting bodies are epigeous, cespitose, up to 24 mm high and 16 mm wide, and are pyriform in outline (Figure 9.9). The persistent peridium is three layered (two-layered exoperidium and one-layered endoperidium), and on the inner exoperidium are orderly arranged warts. There are 2–5 apical pores in the tip of the peridium. None of the specimens, however, has reached sporulation stage, and thus spores, basidia, and hymenium characteristics are not recognized. Poinar (2001) suggests that the colony of *L. tertiarius* originally grew in the rotting fork of a *Hymenaea* tree, and that a large accumulation of resin flowed down and covered the entire series of specimens.

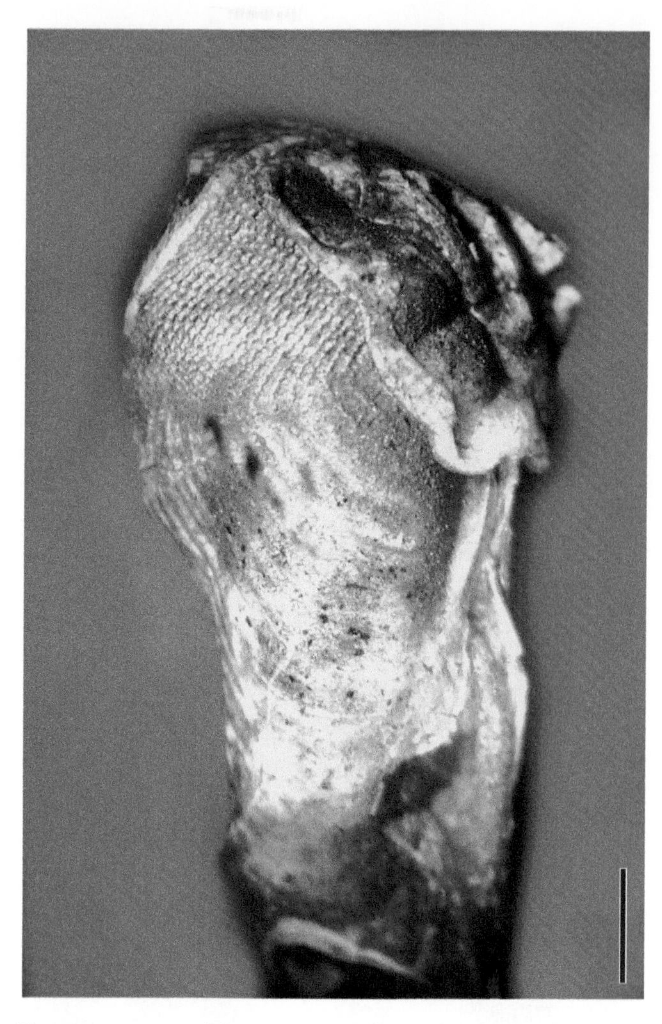

FIGURE 9.9 Partial outer exoperidium of the gasteroid fungus *Lycoperdites tertiarius* preserved in amber. Oligocene–Miocene, Mexico. Bar = 2.0 mm. (Courtesy G.O. Poinar, Jr.)

Geastrum tepexensis is a basidiocarp approximately 2.5 cm in diameter reported from the Miocene–lowermost Pleistocene of Puebla, Mexico (Magallón-Puebla and Cevallos-Ferriz, 1993; Figure 9.10). The fossil consists of a star-shaped exoperidium that is subdivided into 10 non-overlapping segments that are termed rays (Figure 9.11). Each ray may taper to an acute apex or may be blunt at the distal end. The central region is circular, approximately 1.3 cm in diameter, and forms the endoperidium; in the center is a conical protuberance that is interpreted as the ostiole. Although no spores were recovered from the surface of the exoperidium, a few were recovered from the surrounding rock matrix that were globose and approximately 7 μm in diameter. Although morphological features of

FIGURE 9.10 Sergio R.S. Cevallos-Ferriz.

FIGURE 9.11 Star-shaped exoperidium of *Geastrum tepexensis*. Miocene–Pliocene, Mexico. Bar = 5 mm. (Courtesy S.R.S. Cevallos-Ferriz.)

G. tepexensis are like those found in several extant species (e.g. from the genera *Geastrum*, *Astraeus*, and *Myriostroma*), in general the fossil appears most similar to species of *Geastrum*. One enigmatic fossil with possible affinities to the gasteroid fungi comes from the Cretaceous

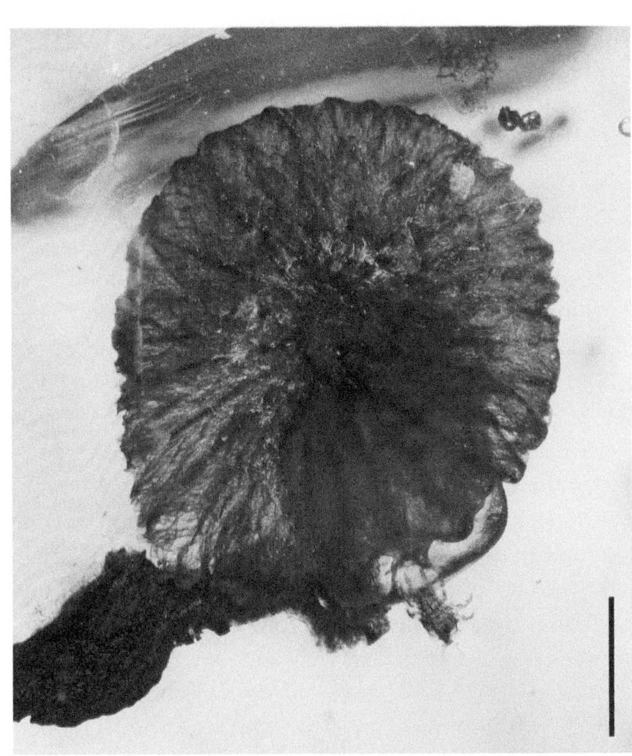

FIGURE 9.12 Pileus of *Coprinites dominicana* preserved in amber. Miocene, Dominican Republic. Bar = 1.0 mm. (Courtesy G.O. Poinar, Jr.)

of Mongolia (Krassilov and Makulbekov, 2003). *Geastroidea lobata* is approximately 2.5 mm in diameter and consists of an exoperidium that shows three lobes each up to 1.0 cm wide; other lobes extend into the matrix. There is a central ostiole and spores up to 10 μm long are reported. Comparison is made with the extant members of *Geastrum*.

A subfossil fungus from the Pleistocene of Alaska has been interpreted as a puffball assignable to the extant species *Bovista plumbea* (paltry puffball; Chaney and Mason, 1936). The specimen is compressed but within the peridium are capillitia; in the associated matrix are spores with an extended pedicle. Extant specimens are spherical to slightly elliptical and up to 10 cm in diameter. At maturity the outer layer of the peridium separates and spores are released through an apical pore.

Gilled Mushrooms

The fossil record of gilled mushrooms is exceedingly meager. All specimens described to date are preserved in amber, which again attests to the significance of amber in documenting the evolutionary history of fungi based on fossils. One of these amber fossils is *Coprinites dominicana* (Figure 9.12), described from the Eocene (now thought to be Miocene; Iturralde-Vinent, 2001) of the Dominican Republic (Poinar and Singer, 1990). The pileus of *C. dominicana* is 3.5 mm in largest diameter, convex with a small depression in the middle, thin-fleshed, and with a squamulose-pectinate surface. The margin is grooved to pleated and incurved. The lamellae are distinct, nondecurrent, and the hymenium bears basidia with spores. Spores are smooth, ellipsoid to oblong, 6 to 7 μm long, and have what appears to be a germ pore. The pileus hyphae are thin-walled, lack clamp connections and sclerified elements, and are broad and elongated at the pileus surface. The stipe diameter is 0.4 mm. As the generic name indicates, this important fossil is believed to have affinities with the modern genus *Coprinus*.

Another fungus from this group preserved in Dominican amber is *Protomycena electra* (Hibbett et al., 1997b). The pileus is 5 mm in diameter, circular, and convex, and the surface is glabrous at the center, becoming striate and translucent (pellucid) toward the margin; the margin is slightly flared and the veil absent. Primary lamellae are distant from each other (6–8 total), moderately broad, and broadly attached to the stipe apex. The stipe is curved, cylindrical, and smooth or minutely textured (gas and liquid-filled bubbles, apparently drawn out of the mushroom during preservation, make it difficult to observe features of the surface). Basal mycelium, rhizoids, or other anchoring structures are absent. The fossil is similar in overall appearance to members in the modern genus *Mycena*, a genus that includes numerous bioluminescent species (Desjardin et al., 2007, 2010). It has been hypothesized that bioluminescence in these white-rot fungi may attract certain arthropods, aiding in spore dispersal and deterring fungivores, or is a side-product of lignin degradation (Desjardin et al., 2008).

Aureofungus yaniguaensis is the third agaric known from Dominican amber (Hibbett et al., 2003). This fossil is characterized by a pileus that is 3 mm broad, convex, and with a broad raised center; the margin is incurved and striated. The lamellae are subdistant and have smooth margins; lamellulae or anastomoses are absent. The stipe is inserted centrally and is cylindrical and smooth; annulus, volva, and rhizoids are absent. Basidiospores are broadly elliptical and appear to be pigmented. The shape and stature of the fruiting body suggest that *A. yaniguaensis* might be related to the smaller pale-spored genera traditionally classified as Tricholomataceae. The first fossils found that morphologically resemble extant bird's nest fungi come from Dominican and Baltic amber (Poinar, 2014). Like modern members of the Nidulariaceae, the funnel-shaped fruiting body of the fossils is characteristic of the group, which appears to have been well established by the Eocene.

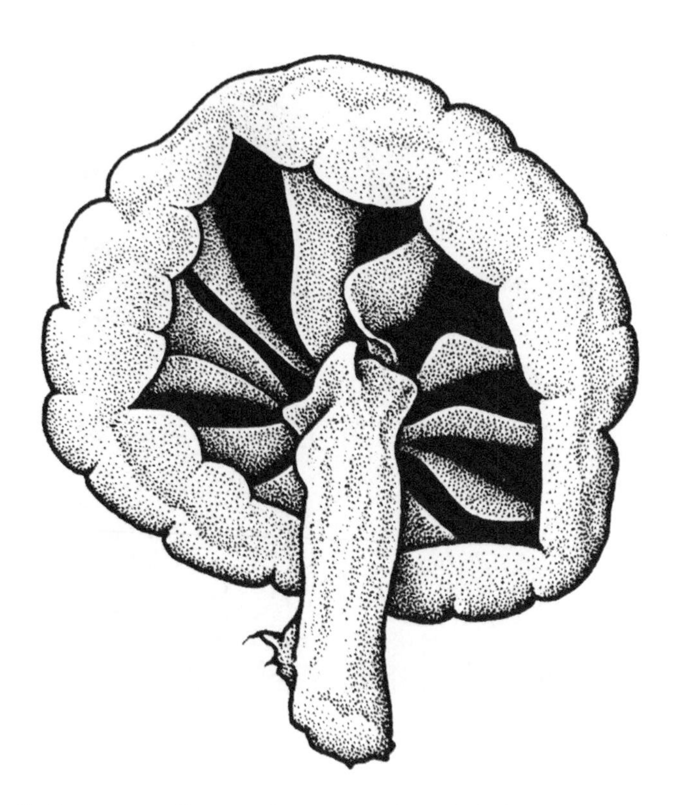

FIGURE 9.13 Suggested reconstruction of *Archaeomarasmius leggetti* based on a fossil from Upper Cretaceous amber, New Jersey, USA. (Courtesy D.S. Hibbett.)

Other fossil agarics are more difficult to accurately place within a phylogenetic framework. Some of these are included tentatively in the Tricholomataceae. One of these is *Archaeomarasmius leggetti* (Hibbett et al., 1997b), which occurs in Atlantic Coastal Plain amber from East Brunswick, New Jersey (USA); the amber is Cretaceous in age (Grimaldi et al., 1989). The basidiocarp is characterized by a pileus that is circular, plano-convex, thin-fleshed, radially sulcate, and up to 6 mm in diameter (Figure 9.13). The margin is incurved and the surface glabrous to minutely textured; a veil is absent and the context (inside flesh of the basidiocarp) is thin. Lamellae are distant to subdistant, without lamellulae, less than 1 mm wide at the widest point, attached to stipe apex, noncollariate, and nonanastomosed; edges are entire. The stipe is centrally inserted, cylindrical, smooth, and exannulate. Basidiospores are broadly ellipsoid to ovoid, smooth or possibly minutely textured, and possess a distinct hilar appendage. They are ellipsoidal and 7.3 × 8.3 μm. The fossil appears similar to the extant genera *Marasmius* and

FIGURE 9.14 David S. Hibbett.

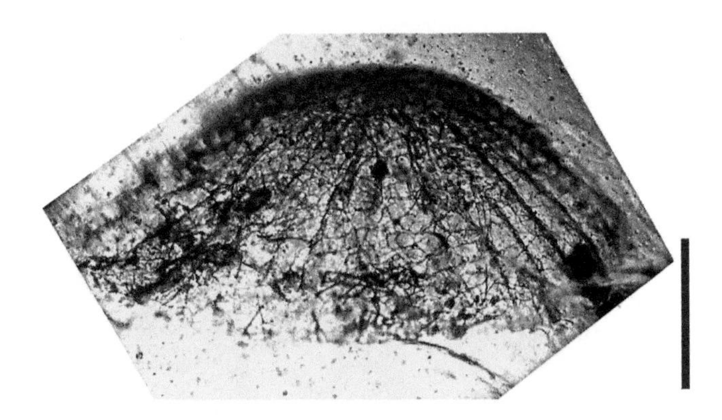

FIGURE 9.15 Pileus of *Palaeoagaricites antiquus* preserved in amber and covered with the mycelium of the mycoparasite *Mycetophagites atrebora.* Cretaceous, Myanmar. Bar = 0.5 mm. (Courtesy G.O. Poinar Jr.)

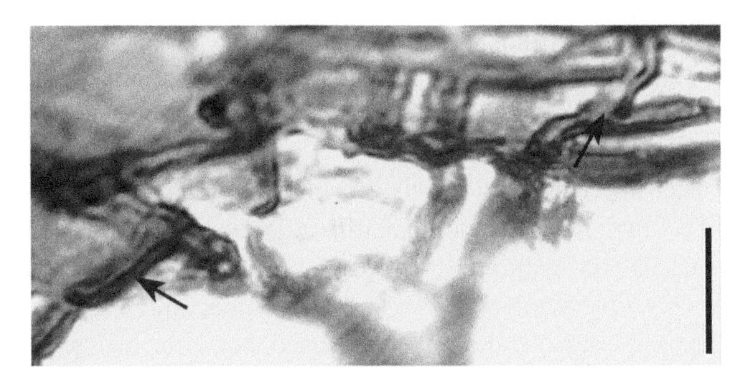

FIGURE 9.16 Mycoparasite *Mycetophagites atrebora* parasitized by the hyperparasite *Entropezites patricii* (*arrows*). Cretaceous amber, Myanmar. Bar = 20 µm. (Courtesy G.O. Poinar, Jr.)

Marasmiellus, two taxa distinguished on basidiome morphology (Mata et al., 2004). Although both morphological and molecular evidence suggest that there has been considerable convergence and parallelism in the evolution of homobasidiomycete fruiting body morphology, the report of *Archaeomarasmius* and *Protomycena* from the Miocene suggests that in at least some lineages, there has been little morphological change (Hibbett et al., 1997b; Figure 9.14).

Palaeocybe striata is an Eocene fungus described from a single specimen from the Bitterfeld amber in Germany (Dörfelt and Striebich, 2000). The delicate, membranaceous cap is approximately 1.6 mm in diameter, hemispherical, and somewhat incurved at the margin. The underside is characterized by radiating ribs that are reminiscent of lamellae. The stalk is up to 0.3 mm wide and constructed of tubes, each 25–35 µm in diameter. More recently, however, additional specimens were found that dismissed the identification of *P. striata* as a fungus. The taxon was subsequently reinterpreted as an indusium of a matoniaceous fern and renamed *Matonia striata* (Schmidt and Dörfelt, 2007).

The oldest fossil evidence to date of gilled mushrooms comes from Burmese amber that is Early Cretaceous in age (Poinar and Buckley, 2007). This fossil, named *Palaeoagaricites antiquus*, consists of a portion of a cap, 2.2 mm in diameter, with 16 radial furrows (Figure 9.15). The lamellae are subdivided into short sections 85–110 µm long. The sides of the lamellae are covered in basidia; the ovoid basidiospores range from light to dark in color, and are 3.0–4.5 µm long. What is especially interesting about this fossil is that the agaric is parasitized by a mycoparasite, *Mycetophagites atrebora*, which in turn is parasitized by a hyperparasitic fungus, *Entropezites patricii* (Figure 9.16). The precise systematic position of *P. antiquus* remains unresolved; however, size, spore shape, pileus features, and likely habitat (conifer trunks) suggest affinities with the genera *Mycena*, *Marasmius*, or *Collybia* of the family Tricholomataceae.

BOLETALES

This is a morphologically diverse group of fungi, which includes stipitate-pileate forms that have tubular or lamellate hymenophores, or intermediates that show transitions between the two hymenophore types, but also includes

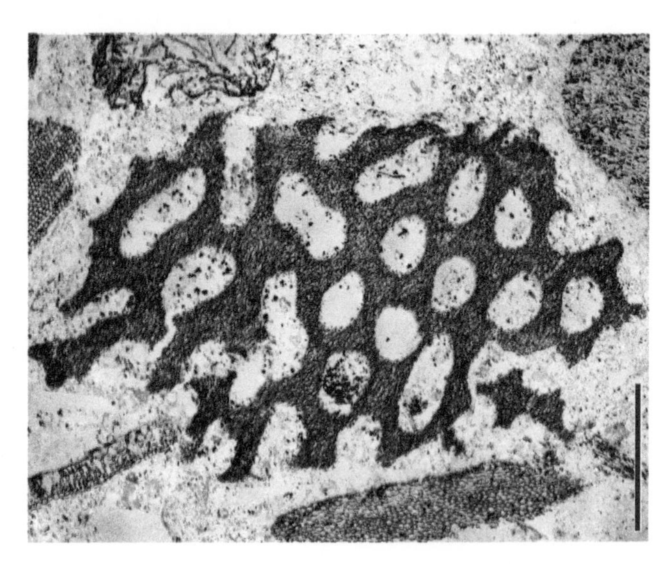

FIGURE 9.17 Partial transverse section of the permineralized poroid hymenophore of *Quatsinoporites cranhamii.* Lower Cretaceous, Canada. Bar = 1.0 mm. (Courtesy S.Y. Smith.)

FIGURE 9.18 Selena Y. Smith.

gasteromycetes (puffball-like forms), resupinate or crust-like fungi that produce smooth, merulioid (wrinkled to warted), or hydnoid (toothed) hymenophores. There is also a single polypore-like species (Binder and Hibbett, 2006). Most Boletales are mycorrhizal, but some forms are saprotrophs, and others mycoparasites. *Scleroderma* is the name of an extant ectomycorrhizal fungus included in this order that has a persistent outer wall (peridium) that splits at maturity. The name has also been used for isolated spores that are small (11–15 µm) and lack a preformed opening. On the surface are delicate spines, except in one region that is believed to be the point of attachment to the sterigma. Specimens of *S. echinosporites* (spores) are reported from the Burrard Formation in British Columbia (Canada), which is Eocene in age (Rouse, 1962).

HYMENOCHAETALES

This group lacks morphological features that are consistent across all forms. In general the group is defined based on molecular characters and usually includes corticioid and poroid types (Larsson et al., 2006).

Two poroid hymenophores preserved in marine calcareous nodules are known from the Cretaceous and Eocene of Vancouver Island, British Columbia, Canada (Smith et al., 2004). *Quatsinoporites cranhamii* is a Cretaceous specimen, at least 5 mm wide, with elliptical to round pores, each up to 540 µm in diameter. The hyphal system is described as monomitic, with simple, septate generative hyphae up to 2.5 µm in diameter (Figure 9.17). Associated hyphae are larger (5 µm in diameter) and characterized by a rugose wall ornament. Within the pores are cystidia, each up to 54 µm long; basidia and basidiospores are not known. Another hymenophore from the same geographic area, but of Eocene age, is *Appianoporites vancouverensis* (Smith et al., 2004; Figure 9.18). The tubes in this fragment are approximately 160 µm in diameter (Figure 9.19).

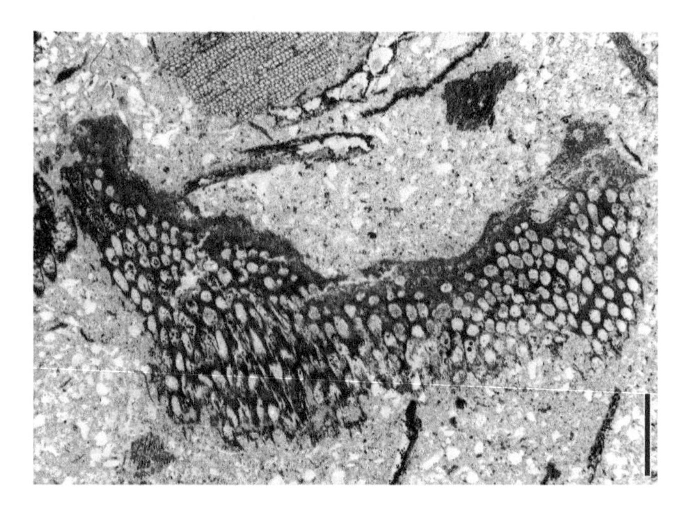

FIGURE 9.19 Oblique transverse section of permineralized poroid hymenophore of *Appianoporites vancouverensis.* Eocene, Canada. Bar = 1.0 mm. (Courtesy S.Y. Smith.)

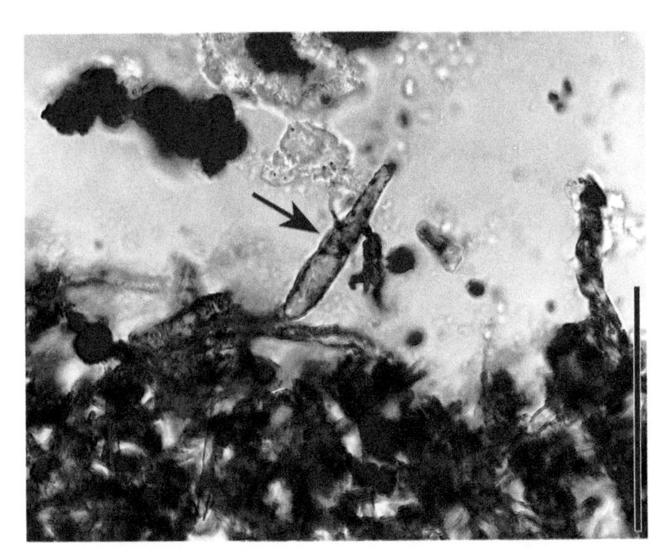

FIGURE 9.20 Detail of *Quatsinoporites cranhamii* ampulliform cystidium (*arrow*). Lower Cretaceous, Canada. Bar = 50 μm. (Courtesy S.Y. Smith.)

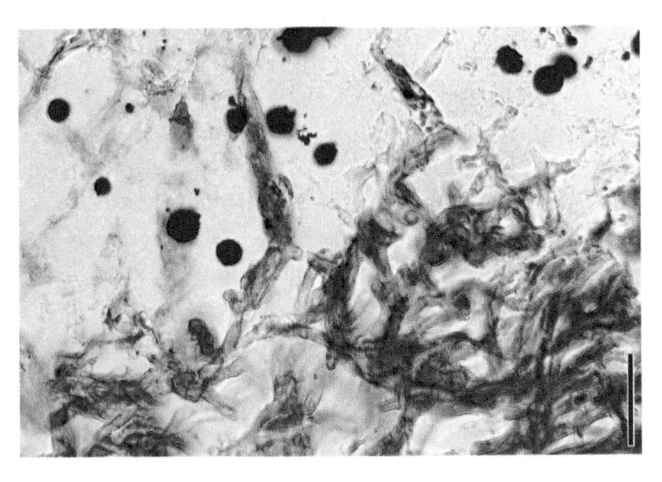

FIGURE 9.21 Cystidia of *Appianoporites vancouverensis.* Eocene, Canada. Bar = 15 μm. (Courtesy S.Y. Smith.)

Although both specimens from Vancouver Island are not complete, the two forms have some differences based on the shape of the pores and size of the cystidia (Figures 9.20, 9.21). It has been suggested that both specimens belong to the Hymenochaetales, although there are modern representatives of the Polyporales (e.g. genus *Phaeolus*) that share similar features.

POLYPORALES

The polypores or polyporous fungi form a large and morphologically heterogeneous group of basidiomycetes that is comprised of over 1000 extant species in many families

FIGURE 9.22 Arthur Hollick.

(e.g. Binder et al., 2013). They exist as saprotrophs (mostly secondary invaders) in partially or extensively decayed tree stumps or logs, or as parasites and perpetrators of mild to severe diseases in conifers and hardwoods. Most are woody and produce basidiospores on the walls of tubes on the lower surface of the hymenophore. The fruiting bodies are sometimes termed conks, shelf, or bracket fungi. The polypores represent an ecologically important element in the nutrient cycles of natural forest ecosystems, due to their ability to degrade different components in wood, including lignin (Ryvarden, 1991). Many polypores show strong antimicrobial activity (Zjawiony, 2004).

Based on a number of Carboniferous fossils, the fossil record of the polypores was historically believed to reach back into the Paleozoic. Forms such as *Dactyloporus archaeus* and *Incolaria securiformis* (Herzer, 1895), *Pseudopolyporus carbonicus* (Hollick, 1910; Figure 9.22), *Polyporites bowmanii* (Lindley and Hutton, 1831–1833; Meschinelli 1892; Figure 9.23), and *Rhizomorpha sigillariae* (Lesquereux, 1877) are species that morphologically were thought to represent some type of polyporous basidiocarp. Almost all these reports of Paleozoic forms, however, were later questioned and the specimens reinterpreted as non-fungal (Pirozynski, 1976a).

FIGURE 9.23 Line drawings of the alleged basidiocarp *Polyporites bowmanii*. Carboniferous, Great Britain. Bar = 1.0 cm. (From Lindley and Hutton, 1831–1833.)

FIGURE 9.24 George R. Wieland.

Likewise, reports of polyporous basidiocarps from the Triassic (Playford et al., 1982; Fohrer and Simon, 2002, 2003), Cretaceous (Wieland, 1934; Brown, 1936; Figures 9.24, 9.25), and Jurassic (Singer and Archangelsky, 1957, 1958) have been questioned (Hibbett et al., 1997c; Kiecksee et al., 2012) or the affinities remain inconclusive to date. Moreover, distinctive coloration patterns in silicified wood from the Triassic of Germany have been interpreted as indirect evidence ("biosignatures") of wood-decaying polypores (Nikel, 2011); however, this interpretation appears to have little basis in fact.

FIGURE 9.25 Roland W. Brown.

Cretaceous and Cenozoic deposits in Europe (e.g. Straus, 1950, 1952a, b, 1956; Dierssen, 1972; Kreisel, 1977; Knobloch and Kotlaba, 1994; Fraaye and Fraaye, 1995; Jansen and Gregor, 1996), North Africa (e.g. Zuffardi-Comerci, 1934; Koeniguer and Locquin, 1979; Locquin and Koeniguer, 1981; Fleischmann et al., 2007), North America (e.g. Mason, 1934; Chaney and Mason, 1936; Buchwald, 1970; Smith et al., 2004), and Japan (Tanai, 1987) have yielded several well-preserved and persuasive specimens of fossil polypores. Moreover, subfossil (Quaternary) basidiocarps have been reported from various archaeological excavation sites and peat-cutting areas in Europe and North America (e.g. Monthoux and Lundstrom-Baudais, 1979; Purdy and Purdy, 1982; Gennard and Hackney, 1989; Chlebicki and Lorenc, 1997; Peintner et al., 1998; Bernicchia et al., 2006; Kreisel and Ansorge, 2009).

Some of these fossils have been formally named, for example *Parapolyporites japonica* (Tanai, 1987) from the Miocene of Japan, and *Fomes mattirolii* from the Miocene of Libya (Zuffardi-Comerci, 1934), while others lack sufficient diagnostic features. In *P. japonica* the carbonaceous compression is attached to a coalified fragment of wood. The basidiocarp is semicircular in outline and gradually thins to the outer margin. The lower surface is smooth and contains numerous pores up to

1.8 mm in diameter; tubes are approximately 1 mm long. Spores were not described.

Trametites eocenicus is the name applied to what is interpreted as a fossil polypore from the Eocene of Bohemia (Knobloch and Kotlaba, 1994). The specimen is semicircular, has an imbricate margin, and is approximately 7.5 cm in the widest dimension. While the overall morphology might suggest affinities with a bracket fungus such as the extant *Trametes*, the fossil is a sandstone impression, and thus detail of what might be the surface, containing pores, is absent. Another extant bracket fungus that is similar to *Trametes*, *Lenzites warnieri* (now considered a synonym of *Trametes*; see Justo and Hibbett, 2011), a white rot pathogen, has been reported from the Pleistocene of Germany (Kreisel, 1977). Other species of *Trametites* have been reported from the Cretaceous and Pliocene (Meschinelli, 1892).

Identifying bracket fungi in the fossil record is difficult because of the variable morphologies and general appearance that can simulate degraded wood. Even when structures that closely resemble pores are preserved, the specimen may not be a polypore (e.g. Brown, 1938). For example, *Polyporites brownii* (Wieland, 1934) and *P. stevensonii* (Brown, 1936), both from the Cretaceous, are no longer considered polyporous fungi. Nevertheless, there are some excellent specimens of polypores that have been described from the fossil record. One of these is *Fomes idahoensis*, a portion of a bracket from the Pliocene of Idaho, USA (Brown, 1940). The partially calcified specimen is approximately 13 cm long and 4.5 cm thick; on the smooth upper surface are convex-rounded ribs that are interpreted as increments of growth. On the lower surface are numerous, closely spaced pores (~750/cm^2); no other features are reported. Although the basidiocarp occurs in sediments that contain fossil wood of several types, there is no indication that the polypore was associated with any of these specimens. The fossil is compared to the extant species *Fomes pinicola* (Buchwald, 1970). Additional specimens of *F. idahoensis* of the same age have been reported that are slightly larger (Andrews, 1948) and show some evidence that the bracket was composed of a mycelium (Andrews and Lenz, 1947). Other *Fomes* fossils have been reported from the Pleistocene of Alaska (Chaney and Mason, 1936). While the specimens were not sectioned, their morphology, including the presence of what are interpreted as growth lines, suggest that they represent bracket fungi. *Lithopolyporales* is the name of a bracket fungus reported from the Cretaceous of central India (Kar et al., 2003). Specimens of *L. zeerabadensis* are up to 19 cm long and 15 cm wide; the largest is 6 cm thick. Although the specimens show some structure, it is difficult to determine whether sections show biologic features or represent crystal or diffraction patterns.

Several fossil basidiocarps have been placed in the extant family Ganodermataceae, a group that contains more than 250 extant taxa (Ryvarden, 1991; Moncalvo and Ryvarden, 1997; Buchanan, 2001). Many species are important white rot fungi that damage forest trees. Although the Ganodermataceae is a cosmopolitan family, most species occur in the tropics. This family represents the most difficult polypore family to identify due primarily to the high intraspecific variability in basidiocarp gross morphology (e.g. Ryvarden, 1991; Seo and Kirk, 2000; Smith and Sivasithamparam, 2003). Nevertheless, most members can be recognized microscopically because they produce distinct ellipsoid basidiospores (i.e. the so-called ganodermatoid spores; e.g. Adaskaveg and Gilbertson, 1988; Ryvarden and Gilbertson, 1993), which are characterized by a double wall consisting of an inner, ornamented endosporium and outer, hyaline exosporium. Endo- and exosporium are separated from one another by delicate inter-wall structures termed pillars.

Several Cenozoic basidiocarps have been directly referred to the Ganodermataceae, including *Ganodermites libycus* and *Archeterobasidium syrtae* from the middle Miocene of northern Libya (Koeniguer and Locquin, 1979; Fleischmann et al. 2007), and a number of *Ganoderma* spp.: *G. adspersum* from the Miocene of the Netherlands (Fraaye and Fraaye, 1995), *G. applanatum* and *G. lucidum* from the Pleistocene of Alaska (Chaney and Mason, 1936), and *G. lucidum* and *G. lipsiense* (= *G. applanatum*) from the Pleistocene and Holocene of Poland (see Kopczyński, 2006). In addition, several Miocene to Pliocene basidiocarps from central Europe are currently included in genera such as *Fomes* and *Trametites* (e.g. Skirgiełło, 1961; Gregor, 1980, 1994; Jansen and Gregor, 1996); however, as these have not been studied in detail, it is not yet possible to determine which, if any, actually belong to the Ganodermataceae. For many of these the description is based on the shape of the structure since most lack other details, including information about the presence of pores.

The most detailed description of a fossil representative of the Ganodermataceae is provided by Fleischmann et al. (2007) for *Ganodermites libycus*, a permineralized (silicified), applanate-ungulate (hoof-shaped), and semicircular-dimidiate basidiocarp from the middle Miocene

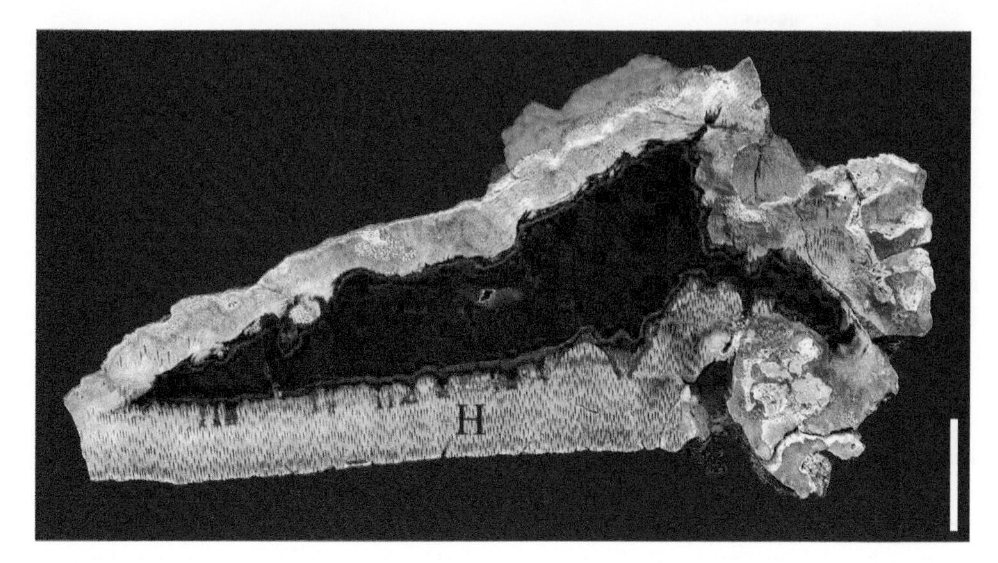

FIGURE 9.26 Longitudinal thick section (polished surface) through basidiocarp of *Ganodermites libycus* showing hymenophoral strata (H). Lower Miocene, Libya. Bar = 1.0 cm.

of North Africa characterized by a stratified (layered) hymenium (Figure 9.26). The individual hymenophoral strata are up to 6.5 mm thick; the context is up to 5 mm thick. Pores are circular to broadly elliptical in outline, generally equidistantly arranged, and with an average of 4 to 6 pores per millimeter. The pores possess thick dissepiments (septa) and contain aggregations of basidiospores that form a dark layer around the inner surface of the pore. The hyphal system is trimitic: (1) Generative hyphae are usually 1.0 μm in diameter (range = 0.2–1.5 μm), colorless, thin walled, septate, rarely branched, with clamps, and are in a parallel arrangement in the hymenium. They taper toward the apex. (2) Skeletal hyphae are thick walled, colorless, non-septate, and characterized by frequent dendritic branching. The main skeletal hyphae are 1–1.5 μm in diameter (range = 0.5–2.0 μm), while the branches are narrower and their tips taper toward an acute apex. (3) Binding hyphae closely resemble the skeletal hyphae (at least in the fossil). They are relatively thick walled, 1–2 μm in diameter, repeatedly branched, and non-septate. In addition, thick-walled clavate pilocystidia (Figure 9.27), some of which are similar to those reported by Moncalvo (2000, fig. 2.1e), regularly occur among the tramal hyphae. Basidia are clavate, 2–4 sterigmate, and possess a basal clamp (Figure 9.28). Most of the basidiospores appear irregular or collapsed, but a few indicate that when mature, spores were ellipsoid in outline and possessed a ganodermatoid wall composed of two layers separated by delicate inter-wall pillars. Fraaye and Fraaye (1995) note that extant *Ganoderma* fungi often contain evidence of insects in the form of

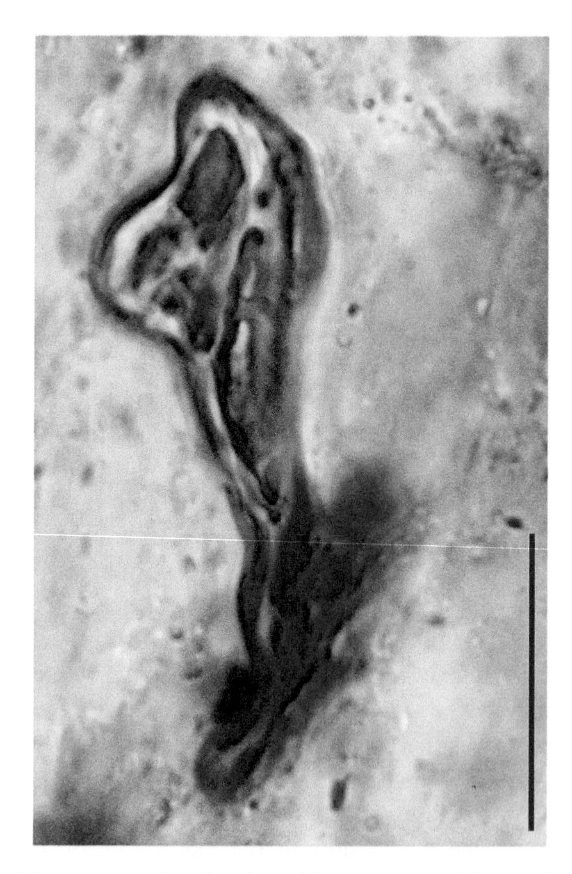

FIGURE 9.27 *Ganodermites libycus* pilocystidium. Lower Miocene, Libya. Bar = 15 μm.

dipteran larvae within the tube layers. Future paleomycological studies of permineralized forms of these fungi might yield evidence of this association in the form of coprolites.

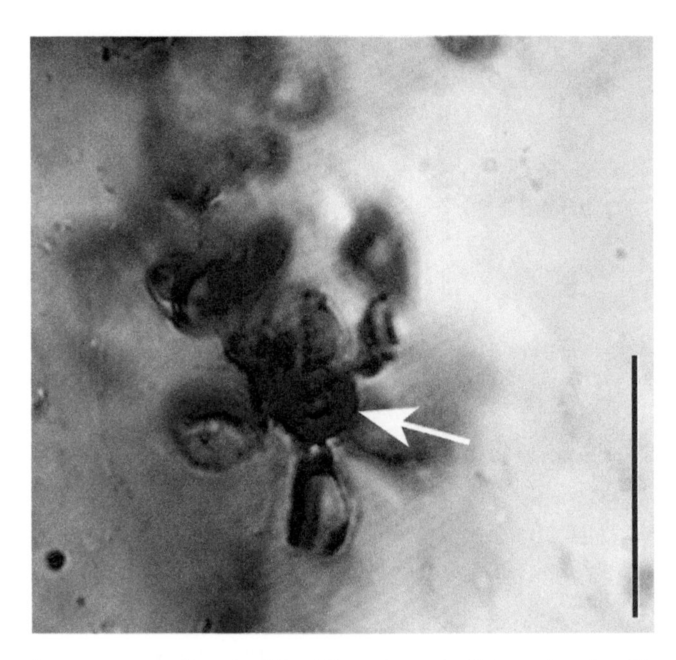

FIGURE 9.28 Detail of basidium (*arrow*) of *Ganodermites libycus.* Lower Miocene, Libya. Bar = 15 μm.

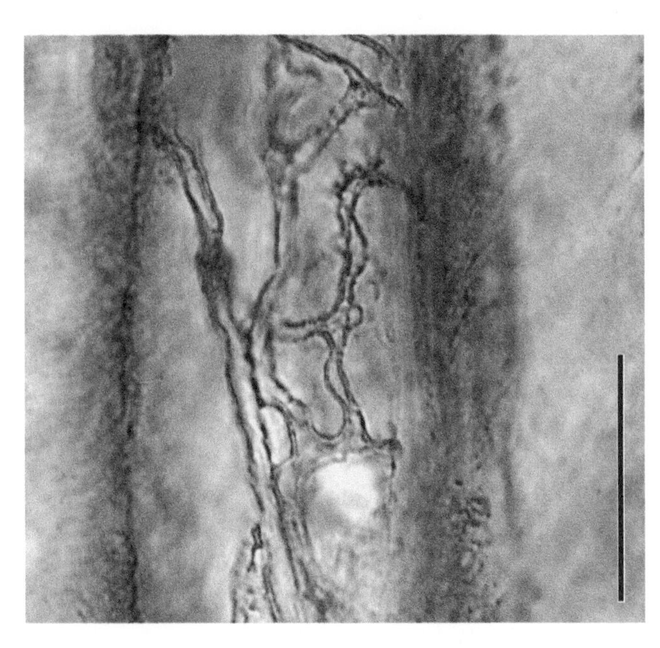

FIGURE 9.29 Longitudinal section of conifer tracheid showing fungal hyphae in lumen. Miocene, Germany. Bar = 30 μm.

Archeterobasidium syrtae is represented by two permineralized basidiocarp fragments from the Miocene of Libya (Koeniguer and Locquin, 1979). These fossils have been assigned to the Ganodermataceae based on basidiospore micromorphology, which corresponds to that seen in extant members of this family. However, Koeniguer and Locquin (1979) also state that the spores of *A. syrtae* display certain similarities to the basidiospores produced by the extant *Heterobasidion annosum*, one of the most important root rot parasites in Northern Hemisphere conifers (Ryvarden, 1991), which causes the disease commonly termed *Fomes* or *Annosus* root rot. The hyphal system of *A. syrtae* is dimitic (Koeniguer and Locquin, 1979). Unfortunately, additional micromorphological details of the *A. syrtae* basidiocarp, such as basidia or cystida, have not been documented.

Three basidiocarps from Miocene deposits of The Netherlands have tentatively been referred to *Ganoderma* (subg. *Elfvingia*) *adspersum* based on macroscopic features, including the absence of a stem and distinct annular rings, a pore density of 3 to 4 per millimeter, and overall habit (Fraaye and Fraaye, 1995). Likewise, basidiocarps from Pleistocene strata near Fairbanks, Alaska (Chaney and Mason, 1936), and Pleistocene and Holocene deposits in Poland (summarized in Kopczyński, 2006) have been assigned to *Ganoderma* based on morphological resemblance to extant *Ganoderma* basidiocarps. None of these fossils has been studied microscopically, however, and features such as composition of the hyphal system and basidiospore micromorphology have not been utilized to substantiate the proposed systematic relationships.

WOOD ROT

The presence of permineralized and petrified wood in the fossil record provides the opportunity to examine wood decay types and the potential fungal groups that were involved. In many instances evidence of the fungi is obvious (Figure 9.29) even though the taxonomic affinities of the wood may be difficult to determine (e.g. Mägdefrau, 1966). Because conducting elements (tracheids, vessels, etc.) are constructed of thick walls, these cells provide a potentially important source of information about fungal decay in fossil wood. In modern woods decay-producing fungi are separated into three primary types according to the pattern of degradation of xylem cells. These include white rot, brown rot, and soft rot (Schwarze et al., 2000; Schwarze, 2007). The reader is referred to the comprehensive volume *Biodiversity in Dead Wood* (Stokland et al., 2012) to better understand the multiple levels of diversity that exist in dead and decaying woody plants.

Brown rot fungi break down cellulose and hemicelluloses at different stages but do not appreciably affect lignin. It is estimated that the residual lignin represents approximately 30% of the forest soil carbon. The majority of the fungi causing brown rot belong to the Polyporaceae, and most attack the wood of conifers (Rayner and Boddy, 1988; Stokland et al., 2012). Wood-decaying polypores are polyphyletic and are identified mainly by spore features; these include color, size and shape (Parmasto et al., 1987). Comparative genomics suggests that brown rot decomposition has evolved from white rot saprotrophy, and that the lack of lignin-degrading enzymes may have been responsible for the evolution of ectomycorrhizae (Eastwood et al., 2011). Interestingly, the divergence times in extant fungal lineages parallel the predicted diversification of conifer families, some time in the Mesozoic (Eckert and Hall, 2006).

White rot fungi are common in angiosperms and are caused by both ascomycetes and basidiomycetes that have the capacity to degrade lignin as well as cellulose and hemicelluloses. In some forms of white rot, channels or erosion troughs are produced close to the hyphae on the lignified wall, while other symptoms include randomly spaced elongate pockets (Figure 9.30) throughout the woody axis and other patterns of decay (e.g. Otjen and Blanchette, 1984; Blanchette, 1992; Figure 9.31). In extant forests with both angiosperms and gymnosperms, Zhou and Dai (2012) showed that over a 10-year period gymnosperms had lower polypore species richness on fallen trunks and in unprotected forests, and a higher proportion of brown rot polypores when compared to angiosperms. The authors suggested that this outcome may be related to the ability of white rot fungi to oxidize lignin in angiosperms more readily than in gymnosperms, owing to the different compositions of lignin, and that this may result in different ecological patterns within the forest.

In soft rot, hyphae extend through the S2 layer of the lignified cell walls and decompose cellulose and lignin, leaving the wood "soft" or spongy. Fungi producing soft rot include both basidiomycetes and ascomycetes (Schwarze et al., 2000).

It is hypothesized that the common ancestor of wood-rotting fungi may have been white-rot types (Floudas et al., 2012), with brown rot perhaps evolving multiple times from white-rot clades (Hibbett and Donoghue, 2001). Some have suggested that lignin-degrading enzymes never evolved in ascomycetes (e.g. Morgenstern et al., 2008), while others point out that some ascomycetes can degrade only certain types of lignin (Lee, Y.S., 2000; Stokland et al., 2012). There may also have been

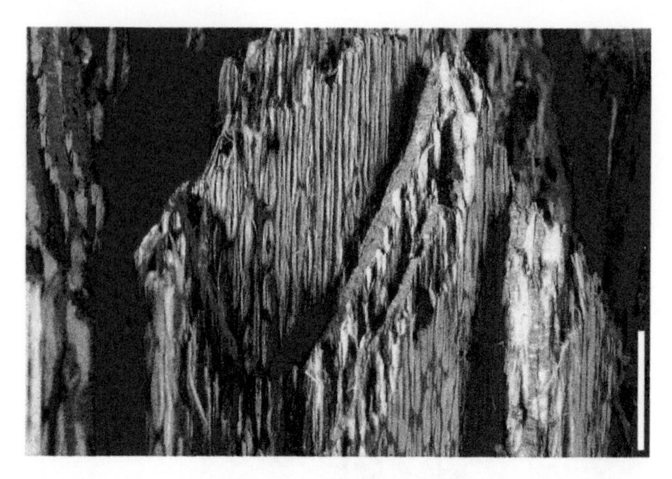

FIGURE 9.30 Fractured piece of modern conifer wood showing longitudinally oriented pockets (*white areas*). Extant. Bar = 1.0 cm.

FIGURE 9.31 Robert A. Blanchette.

wood-decaying ascomycetes that produced symptoms which, in the fossil state, are difficult to distinguish from those produced by basidiomycetes, thus making it more difficult to classify fossil wood-rot fungi. Finally, enzyme systems have also evolved through time within groups of fungi, as have the cell wall types that fungi degrade. All of these factors result in potential uncertainties in defining fossil wood-rot type.

Fossil Wood Rot

The oldest evidence of a wood-rotting fungus occurs in the form of cell wall alterations in progymnosperm

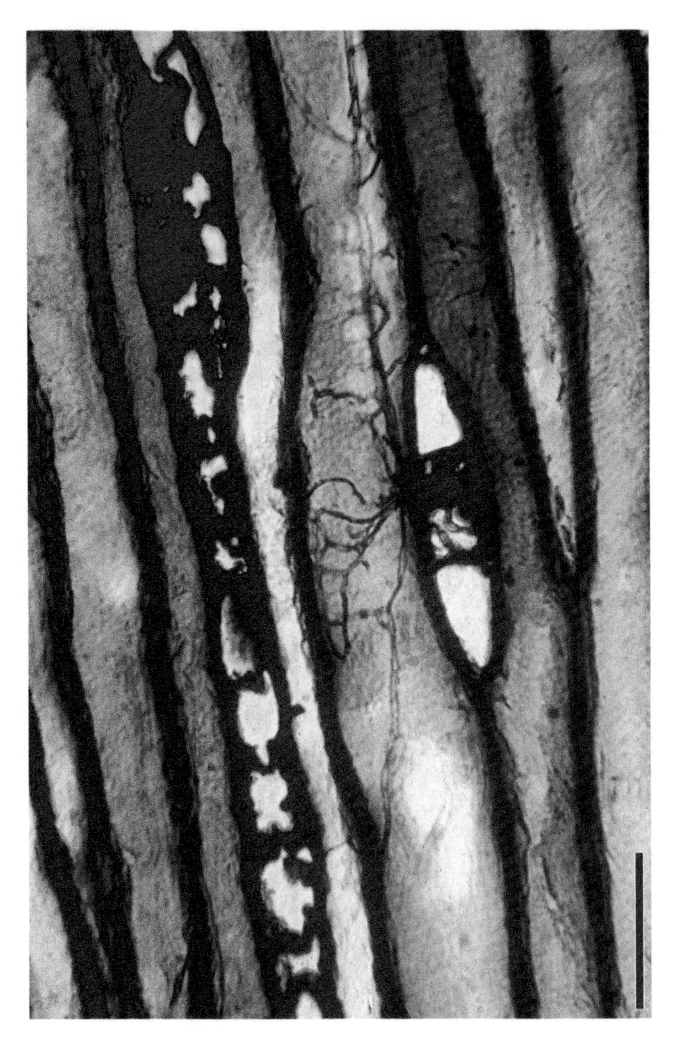

FIGURE 9.32 Longitudinal section of progymnosperm wood *Callixylon newberryi* with fungal hyphae in conducting cell. Devonian, USA. Bar = 35 μm.

wood of *Callixylon newberryi* from the Upper Devonian (~360 Ma) of Indiana, USA (Stubblefield et al., 1985a). Secondary xylem tracheids, decayed in various patterns, show erosion troughs, cavities, and extensive lysis of the tracheid walls that are similar to symptoms caused by modern-day white-rot fungi (Blanchette, 1991). Within the secondary xylem of the tracheids and rays are septate fungal hyphae (Figure 9.32) that range up to 6 μm in diameter. Some hyphae branch and many contain rounded knobs along the surface; in others there are intercalary and terminal swellings. Another interesting feature of the *Callixylon* wood is the presence of variously shaped contents in the ray cells, some of which are opaque and some hollow. Also present are spherical

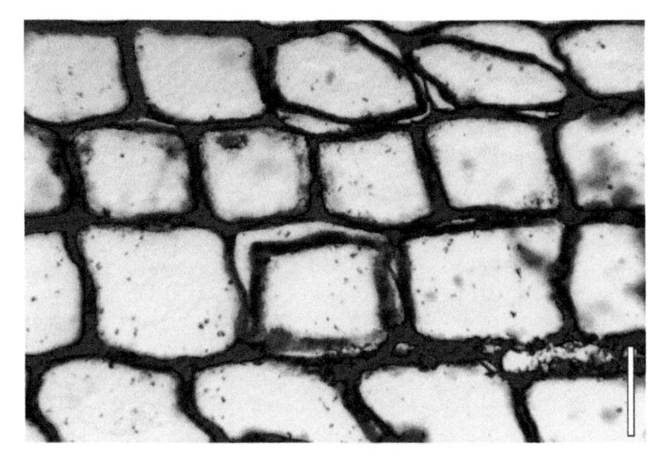

FIGURE 9.33 Transverse section of *Glossopteris* (*Araucarioxylon*) silicified wood with cells showing wall separation. Upper Permian, Antarctica. Bar = 25 μm. (Courtesy C.J. Harper.)

bodies (10–23 μm in diameter), some surrounded by a clear area. No clamp connections were observed in any of the specimens. The contents within the ray cells may represent some type of ergastic substance, perhaps a host response produced as a result of fungal infection (Stubblefield et al., 1985a) and structures with a similar appearance have been reported in extant gymnosperm wood infected by basidiomycetes (e.g. Blanchette, 1991, 1992). In the absence of diagnostic features, either ascomycetes or basidiomycetes may have infected the *Callixylon* wood. This is interesting since molecular data suggest that white rot fungi did not arise until the Late Carboniferous–Early Permian, ~295 Ma (Floudas et al., 2012). However, in this case it is the disease symptoms, rather than evidence from the actual fungus, that has been used to assess the systematic affinities of the fungus.

Several types of fungal symptoms in wood have also been reported from the Triassic and Permian of Antarctica (Figures 9.33, 9.34) (e.g. Maheshwari, 1972; Stubblefield and Taylor, 1986). Pockets of decay in *Araucarioxylon* wood (Figure 9.35) occur randomly in some axes, while in others the decay pattern appears to be associated with growth rings. The decayed areas appear in transverse section as circular areas up to 3.5 mm in diameter; when the wood is sectioned longitudinally the pockets appear spindle shaped and about 3 cm long. Some pockets lack cell wall material while in others there are hyphae (6.5 μm in diameter) that are septate and contain simple and medallion clamp connections (Figure 9.36). Because the fossil wood is so well preserved it is possible

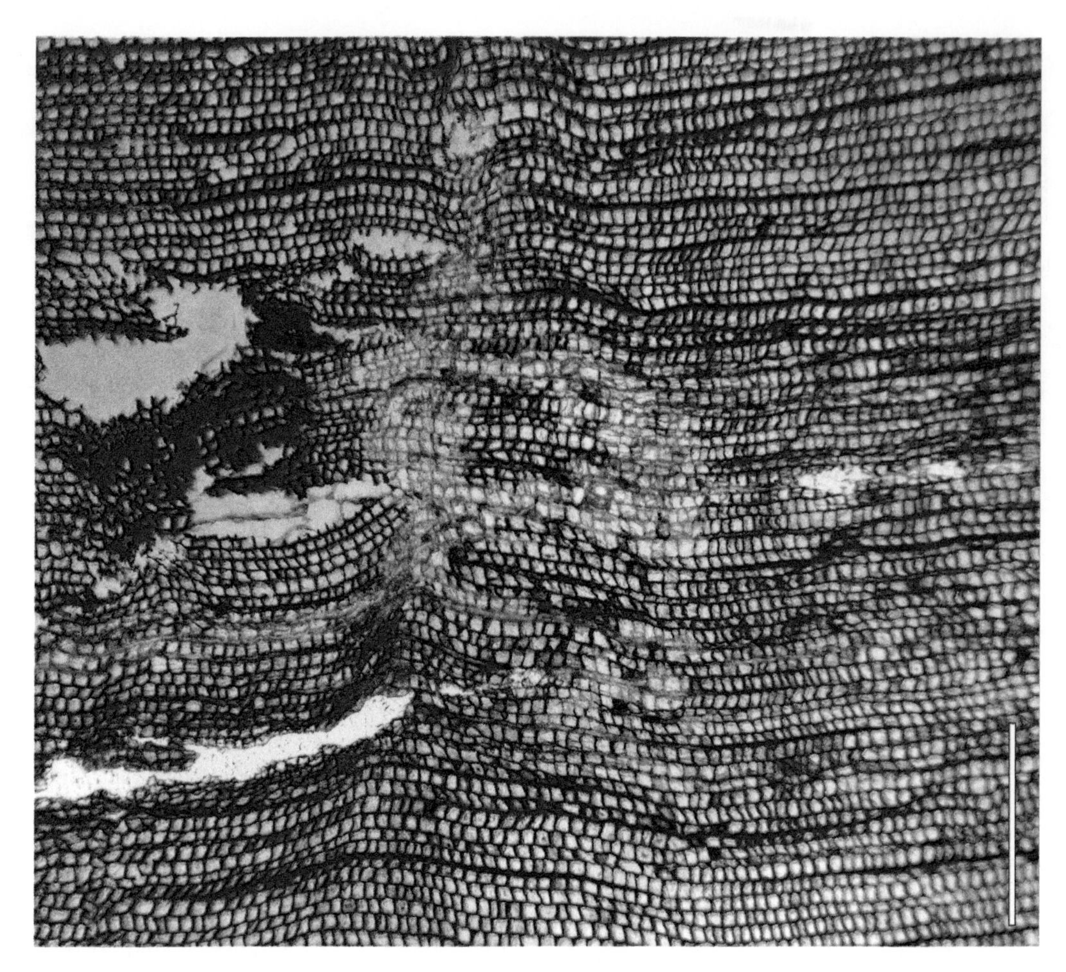

FIGURE 9.34 Transverse section through *Glossopteris* (*Araucarioxylon*) silicified wood showing differential opacity suggesting decay. Upper Permian, Antarctica. Bar = 500 μm. (Courtesy C. J. Harper.)

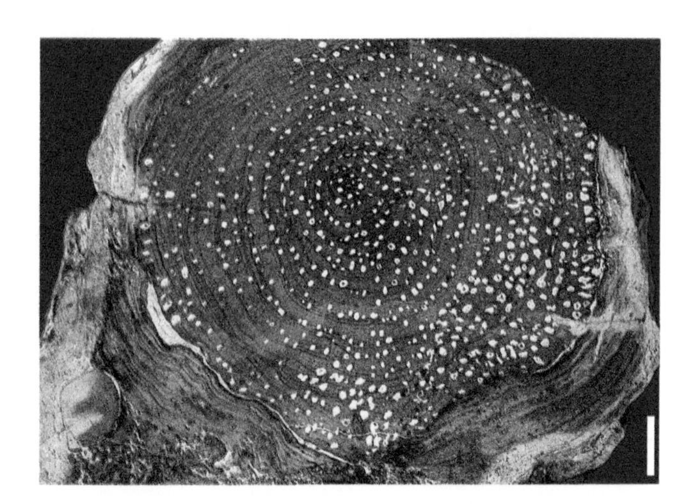

FIGURE 9.35 Transverse section of permineralized (silicified) woody axis of *Araucarioxylon*-type wood showing decay pockets (*white areas*). Middle Triassic, Antarctica. Bar = 2.0 cm.

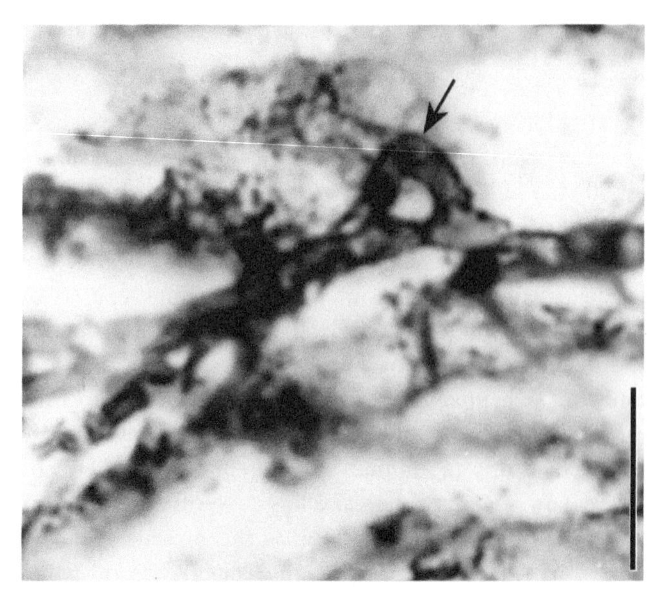

FIGURE 9.36 Basidiomycete clamp connection (*arrow*) in *Araucarioxylon*-type wood. Middle Triassic, Antarctica. Bar = 10 μm.

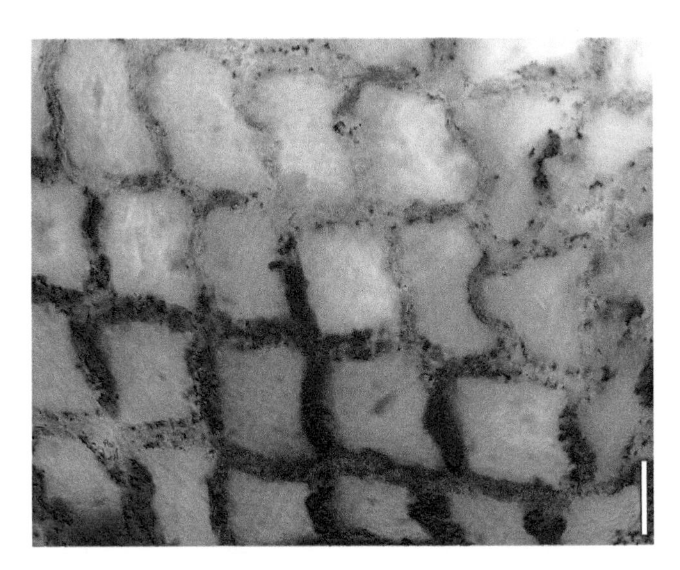

FIGURE 9.37 Transverse section of *Glossopteris* (*Araucarioxylon*) silicified wood showing area where lignin has been degraded. Upper Permian, Antarctica. Bar = 25 μm. (Courtesy C.J. Harper.)

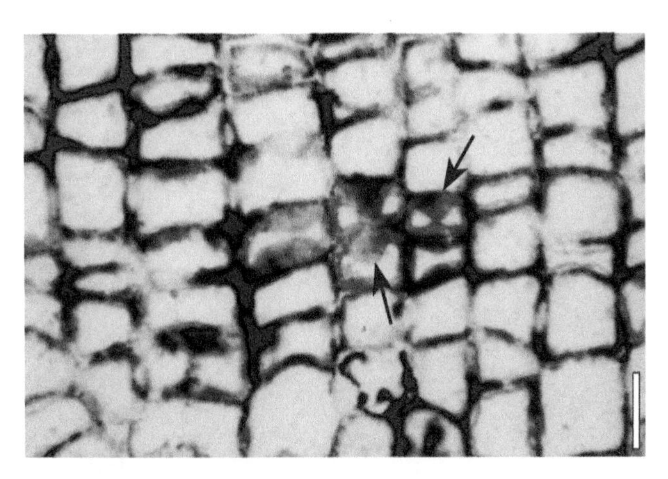

FIGURE 9.38 Transverse section of *Araucarioxylon*-type silicified wood showing thickened areas (*arrows*) suggestive of host response. Middle Triassic, Antarctica. Bar = 25 μm.

to view some of the wall layers and trace the pattern of decay leading to the formation of the pocket based on the degree of opacity (Figure 9.37) and thickness of the cell wall. Similar features in *Glossopteris* root wood (*Vertebraria*) have also been reported (Stubblefield and Taylor, 1986). These decay patterns are consistent with both white rot and white pocket rot, symptoms produced by various basidiomycetes. In both wood types, some specimens show areas of the cell wall that are thickened (Figure 9.38), and these may represent a host response to the infecting fungus.

The symptoms of wood-rotting fungi have also been described from a number of trunks in Late Triassic trees from the Petrified Forest of Arizona (Chinle Formation; Creber and Ash, 1990). In transverse section the wood shows numerous tubes and associated areas in which the cells are partially disrupted. The symptoms within the wood are compared to those that are produced by some modern fungi such as *Heterobasidion* (Otrosina and Garbelotto, 2010). The large number of fossil trunks that show similar fungal symptoms within the Petrified Forest site have been interpreted as evidence of a large scale infection, perhaps comparable to the modern Dutch Elm disease (Creber and Ash, 1990 [Figure 9.39]). It is not possible to determine, however, whether the fossil infections were parasitic or saprobic.

Evidence of a host response to fungal attack has also been reported from Early Jurassic wood from Antarctica

FIGURE 9.39 Geoffrey T. Creber (*left*) and Sidney R. Ash.

(Harper et al., 2012; Figure 9.40). Numerous septate hyphae ranging from 1.5 to 3.0 μm in diameter extend through the tracheids, ray cells, and some elements in the phloem, with some forming knob-like structures and short branches. Within the wood are numerous, inflated (8–25 μm in diameter) ingrowths in tracheids termed tyloses (Figure 9.41). In extant plants these structures originate from parenchyma cells via pits in the cell walls

FIGURE 9.40 Carla J. Harper. (Courtesy G. Mullinix.)

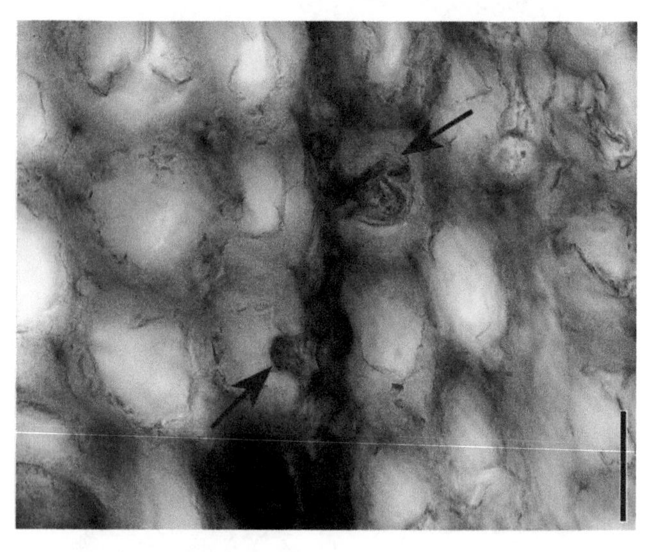

FIGURE 9.41 Transverse section of silicified conifer wood showing tyloses (*arrows*) formed by ray cells and extending into the adjacent tracheids. Lower Jurassic, Antarctica. Bar = 25 μm.

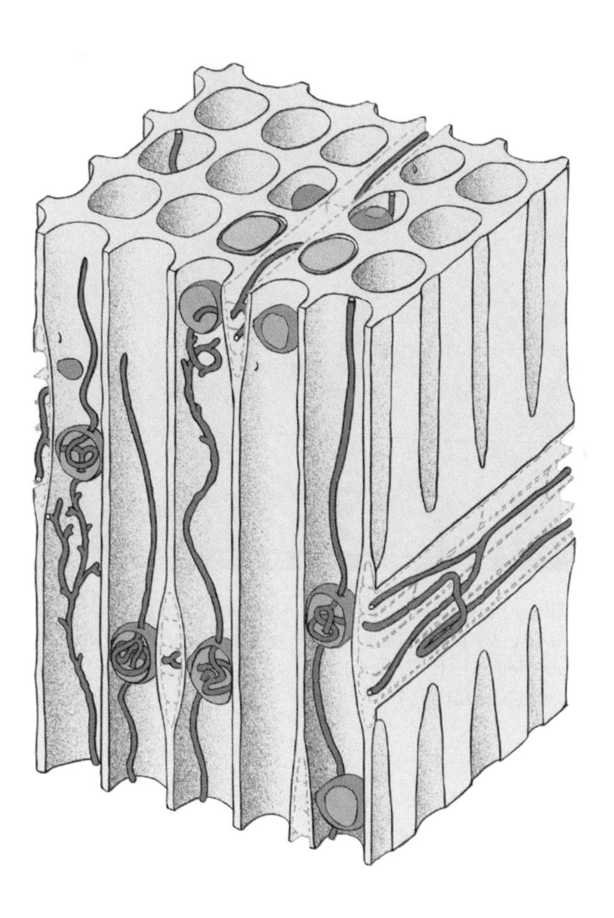

FIGURE 9.42 Diagrammatic representation of the relationship between tylosis formation (globose structures) and fungal distribution in tracheids (*vertical*) and ray cells (*horizontal*) in a block of Early Jurassic wood from Antarctica. (Modified from Harper et al., 2012.)

and may be filled with various substances (tannins, gums, resins, etc.) as a result of physical damage or parasite activity (Collins and Parke, 2008). In the Jurassic wood there are some regions where hyphae penetrate a tylosis, coil inside, and extend to adjacent cells (Figure 9.42). Because fungal infection is not extensive in this fossil wood, it is hypothesized that the specimen may represent an early stage of fungal colonization. The morphology and pattern of colonization suggest that the fossil shares similarities with various extant ascomycetous fungi, including sap-stain, blue-stain, and dark-stain types that are pathogens of various conifers (e.g. Ballard et al., 1982; Hessburg and Hansen, 1987).

There are numerous reports of tyloses in fossil woods that extend back to the late Paleozoic and include progymnosperms (e.g. Scheckler and Galtier, 2003; Figure 9.43), several ferns (e.g. Weiss, 1906; Phillips and Galtier, 2005, 2011), and various types of wood from Mesozoic and Cenozoic permineralized specimens (e.g. Nishida et al., 1990; Privé-Gill et al., 1999; Castañeda-Posadas et al., 2009). It is not known whether the tyloses in any of these reports are the result of fungal infection (Harper et al., 2012); however, the abundance of anatomically preserved wood in the fossil record provides an almost unlimited source of material that can be used to search for the presence of fungi and potential

FIGURE 9.43 Jean Galtier.

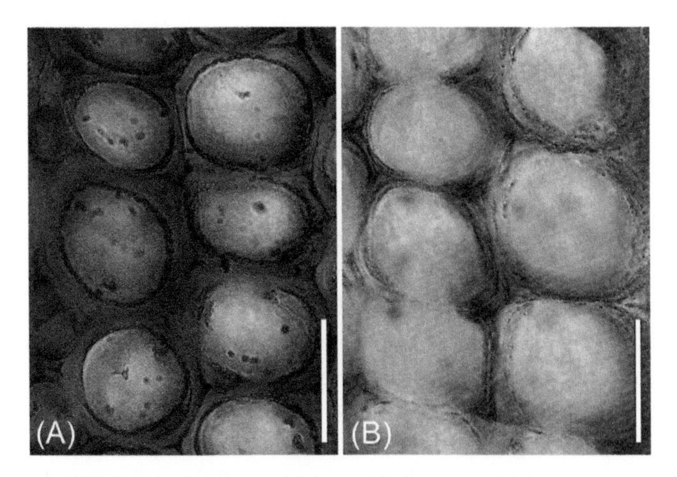

FIGURE 9.44 Transverse section of fossil wood with tracheid walls intact (**A**) and those that have been partially degraded by fungi (**B**). Miocene, Germany. Bar = 25 μm.

host responses, including structures like tyloses. When permineralized woods are examined the primary focus is usually the taxonomy of wood, and, in common with other aspects of paleomycology, little attention is given to fungal host responses or the presence of fungi. Since fungi decompose wood (Figure 9.44), specimens that show the activities of fungi would not necessarily provide the diagnostic features of the wood that would be useful in their illustration or description. As a result, we believe that the abundance of well-curated collections of fossil woods represents a potential source of new information about wood-rotting fungi in the fossil record, including information about hosts.

PUCCINIOMYCOTINA

These fungi are sometimes termed Urediniomycetes or simple-septate basidiomycetes, and include basidiomycetes that almost never form basidiocarps (Bauer et al., 2001). They have simple septal pores that lack membrane-bound caps, and the sugar composition of the group is different from that in the Ustilaginomycotina and Agaricomycotina (Prillinger et al., 2002; Aime et al., 2006). Within this group insect parasitism is restricted to a single family, the Septobasidiaceae (Henk and Vilgalys, 2007). The Pucciniales (=Uredinales) is the largest and economically most important group because they are obligate parasites of vascular plants (Bauer et al., 2001).

PUCCINIALES

Extant members are often termed rusts and cause a variety of economically important plant diseases of angiosperms and gymnosperms, e.g. the cereal rusts (Swann et al., 2001), but some also occur on ferns (e.g. Ando, 1984). These fungi are obligate biotrophic parasites and require a living host to complete their life cycle (Duplessis et al., 2011). The life cycle is complex and may result in the production of up to five different asexual spore types (i.e. pycniospores, aeciospores, urediniospores, teliospores, basidiospores) on different host plants (Petersen, 1974; Bruckart et al., 2010). Some members induce complex perturbations in their host plants to attract insects, thereby aiding spore dispersal and sexual reproduction of the rust fungus. For example, *Puccinia monoica* triggers the formation of flower-like structures (pseudoflowers) in its host plant that mimic co-occurring and unrelated flowers such as buttercups (Cano et al., 2013). Rusts generally do not kill the host but repeated infection may ultimately lead to death. There are approximately 8000 extant species described to date, estimated to represent approximately 30% of the Basidiomycota, and the number may extend to 10,000 (Toome and Aime, 2012). Clamp connections are not produced in this group (Bauer et al., 2001); sexual reproduction includes the production of two mating types of basidiospores that germinate to form monokaryons that eventually fuse.

Fossil Rusts

Although it has been suggested that the evolutionary history of the rusts and their progenitors reaches back into the late Paleozoic (e.g. Arthur, 1924; Anikster and Wahl, 1979), there are hardly any fossils that could be used to document this evolutionary history. Nevertheless, there are a variety of fossil spore taxa that appear morphologically similar to the spores produced by rusts (e.g. Ramanujam and Ranachar, 1963). One of these is *Pucciniasporonites* from the Miocene of India (Ramanujam and Ramachar, 1980). In *P. arcotensis* each of the two-celled teliospores is attached to a single distinct pedicle; the individual spores (~13 μm) and subdivided by a horizontal septum (Kalgutkar and Jansonius, 2000). The fossils are compared to spores of the extant genus *Puccinia*, a common parasite of grasses, and *Gymnosporangium*, which is found on gymnosperms. The literature also contains several references to *Puccinites* (Ettingshausen, 1853; Whitford, 1916) but none of these specimens is conclusive. Hyphae of a putative rust have also been reported from the wood of several Cretaceous conifers (Penny, 1947).

Another interesting report is a cluster of *Puccinia* teliospores in Eocene sediments from Kentucky, USA (Wolf, 1969c), but no details or magnification are provided. Among the various organisms isolated from another Eocene formation (Green River Formation of Colorado and Utah) are several spores described as *Uromyces* (Bradley, 1931). Extant spores of *Uromyces* have been useful in identification of species (e.g. Berndt, 2013).

Aecidites is used for impressions of what are interpreted as reproductive structures on an angiosperm leaf from the Upper Cretaceous of Germany (Debey and Ettingshausen, 1859). These are reproduced at very low magnification and offer nothing about the structure of the unit. Other specimens have been reported from the Eocene (Bureau, 1881; Bureau and Patouillard, 1893), Oligocene (Engelhardt, 1881), and Miocene (Göppert, 1855), but in none of these is there convincing evidence that they represent some type of rust fungus.

Since the fossil evidence of rusts is rare, insight into the phylogenetic position of the group has relied on two basic forms of evidence. One of these suggests that the most primitive rusts occurred on primitive hosts such as ferns (e.g. Savile, 1978, 1979), especially tropical forms (e.g. Leppik, 1955). Using a cladistic approach, the reverse hypothesis appears to be equally plausible in which short life-cycle rusts on angiosperms are interpreted as a basal group, while rusts on conifers and ferns form a nested terminal clade (Hart, 1988). At least some and perhaps all rusts that occur on ferns appear to be of recent origin and closely related to those that appear on flowering plants (e.g. Sjamsuridzal et al., 1999). Earlier, Leppik (1973) hypothesized that rusts occurred principally on conifers in the Northern Hemisphere as a result of continental breakup. Molecular phylogenies of some rust lineages indicate that new hosts are colonized by rusts jumping to taxonomically unrelated but ecologically associated plants (Van der Merwe et al., 2008). Other hypotheses are based on the nature of the life cycle and the simple nature of the teliospores (Ono and Hennen, 1983). Finally, molecular clock estimates suggest that rusts evolved on primitive flowering plants.

USTILAGINOMYCOTINA

Included is this group are approximately 1700 parasitic basidiomycetes that cause smuts of various angiosperms (~95 families), especially cereals (Bauer et al., 2001; Cai et al., 2011). A few smuts have been reported on lycophytes, ferns, and conifers, but the number is less than ten. Members of the Ustilaginomycotina are characterized by septal pores with membrane caps, a particular carbohydrate composition (glucose is dominant; xylose is absent), DNA sequence characters, and host–parasite interaction zones with fungal deposits resulting from exocytosis of primary interactive vesicles. These zones provide ultrastructural characters diagnostic for higher groups in the Ustilaginomycetes (Bauer et al., 2001; Begerow et al., 2006). The life history includes a parasitic dikaryophase and saprotrophic haplophase. The slender septate hyphae are often perennial and infection frequently occurs intercellularly in parts of the flower; some hyphae may grow through cells, typically with no observable host response. Masses of dark, thick-walled teliospores are formed in sori in most plant parts (Piepenbring et al., 1998a, b). It is these spore masses that are the primary dispersal agents. Teliospores may lie dormant for several years before germinating and eventually result in the production of haploid basidiospores. Basidiospores germinate to form hyphae or secondary spores (Bauer et al., 2008).

FOSSIL SMUTS

There is little in the way of a fossil record of this group. *Ustilago deccani* was initially reported from the upper Maastrichtian Deccan Intertrappean beds of Mohgaonkalan, India (Chitaley and Yawale, 1976, 1978), but the fossil has subsequently been included in

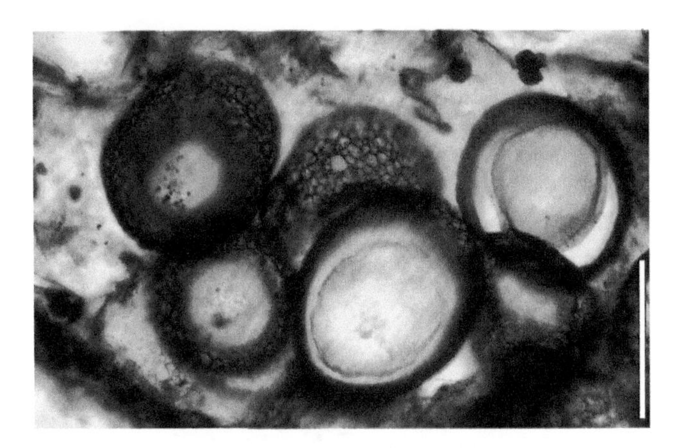

FIGURE 9.45 Several specimens of *Arthroon rochei* in degraded lycophyte tissue. Mississippian, France. Bar = 100 µm.

Inapertisporites (Kalgutkar and Jansonius, 2000). Small spores (6–8 µm in diameter) in angiosperm anthers from the Eocene Princeton chert were initially identified as a smut fungus (Currah and Stockey, 1991; LePage et al., 1994) but are now thought to represent minute pollen grains produced by the flower *Saururus tuckerae* (Smith and Stockey, 2007). There are other reports of what are interpreted as teliospores, but these are problematic. *Arthroon rochei* is an enigmatic fossil that occurs singly or in clusters in partially degraded tissue of arborescent lycophytes from the Carboniferous of France (Renault, 1894, 1896b). Specimens are spheroidal to ovoid (Figure 9.45), up to 160 µm long and 100 µm wide, and characterized by a relatively thick wall and reticulate surface ornament; some specimens possess a short projection with which they appear to be attached to the host tissue. Renault (1896b) interpreted *A. rochei* as eggs of some arthropod, but it has also variously been pointed out that the surface ornament is similar to those found in the teliospores of certain smut fungi (see Piepenbring et al., 1998c); however, *A. rochei* is considerably larger than the spores of any extant smut fungus. Since features of the teliospores, including size, ornamentation, and ultrastructure, are used in identification of extant smuts (e.g. Piepenbring et al., 1998a), palynological preparations as well as permineralized angiosperms may be an important source of fossil data about this group.

ECTOMYCORRHIZAE

Ectomycorrhizae are fungal mycorrhizal symbioses that are generally found associated with various species of woody angiosperms and conifers (e.g. Anderson and Cairney, 2007), as well as with a few shrubs and herbaceous plants (Malloch et al., 1980). It is estimated that perhaps 6000–10,000 species of fungi and about 8000 species of plants are able to form ectomycorrhizae (Baxter and Dighton, 2005; Taylor and Alexander, 2005). This symbiosis is important on a global scale because the dominant trees in most of the temperate and boreal forest ecosystems are ectomycorrhizal (ECM; Horton and Bruns, 2001; Figure 9.46). The ability to form ectomycorrhizae has evolved more than 66 times independently from non-ECM humus and wood saprophytic fungi; ECM-forming fungi include Basidiomycota (Agaricales), Ascomycota, and a few zygomycetous fungi (Hibbett et al., 2000; Hibbett and Matheny, 2009; Tedersoo et al., 2010). The rise of woody angiosperms in the Cretaceous was important in the evolution of ECM-fungal associations in angiosperm and gymnosperm trees, and ECM symbioses have increasingly replaced trees with ancestral arbuscular mycorrhizal associations (Taylor, L.L., et al., 2011). Two hypotheses have been advanced to explain the diversity of basidiomycetous ECM fungi: (1) dual origins, initially with the Pinaceae in the Jurassic and later with angiosperms during the Late Cretaceous (Halling,

FIGURE 9.46 Thomas D. Bruns.

2001), and (2) a simultaneous and convergent radiation of ECM lineages in response to cooling climate during the Paleogene and advancing temperate ECM plant communities (Bruns et al., 1998). A recent study by Ryberg and Matheny (2012), however, indicates that the data currently support neither of these hypotheses.

The large number of ectomycorrhizal species is thought to be directly related to soil nutrient heterogeneity (e.g. Rosling and Rosenstock, 2008) and the interactions between saprotrophic and ECM mycelial systems (Leake et al., 2002). The evolution of ECM symbioses with trees is also seen as the major contributor to the drawdown of atmospheric CO_2 over the last 120 myr through increased silica weathering (Taylor, L.L., et al., 2011; Quirk et al., 2012), and the sporocarps represent important carbon sinks for host trees (Teramoto et al., 2012). Ectomycorrhizae are capable of penetrating mineral grains at the nanoscale level by producing organic acids that allow access to nutrients that are not available to roots (Rosenstock, 2009; Gazzè et al., 2012). These include not only potassium and phosphorus but also nitrogen, believed to be the most important nutrient in boreal forests and the one that cannot be derived from minerals (e.g. Landeweert et al., 2001; Wu, T., 2011). Thus, there is an integrated series of plant and fungal physiological mechanisms involved in the ECM symbiosis that can be demonstrated at the level of the mineral grain surface (Smits et al., 2008, 2012). Fungi with ectomycorrhizae increase resource availability to the host plants by their ability to absorb both inorganic and organic nutrients through a wide range of extracellular enzymes (Tibbett and Sanders, 2002). It is hypothesized that ectomycorrhizae have a greater impact on biogenic weathering than arbuscular mycorrhizae because the latter do not secrete chelating molecules (Hoffland et al., 2004; Taylor, L.L., et al., 2009; McMaster, 2012).

In most instances there is a distinct morphological change in the branching patterns of lateral roots as a result of the ECM symbiosis. Hyphae of ECM fungi generally cover the tips of roots, including the root cap and root hairs, and through repeated branching form a mantle of hyphae up to 100 μm thick that also may include certain bacteria, which are believed to increase the effectiveness of the mycorrhizae (e.g. Garbaye, 1994a, b). This mantle may provide protection for the roots from water loss and various pathogens. Moreover, the mantle represents the transition zone between plant and fungus, as well as between plant and soil, thus functioning as the interface for the exchange of photoassimilates, soil water, and soil nutrients (e.g. Ostonen and Lõhmus, 2003; Kothe et al., 2013). Various specialized cells are produced on the end cells of mantle hyphae that may also offer some defense against fungivory. Some of these cells are pointed while others (e.g. cystidia) are swollen and contain chemical compounds that act as a defensive mechanism (e.g. Marx, D.H., 1972; Agerer et al., 6 vols., 1996–2002). The morphology and organization of mantle cells, together with molecular phylogenetic analyses, have been useful in defining some modern ectomycorrhizae, suggesting the potential for the systematic treatment of fossils when they are found (e.g. Zak, 1973; Wei et al., 2010).

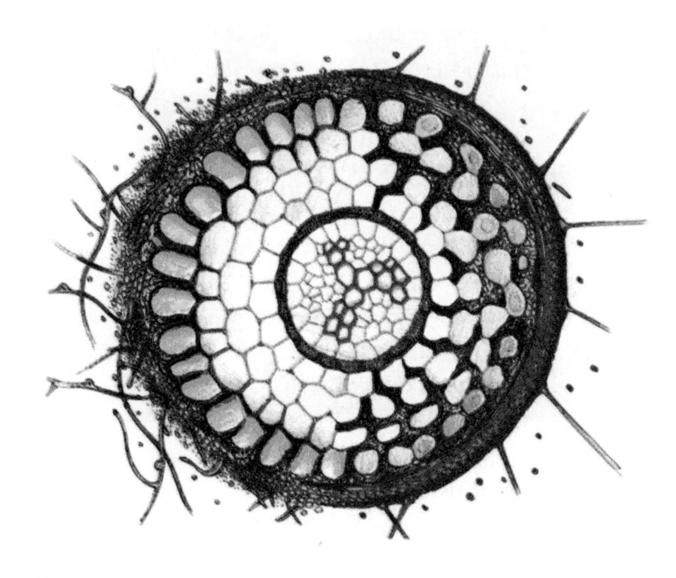

FIGURE 9.47 Transverse section of idealized root showing difference in ectomycorrhizal organization between an angiosperm (*left*) and a conifer (*right*). In angiosperms the Hartig net is typically confined to the epidermis, while in conifers it forms around cells of the epidermis and cortex. (From Peterson et al., 2004, courtesy R.L. Peterson.)

Perhaps the defining feature of ectomycorrhizae is the Hartig net (Figure 9.47), named for the nineteenth-century German forest scientist and mycologist Robert Hartig (1839–1901), which is an aggregation of highly branched hyphae in the inner region of the mantle that develops between cells of the root and may extend into layers of the root cortex (e.g. Kottke and Oberwinkler, 1989). This is the site of nutrient exchange between the fungus and the host, but it is also where carbohydrates and various compounds are stored. In some ectomycorrhizae, water and nutrients can move into the mantle through interhyphal spaces and hyphal walls. The mycelia of ectomycorrhizae can also dissolve certain toxic metal-bearing minerals (e.g. copper, zinc, lead) and phosphates (e.g. Fomina et al., 2004).

Extending from the outer portion of the mantle and into the soil are hyphae that make up the extraradical mycelium, which represents an extensive system of hyphae in the rhizosphere that function in interconnecting individual roots of the same plant as well as in extending out to the roots of neighboring plants. It is also known that ECM mycelia are connected to seedlings of the same species and that this interconnection has a positive effect on seedling growth because of increased nutrient acquisition (e.g. Nara, 2006). Some ECM fungi form rhizomorphs, thick mycelial strands formed of interconnected hyphae that absorb and store nutritive substances (e.g. Agerer, 2001; Schweiger et al., 2002). Some rhizomorphs contain hyphae that contain specialized septae thought to regulate water and nutrients (Cairney, 1992). Moreover, it is now understood that certain regions of the mycelium are specifically associated with nutrient absorption, while others are involved in extracellular enzyme secretion and solute translocation (e.g. Cairney and Burke, 1996). Contact between hyphae increases the mycelial network, which can enlarge the area and amount of nutrient exchange in the ECM system (e.g. Wu et al., 2012).

The formation of basidiocarps or ascocarps, the sexual reproductive structures of the fungus in the ECM association, occurs as a result of branching of the extraradical hyphae. These include different types of sporocarps, including the mushrooms found around trees; other sporocarps are hypogeous. These structures produce basidiospores or ascospores that are dispersed by various methods. Sporocarps contain high concentrations of nitrogen and phosphorus, and it has been suggested that in the absence of a host, their formation would require approximately 1800 km of hyphae to produce a single sporocarp (Taylor and Alexander, 2005). Moreover, sporocarps represent significant carbon sinks for ECM fungi (e.g. Högberg et al., 1999, 2001); they are last in the carbon flux pathway from the source of assimilation and thus particularly sensitive to changes in the availability of fixed carbon (Andrew and Lilleskov, 2009). Some ectomycorrhizae produce sclerotia, a type of asexual resting structure composed of plectenchyma that may be cutinized and from which new vegetative growth may develop (e.g. Grenville et al., 1985a, b).

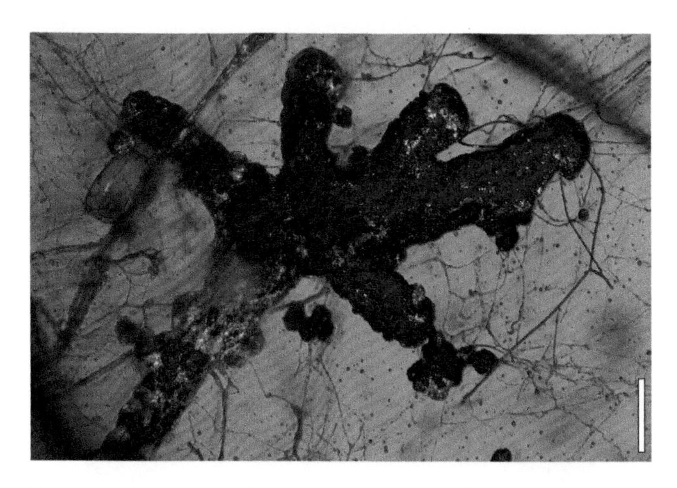

FIGURE 9.48 *Eomelanomyces cenococcoides* showing monopodial-pinnate ectomycorrhizal system forming microsclerotia. Eocene, India. Bar = 100 μm. (Courtesy C. Beimforde and A.R. Schmidt.)

FOSSIL ECTOMYCORRHIZAE

There are a variety of structures produced by ectomycorrhizae that make it possible to identify this symbiosis in the fossil record. These include the presence of a Hartig net around the tips of permineralized roots, specialized cells on hyphae of the mantle, and perhaps ornamented hyphae that contain crystals. The presence of other structural and morphological features indicative of certain ectomycorrhizae (e.g. cell shape, cystidia, organization of certain layers; e.g. Peterson et al., 2004; Wei et al., 2010) may also be useful in the identification of ectomycorrhizae in the fossil record. In spite of these potential characters, there are relatively few conclusive examples of fossil ectomycorrhizae. The oldest of these is reported from lower Eocene amber from western India (Beimforde et al., 2011). The specimens consist of approximately 20 cruciform and monopodially pinnate ectomycorrhizae (Figure 9.48) fossilized adjacent to rootlets and preserved in different developmental stages of the mycorrhiza. Young specimens are characterized by pseudoparenchymatous mantles from which septate hyphae extend. Compounds of melanins are detectable in these hyphae. Clavate or broad fusiform to ovate, chlamydospore-like distal swellings occur on some of the hyphae. Dense hyphal systems extend in all directions into the translucent amber, suggesting that some of the specimens were still alive when initially embedded. Sometimes hyphae form simple rhizomorphs. In developmentally older specimens, spherical to ovoid microsclerotia are

FIGURE 9.49 James F. Basinger.

FIGURE 9.50 Ben A. LePage.

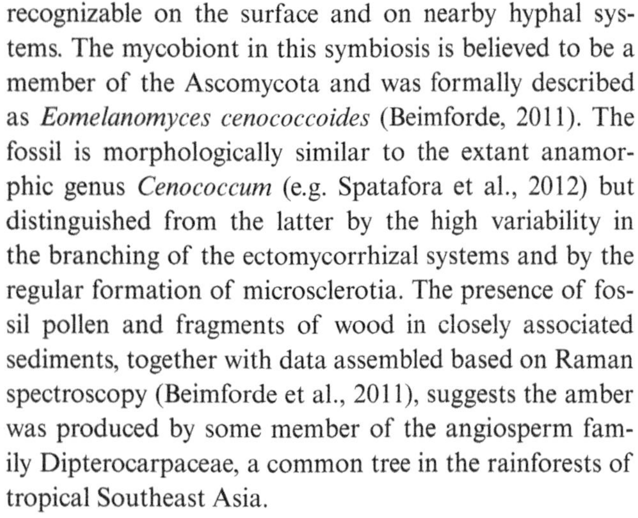

recognizable on the surface and on nearby hyphal systems. The mycobiont in this symbiosis is believed to be a member of the Ascomycota and was formally described as *Eomelanomyces cenococcoides* (Beimforde, 2011). The fossil is morphologically similar to the extant anamorphic genus *Cenococcum* (e.g. Spatafora et al., 2012) but distinguished from the latter by the high variability in the branching of the ectomycorrhizal systems and by the regular formation of microsclerotia. The presence of fossil pollen and fragments of wood in closely associated sediments, together with data assembled based on Raman spectroscopy (Beimforde et al., 2011), suggests the amber was produced by some member of the angiosperm family Dipterocarpaceae, a common tree in the rainforests of tropical Southeast Asia.

To date, the most convincing evidence of fossil ectomycorrhizae in gymnosperms occurs in the lower–middle Eocene Princeton chert of British Columbia (e.g. Basinger and Rothwell, 1977). One of the plants from this site is *Metasequoia milleri*, a conifer that is well known, based on a number of vegetative and reproductive parts (Basinger, 1981, 1984; Figure 9.49). Also in the chert are roots of a conifer that contain coralloid clusters of dichotomously branched small (0.1–0.5 mm in diameter) rootlets lacking

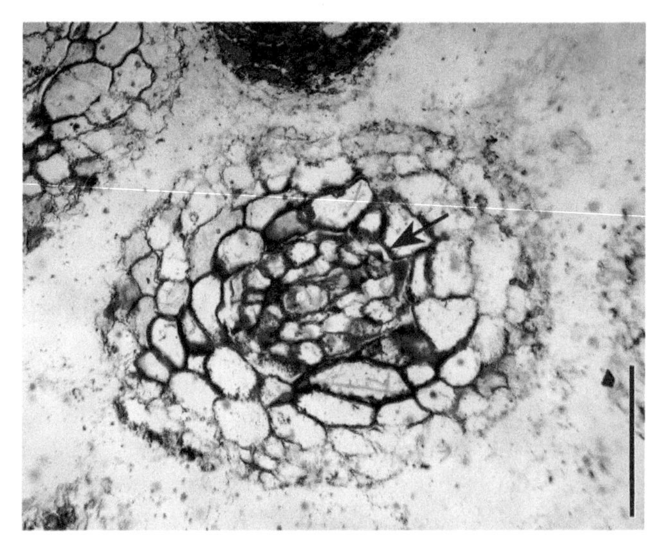

FIGURE 9.51 Transverse section of a permineralized ectomycorrhizal rootlet showing Hartig net (*arrow*) between the cortical cells and endodermis. Bar = 100 μm. Eocene, Canada. (Courtesy R.A. Stockey.)

root hairs (LePage et al., 1997) (Figure 9.50). Within the cortex of these rootlets are densely packed septate hyphae a single cell layer thick that represent the Hartig net (Figure 9.51); there is some evidence of the mantle, including simple-septate hyphae 1–2 μm in diameter. The fungi are compared with the extant basidiomycete taxa *Rhizopogon* and *Suillus* (Bruns et al., 1989; Molina and Trappe, 1994), two genera in the Boletales that today form ectomycorrhizal associations principally with members of the Pinaceae (e.g. Grubisha et al., 2002).

The presence of other structural and morphological features indicative of certain types of ectomycorrhizae, such as cell shape, cystidia, and organization of tissue systems (e.g. Peterson et al., 2004; Wei et al., 2010), may also be useful in the identification of additional ectomycorrhizae in the fossil record.

10

LICHENS

THALLUS MORPHOLOGY AND STRUCTURE 202
LICHEN REPRODUCTION 204
LICHEN EVOLUTION 205
PRECAMBRIAN EVIDENCE OF LICHENS 205
PALEOZOIC LICHENS 206
MESOZOIC LICHENS 210
CENOZOIC LICHENS 211
FOSSILS THAT MIGHT BE LICHENS 214

Lichens are extraordinary life forms; they represent a consortium of organisms that may include representatives from three different kingdoms. Lichens represent an ecologically obligate and stable mutualism between a dominant fungal partner (mycobiont), from which the lichen gets its name, and an inhabitant population of extracellularly located unicells (phycobiont) that may be a green alga, cyanobacterium (cyanobiont), or both (e.g. Grube and Hawksworth, 2007; Henskens et al., 2012). The lichen symbiosis was once portrayed as representing fungi that discovered agriculture. Some would challenge the concept that lichens represent a stable relationship and refer to such an arrangement as an example of controlled parasitism (Ahmadjian, 1993, 1995 [Figure 10.1]; Piercey-Normore and DePriest, 2001). In the lichen association the photobionts transfer sugar alcohols and/or glucose to the mycobiont, while the fungus provides a protective layer around the photobionts that shields them from excessive light, reduces water loss, and helps to absorb mineral nutrients. Many lichens possess multiple photobionts (Hawksworth, 1988). Recent research indicates that highly structured bacterial communities are also an integral component of lichen thalli and serve distinct functional roles within the lichen symbiosis, suggesting that the classical view of this symbiotic relationship should be expanded to include bacteria (e.g. Grube et al., 2009; Bates et al., 2011). Moreover, lichens serve as a habitat for numerous organisms, including other fungi, arthropods, and nematodes (e.g. Bates et al., 2012). There are thought to be approximately 18,500 lichen species made up of more than 20% of all fungi; approximately 40% of these are Ascomycota with a far smaller number of Basidiomycota (e.g. Oberwinkler, 1970; Kirk et al., 2008). Basidiomycete lichens are included in the Agaricomycetes.

The majority of lichen photobionts belong to the green algal class Chlorophyceae, with the most common genera being *Trebouxia* and *Trentepohlia.* In lichens where cyanobacteria are present, *Nostoc* is perhaps the primary photobiont. Lichens grow slowly and are generally found in all types of ecological settings ranging from the Arctic to Antarctica; some have even been termed lichen weeds because they are unnoticed by humans (Purvis, 2000). They are found on all forms of substrate, including bare rock, desert sand, dead wood, and bones – anywhere there is sufficient light. When they colonize rock, lichens have a major influence in the chemical and physical processes connected to weathering (e.g. McIlroy de la Rosa et al., 2012; Zambell et al., 2012) as well as in the composition of microbial metacommunities on the surface

T.N. Taylor, M. Krings & E.L. Taylor: Fossil Fungi.

DOI: http://dx.doi.org/10.1016/B978-0-12-387731-4.00010-4

FIGURE 10.1 Vernon Ahmadjian.

FIGURE 10.2 Robert Lücking.

(e.g. Bjelland et al., 2011). Many lichen species have been used as bioindicators of, among other aspects, habitat disturbance (e.g. Hawksworth et al., 2005), air pollution (e.g. Conti and Cecchetti, 2001; Jovan, 2008), freshwater quality (e.g. Nascimbene et al., 2013), and long term ecological continuity in undisturbed forests (e.g. Tibell, 1992; Lücking, 1997 [Figure 10.2]; Tibell and Koffman, 2002). When combined with climate data, lichens have also been used to date divergence times of certain species (e.g. Printzen and Lumbsch, 2000).

THALLUS MORPHOLOGY AND STRUCTURE

The unique association between a heterotrophic partner and photoautotrophic organisms results in the formation of a new structure, the lichen thallus, where the physiology of the partners is merged and new chemical substances may be synthesized (e.g. Honegger, 1993; Upreti and Chatterjee, 2007). These secondary chemicals, the so-called lichen substances or *Flechtenstoffe*, represent one of the most intriguing aspects of the lichen symbiosis. Although they are produced entirely within the fungal partner, they are not (with a few exceptions) produced by the isolated lichen fungi, only by those that are combined with the appropriate algal host (Lawrey, 1986; Kowalski et al., 2011). More than 1000 different biologically active metabolites have been identified in lichens, many holding promise for use in medicine and in various therapeutic areas (e.g. Stocker-Wörgötter and Elix, 2004; Boustie and Grube, 2005; Molnár and Farkas, 2010), and this number will continue to grow as more lichens are analyzed. Thus far these compounds have been interpreted as functioning to control light, serve as antiherbivore agents, deter microbes, and perhaps negatively influence plants (e.g. Kowalski et al., 2011); a greater understanding of their role and potential value to human health is certain to follow as research continues.

There are two fundamental principles regarding thallus organization in lichens. In homoiomerous (also homomerous or homomeric) thalli there is little differentiation into individual layers or strata; rather, the entire thallus is composed of a loosely arranged mass of fungal hyphae with algal cells, which are equally distributed throughout. On the other hand, heteromerous (also heteromeric) thalli may be constructed of up to four principal layers of interlaced fungal hyphae. The outer (or superior) cortex (1) consists of conglutinated hyphae and provides protection to the layer beneath, the gonidial or algal layer (2), which contains the photobiont. Beneath this layer is the medulla (3), consisting of loosely organized or conglutinated (e.g. central strand in *Usnea*) fungal hyphae where gas exchange takes place. In some lichens there is a cortical zone beneath the medulla, called the lower or inferior cortex (4), from which rhizines may arise. Within the lichen thallus there may also be other ascomycete fungi in the form of endophytes, termed endolichenic fungi. They are distinct from the lichen mycobiont and have been useful in hypotheses regarding trophic transitions in the Ascomycota (Arnold et al.,

FIGURE 10.3 James D. Lawrey.

2009). Although associated with the photobionts in lichen thalli, they do not inflict symptoms of disease in the host. Endolichenic fungi are horizontally transmitted and represent at least five classes of non-lichenized Pezizomycotina (U'Ren et al., 2012).

Other fungi and microlichens live on lichens as host-specific parasites, pathogens, saprotrophs, and commensals (e.g. Lawrey, 1995). These lichenicolous fungi and lichens, which are highly specialized and include ascomycetes and a smaller number of basidiomycetes, are estimated to include upwards of 3000 species (Lawrey and Diederich, 2003). A worldwide checklist of lichenicolous fungi can be found at: http://www.lichenicolous.net (Lawrey and Diederich, 2011; Figure 10.3).

Although within all lichens there are multiple types of thallus morphology based on form and growth patterns (e.g. Ryan et al., 2002), we will mention only four basic macroscopically visible types:

1. Crustose lichens, which are particularly slow growing (e.g. Armstrong and Bradwell, 2010), form crusts that are directly attached to the substrate so tightly that it requires damaging the substrate to remove them. Instead of a lower cortex, they usually have a medulla that is in direct contact with (and often extends into) the substrate. In some crustose forms, distinctive lobes may be formed around the margin (placodioid).

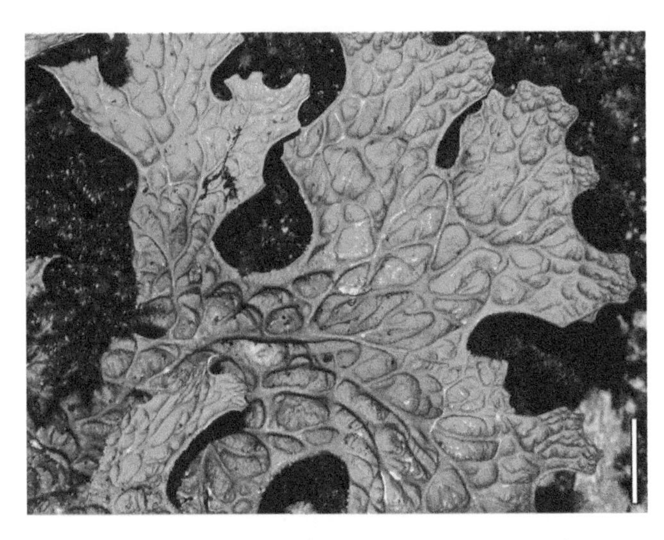

FIGURE 10.4 Thallus of foliose lichen *Lobaria pulmonaria*. Extant, Germany. Bar = 1.0 cm. (Courtesy R. Lücking.)

FIGURE 10.5 Squamulose thallus of *Flakea papillata*. Extant, Costa Rica. Bar = 1.0 mm. (Courtesy R. Lücking.)

2. Foliose lichens, in which the thallus consists of flat lobes that look leaf-like (Figure 10.4) (Armstrong and Bradwell, 2011). They are loosely attached to the substrate, often by rhizines sometimes termed rhizomorphs (Sanders, W.B., 1994), and grow parallel to the surface, with the lobes having an upper and lower surface; the surfaces are often differently colored.
3. Another growth form (squamulose), which is sometimes included with the foliose lichens, is characterized by small scale-like lobes (squamules) that overlap (Figure 10.5).

FIGURE 10.6 Fruticose thallus of *Usnea sanguinea*. Extant, Mexico. Bar = 8.0 mm. (Courtesy R. Lücking.)

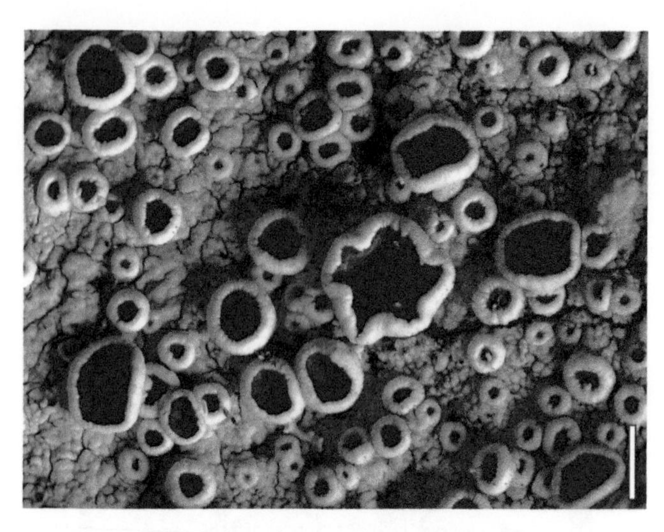

FIGURE 10.7 Apothecia on thallus of *Tephromeia atra*. Extant, Costa Rica. Bar = 1.0 mm. (Courtesy R. Lücking.)

4. Lichens with three-dimensional morphologies that may branch multiple times are termed fruticose (Figure 10.6); these growth forms appear like delicate shrubs and may also hang down (pendant) from the substrate.

Although the traditional thallus morphologies have been important in lichen systematics, especially at the level of the family, the hypothesis has been advanced that changes in hydration rates can also markedly alter thallus form and may have contributed in multiple ways to the evolution of thallus characters (Grube and Hawksworth, 2007).

LICHEN REPRODUCTION

Lichens can reproduce by several asexual methods. In one of these, small (20–50 μm in diameter) clusters of photobiont cells are intermixed with fungal hyphae. These structures, which are a type of diaspore, are termed soredia and develop in special areas of the thallus (soralia) as outgrowths of the medulla before being released through openings in the cortex where they can be dispersed by air currents, water, and animals (e.g. Bailey, 1968; Marshall, 1996). Since they have both components of the lichen – mycobiont and photobiont – they can develop directly into new lichen thalli. A second basic pattern of lichen reproduction is by means of isidia, which are variously shaped outgrowths of the medulla and characterized by a cortical covering (e.g. Lindahl, 1960; Kershaw and Millbank, 1970). These structures are easily ruptured and are distributed as "miniature" lichens. A third type of lichen reproduction involves flaps of tissue (lobules or phyllidia) along the margin of some foliose species, which, if dislodged, can form new lichen thalli. Lichens also reproduce asexually in a nonsymbiotic manner via the production of conidiophores within pycnidia. The conidiospores can give rise to the fungus, or in some instances can function as the male gamete in sexual reproduction.

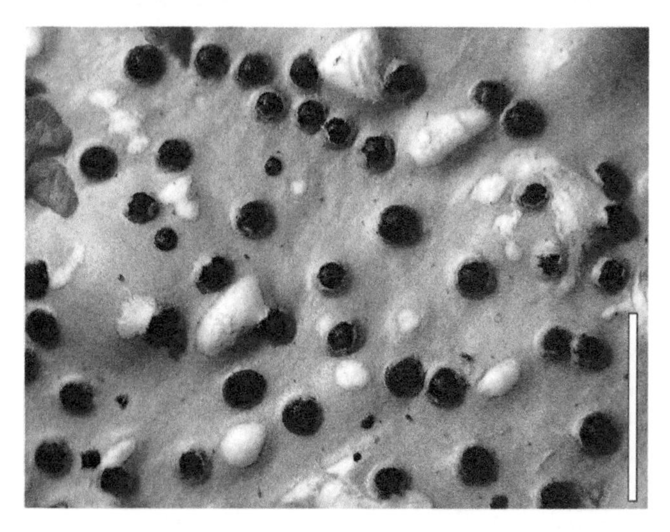

FIGURE 10.8 Perithecia on thallus of *Pyrenula papilligera*. Extant, Costa Rica. Bar = 5.0 mm. (Courtesy R. Lücking.)

Sexual reproduction in lichens only occurs in the fungus; in lichens with ascomycete mycobionts this occurs in mazaedia, apothecia (Figure 10.7), hysterothecia, or perithecia (Figure 10.8), which are often situated on the upper surface of the thallus (Honegger and Scherrer, 2008). Asci and ascospores are produced within the apothecia. Lichenized basidiomycetes form basidiocarps similar to those seen in free-living taxa. If new lichen thalli are to be formed from the sexually produced fungal spores, the spores must find the appropriate photobiont and re-establish the symbiosis at the germination site. This challenge has been overcome by various ways, including the picking up of photobiont cells from the maternal thallus by the spores when released, dispersal of the spores via arthropods (e.g. mites) that also transport photobiont cells, and by survival as saprotrophs or parasites until the right photobiont is found (e.g. Bailey, 1976; Krishnamurthy and Upreti, 2001; Wedin et al., 2004).

LICHEN EVOLUTION

Determining whether a structure in the fossil record is a lichen is not an easy proposition, but it may be possible if the presence of both the myco- and photobiont can be detected. Another morphological character is the presence of a thallus that differs from that of either the photobiont or mycobiont if they were growing independently. Lastly, and certainly the most difficult to determine, is whether there is any evidence that the symbionts interacted in a physiological manner, thus demonstrating the interdependence of the partners. Fossil evidence of this type might include the consistent association of cells of the photobiont with the mycobiont, along with their spatial distribution within a thallus, and the repetition of this same pattern in multiple specimens (Taylor et al., 1997).

Molecular clock assumptions suggest that lichens may have existed by the Precambrian and there is some fossil evidence to support this idea (e.g. Heckman et al, 2001). The evidence, however, is exceedingly scarce, which might be due in part to the structure and organization of lichen associations today and in the past. Ancient lichens may have been different structurally from any of the modern thallus morphologies and therefore difficult to recognize as fossils. The fact that there is currently no evidence to suggest that present-day lichens are direct descendants of primeval lichen-like associations adds support to this hypothesis (e.g. Printzen and Lumbsch, 2000; Schoch et al., 2009a). Because the fungi that form lichens are classified in at least twelve orders of Ascomycota, the majority of which exclusively include lichenized forms, determining the evolutionary origin of lichens continues to be perplexing (Grube and Winka, 2002). Some have postulated five independent origins (e.g. Gargas et al., 1995), while others have suggested twelve (e.g. Aptroot, 1998). In addition, molecular studies suggest that some non-lichenized lineages have arisen from lichen-forming ancestors and that the evolution of biochemical pathways used in forming lichen substances is critical in lichenization (Lutzoni et al., 2001). An interesting study by Schwartzman (2010) suggested that the origin of the lichen symbiosis may have been triggered by declining atmospheric CO_2 levels. It has also been hypothesized that the evolution of land-plant diversity, and the resulting potential of new substrates, may have been an important factor in the radiation of lichenized fungi (Grube and Winka, 2002). Perhaps what is most interesting is that the lichen lifestyle has been gained and lost multiple times among the fungal, green algal, and cyanobacterial partners, and that the coevolution of lichen symbionts is to a large degree based on the optimal symbiont phenotypes (DePriest, 2004).

PRECAMBRIAN EVIDENCE OF LICHENS

One of the earliest structures interpreted as a lichen was *Thuchomyces lichenoides*, a structure found in Precambrian rocks of South Africa (Hallbauer and van Warmelo, 1974; Hallbauer et al., 1977). It was described as a horizontal thallus with small upright columns constructed of branched septate hyphae, each column terminating in a vegetative spore. Evidence of the photobiont was difficult to determine, except for certain biochemical signatures that were ascribed to algae. The occurrence of these structures in rocks that were no doubt strongly heated during diagenesis (a gold-bearing, uranium-lead-oxide conglomerate) makes their assignment as lichens perhaps less convincing. It is now believed by many that the organization of this structure was not biotic, but rather the result of abiotic activities (Cloud, 1976). It is interesting to note that some of these ancient rocks appear to contain biogenic carbon in the form of kerogen and bitumen, which originated in situ from living organisms, perhaps from microbial mats (e.g. Mossman et al., 2008). Others have suggested that the filaments described in *T. lichenoides* are more similar to those produced by various bacteria. Another Precambrian fossil from paleosols in South Africa that has been compared to a lichen is *Diskagma buttonii* (Retallack et al., 2013b). It is an elongate, urn-shaped structure that is up to 18 mm long and apparently attached in groups to a basal structure; along the body are occasional spines. While it remains unclear exactly what *D. buttonii* was, Retallack (2013b) compares these ancient fossils to the modern *Geosiphon pyriformis*, a fungus in the Glomeromycota that is characterized by bladders harboring symbiotic *Nostoc* cyanobacteria (endocytobiosis; Schüßler et al., 1995, 1996). While structures like *Thuchomyces* and *Diskagma* might be interpreted as representing a certain grade of organism evolution, such as a lichen or lichen-like association between different organisms, in most instances there is simply far too little information available with which to hypothesize such affinities.

Other Precambrian fossils have variously been suggested to be lichens or lichen-like symbioses. Foremost among these are the famous Ediacaran fossils or Vendobionta, which represent the oldest assemblages of diverse macroscopic organisms in Earth history to date. Some Ediacaran fossil morphologies have been reported from at least 30 sites worldwide and are preserved as impressions that vary from circular discs a few centimeters in diameter to plant-like fronds up to 1 m long. All Ediacaran fossils come from shallow to deep marine facies; however, one report based on the color and pattern formed on the rocks by weathering hypothesized instead that the fossils were present in paleosols, indicating a terrestrial habitat (Retallack, 2012), although this interpretation has been questioned (Callow et al., 2013). The Vendobionta have historically been interpreted as representing early metazoans (e.g. Erwin et al., 2011) or protist lineages (e.g. Seilacher et al., 2005), some of which may have evolved into cnidarians, poriferans, and certain types of worms, while others became extinct. Retallack (1994) reinterpreted some of these forms as lichens, but in general these ideas have not been accepted (e.g. Waggoner, 1995). It has been suggested that some Ediacaran fossils from the Avalon Peninsula of Newfoundland, Canada are biologically similar to certain fungi (Peterson et al., 2003). To date the accumulated evidence based on features of the Ediacaran organisms and their depositional environments does not offer anything new that would indicate that these organisms had anything in common with lichens, or any other microbial form of life (Xiao, 2013).

An example of a structure that may represent a Neoproterozoic lichen-like symbiotic association was reported from the early Ediacaran upper Doushantuo Formation at Weng'an, South China and is dated around 580–551 Ma (Yuan et al., 2005). The specimens are phosphatized and consist of what are interpreted to be fungal filaments, each up to 0.9 µm wide, and these branch dichotomously to form a net-like organization (Figure 10.9). Many of the filaments end in a slightly inflated pyriform structure, which may be some type of vesicle, resting spore, or some other reproductive unit. Within the lumina of the three-dimensional network are groups of coccoid unicells, each 6–15 µm in diameter, that morphologically resemble those of cyanobacteria or some algae. The fossils are believed to have been deposited in a shallow subtidal environment. It remains impossible to know precisely what these Doushantuo fossils might have represented biologically; however, if the filaments are fungal and the coccoid cells are cyanobacteria, they do demonstrate the consistent close association of structures that morphologically define extant lichens.

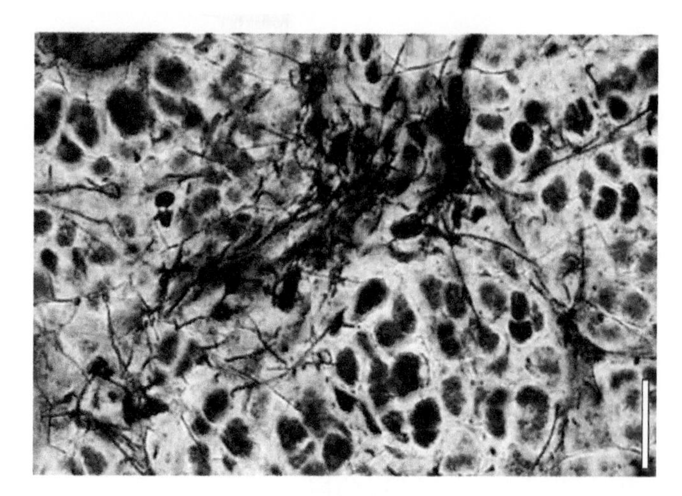

FIGURE 10.9 Hyphal net-like structure containing spherical cells interpreted as cyanobacteria. Neoproterozoic, China. Bar = 20 µm. (Courtesy S. Xiao.)

PALEOZOIC LICHENS

There are several reports from Cambrian rocks that have been interpreted as soil crusts, based on weathering patterns, the presence of certain types of minerals, presumed biological structures, and several other types of evidence (Retallack, 2011; Figure 10.10). Some of these problematic fossils have been interpreted as lichens, based on the morphology of various axial structures; other fossils in this assemblage are thought to represent fungi. While it has been hypothesized that lichens and various groups of fungi were in existence by the Cambrian era, based on several lines of indirect evidence (e.g. Prieto and Wedin, 2013), the various traces that have been offered as evidence of these organisms remain unconvincing. *Farghera robusta* is the name applied to certain compression fossils from Australia that could represent any number of

FIGURE 10.10 Gregory J. Retallack.

biological entities, including lichens (Retallack, 2009). The specimens consist of what appear to be thalli, ~1.4mm wide, composed of filaments that branch dichotomously and monopodially outward from a central segment. Ultimate thallus segments are thin and show a rounded outline defined by thickened margins. Others have argued that this fossil represents an iron oxide weathering pattern on the sandstones (Jago et al., 2012).

Within Ordovician–Silurian rocks from a number of sites in the Appalachian Basin (North America) are various thallus-like structures that have been interpreted as representing a wide range of levels of biological organization, including embryophytes, algae, fungi, cyanobacteria, and lichens. Examination of layering, opacity, and other patterns in sections of the rock/organism interface has revealed that, although there is no consistent internal structure in the thallus-like sheets, there are several organizational types, some of which may represent fungi (Tomescu and Rothwell, 2006). Moreover, experiments conducted on a variety of extant organisms to simulate the effects of pressure and heat during fossilization produced internal structures similar to those of the fossils in some algae, fungi, lichens, and bryophytes (Tomescu et al., 2010). While the results of stable isotope analyses of some of the thalloid fossils suggested that they represent evidence of a terrestrial biota (e.g. Tomescu et al., 2009), they remain enigmatic as to their biological affinities. It is important to note, however, that at least one of these sites includes convincing fossil evidence of bacterial-cyanobacterial consortia (Tomescu et al., 2008).

In younger rocks there are several fossils that suggest the presence of lichens by Early Devonian time. Two of these, *Cyanolichenomycites devonicus* and *Chlorolichenomycites salopensis*, are interpreted as charred lichen thalli from the Lower Devonian (Lochkovian) of Shropshire, West Midlands, UK (Honegger et al., 2013b). In order to make comparisons that reflect preservational characteristics, a series of experiments was undertaken to coalify extant lichen thalli and then examine them with scanning electron microscopy (SEM). The thallus of *Cyanolichenomycites devonicus* is heteromerous, having a cortex several cell layers thick and a medullary layer of loosely organized septate hyphae and presumed colonies of cyanobacterial cells, some with evidence of a gelatinous sheath and bearing a morphological resemblance to *Nostoc*. The fossil contains a central depression surrounded by lobules, and this is interpreted to be the pycnidial conidiomata, which contains conidiophores; conidiogenesis is phialidic. On the other hand, *Chlorolichenomycites salopensis* has been interpreted as a green-algal lichen. The fossil represents a dorsiventral thallus that is a few millimeters in diameter. Fractured sections show septate hyphae intermixed with globose cells (16–21 μm in diameter) that are thought to represent a green alga morphologically similar to the common extant green algal lichen photobiont *Trebouxia*. The upper surface of the thallus is a single cell layer thick; the lower thallus surface consists of interwoven hyphae. Both of these fossil lichens share a number of features with extant forms that can only be compared based on sections or fractured surfaces of the thallus, and using sufficient resolution to detail the diagnostic features. Moreover, bacterial colonies, actinobacterial filaments, and narrow hyphae of presumed endolichenic fungi are associated with the thallus of *Chlorolichenomycites salopensis*, indicating that, as in extant lichens, this fossil served as host to a diverse microbial community (Honegger et al., 2013a).

Although there are few Paleozoic lichens, especially those dated relatively close to the early terrestrialization of the Earth, the study by Honegger et al. (2013b; Figure 10.11) clearly indicates that within Early Devonian ecosystems the lichen symbiosis was already present and showed some level of diversity. Moreover, these fossils suggest that lichen evolution already included multiple types of photobionts by the Early Devonian, a hypothesis that is consistent with fossil evidence on other types of fungal mutualisms, which played a critical role in the development of early ecosystems (e.g. Taylor and

FIGURE 10.11 Rosemarie Honegger.

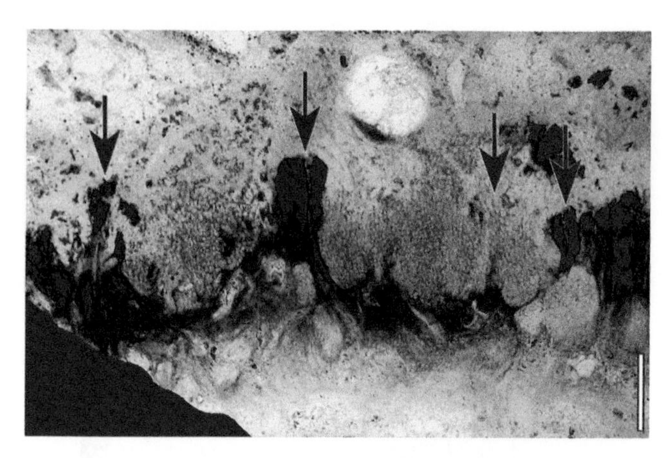

FIGURE 10.12 Longitudinal section through thallus of *Winfrenatia reticulata* showing hyphal ridges (*arrows*). Lower Devonian Rhynie chert, Scotland. Bar = 375 μm.

Osborn, 1996; Selosse and Le Tacon, 1998). Finally, the critical examination of these mesofossil lichens underscores a relatively untapped source of fossil material that will be extremely valuable in deciphering fungal evolution in the future.

A slightly younger fossil from the Lower Devonian Rhynie chert is *Winfrenatia reticulata*, which is assigned to the lichens based on structural features and the consistent association of fungal hyphae and photobiont cells (Taylor et al., 1995b, 1997). This fossil is permineralized and consists of a single, incomplete specimen approximately 10 cm long and up to 2 mm thick. Most of the thallus consists of a thin mycelial mat, 1–2 mm thick. Each mat is formed by two to four superimposed layers of parallel hyphae, and each layer is two to six hyphae thick. The uppermost two layers are vertically oriented and folded into loops that form a pattern of ridges and circular-to-elliptical depressions or indentations on the thallus surface, each 0.5–1.0 mm deep (Figure 10.12). Extending from the walls of the depressions are slightly narrower hyphae of the same type, and these form a three-dimensional network. As a result of hyphal branching each depression along the surface of *W. reticulata* consists of relatively uniform (20–30 μm in diameter) lacunae that are formed by the mycobiont. Perhaps as a result of preservation, or simply because it does not exist, there is no evidence of any layer that might be interpreted as a cortex.

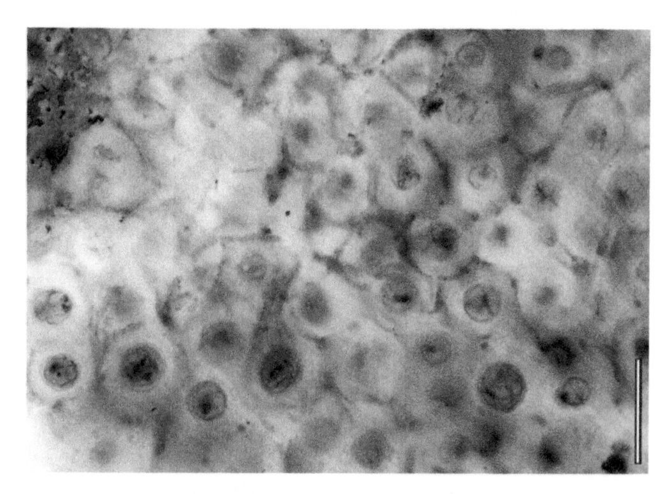

FIGURE 10.13 Hyphal net of *Winfrenatia reticulata* with each lacuna containing a single unicell. Lower Devonian Rhynie chert, Scotland. Bar = 35 μm.

The photobiont in *W. reticulata* consists of coccoid unicells or, in some depressions of the thallus, clusters of unicells that occur within the lacuna of the three-dimensional hyphal net (Figure 10.13). Individual unicells range from 10 to 16 μm in diameter and each is surrounded by a sheath up to 6 μm thick (Figure 10.14). There are multiple sheaths present when the unicells are organized in clusters. One of the most interesting features of *W. reticulata* is the relationship between the stages of development of the photobiont cells within the depressions of the thallus. In the earliest stage there is a gradation of cell size in the depressions, from small ones near the base to larger ones in various stages of cell division toward the top of the depression (Figure 10.15).

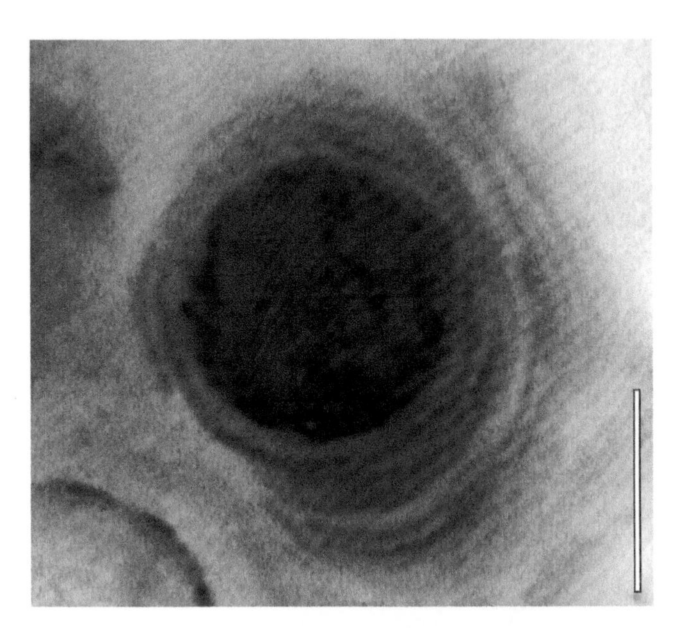

FIGURE 10.14 Multiple sheaths surrounding unicell in hyphal net of *Winfrenatia reticulata*. Lower Devonian Rhynie chert, Scotland. Bar = 10 μm.

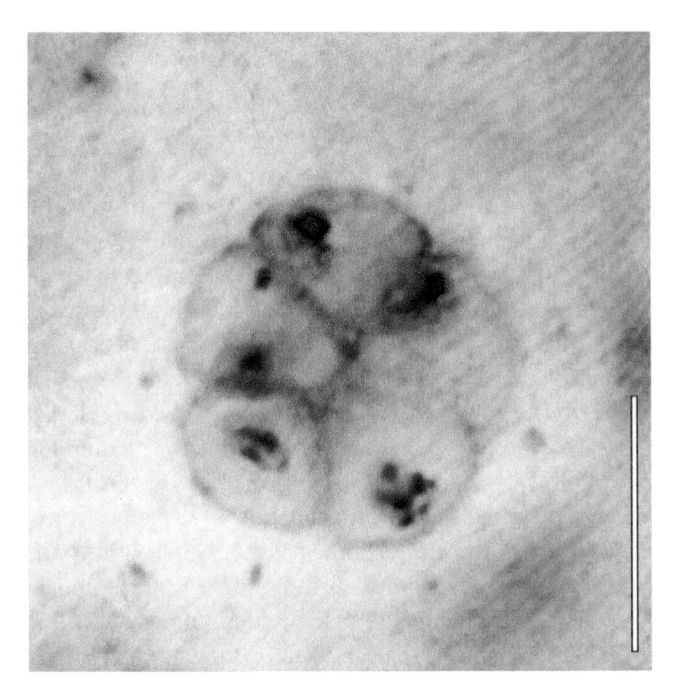

FIGURE 10.15 Multi-celled photobiont of *Winfrenatia reticulata*. Lower Devonian Rhynie chert, Scotland. Bar = 10 μm.

In the next several stages photobiont cells are increasingly enclosed by the hyphal net of the mycobiont, with some of these photobiont-mycobiont clusters detached from the thallus, suggesting that perhaps they represent some type of reproductive unit such as soredia. The consistent relationship between photobiont cells and the hyphal net suggests that in *W. reticulata* there may have been some level of controlled parasitism in which there was a distinct balance between the production of new photobiont cells, production of endospores that may have resulted in additional thallus depressions being formed by the mycobiont, and the physiological interaction between both.

There are, however, several questions that remain about the affinities and organization of *W. reticulata*. One of these regards the affinities of the mycobiont. In some regions of the thallus there are large (~65 μm in diameter), thick-walled spores that have highly sculptured walls and show some similarities to spores produced by certain members of the Mucoromycotina. It has been suggested that the photobiont cells are morphologically similar to modern cyanobacterial coccoid cells of *Gloeocapsa*, *Chroococcus*, and *Chroococcidiopsis* (Taylor et al., 1997). Obviously, the physiological aspects of this lichen are unknown, but the presence of a single coccoid cell in each fungal net space suggests that there might have been some level of interaction between the organisms interpreted as mycobiont and photobiont. In addition, the sheath-like material present in the fossil is quite similar to the gelatinous material surrounding photobiont cells, which represents the site of carbon transfer in certain modern lichens (Honegger, 1991, 1992).

Another specimen of *W. reticulata*, also based on material from the Rhynie chert, provides additional features and offers another interpretation of this interesting organism (Karatygin et al., 2009). In this interpretation, the majority of the thallus contained empty sheaths of filamentous cyanobacteria, like those in the Nostocales, together with hyphae. The presence of multicellular cyanobacterial trichomes within the thallus, in addition to the coccoid cells, is interpreted as indicating that the photobiont in *W. reticulata* included two different cyanobacteria. If this interpretation is accurate, it suggests that perhaps in this lichen symbiosis the structural and morphological dominance that is present in modern lichens was far different and may have relied more on the physiological interrelationships among the organisms. Such a scenario might be expected in the evolution of the lichen symbiosis and would help to explain the absence of findings of fossil lichens that are similar to their modern counterparts. The problem with this interpretation, however, is that the lichens described by Honegger et al. (2013b), from slightly older rocks, are morphologically very similar to modern lichen thalli.

An interesting fossil from the Middle Devonian (Givetian) of Kazakhstan that is thought to be some type of foliose lichen is *Flabellitha elinae* (Jurina and Krassilov, 2002). The specimens were initially described as *Flabellofolium elinae* and interpreted as some type of Paleozoic plant with fan-shaped leaves, similar to a modern *Ginkgo* (Jurina and Putiatina, 2000). The specimens consist of irregularly lobed impressions of a laminar structure, interpreted as a thallus, which is organized into a cortex constructed of broad tubular filaments termed hyphae. Beneath this zone is a layer of narrower (5–7 µm diameter) septate filaments associated with aggregations of spherical cells. Also present are structures thought to be sunken apothecia, each with two, bi-celled elliptical ascospores 40 µm long. The absence of structures that can be interpreted with confidence as photobiont cells, however, makes the suggestion of *Flabellitha elinae* as a type of lichen somewhat equivocal. Because some of the thallus is represented as a ferruginous film, another possibility is that this lichen-like structure is abiotic or some type of biomimetic structure (e.g. Klymiuk et al., 2013a).

FIGURE 10.16 Compressed thalli of *Daohugouthallus ciliiferus*. Middle Jurassic, Inner Mongolia, China. Bar = 5 mm.

MESOZOIC LICHENS

Historically there have been a few fossils described from Mesozoic rocks under the name *Thallites* that have been interpreted as some type of alga, or perhaps a bryophytic level of evolution. It has also been suggested that some have a morphological resemblance to extant foliose lichens (e.g. Webb and Holmes, 1982), although none shows evidence of any features or internal organization that would establish any type of lichen affinity. Some fossils suggestive of a lichen thallus have been described from an Upper Triassic clay from Germany (Ziegler, 1992). The specimens consist of a thallus of filaments and what are interpreted as apothecia and clusters of some type of possible photobiont cells. Although the SEM images and the reconstructions reveal structures in support of the lichen thallus hypothesis, the examination of additional specimens is needed to determine whether these fossils are in fact lichen thalli.

One structure believed to represent a lichen thallus has been reported from Middle Jurassic (Jiulongshan Formation) strata of Inner Mongolia Autonomous Region, China (Wang, X., et al., 2010). While other biological affinities might include some type of thallose liverwort or an alga, it is most likely that *Daohugouthallus ciliiferus* represents a lichen. The fossil is characterized by a highly branched thallus of elongate primary axes, each about 2.0 mm wide, that give rise to lateral and terminal branches, each of which may branch multiple times (Figure 10.16). Some of the ultimate branches terminate in one to several wart-like projections or small lobes that appear to be ruptured and look similar to lichen soralia. The filiform appendages that arise from the thallus are morphologically similar to cilia formed by certain extant lichens; these are believed to function in some moisture-condensation capacity (Hannemann, 1973). While these morphological similarities correspond to those found in modern lichens, there is to date nothing known about the mycobiont or photobiont of *D. ciliiferus.*

Another fossil interpreted as a foliose or squamulose lichen thallus from the Lower Cretaceous of British Columbia, Canada is *Honeggeriella complexa* (Matsunaga et al., 2013). The fossil is a permineralized fragment of a heteromerous thallus slightly larger than a millimeter and approximately 260 µm thick. In section view the thallus consists of distinct layers (Figure 10.17). A slightly thicker upper cortex and a lower cortex, both constructed of thick-walled hyphae that are described as

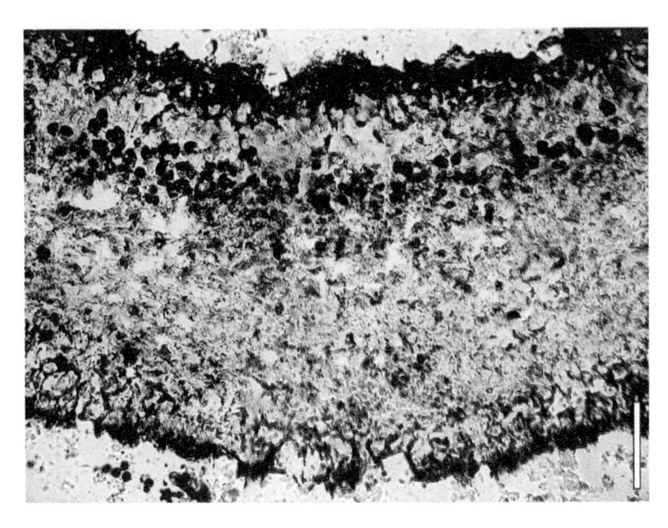

FIGURE 10.17 Partial section showing the organization of the *Honeggeriella complexa* thallus. Lower Cretaceous, Canada. Bar = 50 μm. (Courtesy K.K.S. Matsunaga.)

FIGURE 10.18 Detail of the thallus photobiont layer of *Honeggeriella complexa.* Lower Cretaceous, Canada. Bar = 10 μm. (Courtesy K.K.S. Matsunaga.)

forming a plectenchymatous structure. The inner portion of the cortex is composed of loosely arranged anastomosing hyphae (~1.1–3.5 μm diameter) that form circular lacunae in the thallus. Medullary hyphae are septate, parallel to the cortical layers, and range from 1.4 to 2.4 μm in diameter (Figure 10.18). Cells interpreted as the photobiont are globose and range up to approximately 12 μm in diameter. The intimate physical association between the photobiont and mycobiont is evidenced by narrow (0.5 μm) hyphae and appressoria on the surface of the photobiont cells, with some of the appressoria even slightly penetrating the cells of the photobiont. While the mycobiont is most likely ascomycetous, the nature of the photobiont is more difficult to interpret. Although *H. complexa* is represented only by a small thallus fragment, this report of a permineralized lichen holds great promise for increasing our understanding of the diversity of Mesozoic lichens. Moreover, the report also demonstrates the potential of identifying fossil lichens in ecosystem settings where the primary focus has been on vascular plants.

CENOZOIC LICHENS

While there have been relatively few lichens reported from the Mesozoic, many more are known from the Cenozoic. It is difficult to determine whether this relates to an increasing diversity of lichens in geologically younger strata, their preservational type, or better preservation in younger, less altered rocks (e.g. Green and Lange, 1994). Most of the reports are based on forms preserved in amber (e.g. Garty et al., 1982), but the fossils suggest that by the Cenozoic many modern lichen taxa were already diversified. *Anzia electra* (Lecanorales), reported from Eocene Baltic amber, is a small heteromerous, lobed thallus characterized by prominent cushions of spongiostratum, a layer of anastomosing hyphae on the lower surface of certain lichens that is not effective in attachment (Rikkinen and Poinar, 2002a). The upper and lower cortex are smooth and the medulla is two-layered, with the outer region constructed of loosely interwoven hyphae and an inner zone of longitudinally oriented agglutinated hyphae. The inner portion of the medulla sharply delimits the upper medulla from the spongiostratum. The photobiont of *A. electra* is interpreted as a unicellular green alga. Isidia, soredia, and sexual reproductive structures of the mycobiont are not present. *Anzia* includes 34 extant species (Jayalal et al., 2012), which are sometimes termed the black-foam lichens and are included in the Parmeliaceae, the largest family of lichen-forming fungi (Crespo et al., 2007). The fossil belongs to *Anzia* sect. *Anzia*, the closest living relatives of which are found in East Asia and eastern North America. The presence of the fossil in Europe suggests that *Anzia* sect. *Anzia* once had a circum-Laurasian range but later became extinct in Europe. Rikkinen and Poinar (2002a) noted that the present-day distribution of *Anzia* sect. *Anzia* is paralleled by the relict ranges of gymnosperms like *Cunninghamia*, *Metasequoia*, and *Ginkgo*, and of angiosperms such as *Liriodendron* and *Magnolia*.

Alectoria succini, also from Baltic amber, is thought to represent another lichen (Mägdefrau, 1957). The thallus is hair-like and consists of multiple tangled filaments. Structures interpreted as apothecia are not well illustrated and it is difficult to determine what they represent.

Two other fossil lichens, both identified as members of the Parmeliaceae, occur in Dominican amber (Miocene; Poinar et al., 2000). One of them, *Parmelia ambra*, consists of a branched foliose thallus with dichotomously organized lobes. Arising from the lower surface are uniform, branched rhizines up to 900 μm long and 30–60 μm wide; the upper surface is smooth and a single isidium occurs in the fossil. In section view the thallus of *P. ambra* is constructed of thick-walled, intertwined medullary hyphae each up to 2.3 μm in diameter. The photobiont consists of algal cells (5.8–11.6 μm in diameter), which occur between the medulla and upper cortex. The thallus of the second species, *P. isidiiveteris* (Figure 10.19), is thinner and characterized by sinuate lobes with multiple isidia (300–500 μm tall) that are unevenly scattered on the thallus. There is no evidence of apothecia or pycnidia in these fossils.

The amber fossils of *Anzia* and *Parmelia* have recently been used as calibration points in a study on the origin and diversification of major clades in parmelioid lichens (Amo de Paz et al., 2011) and an analysis of their biogeography. This study concluded that the parmelioid group diversified around the Cretaceous–Paleogene boundary, and the major clades diverged during the Eocene and Oligocene, triggered and fostered by globally changing climatic conditions of the early Oligocene, Miocene, and early Pliocene.

Phyllopsora dominicanus is a lecanoralean lichen preserved in Dominican amber (Rikkinen and Poinar, 2008). Thallus morphology ranges from squamiform to subfoliose, with the overlapping squamules each approximately 1.0 mm wide. The upper cortex is up to 60 μm thick, whereas the lower cortex is described as a net-like pseudocortex. The photobiont includes unicells that are 8–12 μm in diameter, many with opaque inclusions, and fungal hyphae with appressoria in contact. No reproductive structures are known for *P. dominicanus*, although small dorsiventrally constricted thallus sections might represent some type of vegetative reproduction unit. The genus *Phyllopsora* today consists of species growing primarily on the bark of tree trunks in tropical and subtropical humid forests; more than 100 species have been named in the literature (e.g. Timdal, 2008).

FIGURE 10.19 Thallus of *Parmelia isidiiveteris* preserved in amber. Miocene, Dominican Republic. Bar = 5 mm. (Courtesy G.O. Poinar, Jr.)

Fossils assignable to the modern families Caliciaceae (*Calicium*) and Coniocybaceae (*Chaenotheca*) have been reported from Baltic amber (Rikkinen, 2003). Both families are referred to as calicioid lichens, a group that includes both lichenized and non-lichenized species; the systematics of some genera in the group remains enigmatic (Tibell and Wedin, 2000). The lichenized forms have thin-walled, evanescent asci; these produce spores that accumulate on the surface of the ascoma, where they form a mazaedium. The amber fossil that is morphologically similar to *Chaenotheca* is 100–200 μm thick and includes several stalked ascomata preserved in situ, the mature ones with a compact mazaedium, a type of reproductive structure (Figure 10.20). Spores range up to 7 μm in diameter and possess an irregular surface ornament. Nothing is known about the photobiont of the fossil, but extant species of *Chaenotheca* are known to be associated with four different genera of algae (Tibell, 2001). In another piece of amber there is a single, stalked ascoma similar to that found in extant species of *Calicium* (Figure 10.21). The well-preserved spores are ellipsoidal, once-septate, and up to 15 μm long. Although the fossils are not formally named, the close morphological similarity to modern genera indicates that, at least in the features that are preserved, there has been little change over several million years, additional evidence of early divergence within the group.

Other evidence of what is interpreted as a lichen preserved in Baltic amber consists of a multibranched structure that is approximately 2.0 cm wide; individual branches are 1.0 mm in diameter (Garty et al., 1982). On the surface are perforations that are morphologically

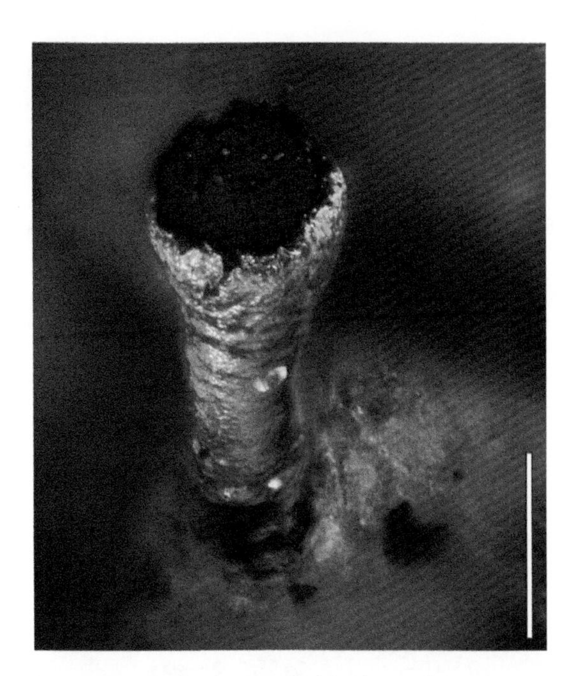

FIGURE 10.20 Stalked ascoma of *Chaenotheca* sp. preserved in amber. Eocene, northern Europe. Bar = 90 μm. (Courtesy A.R. Schmidt.)

similar to aeration pores, larger depressions that contain spore-like structures, and other structures interpreted as apothecia. Elemental analyses on the fossil suggest the presence of sulfur, calcium, iron, silicon, and aluminum. These results have been used to suggest that the lichen accumulated iron and sulfur from the Eocene atmosphere prior to fossilization and thus could be regarded as a bioindicator of air pollution at that time. Although SEM and energy dispersive X-ray analysis were used to characterize this fossil, there is still some uncertainty as to whether the structure does, in fact, represent a lichen.

In other instances certain morphological features present in impression and compression fossils have been used to determine lichen affinities. One example is an impression fossil from the middle Miocene of California (MacGinitie, 1937). The fossil shows a portion of a thallus with a reticulate margin (Peterson, E.B., 2000) that suggests a resemblance to some modern species of *Lobaria* (lung lichen), a genus that is characterized by having both a cyanobacterium and an alga as symbiotic partners. Since the fossil is preserved as an impression, however, it is difficult to determine the precise affinities. The diversity of modern thallus types sometimes makes it difficult to distinguish lichens from other organisms in the fossil record. In his comprehensive study of Eocene

FIGURE 10.21 Stalked ascoma preserved in amber that is morphologically identical to the extant taxon *Calicium*. Eocene, northern Europe. Bar = 160 μm. (Courtesy A.R. Schmidt.)

plant cuticles from Tennessee, USA, Dilcher (1965) suggested that one structure found on *Chrysobalanus* leaves was a member of the Microthyriaceae. *Pelicothallus villosus* consists of a lobed stroma constructed of laterally radiating hyphae and bearing blunt-tipped setae; spores were not found in association with the fossil. Another interpretation is that *P. villosus* might represent an alga (Reynolds and Dilcher, 1984) – one that is morphologically similar to extant *Cephaleuros* (Trentepohliaceae), a genus of green algae that includes forms parasitic on land plants (e.g. Joubert and Rijkenberg, 1971) as well as those that function as lichen phycobionts. Consequently, the fossil was transferred to the genus *Cephaleuros* and named *C. villosus* by Thompson and Wujek (1997). Sherwood-Pike (1985), however, views the fossil as a lichen similar to foliicolous species in the modern genus *Strigula*. These lichens form by lichenization of preexisting *Cephaleuros* thalli; the process occurs by chance through colonization by a compatible fungus (Schubert, 1981). Moreover, what were suggested to be the possible ostioles on the surface of the Eocene fossil are quite similar to pycnidia in *Strigula* (e.g. Sérusiaux, 1998). While the morphological similarity between *Pelicothallos* and *Cephaleuros*/*Strigula* is striking, precisely what type of organism the fossil represents (e.g. microthyriaceous fungus, lichen, or perhaps a parasitic alga) remains undecided.

FOSSILS THAT MIGHT BE LICHENS

There are several interesting organisms from the Paleozoic that have been variously interpreted as some level of lichen evolution (e.g. Taylor, T.N., 1982). While some of these are preserved as different types of compressions, including those that are charcoalified (Glasspool et al., 2006), others are structurally preserved in extraordinary detail (e.g. Hueber, 2001). In spite of this remarkable degree of preservation, the biological affinities of these fossils remain in question. The nematophytes, or Nematophyta (Strother, 1993), is a group of enigmatic Silurian–Devonian organisms that range from millimeter-sized structures to several-meter-long cylindrical objects termed "logs." Their plant bodies are constructed entirely of variously sized and shaped tubes, which are also frequently found isolated in palynological samples from those time periods (e.g. Taylor and Wellman, 2009). Although nematophytes have been studied intensively for more than 150 years, little is known about their systematic affinities, biology, and ecology (Lang, 1937). Perhaps the most famous of the nematophytes is *Prototaxites*, a unique life form that is known from Silurian and Devonian rocks; in addition, there is one equivocal report from the Ordovician (Koeniguer, 1975). One of many interesting features of *Prototaxites* is the size of the organism; some specimens are more than a meter in diameter and more than 8 m long (Figure 10.22) (e.g. Arnold, C.A., 1952; Schweitzer, 2000; Hueber, 2001 [Figure 10.23]). The most definitive study of the structure of *Prototaxites* is that of Hueber (2001), who examined multiple specimens in his detailed analysis. Internally, *Prototaxites* is constructed of a series of interlaced longitudinally oriented tubes of three principal types, which Hueber (2001) refers to as a trimitic hyphal system (Figure 10.24), thereby adopting the terminology used for hyphal types in extant trimitic Basidiomycota. One

FIGURE 10.22 "Log" of *Prototaxites*. Devonian, Saudi Arabia. Bar = 30 cm. (Courtesy F.M. Hueber.)

FIGURE 10.23 Francis M. Hueber.

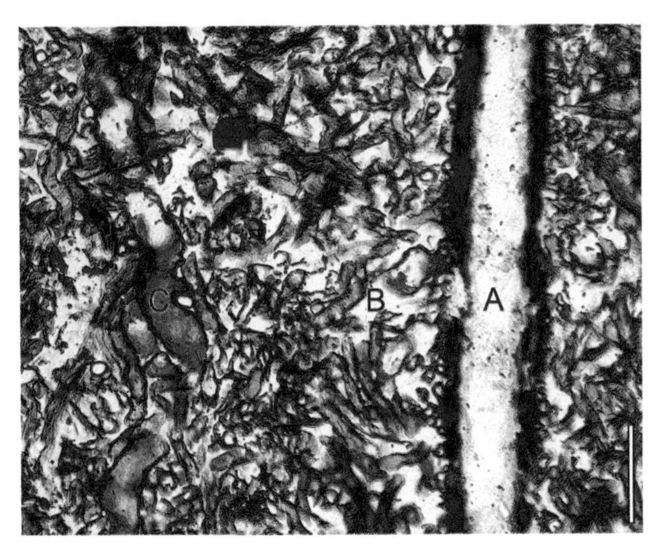

FIGURE 10.24 Longitudinal section of *Prototaxites southworthii* showing three basic types of tubes: **A.** skeletal; **B.** binding; **C.** generative. Devonian, Canada. Bar = 25 μm.

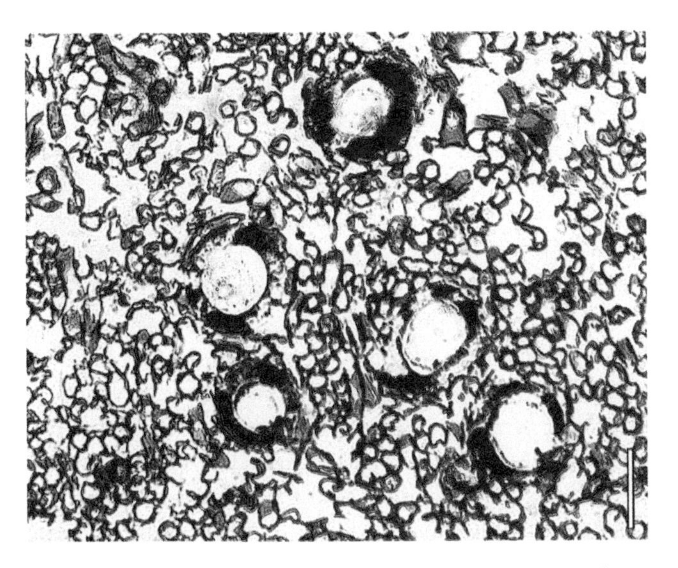

FIGURE 10.25 Transverse section of *Prototaxites southworthii* showing organization of tubes. Devonian, Canada. Bar = 25 μm.

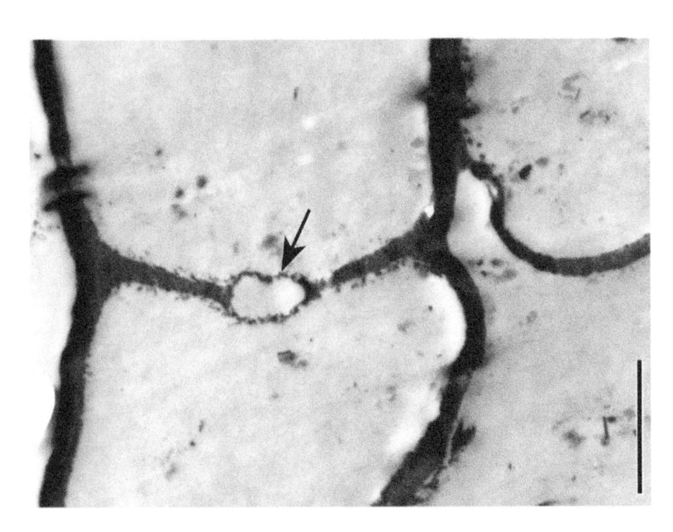

FIGURE 10.26 Section of *Prototaxites southworthii* showing septal pore (*arrow*). Devonian, Canada. Bar = 2 μm. (Courtesy R. Schmid.)

type, termed skeletal hyphae, are thick walled, aseptate, relatively straight, unbranched, and 18–50 μm in diameter (Figure 10.25). Tubes with thinner walls are referred to as generative hyphae. They are highly branched and characterized by distinct septa, often with occluded pores (Figure 10.26), and range from 12 to 42 μm wide (Schmid, 1976). The majority of the *Prototaxites* body is made up of the third type of hyphae, which are referred to as binding hyphae. These are up to 7 μm in diameter, highly branched, thin walled, septate, and have a distinct pore in the septum. The three types of tubes result in the formation of a pseudoparenchymatous or plectenchymatous system that forms the biomass of the large axes. In transverse section *Prototaxites* appears to be constructed of a series of concentric to eccentric increments that have the superficial appearance of growth rings in the secondary xylem of certain woody plants. Although these increments suggest some type of periodicity in growth, the borders of the so-called growth increments are marked by increased density of the hyphae that make up this "tissue," and the increments are not as regular as vascular plant tree rings. In addition to the tubes, numerous lacunae are scattered throughout the pseudoparenchyma. These spaces often contain disintegrated tissue and, occasionally, tubes, and are interpreted as having been formed in response to some type of parasite. Although the outer surface of the fossil is often smooth or with some patterning in the form of longitudinal ribs, it is likely that there may have been additional "tissue" systems on the exterior of the stems as all of the specimens appear to have been transported.

Prototaxites was initially described from the Gaspé region of Canada as some type of gymnosperm wood (i.e. the name refers to the extant conifer genus *Taxus*) because the tubes superficially looked like some type of vascular plant conducting element (Dawson, 1857, 1859; Figure 10.27). Since that time there have been various interpretations of the biological affinities of the organism, including: different morphological types of an

FIGURE 10.27 John William Dawson.

alga (e.g. Kräusel, 1936; Niklas and Smocovitis, 1983; Schweitzer, 2003), a land-plant lineage that became extinct (e.g. Lang, 1937; Abbott et al., 1998), a type of alga that was partially adapted to land (Arbey and Koeniguer, 1979), an alga in some type of lichen association (e.g. Selosse, 2002), a giant terrestrial saprotroph with possible basidiomycetous affinities (Church 1919; Hotton et al., 2001, Hueber, 2001; Moore, D., 2013), or a type of heterotroph that may have used algal-derived organic matter a a nutrient source (Hobbie and Boyce, 2010). The unique organization of the tubes in *Prototaxites* and the presence of pseudo growth increments has led to the interpretation of the organism as representing a rolled up mat of liverworts (e.g. Graham et al., 2010). However, this interpretation has not been supported (e.g. Boyce and Hotton, 2010; Taylor et al., 2010). The use of thin sections to detail the structure of the organism, transmission electron microscopy (TEM) (Schmid, 1976), chemotaxonomy, biomarkers, and isotopic signatures (e.g. Niklas, 1976; Boyce et al., 2007; Graham et al., 2010), and details about sedimentology and taphonomy (Griffing et al., 2000; Taylor et al., 2010) have all been employed to help resolve the affinities of this enigmatic organism.

So what type of an organism is *Prototaxites*? It was certainly large, based on the size of the so-called "logs" but thus far nothing is known about its distal or basal parts. The fossils were all apparently transported into sandstone channels, so external parts of the organism were probably removed in the process (Griffing et al., 2000). Casts approximately 1.0 cm in diameter have been suggested to be possible rooting structures of *Prototaxites* (Hillier et al., 2008), and putative branch scars have also been described (Altmeyer, 1973). Interestingly, the pores in some of the tubes are like the septal pores occurring in certain fungi (e.g. Basidiomycota), but similar structures have been reported in red algae (e.g. Lee, R.E., 1971). The significance of the pores may simply reflect the functional requirements of an organism that evolved a physiological mechanism for long-distance transport within the axis (as seen in large brown algae, for example) rather than being of phylogenetic value.

It has also been reported that small outgrowths near the septa in some specimens resemble (incomplete) clamp connections (Hueber, 2001), suggesting possible affinities with the Basidiomycota. Hyphal anatomy, and the designation of various types of hyphae (i.e. binding, generative, and skeletal), was one of the first features used to distinguish fruiting bodies in the extant Agaricomycotina (Homobasidiomycetes; e.g. Corner, 1966; Pegler, 1996), and there is still some correlation between hyphal types and molecular data (e.g. Hibbett and Donoghue, 1995; Hibbett and Thorn, 2001). Interestingly, the structural organization of *Prototaxites*, with its various types of tubes (hyphae), superficially resembles the structure of a rhizomorph of an agaricoid basidiomycete such as *Ossicaulis lignatilis* (Clémençon, 2005). Although the size range of these two structures is quite different, the general organization of the hyphae is structurally quite similar.

Another method used to identify the potential affinities of *Prototaxites* and its presumed heterotrophic lifestyle is based on isotopic signatures of fungi growing in modern environments that are thought to be similar to the sites where *Prototaxites* once grew (Hobbie and Boyce, 2010). The majority of the geochemical data assembled on *Prototaxites* suggests that it had some level of heterotrophy, and thus a component of the life history may have functioned like a fungus. Whether *Prototaxites* grew as a giant saprotroph is more difficult to envision since the landscape during the Silurian and into the Early Devonian is believed to have supported very small (less than 20 cm tall) terrestrial plants, which would have led to only a relatively small biomass, and thus a meager carbon source for a gigantic organism such as *Prototaxites*. Even

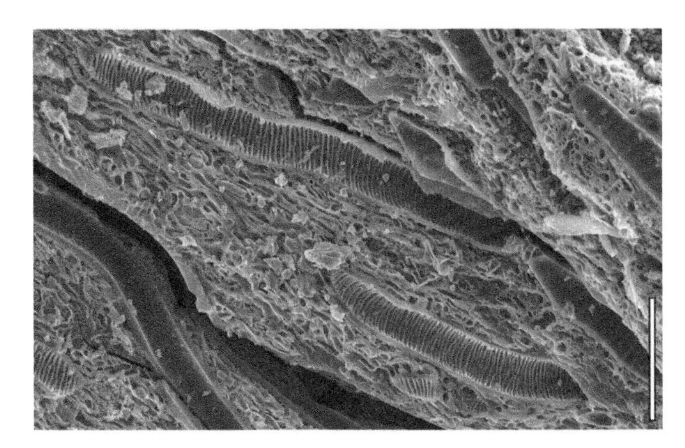

FIGURE 10.28 Fractured surface of *Nematasketum* sp. showing internal thickening of tubes. Devonian, England. Bar = 50 μm. (Courtesy D. Edwards and L. Axe.)

FIGURE 10.29 Dianne Edwards.

including the evidence of various algae and cyanobacterial mats that were present in the ecosystem, one wonders whether there was a sufficient carbon source to be able to support organisms of its size. An alternative hypothesis is to view *Prototaxites* as representing some type of lichen-like association (Selosse, 2002). Although there is nothing directly known about *Prototaxites* that suggests a relationship with a photobiont of some type, the report of both algal and cyanobacterial lichen structures from the Lower Devonian (Taylor et al., 1995b, 1997; Honegger et al., 2013b), indicates that this symbiotic association was no doubt in existence as early as the Silurian. One idea is that the charcoalified fragments of heteromerous lichen thalli reported by Honegger et al. (2013b) may have covered the surface of *Prototaxites* and provided the carbon source for these organisms. The fact that there were other enigmatic organisms constructed of interlacing tubes (e.g. *Nematothallus*) known from coeval rocks adds some support to the suggestion that *Prototaxites* was some type of lichen, albeit a very large one.

Another early land organism found in both marine and freshwater deposits of the Lower Devonian, and constructed of various tubes, is *Nematasketum* (Burgess and Edwards, 1988; Figure 10.28). These specimens are preserved as three-dimensional charcoal fragments, some up to 50 mm long. At least half of the specimens consist of tubes (filaments) up to 25 μm in diameter and of undeterminable length. As in the case of *Prototaxites*, there are narrower (~6 μm in diameter) branched filaments, each with a distinct central swelling in the middle of the septum. In some regions there are highly branched, narrow, smooth-walled tubes that lack a preferred orientation; termed medullary spots, these have been suggested as sites of hyphal generation (Burgess and Edwards, 1988; Edwards et al., 2012). Like *Prototaxites*, *Nematasketum* is thought to perhaps have affinities with some type of fungus (basidiomycete) (Edwards and Axe, 2012; Figures 10.29, 10.30) or perhaps a lichen.

Tubes also characterize the organization of *Nematothallus*, another Early Devonian organism from the same general stratigraphic level as *Nematasketum* (Lang, 1937). The basic organization in this taxon is one of interlacing tubes of two basic types, some of which have been described as internal wall thickenings; however, the thallus is covered by a cuticle-like sheet showing a pseudocellular pattern (Strother, 1988, 1993). The complex anatomy and septate tubes of *Nematothallus* suggest an affinity (Figure 10.31) with lichenized fungi. Limited data support a fungal, rather than an embryophytic chemistry (Edwards et al., 2013). Cuticle-like sheets that appear similar, some with distinct pores (strengthened holes in the cuticle) (Edwards and Rose, 1984), have also been reported from as early as the Ordovician. Some specimens are also found associated with spores, but it is not known whether the cuticle-like sheets, tubes, and spores are parts of the same organism, or even from the same group of organisms. The function of the preformed perforations in the cuticle remains an enigma. Suggested

FIGURE 10.30 Lindsey Axe.

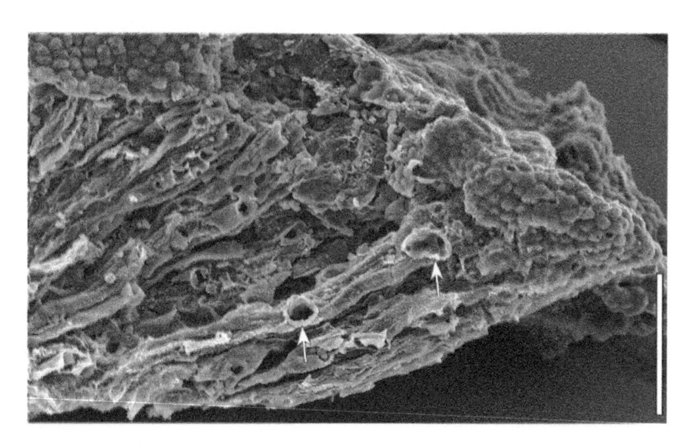

FIGURE 10.31 Fragment of *Nematothallus williamii* showing cortex, palisade zone, and outgrowths (*arrows*). Lower Devonian, Shropshire, England. Bar = 100 μm. (Courtesy D. Edwards and L. Axe.)

functions include some type of aerating structure, or an exit site for spores, gametes, or some type of pathogen (Edwards and Rose, 1984). Coprolites in the form of discoidal structures, which were recovered by acid maceration from Lower Devonian rocks, contain fragments of *Nematasketum* and are interpreted as evidence that millipedes used nematophytes as a food source (Edwards et al., 2012).

New fossils that superficially resemble *Nematothallus* have been described from the Burgsvik Formation (Upper Silurian) of Sweden (Smith and Butterfield, 2013). The specimens, formally assigned to a new genus, *Nematothallopsis*, consist of carbonaceous sheets that have a smooth outer surface and ridges along the inner side; occasional elliptical apertures are found in some regions. A number of specimens of *N. gotlandii* were found to have hollow, septate filaments (1–13 μm in diameter) attached to the inner surface of the cuticle, some of which are closely associated with the apertures. Additional features include organic-walled spheroidal structures ranging in size from 7 to 25 μm and different types of filaments attached only at the edges of the cuticle. *Nematothallopsis* is interpreted as some type of thalloid organism with a pseudoparenchymatous internal organization and reproductive structures that form on the inner surface of the outer cuticle-like sheet. Interestingly, it has been suggested that this fossil represents some type of red alga (Rhodophyta), perhaps related to the Corallinaceae (Smith and Butterfield, 2013).

Another Early Devonian organism that is included among these enigmatic forms is *Nematoplexus rhyniensis*, which, as the species name implies, is from the Rhynie chert (Lyon, 1962). The organization of the thallus is also a system of interlaced tubes of various diameters with and without internal wall thickenings. Septa that are morphologically similar to those in *Prototaxites* occur in some of the narrow tubes. In some regions the tubes are aggregated into dense multibranched clusters (so-called branch knots) (Figure 10.32) and in general, tube branching appears to be restricted to these clusters. It has been suggested that *N. rhyniensis* is a permineralized equivalent of *Nematothallus* (e.g. Taylor, T.N., et al., 2009).

There are other enigmatic fossils that occur in Silurian and Devonian strata in which the biological affinities remain unknown (e.g. Taylor, T.N., 1982). Some of these, such as *Pachytheca* (Silurian–Devonian), are small (~1.0 cm in diameter) and organized into two distinct zones (e.g. Gerrienne, 1991). The inner zone (medulla) consists of densely spaced, intertwined tubes, while the outer layer (cortex) is constructed of tubes that are distinctly radially aligned. Between the cortex and medulla is a narrow area in which the tubes seem to change from random to radial alignment. In some specimens, a narrow canal is visible that appears to connect the medulla with the exterior. Others, like *Spongiophyton* (Devonian),

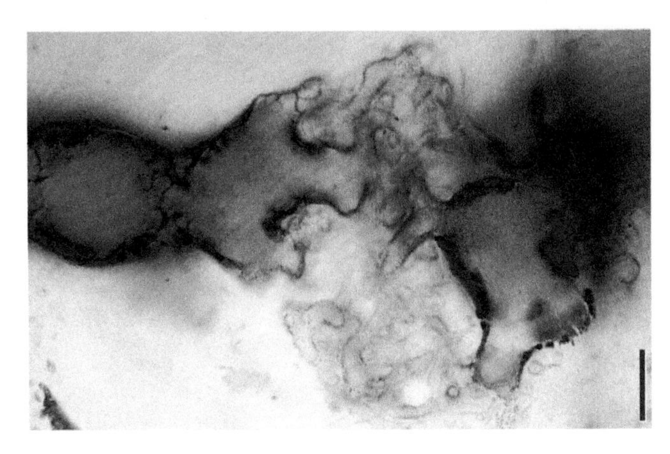

FIGURE 10.32 Developing "branch knot" of *Nematoplexus rhyniensis* tubes. Lower Devonian Rhynie chert, Scotland. Bar = 45 μm. (Courtesy H. Hass.)

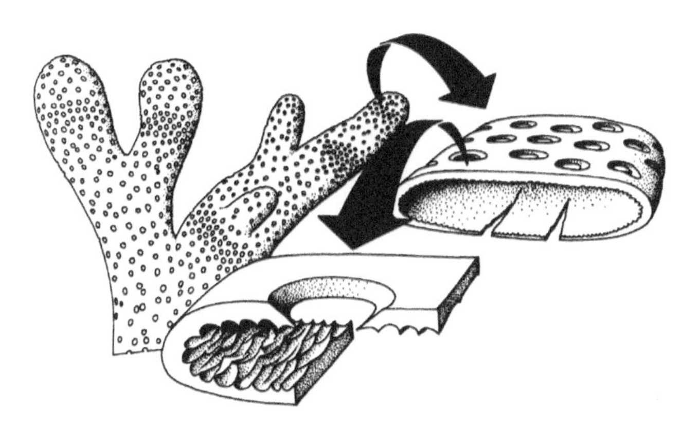

FIGURE 10.33 Diagrammatic reconstruction of *Spongiophyton nanum* thallus showing pores and inner flanges. Devonian. (From Taylor, T.N., 1988.)

are morphologically more complex. This thalloid fossil has variously been interpreted as a colonial animal, alga, vascular plant, and bryophyte. Specimens are small (25 mm wide by 2.5 cm long), branched axes with rounded tips. It has been suggested that *Spongiophyton* grew in a desiccating environment, based on the presence on one surface of the thallus of circular pores ranging from 200 to 300 μm in diameter (Figure 10.33). The pores are believed to have functioned in gas exchange. Additional support for life in a desiccating environment is the presence of a cuticle (or cuticle-like surface layer) covering the *Spongiophyton* thallus. Based on internal tissues, Stein et al. (1993) and Taylor et al. (2004) proposed that *Spongiophyton* represents a lichen, with linear chains of cells in the interior. The lichen hypothesis was initially supported by carbon isotope ratios from the fossil (Jahren et al., 2003). Fletcher et al. (2004), however, found that the range of $\delta^{13}C$ values in *Spongiophyton* is not significantly different from those of lichens, liverworts, or mosses, so carbon isotopes are not useful in identifying this fossil. As a result, the affinities remain conjectural. Continued research is needed to clarify whether this organism possesses a complement of features that are intermediate between algae and land plants.

There are numerous records of isolated spores, cuticle-like sheets, and tubes that have been variously interpreted as originating from algae, fungi, bryophytes, land plants, and lichens. Some of these have also been placed in the nematophytes (Strother, 1988, 1993). Another approach has been used by Burgess and Edwards (1991), who grouped the Ordovician–Early Devonian tubes into an artificial classification system, perhaps generating a broader perspective of the relationships based on patterns of diversity when plotted against other parameters (e.g. time, depositional environment, associated structures).

Despite using various techniques, including TEM and SEM, computer simulation, taphonomic analyses, geochemical profiles, and comparison with various extant organisms that have been physically and chemically altered in an attempt to reproduce similar patterns seen in these fossils, to date there is no definitive answer as to the precise biological affinities of any of the organisms or structures described in the paragraphs above. We are increasingly of the mind to believe that the structural organization seen in these fossils is not of phylogenetic significance, but rather represents structural and physiological adaptations of a diversity of organisms involved in the early move into the new terrestrial niche, perhaps analogous to the diversity of organisms seen in the Ediacaran–Early Cambrian radiation. Our aim in providing information about these enigmatic organisms is simply to increase awareness of their existence and to provide information so that others in the scientific community, who might not be aware of these fossils, will perhaps recognize their systematic affinities. It is possible that these Paleozoic sheets of cuticle-like material represent different types of organisms (e.g. algae, fungi, lichens) that are grouped together because of their general organization, and they may simply represent a particular grade of structural evolution. One of the issues faced by paleobiologists, irrespective of what groups of fossils they may study, relates to interpreting the specimens in the context of extant organisms. No one knows how many groups of organisms existed for a period of time in the history of the Earth and then became extinct, leaving only a minor trace of their existence. The so-called nematophytes may represent one or more lineages for which we have evidence, albeit meager, but no modern analogues with which to make direct comparisons. Whatever these organisms were biologically, they were apparently quite common in the Paleozoic, and it will be interesting to see whether their affinities can be determined in the future. The fact that they are included in the chapter on lichens betrays our bias as to ultimate affinities.

11

FUNGAL SPORES

NAMING FUNGAL SPORES 222
FUNGAL SPORES IN STRATIGRAPHY 223
FUNGAL SPORES IN PALEOECOLOGY 223
FUNGAL SPORE TAXA 224
Amerospores 224
Didymospores 227
Phragmospores 229
Dictyospores 232
Scolecospores 233
Helicospores 233
Staurospores 235
OTHER FUNGAL SPORES AND STRUCTURES 235
Hyphomycetes 236
Agonomycetes 237
Coelomycetes 237

In palynological preparations, fungal spores have traditionally received far less attention than the pollen and spores of vascular plants. Nevertheless, a variety of fungal remains, including spores, hyphae, and various types of reproductive structures and mycelia, are often present in such residues and can be found throughout much of the geologic column. In palynological analyses, these remains are usually treated together with other microfossils of various origins (e.g. cyanobacteria, algae, aquatic invertebrates) and collectively referred to as non-pollen palynomorphs or NPPs (van Geel 2001; Cook et al., 2011). Many NPPs are today used as proxy indicators in paleoecological studies, especially in the Cenozoic. While some of the earliest of these studies recognized fungal spores among the non-pollen/spore remains, they were not dealt with in a systematic manner, although some were simply related to modern forms, a practice that is sometimes followed even today. In addition, it was quickly understood that some modern fungi produce more than one type of spore.

Palynological preparations have been and will continue to be an important source of paleomycological information (e.g. Traverse, 2007; Figure 11.1). In general, when pollen grains and spores are described, representative examples that provide a range of sizes, morphology, and patterns of ornament will be used. Pollen grains with structures adhering to the outer surface or in the body of the grain will probably be discounted or not carefully examined because the extraneous material will be regarded as obscuring important morphological features of the pollen or spore specimen. As in macrofossil research, however, it is important to examine some of these so-called atypical palynomorphs for evidence of: (1) various types of fungi that may have attacked the grains, or (2) infection patterns (e.g. Moore, L.R., 1963a; Huang et al., 1999), as well as for examples of mycoparasites like those seen in modern pollen (e.g. Olivier, 1978; Huang et al., 2003). Such studies, however, must also be certain that structures which appear to adhere to the surface of a pollen grain or spore are in fact fungal and not some structural component of the pollen grain. For example, in some pollen grains the oncus (intine of the aperture region) protrudes through the aperture (Kuang et al., 2012) and might be interpreted as a chytrid zoosporangium on the surface of the grain.

T.N. Taylor, M. Krings & E.L. Taylor: Fossil Fungi.

DOI: http://dx.doi.org/10.1016/B978-0-12-387731-4.00011-6

FIGURE 11.1 Alfred Traverse.

FIGURE 11.2 W. Bryce Kendrick.

NAMING FUNGAL SPORES

One of the first systems for classifying fungal spores by comparing features with those of modern analogues relied on conidial characters of size, shape, color, and degree of septation. This system was proposed by Saccardo et al. (1882–1931), and the multivolume work, *Sylloge fungorum omnium hucusque cognitorum*, in which they assembled details of all the fossil species published up until 1920, was a monumental accomplishment; additions and corrections were later made by Kirk (1985) and in subsequent editions. The initial system was modified by van der Hammen (1954a, b), who assigned groups of pollen and spores to various morphological categories and had the suffix *-sporites* added for fungal spores; later, Clarke (1965) suggested the suffix *-sporonites* be used to indicate that the genus was of a fossil spore. Surface texture, including ornamentation, and color have also been used in defining some fungal spores (Elsik, 1976a, b, 1983, 1996), and there is now some uniformity in the terminology that has been developed and refined for pollen and spores (e.g. Dominguez de Toledo, 1994; Punt et al., 1994). Because the shape of the fungal spores may be variable, Kendrick and Nag Raj (1979) (Figure 11.2) modified the system to remove inconsistencies and provide greater resolution for some features.

Based primarily on Cenozoic fungal spores, Elsik (1976a) suggested a classification based on the two features he regarded as most consistent, which were septation and the presence or absence of an aperture. Within this classification he proposed multiple families within two major orders of Fungi Imperfecti; the Sporae Dispersae included families Sporae Monocellae, Sporae Monodicellae, and Sporae Tetracellae, while the Mycelia Sterilia contained the Cellae and Hyphae. Other classifications of fungal spores used a more descriptive terminology (e.g. monocellate fungal spore, multicellate fungal spore, fruiting bodies; Norris, 1986, 1997), features of the apertures (Song et al., 1999), and both features and number of apertures (Song and Huang, 2002). Today one of the systems that is commonly used was initially proposed by Pirozynski and Weresub (1979), who described fungal spores based on shape and number of cells to identify seven different categories of spores. Kendrick and Raj (1979) also provided a key for mature spores of Fungi Imperfecti.

Irrespective of what classification system is used, one of the major problems that still remains unresolved in studying fossil fungal spores is the fact that some spores may be highly pleomorphic and the different forms may have different names (e.g. Arx, 1987; Sugiyama, 1987). In addition, the teleomorph and the anamorph may each have a specific name, and in studies of both extant and fossil fungi, the biological relationship between the two is unknown (Weresub and Pirozynski, 1979). Despite the desire to relate fossil fungal spores to extant forms, and thus define the geologic history of a taxon or family, the problem still remains that some fossil forms may simply represent types that lack modern analogues because the group that produced them is now extinct. Moreover, with the limited number of spore characters available, it is highly probable that the same spore type was produced by different types of fungi (Stubblefield and Taylor, 1988). As research has continued on fossil fungal spores, imaging systems such as scanning electron microscopy (SEM) have provided greater resolution to some features and allowed others to be interpreted with greater clarity. Interestingly, despite the importance placed on wall thickness and layering, as well as on septal and other characters, transmission electron microscopy (TEM) has not been utilized to any major extent in the study of either extant or fossil fungal spores.

FUNGAL SPORES IN STRATIGRAPHY

Some fungal spores have been used in biostratigraphy (sometimes termed mycostratigraphy), for example when defining continental marine offshore deposits (Elsik, 1980) or delimiting a particular sequence of rocks (Varma and Patil, 1985). As a general rule, however, using fungal palynomorphs as index fossils has not been universally adopted (e.g. Clarke, 1965; Elsik, 1976b, 1996). Fungal spores have been used in association with other microfossils (e.g. algae) to define basin geology (e.g. Zhang, 1980; Ediger, 1981). Helicospores have been used to indicate marshy and swamp-like conditions such as those suggested for the Late Cretaceous of southern Alberta, Canada (e.g. Kalgutkar and Braman, 2008).

Some spore genera and species have been critically examined relative to character plasticity and such variation has been useful in more accurately defining species, and thus increasing their usefulness in biostratigraphy (e.g. Elsik, 1990; Elsik et al., 1990). In general, however, most fungal spores have had a limited role in defining rock boundaries (Pirozynski, 1978; Saxena and Tripathi, 2005). Nevertheless, a few studies have demonstrated the usefulness of fungal spore assemblages to define microenvironments in certain strata (e.g. Paleogene in Parsons and Norris, 1999). In this study seven interval zones based on fungal propagules were recognized in deposits in the Beaufort-Mackenzie Basin (northern Canada), and these in turn were used to hypothesize inferences about the depositional setting of the associated strata. In the Paleozoic, fungal spores have been used less often for biostratigraphy in coal deposits than in those of either the Mesozoic or Cenozoic, perhaps as a consequence of the oxidation process necessary in coal macerations (Elsik, 1996).

FUNGAL SPORES IN PALEOECOLOGY

Fungal spores occurring in various archaeological sites have been used as proxy indicators in reconstructing the type of flora that was present (e.g. van Geel and Aptroot, 2006). For example, in the absence of pollen, or to corroborate the presence of a particular taxon, the occurrence of fungal spores known to be a pathogen of a particular plant has been used to demonstrate the existence of that plant. An example is *Amphisphaerella dispersella*, an ascomycete (Xylariales) that is typically associated with the dicot *Populus* (van Geel and Aptroot, 2006).

Fungal spores have also been used to interpret paleoecological conditions at the time the fungi were fossilized (e.g. van Geel, 1978; Gelorini et al., 2012; Hooghiemstra, 2012; Worobiec and Worobiec, 2013). In Neogene sediments, spores are especially common in coals, lignites, and peat deposits (Ramanujam and Rao, 1978). In some instances where the fungi were preserved on leaf cuticles, the combination of the type of leaf together with the fungus has been used as a broad indicator of climatic conditions. An example of this is the study of Singh and Chauhan (2008), who examined Neogene fungal spores from the Mahuadanr Valley in India and modern relatives that are pathogens of cereal grasses. The high diversity of fungal remains at this site suggests that the area existed in a climate with high humidity. The presence of epiphyllous fungi with their dispersed spores provides additional information that the climate included high temperatures that were optimal for growth. The presence of fish fossils at the site provided details about the depositional setting (a pond).

Fungal spores, together with pollen and spores of vascular plants, have been used to define the environment of deposition, as well as in the interpretation of the precise stratigraphic position of the sample (e.g. Jain, R.K., 1968; Saxena and Trivedi, 2009). In one example, the presence of lignite together with marine fossil fungi similar to those found in modern environments provided data to suggest that this Neogene site represented a fossil mangrove deposit (Kumaran et al., 2004). The identification of different fungal NPPs (i.e. dung-coprophilous and parasitic fungi) in different layers of Holocene peat from Brazil was used to suggest agricultural and domestic activities, and to improve our overall understanding of peat deposition (Medeanic and Silva, 2010).

In other instances fungal spores may be related to specific geographic areas at a particular point in geologic time, such as the Cenozoic equatorial coast of Africa (Salard-Cheboldaeff and Locquin, 1980) or the Miocene of Brazil (e.g. Guimarães et al., 2013). A *Tetraploa*-like fungus that today grows on bamboos was a component of a Miocene site that was a wetland with reed and riparian vegetation (Worobiec et al., 2009). The presence of fungal spores in certain deposits, such as peat, can provide data on the formation of the deposit (e.g. Vishnu-Mittre, 1973). Fungal remains may also provide information on the taphonomy of a site. For example, plant fossils from the Miocene Clarkia site in northern Idaho lacked basidiomycete hyphae with clamp connections (they were found neither associated with leaves nor in the palynological preparations) (e.g. Phipps, 2001). This suggests that the fossils do not represent debris (litter) from a surrounding forest floor where basidiomycetes would be common (Sherwood-Pike and Gray, 1988). This research emphasizes the importance of recording the fungal spores present at a site and where possible, the typical substrate (e.g. Sherwood, 1985).

FIGURE 11.3 Alan Graham.

Another paleoecological source of information for fungal spores is plant resins. These not only provide a substrate for the fungi but also offer some indication of the types of resin-producing plants that may have existed in the ecosystem. This is especially important because some fungi occur exclusively on plant exudates (i.e. resinicolous fungi; see e.g. Rikkinen, 1999). Cretaceous amber inclusions from France containing insect fecal pellets made up entirely of remnants of polypores, including hyphae, setae (spinulae), and basidiospores, indicate that these macrofungi served as a habitat for fungivorous insects in the Mesozoic (Schmidt et al., 2010b).

Various fungal spores have been described from more recent sediments (Quaternary) and compared with those found in modern ecosystems and sediment types (e.g. Wolf and Cavaliere, 1966; Yeloff et al., 2007). In some instances the primary focus of the research with fungal spores was floral succession within a well-defined stratigraphic range and/or geographic setting (van Geel et al., 1986).

FUNGAL SPORE TAXA

Graham (1962; Figure 11.3) provided a list of 204 genera of fossil fungal spores, indicating both their geographic and stratigraphic occurrences, and also noted the importance of fungal spores in documenting the vegetational history of a region. We have adopted this classification, including additions by Kirk (1985) and subsequent modifications that were incorporated in the extraordinarily comprehensive volume by Kalgutkar and Jansonius (2000), and by Saxena and Tripathi (2011). The following include several examples within each fungal spore type (Saxena and Tripathi, 2005). Another especially useful catalogue that documents Cenozoic fungi from India from 1988 to 2005 is found in Saxena (2006). The catalogue provides detailed information regarding the nomenclature of 172 species. As might be expected, a number of species of fossil fungi were not validly published initially or have had diagnoses corrected later and the appropriate repository identified (e.g. Saxena, 2012).

The following descriptions illustrate some of the basic types of fungal spores together with some of the representative species of each of the major types.

AMEROSPORES

Spores in this group are unicellate and may be aperturate or inaperturate; they may have a single pore or hilum, or variable apertures. The length to width ratio in amerospores is <15:1; strongly curved and very long types are excluded.

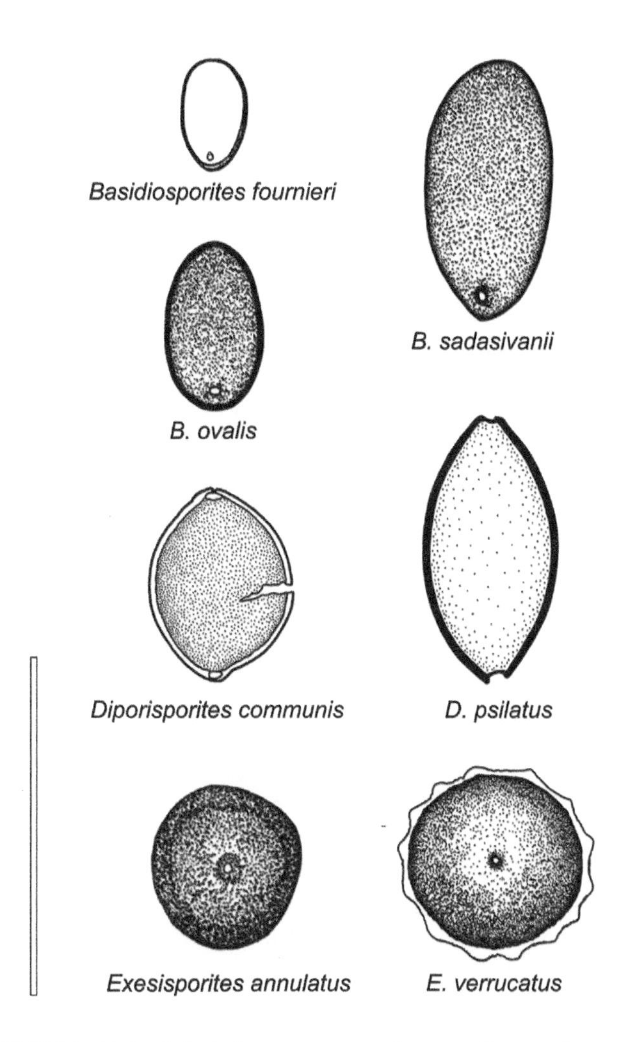

FIGURE 11.4 Selected amerospores. See text for species age and provenance. Bar = 50 μm. (Redrawn from Kalgutkar and Jansonius, 2000.)

Basidiosporites

In *B. fournieri* from the Paleocene of Texas (Figure 11.4), USA, specimens are up to 15 μm long and possess a psilate outer wall (Elsik, 1968). The pore in this form is slightly offset, a condition that is also present in the larger spore *B. ovalis* (=*Monosporiosporites ovalis*) (see Figure 11.4) from the Eocene of Tennessee (Sheffy and Dilcher, 1971). Specimens from late Quaternary sediment cores from the Arabian Sea are biconvex and up to 43 μm long; the offset pore in *B. sadasivanii* (see Figure 11.4) is slightly thickened (Chandra et al., 1984).

Diporisporites

This genus is characterized by its elongate shape and, as the name suggests, a diporate condition (see Figure 11.4), there being one pore on each end of the unicellate spore (van der Hammen, 1954a). Some forms have features of the pores that have been used to define the species, notably *D. communis* (see Figure 11.4) from the Eocene–Oligocene of China (Ke and Shi, 1978) and *D. bhavnagarensis* from the Eocene of India (Saxena, 2009). A number of forms are quite large, as in *D. giganticus* (101–130 μm long) from the Miocene–Pliocene of India (Kar, 1990).

Exesisporites

In *Exesisporites* there is a single pore that is thickened around the margin (Elsik, 1969). In *E. annulatus* (see Figure 11.4) from the upper Pleistocene of Canada, specimens may be monoporate or diporate with a collar surrounding the pore (Kalgutkar, 1993). What is interpreted as a verrucate-like sculpture (see Figure 11.4) occurs along the periphery of the spore wall in *E. verrucatus* (Kumar, 1990). *Exesisporites neogenicus* is distinguished by the absence of a clear annular ring around the pore (Kalgutkar and Braman, 2008). The high number of *Exesisporites* spores together with other microfossils suggests that the climate was predominantly dry, accompanied by a lowered sea level, during the Pliocene–Pleistocene of Nigeria (Durugbo et al., 2010).

Hypoxylonites

This is an oval-elongate spore that is characterized by an elongate scar, slit, or furrow that parallels the long axis (Elsik, 1990; Saxena, 2012). Spores of this type have been interpreted as being similar to those produced by certain living ascomycetes in the family Xylariaceae (e.g. Elsik, 1990; Nandi et al., 2003; Prager et al., 2006). The taxon ranges from the Cretaceous to the recent and is known from deposits around the world (Elsik, 1981), including specimens preserved in amber (Antoine et al., 2006). The wall ornament ranges from psilate to small spines. Specimens of *H. brazosensis* (Figure 11.5) from the upper middle Eocene of Texas (USA) have a furrow that extends to the ends of the spore, while in *H. chaiffetzii* (Neogene, Gulf Coast, USA) the furrow runs approximately three quarters of the length of the spore. Some specimens are described as having two wall layers (see Figure 11.5), but this feature has been determined as a fold along the furrow. In *H. oblongus* the furrow is especially short (Dueñas, 1979). On account of the difficulty in resolving the furrow, some spore types from the Eocene of Tennessee (USA) included in *Inapertisporites* (Sheffy and Dilcher, 1971) have been moved to *Hypoxylonites*

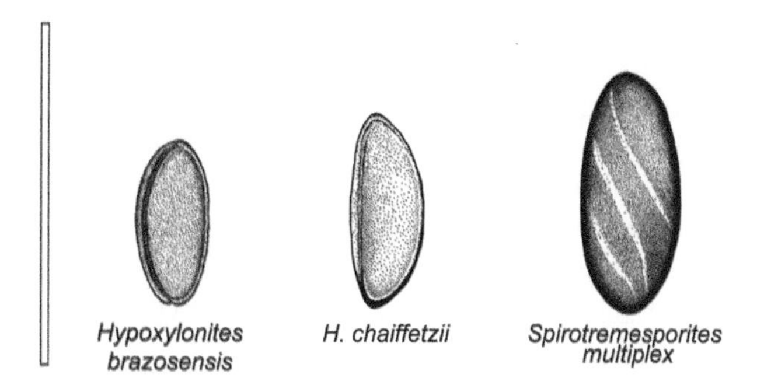

FIGURE 11.5 Selected amerospores. See text for species age and provenance. Bar = 50 μm. (Redrawn from Kalgutkar and Jansonius, 2000.)

(Elsik, 1990). The abundance of *Hypoxylonites* was used to establish an early Oligocene biozone in the northwestern Gulf of Mexico (Elsik and Yancey, 2000).

Another spore type that is morphologically similar to *Hypoxylonites* is *Spirotremesporites* (Dueñas, 1979), which has one to several furrows that may be sigmoidal in outline or positioned at an angle to the long axis of the spore (see Figure 11.5). Elsik (1990) provided a discussion of the difference between the two genera and noted that *Hypoxylonites* is found in rocks that reflect cooling environments.

Inapertisporites

These spores are characterized as unicellular and aseptate, with a highly variable size; the defining character is the lack of a preformed aperture. Specimens of *I. variabilis* (Figure 11.6) from the Upper Cretaceous of Colombia are approximately 31 μm long and psilate (van der Hammen, 1954a). Several spores that clump together appear to be a characteristic of *I. granulatus* (see Figure 11.6) from the Eocene–Oligocene of Shandong Province, China (Ke and Shi, 1978). Specimens of *I. argentinus* (see Figure 11.6), originally described as a species of *Reticulatasporites* (Jain, R.K., 1968) from the Middle Triassic of western Argentina, have an ornament of widely spaced brochi (lumen or mesh reticulum) and a size range of 40 to 61 μm in diameter (Kalgutkar and Jansonius, 2000). Other forms may be highly variable in ornamentation, as in *I. trivedii* (see Figure 11.6) from the lower Eocene of Andhra Pradesh, India, where the wall surface ranges from punctate to variously striate (Ambwani, 1982). The affinities of *I. circularis* (see Figure 11.6) are debatable. The fossil was considered to be related to fungi within the Basidiomycota (Graham, A., 1965), while the spore type may be a conidiospore of a microthyriaceous ascomycete (Dilcher, 1965). In other reports, similar spores have been associated with wood of *Fagus* (Soomro et al., 2010).

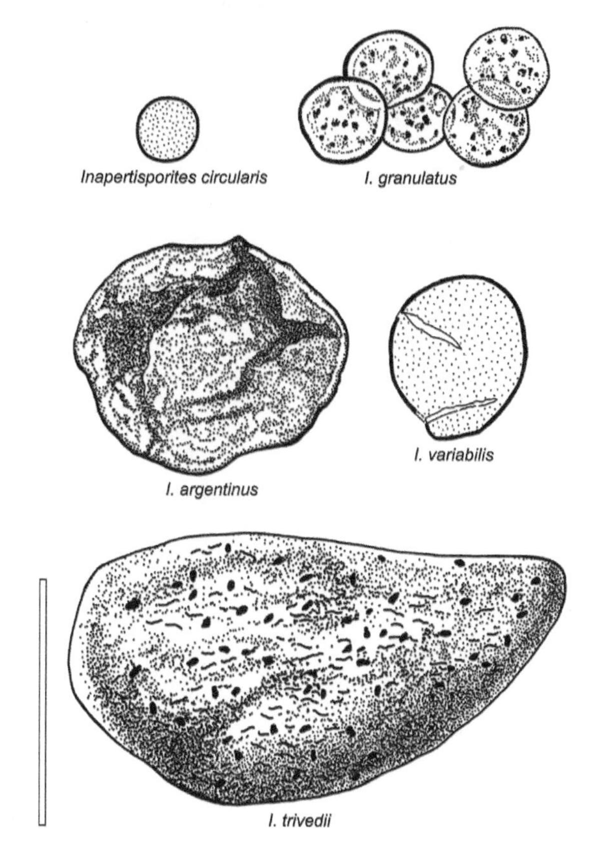

FIGURE 11.6 Selected amerospores. See text for species age and provenance. Bar = 50 μm. (Redrawn from Kalgutkar and Jansonius, 2000.)

Lacrimasporonites

Spores of this genus are monoporate and spatulate to elliptical in outline (Clarke, 1965; Figure 11.7); one species, *L. sondensis* from the Paleocene of Pakistan, has been described as having two germinal apertures (Soomro et al., 2010). Kalgutkar and Jansonius (2000) noted that using spore shape as a diagnostic character has potential problems and is a difficult character to use in practice. The genus has been emended to include specimens with a flat hilar scar at one end and a pore at the other (Kalgutkar and Jansonius, 2000). Some fungal spores that lack the spatulate shape have been transferred to *Monoporisporites* (Kalgutkar and Jansonius, 2000). Spores of *Lacrimasporonites* found in association with mastodon bones from Canada were regarded as similar to ascomycetes in the Sordariales (Pirozynski et al., 1988).

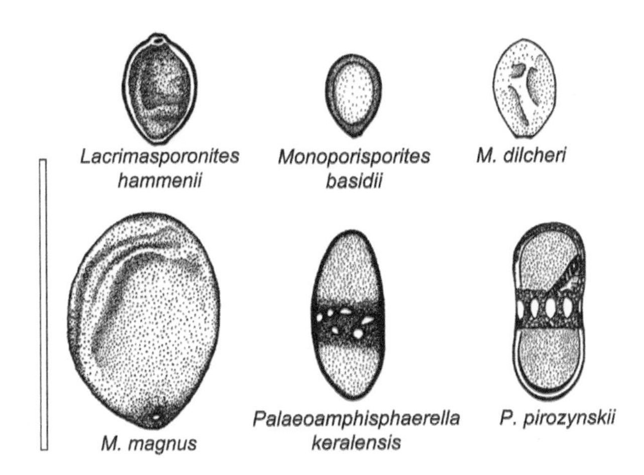

FIGURE 11.7 Selected amerospores. See text for species age and provenance. Bar = 50 μm. (Redrawn from Kalgutkar and Jansonius, 2000.)

MONOPORISPORITES

This genus is used for fungal spores that are monoporate (see Figure 11.7) and generally spherical, with psilate to punctuate wall ornament. The type species, *M. minutus* from the Upper Cretaceous of Colombia, includes small (14 μm) spores (van der Hammen, 1954a). Specimens of *M. basidii* (see Figure 11.7) from the Paleocene of Texas (USA) have a two-layered wall that extends outward in the region of the pore (Elsik, 1968). In *M. dilcheri* (see Figure 11.7) (Upper Cretaceous of Mexico) the pore is 2 μm in diameter and the spore is slightly bilateral (Martínez-Hernández and Tomasini-Ortiz, 1989). Reports of the genus include the Upper Cretaceous of the Antarctic Peninsula (Song and Cao, 1994). Along with other palynomorphs, *Monoporisporites* has been found in large numbers at the Cretaceous–Paleogene boundary (Vajda and McLoughlin, 2004). Some specimens included in *Monoporisporites* have been related to teliospores of certain Uredinales (rust fungi) (Ramanujam and Ramachar, 1980), while others from the Oligocene of equatorial Africa (Cameroon) are aligned simply with the Basidiomycota (Salard-Cheboldaeff and Locquin, 1980).

PALAEOAMPHISPHAERELLA

The occurrence of equatorial pores on an elliptical spore is the primary feature that defines *Palaeoamphisphaerella* (Ramanujam and Srisailam, 1978). Specimens of *P. keralensis* (see Figure 11.7) from the Miocene of India are rhomboidal in outline, 30 μm in the longest axis, and have three to six pores, which are occasionally arranged in an irregular pattern. Surface ornament is scabrate.

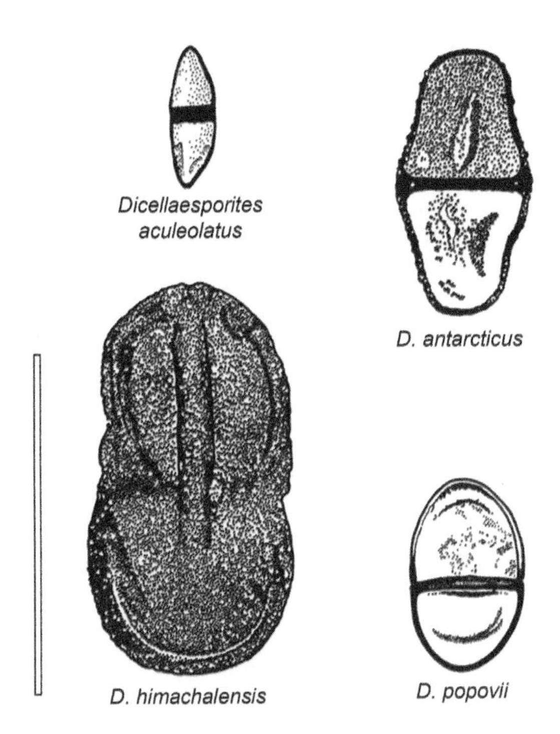

FIGURE 11.8 Selected didymospores. See text for species age and provenance. Bar = 50 μm. (Redrawn from Kalgutkar and Jansonius, 2000.)

Palaeoamphisphaerella pirozynskii from the same stratigraphic level is distinguished by 8–10 equatorial pores (see Figure 11.7). The spores closely compare, morphologically, with those of some members of the Xylariaceae (i.e. *Amphisphaerella*), although it is noted that in the extant forms the pores are more slit-like (Eriksson, O., 1966).

DIDYMOSPORES

As the name implies, spores in this category are two-celled (Figure 11.8). Sometimes one cell is smaller than the other, and the spore is characterized by a pore in the smaller cell; some are described as inaperturate. Sculpture ranges from psilate to slightly punctuate.

DICELLAESPORITES

These spores are isopolar, two-celled, and inaperturate. In some species the individual cells are slightly tapered at the ends (e.g. *D. aculeolatus*; Sheffy and Dilcher, 1971), while in others the two cells are rounded at the ends (see Figure 11.8) (e.g. *D. antarcticus*; Song and Cao, 1994). In *D. himachalensis* it is difficult to see where the cells are constricted (see Figure 11.8; Saxena and Bhattacharyya, 1990). In *D. popovii* from the Paleocene of Texas (USA) each of the cells is about 19 × 29 μm and the septum is two layered (see Figure 11.8; Elsik, 1968). Numerous species have been recorded from Cenozoic sediments of India (Saxena and Tripathi, 2011). Some fossil spores have been compared with those produced by the extant ascomycete *Apiosporina* (Dothideomycetes), the fungus responsible for black knot disease in species of *Prunus* (plums, cherries, peaches, etc.), which results in hyperplasia of woody tissues (e.g. Fernando et al., 2005).

Ascospores with two cells are associated with epiphyllous colonies of *Molinaea asterinoides* on dicot leaves from the Maastrichtian (uppermost Cretaceous) of Colombia (Doubinger and Pons, 1975). Hyphae are long and the lateral margins of some cells are sinuous. The hyphopodia are capitate, with the terminal cell of each hyphopodium (stigmocyst) containing a delicate lateral pore.

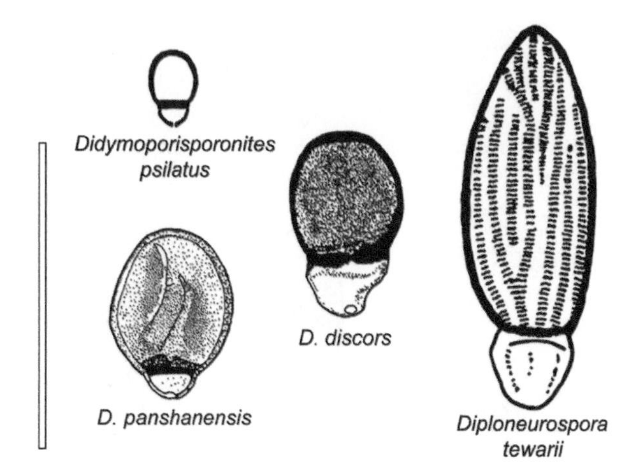

FIGURE 11.9 Selected didymospores. See text for species age and provenance. Bar = 50 μm. (Redrawn from Kalgutkar and Jansonius, 2000.)

DIDYMOPORISPORONITES

Spores in this genus are two-celled, with one of the cells being larger than the other (see Figure 11.9). The type species, *D. psilatus*, is generally oval with the single pore occurring in the smaller cell (Sheffy and Dilcher, 1971). The original specimens come from the middle Eocene of Tennessee (USA) and are distinguished by their size (6.8–11.1 μm; Kalgutkar and Jansonius, 2000). Specimens of *D. discors* (Figure 11.9) from Yukon Territory are slightly larger and have a pore in the septum (Kalgutkar, 1993; name revised by Kalgutkar and Jansonius, 2000). As the name implies, *D. conicus* is characterized by one of the cells being cone-shaped (Kalgutkar, 1997). Specimens have been reported from the upper Paleocene–lower Eocene of Axel Heiberg Island, Canada.

DIPLONEUROSPORA

Specimens of this genus are unusual because the two cells are unequal and there is sculpturing on the upper cell consisting of longitudinal ribs or stripes of thickened wall material; on the lower, smaller cell sculpture is not well defined. Miocene fungal spores of *D. tewarii* from South India have unequal cells in which one cell is approximately one third the size of the other (see Figure 11.9) (Jain and Gupta, 1970). In this species the smaller cell is described as appendage-like. The generic name in part reflects the spore morphology and sculpture found in the extant ascomycete *Neurospora* (see Davis, 2000).

DYADOSPORITES

These spores consist of two cells, each having a single pore at the apex (e.g. van der Hammen, 1954a; Clarke, 1965). Ornamentation is variable. Specimens of *D. cannanorensis* from the Miocene of India have pores that are slightly offset from the long axis of the spore (Ramanujam and Rao, 1978; Figure 11.10). The spore wall and the septum are bilayered. In *D. ellipsus* (see Figure 11.10) from the Upper Cretaceous of Colorado, USA, the wall is finely punctuate (Clarke, 1965). Perhaps the largest species in spore size is *D. grandiporus* from the lower Miocene of Assam (India). Specimens are longer than 100 μm (see Figure 11.10) and contain equally large pores (14–16 μm), each with a thickened rim (Singh et al., 1986). Kalgutkar and Jansonius (2000) commented that the thickened rim may in fact denote pigmentation in the region. In *D. megaporus* (see Figure 11.10) (Paleocene–Pliocene of China) the pore in the septum is disciform (Zhu et al., 1985). *Dyadosporites sahnii* (see Figure 11.10; originally *Granodiporites*; see Kalgutkar and Jansonius, 2000) from the Eocene–Miocene (Assam, India) has its pore area covered by a delicate membrane (Varma and Rawat, 1963). The genus is also known from the Upper Cretaceous of the Fildes Peninsula, King George Island, Antarctic Peninsula (Song and Cao, 1994) and from the Miocene of Taiwan (Huang, 1981). Germinating spores assigned to *Dyadosporites* have been reported in the coal maceral ulminite and in palynological preparations from Miocene coal from Slovakia (O'Keefe et al., 2011b).

Dyadosporites must not be confused with *Dyadospora*, which is a taxon used for certain Paleozoic cryptospores (Strother and Traverse, 1979).

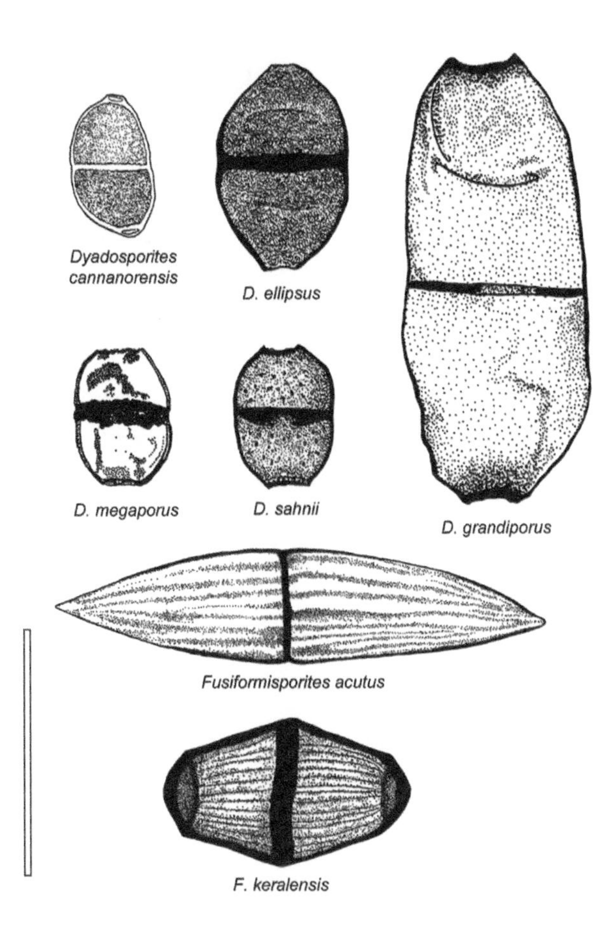

FIGURE 11.10 Selected didymospore types. See text for species age and provenance. Bar = 50 μm. (Redrawn from Kalgutkar and Jansonius, 2000.)

FUSIFORMISPORITES

These spores are fusiform in outline and lack apertures (Rouse, 1962; see Figure 11.10). Specimens range from 20 to 100 μm, possess a two-layered septum, and extend from the Cretaceous to recent (Martínez-Hernández and Tomasini-Ortiz, 1989). Ornamentation includes radiating longitudinal striae that arise from either pole in the pattern of a spindle (Elsik, 1968). In some species the striae may branch near the equator. In *F. keralensis* (see Figure 11.10) the ends of the spores are truncate and the septum is thick (Ramanujam and Rao, 1978). Spores included in *Striadyadosporites* (Dueñas, 1979) have been transferred to *Fusiformisporites.* Based on stratigraphic occurrence, Elsik (1976b) hypothesized that *Fusiformisporites* and other longitudinally striate and ribbed fungal spore types such as *Striatetracellaeites*, *Striadiporites*, *Striasporonites*, and the verrucate *Verrusporonites* are phylogenetically related. Spores that are similar in appearance are found in the extant ascomycete *Cookeina* (Pezizales; Weinstein et al., 2002); *C. tricholoma* spores are also known in the subfossil state from the Holocene of Tanzania, Africa (Wolf, 1967).

PHRAGMOSPORES

Spores of this type have two or more transverse septa resulting in three or more cells. Both aperturate and inaperturate forms are included.

FRACTISPORONITES

Fractisporonites is a common type of phragmospore that has been found at multiple sites ranging from the Jurassic (Clarke, 1965; Traverse and Ash, 1994) to the Eocene (Kalgutkar, 1993). Morphologically similar microfossils from the Precambrian are known as *Arctacellularia* (Hermann and Podkovyrov, 2008).

REDUVIASPORONITES

This is the name used for chains of spores that were once interpreted as a fossil *Penicillium* (Wilson, L.R., 1962). This fossil is the focal point of several influential studies demonstrating a massive accumulation (a so-called "fungal spike" or "fungal abundance event") of *Reduviasporonites* fossils at the end of the Permian, and suggesting that this accumulation is indicative of the destruction of terrestrial vegetation by fungal pathogens that led to the end-Permian collapse of terrestrial ecosystems (Visscher et al., 1996, 2011; Steiner et al., 2003). The conclusion is based on the fact that *Reduviasporonites* is morphologically similar to a resting stage formed by members in the extant basidiomycete genus *Rhizoctonia*, which includes several widespread plant pathogens (e.g. Parmeter, 1970; González García et al., 2006). The authors (Visscher et al., 2011) state that fungal disease was an essential accessory in the destabilization of the vegetation that accelerated widespread tree mortality during the end-Permian crisis. Visscher et al. (2011) dismiss results from a study by Foster et al. (2002), who, based on geochemical evidence, concluded that *Reduviasporonites* may be of algal origin. It is interesting to note that a fungal spike has also been recorded from the Pennsylvanian of Peru (Wood and Elsik, 1999).

BRACHYSPORISPORITES

Spores of this type are organized in three or more cells, with the individual cells decreasing in size from a large dome-shaped apical cell (Figure 11.11) to a smaller attachment cell that is sometime described as hyaline

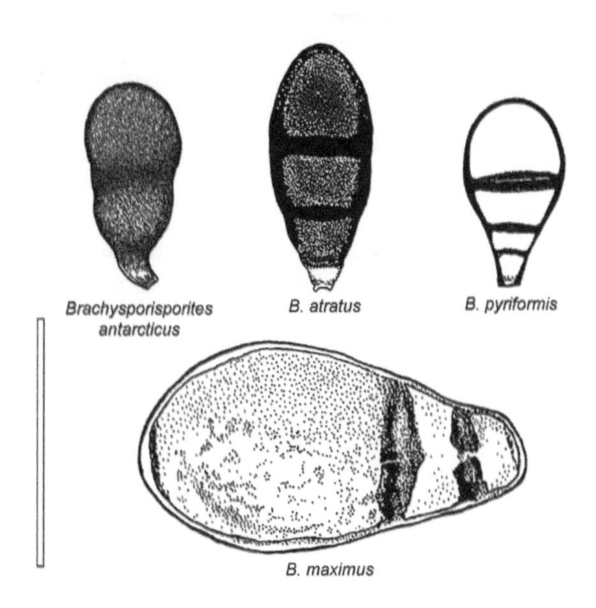

FIGURE 11.11 Selected phragmospore types. See text for species age and provenance. Bar = 50 μm. (Redrawn from Kalgutkar and Jansonius, 2000.)

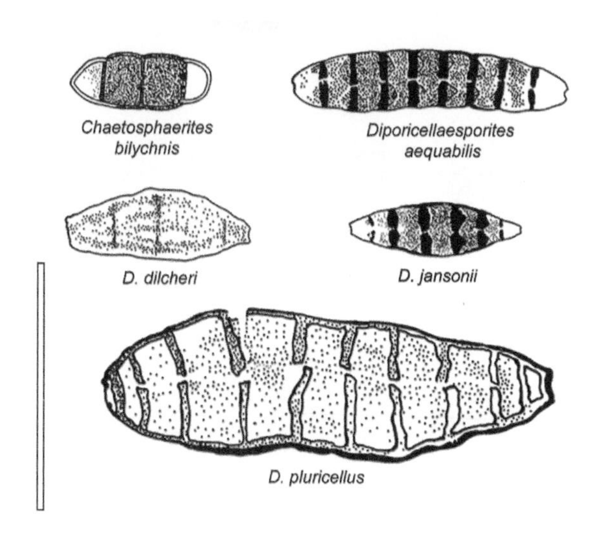

FIGURE 11.12 Selected phragmospore types. See text for species age and provenance. Bar = 50 μm. (Redrawn from Kalgutkar and Jansonius, 2000.)

(Lange and Smith, 1971; Ediger, 1981). A single pore is present at the narrower end of the spore; some species are described as having a pore in the septum. In *B. antarcticus* (see Figure 11.11) from the Cretaceous of King George Island, Antarctica, the spores are up to 45 μm long and often occur in chains of four or five cells (Song and Cao, 1994). Some Cenozoic species (e.g. *B. atratus*; see Figure 11.11) have especially thick septa (Kalgutkar, 1993). In *B. inuvikensis*, a Cenozoic form, the spores are tetracellate, with the two distal cells making up the majority of the spore (Parsons and Norris, 1999). Three-celled forms up to 85 μm long from the Eocene–Oligocene (Bohai, China) are inaperturate (Ke and Shi, 1978). The balloon shape of the basal cell in *B. magnus* (Eocene) of Gujarat, western India, morphologically appears as a head with a tapering cap (Samant, 2000). Fossil spores of the *Brachysporisporites* type are generally compared with the extant saprotrophic, dematiaceous hyphomycete genus *Brachysporium* (Sordariomycetes, Ascomycota; Réblová and Seifert, 2004). A relationship of *Brachysporisporites* to the anamorphic genus *Monotosporella* (Sordariomycetes) has also been considered (Sadowski et al., 2012). Spores assigned to the genus *Anatolinites* are similar to *Brachysporisporites* in size and shape but differ from the latter by having two simple pores (Elsik et al., 1990). Several affinities have been proposed for *Anatolinites*, including the ascomycete plant pathogen *Alternaria* and teliospores of the rust *Puccinia* (Basidiomycota).

Chaetosphaerites

The barrel shape and tetracellate organization are distinctive features of the genus *Chaetosphaerites* (Felix, 1894) (=*Cannanorosporonites*; Ramanujam and Rao, 1978). Terminal cells are smaller than the central cells (Figure 11.12), with the latter typically bulging. The apical cell has a single pore. In the original description of *C. bilychnis* (see Figure 11.12) from the Eocene of Baku (Azerbaijan), the fungus was found on dicot wood believed to have affinities with the Rhamnaceae (Felix, 1894). The fossil specimens are compared to *Torula*, a member of the Pezizomycotina (Schoknecht and Crane, 1977; Kirk et al., 2008), perhaps closely related to extant members of the ascomycete family Sphaeriaceae (Xylariales), which includes parasitic fungi with perithecia. *Felixites* is the name applied to spores that appear similar, but are generally from the Paleozoic (Elsik, 1989). Specimens of *F. playfordii* (formerly *Chaetosphaerites pollenisimilis* Butterworth and Williams, 1958) from the Carboniferous of Spitsbergen are up to 52 μm long and at maturity may split into single cells, so-called half specimens (Elsik, 1990).

Diporicellaesporites

These spores are elongate, multicellular, and have a single pore at each end (Elsik, 1968). The number of cells in *D. aequabilis* (see Figure 11.12) ranges from 9 to 13 with the cells that are more central also larger in diameter (Kalgutkar, 1993). This species from the upper Paleocene–lower Eocene of the Yukon Territory,

Canada, has a two-layered wall and septal flaps. A fungal spore of late Quaternary age that was originally placed in *Inapertisporites* (Chandra et al., 1984) is now designated as *D. dilcheri* (see Figure 11.12; Kalgutkar and Jansonius, 2000). *Diporicellaesporites jansonii* (see Figure 11.12) has between six and nine septa with the pores slightly incurved (Kalgutkar, 1993), while in *D. pluricellus* (see Figure 11.12) the pores protrude slightly (Kar and Saxena, 1976). The genus has been identified in rocks from as early as the Cretaceous (e.g. Kalgutkar and Braman, 2008) and has been recorded extensively from various Cenozoic stratigraphic levels in North America (e.g. Price and Thorne, 1977) and India (Saxena and Tripathi, 2011), as well as from the Oligocene of China (Zhang, 1980). Fossil spores have been morphologically compared to extant members of *Annellophora* (Pezizomycotina), a pathogen common on various palms (e.g. Vann and Taber, 1985).

Foveoletisporonites

These spores are five or more septate, inaperturate, and the two central cells are elongate. In the type species, *F. miocenicus* (Figure 11.13) from the Miocene of India, the spores maybe up to 120 μm long with a foveolate (having small pits) ornament (see Figure 11.13; Ramanujam and Rao, 1978). In some species, such as *F. indicus*, the foveolae may be irregular in arrangement (Ramanujam and Srisailam, 1978). The fossils are generally attributed to the Pezizomycotina.

Multicellaesporites

This genus is characterized by spores constructed of three to five cells with a longitudinal furrow and slight curvature to their shape (Elsik, 1968; Kumar, 1990). Ornamentation may be variable or differentially thickened. In *M. bilobus* (see Figure 11.13) from the upper Pliocene, the septa have a well-defined pore; specimens are reported in the 75 μm size range (Rouse and Mustard, 1997). In *M. dilcheri* the terminal cells are longer (Samant, 2000). Similar fungal spores that are inaperturate are placed in *Multicellites* (Kalgutkar and Jansonius, 2000). Spores of this type were found extensively in the rhizome of the Eocene aquatic angiosperm *Eorhiza arnoldii* (Robison and Person, 1973). In a detailed investigation of the hyphomycetes in the cortical aerenchyma of *E. arnoldii*, Klymiuk et al. (2013b) compared the spores that Robison and Person (1973) attributed to *Multicellaesporites* to the extant *Thielaviopsis* (Pezizomycotina).

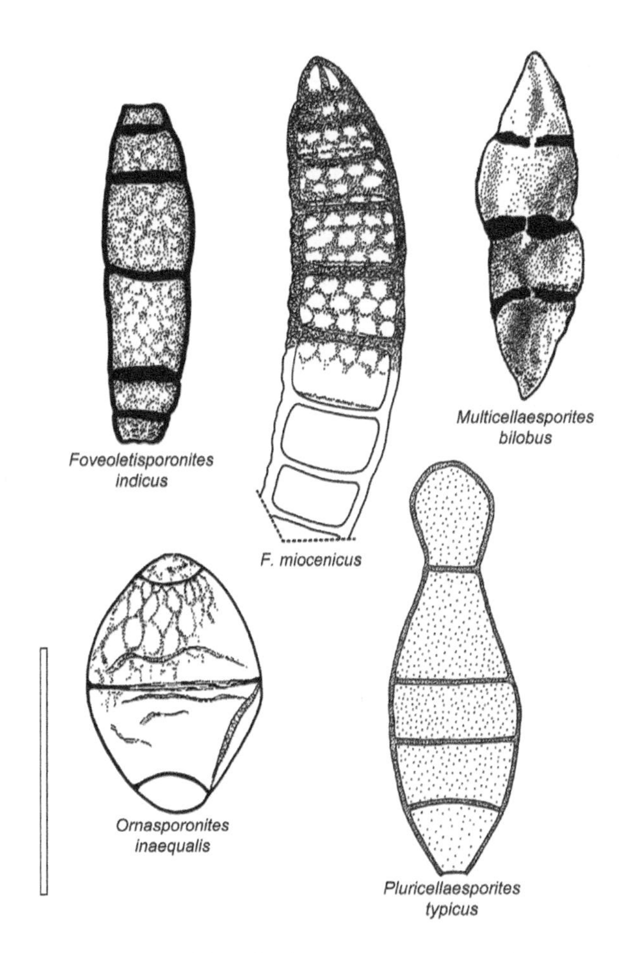

FIGURE 11.13 Selected phragmospore types. See text for species age and provenance. Bar = 50 μm. (Redrawn from Kalgutkar and Jansonius, 2000.)

Ornasporonites

Ornasporonites is the name used for tetracellate spores in which the basal and apical cells are smaller, with each of these cells containing a single, simple pore (Ramanujam and Rao, 1978). In *O. inaequalis* (see Figure 11.13) from the Miocene lignite of Cannanore, India, the ornament is rugulate-reticulate. The fusiform specimens range up to 63 μm long by 42 μm in diameter. Several examples are now known from other Miocene sites in India (Saxena and Tripathi, 2011).

Pluricellaesporites

This dispersed fungal spore is monoporate and constructed of three or more cells (van der Hammen, 1954b). In early discussions of this spore type the genus was sometimes considered as representing algal cells (e.g. Elsik, 1968); however, the fungal nature was subsequently confirmed (Elsik and Jansonius, 1974). In

P. typicus (see Figure 11.13) (Neogene) the septa may be up to 4 μm thick, entire or split, and the spore may be variously ornamented. Other species (e.g. *P. eocenicus*, *P. globatus*) are known from the Eocene of India (Samant and Tapaswi, 2000; Samant, 2000) and Cretaceous of Canada (e.g. Srivastava, 1968); the genus has also been recorded from the Paleocene of Brazil (Pedrão Ferreira et al., 2005). Kalgutkar and Jansonius (2000) provided a history of the possible affinities of *Pluricellaesporites*, including the conidia of the extant genus *Alternaria* (Pleosporaceae), many species of which are plant pathogens found in most environments of the world. An interesting genus of extant hyphomycetes, *Xylomyces*, initially discovered on immersed decaying wood in a tidewater stream in Rhode Island, USA (Goos et al., 1977) but later also reported from several other countries (Goh et al., 1997; Kohlmeyer and Volkmann-Kohlmeyer, 1998), is characterized by large, dematiaceous, thick-walled, multiseptate, and more or less fusiform chlamydospores. Goos et al. (1977) note that the spores of *X. chlamydosporis* bear a striking resemblance to the fossil *Pluricellaesporites psilatus* from the Cretaceous of North America (Clarke, 1965).

Polycellaesporonites

Spores of this type have a large, round basal cell that gives rise to a tubular or beak-like extension (Chandra et al., 1984). The spore is multicellate; cells are arranged in a cluster, and no aperture is present. In some species (*P. acuminatus* from the late Paleocene of British Columbia, Canada; Figure 11.14) there may be up to 12 septa (see Figure 11.14), some of which are longitudinal (Rouse and Mustard, 1997; Jansonius and Kalgutkar, 2000). A form known from the Oligocene–lower Miocene (Saxena and Bhattacharyya, 1990) and Quaternary (Chandra et al., 1984) is *P. bellus* (see Figure 11.14). It is up to 68 μm long and has a tube-like projection that appears to be hyaline. In this species the individual cells are rectangular and not arranged along one axis. Specimens have also been reported from the Paleocene and Eocene of India (Gupta, 2002; Saxena and Ranhotra, 2009). The conical beak in *P. alternariatus* (see Figure 11.14) is thought to represent the region of attachment in the conidial chain (Kalgutkar and Jansonius, 2000).

DICTYOSPORES

Sometimes termed muriform spores, the inaperturate spores in this category are characterized by both longitudinal and transverse septa. Shape is variable.

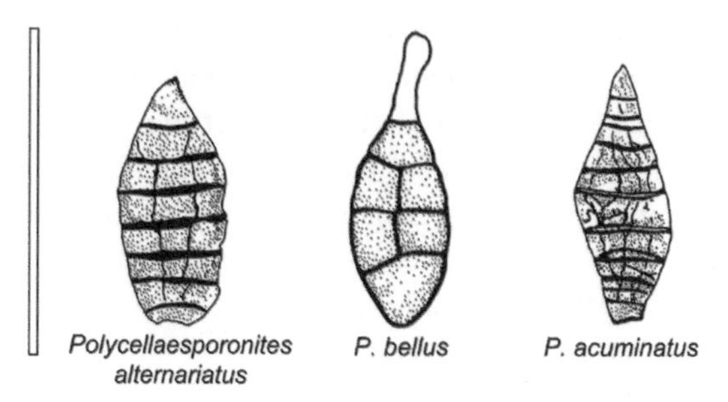

FIGURE 11.14 Selected phragmospore types. See text for species age and provenance. Bar = 50 μm. (Redrawn from Kalgutkar and Jansonius, 2000.)

Dictyosporites

These are muriform, inaperturate spores that are multicellate, and the individual cells are typically spherical (Felix, 1894). Some forms are described in association with a mycelium of septate hyphae called *Arbusculites dicotylophylli* (Figure 11.15; Paradkar, 1974); however, this taxon is now included in *Dictyosporites* (*D. dicotylophylli*) from the Upper Cretaceous (Maastrichtian) of India (Kalgutkar and Jansonius, 2000). *Arbusculites argentea*, an enigmatic fossil from the Carboniferous of Great Britain (Murray, P., 1831), has been interpreted as being of animal origin. In the Oligocene form *D. dictyosus* (see Figure 11.15) from Cameroon, the number of cells is 16 (Salard-Cheboldaeff and Locquin, 1980). Some spores show a size and shape distinction between the central cells and those of the periphery, which may be more rectangular in outline. This can be seen in *D. globimuriformis* (see Figure 11.15; Kalgutkar, 1997). Many of the species can be compared morphologically with the conidia of several modern genera (e.g. *Dictyosporium, Stemphylium, Septosporium, Alternaria*; Kalgutkar and Jansonius, 2000).

Spinosporonites

These inaperturate spores are subcircular and up to 42 μm in diameter; extending from the surface of each cell is a prominent spine that is up to 9 μm long (Saxena and Khare, 1991). The type species, *Spinosporonites indicus* (late Paleocene–middle Eocene of India), has been compared to specimens that are more similar to setose pycnidia found in some coelomycetes (Kalgutkar and Jansonius, 2000).

Staphlosporonites

Spores in this genus are variable in shape, but generally elongate and consisting of more than four irregular

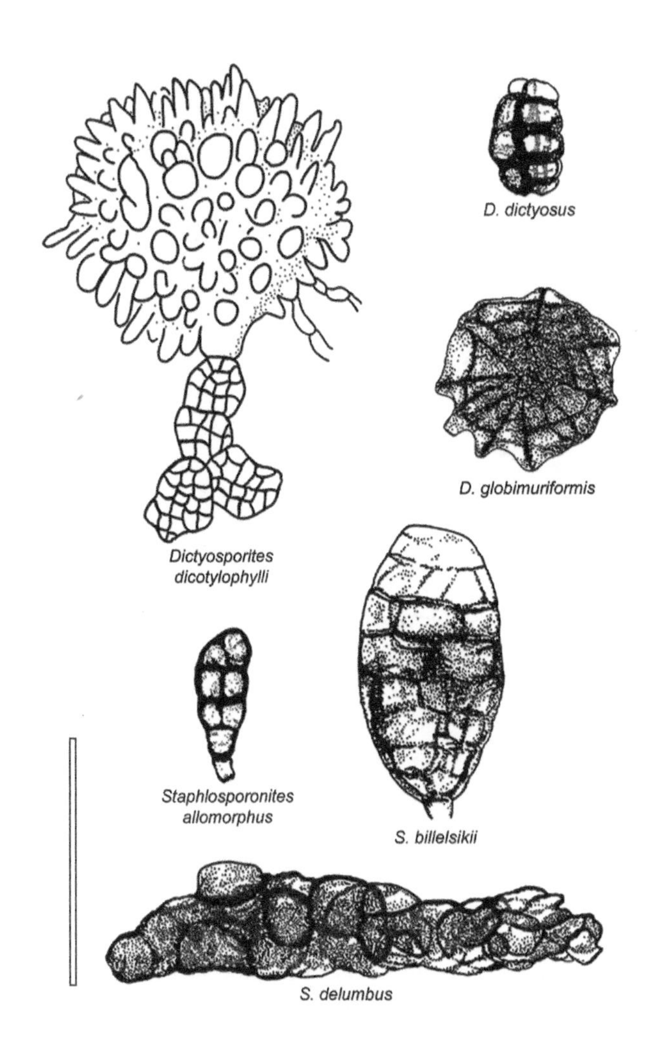

FIGURE 11.15 Selected dictyospore types. See text for species age and provenance. Bar = 50 μm. (Redrawn from Kalgutkar and Jansonius, 2000.)

cells, which are usually arranged in clusters along more than a single axis. They are inaperturate and have a variable ornamentation that ranges from psilate to mildly punctate (Sheffy and Dilcher, 1971). Specimens of *S. allomorphus* (see Figure 11.15) from the Eocene of the USA are oblong and about 30 μm long; each spore is constructed of more than eight irregular cells (see Figure 11.15). Broadly elliptical spores with a rounded apex and protruding hilum are included in *S. billelsikii* from the upper Paleocene–lower Eocene of Canada (Kalgutkar and Jansonius, 2000). Another form initially described from the Eocene–Oligocene of Canada is *S. delumbus* (see Figure 11.15; Norris, 1986). In this species there are multiple cells in the center of the spore that may overlap; specimens range up to 150 μm long. In *S. dichotomus* (Eocene of India) the arrangement of cells can produce a dichotomous pattern (Gupta, 2002). As is the case with a number of dictyospores, several species of *Staphlosporonites* share morphological features with extant members of the Moniliales (Fungi Imperfecti), especially *Alternaria* (e.g. Joly, 1964).

Tricellaesporonites

Spores in this genus have three cells and are inaperturate (Sheffy and Dilcher, 1971). The individual cells are not associated in a single plane in some forms and it has been suggested that they may in fact not be fungal, but rather some type of resinous globules (Kalgutkar and Jansonius, 2000). In *T. semicircularis* the three cells are arranged in a slightly convex plane (Sheffy and Dilcher, 1971). This species from the Eocene of the USA has septa with double walls and ornament ranging from psilate to punctate. Gupta (2002) described *T. granulatus* from the Eocene of India, which, morphologically, looks like three cells of a tetrad.

SCOLECOSPORES

These spores are especially long (length:width ratio of >15:1) and thread-like, and may include both transverse septate and non-septate types. The proximal or distal end may have a hilum or pore.

Scolecosporites

The spores of *S. longus* (Figure 11.16) are up to 35 μm long and consist of 28 cells, each approximately 5 μm long (Song et al., 1999). This late Eocene–middle Oligocene species from China has a thin spore wall and smooth surface. Other specimens, such as *S. maslinensis* (see Figure 11.16) from the lower middle Eocene of Australia, may be 400 μm long and constructed of up to 38 cells (Lange and Smith, 1971). In some descriptions these spores are referred to as scoleco-phragmospores or termed filamentous phragmospores because of the range of curvature, a feature that could be the result of preservation. Scolecospores are a common spore type in the modern family Phyllachoraceae (Ascomycota), a group of obligate biotrophs that produce tar spots on host plants (e.g. Cannon, 1991; Pearce et al., 1999). Similar spores have also been identified as hyperparasites of *Phyllachora* (Parbery and Langdon, 1963).

HELICOSPORES

Spores of this type are easy to recognize because they appear as a rolled up tube (Figure 11.17) or a

FIGURE 11.16 Selected scolecospore types. See text for species age and provenance. Bar = 50 μm. (Redrawn from Kalgutkar and Jansonius, 2000.)

superimposed spiral curved through 180° with a pore at one end. Helicospores may involve multiple revolutions and may coil in a single plane (planispiral) or in three dimensions.

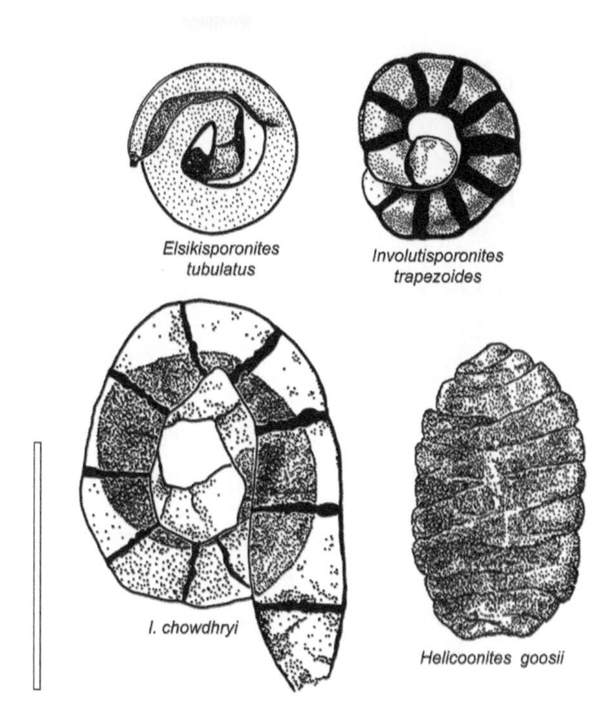

FIGURE 11.17 Selected helicospore types. See text for species age and provenance. Bar = 50 μm. (Redrawn from Kalgutkar and Jansonius, 2000.)

Elsikisporonites

In *E. tubulatus* (Figure 11.17) the tube lacks septa and the cell is widest at the midpoint (Kumar, 1990). The wall is smooth and the pore has a slight extension at the tip. Specimens from the middle Miocene of India are approximately 386 μm long. The wall is generally smooth but may be slightly folded.

Involutisporonites

These spores are planispiral and constructed of multiple cells (Clarke, 1965); in some species the septum contains a pore. Some forms have smaller cells in the center of the spiral (e.g. *I. chowdhryi* (Figure 11.17); Miocene of India) with the cells becoming larger in diameter toward the outside (Jain and Kar, 1979). In *I. trapezoides* (see Figure 11.17) individual cells are trapezoid shaped (the outer wall much longer than the inner) and the septa are thick (Kalgutkar, 1993). Specimens of *Helicoonites goosii* (Figures 11.17, 11.18A) have a three-dimensional morphology in which the spiral of cells forms ellipsoidal to barrel-shaped or beehive-shaped conidia (Kalgutkar and Sigler, 1995).

Many of the spores within the helicospore group are morphologically similar to members of the Dothideales and other ascomycetes and include such genera as

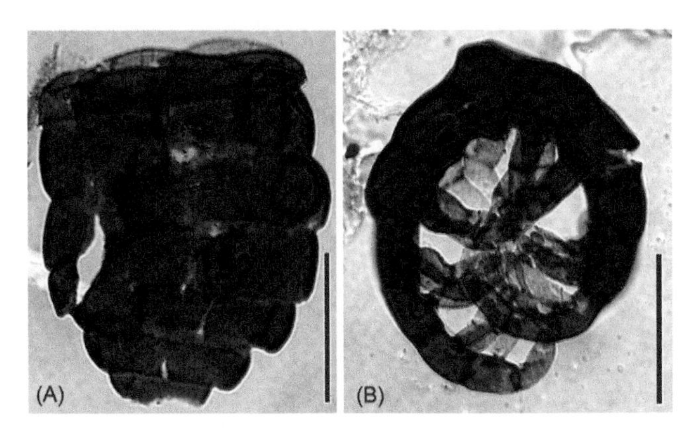

FIGURE 11.18 *Helicoonites goosii* (**A**) and *Pesavis tagluensis* (**B**). Eocene, Canada. Bar = 25 μm. (Courtesy M.G. Parsons.)

Helicoma and *Helicomyces*; however, in these forms conidia are septate. Most occur on plant litter and wood in moist environments (Zhao et al., 2007). Molecular systematics of the extant helicoid forms suggest that most of the species in the anamorphic genera (*Helicoma, Helicomyces*, and *Helicosporium*) form a monophyletic group that corresponds to the teleomorph *Tubefia*, and that the presence of helicoid conidia may represent a good taxonomic indicator for higher taxonomic placement (Tsui et al., 2006).

STAUROSPORES

The spores in this category are unusual in that they are stellate with three or four radiating arms or protuberances. Spores may or may not have septa and have more than a single axis.

Tribolites

This is an Eocene spore with two to six radiating arms that extend 30–45 μm out from a polyhedral central cell (Bradley, 1964; Kalgutkar and Jansonius, 2000). One of the features of *T. tetrastonyx* is that one of the arms is slightly flattened. The fossil is morphologically similar to the conidia of extant species of *Tetrachaetum* and *Lemonniera* (Kalgutkar and Jansonius, 2000).

Frasnacritetrus

This is a genus initially used for Late Devonian (Frasnian) microfossils from northern France; these are rounded to slightly bell shaped at one end, and, at the other, more rectangular in outline with a single process extending from each corner (Taugourdeau, 1968). While most palynologists today use the name *Tetraploa* for geologically younger fossils displaying this morphology, some still prefer to use *Frasnacritetrus* for these fossils.

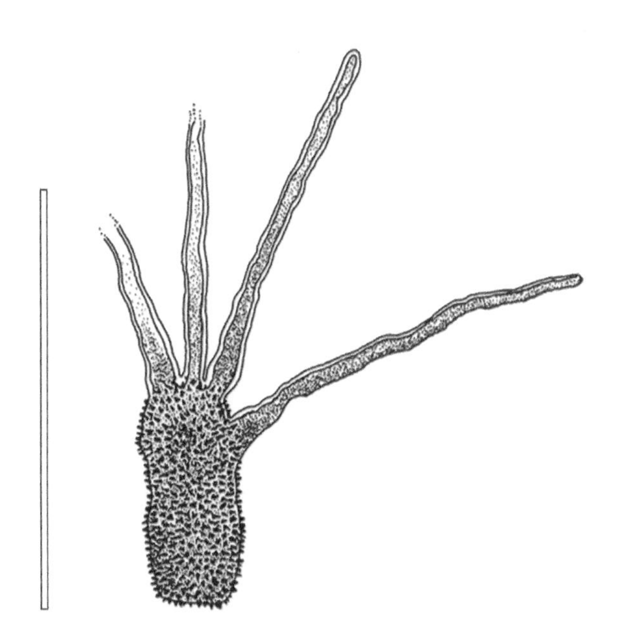

FIGURE 11.19 *Frasnacritetrus conatus*, a type of staurospore. Miocene, Himachal Pradesh, India. Bar = 50 μm. (Redrawn from Saxena and Sarkar, 1986.)

Several studies demonstrated that the number of processes in *Tetraploa*/*Frasnacritetrus* could range from two to four, and that the central cell was divided into multiple chambers (Saxena and Sarkar, 1986). In *F. conatus* (Figure 11.19) from the Miocene of India, the processes are hollow and the surface of the spore is ornamented by small coni (Saxena and Sarkar, 1986). Fungal spores of *Frasnacritetrus*, together with other palynomorphs, have been used to more accurately interpret the sequence of rock layers and climate during short intervals in the latest Cretaceous–Paleocene Intertrappean beds of India (e.g. Tripathi, 2001; Saxena and Ranhotra, 2009).

OTHER FUNGAL SPORES AND STRUCTURES

There are a variety of fungi referred to as anamorphic, mitosporic, asexual, conidial, and Fungi Imperfecti, and, in some literature, Deuteromycotina (Kirk et al., 2008). Sutton (1996) provided an excellent historical record of the attempts to correlate and classify these fungi based mainly on conidiogenesis. In general these fungi are asexual stages of primarily Ascomycota and a few Basidiomycota, and they produce mitotic structures termed conidia. These structures are highly variable in color, size, shape, and septation (e.g. Hennebert and

Sutton, 1994). Some of these fungi appear to have lost sexuality altogether. Most are terrestrial, although some occupy marine and freshwater ecosystems. Nutritionally they are saprotrophs and parasites, as well as lichen mycobionts, endophytes, mycoparasites, and mycorrhiza formers. Historically, several classes have been proposed. Fossil examples of three of these informal classes are described below.

HYPHOMYCETES

Hyphomycetes are anamorphs of a large number of genera and species (Kendrick, 2003; Seifert and Gams, 2011; Seifert et al., 2011). These fungi lack locular fruiting bodies (conidiomata), and sporulation occurs on separate septate hyphae that are exposed rather than within some specialized structure (e.g. Bernadovičová and Ivanová, 2011). The diversity of extant forms is extraordinary (e.g. Ellis, M.B., 1971, 1976). Fungi in this group grow in a wide range of ecological niches and gain carbon from living or dead organic matter; many forms are aquatic (Bärlocher, 1992). One of the features of many members of this group is that they produce passively discharged conidia from hyphae that allow them to move some distance from the substrate, where air currents can move the spores. Other spores can be moved by water or insects. They are regarded as polyphyletic.

These so-called Ingoldian fungi, which are aero-aquatic hyphomycetes and named in honor of C.T. Ingold, are one of a few groups where species can be identified based on the morphology of their long, narrow, branched conidia (e.g. Belliveau and Bärlocher, 2005; Descals, 2005; Prokhorov and Bodyagin, 2007). In this regard they have the potential to be recognized and identified in the fossil record. Some forms are referred to as aero-aquatic hyphomycetes because they form conidia with special flotation devices (e.g. Michaelides and Kendrick, 1982). The increased surface area provided by the stick-like morphology of these fungi provides buoyancy in turbulent streams; when they come to rest on a substrate, adhesive pads or appressoria quickly develop from some of the branch tips. In other instances the arms of the spore become tangled in debris that offers a substrate for the developing germlings. These micro-organisms release several exoenzymes that are primarily responsible for the degradation of cell wall material (e.g. Chamier, 1985). This makes these fungi an important component link in stream food webs because the resulting decomposed plant matter attracts detritus-feeding invertebrates.

An example of a fossil saprophytic hyphomycete is one preserved in the cortical tissues of the Eocene angiosperm *Eorhiza arnoldii* from the Princeton chert of British Columbia, Canada (Robison and Person, 1973; LePage et al., 1994). Additional specimens of these fungi in *E. arnoldii* have been characterized into three types and examined relative to their growth and development in the plant tissues (Klymiuk et al., 2013b). One type includes darkly pigmented holothallic macroconidia up to 125 μm long and with 30–35 transverse septa. Morphologically they appear most similar to the extant aquatic ascomycete *Xylomyces* (e.g. Goh et al., 1997). In the second type, propagation is the result of chains of amerospores, each with a simple septal pore between successive cells. Conidiophore morphology is not known in detail but there is some suggestion that conidia were borne on short branches like those in *Thielaviopsis* (Ellis, M.B., 1971). Pyriform phragmospores, each about 25 μm in diameter, represent the third type of anamorph found in *E. arnoldii*. Conidiogenous cells are approximately 5 μm in diameter and retained at the base of the dispersed conidium. Some hyphae associated with the spores produce curved branches. Features of the three anamorph types suggest affinities with chlamydospores produced by *Culcitalna achraspora*, a member of the Halosphaeriaceae (Sordariomycetes; Seifert et al., 2011). Also present in *E. arnoldii* are several types of sterile mycelia, including monilioid hyphae with intercalary branching, which are in turn associated with distinct morphotypes of microsclerotia (Klymiuk et al., 2013c). This careful study indicates the potential for determining the presence in fossil plants of asymptomatic endophytes and dark-septate endophytes, groups of fungi that can function in modern ecosystems as both pathogens and saprotrophs (Menkis et al., 2004). These studies demonstrate that developmental features of fossil fungi, if preserved in sufficient detail, can be used to determine biological affinities as well as stages in the life history of the fungi. Developmental stages also represent another set of parameters that will be useful in defining the ecological parallels between modern and fossil ecosystems. Moreover, they offer the opportunity to use certain life-history features (e.g. microsclerotia, chlamydospores) as minimum calibration points to document the presence of freshwater lignicolous saprotrophs during the early Eocene (Klymiuk et al., 2013b).

Another unusual fungal fossil, initially described from the upper Paleocene of Britain (Smith and Crane, 1979), is *Pesavis tagluensis* (Figure 11.18B), a fungal spore that subsequently has also been reported from a number of other Cretaceous and Cenozoic sites worldwide (e.g. Elsik and Jansonius, 1974; Jansonius, 1976; Lange, 1978b; Kalgutkar and Sweet, 1988; Lyck and Stemmerik, 2000). In fact, some have noted that it is difficult to determine whether the structure is a spore, fruiting body, or some type of capture device of a parasitic fungus (Kalgutkar and Jansonius, 2000). Each spore consists of a central cell with two incurving primary arms, each of which produces secondary arms (hyphae). Some have suggested that the affinities of *P. tagluensis* may lie with some type of dematiaceous hyphomycete (Pirozynski, 1976b), perhaps morphologically similar to conidia of *Dictyosporium* (Ellis, M.B., 1971). Functionally, the inwardly curving arms may have served to trap a bubble of air (see Michaelides and Kendrick, 1982). Spores with a main "stem" (hypha) of seven cells, with branching on one side, are included in *Ctenosporites* from the upper Eocene–lower Oligocene (Elsik and Jansonius, 1974; Lange and Smith, 1975). In *C. eskerensis* the multicellular structure may be up to 43 μm long (Smith, P.H., 1978); larger (56 μm) species are included in *C. sherwoodiae* from Upper Cretaceous coals in Colorado (Clarke, 1965).

AGONOMYCETES

Hyphae are also common in various matrices that contain fossil fungal spores. Little of this evidence for fossil fungi tends to be described unless there is some unusual feature of the hyphae or evidence of a host response that is somehow related to the fungus. Fungi accommodated in the Agonomycetes consist of sterile mycelia (Mycelia Sterilia) in which conidiation does not occur and the fungus reproduces asexually by hyphal growth and fragmentation (Cole, 1986). Within the group are important plant pathogens, such as *Rhizoctonia solani* of potatoes in which sclerotia overwinter on infected tubers. In this fungus the basidiomycete teleomorph (*Thanatephorus cucumeris*) is known (e.g. Hyakumachi and Ui, 1988). A fossil suggested as a member of *Rhizoctonia* is reported from the Permian of India as *R. nandorii* (Biradar and Bonde, 1976). The fossil occurs on gymnosperm wood and consists of septate hyphae of various sizes, but no spores or any type of reproductive structures are present. Another interesting fossil is *Dendromyceliates splendus*, reported from the Miocene of India (Jain and Kar, 1979). The end of the hypha branches dichotomously several times with the tips sharply pointed. In *D. rajmahalensis* the hyphae are up to 3.5 μm wide and aporate (Tripathi, 2001). In this Cretaceous species some hyphae are ornamented by small baculae.

COELOMYCETES

Coelomycetes are characterized by the production of conidia that form in some type of cavity in the host tissue (Sutton, 1980; Somrithipol et al., 2008). Today coelomycetes are known from temperate and tropical areas and are thought to include at least 1000 genera and more than 7000 species (Kirk et al., 2008). Although most are anamorphs, a few have been linked to teleomorphs, based on molecular studies (e.g. Rungjindamai et al., 2008). New living forms are still being described, especially from the tropics, and some of these possess morphologically unique types of funnel-shaped conidiomata that should be easy to recognize in the fossil record (e.g. Somrithipol et al., 2008).

Some early reports of fossil fungi represent coelomycetes. For example, *Cladosporites fasciculatus* was used for tufts of hyphae, a few of which bore conidia (Berry, 1916). This fungus was abundant in the vessels of lauraceous wood from the Eocene of Texas (USA). What are interpreted as pycnidia of a coelomycete containing one to three septate, ovoid-elliptical spores were described from the Deccan Intertrappean beds as *Deccanosporium eocenum* (Singhai, 1972). Morphologically, the fossil shares a number of features with *Camarosporium*, a plant pathogen historically placed in the Pleosporales.

Fossil coelomycetes have also been described from Miocene Dominican and Mexican amber (Poinar, 2003). Three different species – *Asteromites mexicanus* (Figure 11.20), *Leptostromites ellipticus* (Figure 11.21), and *Leptothyrites dominicanus* (Figure 11.22) – were discovered, one on a petal of a dicot flower. The fossils were compared to modern forms based on pycnidia and other features, and one form was suggested to be parasitic because the leaf it is attached to shows no evidence of decay. This analysis forms a framework that can be used to understand host/parasite relationships for some groups through time.

Permineralized fossil coelomycetes have been reported from the Upper Cretaceous (middle Turonian) of Hokkaido, Japan as *Archephoma cycadeoidellae* and *Meniscoideisporites cretacea* (Watanabe et al., 1999). The specimens occur in a bennettitalean cone, *Cycadeoidella japonica*. In *A. cycadeoidellae* the pycnidia are partially embedded within a pycnidium wall that is two or three layers thick. Aseptate conidia range up to 3 μm in diameter. Several features suggest affinities with the extant

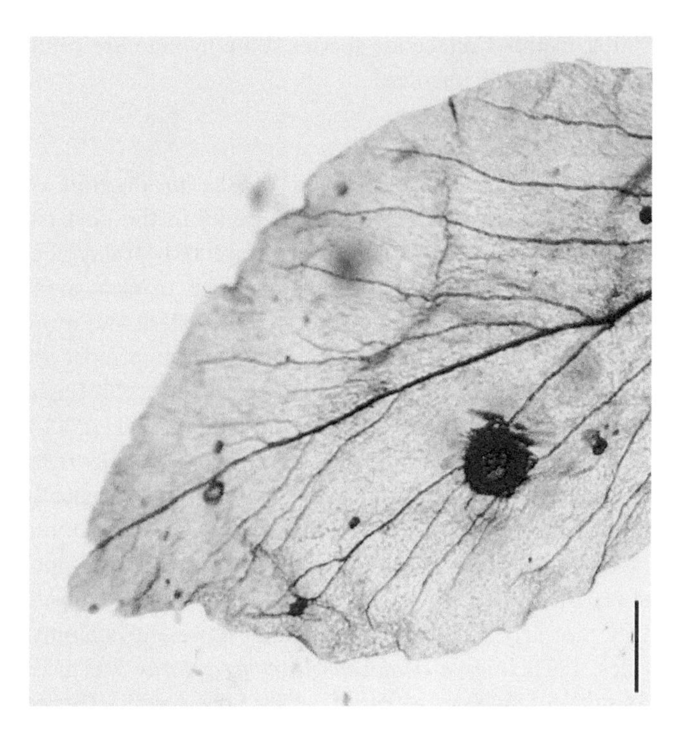

FIGURE 11.20 *Asteromites mexicanus* pycnidium on fossil flower petal preserved in amber. Oligocene–Miocene, Mexico. Bar = 500 μm. (Courtesy G.O. Poinar, Jr.)

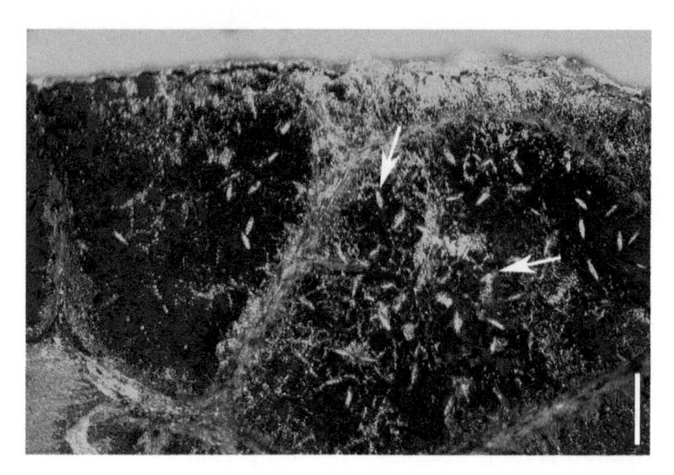

FIGURE 11.21 Pycnidia (*arrows*) of *Leptostromites ellipticus* preserved in amber. Miocene. Dominican Republic. Bar = 1.0 mm. (Courtesy G.O. Poinar, Jr.)

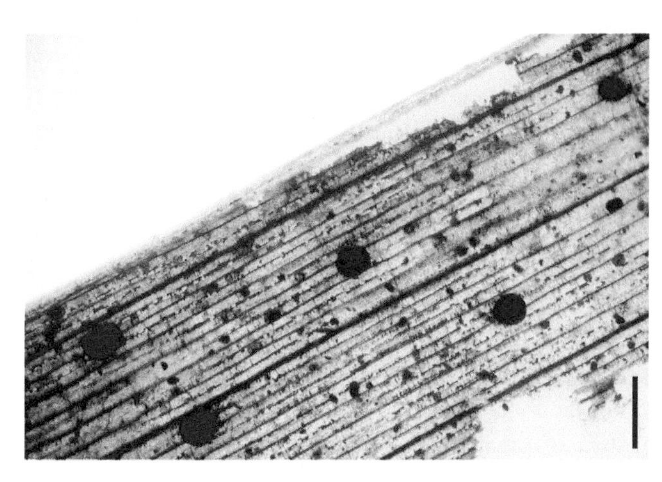

FIGURE 11.22 Several pycnidia of *Leptothyrites dominicanus* on a monocot leaf in amber. Miocene, Dominican Republic. Bar = 1.0 mm. (Courtesy G.O. Poinar, Jr.)

genus *Phoma*, a common soil fungus. Both pycnidial and acervial conidiomata occur in *M. cretacea*, both types producing meniscoid conidia and what are termed hyaline cells.

Another fungus from the Middle Jurassic of Santa Cruz Province, Argentina, is *Palaeopericonia* (Ibañez and Zamuner, 1996). *Palaeopericonia fritzschei* occurs in an araucarian cone and consists of layers of hyphae, each up to 9.6 μm in diameter with infrequent septa. The fossil has micronematous conidiophores with blastic conidiogenesis. Blastoconidia have a verrucose ornament and occur singly or in chains of up to nine elements. Also present in the specimen are terminal chlamydospores. The fossil is compared to several dematiaceous genera and, as the name indicates, is perhaps closest to modern *Periconia*, a common saprophyte (Cantrell et al., 2007). A fossil from the Pleistocene shares some features with the extant genus *Clasterosporium*, an anamorph included in the Magnaporthaceae (Sordariomycetes). The septate hyphae of the fossil are approximately 7 μm in diameter and hyphopodia are conspicuous (Cowley and Colquhoun, 1966).

12

FUNGAL INTERACTIONS

FUNGUS–ANIMAL INTERACTIONS 239
Glomeromycota and Animals 240
Coprolites 241
FUNGI AND ARTHROPODS 243
Fungus Gardens 245
Carnivorous Fungi 247
Trichomycetes 248
FUNGI IN EGGS 250
TRACE FOSSILS (ICHNOFOSSILS) 251
FUNGUS–FUNGUS INTERACTIONS 251
FUNGUS–PLANT INTERACTIONS 251
FUNGUS–GEOSPHERE INTERACTIONS 254
Bioerosion and Rock Weathering 254
Rhizosphere 255
Substrate Boring 256

FUNGUS–ANIMAL INTERACTIONS

Fungal interactions with animals and the geosphere are probably not the initial types of associations that one thinks about when considering the roles fungi play in modern ecosystems, but such interactions are common and quite variable (e.g. Wilding et al., 1989). The primary focus, however, is often on fungal interactions with plants in the form of saprotrophism, parasitism (including various forms of pathogenicity and severe diseases), and varying degrees of mutualism, since these symbioses are so much a part of our daily lives and of such economic importance. Nevertheless, there are a large number of saprotrophic, parasitic, and mutualistic fungal interactions with extant animals, and, certainly, they were present in ancient ecosystems as well. Moreover, there are multiple ways in which fungi are used by animals, including as a food source (mycophagy), as camouflage, and strategies to distribute their spores with the help of animals, especially by various arthropods (e.g. Blackwell, 1994; Moore, 1996; Bultman and Leuchtmann, 2008; Hågvar and Steen, 2013). Some animals also exhibit adaptations to ward off fungal infections; for example, certain rotifers can prevent infection by a parasitic fungus through complete desiccation (anhydrobiosis) and dispersal by wind to establish new populations where the fungus is absent (e.g. Wilson and Sherman, 2010, 2013). Even in some interactions that involve extant fungi and animals, the outcome of the interaction may have multiple dimensions (e.g. Bultman and Mathews, 1996; Silliman and Newell, 2003).

From the perspective of the paleomycologist, the best evidence of interactions among fungi, animals, and other microorganisms is the recycling of carbon during the decay process. The actual players in this process, however, are rarely identifiable based on fossils (e.g. Taylor and Scott, 1983). One principal reason for this relates to the physical collection of fossils. In almost all instances paleontologists search for specimens that are "showy" and/or demonstrate a complete complement of features that define the organism; specimens that are poorly preserved are often discarded. This means that stages in the decay process, including evidence of the decaying organism, are rarely collected or observed. This is sometimes

T.N. Taylor, M. Krings & E.L. Taylor: Fossil Fungi.

DOI: http://dx.doi.org/10.1016/B978-0-12-387731-4.00012-8

not the case when examining permineralized or unaltered materials, such as amber, in which the matrix may show both the decayer and the decaying organism. It is probable that there is no museum in the world that exhibits any type of poorly preserved fossils with a description indicating that these organisms represent an exceptional example of the decaying activities of fungi and other microorganisms! This pattern to some degree parallels that regarding interactions between fossil animals and plants, especially evidence of herbivory. While there are early reports of plant–animal interactions characterized by specific patterns of tissue loss thought to be the result of feeding activities, these patterns have only recently become more widely known from the fossil record and are today established as offering compelling evidence for the existence of certain groups of animals in time and space (e.g. Labandeira, 1998; Wilf et al., 2005; Labandeira et al., 2007; Labandeira and Currano, 2013). They also establish feeding preferences for the animals that have been linked to global-scale activities (e.g. Labandeira and Phillips, 2002; Currano et al., 2010). In other examples where the affinities of the fossil organism remain equivocal (e.g. the Silurian–Devonian organism *Prototaxites*), the presence of coprolites within the tissues has been used to help identify the nutritional mode of the herbivore (e.g. Hotton et al., 1996; Edwards et al., 2012), which can provide ecological data on the host. In certain instances where the focus of the research is on the animals, evidence of microbial activity can sometimes be discovered. For example, *Gigantoscorpio willsi* is a scorpion from the Mississippian of Scotland (Størmer, 1963). Associated with this fossil animal are structures interpreted as bacteria (Moore, L.R., 1963b), but there are also thread-like filaments that may represent hyphae. An interesting side note to this study is that while some of the bacteria were demonstrated to be fossil, others were interpreted as modern contaminants.

It is very likely that many of the fungus–animal interactions that are so common today also existed in ancient ecosystems (Taylor and Osborn, 1996), and there is an increasing body of literature examining animals and microarthropods and what they ate (e.g. Edwards et al., 2012). Some of these interactions today involve various invertebrates and saprophytic Basidiomycota that represent major trophic levels and functional components in soil environments (e.g. Crowther and A'Bear, 2012). Modern ecosystems also contain a large number of invertebrates that consume fungi and have developed a fungivorous or mycophagous food preference (e.g. Shaw, 1992; McGonigle, 1995; Harrington, 2005). These organisms graze not only on fungal hyphae but also on spores and fungal fruiting bodies, and there is evidence that animals have distinct species preferences for certain types of fungi (e.g. Dighton, 2003; Schigel, 2008). In addition, the presence of mycophagous invertebrates can be determined, based on animal diet, since hyphae and spores often pass through the alimentary system of these organisms unscathed. In this way, animals also serve as dispersers for the spores of many saprotrophic and parasitic fungi, but appparently certain soil invertebrates (e.g. oribatid mites) do not disperse mycorrhizal fungi (e.g. Renker et al., 2005). This may be related to spore size since spores of *Acaulospora* have been reported inside oribatid mites (Rabatin and Rhodes, 1982). Interestingly, while there are multiple types of coprolites documented in the Early Devonian Rhynie chert ecosystem food web (e.g. Habgood et al., 2004), none contained evidence of glomeromycotan spores, a common fungal spore type in the Rhynie chert. There are, however, numerous examples of mycophagy, including the dispersal of ectomycorrhizal spores, by multiple groups of extant mammals (e.g. Cázares and Trappe, 1994; Schickmann et al., 2012), and there is evidence that oribatid mites prefer feeding on certain types of ectomycorrhizal fungi rather than saprotrophic fungi (Schneider et al., 2005).

GLOMEROMYCOTA AND ANIMALS

One of the benefits that land plants gain from entering into AM associations, in addition to nutrient uptake, is the apparent protection from various types of fungal pathogens and soil microorganisms like nematodes (Borowicz, 2001; Wehner et al., 2011). The beneficial effect on pathogen suppression, however, is sometimes achieved only in concert with certain saprotrophic fungi, termed plant growth-promoting fungi (PGPF; e.g. Muñoz et al., 2008; Saldajeno et al., 2008, 2012). In modern ecosystems, precisely how the presence of arbuscular mycorrhizae keeps such pathogens in check is not well understood (e.g. Pozo et al., 2002; de la Peña et al., 2006). In one study, the degree to which various plant pathogens were suppressed by one AM fungus was similar, irrespective of the pathogen tested (Veresoglou and Rillig, 2012). When controlling nematodes, however, multiple species of AM fungi were effective. These authors also noted that members in the Acaulosporaceae were the least protective against pathogens, and that nitrogen-fixing plants received more protection as a result of more intense colonization of arbuscular mycorrhizae. A study by Duhamel et al. (2013) indicates that a tritrophic relationship exists between AM fungi, their host plants, and certain fungivores, because the metabolite catalpol was consistently found in AM fungal hyphae in host plants exposed to fungivores but was undetectable in hyphae when fungivores were absent. This suggests that certain secondary metabolites may be transferred from the host plant to the AM fungi to increase their defense against fungivores.

COPROLITES

Arthropod Coprolites

One of the most common, and perhaps oldest examples of fungus–animal interactions occurs in the form of coprolites (fecal pellets) produced by a microarthropod. Included in the macerations of the Burgsvik Sandstone (Upper Silurian of Sweden), together with the evidence of ascomycete anamorphs, are large aggregations of hyphae, some up to 260 μm long. These are interpreted as coprolites, similar to those produced by modern mycophagous microarthropods (Sherwood-Pike and Gray, 1985; Gray and Boucot, 1994). On the other hand, while various types of coprolites (based on size, shape, and content) are frequently encountered in thin sections of the slightly younger Rhynie chert ecosystem, suggesting a diversity of consumers (Habgood et al. 2004), only very few of these are made up of fungal hyphae or spore fragments (Figure 12.1). This may reflect either the types of animals that existed in this food web during the Early Devonian or a preservational bias against coprolites composed entirely of fungal remains. Studies using extant soil microarthropods suggest that fungal hyphae show little degradation after passing through the guts of millipedes and isopods (e.g. Crowther and A'Bear, 2012). There are, however, a few examples of Rhynie chert plants that have disrupted cells and tissues, which suggests that some type of feeding interaction existed with certain types of animals (e.g. Kevan et al., 1975) that may have secondarily increased the potential for fungal attack.

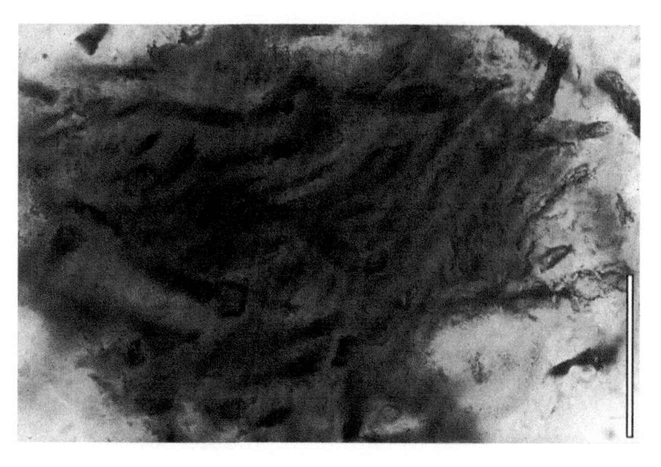

FIGURE 12.1 Coprolite composed of fungal hyphae. Lower Devonian Rhynie chert, Scotland. Bar = 50 μm.

Permineralized Triassic bennettitalean roots from Hopen Island (Svalbard Archipelago) contain coprolites, presumably produced by oribatid mites, which in turn are colonized secondarily by saprotrophic fungi (Strullu-Derrien et al., 2012). Coprolites attributed to oribatid mites have also been reported in permineralized wood from the Permian, Triassic, and Jurassic in Antarctica (Kellogg and Taylor, 2004). This study suggested that the microarthropods within the plant tissues might have been consuming fungi and/or bacteria in addition to, or in place of, plant material. In other reports from Triassic rocks, oribatid mite body fossils were recovered together with various fungal remains, including a diverse assemblage of spores and hyphae (Vijaya and Murthy, 2012). Fungi in the form of variously shaped sclerotia have also been reported in coprolites in the stem of a Late Permian *Psaronius* from Yunnan Province, China (D'Rozario et al., 2011). In this association it is thought that the colonization of the fungi occurred soon after the stem died. Other coprolites containing evidence of fungi have been reported from coal ball permineralizations (e.g. Baxter, 1975; Wu et al., 2007).

Another fossil example of fungivory was reported in silicified peats from the Middle Permian Bainmedart Coal Measures, in the Prince Charles Mountains of Antarctica (Slater et al., 2012). This study examined coprolites from acetate peels and thin sections; other pieces of the peat were macerated with hydrofluoric acid and intact coprolites picked from the residue. Also present in these macerations were parts of arthropod exoskeletons. These coprolites were up to 220 μm long and composed entirely of fungal spores; most occurred in galleries of stem and root wood of glossopterid seed ferns. Since most of the fungal spores were flattened and some split, it is hypothesized that they represented a portion of the diet of some invertebrate; the occurrence of fungal hyphae in some of these coprolites is interpreted as post-depositional colonization of the coprolites by the fungi. The discovery of multiple types of coprolites in *Glossopteris* plants from this Middle Permian site clearly indicates that, in addition to mycophagy, there were multiple feeding strategies being used by invertebrates in this high-latitude paleoecosystem. The increased resolution obtained by using field emission scanning electron microscopy may make it possible to determine the types of fungi that formed the spores.

Albian amber from France includes coprolites containing hyphae, setae, and basidiospores assignable to the Polyporales (Basidiomycota), suggesting that by Early Cretaceous time bracket fungi producing large fruiting bodies may have been a food source for fungivorous beetles (Schmidt et al., 2010b). A large number of fungal remains have also been reported from pieces of Cenomanian (Upper Cretaceous) amber from Ethiopia, where they are found not only in the matrix but also within coprolites (Schmidt et al., 2010a,b). It is suggested that the spores and hyphae occurring in the matrix were attached to the surface of the plant resin and then subsequently covered by a new resin flow; this process was repeated multiple times resulting in a piece of amber containing thousands of spores and other fungal parts. Spores are variable in shape, but generally have tapered ends. Within the population of spores are various degrees of septation, which represent stages of conidial development. The hyphae in the amber are branched and possess short lateral protrusions. This assemblage of fungal remains is described as *Palaeocurvularia variabilis* (Schmidt et al., 2010b), suggesting the morphological resemblance to the extant anamorph *Curvularia*, a hyphomycete that is a plant parasite found mainly in tropical ecosystems.

Another putative nutritional relationship between animals and fungi has been reported from Eocene amber from northern France (Nel et al., 2013). Preserved in this amber are thrips, formally described as *Uzelothrips eocenicus*, to which numerous fungal remains are superficially attached. Since remains of fungi are also frequently attached to the bodies of extant *Uzelothrips* species, these authors concluded that the association of the fossil *U. eocenicus* with fungi was not accidental, but rather a reflection of a long-term interaction, such as fungivory, along with accidental attachment of loose mycelia fragments. These thrips have probably been living near or inside clusters of epiphytic fungi such as sooty molds, their assumed food source, for at least 53 myr.

Large Animal Coprolites

Fungal spores in large animal coprolites have long been used as a proxy record to study animal diet and, thus, the presence of particular herbivores in paleoecosystems (e.g. Pirozynski et al., 1984; Chin, 2007). In some instances coprolites can provide a more accurate picture of the landscape because they reflect environmental variability (e.g. van Geel, 2001). Certain types of fungi, which are sometimes termed dung or coprophilous fungi, have been especially important in assessing past herbivore populations based on selected fungal spores of the *Sporormiella*-, *Sordaria*-, and *Podospora*-types (Baker et al., 2013; Gill, 2013). While many of these studies have focused on Quaternary sediments (e.g. van Geel and Aptroot, 2006; López-Sáez and López-Merino, 2007), the fact that some fungi have restricted ecological ranges may make it possible to link some of these spore morphologies with major groups of herbivores in time and space (e.g. Gill et al., 2013). Another approach that underlines the importance of fossil fungal spores focuses on more accurately understanding the decline of certain megafaunal elements (e.g. Burney et al., 2003). For example, large numbers of fungal spores were present from 200–400 CE in Madagascar and then abruptly dropped off. Since various records, including paleontological ones, indicate that the island country had a considerable diversity of large animals, the presence or absence of certain types of fungal spores suggests that the decline in some types of animals may correlate with habitat destruction (Comandini and Rinaldi, 2004).

Fungal remains have been identified from coprolites of sauropods collected from the Lameta Formation (Maastrichtian) of western India (Kar et al., 2004b; Sharma et al., 2005). Acid maceration yielded a number of plant parts (e.g. leaves, petioles, seeds, pollen) as well as fungi of two basic types. One of these is thought to represent a mycorrhizal fungus and is placed in the genus *Archaeoglomus*, a fossil taxon for oval chlamydospores up to 38 μm in diameter. The other fungi in the samples include various types of epiphyllous ascostromata of varying sizes and shapes. Some of these fruiting bodies have variously shaped appendages used to suggest affinities with extant members of the Erysiphales (Leotiomycetes: Ascomycota); others are compared with microthyriaceous forms that lived on the leaves that were browsed by these dinosaurs. Diverse types of fungal spores have also been reported as occurring in other dinosaur coprolites from the Upper Cretaceous of India (Ghosh et al., 2003; Sander et al., 2010).

An interesting association between fungi and large animals has been suggested by the report of dinosaur coprolites believed to have been produced by *Maiasaura* hadrosaurs from the Upper Cretaceous Two Medicine Formation of Montana (Chin, 2007). The coprolites are irregular calcareous blocks that contain fragmented plant material and are identified by the presence of distinctive, backfilled dung-beetle burrows (Chin and Gill, 1996). An analysis of the coprolite content indicates a high percentage of conifer wood in some, while in others there are poorly preserved cells of plants and a few fragments of wood. Microscopic sections of the coprolites show that the wood was partially decayed, suggesting that the animals were consuming decaying wood. Although the microscopic sections of the coprolites lack unequivocal direct evidence of fungi, there is evidence of tissue degradation by fungi based on which components of the tracheid walls are preserved. Although fragments of wood might be consumed along with foliage by a large herbivore, the high percentage of wood in these coprolites has been used to suggest that rotting wood may have represented a significant component of the diets of these Late Cretaceous animals (Chin, 2007).

Coprolites thought to be deposited by mosasaurs, large fishes, or sharks have been reported from the Bearpaw Formation (Cretaceous) of Saskatchewan, Canada (Mahaney et al., 2013). The specimens were examined using scanning electron microscopy with energy-dispersive X-ray spectroscopy (SEM/EDS), and several forms of bacteria and filamentous fungi were reported, although these were not described in detail. Various organisms have been preserved in mammalian and reptile coprolites from the middle Eocene Bridger Formation of southwestern Wyoming (Bradley, 1946); they include silicified bacteria and freshwater algae of several types, but, interestingly, no fungi. If the hypothesis that the animals drank from highly acidic water is accurate, then this might explain the absence of fungi from the coprolites. Finally, fungal palynomorphs have been reported from coprolites of a late Pleistocene squirrel from Yukon, Canada (Pirozynski et al., 1984). They consisted of saprotrophic and coprophilous fungi, several of which are characteristic of rodent dung.

A different type of fossil matrix containing evidence of fungi occurs in the form of clay balls adhering to some of the bones of a 37 ka skeleton of an American mastodon (Pirozynski et al., 1988). These are globular structures (some greater than 10 cm in diameter) composed principally of calcium carbonates and they contain various fragments of plant material as well as fungal remains. The fungi in the clay balls are very diverse and believed to reflect an ecologically wide range of habitats. What these clay ball structures actually represent relative to the diet of the animal remains a question. Similarly, remains of fungi, including hyphae and several types of well-preserved spores, were macerated from Pleistocene clay sediments that contained the remains of a giant ground sloth (Elsik, 1986). Several of these spores suggest that the site represented a grassland ecosystem.

FUNGI AND ARTHROPODS

An interesting fossil that shows the partial decay of an arachnid cadaver is preserved in amber dated as early–middle Miocene (Iturralde-Vinent and MacPhee, 1996) from the Dominican Republic. *Geotrichites glaesarius* consists of septate, sometimes branching hyphae up to 5 µm wide (Stubblefield et al., 1985c). As the generic name suggests, this fossil shares a number of morphological features with the ascomycete *Geotrichum*, a cosmopolitan extant fungus found in water, sewage, soil, plants, and as a normal component of the human intestinal flora. Holoarthric conidia arise from unspecialized hyphae on the fossil (Figure 12.2). The fact that the fungus was not present within the body cavity of the spider suggests that *G. glaesarius* was a saprotroph rather than a parasite. *Geotrichum*-like inclusions have also been reported from Dominican amber by Rikkinen and Poinar (2002b). These yeast-like cells range from 2 to 5 µm in diameter and many are represented as refractive casts of cells at various stages of development, some undergoing budding. The fossils are interpreted as perhaps early stages after germination of some filamentous Pezizomycotina spores.

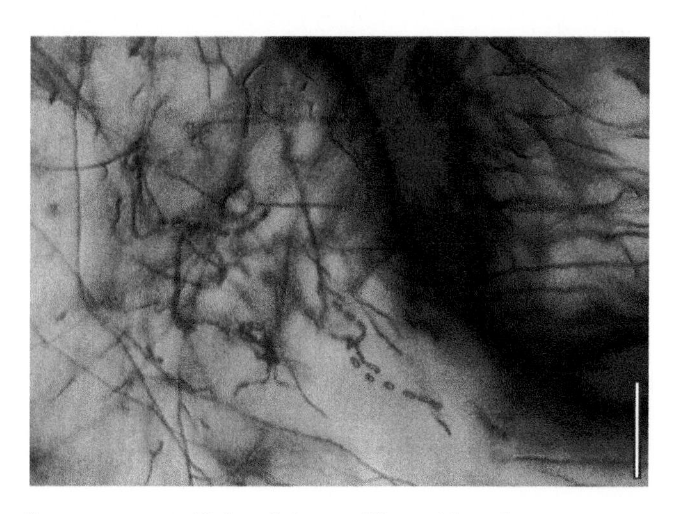

FIGURE 12.2 Holoarthric conidia arising from a mycelium growing on a spider corpse preserved in amber. Miocene, Dominican Republic. Bar = 50 µm.

FIGURE 12.3 Fossil ant in amber covered with conidia that morphologically resemble the ascomycete *Beauveria* sp. Miocene, Dominican Republic. Bar = 0.5 mm. (Courtesy G.O. Poinar, Jr.)

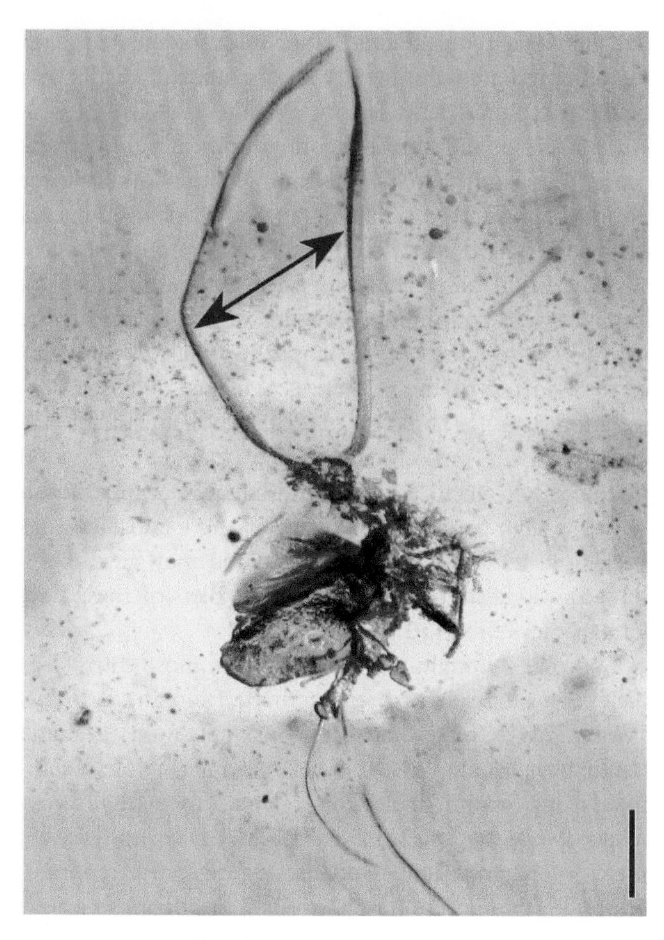

FIGURE 12.4 The fungal parasite *Paleoophiocordyceps coccophagus* showing two synnemata (*arrows*) arising from the head of a scale insect. Lower Cretaceous amber, Myanmar. Bar = 0.65 mm. (From Sung et al., 2008, courtesy G.O. Poinar, Jr.)

Fungi have been reported in association with a worker ant, also preserved in amber from the Dominican Republic (Figure 12.3) (Poinar and Thomas, 1984). Extending from the surface of the ant are conidiophores bearing single-celled, ellipsoidal conidia. The fossil is compared to *Beauveria*, an entomopathogenic ascomycete in the Hypocreales (Rehner et al., 2011), some species of which have been used as biocontrol agents because the fungus can penetrate the cuticle of the insect (Xiao et al., 2012). While there are multiple extant examples of ant–plant symbioses, including those that involve domatia (chambers produced by plants to house arthropods) where fungi are grown as food (e.g. Defossez et al., 2009, 2011; Blatrix et al., 2012), we are unaware of any reports that suggest this pattern from the fossil record (but see Fungus Gardens below).

In ancient ecosystems fungi were also parasites of various insects. *Paleoophiocordyceps coccophagus* (see Chapter 8, p. 142) is a fungal parasite of a scale insect trapped in Lower Cretaceous amber from lignitic seams in the Hukawng Valley of Myanmar (Figure 12.4) (Sung et al., 2008). Arising from the head of a male individual are two synnemata; on the surface of each are elongate cylindrical phialides bearing conidia. The fossil has been used to calibrate a multigene phylogeny within the extant Hypocreales (Sordariomycetes), which suggests that pathogenicity in arthropods associated with this group of fungi first occurred in the Jurassic, and that it is closely tied to the coevolutionary interactions with angiosperm diversification during the Early Cretaceous.

Fungi can also affect phytophagous insects in a variety of ways, only some of which are obvious and direct (e.g. Hatcher, 1995). Examples include insects feeding on fungal mycelia and spores that they cultivate in a host plant and insects being impacted by fungal-induced toxins. Less obvious cases are those where plant pathogenic fungi alter host-plant quality or the number of plants available for the insect; in general this situation results in a reduction in the size of the plant population.

Fungal parasites have also evolved the capacity to modify the behavior of their insect hosts (Moore, J., 2002). One example of this is the so-called death-grip fungus of ants that are infected by the ascomycete *Ophiocordyceps unilateralis* (Hypocreales) (e.g. Hughes et al., 2011a). In this interaction, the fungus manipulates the behavior of the ants (e.g. Mongkolsamrit et al., 2012), which upon infection behave as zombies and display predictable stereotypical behaviors resulting in a final "death grip" on a leaf vein. During the death grip, the mandibles penetrate deeply into vein tissue, accompanied by extensive atrophy of the mandibular muscles. This lock-jaw behavior means the ant will remain attached to the leaf after death, thus assisting parasite reproduction (Hughes et al., 2011a). The fungus then develops a stalked, spore-producing structure, which arises from the head of the ant and serves to disseminate spores, which can then infect additional ants. An extraordinary fossil example of this complex interaction has been reported on leaves from the famous Eocene Messel shale site (~47 Ma) in Hesse, Germany (Hughes et al., 2011b). Ant mandible puncture scars, identical to those formed in modern ant death-grip examples (e.g. Pontoppidan et al., 2009), were found on the surface of a dicot leaf *Byttnertiopsis daphnogenes*. The Messel study indicates that by careful analysis of fossil plant material it is possible not only to document plant–animal interactions but also to characterize behavioral traits that have involved fungi as the major behavioral manipulators.

Microbial interactions as exhibited in fossils may be subtle. For example, the Messel site is one of several Cenozoic localities known to contain insects that still retain distinct metallic colors (McNamara et al., 2012). These colors are the result of ordered nanometer-scale variations in the structure of the tissue, which, in the fossil state, can be influenced by multiple factors, including thermal alteration, physical deformation, degree of dehydration, oxidation, and microbial degradation. On some specimens there are filaments up to 100 µm long and approximately 2 µm in diameter that are interpreted as fossilized bacteria and fungi. Precisely what microbial agents these might represent is not known, but future studies in this area will be important, especially in the preservation and stewardship of modern collections of structurally colored specimens (McNamara et al., 2012).

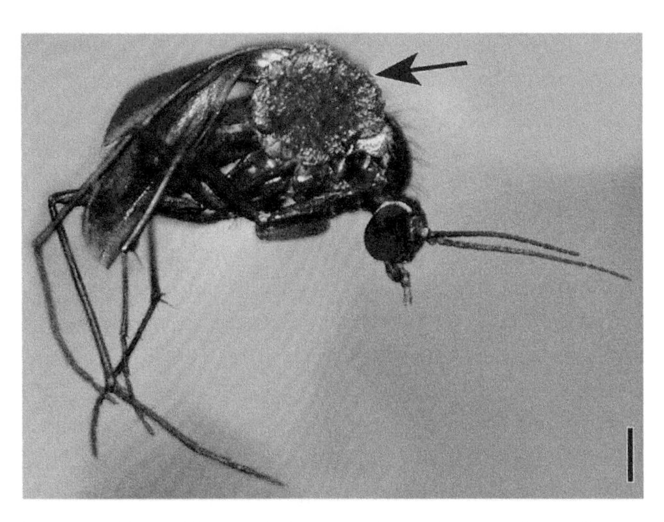

FIGURE 12.5 *Entomophthora* sp. (*arrow*) on surface of gnat. Miocene amber, Dominican Republic. Bar = 300 µm. (From Poinar and Poinar, 2005, courtesy G.O. Poinar Jr.)

Another example of a fossil fungus (ascomycete) that is parasitic on an insect was reported from Baltic amber (Rossi et al., 2005). In this association the fungus consists of thalli, each about 0.3 mm long, that are attached to the thorax of a stalk-eyed fly (*Prosphyracephala succini*) (see Figures 8.59, 8.60). The morphological characters of the fossil are so specific that the specimen can be included in the extant genus *Stigmatomyces*, a member of the Laboulbeniales (Rossi and Weir, 2007). Not only does this discovery represent the oldest record (Eocene) of a fungus that is parasitic on an insect, but it also marks the first occurrence of this fungal family in the fossil record (Rossi et al., 2005). Another fly preserved in Baltic amber associated with a fungus was reported by Göppert and Berendt (1845), but in this association the fungus may be a saprotroph (Poinar and Thomas, 1982). While there are numerous reports of fungi in amber, a significant report documents evidence of insect pathogens from pieces of amber of various ages (Poinar and Poinar, 2005). In this study several types of fungi are reported on the cuticle of a mosquito and the prothorax of a fungus gnat (Figure 12.5).

FUNGUS GARDENS

A highly sophisticated and complex association between insects and fungi occurs in modern ecosystems where certain groups of termites (and ants) cultivate basidiomycetous fungi in special gardens (e.g. Wood and Thomas, 1989). The fungus breaks down food provided by the animals through laccase activity (e.g. Taprab et al., 2005) and also adds nitrogen and other nutrients to the animals' diet. In this association the fungus grows on a sponge-like structure termed a fungus comb, which develops from the feces of the termites (e.g. Arshad and Schnitzer, 1987). The oldest fossil specimens of fungus combs have been reported from the Upper Cretaceous of Texas, USA (Rohr et al., 1986). Complex structures of probable termite origin, including some reminiscent of fungus combs, have been reported from the Lower Jurassic of the Karoo Supergroup of South Africa and Lesotho (Bordy et al., 2009). Fossil fungus combs are also known from Miocene sediments from Chad where they consist of ovoid-shaped structures 6.5 × 2.5 cm that have an alveolar organization (Duringer et al., 2006, 2007). In these fossils, the individual spaces (cells) are empty; however, one of the characteristic features of the fossil comb is the presence of millimeter-sized spheres that are morphologically identical to the mylospheres, a type of fecal pellet that forms the structure of modern fungal combs. Molecular tools suggest that in some extant fungal combs there are filamentous fungi with yeasts as secondary colonizers, but these are not associated with the symbiosis (Guedegbe et al., 2009).

It is believed that fungus-growing termites originated in the African rainforest just before the expansion of the savanna, about 31 Ma (49–19 Ma), based on molecular divergence dating (Nobre et al., 2011). Today approximately 330 species of termites are actively engaged in fungiculture in which they cultivate several species of the genus *Termitomyces* (e.g. Abe et al., 2000; Makonde et al., 2013), and it is hypothesized that growing fungi for food developed only once in termites (Aanen et al., 2002; Aanen and Eggleton, 2005) but multiple times in other animals (e.g. ants and beetles; e.g. Mueller and Gerardo, 2002; Biedermann et al., 2009; Mehdiabadi and Schultz, 2010).). Extant fungi also have evolved the capacity to morphologically and chemically mimic termite eggs in the form of sclerotia within the nests, resulting in the fungi having a competitor-free habitat (Matsuura et al., 2009; Matsuura and Yashiro, 2010). This process also extends to other organisms since there are some flowering plants that mimic the scent of fungal fruiting bodies, and fungi that stimulate the plant to produce non-functional flowers that serve to distribute fungal spores (Kaiser, 2006). There are also numerous extant examples of interactions between animals and fungi in which the animals transmit fungi to the host (e.g. Müller et al., 2002; Rice et al., 2007, 2008). One of these involves the mountain pine beetle that bores through the bark of the host and then inoculates the conducting tissues with a blue-stain fungus, *Grosmannia clavigera*. The fungus colonizes the phloem tissues, which reduces nutrient translocation and eventually kills the tree (e.g. Goodsman et al., 2013). Although either the beetle attack or the fungus can be responsible for the eventual death of the tree, the two organisms together kill much more quickly. In addition, the fungus acts as a food source for the beetle larvae (Mitton and Ferrenberg, 2012). With increasing temperatures in western North America, some beetle populations are now producing two generations per year, exacerbating the epidemic of tree death, which currently extends from Mexico to Canada (e.g. Mitton and Ferrenberg, 2012). While there is currently no evidence of this symbiosis in fossil woods, this type of interaction could be preserved and identified based on evidence of the boring organism (e.g. holes in the bark) and the fungus.

Recognition of fungal combs and potential fungal associates in the fossil record opens up opportunities for multiple areas of inquiry about past ecosystems and the timing of fungiculture in various groups of animals, some of which are well represented in the fossil record. An important future area of research in paleomycology will be the examination of other animal groups engaged in fungus gardening and the discovery of specimens that may indicate these relationships. While there are multiple extant ant–fungus symbioses (Schultz and Brady, 2008; Defossez et al., 2011), there is only one report directly demonstrating this type of association in the fossil record (Genise et al., 2013). Moreover, it is known that fungus-growing ants were common from at least the early Miocene because five attine species from three genera have been found in Dominican amber (Baroni Urbani, 1980; de Andrade, 2003; Schultz, 2007; see also LaPolla et al., 2013). There are numerous additional facets to the complexity of the interaction between ants and fungi. For example, leaf-cutting ants are known to prune their fungi to maintain high productivity (Bass and Cherrett, 1996). On the other hand, the fungal cultivar may produce specialized feeding structures in the form of nutrient-rich swollen hyphal tips (gongylidia), which are fed to the ant larvae and consumed by the adult ants (e.g. Poulsen et al., 2009). In contrast to the structural adaptations relating to the establishment of the association (e.g. growing chambers), the behavioral adaptations necessary to make the association work cannot normally be documented from fossils, and thus their origin and evolutionary history will remain unresolved (but see above: ancient death grip).

CARNIVOROUS FUNGI

Carnivorous fungi are polyphyletic, with approximately 200 species included in multiple extant groups (zygomycetes, Basidiomycota, Ascomycota, and multiple anamorphic stages), and are able to trap and directly obtain nitrogen from various animals, including protozoans, rotifers, springtails, and nematodes (Yang et al., 2007). Perhaps the most common fungi of this type are members of the Orbiliales, a group of Ascomycota that have the ability to form multiple types of trapping devices for nematodes, such as loops, adhesive knobs and columns, and various forms of constricting and non-constricting rings (e.g. Barron, 1977; Swe et al., 2011). A phylogenetic analysis based on nucleotide sequences suggests that the adhesive network was the earliest form of trapping device, at least for nematodes (Yang et al., 2007). Molecular clock assumptions based on two fossil examples hypothesize that the ability of fungi to assimilate nitrogen from prey (carnivory) diverged from saprophytic fungi ~419 Ma, and that active trapping (with constricting rings) diverged from passive carnivory (fungi with adhesive traps) ~246 Ma (Yang et al., 2012). It has been hypothesized that active carnivory evolved from passive forms via two pathways (Li et al., 2005): (1) from adhesive knobs retaining the sticky material by forming simple, two-dimensional networks, and eventually forming complex three-dimensional networks; (2) from adhesive knobs that lost these materials, with their ends meeting to form non-constricting rings, which in turn formed constricting rings with three inflated cells.

Fungi and Nematodes

Nematodes are a common component in most ecosystems today, where they thrive as parasites or feed on bacteria, fungi, other animals (including other nematodes), and/or decomposing organic matter (e.g. Freckman, 1982). There are more than 20,000 species that inhabit freshwater and marine ecosystems and various terrestrial environments, and there is an extensive fossil record of these animals suggesting they evolved about 600–550 Ma (e.g. Poinar et al., 2008), and perhaps were already present in the Precambrian. For additional information, see Poinar (2011), "*The Evolutionary History of Nematodes – As Revealed in Stone, Amber, and Mummies*," which includes descriptions of fossil representatives.

Fossil evidence suggestive of the presence of an interaction between nematodes and a fungus believed to be carnivorous comes from upper Albian amber of southwestern France (Schmidt and Dörfelt, 2007; Schmidt et al., 2008).

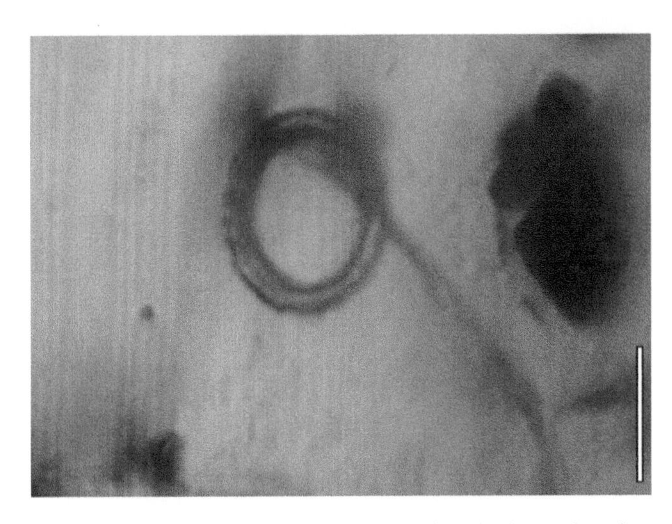

FIGURE 12.6 *Palaeoanellus dimorphus* detached trapping ring. Lower Cretaceous amber, Charente-Maritime, France. Bar = 10 μm. (Courtesy A.R. Schmidt.)

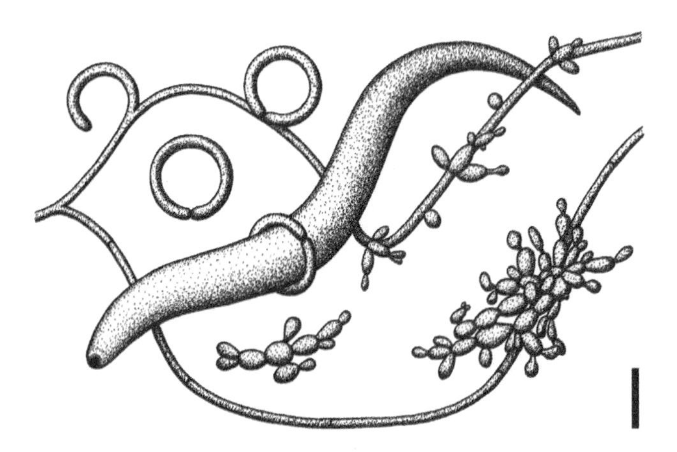

FIGURE 12.7 Suggested reconstruction of carnivorous (nematode) trapping fungus *Palaeoanellus dimorphus* including yeast stage. Bar = 10 μm. (From Schmidt et al., 2007.)

The mycelium of *Palaeoanellus dimorphus* consists of irregularly septate hyphae up to 2 μm in diameter. Short lateral branches of the hyphae occasionally loop, and in this way eventually form rings (Figures 12.6, 12.7). In contrast to modern trapping rings consisting of three cells, these rings are unicellular, with only one septum present at the junction. Mature rings have an inner diameter of 8–10 μm and an outer diameter of up to 15 μm. Adhering particles suggest that the rings may have secreted a substance which added to the efficiency of the trapping device. Blastospores and a yeast stage (Figure 12.8) are also a component of this Early Cretaceous fungus. While nematodes have been found in this amber, none has been

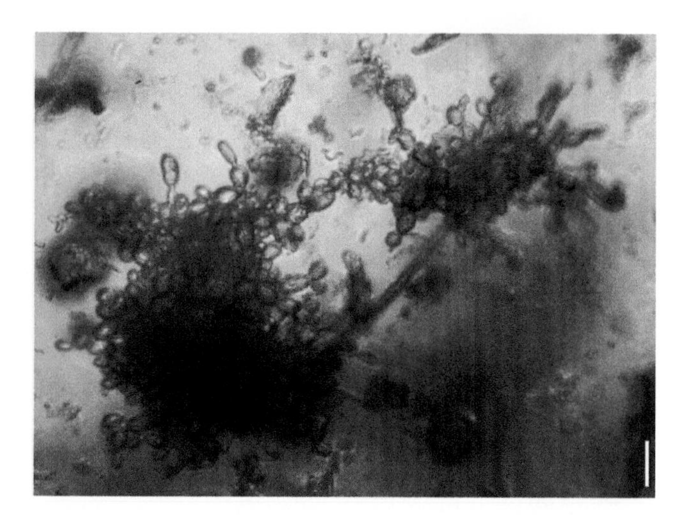

FIGURE 12.8 Yeast stage of *Palaeoanellus dimorphus.* Lower Cretaceous amber, Charente-Maritime, France. Bar = 10 μm. (Courtesy A.R. Schmidt.)

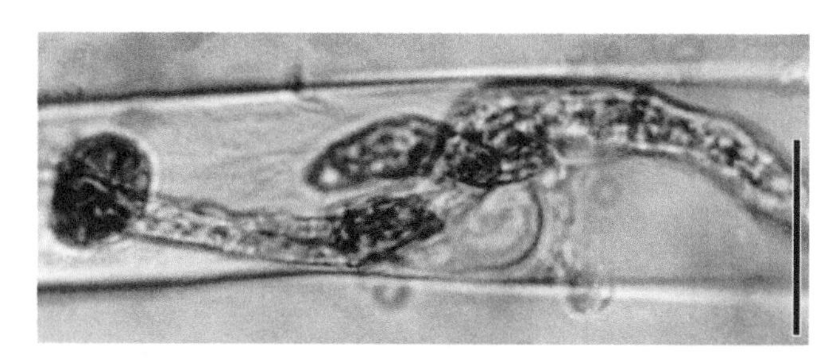

FIGURE 12.9 Fossil nematode containing a fungus. Oligocene–Miocene amber, Mexico. Bar = 15 μm. (From Jansson and Poinar, 1986.)

demonstrated to be directly associated (i.e. trapped in the rings) with *P. dimorphus*. While some have challenged the nature of this fossil fungus because it cannot be assigned to a modern group (e.g. Thorn et al., 2008), the fossil record is replete with examples of organisms that are now extinct and possess combinations of characters that do not fit modern systems of classification.

Nematodes have also been reported in Oligocene–Miocene amber from Mexico dated at approximately 20–15 Ma (Jansson and Poinar, 1986). Present in this amber are three different types of nematophagous fungi, including spores, structures possibly representing adhesive spores or some type of trapping device, and evidence of aseptate hyphae in the body of the nematode (Figure 12.9). Although the precise affinities of the fungi could not be determined, the study is an important one in expanding details about potential carnivorous fungi in time and space.

FIGURE 12.10 Robert W. Lichtwardt.

TRICHOMYCETES

While fungal parts can pass through the digestive tract of various invertebrates and are documented in the fossil record in the form of coprolites, there are other extant fungi that remain in the hind gut of certain insects and other arthropods with chewing mouthparts (Nematocera, Plecoptera, Ephemeroptera, Trichoptera, Collembola, Isopoda; Lichtwardt, 2012; Figure 12.10). These are the gut fungi or trichomycetes, which have hosts that are globally distributed and live in marine, freshwater, and various terrestrial habitats (Lichtwardt, 1986; Hernández Roa et al., 2009).

Trichomycetes are a polyphyletic class within the zygomycetous fungi and, as a result of several molecular studies (e.g. Benny and O'Donnell, 2000; Cafaro, 2005 [Figure 12.11]; White et al., 2006), also include various protists. Today the trichomycetes are regarded as an ecological group rather than a taxonomic class (e.g. Hibbett et al., 2007). Trichomycetes are obligate commensals, but in some instances provide the host with organic nutrients and affect host fitness (e.g. McCreadie et al., 2005; Nelder et al., 2006). These fungi generally lack a mycelium but produce septate, branched thalli that are attached to the lining of the host by specialized cells termed holdfasts

FIGURE 12.11 Matías J. Cafaro.

(e.g. Misra and Lichtward, 2000). Asexual reproduction occurs by means of trichospores that are expelled from the host and germinate after ingestion by a new host. One species, *Spartiella aurensis*, is capable of growing back into the midgut of the host (White and Strongman, 2012b). In some interactions the fungus can switch from being commensal to mutualistic to parasitic (e.g. Nelder et al., 2006); some forms infect the ovaries of adult insects, thereby reducing the overall fitness of the population (e.g. Garms, 1975; Labeyrie et al., 1996). Today the group consists of the fungal orders Harpellales and Asellariales (e.g. Cafaro, 2005; Misra, 2012; Wang et al., 2013). Two other orders once included in the former class Trichomycetes, that is the Eccrinales and Amoebidiales (Lichtwardt, 1986), have been transferred to the Mesomycetozoea, a diverse group of fungus-like organisms that arose near the animal–fungus divergence and is basal to the animal lineage in the Opisthokonta supergroup. They occur in marine, freshwater, and terrestrial environments where they enter into symbioses with a wide range of other organisms, including invertebrates, fish, amphibians, birds, and mammals (Mendoza et al., 2002; Glockling et al., 2013). Additional details, including an extensive bibliography on the trichomycetes, can be found at www.nhm.ku.edu/~fungi.

To date there is one report of a fossil that was interpreted as a trichomycete (White and Taylor, 1989c). The single specimen comes from silicified plant material from the Fremouw Formation (Triassic) in Antarctica, the site of multiple reports of fossil fungi (e.g. White and Taylor, 1988; Osborn et al., 1989). In section view, the fossil consists of several elongate, cylindrical tubes (Figure 12.12), which are about 55 μm long and arise from the inner surface of what is interpreted as the inner hindgut cuticle of an arthropod. Each of the aseptate thalli is attached by an isodiametric cell that is morphologically and topographically like a holdfast cell in some extant trichomycetes. The distal end of the thallus (directed toward the inside of the cavity) is truncated and contains what is described as a septal plug (Figure 12.13). Associated

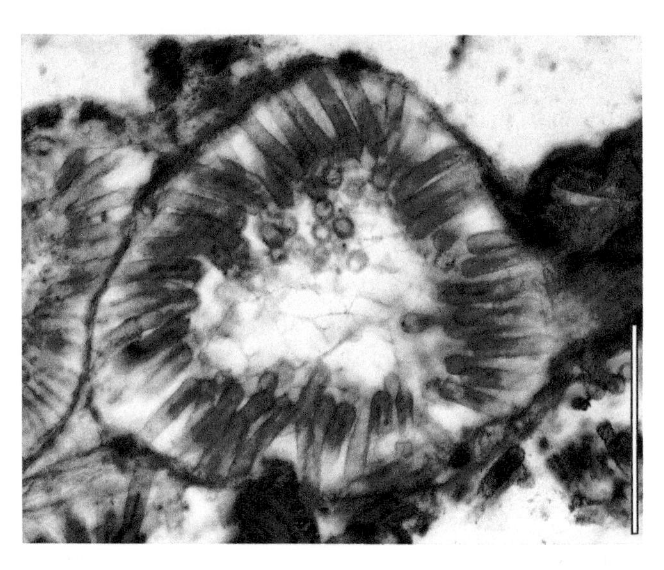

FIGURE 12.12 Elongate thalli arising from inner surface of putative arthropod cuticle. Triassic, Antarctica. Bar = 100 μm.

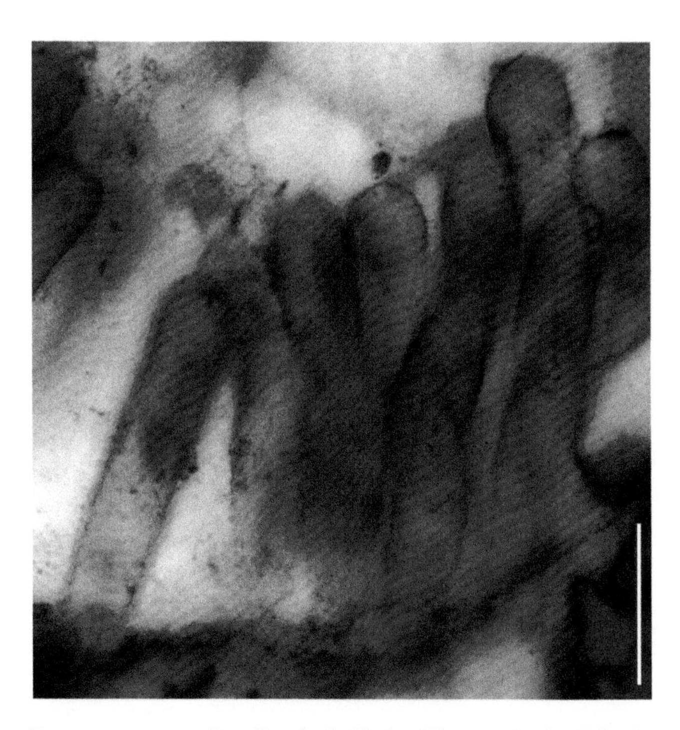

FIGURE 12.13 Detail of thalli in Figure 12.12. Triassic, Antarctica. Bar = 25 μm.

with the thalli are ellipsoidal spores, the dimensions of which are approximately 10 ×14 μm. The fossils are compared morphologically to the extant trichomycete protist *Enterobryus* (Eccrinales), based on the presence of septate thalli, holdfast, spores, and septal plugs (White and Taylor, 1989c). Cafaro (2005), however, discounted this relationship, noting that septal plugs are not present in members of the Eccrinales, nor is there conclusive evidence that the fossil is in fact attached to the cuticle of an arthropod. Lichtwardt (2012) suggested that the wide geographic distribution and primitive nature of trichomycete hosts suggests that in general the group represents an ancient symbiosis. This hypothesis is strengthened by the fact that some members of the Harpellales occur on hosts whose ancestors are believed to have evolved by perhaps 250 Ma (earliest Triassic; Grimaldi and Engel, 2005). This timeline fits well with the putative Triassic trichomycete specimen, which is far younger than the beginning of the Cambrian, the time the stem group of arthropods are believed to have evolved (Budd and Telford, 2009).

One might ask why there is not more evidence of trichomycetes since the fossil record is rich in arthropod remains. Circumstantial evidence based on the global distribution and occurrence of some primitive groups of insects would suggest that this fungus–animal interaction is indeed ancient. With a rapidly expanding database of fossil arthropods (especially those found in amber and other matrices offering extraordinary preservation), including complete organisms, it is surprising that trichomycetes have not been recorded from any of these fossils. Moreover, with new imaging techniques (e.g. X-ray microtomography) it may be possible to identify evidence of fossil trichomycetes, including forms that are at different stages of the life history (e.g. Garwood et al., 2012).

FUNGI IN EGGS

Discoveries in paleontology are often serendipitous. One of these is the preservation of fungi from an Early Cretaceous turtle egg from China (Jackson et al., 2009). The fungal remains preserved together with this egg include branching septate hyphae, conidiophores supporting multiple phialides, and chains of up to five basipetal conidia. The morphology of the fossil fungus is similar to extant taxa within the genus *Penicillium*. This association is believed to provide clues to clutch-related paleoecological interactions. The presence of varied mycobiota have also been reported in modern turtle eggs and nests (e.g. Phillott et al., 2004; Güçlü et al., 2010).

Structures interpreted as fungal mycelia have also been found in sections of dinosaur egg shells from the Upper Cretaceous of central China (Gong et al., 2008). Using environmental SEM/EDS microanalysis, the fungal hyphae appear as needle-, ribbon-, and silk-like filaments up to 18 μm long; all are unbranched and have pointed tips. As a result of the morphological similarity and the areas in the shells in which the fungi occur, it is hypothesized that the fungi were parasitic and invaded the dinosaur shells before they became lithified. These authors also suggested that the fungi were pathogens that contributed to the demise of dinosaurs at the end of the Cretaceous (Casadevall, 2005; Gong et al., 2008). While an interesting idea, this rather sweeping hypothesis will require a great many more studies of egg shells to determine whether the co-occurrence of fungi with these dinosaur embryos was more than a single interaction. Trace fossil evidence believed to document the presence of endolithic fungi and algae in various types of fossil (and extant) hard parts (e.g. shells, teeth [Figure 12.14], bones) occurs in the form of narrow canaliculi that have been collectively described as *Mycelites ossifragus* (Roux, 1887; Bernhauser, 1953). Schmidt (1964) reported *M. ossifragus* from a mid-Pleistocene swan eggshell from Germany.

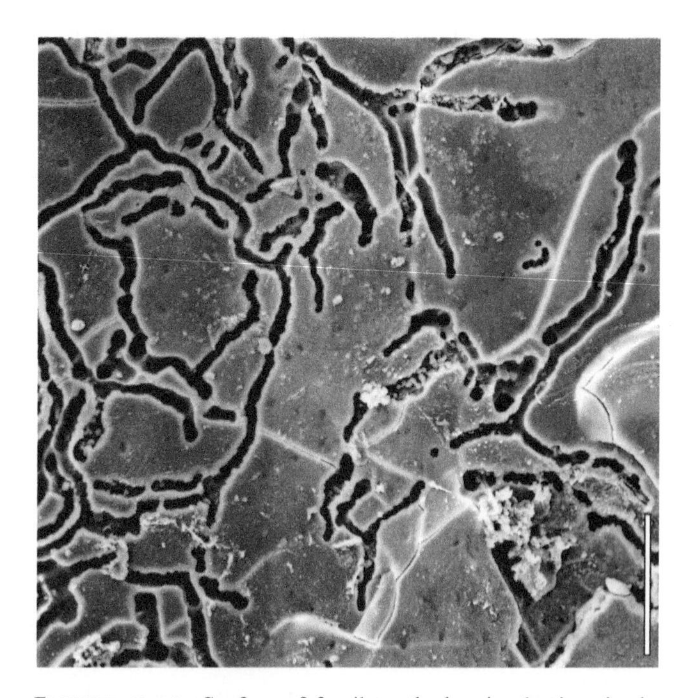

FIGURE 12.14 Surface of fossil tooth showing borings in the enamel of *Mycelites* sp. Jurassic, Great Britain. Bar = 50 μm. (From Martill, 1989.)

TRACE FOSSILS (ICHNOFOSSILS)

A trace fossil is a morphologically recurrent structure resulting from the activity of an organism that was modifying the substrate during life (sediment, rock, dead or living organic matter; Bertling et al., 2006). Some trace fossils have presented special problems in interpretation. *Asthenopodichnium xylobiontum* is an ichnospecies used for a wood-boring organism that produced U-shaped traces thought to represent some Cretaceous mayfly nymphs (e.g. Moran et al., 2010), while *A. lithuanicum*, a Neogene trace believed to have also been produced by mayfly nymphs, is distinguished by a pouch-like shape with a J-shaped limb that is wider than the remaining part of the pouch (Uchman et al., 2007). Traces of *A. lignorum* found in Miocene wood from New Caledonia, however, are interpreted as evidence of fungal activity (e.g. Genise et al., 2012). The traces occur in the form of elongate, and ellipsoidal or almond-shaped scoops (Figure 12.15) that are typically oriented parallel with the fibers of the wood. Similar traces have been reported in fossil wood from Argentina, Egypt, Czech Republic, Lithuania, and the USA (Genise, 2004).

FIGURE 12.15 Ichnofossil *Asthenopodichnium lignorum* consisting of traces formed in wood. Miocene, New Caledonia. Bar = 1.0 cm. (Courtesy J.F. Genise and A. Nei.)

Coprolites and evidence of wood boring have also been reported in several Permian woods from China (e.g. Feng et al., 2010; Feng, 2012). Moreover, permineralized wood from the Upper Triassic Petrified Forest Formation in Arizona, USA, displays various types of heartwood degradation (Creber and Ash, 1990; Tanner and Lucas, 2013). One type occurs in the form of elongated cavities, several millimeters long, that are oriented parallel to the wood grain and may merge to form channels extending longitudinally through the wood, while another consists of circular to elliptical cavities that cut across the wood grain. Tanner and Lucas (2013) interpret both trace types as reflecting preburial biotic activity, probably by pathogenic fungi; however, the size and shape of the longitudinal channels may also be consistent with some forms of arthropod burrowing.

FUNGUS-FUNGUS INTERACTIONS

An excellent fossil example of complex interactions that can be deciphered between different fungi is the hyperparasitism reported from Lower Cretaceous amber (Poinar and Buckley, 2007). In this detailed study the pileus of the basidiomycete, *Palaeoagaricites antiquus* (see Chapter 9), is parasitized by another fungus (*Mycetophagites atrebora*), which consists of hyphae 4–6 um in diameter and septate conidiophores producing short chains of conidia (see Figure 9.15). Individual conidia (aleurospores) are basipetally produced. There is a third level of interaction present in the form of Entropezites patricii, which is interpreted as a necrotrophic hyperparasite because the hyphae invade and apparently destroy the mycelium of *M. atrebora* (see Figure 9.16). The hyperparasite shows some similarities to extant necrotrophic parasites (e.g. *Penicillium, Trichoderma*) which have extensive intrahyphal growth (e.g. Jeffries and Young, 1994; Poinar and Buckley, 2007). Research has also demonstrated that extant species of *Trichoderma* increase nutrient uptake in plants, elicit plant defense reactions against pathogens, and can serve as opportunistic plant symbionts (e.g. Harman et al., 2004; Druzhinina et al., 2011).

FUNGUS-PLANT INTERACTIONS

In addition to the obvious interactions between plants and fungi that are included in this volume, there are also a variety of abnormal plant growths, including galls and witches' brooms (Dreger-Jauffert, 1980), that are the result of interactions between fungi and plants in which the cells of the host are stimulated to produce additional tissues and/or increase in cell size. These morphological and anatomical abnormalities and modifications are typically the result of hypertrophy (increase in cell size), hyperplasia (increase in cell number), or hypoplasia (decrease in number of cells; Preece and Hick, 1994), and may be caused by various types of fungi (e.g. Apt, 1988; Hernández and Hennen, 2003). One of the problems in interpreting such phenomena in the fossil record is that the abnormalities can also be formed as a result of interactions with other organisms (e.g. bacteria, nematodes, insects, viruses), as well as abiotic agents like radiation and chemicals (herbicides). In some fossil specimens there is evidence of several stages of fungi, together with specific host responses, indicating that the fungus was a parasite (e.g. Stubblefield et al., 1984). For example, in the pollen cone *Lasiostrobus polysacci* from the Upper Pennsylvanian of Illinois (USA), septate fungal hyphae occur in the cortex and microsporophylls. Hyphae are up to 4 μm in diameter, highly branched, and produce terminal and intercalary swellings. On the inner surface of the host cells are swellings that may represent some type of wall apposition; these cells are sometimes accompanied by copious amounts of material interpreted as resin. Despite the hyphae being relatively widespread in the cone, the biological affinities of the fungus remain unknown. In some extant gymnosperms including *Pinus*, there is an increase in the number of resin ducts in the xylem as a result of a fungal infection (Martín-Rodrigues et al., 2013). This is certainly an anatomical feature that can be considered when examining fossil wood. However, although tissue disruption can result from fungal infection, it can also be caused by parasitic plants that invade stem tissue (e.g. *Pilostyles*; Gomes and Fernandes, 1994; do Amaral and Ceccantini, 2011). While there is no fossil evidence of this type of plant–plant association, permineralized plant remains may be a potential source of evidence of this type of interaction, especially in angiosperm wood.

An excellent example of hyperplasia that has been detected in fossils is the production of galls. Galls are abnormal growths in plants caused by some type of parasitic organism, ranging from fungi to insects (e.g. Mani, 1964). The gall makers (sometimes termed cecidozoa, or animals that produce cecidia or galls) typically produce enzymes or plant hormones that stimulate hypertrophy and/or hyperplasia. Galls are defined in various ways depending on the organism responsible, where growth takes place, and the general organization of the tissue (Meyer, 1987). Galls are formed by fungi and fungus-like organisms in most of the major groups, and the galls have sometimes been subdivided based on whether they develop as a result of hypertrophy or hyperplasia (Akai, 1950). The fact that unrelated groups of organisms can elicit the formation of galls renders the accurate interpretation of such structures from fossils difficult. In addition, plant galls caused by insects may harbor fungi (e.g. Batra and Lichtwardt, 1963) and, in some instances, rather than being simple endophytes, these fungi are essential for the development of the gall-producing insect, which further aggravates the interpretation problem. For example, certain gall midges obtain their food from fungi that form mycelia on the interior surface of the gall chamber (e.g. Rohfritsch, 2008; Stireman et al., 2010). While direct fossil evidence of this complex symbiosis is lacking to date, gall midges (Cecidomyiidae) have been described as fossils preserved in amber dating back to the Eocene (e.g. Nel and Prokop, 2006). It is hypothesized that gall-forming Cecidomyiidae may have evolved from opportunistic colonization of pathogenic fungi in galls during oviposition or larval invasion on plant tissue (Stireman et al., 2010).

There are numerous reports for fossil galls that date back to the Paleozoic and provide an interesting pattern of both the diversity and evolution of this type of interaction (e.g. Larew, 1986; Scott et al., 1994; Labandeira, 2002; Labandeira and Phillips, 2002; Krassilov, 2007, 2008; Labandeira et al., 2007; Srivastava, 2007). One interesting late Paleozoic example of a gall occurs in the seed fern *Trivena arkansana* from the Mississippian of Arkansas (Dunn et al., 2003). Within the cortex are clusters of thick-walled cells that are suggested to be the wound response to some arthropod vector, which perhaps carried some type of virus. The presence of galleries in the phloem of the stem supports this hypothesis, but aggregations of differently shaped cells could also represent a response to a fungus. Similar histological responses have been noted in other fossil vascular plants (e.g. Beck and Stein, 1987), especially in wood (e.g. Conwentz, 1890; Jeffrey, 1906). An excellent example of fossil galls are those that occur in the Paleogene of Spitsbergen in which there are multiple examples on angiosperm leaves (Figures 12.16, 12.17), and at some sites galls were the most common type of host damage (Wappler and Denk, 2011). The presence of certain types of insect damage on the leaves is an important data source since there

FIGURE 12.16 Fossil angiosperm leaf with conspicuous galls. Paleocene, Spitsbergen. Bar = 1.0 mm. (Courtesy T. Wappler.)

FIGURE 12.17 Galls on *Leguminosites* sp. leaf. Miocene, Czech Republic. Bar = 1.0 mm. (Courtesy T. Wappler.)

is no modern forest equivalent to the High Arctic forest ecosystems that existed in the Cretaceous–Cenozoic. Distinctive galls on oak leaves (*Quercus*) have been reported from the Miocene of Nevada, USA (Waggoner and Poteet, 1996), Eocene of Tennessee (Brooks, 1955; Wittlake, 1969) and Pleistocene of The Netherlands (Stone et al., 2008). The degree of interaction between the gall producers and the host plants, and whether the galls were initially formed by a fungus, will represent difficult obstacles to overcome in the fossil record. However, the documentation and description of abnormalities in the general morphology and structure of cells and tissue systems in various plant parts will be an important first step in defining the breadth of gall formation in the

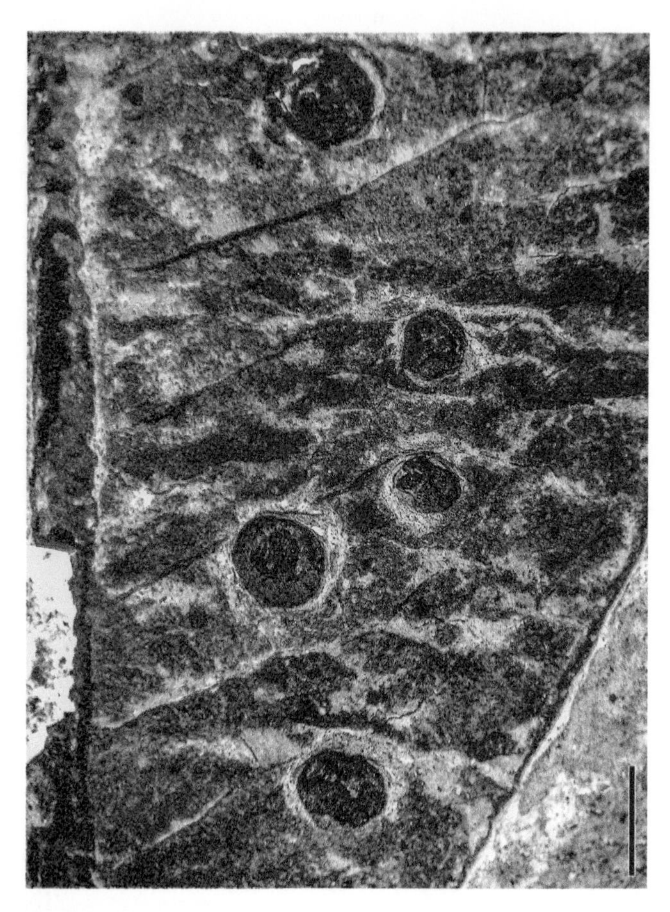

FIGURE 12.18 Five galls located between midvein and leaf margin. Darker inner region of each gall interpreted as a single, inner chamber. Eocene, Germany. Bar = 2.0 mm. (Courtesy T. Wappler.)

fossil record (Figure 12.18). Even if the gall contains evidence of fungal activity, it may be difficult to determine the activities of the fungi; some may represent pathogens, while others may be saprotrophs invading dead galls, or parasites of the gall maker (Wilson, L.R., 1995).

FUNGUS–GEOSPHERE INTERACTIONS

There are multiple ways in which fungi directly and indirectly interact with the rocks and sediments on Earth. For example, microbialites are various organosedimentary deposits that have formed as a result of the development of prokaryotic and eukaryotic communities which trap and bind sediment or serve as the nucleus of mineral precipitation (e.g. Couradeau et al., 2011). The most common of these structures are stromatolites, which have a rich geologic history extending far back into the Proterozoic and Archean (e.g. Allwood et al., 2009). Degraded sheaths of calcifying cyanobacteria are a major component of many stromatolites. During the initial phases of the study of stromatolites the focus centered on the role of prokaryotes in these interactions; however, today fungi are increasingly being examined for their role in the formation of these structures (e.g. Jones et al., 2000). When studying the impact that fungi have on rocks, an important aspect of study is biomineralization, the collective processes by which organisms form minerals (calcium carbonates, silicates, iron oxides, sulfides, etc.) and which may involve bacteria, protists, fungi, plants, and animals (Kolo et al., 2007; Gadd, 2010). On the other hand, the destruction of rocks and minerals by biological activities is termed bioerosion and may include both mechanical and chemical effects and involve a large group of different types of organisms (e.g. Wisshak, 2006a; Tribollet et al., 2011). Fungi are powerful geobiological agents in terrestrial environments, where they promote mineral weathering and the decomposition of organic matter. They also serve a similar role in the oceanic crust, some existing in extreme marine ecosystems where they can be found in deep-sea hydrothermal vents, deep sea water, and methane hydrate-bearing sediments (e.g. Richards et al., 2012; papers cited in Jones and Pang, 2012). Determining the diversity of these modern organisms is based on both culture-dependent and culture-independent methods (Singh et al., 2012).

BIOEROSION AND ROCK WEATHERING

Various types of microorganisms develop in and on the surface of rocks and minerals where they may obtain critical elements and compounds needed to produce cellular energy (Hutchens, 2009). Mineral weathering, once thought to be solely an inorganic process, is now known to involve fungi (and bacteria) that transfer and cycle organic and inorganic nutrients, form biogenic minerals, and transform and accumulate metals (Burford et al., 2003a; Bonneville et al., 2009; Finlay et al., 2009; Rosling et al., 2009); some fungal hyphae are so sensitive they can discriminate different mineral grain sizes (Leake et al., 2008b). Moreover, the chemical composition of the mineral may be a major factor in the development of microbial and fungal communities on the mineral surface, together with the type of microbial exopolymer, pH microenvironment, electrostatic effects, surface properties, and so on (Burford et al., 2003b).

Today fungi are major geologic agents of bioerosion just as they were in the geologic past (e.g. Sterflinger,

2000). Evidence of bioerosion can be documented from as early as the Precambrian (Taylor and Wilson, 2003). For example, in temperate carbonate-rich areas (e.g. reef ecosystems) there is an extraordinarily high diversity of bioeroding organisms, including fungi, some of which may be opportunistic pathogens (e.g. Bentis et al., 2000; Wisshak et al., 2011). Bioerosion also takes place in other environments, including those that are extremely cold (e.g. Etienne and Dupont, 2002). These fungi possess metabolic systems, life histories, and sometimes structural features that allow them to survive in especially harsh environments (e.g. Selbmann et al., 2005). There are also non-lichenized, rock-inhabiting (i.e. endolithic) fungi that have adaptations allowing them to exist in extreme abiotic conditions, including high (or low) temperatures and high UV (Gueidan et al., 2011). These ascomycetes may have also been able to exist on rock surfaces that were initially nutrient poor. Some endolithic fungi, however, can have a protective effect on buildings made of oolitic limestone (Concha-Lozano et al., 2012).

Saprotrophic, mutualistic, and lichen-forming fungi contribute to mineral weathering by penetrating rock by both physical and chemical means, and fungal weathering is no doubt important on evolutionary timescales (Hoffland et al., 2004). The process of fungal weathering has been going on for hundreds of millions of years and there is no rock on the surface of the Earth that escapes some degree of weathering. Mineral weathering by fungi increases nutrients for the fungi and thus supports their fitness. As a result of the mycorrhizal relationships with plant roots, various nutrients are removed from minerals, while at the same time carbon is provided, which can become fixed in soil carbonates (e.g. Balogh-Brunstad et al., 2008).

Fungi are also involved in the production of various metabolites that enhance mineral weathering. They are active agents in lithification, diagenesis, and other alterations of various rocks, especially limestones (e.g. Verrecchia, 2000; Burford et al., 2006). Of particular importance are the activities of ectomycorrhizal fungi that have the capability of absorbing both organic and inorganic nutrients from rocks (Adeleke et al., 2012). Among others, Quirk et al. (2012) documented active weathering of minerals by AM fungi. They suggested that mycorrhiza-driven weathering may have originated more than 350 Ma, and that it subsequently intensified with the evolution of trees and mycorrhizae to affect the Earth's long-term atmospheric and climate history. Controlled experiments using extant ectomycorrhizal fungi grown over a mineral surface indicate that fungal weathering alters the lattice structure of the mineral at the nanometer scale, thus enabling the formation of various microcavities that can be mined by the fungus for essential nutrients (Bonneville et al., 2009). Fungi selectively colonize particular mineral types in the soil (Smits, 2009). Atomic force microscopy has been used to demonstrate the fungal–mineral interface in extant ecosystems (McMaster, 2012), and it would be interesting to see if such features could be identified in fossil mineral surfaces using similar techniques.

RHIZOSPHERE

The rhizosphere, that portion of the soil where microorganism-mediated processes are under the influence of the root system, involves numerous biotic and abiotic factors including exudates produced by the roots (e.g. Berg and Smalla, 2009). Soil microorganisms associated with plant roots also control the weathering of rock-forming minerals (Smits et al., 2009). For fungi, root exudates provide the signaling mechanisms that lead to fungus–plant interactions. The presence of mycorrhizal fungi results in changes in the chemical and physical characteristics of the rhizosphere, and that portion occupied by the fungi is sometimes termed the mycorrhizosphere (Sen, 2005).

There are other structures in fossil soils that may represent the activities of fungi. One of these is termed rhizoliths, organosedimentary structures consisting of concentrations of calcium carbonate that initially formed around roots (e.g. Klappa, 1980; Cramer and Hawkins, 2009; Figure 12.19). Some discovered in the Paleogene of Uruguay are up to a meter in diameter and have been termed ferruginized paleorhizospheres (Genise et al., 2011). Other structures (termed rhizolith balls) reported from the Cretaceous of Patagonia may represent the

FIGURE 12.19 Cluster of rhizoliths exposed by wind erosion. Quaternary, Yucatan, Mexico. Bar = 1.0 cm. (Courtesy J. Foote.)

biomineralization of fungal hyphae, although they may also record the biological activities of other organisms including crayfish, termites, or ants (Genise et al., 2010).

Evidence of Biogenic Activity

Fungi are involved in the indirect precipitation of certain types of calcite in fossil soils around decaying roots (e.g. Wright, 1984). One of these is needle-fiber calcite, a secondary calcium carbonate mineral crystal that has multiple morphologies and origins (Phillips and Self, 1987). Some of these calcites are believed to be inorganically formed from supersaturated solutions that rapidly evaporated to produce certain types of crystals (Borsato et al., 2000). Others, however, represent calcified bacteria or are related to biomineralization associated with fungal hyphae (e.g. Verrecchia and Verrecchia, 1994). Today most evidence indicates that these calcite forms are biogenic – the result of bacterial and fungal activity or in association with basidiomycete fungal hyphae (e.g. Janssen et al., 1999; Cailleau et al., 2009; Bindschedler et al., 2010, 2012). Similar needle-fiber calcite is known from paleosols in Mississippian limestones of South Wales, where it occurs as coating on sediment grains and as rhizoliths (Wright, 1986). Individual needles are a few micrometers wide and up to several hundred micrometers long. They can occur in desiccation cracks, fractures, and other spaces in modern and fossil soils. Stable isotopic composition of needle-fiber calcite from Quaternary paleosols in travertine deposits from Hungary provides an additional source of data that confirms that these needles were of biological origin (e.g. Bajnóczi and Kovács-Kis, 2006). Whewellite, a major mineral of calcium oxalate in the plant kingdom (also found in mammalian kidney stones) occurs in multiple morphologies including prisms and radially aligned crystals known as druses. Such crystals are also produced by fungi and lichens (Stephens, 2012).

Microcodium is a name used for a calcitic feature of solitary or aggregated elongate grains with rounded ends, some up to 1 mm in size, which were once interpreted as some type of alga (Figure 12.20) (e.g. Wood and Basson, 1972; Mamet et al., 1987). These calcification structures are known from the Paleozoic to the Neogene and it has also been suggested that they represent various forms of fungi or fungal interactions (e.g. Durand, 1962; Richard, 1967; Klappa, 1978; Košir, 2004). A number of specimens appear to have thin filaments up to 3 μm in diameter associated with some of the aggregates. A more recent interpretation suggests that *Microcodium* represents some type of biologically induced mineralization, the result of a saprotrophic microorganism or some type of microbial association (Kabanov et al., 2008).

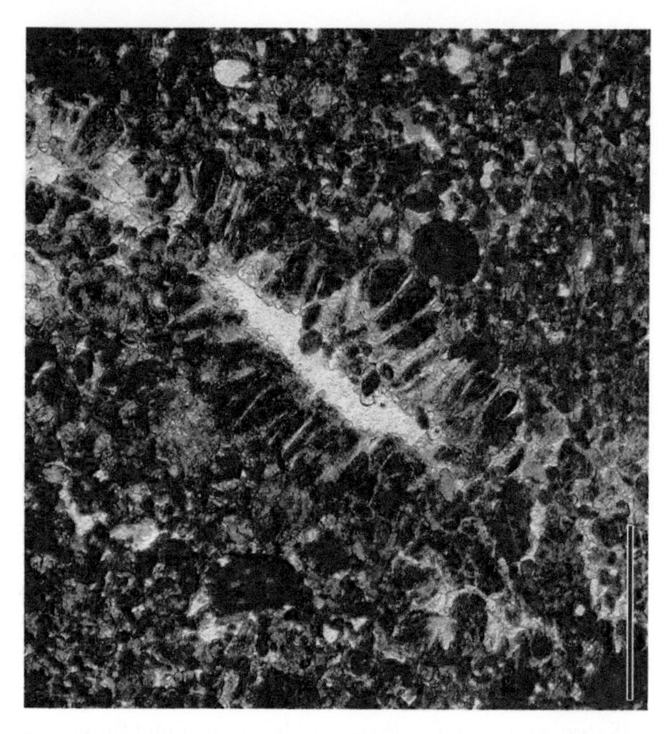

FIGURE 12.20 Thin section of *Microcodium* pattern, interpreted as being the result of some type of microbial or fungal association. Pennsylvanian, Russia. Bar = 1.0 mm. (Courtesy P. Kabanov.)

SUBSTRATE BORING

Fungi and other organisms can be physically associated with rocks and biogenic hard parts (e.g. shells, bones) in multiple ways. If they are attached to the external surface of a hard substrate they are termed epiliths, while organisms like fungi and cyanobacteria that actively penetrate rocks, principally carbonates and phosphates, are termed euendoliths (Golubic et al., 1981). In addition, there are other types of endoliths (e.g. cryptoendolith, chasmoendolith) that are defined by the type of space in the substrate or rock they colonize (Golubic et al., 1981; Ekdale et al., 1984). In this capacity they may function as parasites or various kinds of mutualists. They possess organic acids that allow for penetration of the substrate, including those substrates with organic lamellae, such as shells (Golubic et al., 2005). Today they are found in multiple types of environments ranging from shallow to deep water, and within terrestrial and oceanic crusts; together with algae, they are major inhabitants of corals (Bentis et al., 2000; Kolodziej et al., 2008). Endolithic microorganisms have an extensive geologic history ranging from the Archean (Furnes et al., 2004) onwards and geologically have played a major role in the destruction of rocks (e.g. Jehu, 1918; Knoll et al., 1986), especially carbonates (Zhang and Golubic, 1987). As to exactly why these organisms evolved to inhabit hard substrates remains an interesting question, with some speculating that it is related to protection against sunlight, grazers, and/or desiccation (e.g. Kobluk and Kahle, 1978; Tribollet, 2007).

There are estimated to be more than 50 different species of fungi and fungus-like organisms that have the capacity to penetrate rock, and some of these are regarded as living fossils because of their long stratigraphic ranges. The oldest evidence of boring is etching of glass found in Mesoarchean pillow lavas from the Barberton Greenstone Belt in South Africa (e.g. Furnes et al., 2004, 2008); both fungi and cyanobacteria are known to etch glass today (e.g. Gorbushina and Palinska 1999; Furnes et al., 2008 and papers cited therein). It is believed that the first microborings were produced by cyanobacteria sometime in the Precambrian, with rock-penetrating chlorophytes and fungi evolving by the Ordovician, and microboring members of the Rhodophyta in the Ordovician or Silurian (Glaub and Vogel, 2004). Evidence suggestive of Peronosporomycetes (formerly Oomycetes) having the capacity to penetrate hard substrates appears first in the Triassic. Various types of endoliths, including fungi, are known from marine shells in the Late Ordovician and have been used to reconstruct multiple zonation levels in the substrate related to light levels (Vogel and Brett, 2009). Microborings have also been described from early Cambrian phosphatic and phosphatized fossils (e.g. Zhang and Pratt, 2008; Figures 12.21, 12.22). Other endoliths are known from Middle Devonian volcanic rocks, where they apparently lived in water-filled cavities (Peckmann et al., 2008).

Many types of microborings can be precisely characterized, identified, and assigned to a particular group of producers based on size, type of branching, and growth direction, among other structural features. Some of these microborings represent valuable ecological indicators (e.g. Zeff and Perkins, 1979; Chazottes et al., 2009). As a result, microborings have been identified in the fossil record, described as trace fossils, and used as proxy indicators in paleoecology (see below); however, it is important to point out that there are also abiotic microtunnels and contamination features present in various rock substrates that can be easily mistaken for some type of endolithic microorganism (McLoughlin et al., 2010). Since they were first reported (Bornet and Flahault, 1889), there have been numerous accounts of fungus-like filaments. They have been noted in a variety of substrates such as sponge spicules (e.g. Kölliker, 1860), fragments of mollusk shells (e.g. Zebrowski, 1936), algae (e.g. Chapman, 1913), corals (e.g. Kobluk and Risk, 1974; Raghukumar and Raghukumar, 1991, 1998), foraminifera (Nielsen et al., 2003), and ostracodes (Porter and Zebrowski, 1937).

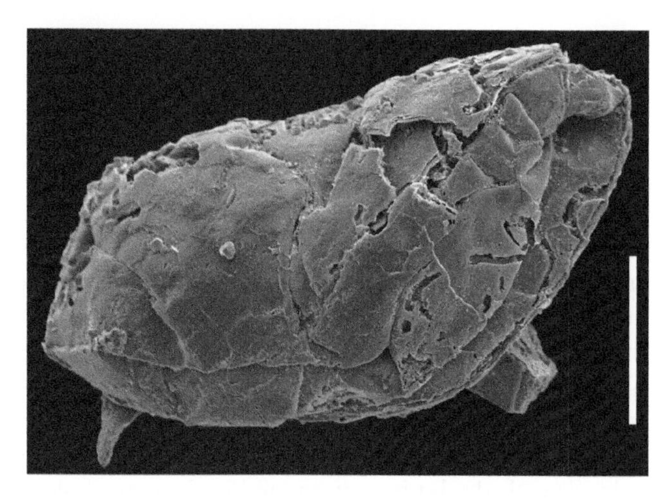

FIGURE 12.21 Microborings on fecal pellet. Bar = 100 μm. Lower Cambrian, South China. (From Zhang and Pratt, 2008.)

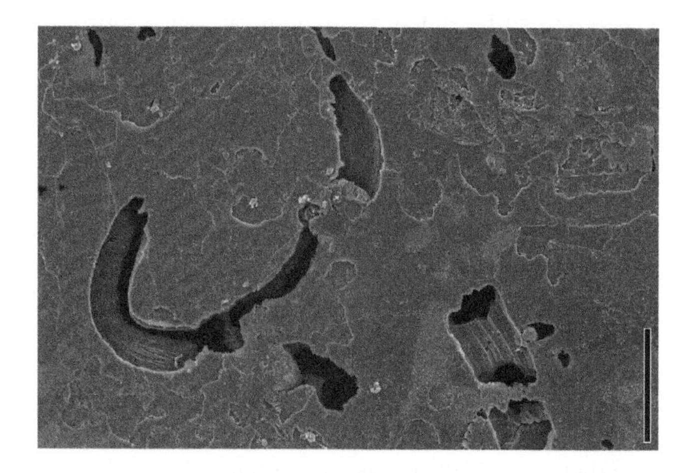

FIGURE 12.22 Microborings on phosphatized arthropod (bradoriid) carapace. Bar = 100 μm. Lower Cambrian, South China. (From Zhang and Pratt, 2008.)

An important issue when dealing with certain types of microborings is whether some of the so-called fungi in shell fragments are in fact fossils or simply represent extant organisms using a fossil shell as a substrate (Cavaliere and Alberte, 1970). Another major problem, which has been perplexing for a long time, is in determining whether the endolithic filaments were produced by an alga or a fungus (e.g. Bornet and Flahault, 1889; Etheridge, 1899). One suggestion is that the boring patterns produced by algae and fungi are significantly different, with fungi more common within the organic lamellae of certain shells, and algae tending to be most conspicuous in the mineral portion of the shell (e.g. Schneider 1976). It also is important to understand that various environmental factors (such as water temperature, light intensity, water depth, and whether oxic or anoxic) affect the organisms that are boring into shells, and that these physical factors may alter organism behavior and the ultimate evidence used to demonstrate their existence (e.g. Gaspard, 2011).

In general, the borings of endolithic algae are 1–4 µm in size, are typically straight or curve slightly, and show evidence of reproductive cells. Traces that branch dichotomously and have a tapered end are consistent with borings produced by endolithic fungi in carbonate substrates (Golubic et al., 2005). Moreover, fungi proliferate within organic lamellae in various directions, whereas endolithic algae extend through the mineral portion of shells and are confined to certain crystallites. Golubic et al. (2005) suggested that endolithic fungi possess enzymes capable of digesting organic matter incorporated in certain shells that are absent in the case of endolithic phototrophs. Evidence from modern corals indicates that the morphology and size of the tunnels differs among euendolithic, cryptoendolithic, and reproductive phases and, if dichotomies are present, whether they are parallel or three dimensional (Priess et al., 2000). An excellent example is *Abeliella*, a microborer that was initially interpreted as fungal in origin based on the dichotomous branching present (Mägdefrau, 1937). It is now known, however, that dichotomous branching can also result from microboring cyanobacteria and certain chlorophytes (Radtke et al., 2010). Another research avenue is to use the water depth at which the organisms were fossilized as a measure to determine whether the borings were produced by algae or fungi. Algae have a limited depth at which they can grow, since they need light to photosynthesize, whereas the heterotrophic fungi do not have this limitation. In other instances morphological features alone do not provide sufficient resolution to characterize various types of putative microorganisms in rocks. For example, pillow basalts collected from the eastern North Atlantic Ocean contain numerous microorganisms that were detected using optical microscopy, confocal laser scanning microscopy, and confocal Raman microscopy; however, the precise affinities of the microorganisms could not be determined (e.g. Cavalazzi et al., 2011).

One of the major obstacles in studying endolithic organisms is the difficulty in preserving the integrity of the organism and spatial relationships between parts of the same organism, in this case fungal hyphae. Using casting and embedding techniques for cases in which the voids formed by the fungus are hollow, resins are allowed to infiltrate the borings and solidify (Golubic et al., 1970). Increasing resin penetration in the borings can be enhanced under vacuum. Once the resin solidifies, the pieces of rock can be dissolved in the appropriate acid to reveal casts of the fungal borings. Details of these casts can then be examined via SEM. In other instances the activities of the boring organisms can be detailed using thin-section preparations, which in some cases may contain infillings of secondary mineral(s).

Taxa

Generic names have been applied to some of the filaments and filament traces that extend through bivalve shells. One of these is *Stichus*, from the Cretaceous of Australia, a taxon that consists of chains of opaque spheres extending through the growth layers of the shell (Etheridge, 1904). Others, like *Dodgella priscus*, consist of what are interpreted as sporangia that range up to 34 µm in diameter. Pores open at the shell surface, where there are also multiple hyphae (1.5 µm in diameter; Zebrowski, 1936). Although the biological affinities of *Dodgella* remain equivocal (endobiotic chytrid?), it is probable that these structures represent a heterogeneous group of organisms rather than a single taxon (Vogel et al., 1987). Fossil endolithic borings in ostracode valves are known from the Cambrian onwards. In the Ordovician of Poland (Olempska, 1986), these microborings are tubes 9–13 µm in diameter, some with terminal swellings up to 40 µm wide. Interestingly, the microborings in these fossils fall outside the diameters generally attributed to endolithic fungi and may have been produced by some type of alga (e.g. Golubic et al., 1975). Fungal traces have also been reported from Pliocene gastropod shells from the seafloor of the Atlantic Ocean (Kaim et al., 2012). In certain areas of these shells are fungal traces that are attributed to *Saccomorpha clava* (Radtke, 1991), an ichnospecies (Ordovician) of borings that are 10–30 µm in diameter and terminate in club-shaped structures 10–30 µm long. Additional specimens are recorded from the Triassic (Schmidt, 1992), Jurassic (Glaub, 1994), and Cretaceous (Hofmann, 1996). According to Wisshak (2006b), *Dodgella priscus* is the tracemaker of *S. clava* traces (Figure 12.23). *Dodgella*, a putative member of the

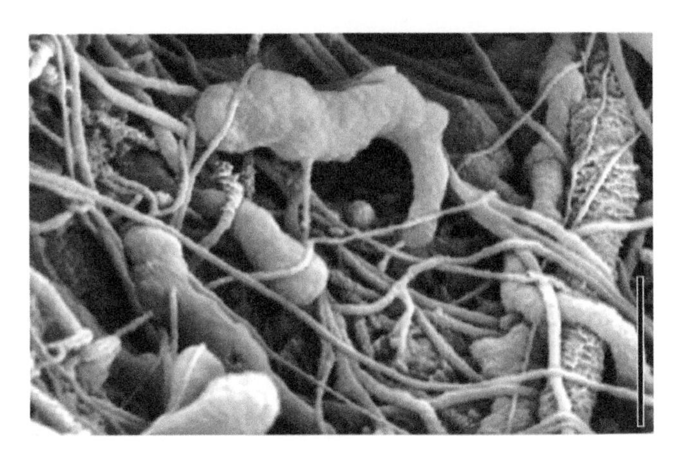

FIGURE 12.23 Ichnospecies *Saccomorpha clava* produced by the fungus *Dodgella priscus*. Extant. Bar = 20 µm. (From Chazottes et al., 2009; courtesy V. Chazottes.)

Chytridiomycota (Zebrowski, 1936), is used as an index taxon of the *Saccomorpha clava–Orthogonum lineare* ichnocoenosis, which is regarded as indicative of fossil and extant open marine aphotic environments (Müller and Löffler, 1992; Glaub, 1999; Wisshak et al., 2005). The tracemaker for *Orthogonum lineare*, an ichnotaxon based on the pattern of gallery branching, is not known.

In some cases the presence of multiple types of ichnotaxa provides support for the idea that some of the morphotypes represent different ontogenetic stages of a single fungal colony or evidence of multiple types of organisms that bioeroded a single type of substrate (e.g. Bromley, 2005). Others extending from the Cambrian are thought to represent members of the Chytridiomycota. Unnamed forms believed to have been produced by fungi have also been reported from the Ordovician of northern Estonia (Podhalańska and Nõlvak, 1995).

Palaeachlya perforane is the name initially used for microborings in Silurian and Devonian brachiopod shells and certain Cenozoic corals (Duncan, 1876). Similar borings have been reported from corals from Carboniferous–Permian rocks from New South Wales, Australia, which consist of ramifying tubes about 20 µm in diameter (Etheridge, 1891). Some of the microborings dichotomize, and some bear small swellings at their tips. At the end of some of the tubes are spheroidal structures that were thought to be spores, but were subsequently determined to be iron pyrite (Etheridge, 1891). The fossils were cautiously compared with the Peronosporomycetes (Smith, W.G., 1877, 1884 but see Chapter 13, p. 272). Tubes of various types that are preserved in thin sections of the fossil bryozoan *Nicholsonella* are given the name *Ordovicimyces* (Elias, 1966). Rather than being fungal, these structures are interpreted as intrazooecial spines (Corneliussen and Perry, 1970).

Paleoecology

Borings in fossil invertebrate shells have been useful as biostratigraphic and paleoecological proxy records to understand environmental conditions. For example, numerous belemnites, an extinct order of Mesozoic cephalopods, were studied using thin sections and cathodoluminescence microscopy (Reolid and Benito, 2012). Various morphological microborings were encountered in the Jurassic specimens, including non-branching traces that morphologically resemble fungal hyphae. The presence of these fungal borings was used to suggest the order of colonization for the various microborers, with fungi being the dominant type in the mid-shelf environments.

Comparisons between associations of endolithic microorganisms today and those found as fossils (so-called ichnocoenoses) can be used to reconstruct a variety of paleoecosystem parameters such as paleobathymetry (the study of water depths in the geologic past; e.g. Glaub 1999, 2004; Chazottes et al. 2009; Vogel and Brett, 2009), turbidity (e.g. Vogel et al. 1987), and organic erosion (e.g. Perkins and Halsey 1971; Vogel et al. 2000). Characterizing the types of microborers and their vertical distribution in the water column is becoming an increasingly important proxy record of paleobathymetry since heterotrophic borers, unlike photosynthetic microorganisms, do not require light and thus the depth can be gauged (e.g. Glaub and Bundschuh, 1997; Golubic et al., 1984; Vogel et al., 1995). The study of bioerosion is especially important as it can provide a useful tool to monitor environmental conditions of deep water corals (e.g. Beuck and Freiwald, 2005).

Last but not least, it should also be noted here that some of the holes in various types of rocks can be caused by other means that are of abiotic origin, including chemical surface reactions along crystal deficiencies (e.g. Sverdrup, 2009).

13

BACTERIA AND FUNGUS-LIKE ORGANISMS

BACTERIA ... 261
Fossil Bacteria ... 262
ACTINOMYCETES ... 265
Fossil Actinomycetes ... 265
MYCETOZOA ... 267
Fossil Mycetozoa ... 267
PERONOSPOROMYCETES ... 268
Fossil Peronosporomycetes ... 269
PERONOSPOROMYCETES: CONCLUSIONS ... 282

Numerous microorganisms resemble fungi to varying degrees, with regard to both morphology and biology, but yet do not belong to that kingdom. Some of these have a fossil record and thus are briefly discussed here.

BACTERIA

Despite their small size, the various types of organisms that are classified as bacteria are the dominant life form of the biosphere in terms of numbers and physiological diversity, and the oldest organisms on Earth. It is estimated that fewer than 10,000 have been identified based on molecular tools and there are many more yet to be discovered. Bacteria occupy every conceivable ecosystem and habitat, including soil and water, and are found within the subsurface of the continents and oceans. They can live in extreme environments and be preserved as fluid inclusions in various types of hydrothermal environments (Ivarsson et al., 2009, 2010; Sankaranarayanan et al., 2011); some are even found in radioactive waste (e.g. Cologgi et al., 2011; Orellana et al., 2013). In certain types of aquatic ecosystems the surface of the water is covered by bacteria that form biofilms, a combination of the bacteria, their mucilaginous secretions, and other organisms that become associated with the biofilm (e.g. Donlan, 2002). Some of these, especially certain types of cyanobacteria, may grow more extensive in thickness and develop into mats. Bacteria are capable of degrading organic matter as well as assimilating inorganic nutrients from the environment, and they can obtain nutrients from living fungi, a strategy termed bacterial mycophagy (Leveau and Preston, 2008). They also play an important role in mineral precipitation during the preservation of structures such as bones (e.g. Daniel and Chin, 2010). Bacteria today are critical in many forms of nutrient cycling and are the major agents of biogeochemical transformations. Some of these include metabolizing organic matter and oxidization of methane and ammonia. Because bacteria require liquid, they have evolved various adaptations to guard against desiccation, including the production of certain types of spores, a mycelial type of vegetative phase, and special types of life histories. Bacteria are involved in various symbiotic associations (e.g. nitrogen fixation and chemolithotrophy).

T.N. Taylor, M. Krings & E.L. Taylor: Fossil Fungi.

DOI: http://dx.doi.org/10.1016/B978-0-12-387731-4.00013-X

FOSSIL BACTERIA

Although modern bacteria are classified using a variety of molecular tools, the group has an excellent fossil record that extends back to the Archean and includes structurally preserved cyanobacteria that are dated at 3.7 Ga. The major morphological forms of extant bacteria, including coccoid, bacilloid, rod-shaped, spiral, and filamentous, are known from the fossil record (e.g. Westall et al., 2006). When detailing the fossil record of bacteria in the following sections, we have excluded members of the cyanobacterial phylum that encompasses the oxygenic photosynthetic prokaryotes (Shih et al., 2013), which have an extensive record (Taylor, T.N. et al., 2009).

Most bacteria reported from the fossil record occur in association with plants and plant parts, various types of animals, and coprolites, and in mineral matrices. Bacteria were one of the main types of microbial life form interpreted in a variety of rocks, especially Carboniferous coals (e.g. Van Tieghem, 1879; Renault, 1894b, 1895c, 1896a–d, 1897, 1899, 1900, 1901; Figure 13.1). Many of these microfossil structures were given generic and specific names (e.g. *Bacillus agilis* and *Micrococcus paludis*; see Renault, 1895c, 1900) and compared with extant bacteria of various types (e.g. Renault, 1894b, 1895c, 1896b, 1900). Although some of these structures probably represent crystals and metamorphosed organic matter, and some are modern bacteria that may have contaminated samples (e.g. Farrell and Turner, 1932), others may in fact represent examples of fossilized bacteria. A brine inclusion within a 250-Ma-old salt crystal from the Permian Salado Formation, New Mexico, USA, has yielded a halotolerant, spore-forming bacterium, which was named *Bacillus permians* and believed to represent the oldest evidence of viable microorganisms in geological salt formations (Vreeland et al., 2000, 2006). Other researchers, however, have contested the antiquity of *B. permians* (Graur and Pupko, 2001). A number of other structures that have been described as fossil bacteria may represent propagules (e.g. spores, conidia) of organisms such as fungi or algae (e.g. Lyons, 1991). With the emergence of the discipline of geobiology, it has been demonstrated that some of these bacteria-like forms represent biomimetic structures. Although paleomycological research generally does not focus on bacteria, increased attention to these organisms in the fossil record represents an important convergence with several disciplines in geomicrobiology, especially in those cases where there are symptoms of disease, decay, and decomposition (e.g. Edwards et al., 2006). There are various reports of bacteria in coal, pyrite, and permineralizations (coal balls) that remain equivocal (e.g. Ehlers et al., 1965). They typically consist of spherical bodies, sometimes directly associated with plant material, such as within the lumen of a vascular plant conducting element (e.g. Lyons, 1991). Damage by bacteria has also been reported on plant cuticles from the Permian of India (Srivastava and Tewari, 1994).

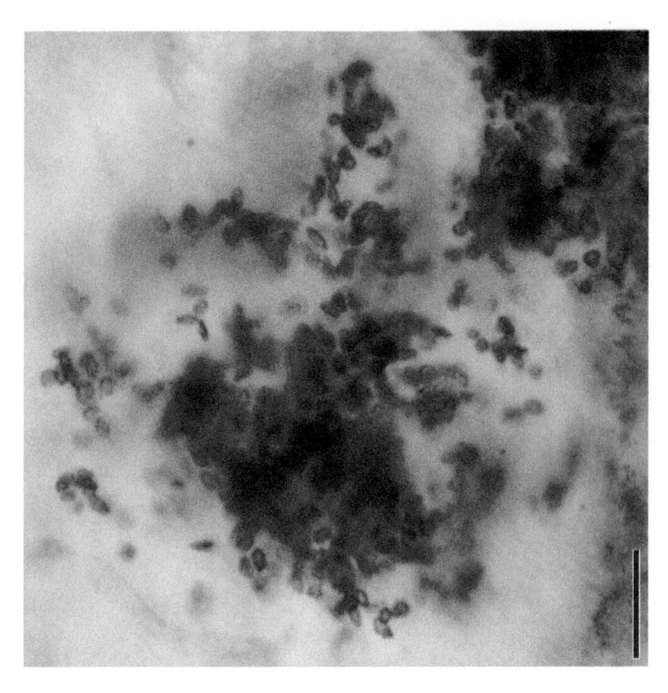

FIGURE 13.1 Cluster of bacteria in *Lepidodendron* tracheid. Mississippian, France. Bar = 10 μm. (Courtesy N. Dotzler.)

Tonsteins are thin, distinctive bands of volcanic ash that have been altered to kaolinite (Spears, 2012) and are associated with some coal seams. Because tonsteins were initially used to date such seams, they were examined both for pollen grains and spores and for other evidence of the coal-forming flora of the time. The presence of tonsteins is still used in the exploration for coal deposits but is now used in facies analysis and sequence stratigraphy as well. Various microorganisms have been reported from these deposits, including filaments, clusters of coccoid cells interpreted as bacteria, and other structures thought to be fungal (Moore, L.R., 1964). A number of organisms considered to be bacteria have been reported from oil shales of Scotland (e.g. Moore, L.R., 1963a). Some of these have been placed in *Micrococcus*, a genus initially established by Cohn (1872; see Stackebrandt et al., 1995) for certain extant bacteria but also used by Renault (1895c, 1896b) for fossilized coccoid to ovoid-shaped cells that were

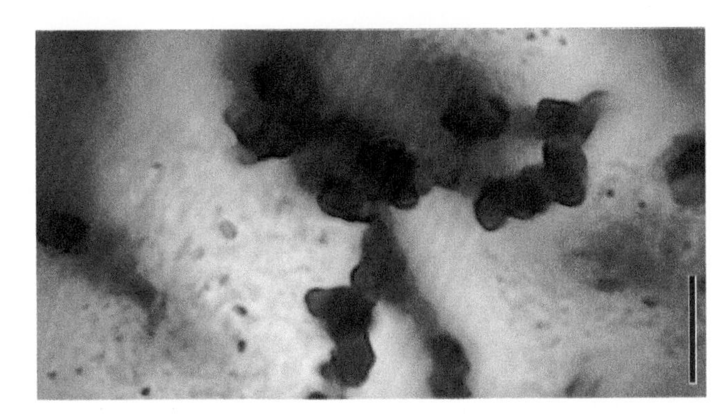

FIGURE 13.2 Chains of spheroidal cell contents, which represent abiotic biomimetic structures, in conducting cells of a Carboniferous fern (*Botryopteris tridentata*). Middle Pennsylvanian, Kentucky, USA. Bar = 5 µm. (Courtesy A.A. Klymiuk.)

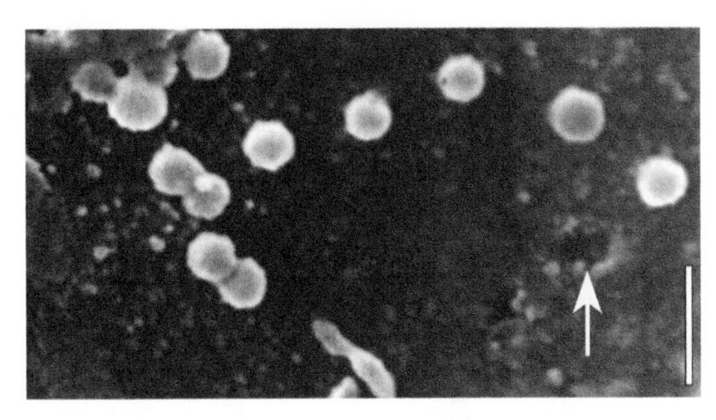

FIGURE 13.3 Spheres of *Annella capitata* interpreted as bacteria (?) on the surface of a pollen grain (*Exesipollenites*). *Arrow* is potential infection cavity. Upper Jurassic, Dorset, Great Britain. Bar = 2 µm. (From Srivastava, 1976.)

interpreted as bacteria, while other types were included in *Polymorphyces*. Today a number of these are known to represent biomimetic structures (Figure 13.2).

While there have been several reports of various types of fossil bacteria (e.g. Blunck, 1929; Katinas, 1983; Poinar, 1992; Waggoner, 1993; Wainwright et al., 2009), the challenge will always be to determine the fossil nature of these organisms (Richter, 1994) or whether some represent modern contaminants. Confirming the biological identity of structures like fossil bacteria is even more challenging because the features suggestive of organic decay are similar in both bacteria and fungi. Nevertheless, it is important to report such structures in the fossil record regardless of their association with other fossils (e.g. Schmitz-Münker and Franzen, 1988) and to devise additional techniques and protocols to determine precisely what they represent. Quite a number of fossil bacteria have been reported from amber of various ages (e.g. Czeczott, 1961; Waggoner, 1996; Schmidt and Schäfer, 2005; Schmidt et al., 2010a). For example, bacterial isolates that biochemically resemble *Staphylococcus* have been extracted from Dominican amber (Lambert et al., 1998). One study even reports that a bacterium isolated from the abdominal contents of an extinct bee (*Proplebeia dominicana*) preserved in Dominican amber was successfully revitalized and cultured (Cano and Borucki, 1995). Enzymatic, biochemical, and DNA profiles have been used to infer that this Miocene bacterium is closely related to the extant *Bacillus sphaericus*. Other studies, however, have questioned the results published by Cano and Borucki (e.g. Yousten and Rippere, 1997). A subsequent study reports on *luxS* sequences in bacteria isolated from Dominican amber that are distinct from *luxS* sequences in modern bacteria (Santiago-Rodriguez et al., 2013). There are also several other accounts of successful isolation of bacteria from amber and subsequent detailed characterization of these fossils by molecular techniques (e.g. Hamamoto and Horikoshi, 1994; Greenblatt et al., 2004). One important area of research in paleomicrobiology will involve the examination of these structures with sufficient resolving power or other applications to determine whether they do in fact represent bacteria or are some type of crystal that formed during the process between resin flow and fossilization.

Evidence of bacterial degradation is known from pollen grains and spores in which the surface morphology and sculpture are altered in various patterns. In some examples the central body of some saccate pollen grains is partially destroyed, while in others the surface is highly degraded, often represented only by the thicker ornament (e.g. Moore, L.R., 1963b). Examination of some Jurassic grains revealed spherules (0.5–1.5 µm in diameter) on the surface (Figure 13.3), each individually attached and believed to have developed from hyphae (Srivastava, 1976). The study described these forms under the name *Annella* and interpreted them as evidence of either bacterial or fungal infection. Similar structures have been noted on pollen from the Lower Cretaceous (Elsik, 1971) and megaspores from the Jurassic of Poland (Marcinkiewicz, 1979). One of these, *Reymanella globosa*, is spherical to oval and up to 80 µm in diameter, constricted at the apex and base, and attached to the surface of the spore wall. The possibility exists, however, that some of these structures represent artifacts caused by preparation procedures used in the scanning electron microscopy analysis.

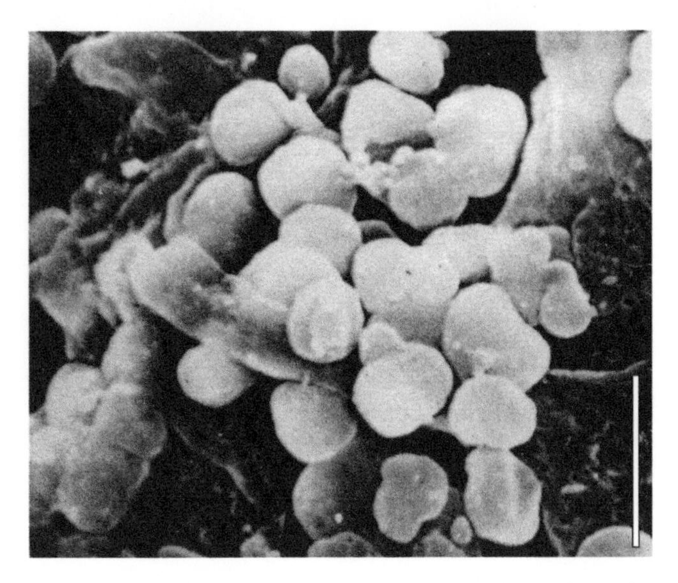

FIGURE 13.4 Bacterial or fungal spores on the surface of a lycopsid megaspore. Triassic, Shaanxi, China. Bar = 5 μm. (From Wang, 1991.)

There have also been reports of fossil pollen grains and spores in which the wall of the grain is differentially thinned (e.g. Havinga, 1971, 1984; Tipping, 1987). Several types of exine disruption in the form of microbial degradation have been reported from pollen and spores from the Carboniferous (e.g. Elsik, 1971) as well as from Cretaceous pollen grains (Skarby and Rowley, 1993). Using transmission electron microscopy, sections of pollen wall show various microchannels extending through the exine, some branching and in the range of 15–600 nm in diameter. Whether this tunneling in pollen and spore walls represents a chemical change of the wall layer or the result of some bacterial or other microbial agent remains unknown (e.g. Rowley et al., 1990), although Elsik (1966a, b) attributed these features in Cretaceous and Cenozoic pollen and spores to fungi or fungus-like organisms. Similar channels have been observed in Cretaceous cuticles and interpreted as evidence of fungal enzymes on cutin (Bajpai, 1997). Degradation of the spore wall has also been reported in Triassic megaspores of the lycopsid *Isoetes* (Wang, 1991). In these spores, small coccoid structures occur on the outer surface of the megaspore wall (Figure 13.4) and there are also microchannels within various layers of the sporoderm.

Fossil Bacteria as Endosymbionts

Bacteria represent a critical component of modern mycorrhizal systems, where they promote development (e.g. Garbaye, 1994a,b) and play a significant role in the functioning of symbioses (e.g. Artursson et al., 2006; Castillo and Pawlowska, 2010). There remain many questions, however, about the interrelationships between the bacteria, fungus, and host plant, and whether the rhizosphere microflora has a positive or negative impact on the total mycorrhizal symbiosis (Frey-Klett and Garbaye, 2005). Some of these bacteria have been termed mycorrhiza helper bacteria (MHB; Garbaye, 1994b; Tarkka and Frey-Klett, 2008) and it has been suggested that their role may be in stimulating mycelial growth, increasing root permeability, inhibiting pathogenic fungi, increasing root branching, altering physiochemical properties of the soil, and aiding in spore germination (Frey-Klett et al., 2007).

In other extant arbuscular mycorrhizal (AM) fungi, there are certain endosymbiotic, Gram-positive bacteria (initially termed bacteria-like organisms, BLOs) that are now known to play a major role in mycorrhizal functioning (e.g. Minerdi et al., 2002; Naumann et al., 2010). These bacteria are located in the fungal cytoplasm and can alter fungal interactions with host plants in many ways because they are vertically transmitted through generations of the fungus. Some are widespread, while others are specific to particular AM fungi (e.g. Bianciotto et al., 2000; Bonfante and Anca, 2009). There is also evidence that it is the fungus that selects for the most efficient bacterial strain (Pivato et al., 2009). One Gram-negative type, named "*Candidatus* Glomeribacter gigasporarum," an endocellular bacterium of the *Burkholderia* lineage, may contribute to host fitness through its ability to synthesize vitamin B12, antibiotics, and certain toxin-resistant molecules (Ghignone et al., 2012). For example, in one set of experiments, the presence of the bacterium *Paenibacillus validus* stimulated growth in the form of highly branched hyphae, coiled structures, and smaller spores in the AM fungus *Glomus intraradices*. These changes did not, however, result in the germination of the newly formed spores in the absence of the host (Hildebrandt et al., 2002, 2006). In addition, it has been demonstrated that the presence of "*Ca.* Glomeribacter gigasporarum," improves hyphal growth after spore germination and prior to root colonization (Lumini et al., 2007; Anca et al., 2009). In this capacity the endobacteria modify lipid metabolism within the spores and increase germination and hyphal development. *Glomeribacter* have been reported as surviving for several weeks outside the fungal host (Jargeat et al., 2004).

Fossil Bacteria in the Rhynie Chert

Molecular phylogenetics suggests that the BLO–AM fungus symbiosis is an ancient one and that this tripartite

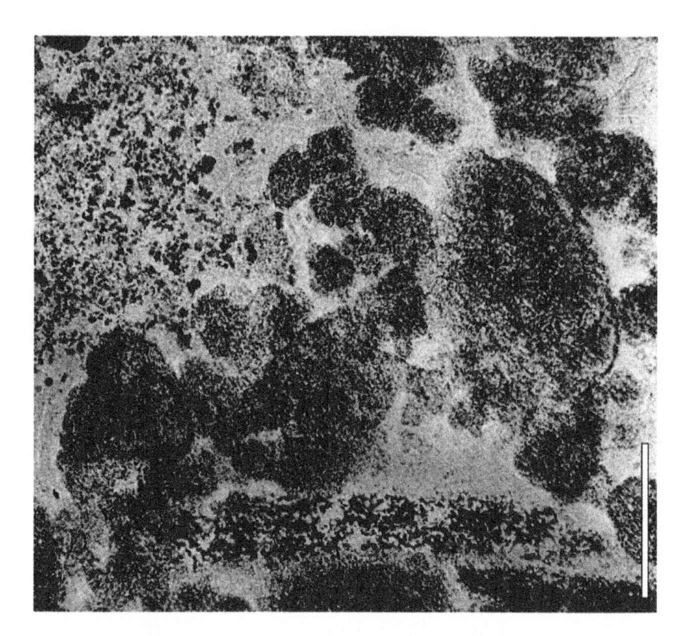

FIGURE 13.5 Thin section of degraded plant material in Rhynie chert showing colonial masses of what are interpreted as dense aggregations of unicellular bacteria. Lower Devonian, Scotland. Bar = 25 μm. (From Kidston and Lang, 1921b.)

association dates back to the time when AM fungi first colonized land plants. As research dealing with this multi-phylum association proceeds with living mycorrhizal systems, it will be worthwhile examining some of the fossil glomeromycotan fungi to see if there are any consistent remains that might be demonstrated as bacterial. It is estimated that extant spores of *Gigaspora margarita* contain approximately 20,000 bacteria, which perhaps increases the likelihood of locating them in fossil spores (e.g. Bianciotto et al., 2004; Cruz and Ishii, 2011). Kidston and Lang (1921b) described some masses of unicellular organisms in the Rhynie chert, which they interpreted as some type of bacterium (Figure 13.5). They were present as clusters in the matrix, rather than in any specific plant or fungus. In may be impossible to resolve whether these structures are in fact some type of endobacterium and, if so, whether they are representatives of the Glomeromycota–*Glomeribacter* symbiosis that is hypothesized to be at least 400 million years old (Mondo et al., 2012).

To date bacteria have not been the focus of Rhynie chert studies or any other silicified ancient ecosystem research; however, the exceptional preservation and the presence of structures that (at least morphologically) resemble bacteria may hold some promise of detecting these organisms within AM fungi (Hass et al., 1994). While conventional transmitted optical systems may not provide sufficient resolution to unequivocally determine the presence of bacteria, there may be other means to confirm whether bacteria of some type were involved in the early evolution of mycorrhizal symbioses (e.g. Bonfante and Anca, 2009).

ACTINOMYCETES

Actinomycetes are high GC (guanine–cytosine) Gram-positive bacteria (e.g. Fox and Stackebrandt, 1987). They may be aquatic or terrestrial and today represent one of the major decomposers of organic matter; some may be pathogens or parasites of various organisms including humans (e.g. Gao and Gupta, 2012). Other actinomycetes are mutualists, including *Frankia*, a nitrogen-fixing form that lives symbiotically with certain living vascular plants (e.g. Sellstedt et al., 1986; Benson, 1988); there is at least one dubious report of *Frankia* from the Pleistocene of Vermont, USA (Baker and Miller, 1980). Some members of this group produce branching, septate filaments that are morphologically like fungal hyphae, although actinomycete filaments are generally smaller. As a result of morphological similarities to some anamorphic fungi, actinomycetes were originally thought to have affinities with the fungi. The use of molecular tools (16S rRNA) has made it possible to place the class Actinobacteria more accurately within a phylogenetic context (e.g. Zhi et al., 2009).

FOSSIL ACTINOMYCETES

The filamentous organization of a number of fossils led to their initial designation as actinomycetes. One of these was described from a middle Precambrian glacial varve (Jackson, 1967) and another from phloem cells of the Pennsylvanian fern *Botryopteris* (Smoot and Taylor, 1983) in which some of the filaments range from 0.5 to 2.0 μm in diameter and appear to be attached to the inner surface of the cell wall. Many branch repeatedly and some have knobs along the surface (Figure 13.6). A reinvestigation of these structures using multiple imaging systems (spinning disc confocal microscopy) has indicated that, rather than actinomycetes, the material in the cell lumina consists of biomimetic structures composed of authigenic carbonate minerals (Klymiuk et al., 2013a). Luminescence mapping suggests that the structures may represent disordered iron-rich dolomites that may have formed as a result of sulfur-reducing bacteria. Also present in some of the cells are spheroidal unicells, 5–8 μm

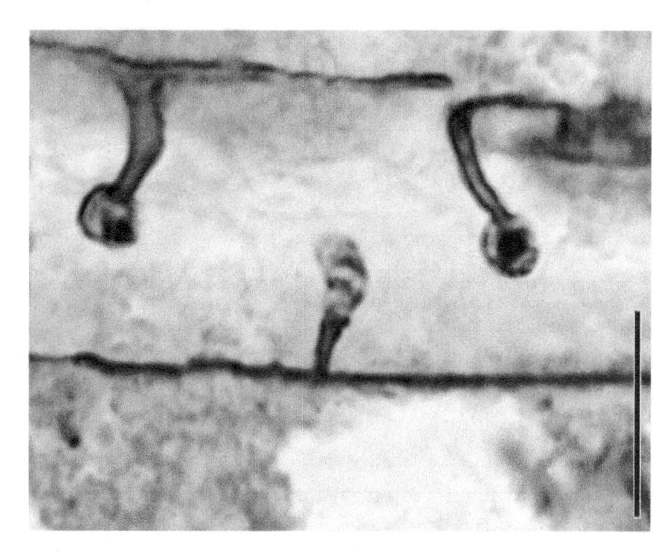

FIGURE 13.6 Intercellular pectic protuberances extending from the inner surface of a cell wall into the cell lumen, Carboniferous fern (*Botryopteris tridentata*). Middle Pennsylvanian, Kentucky, USA. Bar = 10 µm.

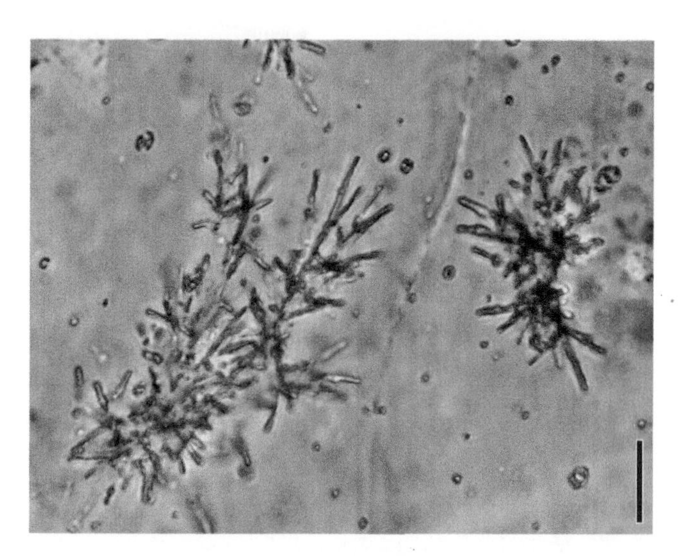

FIGURE 13.7 Mycelia of *Cardonia stellata* showing terminal conidiospores and lateral aleurospores. Upper Cretaceous amber, Ariège, France. Bar = 10 µm. (Courtesy G. Breton).

in diameter, that are now known to represent chytrid zoosporangia (Taylor, T.N. et al., 2009). This study (Klymiuk et al., 2013a) is important for several reasons: first, it demonstrates the potential of confusing biomimetic structures in the fossil record with those thought to be biological in origin, and secondly, it provides an excellent example of the necessity of integrating techniques and aspects of geomicrobiology with studies that define the taphonomy of the fossil.

Several genera (e.g. *Actinomycodium, Mucorodium*) from the Paleozoic of Russia have been suggested as actinomycetes (Zalessky, 1915); however, these may represent anamorphic stages of some fungus. Even in preservation modes that are thought to be extraordinary, the possibility always exists that what are interpreted as fossilized actinomycetes are in fact modern contaminants (e.g. Larsson, 1978).

Nocardioformis dominicanus (Dominican amber) is the name applied to a loose reticulum of filaments like those in modern actinomycetes (Waggoner, 1994). The filaments (1–4 µm wide) are septate and occasionally branch; some assume various shapes and may represent chains of spores. It is hypothesized that the soil-borne spores of these actinomycetes may have been blown onto the fresh surface of oozing amber and actually grew in the resin prior to its being hardened (Waggoner, 1994). *Cardonia stellata* (Figure 13.7) is an actinomycete from Cretaceous amber in France consisting of star-shaped clusters of mycelia in three or four whorls that produce short chains of terminal conidiospores and lateral aleuriospores (Breton et al., 2013). Various morphologies of actinomycetes, together with filaments interpreted as fungal, have also been reported in grains of amber associated with lignite debris from a Cretaceous marine series (Saint Martin et al., 2013).

Fossil actinomycetes have also been reported from vessel elements of dicot wood from the Eocene London Clay Formation (Wilkinson, H.P., 2003). By using scanning electron microscopy on fractured specimens it was possible to identify narrow (<1.0 µm) branching filaments, some of which were coiled. The filaments ramify within the cell lumina and in some instances extend onto the surface of pyrite crystals. Within the ray cells of the wood or on vessel walls are wider filaments that have been interpreted as fungal. These hyphae are up to 2 µm wide, extend through the cell wall and often branch; some specimens show structures that morphologically resemble appressoria. While spore morphology and sculpture, filament morphology and chemistry, coloration of the infected wood, and stages of cell wall degradation are all features that can be used to identify extant actinomycetes and lignin-degrading fungi, these characters are not typically available in fossil woods like those from the London Clay, and thus precise biological identification of the fossils is often impossible. This will be a continuing problem when attempting to determine the biological affinities of similar fossils (Wilkinson, H.P., 2003).

Actinomycites is the name of a putative actinomycete described from Jurassic oolitic limestone from Scotland

(Ellis, 1915). The specimens consist of delicate threads that are highly branched and up to 1.0 μm in diameter. The specimens are poorly preserved and it is difficult to determine their precise affinities. Lower Cretaceous lake sediments have also been described as containing various types of bacteria and actinomycetes (Bradley, 1963). These specimens were interpreted as chains of 2–15 cells; however, after re-examination they were interpreted as fluorite artifacts formed during the preparation procedures and not actinomycete filaments (Bradley, 1968).

Actinomycetes have also been reported from Eocene amber (Waggoner, 1993). The specimens consist of masses of filaments that make up a hyphal mat; at the ends of some filaments are spheroids (spores) that are up to 2 μm in diameter. The fossils are morphologically similar to *Streptomyces*, the largest genus of Actinobacteria that are chiefly found in soil and decaying plant material. Another fossil in amber is *Leptotrichites resinatus* (Schmidt and Schäfer, 2005). It consists of branched filaments 7–10 μm in diameter that are generally swollen at the site of branch formation. In section view the filament lumen is narrow (~ 1 μm) and occasionally filled with chains of rod-shaped cells, some extending from the lumen of the filament. Examination of several morphological features of *L. resinatus* has led to the suggestion that it has affinities with Gram-negative bacteria such as *Leptothrix* and *Sphaerotilus* (e.g. Spring, 2002). Similar structures have also been reported from Albian–Cenomanian (Cretaceous) amber from Kansas (USA) and placed in the modern genus *Leptothrix* (Waggoner, 1996). None of these Cretaceous specimens contained spores within the filaments.

MYCETOZOA

The Mycetozoa (sometimes also called Eumycetozoa) include the cellular (dictyostelid), acellular, syncytial or plasmodial (myxogastrid) slime molds, as well as the protostelids (Stephenson and Stempen, 1994; Baldauf and Doolittle, 1997; Fiore-Donno et al., 2010). Cellular and plasmodial (true) slime molds are characterized by multicellular fruiting bodies that may contain hundreds of spores, while the protostelid forms produce rather simple fruiting bodies consisting of a stalk and one to a few spores. Approximately 900 species of slime molds have been described to date; they are saprotrophs, occur worldwide, and are more or less ubiquitous. Most forms occur in soil or in dead plant materials and contribute to decomposition. As long as there is plenty of food available, slime molds exist as single-celled organisms. When conditions deteriorate, however, they congregate and function as a single organism. In this state they may also reproduce by forming fruiting bodies (sporocarps).

FOSSIL MYCETOZOA

The fossil record of Mycetozoa is exceedingly meager (Keller and Everhart, 2008; Stephenson et al., 2008). A variety of enigmatic Precambrian trace fossils from Australia, Russia, Uruguay, India, and China have been interpreted as traces of the migratory slug (grex or pseudoplasmodium) phase of cellular slime molds. Some of the traces show features such as bulbous ends, interruptions, variable width, narrow ends, and beaded levees that are believed to structurally resemble the slug phase of modern cellular slime molds (Zhuravlev et al., 2009; Chen et al., 2013; Retallack, 2013b). Equally enigmatic, and thus dubious, are fossils from the Carboniferous of France described as *Bretonia hardingheni* and *Myxomycites mangini* and interpreted as slime molds (Renault, 1900).

More persuasive are a variety of fossil spores assigned to the slime molds (Graham, A., 1971), as well as several body fossils of Myxomycetes (true slime molds) preserved in amber. One of the latter is a plasmodium preserved in amber from the Dominican Republic (Waggoner and Poinar, 1992), while another record, formally described as *Stemonitis splendens*, consists of a sporangium and capillitium in Baltic amber (Domke, 1952). While *S. splendens* likely belongs to the subclass Stemonitomycetidae of the Myxomycetes, which are characterized by interiorly stalked sporocarps, two other slime mold sporocarps preserved in Baltic amber can be affiliated with the subclass Myxogasteromycetidae. One of these is an isolated sporocarp that has been described as *Arcyria sulcata* (Dörfelt et al., 2003; Figure 13.8). This fossil is ~0.9 mm high and consists of a stalk and cupuliform base of the peridium (cup), a part of the elastic capillitium with capillitial threads forming net-like anastomoses, and a few spores. The spores are globose, 6–7 μm in diameter, and ornamented. The second fossil consists of several exquisitely preserved sporocarps attached to a plant fragment. The sporocarps have been described as *Protophysarum balticum* (Dörfelt and Schmidt, 2006); each consists of an apically tapering stalk (160–180 μm long) and a nearly globose spore-bearing capitulum (sporotheca) that is up to 105 μm in diameter (Figure 13.9). Spores are recognizable within the sporocarps, whereas evidence of a capillitium and a

FIGURE 13.8 Heinrich Dörfelt.

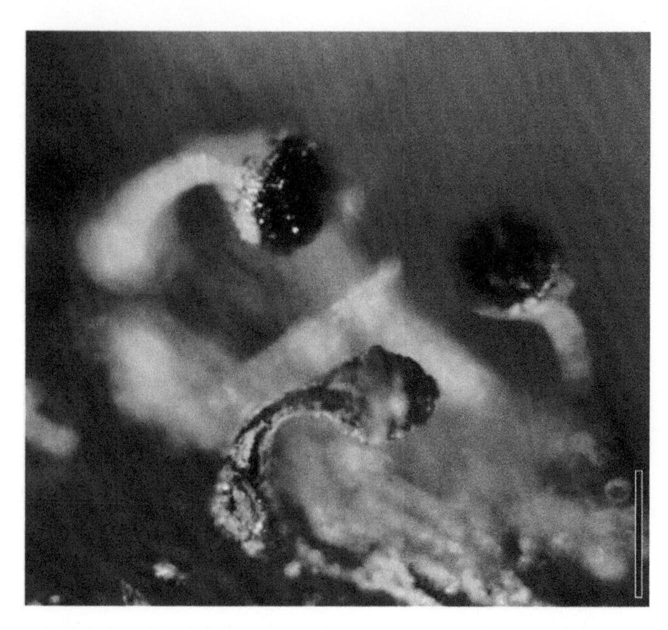

FIGURE 13.9 Three sporocarps of *Protophysarum balticum*. Eocene, Baltic. Bar = 100 μm. (Courtesy A.R. Schmidt.)

central columella is lacking. Spores are globose, 6–13 μm in diameter, and possess a finely punctate, rough surface. *Protophysarum balticum* is very similar morphologically to the extant *P. phloiogenum* (Physarales).

Identifying microorganisms such as bacteria, actinobacteria, and slime molds in the fossil record will always pose a challenge until sufficient techniques have developed to record some type of biomarker or other molecular signature. Nevertheless, at some point it may be possible to identify patterns within fossil ecosystems, or indirect evidence of association with organisms and/or various structures, that signal the presence of these organisms in geologic time. They were certainly present in past ecosystems, and it will become the responsibility of the next generation(s) of paleomycologists and paleobiologists to develop the appropriate methods to reveal their existence.

PERONOSPOROMYCETES

The Peronosporomycetes (also called Peronosporomycota, or formerly Oomycota or Oomycetes; see David, 2002) are heterotrophic eukaryotes that were at one time included in the Fungi, on the basis of similarities in overall morphology and physiology (Dick, 2001). Today, however, they are thought to represent close relatives of the chromophyte algae such as the Bacillariophyta (diatoms), Phaeophyceae (brown algae), and other heterokont protists including the Hyphochytridiomycota, among other groups (Gunderson et al., 1987; Porter, 2006; Rossman and Palm, 2006; Beakes and Sekimoto, 2009). Adl et al. (2005) place the Peronosporomycetes in the Chromalveolata and, within this supergroup, the kingdom Stramenopila or stramenopiles (alternative names in use are Straminipila and Chromista; see Beakes and Sekimoto, 2009).

Peronosporomycetes are distinguished from the true Fungi by the possession of motile zoospores with a flagellar apparatus composed of two different flagella (biflagellate, with the flagella of unequal length [anisokont] and of different morphologies [heterokont]), with basal bodies or kinetosomes, a transitional zone between these regions, and a root that anchors the flagella (Lévesque, 2011). Moreover, peronosporomycetes have oogamous sexual reproduction with gametangial meiosis (except the basal taxa; see p. 271), tubular mitochondrial cristae, and cell walls containing cellulose and the amino acid hydroxyproline (Dick, 2001).

The Peronosporomycetes today comprise approximately 500–800 species that thrive in both aquatic (marine and freshwater) and terrestrial environments, where they function as saprotrophs and disease causative agents in plants (including algae), animals, and even humans (e.g. Margulis and Schwartz, 1998; Jiang and Tyler, 2012). Within the group are economically important and biologically sophisticated phytopathogens, including the root-rotting fungi and downy mildews (e.g. Kamoun, 2003; Govers and Gijzen, 2006; Bozkurt et al., 2012). One form, *Phytophthora infestans*, significantly influenced human history as the causative agent of the disease called late blight or potato blight. An epidemic of late blight was responsible for the Great Famine in Ireland (1845–1852) and the resultant major decline in the population through starvation and emigration (Gregory, 1983; Mizubuti and Fry, 2006).

Although phylogenetically distant, plant-parasitizing peronosporomycetes and fungi share a unique capability compared with other microbial pathogens; that is, they are able to breach the cuticle of the host plant and rapidly establish infection (Soanes et al., 2007; Meng et al., 2009). It has been suggested that the success of some peronosporomycetes as plant pathogens is due in part to acquisition of genes from fungi through horizontal gene transfer (e.g. Richards et al., 2011).

FIGURE 13.10 Michael W. Dick.

FOSSIL PERONOSPOROMYCETES

The Peronosporomycetes are hypothesized to have evolved in the marine realm from a photosynthetic ancestor through the loss of plastids and acquisition of a parasitic or saprotrophic life history biology (Beakes and Sekimoto, 2009; Grenville-Briggs et al., 2011; Beakes et al., 2012; Seidl et al., 2012). However, the evolutionary patterns leading to the present-day diversity of this group remain incompletely understood, in part because molecular markers that might reveal deep phylogenetic patterns are lacking (Lara and Belbahri, 2011). The earliest diverging genera are thought to be *Eurychasma* and *Haptoglossa*, both obligate parasites (e.g. Johnson and Sparrow, 1961; Davidson and Barron, 1973; Küpper et al., 2006; Strittmatter et al., 2009; Grenville-Briggs et al., 2011). Oogamous sexual reproduction, regarded as one of the major evolutionary innovations that defines crown oomycetes, may have developed after migration from the sea to land (Beakes et al., 2012). The basal clade taxa are non-oogamous.

Molecular clock assumptions suggest that the first Peronosporomycetes evolved during the early Neoproterozoic, approximately 1.0 Ga (Bhattacharya et al., 2009); it has also been speculated that Peronosporomycetes may have been among the first eukaryotes on Earth (Pirozynski, 1976a, b). Dick (2001; Figure 13.10) notes that the environment and substrates in ancient terrestrial ecosystems were certainly suitable and could have supported these organisms by Paleozoic time. Efforts to reconstruct the evolutionary history and phylogeny of the Peronosporomycetes or of lineages within this group (e.g. Hudspeth et al., 2000; Petersen and Rosendahl, 2000; Thines and Kamoun, 2010; Uzuhashi et al., 2010) have to date been based exclusively on the analysis of extant members, because the fossil record of the group, until recently, has been almost nonexistent.

The incompleteness of the fossil record includes preservation of parts of individuals and/or isolated stages of a life cycle. The life cycle of Peronosporomycetes includes both an asexual and a sexual phase (Dick, 2001). Most of the structures that constitute the life cycle, however, are not diagnostic at the level of resolution available with transmitted light, rendering it difficult, if not impossible, to identify a fossil peronosporomycete based on these structures alone. One of the features that could be used to positively identify a fossil peronosporomycete are the biflagellate zoospores; however, these structures are too small to be observed in sufficient detail in transmitted light, and the flagella would not normally be preserved (Taylor et al., 1992a). As a result, the only life-cycle stage that both lends itself to preservation and can be used to positively identify fossil Peronosporomycetes is the characteristic oogonium–antheridium complex, a structure that forms during sexual reproduction (Dick, 2001; Judelson, 2009). While the presence of such complexes makes identification of fossil Peronosporomycetes possible (see below), determining their exact systematic position remains difficult – because essential features used today in peronosporomycete taxonomy, especially molecular data, cannot be obtained from the fossils.

The extraordinary level of preservation necessary for Peronosporomycetes to be recognized explains why the fossil record of these organisms is largely limited to the Devonian and Carboniferous. Chert deposits and permineralizations from these time periods represent the only sources of evidence to date for fossil peronosporomycete oogonium–antheridium complexes. Although various types of body fossils of microorganisms and/or indirect evidence of their activities have been preserved by other modes, peronosporomycetes have not yet been documented from these preservation states, with the possible exception of a few amber fossils (Smith, J., 1896; Berry, 1916; Ting and Nissenbaum, 1986; Nissenbaum and Horowitz, 1992; Rikkinen and Poinar, 2001).

Petsamomyces polymorphus from the Paleoproterozoic of the Kola Peninsula is perhaps the oldest Precambrian microfossil that has been suggested to be morphologically similar to structures seen in Peronosporomycetes (see below) (Belova and Akhmedov, 2006); however, similarities also exist with Chytridiomycota and other groups of basal fungi. On the other hand, Pirozynski (1976a) proposed that, among the abundant microfossils collectively termed acritarchs, which are common as early as the Proterozoic, some may represent either the oogonia or oospores of peronosporomycetes. This hypothesis, which he based on morphological similarities, has been largely overlooked by other researchers, and when addressed, has usually been dismissed. Colbath and Grenfell (1995) considered Pirozynski's morphological comparisons to represent superficial similarities rather than common origins, but they also noted that possible fungal affinities for at least some acritarchs could not be completely discounted.

Sporadic reports of ichnofossils (trace fossils or structures based on the fossilized activities of organisms) in the form of microborings (Figure 13.11) have been interpreted as having been produced by peronosporomycetes. One of these is *Palaeachlya,* a microboring that occurs in Ordovician to Miocene corals and other animal hard parts (e.g. Bertling, 1987) and which was historically believed to have been formed by a peronosporomycete (Duncan, 1876; Etheridge, 1891). Although this microboring has some morphological similarities to structures seen in modern peronosporomycetes, diagnostic features that could be used to positively identify the producer(s) are lacking. More recently, *Palaeachlya* has been reinterpreted as representing the activities of endolithic algae (Elias and Lee, 1993). There are several extant examples of endolithic and/or bioeroding marine Peronosporomycetes (e.g. Beuck and Freiwald, 2005; Försterra et al., 2005), and a few ichnotaxa of fossil microborings have been attributed to the same class, based on structural similarities among the borings (e.g. Glaub, 1994). One of these is *Saccomorpha terminalis* (Late Ordovician to recent), an ichnotaxon composed of narrow tunnels from which arise spherical structures on short pedicels that are similar to a *Phytophthora*-like endolithic peronosporomycete (Radtke, 1991; Vogel and Brett, 2009). Others regard the biological affinities of *S. terminalis* as problematic (Wisshak et al., 2008).

An enigmatic body fossil interpreted as a peronosporomycete is *Palaeophthora mohgaonensis* from the uppermost Maastrichtian (Cretaceous) Deccan Intertrappean beds of India (Singhai, 1975, 1978). This fungus consists of a coenocytic mycelium that is endophytic in plant tissue. Some hyphal tips produce spherical structures referred to as oogonia; tubular hyphal tips occurring in the same host cells as the oogonia have been interpreted as antheridia.

Another interesting structurally preserved fossil organism interpreted as a peronosporomycete from

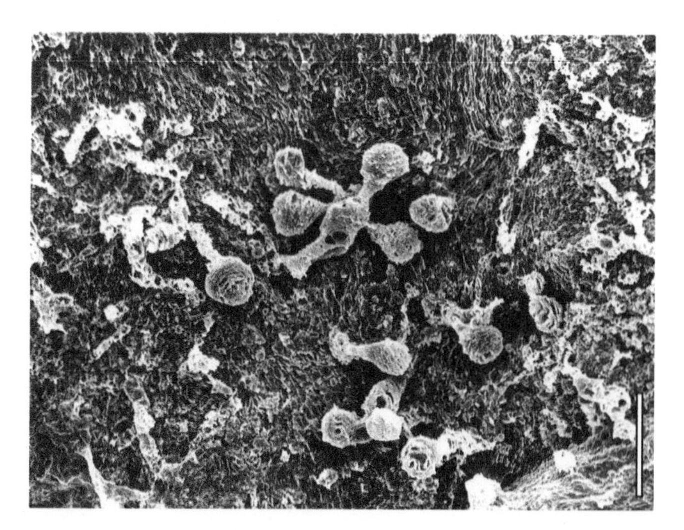

FIGURE 13.11 Casts of *Saccomorpha clava* microborings. Upper Ordovician, Ohio, USA. Bar = 30 μm. (From Vogel and Brett, 2009.)

Carboniferous coal balls of Great Britain was named *Peronosporites antiquarius* (Smith, W.G., 1877, 1878). It consists of terminal and intercalary swellings produced on septate hyphae; smaller spherules within the swellings were interpreted as oospores (Figure 13.12). Williamson (1881), however, concluded that affinities with the Peronosporomycetes were questionable. The fact that the fossils described in this and several other accounts (e.g. in Cash and Hick, 1879 [Figures 13.13, 13.14]; Williamson, 1881; Loomis, 1900; Pampaloni, 1902; Elias, 1966; Agashe and Tilak, 1970) cannot be assigned with confidence because of the absence of diagnostic features is the major reason that there has been no systematic study or comprehensive treatment of the fossil history of the group. Unfortunately, neither *Peronosporites antiquarius* nor *Palaeophthora mohgaonensis* display features of sufficient clarity to enable their assignment to any group of fungi or fungus-like organisms to be made with confidence. Johnson et al. (2002) regarded the entire fossil record published prior to 2002 as inconclusive. This assumption was further supported by Dick (1988, 2001, 2002), who expressed the opinion that no definite fossil Peronosporomycetes had yet been reported, and that fossil genera such as *Achlyites*, *Ordovicimyces*, *Palaeoperone*, *Peronosporoides*, *Palaeachlya*, *Propythium*, and *Pythites* could all be discounted (see Table 13.1 for references on these taxa).

A slightly different opinion was offered by Blackwell and Powell (2000), who considered at least some of the historical records of fossils interpreted as Peronosporomycetes as persuasive. For example, *Peronosporoides palmi* (later renamed *Peronosporites palmae* [Kalgutkar and Jansonius,

FIGURE 13.12 Hyphae and swellings (?oogonia) of *Peronosporites antiquarius*. Pennsylvanian, Great Britain. Bar = 50 μm. (From Smith, W.G., 1884.)

FIGURE 13.13 William Cash.

FIGURE 13.14 Thomas Hick.

TABLE 13.1 Synopsis of fossil taxa (historically) assigned or related to the Peronosporomycetes, arranged according to their appearance in the text.

The summary opinion of the taxonomic assignment (right column), is dealt with in three categories: (1) accepted, indicating that the fossils have been identified as peronosporomycetes based on oogonium-antheridium complexes or other persuasive structural correspondences to living forms; (2) inconclusive, indicating that the fossils are (somewhat) reminiscent of structures seen in modern peronosporomycetes, but lack features of sufficient clarity to assign them to that group with confidence; and (3) dubious, indicating that the fossils in all probability do not belong to the Peronosporomycetes (based on Krings et al., 2011b).

Taxon	References	Geologic age	Occurrence	Our Opinion
Achlyites penetrans	Duncan (1876), James (1893a,b), Meschinelli (1898)	Silurian and Tertiary	endozoic in corals and foraminifers	inconclusive
Albugo-like organism	Stidd and Cosentino (1975)	Pennsylvanian	endophytic in seed preserved in coal ball	inconclusive
Combresomyces	Dotzler et al. (2008), Schwendemann et al. (2009, 2010), Strullu-Derrien et al. (2011), Slater et al. (2013)	several species; Mississippian to Triassic	endophytic in plant tissue preserved in chert and coal balls (Carboniferous, Triassic); isolated in chert matrix (Permian, Triassic)	accepted
Frankbaronia	Krings et al. (2012c, 2013b)	two species; Devonian	in plant debris and microbial mats preserved in chert	accepted
Galtierella biscalithecae	Krings et al. (2010d)	Pennsylvanian	endophytic in sporangia preserved in chert	accepted
Hassiella monospora	Taylor et al. (2006)	Devonian	in chert matrix	accepted
Ordovicimyces	Elias (1966)	two species; Ordovician	endozoic in bryozoans	inconclusive
Palaeachlya	Duncan (1876), Etheridge (1891); Chapman (1911), Elias and Lee (1993)	four species; Ordovician to Miocene	endozoic in corals and other animal hard parts	inconclusive
Palaeoperone endophytica	Etheridge (1891), Meschinelli (1898)	Pennsylvanian	endozoic in corals	inconclusive
Palaeophthora mohgaonensis	Singhai (1975, 1978)	Cretaceous	endophytic in plant tissue preserved in chert	inconclusive
Peronosporites	Smith, J. (1877, 1878), Loomis (1900), Ellis (1918), Pampaloni (1902), Ting and Nissenbaum (1986)	several species; Silurian, Pennsylvanian, Cretaceous, and Miocene	endozoic in animal hard parts (Silurian); in coal ball matrix and endophytic in lycophyte tissue (Pennsylvanian); in amber from Israel (Cretaceous); isolated in sediment (Miocene)	inconclusive
Peronosporoides	Smith, J. (1896), Berry (1916)	two species[1]; Pennsylvanian and Oligocene	in amber [called middletonite] from Great Britain (Pennsylvanian); in silicified palm stem (Oligocene)	inconclusive
Petsamomyces polymorphus	Belova and Akhmedov (2006)	Paleoproterozoic	isolated in black shale	dubious
Propythium carbonarium	Elias (1966)	Pennsylvanian	endozoic in bryozoans	inconclusive
Pythites disodilis	Pampaloni (1902)	Miocene	isolated in sediment	inconclusive
Saccomorpha terminalis	Radtke (1991), Wisshak et al. (2008), Vogel and Brett (2009)	Ordovician to recent	endolithic in rocks and animal hard parts	inconclusive
Unnamed 2 (Figures 13.27–13.29)	Krings et al. (2010c)	Mississippian	in chert matrix	accepted
Unnamed 3 (Figure 13.32)	Krings et al. (2010b)	Pennsylvanian	endophytic in lycophyte periderm preserved in coal ball	accepted

[1]The Oligocene species (*P. palmii* Berry 1916) was later renamed *Peronosporites palmae* by Kalgutkar and Jansonius (2000).

2000]) reported from structurally preserved wood from the lower Oligocene of North America (Berry, 1916; Figure 13.15) consists of spherical structures interpreted as oogonia, each approximately 55 μm in diameter and borne on narrow, septate hyphae (Figure 13.16). Some oogonia contain structures that appear to be oospores. What is especially interesting about these fossils is the fact that there are collapsed sacs attached to the oogonia that have been interpreted as antheridia.

It has also been suggested that the genus *Albugo* was present at least by the Carboniferous (Stidd and Cosentino, 1975). An *Albugo*-like organism, which occurs in a Middle Pennsylvanian seed in a coal ball from Iowa, consists of irregular masses of tissue that extend into the region of the megagametophyte. Oogonia, each about 100 μm in diameter, are reported just beneath the integument (Figure 13.17). They possess a thick wall and may have an opaque central inclusion, which may represent the oosphere or developing oospore. The fossil oogonia are in various stages of development, with some associated with structures suggestive of antheridia.

Oogonium–Antheridium Complexes

The oldest fossil evidence of the Peronosporomycetes that has been identified based on oogonium–antheridium complexes is *Hassiella monospora* from the Lower Devonian Rhynie chert (Taylor et al., 2006). This fossil occurs within degrading plant material and consists of aseptate hyphae 4–6 μm wide (Figure 13.18). Branching is relatively frequent and includes numerous small, sometimes closely spaced papilla-like protuberances. None of these protuberances shows a separation from the vegetative hypha in the form of a septum, although several larger ones, presumably representing later developmental stages, do show separation from the parental hyphae by a simple septum. Also arising from the vegetative hyphae are globose spheres, which measure up to 28 μm in diameter, are typically opaque, possess a thick verrucate outer wall (Figure 13.19A,B), and are characterized by a prominent, funnel-shaped appendage (Figure 13.20). The

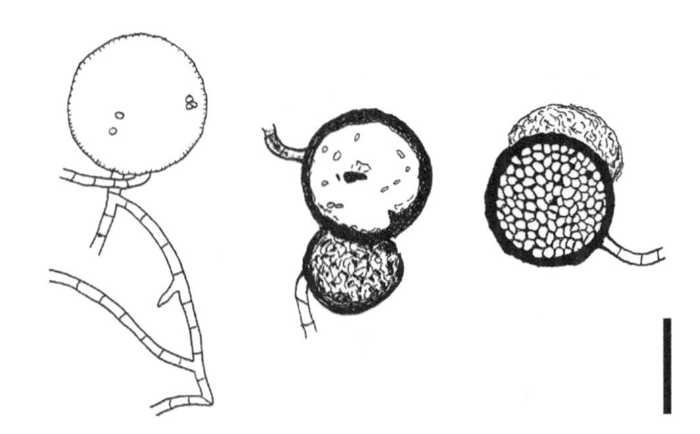

FIGURE 13.16 Several oogonia (?) with attached antheridia (?) of *Peronosporites palmae*. Oligocene, USA. Bar = 50 μm. (From Berry, 1916.)

FIGURE 13.15 Edward W. Berry.

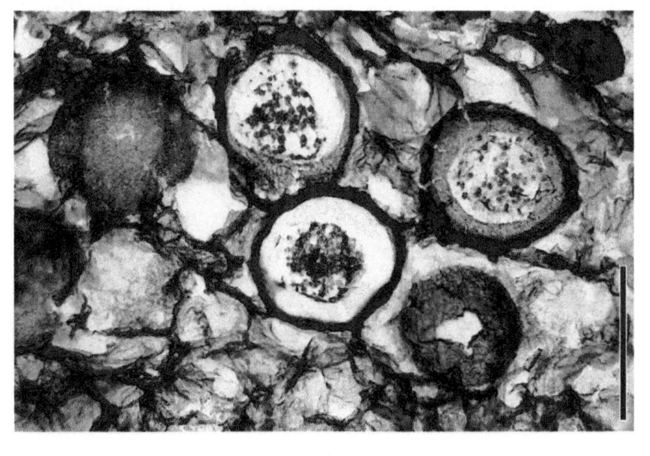

FIGURE 13.17 Several *Albugo*-like oogonia embedded in fossil seed megagametophyte. Middle Pennsylvanian, Iowa, USA. Bar = 150 μm.

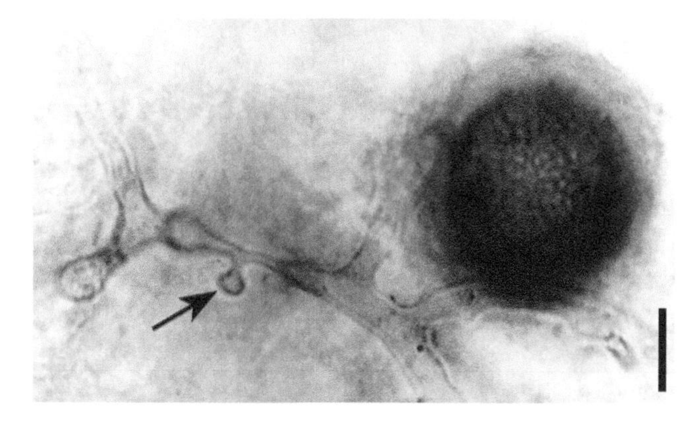

FIGURE 13.18 *Hassiella monospora* showing branching coenocytic hyphae with swelling (*arrow*) and thick-walled oogonium. Lower Devonian Rhynie chert, Scotland. Bar = 10 μm. (From Taylor et al., 2006.)

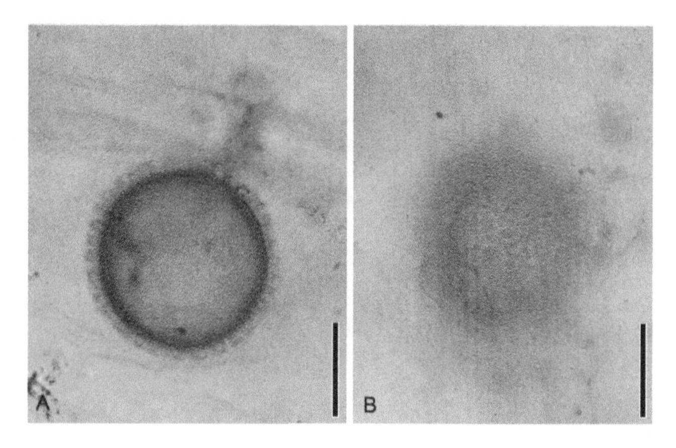

FIGURE 13.19 *Hassiella monospora* showing ornament of oogonium at different focal planes (**A**, **B**). Lower Devonian Rhynie chert, Scotland. Bar = 10 μm.

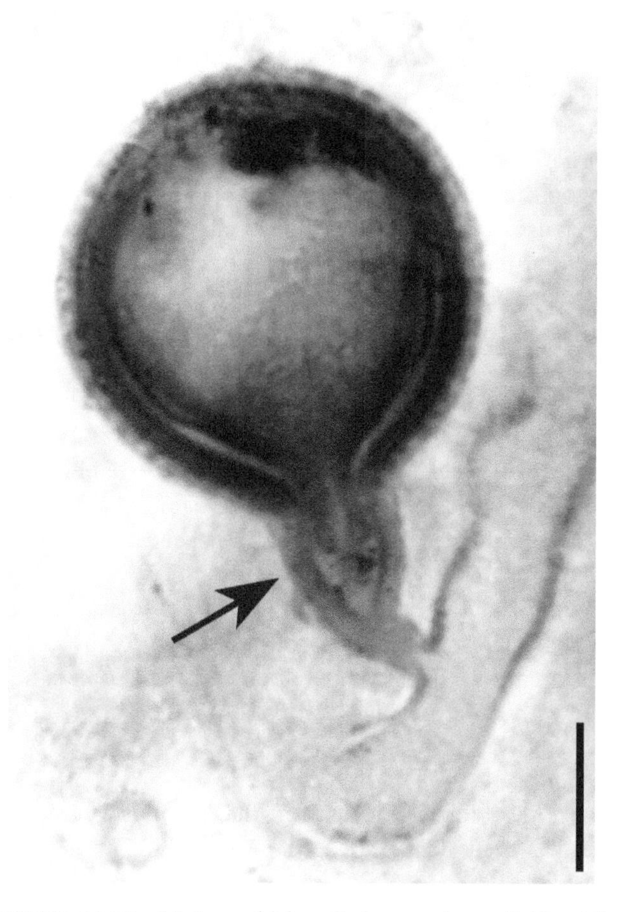

FIGURE 13.20 Mature, thick-walled oogonium of *Hassiella monospora* attached to hypha by funnel-shaped appendage interpreted as an amphigynous antheridium (*arrow*). Lower Devonian Rhynie chert, Scotland. Bar = 10 μm. (From Taylor et al., 2006.)

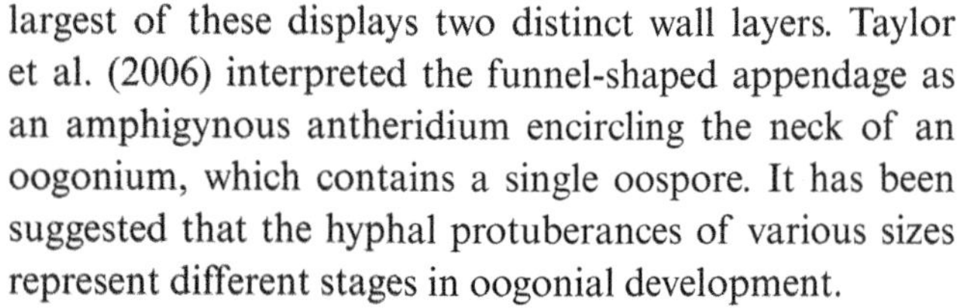

largest of these displays two distinct wall layers. Taylor et al. (2006) interpreted the funnel-shaped appendage as an amphigynous antheridium encircling the neck of an oogonium, which contains a single oospore. It has been suggested that the hyphal protuberances of various sizes represent different stages in oogonial development.

Also present in the Rhynie chert are two distinct ovoid to slightly elongate structures, ranging from ~50 μm to >100 μm in diameter, that have been formally described under the generic name *Frankbaronia* (Krings et al., 2012c, 2013b). They typically occur associated with microbial mats and plant debris, and may sometimes be solitary but more frequently occur in loose clusters. One species, *F. polyspora* (Figure 13.21), is characterized by irregularly distributed prominent wall projections

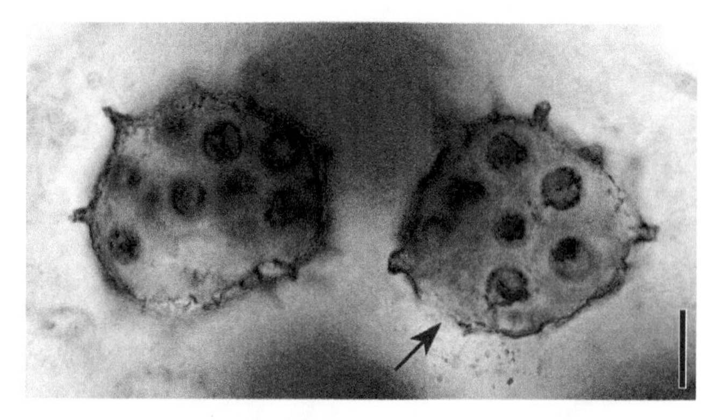

FIGURE 13.21 Two oogonia of *Frankbaronia polyspora* with prominent wall projections. Arrow indicates possible point of attachment. Lower Devonian Rhynie chert, Scotland. Bar = 10 μm.

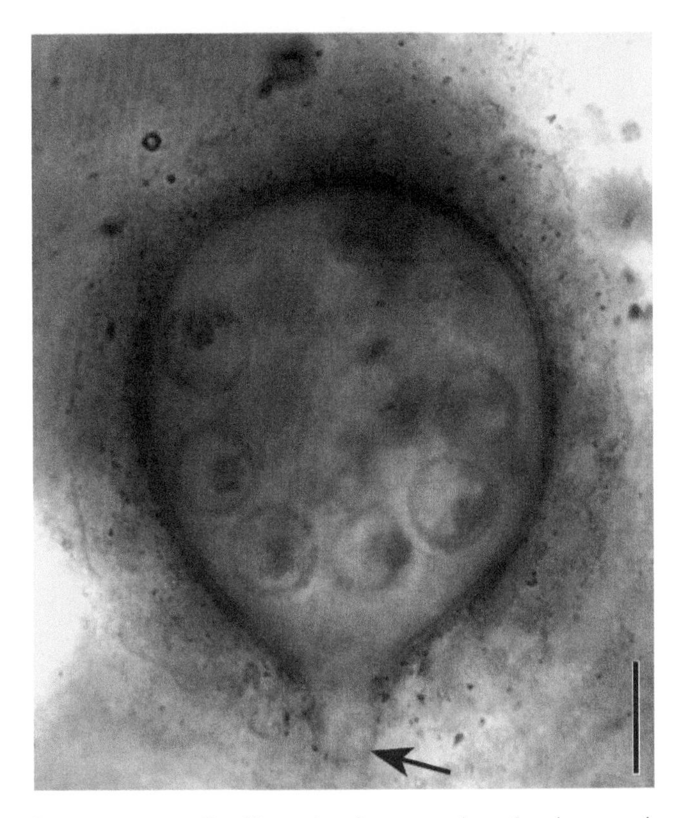

FIGURE 13.22 *Frankbaronia velata* oogonium showing prominent mucilage sheath. Arrow indicates attachment site to parental hypha. Lower Devonian Rhynie chert, Scotland. Bar = 20 µm.

(Krings et al., 2012c). Most of the specimens contain a number of small spheres, the average being between 10 and 40, but there may be only 3–5 and or as many as 80+ in the smallest and largest specimens, respectively. These spheres are smooth and bounded by a thin but distinct wall, and each contains a (sub)centric opaque inclusion. The second type, *F. velata* (Figure 13.22), lacks projections but possesses a prominent truncated extension, which most likely represents the attachment point of a stalk or parental hypha. Each specimen is surrounded by a translucent sheath, in some areas up to 35 µm thick, which may be uniform in texture or stratified, and which is bounded on the outside by a narrow but distinct boundary layer. Specimens usually contain anything from 6 to >150 small spheres, which are smooth, have a distinctive wall, and contain a subcentric opaque inclusion. Although gametangial fusion cannot be demonstrated in *Frankbaronia*, the morphology of the fossils, including their contents, is strikingly similar to that seen in the polyoosporous oogonia of certain extant members of

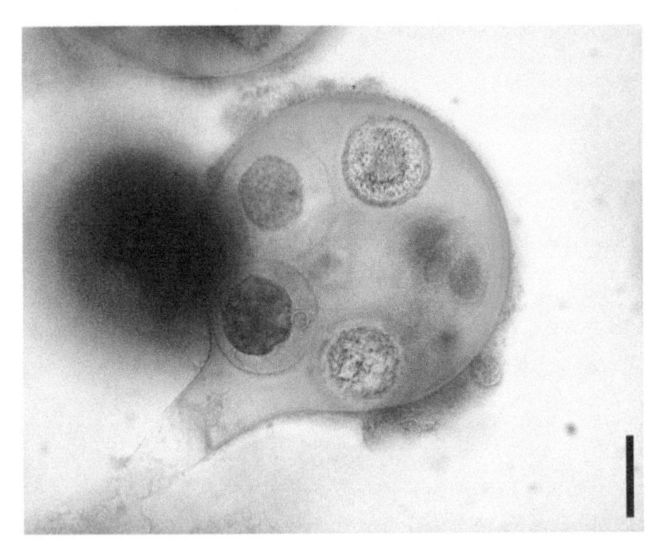

FIGURE 13.23 Oogonium of extant *Saprolegnia* sp. containing several oospores. Extant. Bar = 10 µm.

the Saprolegniaceae (Figure 13.23), such as *Saprolegnia multispora*, in which oogonia may contain up to 100 oospores (Paul and Steciow, 2004).

One of the most interesting aspects of *Frankbaronia polyspora* is demonstrated in specimens that reveal the different stages of development of the small interior spheres. For example, early developmental stages consist of a large mass of opaque material (not bounded by a wall) in the center of the parental structure; the opaque material subsequently becomes apportioned radially and redistributed to the individual small spheres (Figure 13.24). Later in development the inclusions in the individual small spheres remain connected to the center by a narrow, thread-like structure (Figure 13.25). At some stage the inclusions become separated and the walls are closed. Based on the developmental sequence demonstrated in the fossils, Krings et al. (2012c) suggested that *F. polyspora* produced a single oosphere- or coenocentrum-like structure in the center of the oogonium that became apportioned after fertilization (i.e. in a zygotic state) to form several walled oospores (Figure 13.26). This pattern of oosporogenesis is different from that seen in any modern polyoosporous peronosporomycete.

An interesting structural feature of *Frankbaronia velata* is the sheath enveloping the oogonium. No sheaths of similar appearance are known in any other fossil peronosporomycete; however, in several extant peronosporomycetes, development of the oogonial initial is accompanied by a secretion of mucilage, which

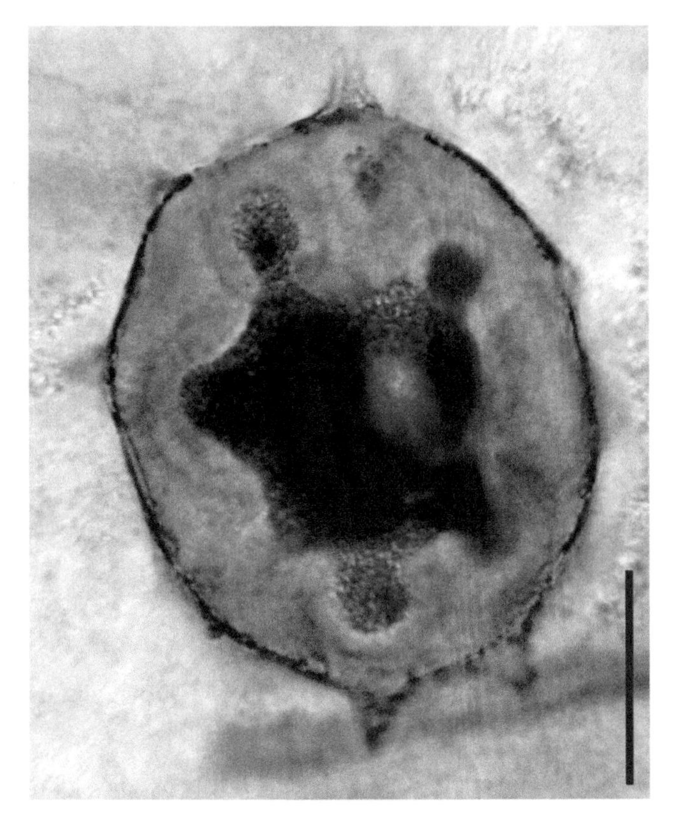

FIGURE 13.24 Oospore development in *Frankbaronia polyspora* showing redistribution of opaque material from center of parental structure to individual oospores. Lower Devonian Rhynie chert, Scotland. Bar = 20 μm.

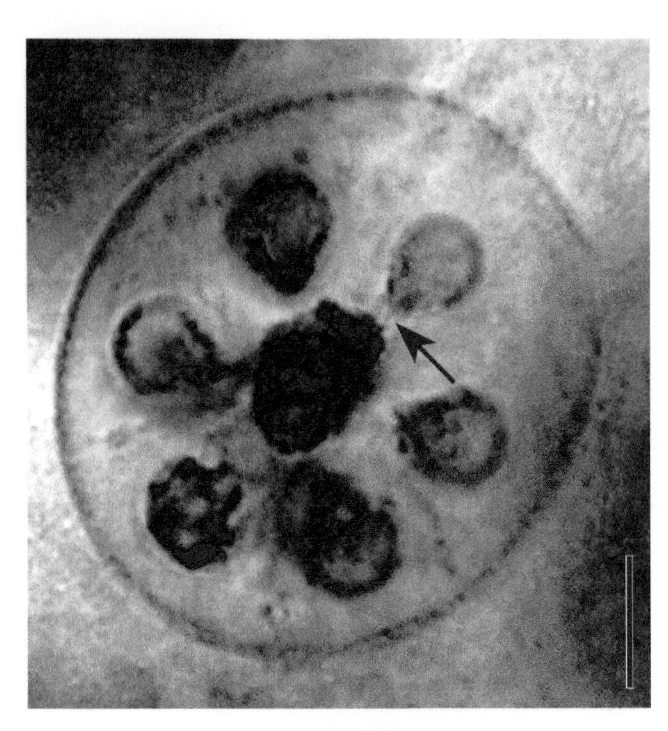

FIGURE 13.25 Later developmental stage than that shown in Figure 13.24, showing oospore inclusions still connected to center of oogonium by thread-like structure (*arrow*). Lower Devonian Rhynie chert, Scotland. Bar = 10 μm.

initially forms a spherical envelope with a sharply defined outer surface around the developing oogonium (e.g. Dick, 2001). As development of this oogonium continues, the envelope becomes irregular, and finally collapses to some degree on the oogonium wall proper (Dick, 1969; Johnson et al., 2002; Spencer et al., 2002). Krings et al. (2013b) speculated that the developing oogonium of *F. velata* may have produced a mucilaginous secretion through the oogonial wall that gradually accumulated on the outer surface of the oogonium, eventually consolidating, but not collapsing, and thus forming a continuous sheath.

Carboniferous–Permian

There are several examples of fossil Peronosporomycetes from the Carboniferous. One of these, from Middle Mississippian cherts of central France, consists of a single spherical oogonium (Figure 13.27) approximately 20 μm in diameter (Krings et al., 2010e). Extending from the surface is a conspicuous ornamentation composed of elongate subtle extensions up to 14 μm long, which are

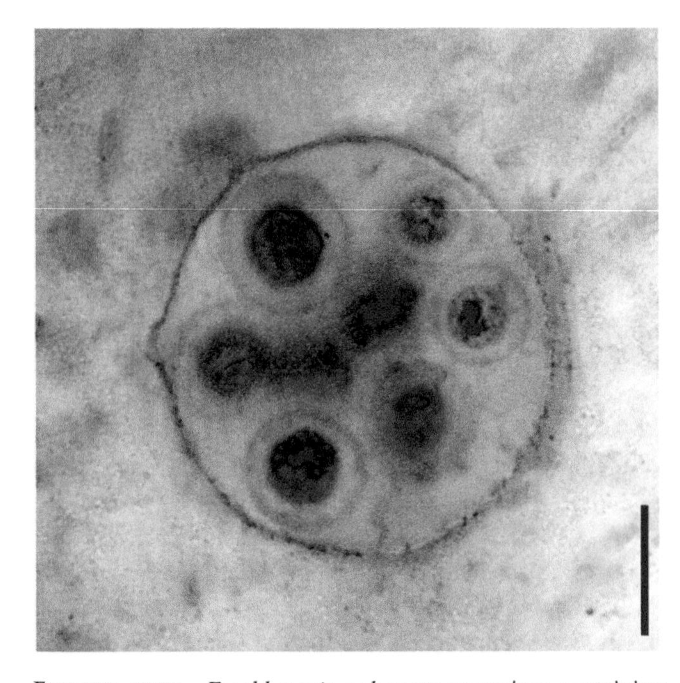

FIGURE 13.26 *Frankbaronia polyspora* oogonium containing well-developed walled oospores. Lower Devonian Rhynie chert, Scotland. Bar = 10 μm.

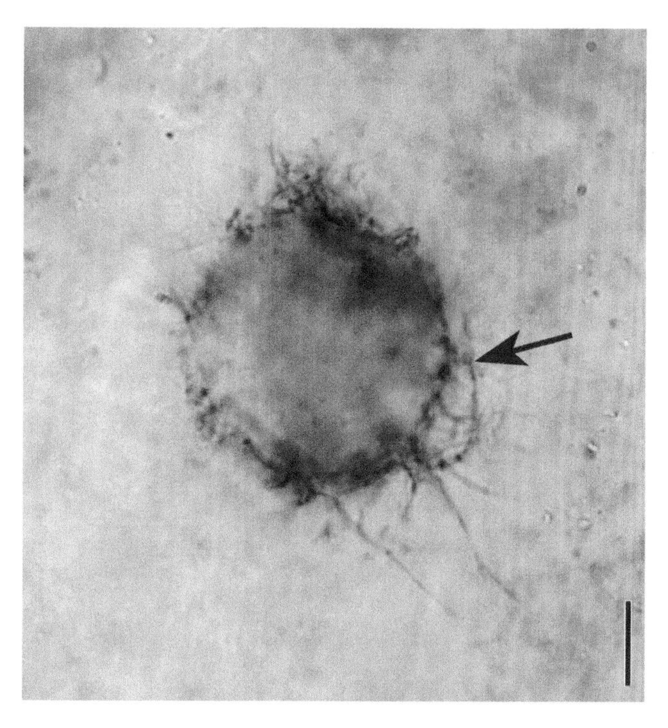

FIGURE 13.27 Peronosporomycete oogonium with conspicuous ornamentation (*arrow*) and subtending hypha. Mississippian, France. Bar = 10 μm.

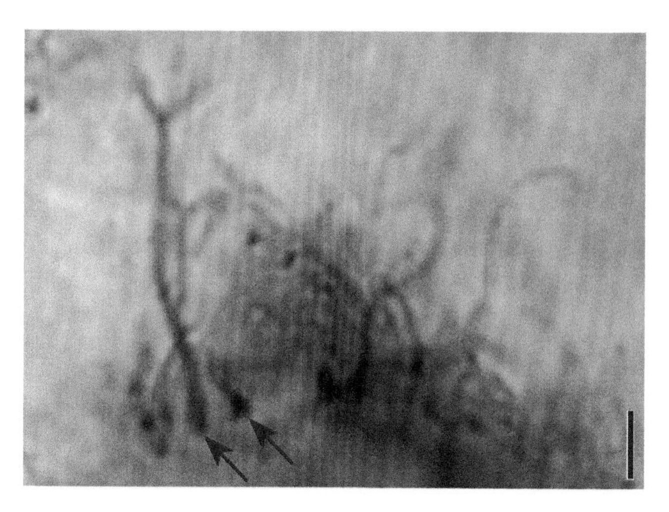

FIGURE 13.28 Detail of surface ornament (*arrows*) of oogonium in Figure 13.27. Mississippian, France. Bar = 2 μm.

densely spaced and regularly distributed over the surface (Figure 13.28). Most of the extensions are one to several times furcate, and attached to a small, bulb-like swelling at the base. Two clavate, thin-walled antheridia, up to 5.5 μm high and 10 μm long, are attached to the oogonium (Figures 13.27 and 13.29). One of the antheridia is still connected to a fragment of the parental hypha or antheridial stalk, and what appears to be a septum occurs between the antheridium and stalk.

Another putative peronosporomycete with ornamented oogonia occurs as an intracellular endophyte in the sporangium wall of the fern *Biscalitheca* cf. *musata* preserved in the Upper Pennsylvanian Grand-Croix cherts from France (Krings et al., 2010d). This organism, *Galtierella biscalithecae*, is composed of wide aseptate hyphae, which are more or less Ω-shaped in cross section, relatively thin-walled, and extend along the inner surface of the host cell walls (Figure 13.30) to form spherical and typically terminal oogonia. Several oogonia possess what are interpreted as amphigynous antheridia (Figure 13.31) at the neck region of the oogonial stalk. Some of these structures are similar to amphigynous antheridia seen in certain extant species of *Phytophthora*. Clusters of small (3.5–11 μm in diameter) dome-shaped structures with

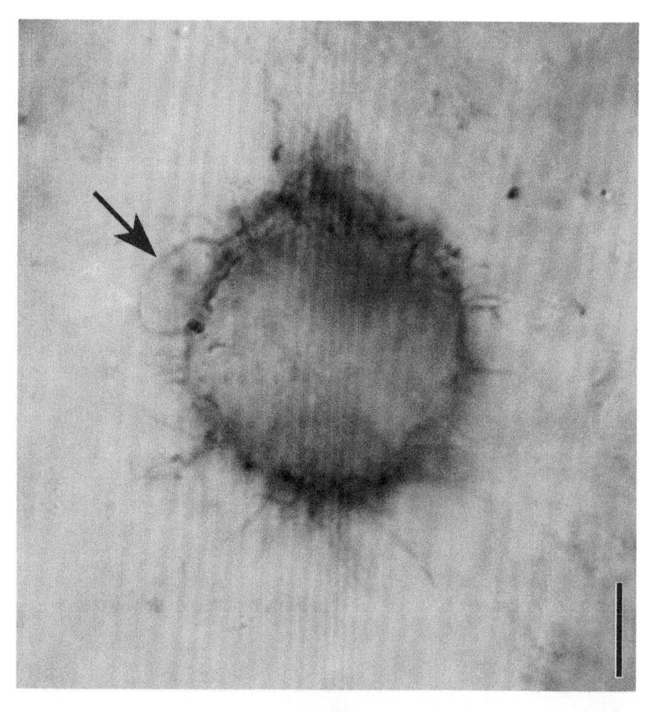

FIGURE 13.29 Same oogonium as in Fig. 13.27, different focal plane showing the second club-shaped antheridium (*arrow*). Mississippian, France. Bar = 10 μm.

thick-walls are attached to the internal walls of many infected host cells; some are distally open. These structures may represent encysted zoospores.

Perhaps the best fossil interpreted as a peronosporomycete is preserved in a coal ball from the Pennine Lower Coal Measures (Lower Pennsylvanian) of Great Britain.

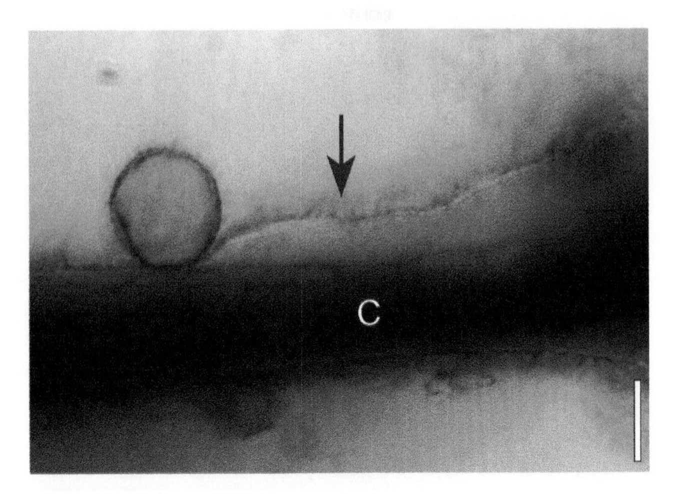

FIGURE 13.30 Host cell wall (C) with oogonium of *Galtierella biscalithecae* and hypha (*arrow*). Upper Pennsylvanian, France. Bar = 10 μm.

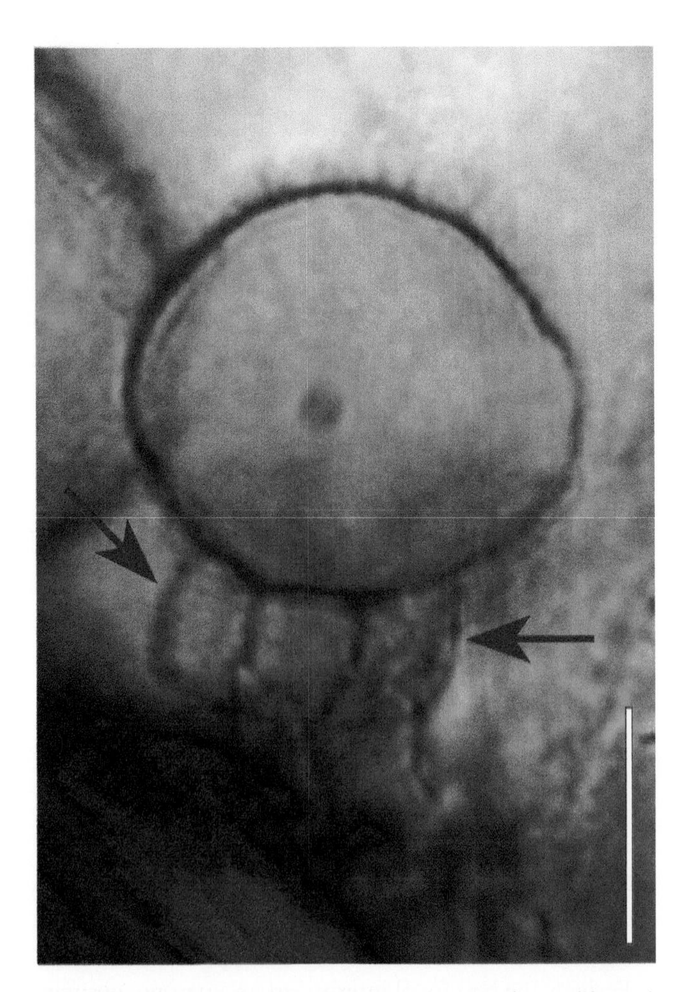

FIGURE 13.31 *Galtierella biscalithecae* oogonium with neck surrounded by what is interpreted as an amphigynous antheridium (*arrows*). Upper Pennsylvanian, France. Bar = 10 μm.

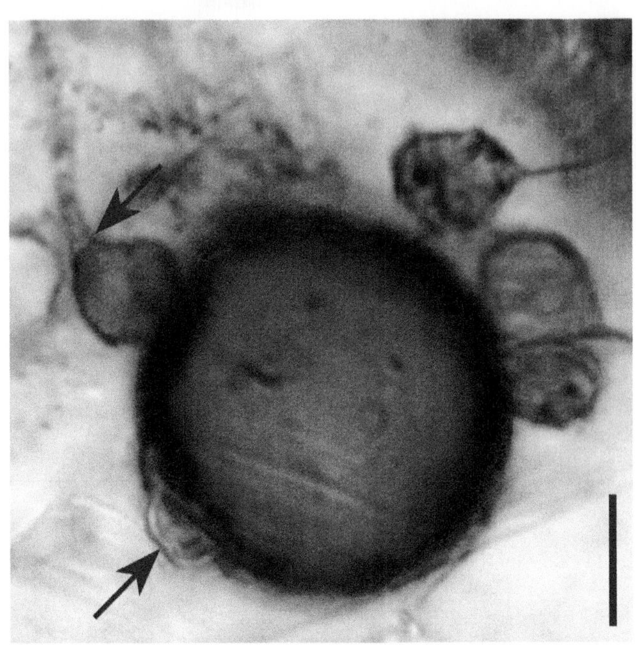

FIGURE 13.32 Oogonium with four attached paragynous antheridia; *lower arrow* indicates attachment of oogonial stalk, *upper arrow* shows septum between antheridium and antheridial hypha. Pennsylvanian, Great Britain. Bar = 5 μm.

This extraordinary fossil occurs within the periderm of an arborescent lycophyte and is represented by an unornamented oogonium approximately 15 μm in diameter to which are attached four paragynous and diclinous, club-shaped antheridia (Figure 13.32) (Krings et al., 2010a). Three antheridia are still in attachment with their subtending hyphae or antheridial stalks. A septum occurs between the antheridium and parental hypha.

An interesting aspect of the geologic history of the Peronosporomycetes concerns the occurrence of *Combresomyces cornifer* in lycophyte periderm cells from the Middle Mississippian cherts of central France (Krings et al., 2007c; Dotzler et al., 2008). This fossil is represented by stalked oogonia (Figure 13.33) with compound surface ornaments of hollow papillations of the oogonial wall. At the tip of each are antler-like extensions (Figure 13.34), which are believed to have formed as a result of the condensation of some mucilaginous extra-oogonial wall secretion (Dotzler et al., 2008). Within some of the oogonia are single spheres interpreted as aplerotic oospores. Other specimens have large clavate and paragynous antheridia adpressed to the oogonial wall (Figure 13.35). No vegetative hyphae have been found attached to *C. cornifer* to date; however, hyphae in some host cells, corresponding in diameter to the base of

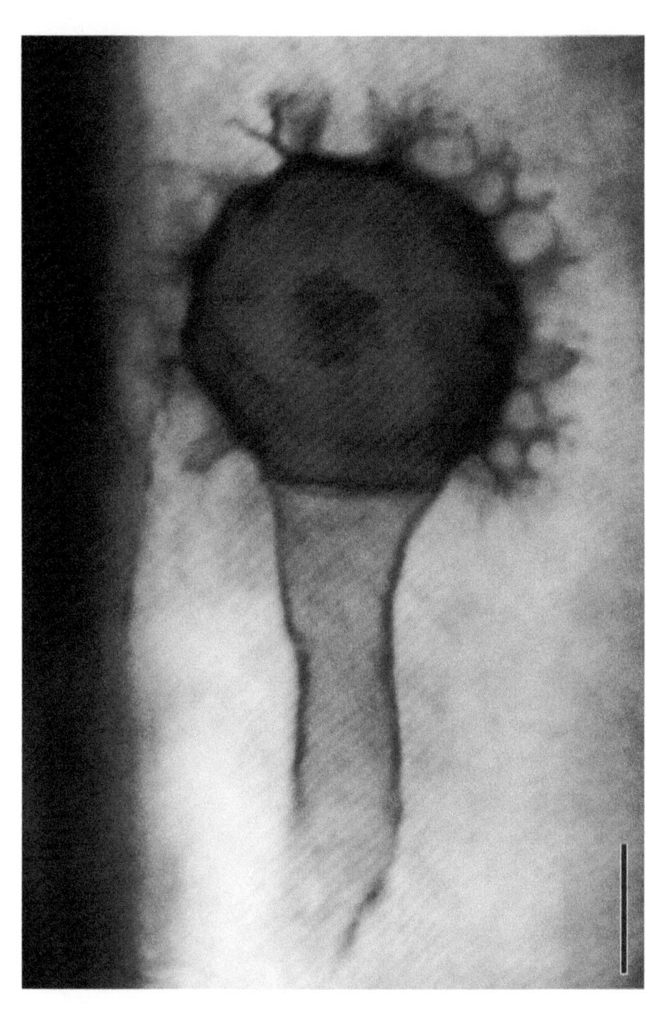

FIGURE 13.33 Stalked oogonium of *Combresomyces cornifer* containing a single oospore. Mississippian, France. Bar = 10 μm.

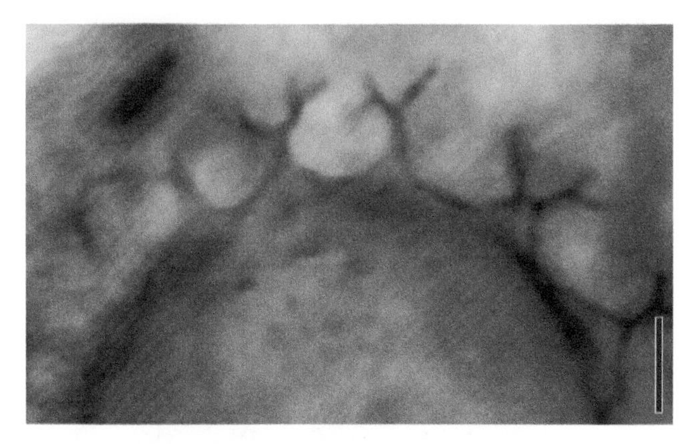

FIGURE 13.34 Surface ornament on *Combresomyces cornifer* oogonium. Mississippian, France. Bar = 5 μm.

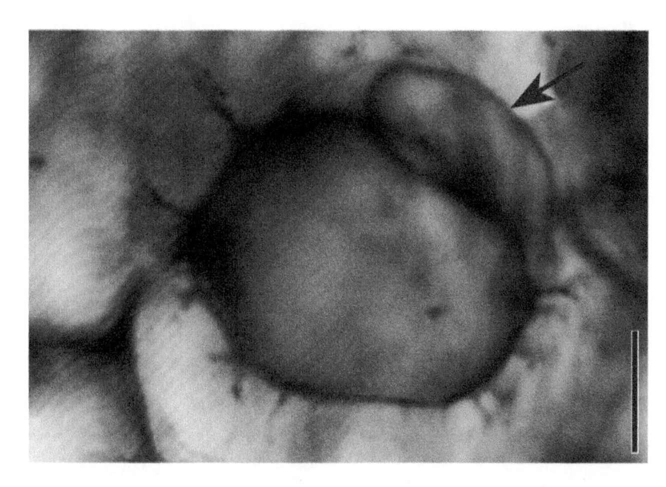

FIGURE 13.35 *Combresomyces cornifer* oogonium with club-shaped paragynous antheridium (*arrow*). Mississippian, France. Bar = 10 μm.

the oogonial stalk, may represent the vegetative part of this organism. Another report of *C. cornifer* comes from the Middle Triassic of Antarctica (Schwendemann et al., 2009; Figure 13.36). While the Triassic oogonia are morphologically identical to those from the Mississippian of France, they are appreciably larger (~100 μm in diameter) and do not occur endophytically, but instead are found isolated within degrading, silicified plant material. Spherical structures inside a conifer seed (Figure 13.37) that are morphologically similar to *Combresomyces* probably represent "sporocarps" (Schwendemann et al., 2010).

A similar type of oogonium, *Combresomyces williamsonii* (Figure 13.38), has been reported in the cortical tissues of a seed fern from the Pennine Lower Coal Measures of Great Britain (Strullu-Derrien et al., 2011; Figure 13.39). These specimens, some of which occur in organic attachment to hyphae, differ from *C. cornifer* in size (they are up to four times larger), general organization of the surface ornament, and the presence of both paragynous and hypogynous antheridia. Strullu-Derrien et al. (2011) recognized certain features in the vegetative hyphae, including hyphal knots and haustoria, and put forward the hypothesis that *C. williamsonii* may have been a hemibiotrophic parasite.

While *Combresomyces cornifer* and *C. williamsonii* have been discovered within plant tissue, several types of prominently ornamented spherical structures closely resembling these two forms also occur dispersed in coal balls from the Pennine Lower Coal Measures of Great Britain (Williamson, 1878, 1880, 1883), as well as in Carboniferous and Permian coal balls and chert

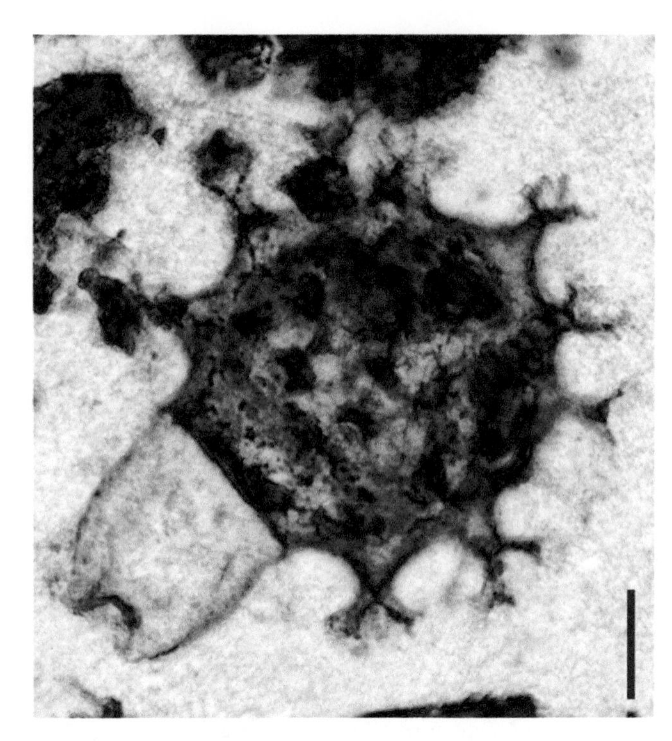

FIGURE 13.36 Oogonium of *Combresomyces cornifer* from the Triassic, Antarctica. Bar = 20 µm. (Courtesy A.B. Schwendemann.)

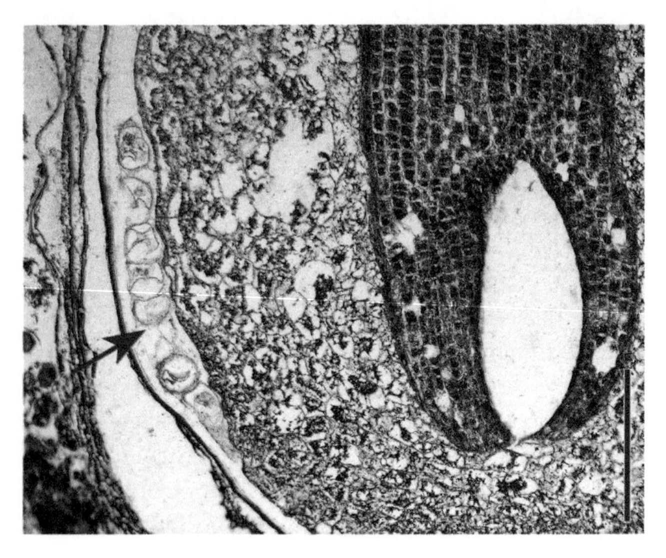

FIGURE 13.37 Partial section of seed showing several "sporocarps" (?), between seed coat and megagametophyte. Triassic, Antarctica. Bar = 100 µm. (Courtesy A.B. Schwendemann.)

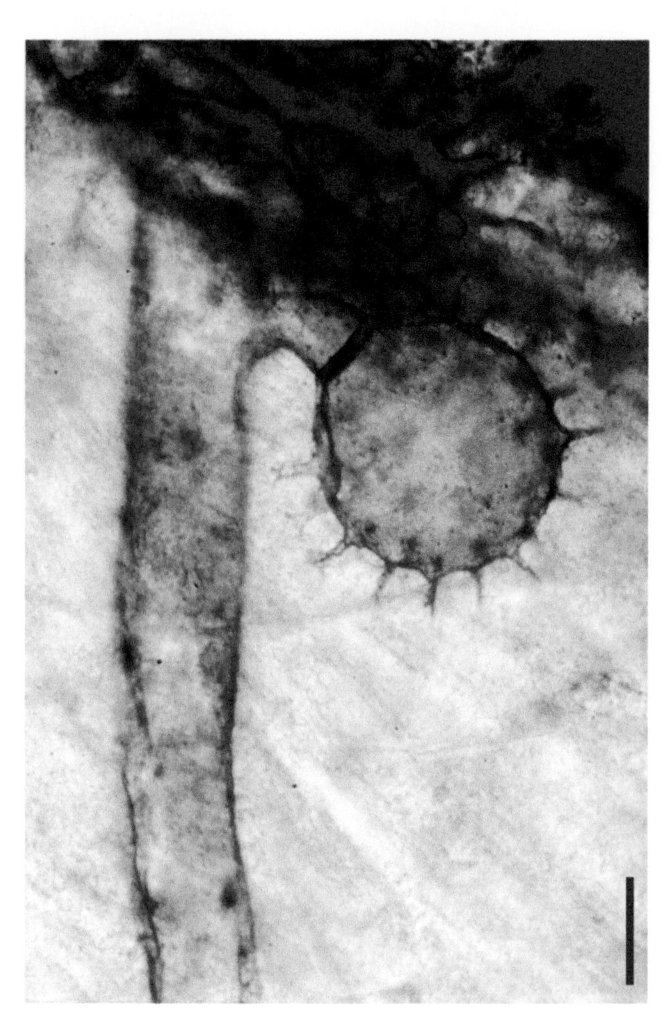

FIGURE 13.38 Oogonium of *Combresomyces williamsonii* attached to wide aseptate hypha. Pennsylvanian, Great Britain. Bar = 30 µm. (Courtesy C. Strullu-Derrien.)

deposits from elsewhere (e.g. Krings et al., 2009e; Slater et al., 2013). One of the more common types was initially named *Zygosporites* by Williamson (1878, 1880), a genus that was interpreted as some type of land plant spore because some of the specimens occurred clustered within a sporangium (Figure 13.40). Later, Williamson (1883) determined that the name *Zygosporites* was superfluous and suggested that the genus be deleted from the record. Interestingly, some of the fossils that might have been included in *Zygosporites* are characterized by surface ornamentation patterns that closely resemble those found on some *Combresomyces* oogonia (Figure 13.41).

Slater et al. (2013) described two prominently ornamented spherical microfossils found in Middle Permian permineralized peat from the Prince Charles Mountains in East Antarctica as additional species of *Combresomyces*. One of these, *C. caespitosus*, has long conical papillae that are densely spaced, hollow, and slender, with multiple, sharply pointed apical branches that commonly form a pseudoreticulate pattern. The

FIGURE 13.39 Christine Strullu-Derrien.

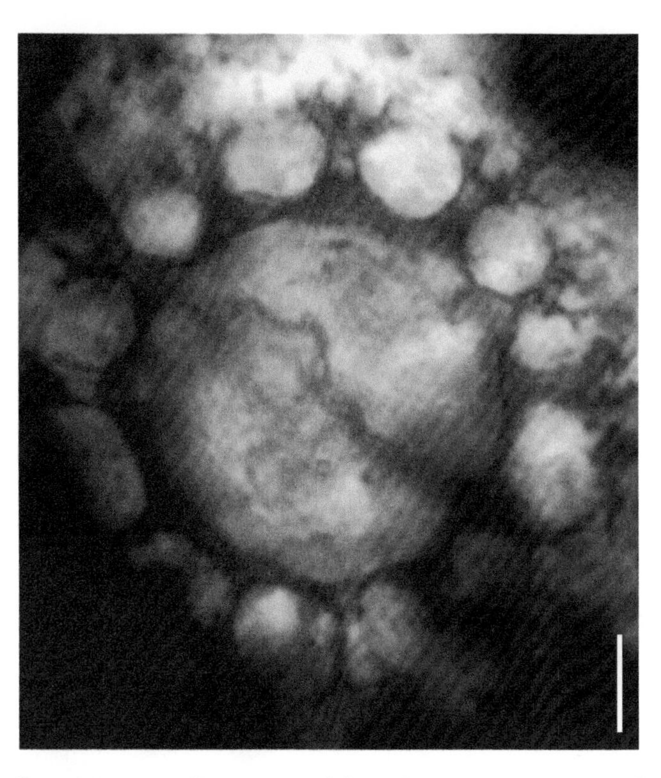

FIGURE 13.41 *Zygosporites* (=?*Combresomyces*) ornamented sphere. Pennsylvanian, Great Britain. Bar = 10 μm.

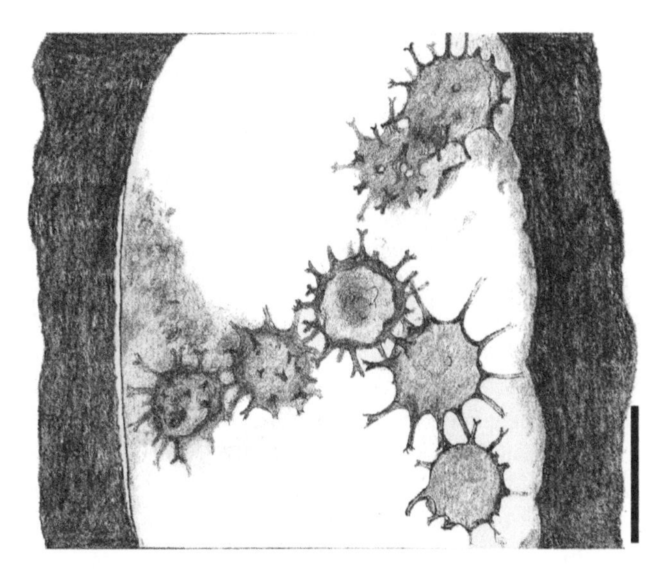

FIGURE 13.40 Line drawing of sporangium with *Zygosporites* (=?*Combresomyces*) ornamented spheres. Pennsylvanian, Great Britain. Bar = 100 μm. (From Williamson, 1883.)

second species, *C. rarus* (Figure 13.42), is characterized by widely spaced conical papillae that are hollow and broad and terminate in a single bifurcation that produces a pair of acutely divergent sharply pointed branches. Neither species has been found with attached antheridia, but some of the specimens display the typical truncated, collar-like extension of the oogonium where the parental hypha was originally attached.

It has been suggested that certain morphogenetic patterns of oogonial wall ornamentation (Figure 13.43) in Peronosporomycetes may have phylogenetic value (Dick, 2001). Dotzler et al. (2008) and Krings et al. (2010e) noted that *Combresomyces* possesses a rather complex oogonial ornament that morphologically differs from that in extant Peronosporomycetes. While acknowledging the uncertain value of this feature with regard to peronosporomycete phylogeny, these authors have hypothesized that fossil groups or lineages of Peronosporomycetes may have existed which possessed features (and developmental patterns) or combinations of features (and developmental patterns) unknown in any extant groups. Based on these speculations, Slater et al. (2013) introduced the family Combresomycetaceae within the order Combresomycetales to accommodate such fossil forms, which they suggest represent an extinct but once cosmopolitan Paleozoic to early Mesozoic branch of the peronosporomycete clade.

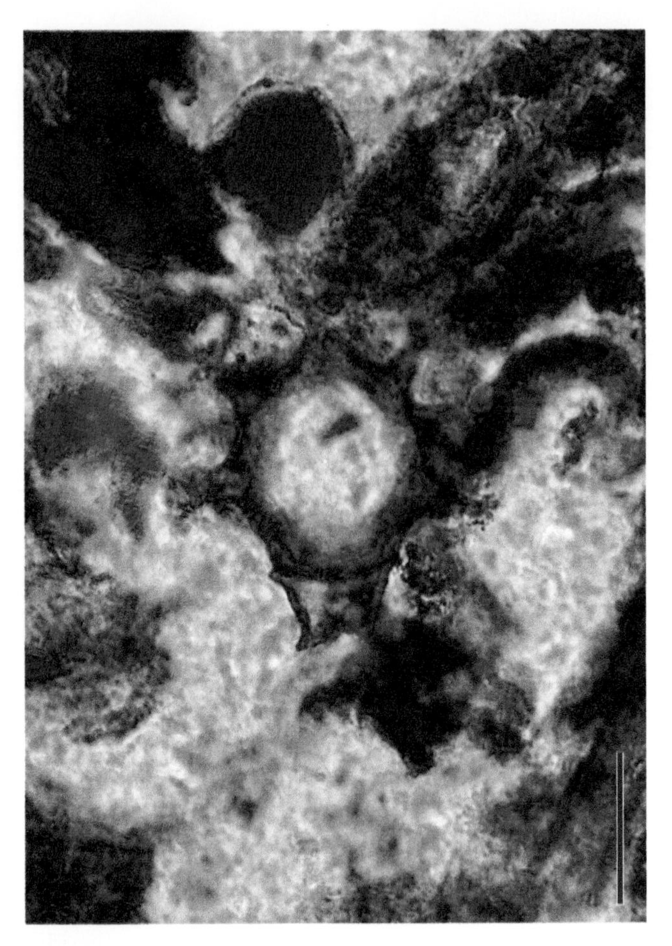

FIGURE 13.42 *Combresomyces rarus* oogonium showing widely spaced ornament. Permian, Antarctica. Bar = 50 µm. (Courtesy B.J. Slater.)

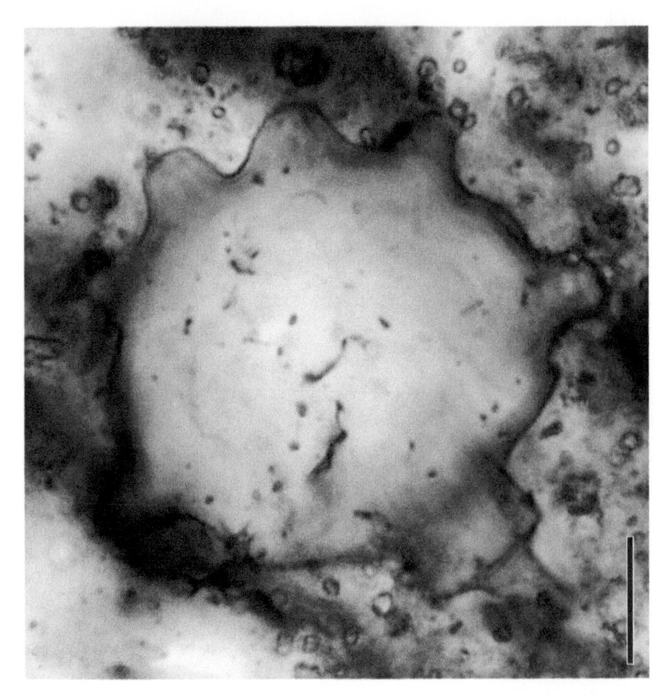

FIGURE 13.43 Putative peronosporomycete oogonium with coarse ornament. Mississippian, France. Bar = 15 µm.

PERONOSPOROMYCETES: CONCLUSIONS

It is perhaps not surprising that the fossil record of a group of organisms like the Peronosporomycetes has been slow to accumulate, especially because the primary diagnostic features are difficult to recognize in fossils. There are several reasons for this that parallel those in much of paleomycology. One concerns the idea that the preservation potential of these organisms is far too low to provide a reliably identified fossil record. A second reason is the preparation technique used to study fossil organisms of this type. The cellulose acetate peel, the standard procedure used in examining Carboniferous coal ball plants, is inadequate compared to the more traditional but time-consuming preparation of thin sections (Taylor et al., 2011; see Chapter 2, p. 22). And finally, there is the continuing issue that the paleobotanical community has the fossils, while the mycological community has the expertise to readily identify them. We hope that in continuing to point out these issues, not only will increased communication be beneficial to the scientific community at large, but there will be an increased realization that groups like the Peronosporomycetes do in fact have a fossil record (see Table 13.1) and thus are important in any discussion regarding the evolution of heterotrophic organisms.

GLOSSARY

Below are short definitions for some of the specialized terms used in this volume. For an extensive glossary of mycological terms, please see Kirk et al. (2008), and Ulloa and Hanlin (2012); for geologic terms, see Neuendorf et al. (2005), and for botanical and paleobotanical terms, Taylor, T.N., et al. (2009).

abaxial The lower surface; on the side away from the axis of the plant (contrast adaxial).

absolute dating Quantitative dating of rocks utilizing decay of radioactive elements (see radiometric dating; contrast relative dating).

acanthomorphic acritarch An acritarch with a clear distinction between the central body and the radially oriented processes attached to the central body.

acaulosporoid spore A developmental pattern in Glomeromycota in which asexual spores are produced within a blastic saccule either laterally or centrally.

acervulus (pl. acervuli) Aggregation of hyphae in the form of a small cushion that ruptures the host epidermis and on which are formed crowded conidiophores.

achlorophyllous Lacking chlorophyll.

acritarch A generalized term for a unicellular, morphologically variable microfossil whose affinities cannot be determined; many acritarchs represent cysts or phycomata of planktonic algae.

Actinobacteria A group of Gram-positive bacteria found in fresh water and, especially, soil.

acuminate Tapering to a slender point.

adaptation A structure or feature that performs a particular function and which presumably results in increased survival or reproduction.

adaxial Upper surface; the side toward the axis; facing the stem (contrast abaxial).

adventitious A structure arising from an unusual or abnormal position (e.g. roots growing from leaves).

aeciospore A binucleate spore produced in an aecium.

aerial Living above the surface of the ground or water (as in aerial roots).

aeroaquatic fungi Hyphomycetous fungi found in still and stagnant water, often having conidia with three-dimensional surface ornaments (e.g. coils, cage-like coverings), which aid in flotation and dispersal.

aerobes Organisms that require or can grow in the presence of free oxygen (O_2).

aerobic Requiring free oxygen (O_2).

agaric A member of the Agaricales, one of the largest groups of basidiomycetes.

Agonomycetes A group of unrelated fungi consisting of sterile mycelia lacking sexual or asexual reproduction by spores.

aleuriospore An asexual spore that develops by the swelling of a terminal or lateral cell of a hypha; found in certain conidial fungi.

alginite An organic component of coal that consists of algal matter.

allochthonous Sediment formed or produced some distance from its present location.

alluvial plain A level or slightly sloping land surface formed by extensive deposition of unconsolidated detrital material by a stream or other body of water.

alternation of generations A type of reproductive cycle in which a haploid or gametophyte (n) phase alternates with a diploid or sporophyte ($2n$) phase.

amb The outline or shape of a spore (or pollen grain) as viewed in the proximo-distal plane (i.e. from one of the poles).

amber Solidified fossilized plant resin, usually yellow-brown in color; may be classified as to origin and color.

amerospore A unicellular spore not divided by a septum.

amphigynous antheridium In peronosporomycetes, an antheridium that encircles the stalk of the oogonium, formed when the oogonium initial grows through the antheridial initial.

anaerobes Organisms that can live in environments lacking free oxygen (O_2).

anaerobic Occurring in the absence of free oxygen (O_2).

analogous Structures with a similar function that do not have a common evolutionary history (i.e. structures resulting from convergent evolution) (contrast homologous).

anamorph The asexual, conidial, or the so-called imperfect state of a fungus in which the spores are produced by mitosis (see also teleomorph and holomorph).

ancestral A preexisting condition or character state.

anhydrobiosis The ability to tolerate extreme desiccation.

anoxic Depleted of oxygen.

anoxygenic photosynthesis Photosynthesis that does not produce oxygen.

antheridium (pl. antheridia) A unicellular (in most fungi and algae) or multicellular (in land plants) reproductive structure that produces sperm (male gametes); a male gametangium.

anthracite A hard black metamorphic coal with a high percentage of fixed carbon (92–98%) and a low percentage of volatile matter.

anticlinal Perpendicular to the surface (contrast periclinal).

aperture A site through which the contents of a spore or pollen grain exit (e.g. laesura, pore, colpus).

apiculate Terminating in a sharp, flexible point.

aplanospore An asexual, immotile spore produced singly or in clusters within a sporangium.

aplerotic oospore A condition in which the oospore does not completely fill the oogonium.
apomorphy A derived feature that provides information on evolutionary relationships (contrast plesiomorphy).
apophysis A natural swelling or enlargement: (1) swelling found beneath the zoosporangium in some aquatic fungi (e.g. chytrids); (2) the wide portion of some mucoralean sporangiophores; (3) the endoperidium of some members of lycoperdalean fungi.
apothecium (pl. apothecia) An open ascocarp.
applanate Shelf-like.
apposition An increase in thickness of a cell wall through the addition of new wall material.
apressorium A slight swelling that forms at the end of a germ tube or vegetative hypha and which attaches to the surface of a host before penetration with some type of infection hypha.
arborescent Treelike.
arbuscule Shrubby and highly branched hyphae that occur inside the host root and rhizome cells of endomycorrhizal fungal associations.
arbutoid mycorrhiza A mycorrhizal type characterized by a mantle, Hartig net, and intracellular hyphae that form complexes in the epidermal cells of the host root.
archegonium (pl. archegonia) Multicellular structure that produces an egg; in land plants, it consists of a swollen venter, which surrounds the egg, and a narrow neck.
arthroconidium A type of asexual fungal spore produced by the segmentation of an existing hypha.
arthrospore A spore that results when a hypha fragments.
artifact In biology, a structure or material not normally present, not naturally occurring; for example, produced by preparation techniques (British spelling = artefact).
***Arum*-type** Pattern of arbuscular mycorrhizal colonization in which hyphae are intercellular with short branches giving rise to arbuscules; common in crop plants (contrast *Paris*-type).
ascocarp A fungal fruiting body containing asci.
ascogenous Pertaining to the production of asci.
ascogenous hyphae In Ascomycota, hyphae that give rise to the ascus.
ascohymenial A type of ascocarp development that is initiated by fertilization (dikaryon condition), followed by differentiation of a hymenium from ascogenous cells and subsequent development of the ascocarp; results in the formation of true apothecia, perithecia, or cleistothecia.
ascolocular A type of ascocarp development that is initiated by formation of a small stroma in which ascogenous cells subsequently differentiate; ascogenous hyphae then develop into cavities or locules where asci form.
ascoma (pl. ascomata) A fruiting body or sporocarp with asci (sing. ascus).
ascospore A sexual haploid spore produced within an ascus, usually as a result of meiosis.
ascostroma (pl. ascostromata) A fungal fruiting body of the Ascomycota that consists of a stroma that forms asci.
ascus (pl. asci) The saclike reproductive structure of the Ascomycota.
aseptate Lacking septa (= nonseptate) (contrast septate).
astomate A closed ascoma that lacks an ostiole; in plants, lacking stomata.
atomic force microscopy (AMF) High-resolution scanning probe microscopy with a resolution of fractions of nanometers of almost any surface.
authigenic A mineral or rock that was formed where it is found.
authigenic cementation A type of preservation that involves soft sediment cementation by iron and carbonate compounds; it results in replicas of surface features including molds and casts (e.g. ironstone concretions from the Mazon Creek area of Illinois, USA).
autochthonous Sediments or fossils which formed or lived in the place where now found.
autotrophy A form of metabolism in which the organism (= autotroph) synthesizes its own food from inorganic compounds; examples include photosynthetic and chemosynthetic organisms.
auxiliary cell In Gigasporaceae, a thin-walled, spore-like structure formed from hyphae after spore germination.
azygospore A spore that is morphologically identical to a zoospore, but which forms without prior gametangial fusion (asexual), as in some species of the Mucorales.
baculae Small rods.
baculate A type of ornamentation in pollen and spores consisting of small rods.
ballistospore In Basidiomycota, a spore that is forcibly discharged from a sterigma of a basidium. In Entomophthorales, a spore forcibly discharged from a conidium.
bark The complement of tissues external to the vascular cambium in plants.
Bayesian methods A field of statistics that expresses degrees of probability based on observed distribution.
basidiocarp A fruiting body or sporophore in/on which basidia and basidiospores are produced.
basidiome Any structure that produces basidia (= basidioma, basidiocarp)
basidiospore A sexual spore formed exogenously on a basidium.
basidium (pl. basidia) A specialized cell on which are formed basidiospores that are supported by the sterigmata.
basipetal Toward the base.
Bennettitales An extinct group of gymnospermous plants characterized by pinnate leaves and reproductive organs in the form of open cones (also called cycadeoids)
benthic Living on or attached to the bottom in aquatic habitats (contrast pelagic, planktonic).
bifid Divided into two parts.
bifurcate To divide into two halves; dichotomize.
binding hyphae Highly branched, sinuous hyphae that interlace and provide flexibility to some basidiomycete sporophores.
biodictyon A three-dimensional network of filamentous structures embedded in soil, sediment, or rock that can trap microorganisms, minerals, sediments, and other materials percolating through the matrix.
bioerosion The erosion of hard marine (e.g. calcium carbonate) and terrestrial substrates by living organisms.
biofilm A community of microorganisms embedded in extracellular biopolymers that form a thin film.
biogenic Of biological origin; formed by organisms.
bioherm A mound or reef-like mass of rock built up by organisms (e.g. coralline algae) and composed primarily of their calcareous remains.
bioinformatics A field that involves storing, retrieving, organizing, and analyzing large amounts of biological data.
biomarker A chemical signature of an organism in the sedimentary record; a product derived from biochemical precursors that provides evidence of the past presence of an organism or group of organisms.
biomimetic A structure or substance that morphologically appears to be biogenic or biological but is not.
biomineralization The process in which living organisms accumulate minerals or cause them to precipitate.
biominerals Crystals precipitated as a result of complex interactions between organic and inorganic ions (e.g. calcium oxalate crystals precipitated on fungal hyphae).

biosignature An element, molecule, isotope or other substance that provides evidence of past or current life.

biostratigraphy Relative dating and/or correlation of rocks by the type of fossil organisms (plants, animals, and/or microfossils) that they contain.

biotroph An organism that lives and multiplies on another living organism; a condition in which only living cells are used as a nutrient source.

biozone Stratigraphic unit(s) defined by the fossil taxa they contain.

bitunicate A type of ascus that has two functional layers at the time of spore release; they are termed the endoascus and exoascus (contrast unitunicate and prototunicate).

blastic conidiogenesis One of two basic types of conidial development in which the conidial primordium undergoes enlargement before it is delimited by a septum. When all wall layers of the conidium are involved in development, the pattern is termed holoblastic; if only the inner wall layer is involved, the pattern is termed enteroblastic (contrast thallic conidiogenesis).

blastospore An asexual spore that develops by budding.

brown coal A brown or yellowish material that is intermediate in rank between peat and sub-bituminous coal and in which some original plant structures can be identified (also called lignite).

brown rot A fungal symptom in wood in which hemicellulose and cellulose are broken down.

bulbous base In gigasporoid spores (Glomeromycota), the initial stage in spore production that is followed by blastic expansion and the production of a large spore.

burl A rounded woody outgrowth of a tree trunk, branch or root in which there are many knots from dormant buds and the wood grain is deformed.

calamite An extinct, spore-producing arborescent fossil plant of the family Calamitaceae, distantly related to the modern horsetails (*Equisetum*) and characterized by whorled branches and leaves.

calcareous nodule Rounded fragment or nodule of calcium carbonate (calcite).

calcification A type of fossilization in which calcium carbonate ($CaCO_3$) either infills intercellular spaces and other voids (permineralization) or replaces the original organic material (petrifaction) of an organism (see also coal ball).

callosity (pl. callosities) A conical structure that develops from the host cell in response to some external stimulus, such as penetrating fungal hyphae (also called apposition or papilla).

canaliculus (pl. canaliculi) A narrow channel, groove, or furrow.

capitate hyphopodium A type of hyphopodium (short lateral branch of a hypha with one or two cells) with a rounded end.

capitulum (pl. capitula) A globose, apical apothecium in certain lichens.

Carboniferous A geologic time period that extends from 359 to 299 Ma. Divided into the Mississippian (Early Carboniferous) and Pennsylvanian (Late Carboniferous) and characterized by widespread coal deposits in Europe and North America.

cast A preservation type that forms within a mold; casts are usually three-dimensional representations of the fossil organism.

catapol A monoterpene with a glucose molecule attached; an iridoid glucoside.

catenulate Forming a chain or chain-like organization, as in some conidia.

cecidozoa Informal designation for animals that produce cecidia or galls (e.g. insects, mites, nematodes).

centrum A complex of fertile and sterile structures that occur in the interior of an ascocarp.

chalaza The proximal or basal region of an ovule (contrast micropyle).

charcoal A black, porous, impure carbon residue resulting from the burning of wood or other organic material in the absence of air.

chasmoendolith A type of endolith that colonizes fissures and cracks in rocks (contrast cryptoendolith and euendolith).

chemolithotrophic A organism that can obtain its energy from the oxidation of inorganic compounds.

chert A hard, brittle sedimentary rock consisting of microcrystalline quartz.

Chitinozoa, chitinozoans Vase-shaped microfossils of an extinct marine group (Cambrian–Devonian) with uncertain affinities.

chlamydospore A thick-walled, asexual fungal resting spore.

Chlorophyceae A large class of freshwater green algae (Chlorophyta) that are characterized in part based on ultrastructural features.

Chromista (Chromalveolata) A supergroup of crown eukaryotic, polyphyletic organisms (brown algae, diatoms, chrysophytes, cryptomonads, hyphochytridiomycetes, oomycetes, Labyrinthulomycota); some contain chlorophyll c and do not store energy in the form of starch (see also stramenopiles, heterokonts).

chytridiomycosis A disease of mostly amphibians, caused by certain parasitic chytrids.

cilium (pl. cilia) A hair-like process found on the surface of some eukaryotic cells; in lichens, filiform thallus appendages.

clade A monophyletic group or lineage.

cladistic A type of methodology for inferring the evolutionary history of a group of organisms by grouping taxa based on shared, derived characters (synapomorphies).

cladogram A tree-like graphic representation that indicates the relative degrees of phylogenetic relationships of the taxa analyzed.

clamp connection A specialized connection between two adjacent hyphal cells in a member of the Basidiomycota; a specialized bridge over a septum that functions to maintain the dikaryon condition.

clavate Club shaped.

cleistothecium (pl. cleistothecia) A type of ascocarp in which the asci are completely enclosed.

coal ball A type of permineralization known from Carboniferous and Permian coal measures in which the plants are preserved by calcium carbonate and other minerals (e.g. pyrite).

coalification The alteration of plant material into coal.

coccoid Spherical; used to describe the shape of certain bacterial cells.

coelomycetes An informal name for asexual fungi that produce conidia on conidiophores contained in an acervulus or pycnidium.

coenocytic A cell in which the nuclei are contained in a common cytoplasm without being separated by septa or cross walls; a multinucleate cell.

collarium (pl. collaria) A ring of tissue that is attached to the proximal edge of the gills on the lower side of a basidiocarp.

compression A preservation type in which the organic matter consists of a thin film of carbon.

concretion A nodule that forms in an accretionary manner around a nucleus or center. These nodules generally differ chemically from the rocks in which they are found and many are known to contain plant and animal fossils, such as the ironstone nodules from Mazon Creek, Illinois.

coni (sing. conus) An ornamentation pattern on spores consisting of cone-shaped elements.

conidiogenous cells Cells that produce conidia.

conidioma (pl. conidiomata) A more complex structure formed by hyphae that produces conidia.

conidiophore A hypha that is morphologically and/or physiologically differentiated from somatic hyphae and on which conidia are produced; a stalk that bears conidia.

conidium (pl. conidia) In fungi, a nonmotile, asexual spore usually formed at the tip or side of a conidiogenous cell.
conk Another term for a shelf or bracket fungus basidiocarp.
context The inside flesh of a basidiocarp.
conus (pl. coni) A type of sculpture in pollen and spores consisting of small, cone-shaped projections.
convergent evolution Evolution that results in similar structures in organisms that do not share a common evolutionary history.
coprolite Fossilized fecal material (see also frass).
coprophilous Growing on or living in animal dung.
coralloid roots In cycads, a cluster of tightly formed dichotomously branched roots that are at the soil level; some may contain cyanobacteria.
cordaite (Cordaitales) A group of extinct gymnosperms that lived during the Carboniferous and Permian and that are thought to be related to modern conifers.
cork A type of secondary tissue produced by a cork cambium in plants; cork cells are nonliving at maturity and have waxy or fatty deposits in their walls.
corm An upright, thickened, underground stem that functions in food storage.
cortex Primary tissue system of parenchymatous cells between the epidermis and conducting tissues of many plant organs.
cross-polarized light In a polarizing microscope, when both the polarizer and analyzer are inserted so that the light vibrates in both an E–W and N–S direction. Sometimes abbreviated to XPL.
crown group In phylogenetics, a group that includes the last common ancestor and all the living members of the clade.
cruciate Four-armed, having the form of a cross.
crustose The thallus of a lichen that is tightly attached to a substrate and lacks a lower cortex; the surface may be powdery, areolate, granulose, or with warts.
cryptoendolith A type of endolith that colonizes structural cavities in porous rocks, including spaces evacuated by euendoliths (contrast chasmoendolith and euendolith).
cryptogam A seedless plant.
Cryptomycota A clade of endoparasitic microorganisms that are now thought to be fungi.
cryptospore A spore type for which the source plant is not known; mostly used for Paleozoic spores older than the earliest land plant megafossils.
cuticle The amorphous, waxy layer that covers and impregnates the walls of the epidermal cells on the aerial parts of plants; made of cutin and waxes.
cutin The waxy polymeric substance that makes up the cuticle.
cutinite A type of coal maceral composed of plant cuticle.
cyanobacterial mat A multilayered sheet or film of bacteria, archaea, and other microorganisms that form a community that is determined by the dominant species, nature of the sediment, and the environment in which the mat develops.
cystidium (pl. cystidia) A large, inflated sterile cell that occurs among basidia and paraphyses in Basidiomycota.
Deccan Intertrappean Beds A series of sedimentary beds intercalated with volcanic flows (the Deccan Traps) that occur in west-central India. Although dating of the Intertraps has been controversial, they are now thought to be latest Cretaceous to Paleogene.
decortication The stripping away of bark, outer cortex, or periderm in a vascular plant.
deliquescent Dissolving or liquefying at maturity.
dentate Having teeth or with a toothed margin.
Deuteromycetes Historical group (i.e. a form group) name used to denote conidial fungi that lack a method of sexual reproduction. Also called asexual fungi, conidial fungi, mitosporic fungi, or Fungi Imperfecti.
dextrorse Turned to the right; clockwise.
diagenesis The physical and chemical changes that occur to sedimentary deposits (and the fossils they contain) after deposition.
dichotomize To divide into two equal parts; bifurcate.
dichotomous Divided into two equal parts.
diclinous A condition in which an antheridium is produced on one hypha or branch and an oogonium on another.
dictyospore A spore with transverse and longitudinal septa (= muriform).
didymospore A two-celled spore.
dikaryon Two nuclei from different mating types that exist in the same cell without karyogamy.
dimidiate An ascocarp in the shape of a shield that appears to lack one half; a basidiocarp in which one side is more developed than the other.
diporate Having a pore at each end of a unicellate spore.
diploid Having two sets of chromosomes in each body cell ($2n$) (contrast polyploid and haploid).
disciform Disc-shaped (= discoidal).
Discomycetes A class of ascomycetes with apothecial ascocarps that are exposed at maturity.
dissepiment Tissue (plectenchyma) between the pores of a poroid fungus; a septum.
distal Away from the axis, center, or point of attachment.
distal surface The surface of a spore or pollen grain away from the center of a tetrad (contrast proximal surface).
dolipore septum (septum with dolipores) A type of septum found in basidiomycetes that is characterized by a central pore covered on both sides by a septal pore cap (parenthesome).
dolomite A mineral or sedimentary rock composed of calcium magnesium carbonate.
domatium (pl. domatia) A small cavity produced by plants that is the habitat for certain arthropods.
dubiofossil A fossil whose biogenic origin (i.e. resulting from a once-living organism) is in question.
Earthstars A group of basidiomycetes in which the exoperidium splits into more or less triangular segments that give a star-like appearance to the basidioma.
ectendomycorrhiza An association of certain ascomycete fungi with a few conifers in which the fungus initially forms a mantle and Hartig net (see ectomycorrhiza below), but later develops intracellular hyphae that extend into the epidermal and cortical cells.
ecto- Prefix meaning outside, outer, external.
ectomycorrhiza A type of mycorrhiza in which the fungus surrounds the root tip with a sheath (mantle) and penetrates into intercellular spaces (Hartig net); roots with ectomycorrhizal infections are typically short, swollen, and branched; common in modern pines (contrast endomycorrhiza).
ectoparasite A parasite that thrives on the exterior of another organism and which may be nourished by haustoria, special structures that penetrate host cells.
ectopic Arising in an unusual or abnormal place.
ectotrophic Deriving nourishment from the surface; a mycorrhiza in which the mycelium forms an external covering on the root but does not penetrate the host cells (i.e. ectomycorrhizae).
edaphic conditions The physical and chemical nature of the soil and/or water in which a plant is growing.
Ediacaran The youngest period of the Neoproterozoic Era (late Precambrian) characterized by the earliest soft-bodied forms of metazoan life.

enation A nonvascularized, epidermal outgrowth found in some early land plants.
endexine The inner, typically homogeneous layer of the exine in pollen and spores that normally stains less deeply.
endo- Prefix meaning inside, within, internal.
endobiotic Living in the interior of another organism.
endodermis A tissue that separates the vascular tissue from the cortex in some vascular plant organs; Casparian strips or thickenings may occur on the anticlinal walls.
endolith An organism that grows into and colonizes the interior of a rock (see also chasmoendolith, cryptoendolith, and euendolith)
endomycorrhiza The most common type of mycorrhizal infection in which the fungus grows intracellularly within the cortex of the root; the root is not conspicuously different in morphology from uninfected roots (contrast ectomycorrhiza).
endoparasite A parasitic organism that develops in the interior of its host.
endoperidium The inner layer of a peridium.
endophyte, endophytic Microorganisms and fungi that live within the tissues of a host in a variety of relationships ranging from symbiotic to pathogenic.
endospore (1) The innermost layer that forms the wall of a complex basidiospore; (2) asexual spore formed from internal budding in some yeasts (= endosporium); (3) a resistant spore that forms inside some bacterial cells.
endosporium The inner layer of a basidiospore; also an asexual spore formed by internal budding of the hyphae of certain yeasts, or endospore.
enigmatic see problematic.
entomopathogenic Fungi that cause disease in insects and other arthropods.
entomophilous Pollination by insects.
epibiotic An organism that lives on the surface of the host.
epilithic Organisms that live on or are attached to a rock surface.
epiphyllous Upon a leaf; an organism that lives on leaves; some epiphylls may also be parasitic on the leaf itself.
epiphyte Growing upon another plant and not attached to the soil.
ergastic deposits Cellular waste products, either crystalline or noncrystalline.
ericoid mycorrhiza An ascomycete mycorrhizal type confined to a few families in the angiosperm group Ericales, in which the fungus forms a branched hyphal complex in each epidermal cell.
erosion trough Areas of thinning or fragmentation in the wall of xylem elements caused by fungal enzymes.
Euascomycetes Filamentous ascomycetes in some classification systems.
eucarpic A chytrid in which only a portion of the thallus is transformed into a reproductive structure(s) (contrast holocarpic).
euendolith A type of endolith that actively penetrates into the interior of rocks, forming tunnels that are the shape of its body; a rock borer (contrast chasmoendolith and cryptoendolith).
eukaryote An organism that possesses a membrane-bound nucleus and other membrane-bound organelles (contrast prokaryote).
eumycorrhizae Mycorrhizae that develop in root systems within the soil.
eurytopic A plant that is able to adapt to a wide range of environmental conditions and so is geographically widespread.
evanescent Short lived.
exine The outermost wall layer of the sporoderm of pollen and spores, outside the intine; made of sporopollenin.
exocytosis A process that results in the contents of secretory vesicles moving out of the cell and into the extracellular space.
exoperidium The outer layer of a peridium.
exosporium The second layer of the wall of a complex basidiospore that is composed of five layers.
extraradical hyphae The portion of a mycorrhizal fungus that occurs outside of the host root.
extraxylary tissues All tissue systems that are external to the secondary xylem; may include phloem and various types of cortical and epidermal tissues.
facies The appearance and all of the characteristics of a rock unit that typically reflect its condition of origin.
facultative parasite An organism that has the ability to invade and live on another organism, but can live on dead organic material depending on the circumstances (contrast obligate parasite).
falcate Sickle shaped.
fasciculate asci Bundles or clusters of asci.
filiform Threadlike or filamentous shape.
fissitunicate A type of bitunicate ascus in which the inner layer is separate from the outer layer and extends out at the time of spore liberation.
foliose Leaflike in appearance, as in foliose lichens, which are loosely attached to the substrate.
Foraminifera A group of heterotrophic protists mostly with shells of calcium carbonate; some are planktonic and some benthic; also termed forams.
fossil taxon See morphotaxon.
foveolate Ornamented with small pits; pitted.
frass Fossilized fecal remains, typically of a small, burrowing arthropod (see also coprolites).
fruticose A lichen thallus attached to the substrate at a single point. It develops branches that result in a three-dimensional morphology and look shrub-like.
fungiculture The cultivation of fungi by certain animals (ants, termites, etc.).
funginite A type of coal maceral made up of fungal spores, sclerotia, and hyphae.
fungivory Feeding on fungi.
fungus comb Complex, convoluted structures formed by certain termites from macerated wood pseudofeces. They are used to grow fungi (see fungiculture) and consist of numerous chambers; the architectural organization is distinct for the particular species.
furcate Forked or bifurcate.
fusain A component of coal characterized by its silky luster, fibrous structure, and black color; fusinized (or fusainized) refers to fossil plants preserved in fusain (i.e. preserved as charcoal or charcoalified).
fusinite A type of inertinite with some cellular structure, a reflectance above vitrinite, and a particle size of 50 μm.
gall Abnormal outgrowths on the surface of plants that can be caused by fungi, other microorganisms, or arthropods; often provide food and habitat for insects.
gametangial fusion A type of sexual reproduction in which two different gametangia or their protoplasts fuse to give rise to a zygote.
gametangium (pl. gametangia) A structure that bears gametes or gametic nuclei sex cells.
gamete A haploid (n) reproductive cell or sex cell; sperm and egg are two types of gametes.
gametophyte The haploid (n) phase of a plant life cycle on which gametes are produced (contrast sporophyte).
ganodermatoid Fungal spore type in which two wall layers are separated by delicate inner wall pillars, as in spores of most members of the polypore family Ganodermataceae.

gasteromycetes (gasteroid fungi) An artificial grouping of basidiomycetes not closely related to each other in which the spores are contained in the interior of the spore-producing tissue.

GC–MS Gas chromatography–mass spectrometry.

generative hyphae Thin-walled basidiomycete hyphae that are multi-branched, regularly septate, and capable of producing basidia.

geobiology The interdisciplinary study of interactions between the biosphere and lithosphere.

geochronology The science of dating rocks.

geomicrobiology The interactions of microorganisms in geological and geochemical processes.

geomycology The role that fungi play in fundamental geological processes, including bioweathering, organic and inorganic transformations, element cycling, and rock and mineral transformations.

geothermal Heat of the interior of the Earth.

germ tube A short hypha that extends from a pore of a spore during germination and that can develop into a larger hypha.

germinal wall Semiflexible layer of the asexual spore of Glomeromycota through which the germ tube emerges.

germination shield A coiled or shield-like, infolded or lobed structure between individual wall layers in gigasporoid and acaulosporoid glomeromycotan spores, which develops prior to germ tube formation.

germling A bud or newly developed propagule with the capacity to grow into a mature organism.

gleba The internal portion of a gasteromycete fruiting body composed of a hymenium that forms basidiospores and a sterile zone; hyphal network in zygomycetous and glomeromycotan sporocarps from which the gametes or spores develop.

gongylidium (pl. gongylidia) A bulbous structure that develops on hyphae of fungi cultivated by termites; the termites feed on these structures.

grana An ornament pattern on spores composed of small rounded elements.

granulose Granular, finely roughened.

halophyte A plant that grows in a saline environment.

haplobiontic life cycle A life cycle in which there is only one free-living organism; if zygotic meiosis is present, the free-living organisms are haploid and the zygote is the only diploid cell.

haploid Having a single set of chromosomes in a cell (n) (contrast polyploid and diploid).

haptotypic mark A characteristic mark or marks on spores or pollen grains that results from contact within a tetrad, for example, the laesurae and contact areas, or a trilete mark; the presence of a haptotypic mark is sometimes used in fossils as evidence for the occurrence of meiosis.

Hartig net A hyphal network that extends into the root between epidermal and cortical cells and is the site of nutrient exchange in ectomycorrhizae.

haustorium (pl. haustoria) The structure that penetrates a host and absorbs nutrients in a parasitic relationship.

heartwood The inner, often darkly colored layers of wood which have ceased to conduct and contain no living cells. Heartwood is generally more resistant to decay.

helicospore A cylindrical fungal spore that is coiled into a spiral or helix.

hemibiotrophic parasite An organism that parasitizes a living host for some time and then thrives on the dead host. For example, in entomogenous fungi there is a parasitic phase of yeast-like cells that inhabit the living insect, as well as a saprobiotic mycelial phase in which the insect becomes hardened, and in which the fungus can survive until reproductive stromata can be produced.

herbaceous A plant having little or no secondary development; nonwoody.

herbivory A type of predation in which animals consume autotrophs (e.g. plants, algae, photosynthetic bacteria).

Heterobasidiomycetes In some classifications, a group of fungi that have heterobasidia and basidiospores that germinate by forming secondary spores instead of mycelia (contrast Holobasidiomycetes).

heterokonts see stramenopiles.

heteromerous A type of lichen thallus in which the mycobiont and photobiont are arranged in distinct layers with the hyphae and photosynthetic cells well delimited.

heterothallic Fungi in which mating requires two different strains, each from a different mycelium or different structures on the same thallus (contrast homothallic).

heterotroph An organism that obtains its food by ingesting organic materials. Heterotrophs are incapable of producing their own food. See also saprotrophism, mixotrophy, osmotrophy, and phagotrophy, which are types of heterotrophy (contrast autotrophy).

hilum (pl. hila) A scar or mark on a spore wall that indicates the point of attachment. In angiosperms, the hilum is the scar on a seed.

holdfast The basal structure in trichomycetes that attaches the thallus to the cuticle of the host.

holoarthic conidiogenesis A pattern of conidial development in which all the wall layers of the conidiogenous cell participate in the formation of the conidium wall.

Holobasidiomycetes In some classifications, a group of fungi that have a homobasidium and basidiospores that germinate to form a mycelium instead of secondary spores (= Homobasidiomycetes) (contrast Heterobasidiomycetes).

holocarpic A chytrid whose entire thallus is converted into one or more reproductive organs (contrast eucarpic).

holomorph The fungus as a total organism that includes both the sexual (teleomorph) and asexual (anamorph) stages.

holozoic Feeding entirely in the manner of an animal by ingesting complex organic matter.

Homobasidiomycetes another name for Holobasidiomycetes.

homologous Having a shared evolutionary history (contrast analogous).

homomerous thallus Lichen thalli in which photo- and mycobiont are uniformly distributed throughout a gelatinous matrix.

homothallic Fungi in which sexual reproduction can occur on a single mycelium (contrast heterothallic).

host A living organism that provides food and habitat to a parasite or a symbiont.

hydromorphic A soil developed in the presence of excess moisture or poor drainage (e.g. in a marsh).

hymenium A layer composed of specialized hyphae involved in the production of spores and which is organized into a palisade and often mixed with specialized sterile elements (e.g. paraphyses or cystidia).

hymenophore The sterile tissue on which the hymenium is borne.

hyperparasite A parasite that lives on or within another parasite.

hyperplasia An abnormal increase in the number of cells, which often increases the size of an organ.

hypertrophy An abnormal increase in the size of cells.

hypha (pl. hyphae) A single tubular filament that represents the structural entity of most fungi. A mass of hyphae makes up the mycelium.

hyphal mantle or investment Ancillary covering of fungal spores and sporangia, e.g. in certain extant and fossil Glomeromycota and zygomycetes, and in fossil fungal "sporocarps."

Hyphochytridiomycetes A class of fungus-like organisms characterized by an anterior tinsel flagellum.

hyphomycetes In some classifications, another name for conidial fungi that lack a fruiting body.

hyphopodium (pl. hyphopodia) A short, lateral hyphal branch represented by one or two cells, and characteristic of certain obligately parasitic fungi.
hypodermis The outer cortex immediately beneath the epidermis in vascular plants; often serves a support function and may include collenchyma or sclerenchyma.
hypogeous Growing below the surface (contrast epigeal).
hypoplasia An abnormal reduction in the number of cells in tissues.
ichnocoenoses An association of trace fossils.
ichnofossil A trace fossil; evidence of past life without the organism itself being preserved (e.g. burrows or trackways).
ichnogenus, ichnospecies Names given to morphotaxa of trace fossils.
imbricate Overlapping.
impression A preservation type that represents a negative imprint of an organism; no organic material remains.
in situ In place, as fossil pollen grains and spores found within the sporangium that produced them (contrast sporae dispersae), or fossils found in growth position.
inaperturate Of spores and pollen, having no aperture or other opening.
index fossil A geographically widespread fossil that is diagnostic of a particular restricted time period and therefore useful in biostratigraphy.
inertinite A coal maceral group characterized by a high carbon content and reflectance higher than vitrinite.
inertodetrinite A type of inertinite with a reflectance greater than associated vitrinite; occurs as angular fragments in coal.
integument Seed coat or testa; tissue that covers the megasporangium (nucellus) in a seed plant.
intercellular Extending or occurring between cells.
intine Generally the innermost layer of the sporoderm in pollen and spores, inside the exine; made up of cellulose and pectates and not present in many fossils.
intracellular Inside the cell, as in hyphae extending through a cell.
intrahyphal Inside a hypha.
intraradical hyphae Portion of a mycorrhizal fungus that lives within a host root.
isidium (pl. isidia) A generally cylindrical or globose outgrowth of a lichen medulla that consists of a cortical covering, the mycobiont, and photobiont; they serve as asexual propagules.
isogametes Gametes of the same size and appearance.
isomorphic A type of alternation of generations in which the haploid and diploid generations are morphologically similar.
isotomous Dividing equally and repeatedly.
kelp A large marine alga of the Laminariales (Phaeophyceae, brown algae).
kerogen The insoluble, amorphous organic matter present in sediments.
lacuna (pl. lacunae) A cavity or space within a plant; lacunate = containing lacunae.
lacustrine deposits Lake bed sedimentary deposits.
laesura (pl. laesurae) A haptotypic mark.
laevigate With a smooth surface (e.g. of pollen and spores).
Lagerstätte (pl. Lagerstätten) An exceptionally well-preserved or taxonomically rich fossil deposit.
lamina (pl. laminae) Flattened blade portion of a leaf or thallus.
lanceolate Longer than wide and tapering to a point.
lax Drooping or relaxed.
liana, liane A vine or woody climbing plant.
lianescent Growing like a liana or vine.
lichen A composite organism consisting of a fungus in a symbiotic relationship with a photobiont, which may be an alga or cyanobacterium, or both.
lichenicolous fungi Fungi that grow in or on lichen thalli.
lignin A biopolymer of phenylpropanoid units that surrounds the cellulose microfibrils in the secondary wall of many plant cells; it occurs in all plants that possess vascular tissue and provides structural strength, resistance to decay, and a barrier to water permeability.
lignite A brownish-black coaly deposit in which the plant material is not highly consolidated; intermediate in coalification between peat and sub-bituminous coal (also called brown coal).
lignophyte A clade composed of the seed plants and progymnosperms and characterized by a bifacial vascular cambium, among other features.
liguliform Strap shaped, flat and narrow.
limonite Hydrated iron oxides; limonitized fossil plants are preserved in limonite.
liptinite A coal maceral group that may include sporinite, cutinite, alginite, resinite, leptodetrinite, cuticular materials, resins and waxes. Sometimes termed exinite.
lithification The conversion of unconsolidated sediment to a coherent solid rock through compaction and cementation.
lithofacies A laterally continuous, mappable subdivision of a designated stratigraphic unit distinguished on the basis of lithology.
lycophyte A group of living and fossil spore-producing vascular plants with dichotomous branching and microphyllous leaves.
lysigenous Formed from the breakdown or lysis of cells, as in lysigenous lacunae.
maceral One of the organic constituents that make up the coal mass.
manginuloid hyphae Hyphae of some epiphyllous fungi that are characterized by the invisibility or evanescence of external walls, resulting in display of the pronounced septa as bars on the surface of the host leaf.
mantle hyphae A compact layer of hyphae that surrounds feeder roots of ectomycorrhizal plants; it is connected to the Hartig net on the inside and extramatrical hyphae on the outside.
mating arena In insects, a well defined area in which most mating occurs.
matrotrophy A developmental pattern in which the embryo is provided with additional nutrition from the parent generation and which is thought to be critical in the origin of embryophytes.
mazaedium (pl. mazaedia) Mucilaginous and powdery mass in certain ascomycetes, composed of ascospores and secretions of paraphyses; type of reproductive structure in certain lichens.
medulla In lichens with a heteromerous thallus, this is the central layer composed of loose hyphae; in algae, the central portion of the thallus.
megaphyll A leaf with more than one vein and a trace associated with a leaf gap in the stele; it may or may not be large in size; characteristic of vascular plants other than lycophytes.
megaspore A large, haploid (n) spore of a heterosporous plant that produces a megagametophyte, or female gametophyte; the single, functional spore in seed plants; in sporae dispersae, megaspores are arbitrarily defined as spores $>200\,\mu m$ in diameter.
meiosis Reduction division; two successive nuclear divisions that reduce the ploidy level of a cell from diploid ($2n$) to haploid (n); in plants, spores are the products of meiosis.
meiosporangium A sporangium in which meiosis occurs, especially those formed on sporothalli of certain members of the Blastocladiomycota.
meiospores Haploid zoospores in some members of the Blastocladiomycota.
melanized Certain fungi that have dark hyphae and spores.
merosporangium A sporangiolum with spores in a row, produced at the swollen end of a sporangiophore.

mesofossil Plant fossils that are intermediate in size (0.25 mm to several millimeters) and require microscopic study (e.g. megaspores, small seeds, flowers).
mesophyll The parenchymatous, chlorophyllous tissue in a leaf located between the upper and lower epidermis; may consist of palisade and spongy parenchyma.
metaphyte A multicellular plant with some degree of organ differentiation.
metazoa (sing. metazoon) Multicellular animals; also metazoan.
microbial mats An multilayered accumulation of films of communities of microorganisms.
microbialites An organosedimentary deposit formed from the trapping, binding, or precipitation of sediments by benthic microbial communities. Stromatolites are a type of microbialite.
microborings General term for micrometer-sized cavities and channels created by certain rock-etching endolithic microorganisms.
microgametophyte The haploid (*n*) plant body produced by the microspore in heterosporous and seed plants; the male gametophyte.
micronematous (1) Having hyphae of small diameter; (2) a conidiophore that appears morphologically identical to vegetative hyphae of the mycelium.
microphyll A leaf vascularized by a single (and undivided) vascular bundle that is not associated with a gap in the stele; typical leaf type of the lycophytes. Other vascular plants have megaphylls.
micropyle A small opening in the integument at the apex of a seed through which either the pollen (in gymnosperms) or the pollen tube (in angiosperms) enters (contrast chalaza).
microsclerotium (pl. microsclerotia) A small cluster of thick-walled dark cells, each of which can form a hypha or germ tube.
microsporangium (pl. microsporangia) A sporangium in which microspores are produced.
microspore In heterosporous plants, a small spore that produces the microgametophyte.
microsporophyll A leaf that bears one or more microsporangia.
miospore A neutral term for fossil spores <200 μm in diameter whose biological function (as micro- or megaspore) is not necessarily known.
mitosporangium A fungal structure in which spores develop by mitosis; for example, the thin-walled zoosporangia formed on the sporothallus of some Blastocladiales in which the nuclei are diploid and result in the formation of mitospores that germinate into new diploid sporothalli.
mixotrophy Obtaining nutrients by both photosynthesis (autotrophy) and ingesting organics or prey (heterotrophy).
mold A three-dimensional preservation type that represents a negative imprint of the plant (compare cast).
molecular clock A technique used to relate the divergence time of two species, based on DNA sequences or proteins measured against a reliable fossil record.
moniliform A series of more or less rounded segments in a chain.
monoblastic A type of holoblastic conidiogenous cell that produces a conidium at a single point or location.
monocentric A thallus with a single center of growth in which is formed the sporangium or resting spore (contrast polycentric).
monokaryon A haploid cell nucleus.
monolete A spore with a single, straight laesura on the proximal surface (contrast trilete).
monomitic A fructification with only generative hyphae which is thin walled, septate, multibranched, and capable of producing basidia.
mononematous With conidiophores borne singly (contrast synnematous).
monophyletic A group of organisms (clade) that consists of a common ancestor and all of its descendants.
monopodial Having a single main trunk or axis.
monopodially pinnate A system in which side branches are formed in two opposite rows and typically are oriented in a single plane.
monotropoid mycorrhizae Mycorrhizal type in some Ericales that form a thick mantle and Hartig net and which develop short hyphae (fungal pegs) that penetrate the epidermal cells.
morphotaxon (e.g. morphospecies and morphogenus) Artificial taxonomic units defined based on macromorphological features. They are used for, and assigned to, isolated parts of fossil plants (e.g. leaves, cones, stems), and indicate structural similarity but not necessarily biological relationship. Although still in use, morphotaxa were replaced by fossil taxa in the International Code of Nomenclature for Algae, Fungi, and Plants (ICN), which was adopted in 2011 (http://www.iapt-taxon.org/nomen/main.php).
mosasaurs A group of large, extinct marine reptiles.
motile Having the ability to swim actively (e.g. zoospores, some algal spores or gametes).
mucronate Ending in a sharp point.
mucronate hyphopodium Lateral hyphal branch that is conidiogenous and terminates in a sharp point.
multicellate A fungal palynomorph having six or more connected cells or chambers separated by septa.
murus (pl. muri) A pollen or algal cyst ornamentation type characterized by a series of ridges or walls that make up a reticulate sculpture pattern.
mutualism A symbiotic relationship between two organisms that is not detrimental to either organism involved; a system that provides mutual benefits to both partners.
mycelial strand See rhizomorph or rhizine.
mycelium (pl. mycelia) The body of a fungus, made up of hyphae which may be aggregated into different morphologies.
mycobiont The fungal component of a lichen.
mycobiota The assemblage of fungi indigenous to a habitat or place.
mycoheterotrophic Parasitic plants that are non-photosynthetic and obtain carbon from the fungi with which they are associated.
mycoparasite A fungus parasitic on another fungus.
mycophagy To eat fungi.
mycorrhiza (pl. mycorrhizae or mycorrhizas) A mutually beneficial association between the hyphae of certain fungi and the roots of vascular plants and certain bryophytes (see also ectomycorrhiza and endomycorrhiza).
mycostratigraphy The organization of rock strata on the basis of the fossil fungal spores they contain.
mylospheres Millimeter-sized pellets of fecal material used to form fungus combs.
nannoplankton Nannofossils; very small (usually <35 μm), planktonic marine fossils.
necrosis Partial or total death of cells or a tissue system.
necrotrophic A parasite that kills the host and then feeds on the dead host.
nematode A nonsegmented worm of the phylum Nematoda; many are important agents of plant disease; a number are parasitic on the roots of higher plants.
nematophagous fungi Fungi that are adapted to trapping and digesting nematodes.
Neotropics The tropical part of the Americas.
nexine The inner, unsculptured layer of the sporoderm; often lamellate in gymnosperms (contrast sexine).
node Position on an axis where appendages are attached.
nomenclature A system of names, terms, and/or the rules used to name organisms.

nonseptate Without cross walls; said of fungal hyphae and rhizoids (= aseptate) (contrast septate).

nucellus The megasporangium in a seed plant; covered by the integument.

obligate commensal A relationship between organisms where one benefits and the other is neither helped nor harmed.

obligate parasite An organism that can only live as a parasite of another organism.

obovate Egg shaped, with the large end facing outward.

oligotrophic An environment with low levels of nutrients, often nitrogen.

ontogeny The development of an organism (contrast phylogeny).

oogamy Sexual fusion of a small, motile sperm and a large, nonmotile egg.

oogonium (pl. oogonia) A unicellular gametangium that contains an egg; also the structure containing the oosphere(s) in peronosporomycetes.

oolitic limestone A rock made up of ovate accretionary bodies of calcium carbonate formed by inorganic precipitation.

ooplast A membrane-bound inclusion that forms from the coalescence of vesicles in peronosporomycete oospores.

oosphere A large, nonmotile gamete (= egg) in water molds.

oospore Sexual spore that develops within an oogonium from a fertilized oosphere upon fusion with an antheridium in some algae and fungal-like organisms.

operculate Having a lid-like opening in which dehiscence is circumscissile, as in operculate zoosporangia or asci.

operculum A lid covering the sporangium in some chytrids.

orchid mycorrhiza Mycorrhizal type characterized by hyphal coils (pelotons) within host cells that are essential for seed germination and seedling establishment in the Orchidaceae.

organotaxis The arrangement of organs on an axis.

oribatid mites (Oribatida) A group of small (~0.2–1.5 mm) arthropods with paired chelicerae, four pairs of legs, and the division of the body into an anterior and posterior region.

orthotropic Growing vertically or upright (contrast plagiotropic or prostrate).

osmotrophy Obtaining nutrition by the osmotic uptake of dissolved organics; a type of heterotrophy.

ostiolate perithecium A perithecium with a small pore.

ostiolate pycnidium A pycnidium with a small pore.

ostiole A small opening or pore, especially in a reproductive body.

paleobathymetry The study of past water depths.

paleoecology The study of the relationships between ancient organisms and their environment.

paleoenvironment The conditions under which ancient organisms lived.

paleomicrobiology The detection, identification, and characterization of microorganisms in ancient remains, generally in human populations.

paleomycology The study of the accumulated evidence about fungi and their activities measured in geologic time.

paleopathology The study of ancient diseases, generally based on skeletal or other hard part remains.

paleopedology The study of fossil soils.

paleophytogeography The past distributions and migrations of plants.

paleosol A fossil soil.

palisade parenchyma A layer or layers of column-shaped cells just below the upper epidermis in leaves; the cells contain chloroplasts.

palynology The study of pollen grains and spores; more generally, the study of organic microfossils including those produced by fungi.

palynomorph A general term for organic-walled microfossils with resistant walls; includes pollen, spores, acritarch, cysts, etc.

palynostratigraphy Application of palynological methods to defining sequences of rocks.

paper coal A sedimentary organic deposit that resembles coal but consists almost entirely of compressed sheets of plant cuticles.

papilla (pl. papillae) A small, slightly elevated thin area of a sporangium or gametangium through which zoospores or planogametes exit; a short protuberance; a very simple, unicellular hair. A host response to fungal invasion; also called a callosity or apposition.

papilliform Shaped like a papilla.

paragynous antheridium An antheridium that attaches to the side of the oogonium.

paramycorrhiza Mycorrhizae that occur in rootless thalli and shoot systems above the substrate in which they are growing (from Strullu-Derrien and Strullu, 2007).

paraphyletic A group that includes a common ancestor plus some, but not all, of the descendants of that common ancestor (see also polyphyletic and monophyletic).

paraphysis (pl. paraphyses) Sterile, elongate, basally attached structures present between asci in the hymenium of certain fungi (e.g. ascomycetes).

parasite An organism that derives its nourishment from another living organism.

parasitism A symbiotic relationship that is detrimental to one of the symbionts and beneficial to the other (contrast mutualism or saprotrophism).

parenchyma The basic cell and tissue type in plants; parenchyma cells are thin walled, living at maturity and variable in size and shape (contrast sclerenchyma).

parenthesome Parenthesis-shaped structures occurring on either side of the pores in the dolipore septum in Basidiomycota.

***Paris*-type** Pattern of arbuscular mycorrhizal colonization in which the hyphae are intracellular with arbuscules forming along coils (contrast *Arum*-type).

parsimony An axiom in data analysis that assumes that the simplest or most economical data set is the most accurate.

peat Partially decayed plant material, consisting of fragments of mosses, vascular plants, fungi, and algae that grew in a wet, swampy habitat; the moisture content is high (75%) and the carbon content is about 60% (contrast lignite and coal).

pelagic Pertaining to marine organisms that are free swimming (nektonic) or free floating (planktonic) (contrast benthic).

pellucid Translucent or transparent.

peloton Complex hyphal coils found in host cells of orchid mycorrhizae.

peri-abuscular membrane A plant-derived membrane that develops around the branching hyphae of the arbuscule, separating the fungus from the plant cell cytoplasm.

periclinal Parallel to the surface, as in periclinal cell walls (contrast anticlinal).

periderm Outer tissue in an axis consisting of a cork cambium and its derivatives.

peridium A protective structure or wall composed of branching and thick-walled anastomosing hyphae.

perine See perispore.

periphysis (pl. periphyses) Sterile hyphae that line the interior wall of the ostiolar canal of a perithecium.

perispore An extraexinous wall layer found in some pollen and spores; made of sporopollenin; also called perine.

perithecium (pl. perithecia) An ascocarp that is closed at maturity except for a small pore or ostiole at the top.

permineralization A preservation type in which mineral matter has filled in the cell lumens and intercellular spaces but has not replaced the cell walls; permineralizations can be studied by the peel technique or by thin sections.

perylene A polycyclic aromatic hydrocarbon that has been suggested as a fossil fungal biomarker.

petrifaction A type of preservation in which mineral matter not only fills in the intercellular spaces and cell lumens but also replaces the cell walls; petrifactions can be studied by making thin sections of the specimen.

phagotrophy Obtaining nutrition by engulfing other organisms, a type of heterotrophy.

phenetic system A classification system that groups organisms based on the overall similarity of characters, whether those characters are primitive or derived.

phialide A short, flask-shaped conidiogenous cell that produces blastic conidia in basipetal succession.

phialidic conidiogenesis A type of conidiogenesis in which each conidium develops basipetally by the formation of new cell wall material and not from cell walls that already exist.

photoautotrophs Organisms that produce their own food via photosynthesis.

photobiont The photosynthetic symbiont of a lichen.

phototactic Showing a response to light.

phototroph An organism that uses photons to acquire energy.

phragmospore Multicellular spore divided by more than two septa.

phycomycetes A class name once used to represent a single taxonomic group of so-called lower fungi that are now included in the Chromista.

phylloplane All biological and physical aspects associated with the surface of a leaf.

phylogenetic systematics A methodology used for the reconstruction of phylogenetic trees and the discovery of monophyletic groups by the use of shared (homologous), derived character states.

phylogenetics The study of the evolutionary relationships among groups of organisms which are based on morphological characters and molecular sequence data sets.

phylogeny The evolution or history of a particular group or lineage (contrast ontogeny).

phytolith A small, stony structure that represents the remains of a plant secretion, often consisting of calcium oxalate or opaline silica.

phytopathogenic Microorganisms that can induce disease in plants.

phytophagous An organism that feeds on plants; an herbivorous organism.

pilae A spore ornamentation pattern consisting of clavate-shaped units.

pileus (pl. pilei) The expanded upper cap portion of certain types of ascocarps and basidiocarps that contains the hymenium.

pillar A column-like structural element in the wall of a spore.

pilocystidium A large cell that extends from the sterile surface of the pileus of some basidiomycetes.

plagiotropic Having the longer axis inclined away from the vertical line; growing horizontally or prostrate (contrast orthotropic).

planispiral A pattern of coiling in which growth is in a single flat plane.

plankton Free-floating microscopic aquatic organisms.

planktonic Free floating (contrast benthic).

plectenchymatous Tissue composed of interwoven hyphae (see also pseudoparenchyma).

plerotic oospore An oospore completely filling the available space in the oogonium (contrast aplerotic oospore).

plesiomorphy In cladistics, a primitive feature (contrast apomorphy).

pollen The developed microspore of seed plants that contains the microgametophyte or male gametophyte, including two gametes.

pollen chamber Cavity formed at the distal end of a megasporangium (nucellus) in gymnosperm seeds.

polycentric A thallus that has multiple interconnected centers of growth in which are formed sporangia or resting spores (contrast monocentric).

polymorphism Natural variations in a gene, DNA sequence, or chromosome; two or more different phenotypes within a population.

polyoosporous Containing multiple oospores.

polyphyletic Evolutionary descent based on shared, but not homologous, character states. A polyphyletic taxon is made up of organisms descended from more than one ancestor and so not related evolutionarily.

polyploid An organism, cell, or tissue that has more than two sets of chromosomes (contrast haploid, diploid).

polypore General name of certain basidiomycete fungi with a tubular (porose) hymenophore; also called bracket fungi.

Precambrian Approximately 90% of all Earth history, including the rocks that were formed before the Paleozoic (i.e. from the time of the formation of the Earth until the beginning of the Cambrian period, currently 541 Ma). The Precambrian is divided into the Hadean, Archean, and Proterozoic.

primary growth Growth in the height or length of a stem or root brought about by cells of the apical meristem.

problematic/enigmatic Describes organisms (including fossils) whose systematic affinities are uncertain or unknown.

prokaryote An organism that lacks a membrane-bound nucleus and other membrane-bound organelles (i.e. Bacteria and Archaea) (contrast eukaryote).

prolate A shape characterized by a polar diameter that is longer than the equatorial diameter (a ratio of 1.25–2.0 to 1); drawn out toward the poles.

prostrate Growing flat or trailing; plagiotropic (contrast orthotropic).

Protista A heterogeneous group of heterotrophic and autotrophic organisms; describes: (1) a eukaryote that is neither animal, nor (land) plant, nor Fungus; or (2) a unicellular eukaryote that never becomes multicellular.

prototunicate An ascus type with a thin wall that liberates spores via deliquescence (contrast unitunicate and bitunicate).

proximal Near; closest to the axis or point of attachment (contrast distal).

proximal surface The surface of a spore or pollen grain that faces toward the center of a tetrad (contrast distal surface).

pseudofossil A natural object, structure, or mineral of inorganic origin that may resemble a fossil.

pseudomycelium A grouping of cells arranged in loosely organized chains as a result of budding.

pseudoparaphysis (pl. pseudoparaphyses) Sterile hyphae that develop at the top of the centrum of an ascostroma and grow downward to the base, where they become associated with the ascogenous system.

pseudoparenchyma A “tissue” in some fungi that consists of numerous closely united isodiametric cells in which the interwoven hyphae can no longer be identified; resembling plant parenchyma in section view

pseudoperidium A membranous cup enclosing spores; seen in some rusts.

psilate Having smooth walls and lacking a conspicuous ornamentation.

pteridosperms (seed ferns) An extinct, paraphyletic group of seed-producing vascular plants, many with fern-like foliage.

pubescent Having trichomes.
puff balls The common name of a group of basidiomycetes (Lycoperdales) that produce clouds of spores when the peridium is compressed.
punctae An ornamentation pattern of small pits; punctate = ornamented with punctae.
pycnidiospore A conidium (asexual spore) produced in a pycnidium.
pycnidium (pl. pycnidia) A spherical fungal fruiting body that resembles a perithecium but produces only asexual spores.
Pyrenomycetes Another name for the Sordariomycetes, a group of ascomycetes that have perithecial ascomata.
pyriform Pear shaped.
pyrite An iron-sulfide mineral (FeS_2), also known as iron pyrite, and commonly called fool's gold.
pyritized Preserved via deposition of iron pyrite.
quorum sensing In bacteria, the regulation of gene expression in response to changes in the density of a cell population.
radiolarian A type of marine zooplankton (protozoa) possessing a mineral skeleton. They are important as index fossils.
radiometric dating Quantitative dating of rocks by the decay of radioactive isotopes (see absolute dating).
Raman spectroscopy A spectroscopic laser technique used to observe vibrational, rotational, and other low frequency modes.
rbcL Large-chain protein subunit of ribulose-1,5-bisphosphate carboxylase/oxygenase (RuBisCO), an enzyme important in photosynthesis.
relative dating Determining the age of rocks in relationship to other rocks, either based on the fossils contained in them (biostratigraphy) or on lithologic (i.e. rock) relationships (contrast absolute dating).
relaxed molecular clock Statistical models that take into consideration rate variation of species divergence times.
reniform Kidney shaped.
resinicolous fungi Fungi that live primarily on plant exudates such as resins.
resinite A type of coal maceral composed of resinous compounds that were formed from material in cells or specialized groups of cells (e.g. resin rodlets).
resupinate Any part of an organ that is inverted with respect to the normal position.
reticulate A type of pollen or spore sculpture consisting of a series of ridges or muri; netlike; a type of netlike wall-thickening pattern found in some conducting cells.
rhizine A thick, resistant mycelial cord with a growing point that penetrates the substrate surface and functions in absorption and conduction (= rhizomorph or mycelial strand).
rhizoid A small, usually unicellular hairlike structure; any root-like structure that functions like a root (i.e. to anchor the plant and absorb water and minerals from the soil) but lacks vascular tissue.
rhizoliths A trace fossil that forms through the encrustation of root systems by mineral materials through the processes of chemical weathering, decomposition, reprecipitation, and cementation; rhizoliths may or may not contain remains of the root and they may be filled with sediment, thus forming a mold of the root (see also Kraus and Hasiotis, 2006).
rhizome Horizontal shoot axis or stem, occasionally underground.
rhizomorph A bundle of hyphae with a growing tip which allows penetration of the substrate and functions in absorption and conduction of nutritive substances (= mycelial strand or cord; rhizine); also a rootlike organ, especially the lower portion of living and many fossil lycopsids; may be stigmarioid (i.e. branched and extending out from the base of the stem) or cormose (i.e. unbranched, club-shaped, and not extensive).
rhizomycelium A rudimentary rhizoidal system that superficially resembles a mycelium but in which the branches lack nuclei. Found in many chytrids.
rhizosphere The area of the soil that includes the roots of plants and surrounding soil microorganisms.
riparian Growing by rivers or streams.
RNA Ribonucleic acid.
rotifers Bilaterally symmetrical, multicellular microscopic animals that inhabit fresh water and move by means of a crown of cilia.
rRNA Ribosomal RNA.
ruderal Plants growing in disturbed habitats (e.g. along streams).
rugulate In pollen and spores, a type of ornamentation consisting of wrinkles that may irregularly anastomose.
rusts Plant diseases caused by specialized plant parasites included in the Pucciniales; the fungi that cause those diseases (i.e. rust fungi).
saccate A pollen grain that has a saccus; saccate grains may be protosaccate or eusaccate.
saccus (pl. sacci) In pollen grains, a winglike continuation of the exine that extends beyond the central body or corpus of the grain and contains internal ornamentation; characteristic of many conifers and other plants with wind-borne pollen.
sagittate Shaped like an arrowhead.
sapromyxite A coal type composed primarily of algal matter (also called boghead coal).
saprotroph An organism that obtains food from nonliving (i.e. dead or decaying) organic matter (contrast parasite).
saprotrophism A nutritional mode in which an organism obtains its food from nonliving organic matter (contrast parasitism).
sclerenchyma A cell type in vascular plants with thickened secondary walls that is important in support; cells are dead at maturity and include fibers, fiber-sclereids, and sclereids.
sclerotium (pl. sclerotia) A hard, often spherical fungal resting body that may give rise to a mycelium, sporocarp, or stroma.
scolecospore An elongate, thread-like or vermiform spore that may have septa.
secondary tissues In vascular plants, tissues produced by a lateral meristem, such as the vascular cambium or the cork cambium (e.g. wood, bark, or periderm).
seed A fertilized ovule; an integumented megasporangium that contains an embryo.
seed ferns (pteridosperms) An extinct, paraphyletic group of seed-producing vascular plants, many with fern-like foliage.
self-incompatibility Inability for fertilization to occur between gametes derived from a single genotype.
septate With cross walls or septa (e.g. certain fungal hyphae) (contrast aseptate or nonseptate).
septum (pl. septa) A cross wall.
serrate Saw-toothed.
sessile Without a stalk; borne directly on an axis or a leaf.
seta (pl. setae) In fungi, pointed stiff hair or bristle that occurs on the fruiting bodies of various fungi.
sexine The outer, sculptured portion of the exine of pollen and spores (contrast nexine).
silicification A type of fossilization in which silica either infills intercellular spaces and other voids (permineralization) or replaces the original organic material (petrifaction); the silica is usually in the form of chalcedony, quartz, or opal.
skeletal hyphae Thick-walled, unbranched, aseptate hyphae in some basidiomycete fruiting bodies.
soft rot A fungal symptom in wood where cellulose is broken down.
sooty mold A black powdery coating or crust formed from the mycelia of certain ascomycetes that grow on plant exudates and the sugar secretions of some insects that feed on the plant.

soralium (pl. soralia) In lichens, a cluster of densely spaced soredia.
soredium (pl. soredia) Lichen asexual propagule composed of fungal hyphae (mycobiont) intertwined with the photosynthetic cells of a green alga or cyanobacterium (photobiont).
spongiostratum A layer of anastomosing hyphae on the lower surface of certain lichens that is not effective in attachment.
spongy parenchyma A leaf mesophyll tissue that consists of loosely arranged parenchymatous cells, which contain chloroplasts.
sporae dispersae Spores and pollen grains that are present in the rocks or sediments and can be obtained by bulk dissolution of the rock in acid (contrast in situ); palynomorphs that are not within reproductive organs.
sporangiolum (pl. sporangiola) A small sporangium that contains a few spores; sometimes a single spore.
sporangiophore A specialized hypha that produces one or more sporangia.
sporangiothecium Term used for the zygosporangial wall.
sporangium (pl. sporangia) A structure in which spores are produced.
sporiferous saccule A pyriform structure that develops at the tip of a hypha and is associated with the development of the spore in certain arbuscular mycorrhizae.
sporocarp A generalized term for a fungal reproductive structure that produces spores; in fossils the term is sometimes included in quotation marks to underscore the fact that they may not all be homologous or produced by the same group of fungi as extant sporocarps.
sporoderm A general term for the entire wall of a pollen grain or spore; includes the inner intine, the outer exine, and the external perine, if present.
sporophore A structure or stalk that bears spores or sporangia.
sporophyll A modified leaf that bears a sporangium or sporangia.
sporophyte The diploid ($2n$) phase of a plant life cycle, which produces spores (contrast gametophyte).
sporopollenin A high-molecular-weight polymer of C–H–O (possibly a carotenoid ester) that makes up the exine and perine of pollen grains and spores; it is very resistant to decay.
squamules Scale-like lobes with a distinct lower surface that are found on squamulose lichens.
squamulose A lichen thallus composed of multiple small scale-like lobes that overlap.
staurospores Stellate spores with three or four radiating arms.
stellate Having several arms arising from a common point; star-shaped.
stem group All fossils more closely related to the crown group than to any other.
sterigma (pl. sterigmata) A small outgrowth on a basidium on which a basidiospore forms.
stigmatocyst The terminal cell of a capitate hyphopodium.
stipe A stalk-like feature supporting a basidiocarp or ascocarp.
stipitate Formed on a stipe.
stoma (pl. stomata) A specialized opening in the epidermis of plants that is surrounded by guard cells and through which gases pass; it also enables the plant to regulate transpiration.
Stramenopila, stramenopiles Eukaryotic organisms that include diatoms, chrysophytes, brown algae, peronosporomycetes, and some protozoa characterized by a single anterior tinsel flagellum with two rows of tubular hairs (= heterokonts).
stroma (pl. stromata) In ascomycetes, an undifferentiated mass of fungal tissue on which asci are borne.
stromatolite A layered, usually calcareous, organosedimentary structure that results from the accretion of detrital and precipitated minerals by microbial mats; these mats may include bacteria, cyanobacteria, and algae; a type of microbialite.
suberin A waxy substance in vascular plants, which, being hydrophobic, inhibits water from entering plant tissues.
suberinite A maceral pf brown coal/lignite formed from the suberin of certain cortical cells.
subfossil Remains that are not as old as a typical fossil, having been preserved for approximately 6000 years.
substomatal chamber A cavity that is immediately proximal to the stoma in green plants.
subtending hypha Hypha giving rise to a reproductive structure or propagule (= parental hypha).
suspensor In fungi, a special hyphal branch that subtends the gametangia in zygomycetous fungi.
suture A line of fusion or a line where dehiscence may occur.
symbiosis, symbiotic relationship A permanent or long-lasting association between different species of organisms without respect to the outcome of the association; parasitism and mutualism are two types of such relationships.
symplesiomorphy An ancestral trait shared by two or more taxa.
synapomorphy In cladistics, a shared, derived character.
synnema (pl. synnemata) A bundle of conidiophores that produce conidia on the apex.
synnematous Bearing conidiophores in clusters (contrast mononematous).
taphocoenosis (pl. taphocoenoses) A set of fossils brought together after death by sedimentary processes rather than by having originally lived together.
taphonomy The branch of paleoecology that deals with all of the processes occurring after the death of an organism until it is discovered.
taxon (pl. taxa) A general term for a taxonomic unit at any level (e.g. a species or genus).
teleomorph The sexual or the so-called perfect state of a fungus in which the spores are produced by meiosis (see also anamorph and holomorph).
teliospore A thick-walled binucleate spore in rusts and smuts.
tetrad A group of four spores formed from a spore mother cell by meiosis.
tetrahedral tetrad A three-dimensional arrangement of four cells, especially spores, such that three cells are in one plane, the fourth is in an adjacent plane, and each cell is in contact with its three neighbors.
thallic conidiogenesis One of two basic types of conidial ontogeny characterized by the transformation of an existing cell in a hypha by formation of a septum before the conidium is initiated. If all layers of the conidiogenous cell are involved in formation of the wall the process is termed holothallic; if only the inner wall is involved the process is termed enterothallic (contrast blastic conidiogenesis).
thallophyte A plant with a body that is undifferentiated (i.e. not subdivided into roots, shoot axis, and leaves).
thallus (pl. thalli) The entire body of a fungus; a generalized term for the simple plant body of nonvascular plants; thalli are not differentiated into roots, shoot axis, and leaves.
thin section A technique to prepare rocks and fossils for microscopic examination. A piece of rock is cemented to a glass microscope slide and ground thin enough to transmit light; used to examine fossil permineralizations and petrifactions.
thrips Small, cylindrical hemimetabolic insects approximately 1 mm long with asymmetrical mouthparts and legs that end in tarsal segments.
thyriothecium (pl. thyriothecia) A shield-shaped ascocarp that occurs in epiphyllous fungi.
tonstein A compact rock representing the alteration products of volcanic ash and which occurs as a thin band in coal seams.

tracheid A type of tracheary element (specialized water-conducting cells in the xylem of vascular plants) that is elongate, thick-walled, and nonliving at maturity; tracheids are found in all vascular plants and function in water conduction and support.

tramal hyphae Hyphae that make up the tissue (trama) supporting the hymenium in certain basidiomycetes.

transverse section A cross section; a section perpendicular to the longitudinal axis of the plant organ or organism.

trichome In plants, an epidermal hair; in cyanobacteria, a string of cells (excluding the mucilaginous sheath).

trichospore A unispored sporangium in the trichomycetes with one to several filamentous or piliform appendages.

trichotomous Dividing into three equal branches at the same point (= trifurcate).

trifurcate With three forks or branches (= trichotomous).

trilete A three-armed, Y-shaped mark found commonly on the proximal surface of some spores and pollen resulting from the configuration of the spores within a tetrahedral tetrad.

trimitic Having three types of hyphae: skeletal, generative, and binding hyphae; used for some extant basidiomycetes and for the fossil *Prototaxites*.

triradiate mark See trilete.

tubercle A wartlike or knoblike outgrowth.

tuff A general term for consolidated volcanic ash.

tylosis (pl. tyloses) Outgrowths of xylem parenchyma cells through pits that occlude the conducting elements.

ungulate Hoof-like.

unipapillose Bearing a single hair or papilla.

unistratose Consisting of a single layer of cells, as in moss leaves.

unitunicate An ascus type in which the two wall layers do not separate during the release of spores (contrast bitunicate and prototunicate).

Urediniomycetes In older classifications, the name given to certain basidiomycetes including plant pathogens that form rusts. Same as Pucciniomycetes.

urediniospore The primary reinfecting spore type of rust fungi, in the form of a binucleate spore that is rust in color and produced in a uredium.

varnish A thin, dark film of iron oxide and traces of organic matter on rocks or soils formed as a result of the evaporation of mineralized solutions.

vascular cambium A type of lateral meristem that gives rise to secondary vascular tissue (i.e. secondary xylem/wood and phloem).

vascular cryptogam A seedless vascular plant.

vascular rays A series of cells, usually parenchymatous, that extend in radial series through secondary xylem and phloem; produced by the ray initials of the vascular cambium.

venation The arrangement of veins (vascular bundles) in a leaf.

ventral On the inner face of an organ; on the upper surface of a leaf.

vermiculate A type of pollen and spore ornamentation consisting of scattered, elongated depressions.

verruca (pl. verrucae) A type of pollen and spore ornamentation consisting of wartlike projections.

verrucate Ornamented with verrucae.

vesicle Thick-walled structures in certain fungi that may function in storage.

vesiculus (pl. vesiculi) Small bladder or vesicle.

vessel A tubelike structure, consisting of vessel members, a type of tracheary element (specialized water-conducting cells in the xylem of vascular plants); perforation plates on the end walls of individual vessel members allow for the free movement of water through the vessel.

vitrinite A type of maceral (organic constituent of coal) that is made of humic material and has a mid-level reflectance; it is black and has a vitreous luster.

volva (pl. volvae) A cup-like structure at the base of a mushroom; it is a remnant of the membranous tissue (universal veil) that envelops immature fruiting bodies in certain gilled fungi.

white pocket rot A symptom of fungal decay in wood in the form of small pits or canals.

white rot A fungus symptom in wood in which lignin and polysaccharides are broken down.

whorl Three or more appendages attached at a single node.

wood Secondary xylem (i.e. xylem produced by a vascular cambium).

Woronin bodies A cytoplasmic organelle surrounded by a simple membrane and found near the septa of ascomycetes and their conidial states.

X-ray synchrotron A non-destructive method that uses diffraction patterns to examine opaque specimens.

zoosporangium A sporangium in which zoospores are produced.

zoosporic fungi Pertaining to zoospores; sometimes used to identify certain aquatic fungi.

Zosterophyllophyta A phylum of extinct seedless vascular plants known from the Silurian and Devonian. They lacked true roots and leaves and had sporangia borne on reduced stalks.

zygosporangium In zygomycetes, a sporangium that develops after the fusion of two gametangia and which contains a zygospore.

zygote The diploid ($2n$) cell that results from the fusion of two gametes and is the first cell of the new sporophyte generation.

zygotic meiosis Meiosis that occurs in the zygote; a life cycle in which the zygote is the only diploid ($2n$) cell.

tegmen [illegible] and consisting of [illegible] ... [illegible]

[illegible]

vexillum The arrangement of [illegible] in a leaf; [illegible] the upper surface [illegible].

verruciform A type of pollen and spore ornamentation consisting of elongated [illegible].

verrucate A type of pollen and spore ornamentation consisting of wartlike protuberances.

vermiculate Ornamented with wormlike [illegible].

vessels [illegible]

[illegible]

REFERENCES

Aanen, D.K., and P. Eggleton. 2005. Fungus-growing termites originated in African rain forest. Current Biology 15: 851–855.

Aanen, D.K., P. Eggleton, C. Rouland-Lefèvre, T. Guldberg-Frøslev, S. Rosendahl, and J.J. Boomsma. 2002. The evolution of fungus-growing termites and their mutualistic fungal symbionts. Proceedings of the National Academy of Sciences, USA 99: 14887–14892.

Aassoumi, H., J. Broutin, M. El Wartiti, P. Freytet, J.-C. Koeniguer, C. Quesada, F. Simancas, and N. Toutin-Morin. 1992. Pedological nodules with cone in cone structure in the Permian of Sierra Morena (Spain) and central Morocco. Carbonates and Evaporites 7: 140–149.

Abbott, G.D., G. Ewbank, D. Edwards, and G.-Y. Wang. 1998. Molecular characterization of some enigmatic Lower Devonian fossils. Geochimica et Cosmochimica Acta 62: 1407–1418.

Abe, T., D.E. Bignell, and M. Higashi (Eds). 2000. Termites: Evolution, Sociality, Symbioses, Ecology. Kluwer Academic Publishers, Dordrecht, The Netherlands. 466 pp.

Abrouk, M., F. Murat, C. Pont, J. Messing, S. Jackson, T. Faraut, E. Tannier, C. Plomion, R. Cooke, C. Feuillet, and J. Salse. 2010. Palaeogenomics of plants: Synteny-based modelling of extinct ancestors. Trends in Plant Science 15: 479–487.

Abu Hamad, A., H. Kerp, B. Vörding, and K. Bandel. 2008. A Late Permian flora with *Dicroidium* from the Dead Sea region, Jordan. Review of Palaeobotany and Palynology 149: 85–130.

Adams, D.G. 2002. Cyanobacteria in symbiosis with hornworts and liverworts. In A.N. Rai, B. Bergman, and U. Rasmussen (Eds), Cyanobacteria in Symbiosis, pp. 117–135. Kluwer Academic Publishers, Dordrecht, The Netherlands.

Adams, D.G., and P.S. Duggan. 2008. Cyanobacteria-bryophyte symbioses. Journal of Experimental Botany 59: 1047–1058.

Adaskaveg, J.E., and R.L. Gilbertson. 1988. Basidiospores, pilocystidia and other basidiocarp characters in several species of the *Ganoderma lucidum* complex. Mycologia 80: 493–507.

Addy, H.D., M.M. Piercey, and R.S. Currah. 2005. Microfungal endophytes in roots. Canadian Journal of Botany 83: 1–13.

Adeleke, R.A., T.E. Cloete, A. Bertrand, and D.P. Khasa. 2012. Iron ore weathering potentials of ectomycorrhizal plants. Mycorrhiza 22: 535–544.

Adl, S.M. 2003. The Ecology of Soil Decomposition. CABI Publishing, Wallingford, Oxon, UK. 368 pp.

Adl, S.M., A.G.B. Simpson, M.A. Farmer, R.A. Andersen, O.R. Anderson, J.R. Barta, S.S. Bowser, G. Brugerolle, R.A. Fensome, S. Fredericq, T.Y. James, S. Karpov, P. Kugrens, J. Krug, C.E. Lane, L.A. Lewis, J. Lodge, D.H. Lynn, D.G. Mann, R.M. McCourt, L. Mendoza, Ø. Moestrup, S.E. Mozley-Standridge, T.A. Nerad, C.A. Shearer, A.V. Smirnov, F.W. Spiegel, and M.F.J.R. Taylor. 2005. The new higher level classification of eukaryotes with emphasis on the taxonomy of protists. Journal of Eukaryotic Microbiology 52: 399–451.

Adl, S.M, V. Girard, G. Breton, M. Lak, A. Maharning, A. Mills, V. Perrichot, M. Trionnaire, R. Vullo, and D. Néraudeau. 2011. Reconstructing the soil food web of a 100 million-year-old forest: The case of the mid-Cretaceous fossils in the amber of Charentes (SW France). Soil Biology & Biochemistry 43: 726–735.

Agashe, S.N., and S.T. Tilak. 1970. Occurrence of fungal elements in the bark of arborescent calamite roots from the American Carboniferous. Bulletin of the Torrey Botanical Club 97: 216–218.

Agerer, R. 2001. Exploration types of ectomycorrhizae. Mycorrhiza 11: 107–114.

Agerer, R., R.M. Danielson, S. Egli, K. Ingleby, D. Luoma, and R. Treu. 1996–2000. Descriptions of Ectomycorrhizae. Einhorn-Verlag, Schwäbisch-Gmünd, Germany.

Ahmadjian, V. 1993. The Lichen Symbiosis. John Wiley & Sons, Inc., New York, NY. 250 pp.

Ahmadjian, V. 1995. Lichens – specialized groups of parasitic fungi. In U.S. Singh, K. Kohmoto, and R.P. Singh (Eds), Pathogenesis and Host Specificity in Plant Diseases: Histopathological, Biochemical, Genetic and Molecular Bases, Volume II: Eukaryotes, pp. 277–288. Pergamon Press, Oxford, UK.

Aide, M. 2005. Elemental composition of soil nodules from two alfisols on an alluvial terrace in Missouri. Soil Science 170: 1022–1033.

Aime, M.C., P.B. Matheny, D.A. Henk, E.M. Frieders, R.H. Nilsson, M. Piepenbring, D.J. McLaughlin, L.J. Szabo, D. Begerow, J.P. Sampaio, R. Bauer, M. Weiß, F. Oberwinkler, and D. Hibbett. 2006. An overview of the higher level classification of Pucciniomycotina based on combined analyses of nuclear large and small subunit rDNA sequences. Mycologia 98: 896–905.

Aist, J.R. 1977. Mechanically induced wall appositions of plant cells can prevent penetration by a parasitic fungus. Science 197: 568–571.

Akai, S. 1950. Studies on the pathological anatomy of fungus galls of plants. Memoirs of the College of Agriculture, Kyoto University 58 (Phytopathological Series 10): 1–60.

Akiyama, K., and H. Hayashi. 2006. Strigolactones: Chemical signals for fungal symbionts and parasitic weeds in plant roots. Annals of Botany 97: 925–931.

Albertin, W., and P. Marullo. 2012. Polyploidy in fungi: Evolution after whole-genome duplication. Proceedings of the Royal Society 279B: 2497–2509.

Alberton, O., T.W. Kuyper, and R.C. Summerbell. 2010. Dark septate root endophytic fungi increase growth of Scots pine seedlings under elevated CO_2 through enhanced nitrogen use efficiency. Plant and Soil 328: 459–470.

Alexander, T., R. Toth, R. Meier, and H.C. Weber. 1989. Dynamics of arbuscule development and degeneration in onion, bean, and tomato with reference to vesicular-arbuscular mycorrhizae in grasses. Canadian Journal of Botany 67: 2505–2513.

Alexopoulos, C.J., C.W. Mims, and M. Blackwell. 1996. Introductory Mycology (Fourth Edition). John Wiley & Sons., Inc., New York, NY. 868 pp.

Algeo, T.J., and S.E. Scheckler. 1998. Terrestrial-marine teleconnections in the Devonian: Links between the evolution of land plants, weathering processes, and marine anoxic events. Philosophical Transactions of the Royal Society 353B: 113–130.

Allison, C.W. 1988. Paleontology of Late Proterozoic and Early Cambrian rocks of east-central Alaska. United States Geological Survey Professional Paper 1449: 1–50.

Allwood, A.C., J.P. Grotzinger, A.H. Knoll, I.W. Burch, M.S. Anderson, M.L. Coleman, and I. Kanik. 2009. Controls on development and diversity of Early Archaean stromatolites. Proceedings of the National Academy of Sciences, USA 106: 9548–9555.

Almeida, R.T., and N.C. Schenck. 1990. A revision of the genus *Sclerocystis* (Glomaceae, Glomales). Mycologia 82: 703–714.

Altehenger, A. 1959. Floristisch belegte Klimaschwankungen im mitteleuropäischen Pliozän der Reuver-Stufe. Palaeontographica 106B: 11–70.

Altmeyer, H. 1973. Astnarben an Prototaxiten? Aufschluss 24: 350–356.

Alvin, K.L., and M.D. Muir. 1970. An epiphyllous fungus from the Lower Cretaceous. Biological Journal of the Linnean Society 2: 55–59.

Amano, K. 1986. Host Range and Geographical Distribution of the Powdery Mildew Fungi. Japan Scientific Societies Press, Tokyo, Japan. 741 pp.

Ambwani, K. 1982. Palynology of the Deccan Intertrappean beds of Rajahmundry District, Andhra Pradesh. Palaeobotanist 30: 28–33.

Amo de Paz, G., P. Cubas, P.K. Divakar, H.T. Lumbsch, and A. Crespo. 2011. Origin and diversification of major clades in parmelioid lichens (Parmeliaceae, Ascomycota) during the Paleogene inferred by Bayesian analysis. PLoS ONE 6: e28161.

Amon, J.P. 1984. *Rhizophydium littoreum*: A chytrid from siphonaceous marine algae – an ultrastructural examination. Mycologia 76: 132–139.

Anca, I.-A., E. Lumini, S. Ghignone, A. Salvioli, V. Bianciotto, and P. Bonfante. 2009. The *ftsZ* gene of the endocellular bacterium '*Candidatus* Glomeribacter gigasporarum' is preferentially expressed during the symbiotic phases of its host mycorrhizal fungus. Molecular Plant-Microbe Interactions 22: 302–310.

Anderson, I.C., and J.W.G. Cairney. 2007. Ectomycorrhizal fungi: Exploring the mycelial frontier. FEMS Microbiology Reviews 31: 388–406.

Anderson, J.B, and L.M. Kohn. 2007. Dikaryons, diploids, and evolution. In J. Heitmann, J.W. Kronstad, J.W. Taylor, and L.A. Casselton (Eds), Sex in Fungi: Molecular Determination and Evolutionary Implications, pp. 333–348. ASM Press, Washington, DC.

Anderson, J.P., C.A. Gleason, R.C. Foley, P.H. Thrall, J.B. Burdon, and K.B. Singh. 2010. Plants versus pathogens: An evolutionary arms race. Functional Plant Biology 37: 499–512.

Anderson, R.S., R.L. Homola, R.B. Davis, and G.L. Jacobson., Jr. 1984. Fossil remains of the mycorrhizal fungal *Glomus fasciculatum* complex in postglacial lake sediments from Maine. Canadian Journal of Botany 62: 2325–2328.

Ando, K. 1984. Phylogeny of the fern rusts (*Uredinopsis, Milesina* and *Hyalopsora*). Transactions of the Mycological Society of Japan 25: 295–304.

Andrew, C., and E.A. Lilleskov. 2009. Productivity and community structure of ectomycorrhizal fungal sporocarps under increased atmospheric CO_2 and O_3. Ecology Letters 12: 813–822.

Andrews, H.N. 1948. A note on *Fomes idahoensis* Brown. Annals of the Missouri Botanical Garden 35: 207.

Andrews, H.N., and L.W. Lenz. 1943. A mycorrhizome from the Carboniferous of Illinois. Bulletin of the Torrey Botanical Club 70: 120–125.

Andrews, H.N., and L.W. Lenz. 1947. Fossil polypores from Idaho. Annals of the Missouri Botanical Garden 34: 113–114.

Andújar, C., J. Serrano, and J. Gómez-Zurita. 2012. Winding up the molecular clock in the genus *Carabus* (Coleoptera: Carabidae): Assessment of methodological decisions on rate and node age estimation. BMC. Evolutionary Biology 12: 40.

Angier, N. 1992. Twin crowns for 30-acre fungus: World's biggest, oldest organism. The New York Times, April 2: A1.

Anikster, Y., and I. Wahl. 1979. Coevolution of the rust fungi on Gramineae and Liliaceae and their hosts. Annual Reviews of Phytopathology 17: 367–403.

Antoine, P.-O., D. De Franceschi, J.J. Flynn, A. Nel, P. Baby, M. Benammi, Y. Calderón, N. Espurt, A. Goswami, and R. Salas-Gismondi. 2006. Amber from western Amazonia reveals Neotropical diversity during the Middle Miocene. Proceedings of the National Academy of Sciences, USA 103: 13595–13600.

Antunes, P.M., J. Miller, L.M. Carvalho, J.N. Klironomos, and J.A. Newman. 2008. Even after death the endophytic fungus of *Schedonorus phoenix* reduces the arbuscular mycorrhizas of other plants. Functional Ecology 22: 912–918.

Appoloni, S., Y. Lekberg, M.T. Tercek, C.A. Zabinski, and D. Redecker. 2008. Molecular community analysis of arbuscular mycorrhizal fungi in roots of geothermal soils in Yellowstone National Park (USA). Microbial Ecology 56: 649–659.

Apt, K.E. 1988. Morphology and development of hyperplasia on *Cystoseira osmundacea* (Phaeophyta) associated with *Halgoguingnardia irritans* (Ascomycotina). American Journal of Botany 75: 979–984.

Aptroot, A. 1995. A monograph of *Didymosphaeria*. Studies in Mycology 37: 1–160.

Aptroot, A. 1998. Aspects of the integration of the taxonomy of lichenized and non-lichenized pyrenocarpous ascomycetes. Lichenologist 30: 501–514.

Arbey, F., and J.-C. Koeniguer. 1979. Les Nématophytes et les algueraies de l'Ordovicien et du Dévonien saharien (Ordovician and Devonian nematophytes and seaweeds in the Sahara). Bulletin des Centres de Recherches Exploration-Production Elf-Aquitaine 3: 409–418.

Archer, K.J., and A.L.J. Cole. 1986. Cuticle, cell wall ultrastructure and disease resistance in maidenhair fern. New Phytologist 103: 341–348.

Arditti, J. 1967. Factors affecting the germination of orchid seeds. Botanical Review 33: 1–97.

Armstrong, L., and R.L. Peterson. 2002. The interface between the arbuscular mycorrhizal fungus *Glomus intraradices* and root cells of *Panax quinquefolius*: A *Paris*-type mycorrhizal association. Mycologia 94: 587–595.

Armstrong, R.A., and T. Bradwell. 2010. Growth of crustose lichens: A review. Geografiska Annaler, Series A. Physical Geography 92: 3–17.

Armstrong, R.A., and T. Bradwell. 2011. Growth of foliose lichens: A review. Symbiosis 53: 1–16.

Arnaud, G. 1918. Les Astérinées. Annales de l'École Nationale d'Agriculture de Montpellier 16: 1–288.

Arnold, A.E. 2007. Understanding the diversity of foliar endophytic fungi: Progress, challenges, and frontiers. Fungal Biology Reviews 21: 51–66.

Arnold, A.E. 2008. Endophytic fungi: Hidden components of tropical community ecology. In W.P. Carson, and S.A. Schnitzer (Eds), Tropical Forest Community Ecology, pp. 254–271. Blackwell Publishing Ltd., Hoboken, NJ.

Arnold, A.E., and B.M.J. Engelbrecht. 2007. Fungal endophytes nearly double minimum leaf conductance in seedlings of a neotropical tree species. Journal of Tropical Ecology 23: 369–372.

Arnold, E.A., and F. Lutzoni. 2007. Diversity and host range of foliar fungal endophytes: are tropical leaves biodiversity hotspots? Ecology 88: 541–549.

Arnold, A.E., Z. Maynard, G.S. Gilbert, P.D. Coley, and T.A. Kursar. 2000. Are tropical fungal endophytes hyperdiverse? Ecology Letters 3: 267–274.

Arnold, A.E., L.C. Mejía, D. Kyllo, E.I. Rojas, Z. Maynard, N. Robbins, and E.A. Herre. 2003. Fungal endophytes limit pathogen damage in a tropical tree. Proceedings of the National Academy of Sciences, USA 100: 15649–15654.

Arnold, A.E., D.A. Henk, R.L. Eells, F. Lutzoni, and R. Vilgalys. 2007. Diversity and phylogenetic affinities of foliar fungal endophytes in loblolly pine inferred by culturing and environmental PCR. Mycologia 99: 185–206.

Arnold, A.E., J. Miadlikowska, K.L. Higgins, S.D. Sarvate, P. Gugger, A. Way, V. Hofstetter, F. Kauff, and F. Lutzoni. 2009. A phylogenetic estimation of trophic transition networks for ascomycetous fungi: Are lichens cradles of symbiotrophic fungal diversification? Systematic Biology 58: 283–297.

Arnold, C.A. 1931. On *Callixylon newberryi* (Dawson) Elkins et Wieland. Contributions from the Museum of Paleontology. University of Michigan 3: 207–232.

Arnold, C.A. 1952. A specimen of *Prototaxites* from the Kettle Point black shale of Ontario. Palaeontographica 93B: 45–56.

Arshad, M.A., and M. Schnitzer. 1987. The chemistry of a termite fungus comb. Plant and Soil 98: 247–256.

Arthur, J.C. 1924. Fern rusts and their aecia. Mycologia 16: 245–251.

Artursson, V., R.D. Finlay, and J.K. Jansson. 2006. Interactions between arbuscular mycorrhizal fungi and bacteria and their potential for stimulating plant growth. Environmental Microbiology 8: 1–10.

Arx, J.A. von. 1987. Plant Pathogenic Fungi. Gebr. Borntraeger Verlagsbuchhandlung, Berlin, Germany. 288 pp.

Arx, J.A. von, and E. Müller. 1975. A Re-evaluation of the Bitunicate Ascomycetes with Keys to Families and Genera. Centraalbureau voor Schimmelcultures, Baarn, The Netherlands. 159 pp.

Ascaso, C., J. Wierzchos, M. Speranza, J.C. Gutiérrez, A.M. González, A. de los Ríos, and J. Alonso. 2005. Fossil protists and fungi in amber and rock substrates. Micropaleontology 51: 59–72.

Atsatt, P.R. 1988. Are vascular plants "inside-out" lichens? Ecology 69: 17–23.

Atsatt, P.R. 1991. Fungi and the origin of land plants. In L. Margulis, and R. Fester (Eds), Symbiosis as a Source of Evolutionary Innovation, pp. 301–315. MIT Press, Cambridge, MA.

Aufderheide, A.C., and C. Rodríguez-Martín. 1998. Cambridge Encyclopedia of Human Paleopathology. Cambridge University Press, New York, NY. 496 pp.

Augé, R.M. 2004. Arbuscular mycorrhizae and soil/plant water relations. Canadian Journal of Soil Science 84: 373–381.

Baar, J., I. Paradi, E.C.H.E.T. Lucassen, K.A. Hudson-Edwards, D. Redecker, J.G.M. Roelofs, and A.J.P. Smolders. 2011. Molecular analysis of AMF diversity in aquatic macrophytes: A comparison of oligotrophic and utra-oligotrophic lakes. Aquatic Botany 94: 53–61.

Baayen, R.P., G. Cochius, H. Hendriks, J.P. Meffert, J. Bakker, M. Bekker, P.H.J.F. van den Boogert, H. Stachewicz, and G.C.M. van Leeuwen. 2006. History of potato wart disease in Europe – A proposal for harmonisation in defining pathotypes. European Journal of Plant Pathology 116: 21–31.

Baele, J.-M. 1999. Karst et silicification en milieu continental et en milieu margino-littoral: deux exemples dans le Méso-Cénozoïque de la Belgique méridionale. In: P. Audra (ed.), Karst 99, Colloque européen 10–15 Septembre, 1999, Université de Provence, France, Etudes de Géographie Physique, Travaux 1999, Sup. 28: 49–54.

Bailey, R.H. 1968. Studies on the dispersal of lichen soredia. Journal of the Linnean Society of London, Botany 59: 479–490.

Bailey, R.H. 1976. Ecological aspects of dispersal and establishment in lichens. In D.H. Brown, D.L. Hawksworth, and R.H. Bailey (Eds), Lichenology: Progress and Problems, pp. 215–247. Academic Press, London, UK.

Bajnóczi, B., and V. Kovács-Kis. 2006. Origin of pedogenic needle-fiber calcite revealed by micromorphology and stable isotope composition – A case study of a Quaternary paleosol from Hungary. Chemie der Erde 66: 203–212.

Bajpai, U. 1997. Taphonomic constraints on preservation of cuticles in compression fossils: Fungi induced ultrastructural changes in cuticular membranes. Palaeobotanist 46: 31–34.

Bajpai, U., and H.K. Maheshwari. 1987. Epiphyllous fungi from the Gondwana. Palaeobotanist 36: 210–213.

Baker, A.G., S.A. Bhagwat, and K.J. Willis. 2013. Do dung fungal spores make a good proxy for past distribution of large herbivores? Quaternary Science Reviews 62: 21–31.

Baker, D., and N.G. Miller. 1980. Ultrastructural evidence for the existence of actinorhizal symbioses in the Late Pleistocene. Canadian Journal of Botany 58: 1612–1620.

Baldauf, S.L., and W.F. Doolittle. 1997. Origin and evolution of the slime molds (Mycetozoa). Proceedings of the National Academy of Sciences, USA 94: 12007–12012.

Ballard, R.G., M.A. Walsh, and W.E. Cole. 1982. Blue-stain fungi in xylem of lodgepole pine: A light-microscope study on extent of hyphal distribution. Canadian Journal of Botany 60: 2334–2341.

Ballhaus, C., C.T. Gee, C. Bockrath, K. Greef, T. Mansfeldt, and D. Rhede. 2012. The silicification of trees in volcanic ash – An experimental study. Geochimica et Cosmochimica Acta 84: 62–74.

Balogh-Brunstad, Z., C.K. Keller, R.A. Gill, B.T. Bormann, and C.Y. Li. 2008. The effect of bacteria and fungi on chemical weathering and chemical denudation fluxes in pine growth experiments. Biogeochemistry 88: 153–167.

Banerji, J., and A.K. Ghosh. 2002. Mutualism/symbiosis from the Early Cretaceous (intertrappeans) of Rajmahal Basin, Jharkhand, India. Current Science 83: 1073–1074.

Banks, H.P., and B.J. Colthart. 1993. Plant-animal-fungal interactions in Early Devonian trimerophytes from Gaspé, Canada. American Journal of Botany 80: 992–1001.

Barlinge, S.G., and S.A. Paradkar. 1982. Record of new fossil algal and fungal forms from the Deccan Intertrappean of Mohgaon-Kalan, M.P., India. Botanique (Nagpur) 10: 163–175.

Bärlocher, F. 1992. The Ecology of Aquatic Hyphomycetes. Springer-Verlag, Berlin, Germany. 225 pp.

Baroni Urbani, C. 1980. First description of fossil gardening ants (Amber Collection Stuttgart and Natural History Museum Basel; Hymenoptera: Formicidae. I. Attini). Stuttgarter Beiträge zur Naturkunde 54B: 1–13.

Barr, D.J.S. 1970. *Hyphochytrium catenoides*: A morphological and physiological study of North American isolates. Mycologia 62: 492–503.

Barr, D.J.S. 1971. Morphology and taxonomy of *Entophlyctis confervae-glomeratae* (Chytridiales) and related species. Canadian Journal of Botany 49: 2215–2222.

Barr, D.J.S. 1980. An outline for the reclassification of the Chytridiales, and for a new order, the Spizellomycetales. Canadian Journal of Botany 58: 2380–2394.

Barr, D.J.S. 1984. The classification of *Spizellomyces, Gaertneriomyces, Triparticalcar*, and *Kochiomyces* (Spizellomycetales, Chytridiomycetes). Canadian Journal of Botany 62: 1171–1201.

Barr, D.J.S. 1990. Phylum Chytridiomycota. In L. Margulis, J.O. Corliss, M. Melkonian, and D.J. Chapman (Eds), Handbook of Protoctista, pp. 454–466. Jones & Bartlett, Boston, MA.

Barr, D.J.S. 2001. Chytridiomycota. In D.J. McLaughlin, E.G. McLaughlin, and P.A. Lemke (Eds), The Mycota VIIA: Systematics and Evolution, pp. 93–112. Springer-Verlag, Berlin, Germany.

Barr, D.J.S., N.L. Désaulniers, and J.S. Knox. 1987. *Catenochytridium hemicysti* n. sp.: Morphology, physiology and zoospore ultrastructure. Mycologia 79: 587–594.

Barr, M.E. 1987. Prodromus to Class Loculoascomycetes. Hamilton I. Newell, Inc, Amherst, MA. 168 pp.

Barr, M.E., H.D Ohr, and M.K. Murphy. 1989. The genus *Serenomyces* on palms. Mycologia 81: 47–51.

Barron, G.L. 1977. The nematode-destroying fungi. Topics in Mycobiology I. Canadian Biological Publications, Guelph, Ontario, Canada. 140 pp.

Barrow, J.R. 2003. Atypical morphology of dark septate fungal root endophytes of *Bouteloua* in arid southwestern USA rangelands. Mycorrhiza 13: 239–247.

Barthel, M. 1961. Ein Pilzrest aus dem Saarkarbon. Geologie 10: 856–857.

Barthel, M., M. Krings, and R. Rössler. 2010. Die schwarzen Psaronien von Manebach, ihre Epiphyten, Parasiten und Pilze. Semana 25: 41–60.

Bartnicki-Garcia, S. 1970. Cell wall composition and other biochemical markers in fungal phylogeny. In J.B. Harborne (Ed.), Phytochemical Phylogeny, pp. 81–103. Academic Press, London.

Basinger, J.F. 1981. The vegetative body of *Metasequoia milleri* from the Middle Eocene of southern British Columbia. Canadian Journal of Botany 59: 2379–2410.

Basinger, J.F. 1984. Seed cones of *Metasequoia milleri* from the Middle Eocene of southern British Columbia. Canadian Journal of Botany 62: 281–289.

Basinger, J.F., and G.W. Rothwell. 1977. Anatomically preserved plants from the Middle Eocene (Allenby Formation) of British Columbia. Canadian Journal of Botany 55: 1984–1990.

Bass, M., and J.M. Cherrett. 1996. Leaf-cutting ants (Formicidae, Attini) prune their fungus to increase and direct its productivity. Functional Ecology 10: 55–61.

Bates, S.T., G.W.G. Cropsey, J.G. Caporaso, R. Knight, and N. Fierer. 2011. Bacterial communities associated with the lichen symbiosis. Applied and Environmental Microbiology 77: 1309–1314.

Bates, S.T., D. Berg-Lyons, C.L. Lauber, W.A. Walters, R. Knight, and N. Fierer. 2012. A prelimilary survey of lichen associated eukaryotes using pyrosequencing. Lichenologist 44: 137–146.

Batra, L.R., and R.W. Lichtwardt. 1963. Association of fungi with some insect galls. Journal of the Kansas Entomological Society 36: 262–278.

Batra, L.R., R.H. Segal, and R.W. Baxter. 1964. A new Middle Pennsylvanian fossil fungus. American Journal of Botany 51: 991–995.

Battin, J. 2010. Le feu saint-Antoine ou ergotisme gangreneux et son iconographie médiévale. Histoire des Sciences Medicales 4: 373–382.

Bauer, R., D. Begerow, F. Oberwinkler, M. Piepenbring, and M.L. Berbee. 2001. Ustilaginomycetes. In D.J. McLaughlin, E.G. McLaughlin, and P.A. Lemke (Eds), The Mycota VIIB: Systematics and Evolution, pp. 57–83. Springer, Berlin, Germany.

Bauer, R., D. Begerow, and F. Oberwinkler. 2008. Ustilaginomycotina. The true smut fungi. Version 23 January 2008 (under construction). http://tolweb.org/Ustilaginomycotina/20530/2008.01.23 in The Tree of Life Web Project, http://tolweb.org/.

Baxter, J.W., and J. Dighton. 2005. Diversity-functioning relationships in ectomycorrhizal fungal communities. In J. Dighton, J.F. White, and P. Oudemans (Eds), The Fungal Community: Its Organization and Role in the Ecosystem, pp. 383–398 (Third Edition). CRC Press, Boca Raton, FL.

Baxter, R.W. 1960. *Sporocarpon* and allied genera from the American Pennsylvanian. Phytomorphology 10: 19–25.

Baxter, R.W. 1975. Fossil fungi from American Pennsylvanian coal balls. University of Kansas Paleontological Contributions Paper 77: 1–6.

Baylis, G.T.S., R.F.R. McNabb, and T.M. Morrison. 1963. The mycorrhizal nodules of podocarps. Transactions of the British Mycological Society 46: 378–384.

Bayman, P. 2007. Fungal endophytes. In C.P. Kubicek, and I.S. Druzhinina (Eds), Environmental and Microbial Relationships, pp. 213–227. Springer-Verlag, Berlin, Germany.

Beakes, G.W., and S. Sekimoto. 2009. The evolutionary phylogeny of Oomycetes – insights gained from studies of holocarpic parasites of algae and invertebrates. In K. Lamour, and S. Kamoun (Eds), Oomycete Genetics and Genomics: Diversity, Interactions, and Research Tools, pp. 1–24. John Wiley & Sons, Inc., Hoboken, NJ.

Beakes, G.W., S.L. Glockling, and S. Sekimoto. 2012. The evolutionary phylogeny of the oomycete "fungi". Protoplasma 249: 3–19.

Beck, C.B., and W.E. Stein, Jr. 1987. *Galtiera bostonensis*, gen. et sp. nov., a protostelic calamopityacean from the New Albany Shale of Kentucky. Canadian Journal of Botany 65: 348–361.

Bedini, S., A. Maremmani, and M. Giovannetti. 2000. *Paris*-type mycorrhizas in *Smilax aspera* L. growing in a Mediterranean sclerophyllous wood. Mycorrhiza 10: 9–13.

Beeli, M. 1920. Note sur le genre *Meliola* Fr. Espèces et variétés nouvelles récoltées au Congo. Bulletin du Jardin botanique de l'État a Bruxelles 7: 89–160.

Beerling, D.J. 2009. Coevolution of photosynthetic organisms and the environment. Geobiology 7: 97–99.

Begerow, D., M. Stoll, and R. Bauer. 2006. A phylogenetic hypothesis of Ustilaginomycotina based on multiple gene analyses and morphological data. Mycologia 98: 906–916.

Behie, S.W., P.M. Zelisko, and M.J. Bidochka. 2012. Endophytic insect-parasitic fungi translocate nitrogen directly from insects to plants. Science 336: 1576–1577.

Beimforde, C., N. Schäfer, H. Dörfelt, P.C. Nascimbene, H. Singh, J. Heinrichs, J. Reitner, R.S. Rana, and A.R. Schmidt. 2011. Ectomycorrhizas from a Lower Eocene angiosperm forest. New Phytologist 192: 988–996.

Bélanger, R.R., and T.J. Avis. 2002. Ecological processes and interactions occurring in leaf surface fungi. In S.E. Lindow, E.I Hecht-Poinar, and V.J. Elliott (Eds), Phyllosphere Microbiology, pp. 193–207. APS Press, St. Paul, MN.

Belkin, H.E., S.J. Tewalt, J.C. Hower, J.D. Stucker, and J.M.K. O'Keefe. 2009. Geochemistry and petrology of selected coal samples from Sumatra, Kalimantan, Sulawesi, and Papua, Indonesia. International Journal of Coal Geology 77: 260–268.

Belkin, H.E., S.J. Tewalt, J.C. Hower, J.D. Stucker, J.M.K. O'Keefe, C.A. Tatu, and G. Buia. 2010. Petrography and geochemistry of Oligocene bituminous coal from the Jiu Valley, Petroşani basin (southern Carpathian Mountains), Romania. International Journal of Coal Geology 82: 68–80.

Bellgard, S.E., and S.E. Williams. 2011. Response of mycorrhizal diversity to current climatic changes. Diversity 3: 8–90.

Belliveau, M.J.-R., and F. Bärlocher. 2005. Molecular evidence confirms multiple origins of aquatic hyphomycetes. Mycological Research 109: 1407–1417.

Belova, M.Y., and A.M. Akhmedov. 2006. *Petsamomyces*, a new genus of organic-walled microfossils from the coal-bearing deposits of the Early Proterozoic, Kola Peninsula. Paleontological Journal 40: 465–475.

Beneš, K. 1956. Neue Erkenntnisse aus dem Gebiet der Paläomykologie der Kohle. Freiberger Forschungshefte C 30: 49–56.

Beneš, K. 1960. Paleomycology – a new trend in microscopic study of coals. Akademiia Nauk SSSR, Izvestiya, Seriya Geologicheskaya 25: 32–36.

Beneš, K. 1961. Príspevek k poznání sporogenních orgánů nekterých askomycetních hub z uhlí hornoslezské pánve. Přírodovedný časopis Slezský: Acta Rerum Naturalium Districtus Silesiae 22: 543–547.

Beneš, K. 1978. Fungal spores from the Carboniferous and Tertiary coals and the role of coal facies in fungal distributions. In: D.C. Bharadwaj et al. (eds), Proceedings of the 4th International Palynology Conference, 1976–1977, pp.

312–316. Birbal Sahni Institute of Palaeobotany, Lucknow, India.

Beneš, K., and J. Kraussová. 1964. Carboniferous fossil fungi from the Upper Silesian Basin (Ostrava-Karviná Coal District). Sborník Geologických věd Paleontologie 4: 65–89.

Beneš, K., and J. Kraussová. 1965. Paleomycological investigation of the Tertiary coals of some basins in Czechoslovakia. Sborník Geologických věd Paleontologie 6: 149–168.

Benjamin, R.K. 1979. Zygomycetes and their spores. In B. Kendrick (Ed.), The Whole Fungus: The Sexual-Asexual Synthesis (Volume 2, pp. 573–621). National Museum of Natural Sciences, National Museums of Canada, Ottawa, and The Kananaskis Foundation, Ottawa, Canada.

Benjamin, R.K., and B.S. Mehrotra. 1963. Obligate azygospore formation in two species of *Mucor* (Mucorales). Aliso 5: 235–245.

Benning, L.G., V. Phoenix, and B.W. Mountain. 2005. Biosilicification: The role of cyanobacteria in silica sinter deposition. In G.M. Gadd, K.T. Semple, and H.M. Lappin-Scott (Eds), Society of General Microbiology Symposium 65: Micro-organisms and Earth Systems – Advances in Geomicrobiology, pp. 131–150. Cambridge University Press, Cambridge, UK.

Benny, G.L. 2012. Current systematics of Zygomycota with a brief review of their biology. In J.K. Misra, J.P. Tewari, and S.K. Deshmukh (Eds), Systematics and Evolution of Fungi, pp. 55–104. Science Publishers, St. Helier, Jersey, British Channel Islands.

Benny, G.L., and K. O'Donnell. 2000. *Amoebidium parasiticum* is a protozoan, not a Trichomycete. Mycologia 92: 1133–1137.

Benny, G.L., R.A. Humber, and J.B. Morton. 2001. Zygomycota: Zygomycetes. In D.J. McLaughlin, E.G. McLaughlin, and P.E. Lemke (Eds), The Mycota VIIA: Systematics and Evolution, pp. 113–146. Springer-Verlag, Berlin, Germany.

Bensch, K., J.Z. Groenewald, J. Dijksterhuis, M. Starink-Willemse, B. Andersen, B.A. Summerell, H.-D. Shin, F.M. Dugan, H.-J. Schroers, U. Braun, and P.W. Crous. 2010. Species and ecological diversity within the *Cladosporium cladosporioides* complex (Davidiellaceae, Capnodiales). Studies in Mycology 67: 1–94.

Benson, D.R. 1988. The genus *Frankia*: Actinomycete symbionts of plants. Microbiological Sciences 5: 9–12.

Bentis, C.J., L. Kaufman, and S. Golubic. 2000. Endolithic fungi in reef-building corals (Order: Scleractinia) are common, cosmopolitan, and potentially pathogenic. Biological Bulletin 198: 254–260.

Benton, M.J., and F.J. Ayala. 2003. Dating the tree of life. Science 300: 1698–1700.

Beraldi-Campesi, H. 2013. Early life on land and the first terrestrial ecosystems. Ecological Processes 2: 1.

Berbee, M.L., and J.W. Taylor. 1992. Two Ascomycete classes based on fruiting-body characters and ribosomal DNA sequence. Molecular Biology and Evolution 9: 278–284.

Berbee, M.L., and J.W. Taylor. 1993. Dating the evolutionary radiations of the true fungi. Canadian Journal of Botany 71: 1114–1127.

Berbee, M.L., and J.W. Taylor. 2001. Fungal molecular evolution: Gene trees and geologic time. In D.J. McLaughlin, E.G. McLaughlin, and P.A. Lemke (Eds), The Mycota VIIB: Systematics and Evolution, pp. 229–245. Springer-Verlag, Berlin, Germany.

Berbee, M.L., and J.W. Taylor. 2006. Dating divergences in the Fungal Tree of Life: review and new analyses. Mycologia 98: 838–849.

Berbee, M.L., and J.W. Taylor. 2007. Rhynie chert: A window into a lost world of complex plant-fungus interactions. New Phytologist 174: 475–479.

Berbee, M.L., and J.W. Taylor. 2010. Dating the molecular clock in fungi – How close are we? Fungal Biology Reviews 24: 1–16.

Berch, S.M. 1985. The nomenclatural fate of *Rhizophagites acinus*. Mycotaxon 23: 405–407.

Berch, S.M., and B. Kendrick. 1982. Vesicular-arbuscular mycorrhizae of southern Ontario ferns and fern-allies. Mycologia 74: 769–776.

Berch, S.M., and B.G. Warner. 1985. Fossil vesicular-arbuscular mycorrhizal fungi: Two *Glomus* species (Endogonaceae, Zygomycetes) from late Quartenary deposits in Ontario, Canada. Review of Palaeobotany and Palynology 45: 229–237.

Berch, S.M., O.K. Miller, and H.D. Thiers. 1984. Evolution of mycorrhizae. In: R. Molina (Ed.), Proceedings of the 6th North American Conference on Mycorrhizae, pp. 189–192. Forest Research Lab, Corvallis, Oregon.

Berdan, H.B. 1941. A developmental study of three saprophytic chytrids. II. *Catenochytridium carolinianum* Berdan. American Journal of Botany 28: 901–911.

Berg, G., and K. Smalla. 2009. Plant species and soil type cooperatively shape the structure and function of microbial communities in the rhizosphere. FEMS Microbiology Ecology 68: 1–13.

Berger, L., R. Speare, P. Daszak, D.E. Green, A.A. Cunningham, C.L. Goggin, R. Slocombe, M.A. Ragan, A.D. Hyatt, K.R. McDonald, H.B. Hines, K.R. Lips, G. Marantelli, and H. Parkes. 1998. Chytridiomycosis causes amphibian mortality associated with population declines in the rain forests of Australia and Central America. Proceedings of the National Academy of Sciences, USA 95: 9031–9036.

Berger, L., A.D. Hyatt, R. Speare, and J.E. Longcore. 2005. Life cycle stages of the amphibian chytrid *Batrachochytrium dendrobatidis*. Diseases of Aquatic Organisms 68: 51–63.

Berkeley, M.J. 1848. On three species of Mould detected by Dr. Thomas in the amber of East Prussia. Annals and Magazine of Natural History 2: 380–383.

Bernadovičová, S., and H. Ivanová. 2011. Hyphomycetes and Coelomycetes fungi isolated from affected leaves and twigs of cherry laurel trees. Folia Oecologica 38: 137–145.

Bernard, F., I. Sache, F. Suffert, and M. Chelle. 2013. The development of a foliar fungal pathogen does react to leaf temperature! New Phytologist 198: 232–240.

Berndt, R. 2013. Revision of the rust genus *Uromyces* on Cucurbitaceae. Mycologia 105: 760–780.

Berner, R.A., J.M. VandenBrooks, and P.D. Ward. 2007. Oxygen and evolution. Science 316: 557–558.

Bernhauser, A. 1953. Über *Mycelites ossifragus* Roux. Auftreten und Formen im Tertiär des Wiener Beckens. Sitzungsberichte der Akademie der Wissenschaften, mathematisch-naturwissenschaftliche Klasse 162: 119–127.

Bernicchia, A., M.A. Fugazzola, V. Gemelli, B. Mantovani, A. Lucchetti, M. Cesari, and E. Speroni. 2006. DNA recovered and sequenced from an almost 7000 y-old Neolithic polypore, *Daedalopsis tricolor*. Mycological Research 110: 14–17.

Berry, E.W. 1916. Remarkable fossil fungi. Mycologia 8: 73–79.

Bertling, M. 1987. Ein hardground am Top eines kalkigen Tempestits im Mittleren Kimmeridge auf dem Kalkrieser Berg (Nordwestdeutschland). Osnabrücker Naturwissenschaftliche Mitteilungen 13: 7–22.

Bertling, M., S.J. Braddy, R.G. Bromley, G.R. Demathieu, J. Genise, R. Mikulás, J.K. Nielsen, K.S.S. Nielsen, A.K. Rindsberg, M. Schlirf, and A. Uchman. 2006. Names for trace fossils: A uniform approach. Lethaia 39: 265–286.

Beuck, L., and A. Freiwald. 2005. Bioerosion patterns in a deep-water *Lophelia pertusa* (Scleractinia) thicket (Propeller Mound, northern Porcupine Seabight). In A. Freiwald, and J.M. Roberts (Eds), Cold-water Corals and Ecosystems, pp. 915–936. Springer-Verlag, Berlin, Germany.

Bever, J.D. 2002. Host-specificity of AM fungal population growth rates can generate feedback on plant growth. Plant and Soil 244: 281–290.

Bever, J.D., P.A. Schultz, A. Pringle, and J.M. Morton. 2001. Arbuscular mycorrhizal fungi: More diverse than meets the eye, and the ecological tale of why. BioScience 51: 923–931.

Bhattacharya, D., H.S. Yoon, S.B. Hedges, and J.D. Hackett. 2009. Eukaryotes (Eukaryota). In S.B. Hedges, and S. Kumar (Eds), The Timetree of Life, pp. 116–120. Oxford University Press, New York, NY.

Bianciotto, V., E. Lumini, L. Lanfranco, D. Minerdi, P. Bonfante, and S. Perotto. 2000. Detection and identification of bacterial endosymbionts in arbuscular mycorrhizal fungi belonging to the family Gigasporaceae. Applied and Environmental Microbiology 66: 4503–4509.

Bianciotto, V., A. Genre, P. Jargeat, E. Lumini, G. Bécard, and P. Bonfante. 2004. Vertical transmission of endobacteria in the arbuscular mycorrhizal fungus *Gigaspora margarita* through generation of vegetative spores. Applied and Environmental Microbiology 70: 3600–3608.

Bidartondo, M.I. 2005. The evolutionary ecology of myco-heterotrophy. New Phytologist 167: 335–352.

Bidartondo, M.I, D.J. Read, J.M. Trappe, V. Merckx, R. Ligrone, and J.G. Duckett. 2011. The dawn of symbiosis between plants and fungi. Biology Letters 7: 574–577.

Biddle, J.F., C.H. House, and J.E. Brenchley. 2005. Microbial stratification in deeply buried marine sediment reflects changes in sulfate/methane profiles. Geobiology 3: 287–295.

Biedermann, P.H.W., K.D. Klepzig, and M. Taborsky. 2009. Fungus cultivation by ambrosia beetles: Behavior and laboratory breeding success in three xyleborine species. Environmental Entomology 38: 1096–1105.

Bigall, N.C., and A. Eychmüller. 2010. Synthesis of noble metal nanoparticles and their non-ordered superstructures. Philosophical Transactions of the Royal Society 368A: 1385–1404.

Binder, M., and D.S. Hibbett. 2006. Molecular systematics and biological diversification of Boletales. Mycologia 98: 971–981.

Binder, M., A. Justo, R. Riley, A. Salamov, F. Lopez-Giraldez, E. Sjökvist, A. Copeland, B. Foster, H. Sun, E. Larsson, K.-H. Larsson, J. Townsend, I.V. Grigoriev, and D.S. Hibbett. 2013. Phylogenetic and phylogenomic overview of the Polyporales. Mycologia 105: 1350–1373.

Bindschedler, S., L. Millière, G. Cailleau, D. Job, and E.P. Verrecchia. 2010. Calcitic nanofibres in soils and caves: A putative fungal contribution to carbonatogenesis. In H.M. Pedley, and M. Rogerson (Eds), Tufas and Speleothems: Unravelling the Microbial and Physical Controls. Geological Society of London Special Publication 336, pp. 225–238. Geological Society Publishing House, Bath, UK.

Bindschedler, S., L. Millière, G. Cailleau, D. Job, and E.P. Verrecchia. 2012. An ultrastructural approach to analogies between fungal structures and needle fiber calcite. Geomicrobiology Journal 29: 301–313.

Biradar, N.V., and S.D. Bonde. 1976. On a fossil species of imperfect fungus *Rhizoctonia* DC. from the Kamthi Beds of India. Maharashtra Vidnyan Mandir Patrika 11: 12–19.

Bischoff, J.F., and J.F. White, Jr. 2005. Evolutionary development of the Clavicipitaceae. In J. Dighton, J.F. White Jr., and P. Oudemans (Eds), The Fungal Community: Its Organization and Role in the Ecosystem, pp. 505–518 (Third Edition). CRC Press, Boca Raton, FL.

Bjelland, T., M. Grube, S. Hoem, S.L. Jorgensen, F.L. Daae, I.H. Thorseth, and L. Øvreås. 2011. Microbial metacommunities in the lichen-rock habitat. Environmental Microbiology Reports 3: 434–442.

Blaauw, M., and D. Mauquoy. 2012. Signal and variability within a Holocene peat bog – Chronological uncertainties of pollen, macrofossil and fungal proxies. Review of Palaeobotany and Palynology 186: 5–15.

Blackwell, M. 1994. Minute mycological mysteries: The influence of arthropods on the lives of fungi. Mycologia 86: 1–17.

Blackwell, M. 2000. Terrestrial life – Fungal from the start? Science 289: 1884–1885.

Blackwell, M. 2011. The fungi: 1, 2, 3 … 5.1 million species? American Journal of Botany 98: 426–438.

Blackwell, M., and J.W. Spatafora. 2004. Fungi and their allies. In G.M. Mueller, G.F. Bills, and M.S. Foster (Eds), Biodiversity of Fungi: Inventory and Monitoring Methods, pp. 7–22. Elsevier Academic Press, Burlington, MA.

Blackwell, M., D.S. Hibbett, J.W. Taylor, and J.W. Spatafora. 2006. Research Coordination Networks: A phylogeny for kingdom Fungi (Deep Hypha). Mycologia 98: 829–837.

Blackwell, W.H., and M.J. Powell. 2000. A review of group filiation of stramenopiles, additional approaches to the question. Evolutionary Theory 12: 49–88.

Blackwell, W.H., P.M. Letcher, and M.J. Powell. 2004. Synopsis and systematic reconsideration of *Karlingiomyces* (Chytridiomycota). Mycotaxon 89: 259–276.

Blair, J.E. 2009. Fungi. In S.B. Hedges, and S. Kumar (Eds), The Timetree of Life, pp. 215–219. Oxford University Press, New York, NY.

Blanchette, R.A. 1991. Delignification by wood-decay fungi. Annual Review of Phytopathology 29: 381–398.

Blanchette, R.A. 1992. Anatomical responses of xylem to injury and invasion by fungi. In R.A. Blanchette, and R.A. Biggs (Eds), Defense Mechanisms of Woody Plants Against Fungi, pp. 76–95. Springer, Berlin, Germany.

Błaszkowski, J. 2012. Glomeromycota. W. Szafer Institute of Botany, Polish Academy of Sciences. Kraków, Poland. 303 pp.

Błaszkowski, J., G.M. Kovács, B.K. Gáspár, T.K. Balázs, F. Buscot, and P. Ryszka. 2012. The arbuscular mycorrhizal *Paraglomus majewskii* sp. nov. represents a distinct basal lineage in Glomeromycota. Mycologia 104: 148–156.

Blatrix, R., C. Djiéto-Lordon, L. Mondolot, P. La Fisca, H. Voglmayr, and D. McKey. 2012. Plant-ants use symbiotic fungi as a food source: New insight into the nutritional ecology of ant-plant interactions. Proceedings of the Royal Society 279B: 3940–3947.

Blunck, G. 1929. Bakterieneinschlüsse im Bernstein. Centralblatt für Mineralogie, Geologie und Paläontologie 11: 554–555.

Böcher, T.W. 1964. Morphology of the vegetative body of *Metasequoia glyptostroboides*. Dansk Botanisk Arkiv 24: 1–70.

Bock, W. 1969. The American Triassic Flora and Global Distribution. Geological Center Research Series, Volume 3 and 4. Geological Center, North Wales, PA. 406 pp.

Boedijn, K.B. 1958. Notes of the Mucorales of Indonesia. Sydowia 12: 321–362.

Boehm, E.W.A., G.K. Mugambi, A.N. Miller, S.M. Huhndorf, S. Marincowitz, J.W. Spatafora, and C.L. Schoch. 2009. A molecular phylogenetic reappraisal of the Hysteriaceae, Mytilinidiaceae and Gloniaceae (Pleosporomycetidae, Dothideomycetes) with keys to world species. Studies in Mycology 64: 49–83.

Bonfante, P., and I.-A. Anca. 2009. Plants, mycorrhizal fungi, and bacteria: A network of interactions. Annual Review of Microbiology 63: 363–383.

Bonfante, P., and A. Genre. 2008. Plants and arbuscular mycorrhizal fungi: An evolutionary-developmental perspective. Trends in Plant Science 13: 492–498.

Bonfante, P., and A. Genre. 2010. Mechanisms underlying beneficial plant-fungus interactions in mycorrhizal symbiosis. Nature Communications 1(4): 1–11.

Bonfante-Fasolo, P. 1984. Anatomy and morphology of VA mycorrhizae. In D. Powell, and D.J. Bagyaraj (Eds), VA Mycorrhiza, pp. 5–33. CRC Press, Boca Raton, FL.

Bonfante-Fasolo, P., and S. Scannerini. 1976. The ultrastructure of the zygospore in *Endogone flammicorona* Trappe & Gerdemann. Mycopathologia 59: 117–123.

Bonneville, S., M.M. Smits, A. Brown, J. Harrington, J.R. Leake, R. Brydson, and L.G. Benning. 2009. Plant-driven fungal weathering: Early stages of mineral alteration at the nanometer scale. Geology 37: 615–618.

Bordy, E.M., A.J. Bumby, O. Catuneanu, and P.G. Erkisson. 2009. Possible trace fossils of putative termite origin in the Lower Jurassic (Karoo Supergroup) of South African and Lesotho. South Africa Journal of Science 105: 356–362.

Bornet, E., and C. Flahault. 1889. Sur quelques plantes vivant dans le test calcaire des mollusques. Bulletin de la Société Botanique de France 36: 147–176.

Bornet, M.E.. 1891. Note sur l'Ostracoblabe implexa Bornet and Flahault. Journal de Botanique 5: 397–400.

Borowicz, V.A. 2001. Do arbuscular mycorrhizal fungi alter plant-pathogen relations? Ecology 82: 3057–3068.

Borsato, A., S. Frisia, B. Jones, and K. Van der Borg. 2000. Calcite moonmilk: Crystal morphology and environment of formation in caves in the Italian Alps. Journal of Sedimentary Research 70: 1179–1190.

Boullard, B. 1979. Considerations sur la symbiose fongique chez les Pteridophytes. Syllogeus 19: 4–58.

Boullard, B. 1988. Observations on the coevolution of fungi with hepatics. In K.A. Pirozynski, and D.L. Hawksworth (Eds), Coevolution of Fungi with Plants and Animals, pp. 107–124. Academic Press Limited, London.

Boullard, B., and Y. Lemoigne. 1971. Les champignons endophytes du *Rhynia gwynne-vaughanii* K. et L. Étude morphologique et déductions sur leur biologie. Le Botaniste 54: 49–89.

Boustie, J., and M. Grube. 2005. Lichens – a promising source of bioactive secondary metabolites. Plant Genetic Resources 3: 273–287.

Boyce, C.K. 2008. How green was *Cooksonia*? The importance of size in understanding the early evolution of physiology in the vascular plant lineage. Paleobiology 34: 179–194.

Boyce, C.K., and C.L. Hotton. 2010. "*Prototaxites* was not a taphonomic artifact". American Journal of Botany 97: 1073.

Boyce, C.K., C.L. Hotton, M.L. Fogel, G.D. Cody, R.M. Hazen, A.H. Knoll, and F.M. Hueber. 2007. Devonian landscape heterogeneity recorded by a giant fungus. Geology 35: 399–402.

Boyetchko, S.M., and J.P. Tewari. 1991. Parasitism of spores of the vesicular-arbuscular mycorrhizal fungus, *Glomus dimorphicum*. Phytoprotection 72: 27–32.

Bozkurt, T.O., S. Schornack, M.J. Banfield, and S. Kamoun. 2012. Oomycetes, effectors, and all that jazz. Current Opinion in Plant Biology 15: 483–492.

Brabenec, B. 1909. Souborná květena českého útvaru třetihorního. Část I-II. – Archiv pro přírodovědecké prozkoumání Čech 14: 146–374.

Bradley, W.H. 1931. Origin and Microfossils of the Oil Shale of the Green River Formation of Colorado and Utah. United States Geological Survey Professional Paper 168: 58.

Bradley, W.H. 1946. Coprolites from the Bridger Formation of Wyoming: Their composition and microorganisms. American Journal of Science 244: 215–239.

Bradley, W.H. 1963. Unmineralized fossil bacteria. Science 141: 919–921.

Bradley, W.H. 1964. Aquatic fungi from the Green River Formation of Wyoming. American Journal of Science 262: 413–416.

Bradley, W.H. 1967. Two aquatic fungi (Chytridiales) of Eocene age from the Green River Formation of Wyoming. American Journal of Botany 54: 577–582.

Bradley, W.H. 1968. Unmineralized fossil bacteria: A retraction. Science 160: 437.

Braun, E.J., and R.J. Howard. 1994. Adhesion of fungal spores and germlings to host plant surfaces. Protoplasma 181: 202–212.

Braun, U. 2011. The current systematics and taxonomy of the powdery mildews (Erysiphales): An overview. Mycoscience 52: 210–212.

Braun, U., and R.T.A. Cook. 2012. Taxonomic Manual of the Erysiphales (Powdery Mildews). Centraalbureau voor Schimmelcultures (CBS), Biodiversity Series 11, Utrecht, The Netherlands. 707 pp.

Braun, U., P.W. Crous, F. Dugan, J.Z. Groenewald, and G.S. de Hoog. 2003. Phylogeny and taxonomy of cladosporium-like hyphomycetes, including *Davidiella* gen. nov., the teleomorph of *Cladosporium* s. str. Mycological Progress 2: 3–18.

Bray, P.S., and K.B. Anderson. 2009. Identification of Carboniferous (320 Million Years Old) Class Ic Amber. Science 326: 132–134.

Breton, G., M. Bilotte, and G. Eychenne. 2013. L'ambre campanien du Mas d'Azil (Ariège, France): Gisement, micro-inclusions, taphonomie. Annales de Paléontologie 99: 317–337.

Brochu, C.A., C.D. Sumrall, and J.M. Theodor. 2004. When clocks (and communities) collide: Estimating divergence time from molecules and the fossil record. Journal of Paleontology 78: 1–6.

Bromham, L., and D. Penny. 2003. The modern molecular clock. Nature Reviews Genetics 4: 216–224.

Bromley, R.G. 2005. Preliminary study of bioerosion in the deep-water coral *Lophelia*, Pleistocene, Rhodes, Greece. In A. Freiwald, and J.M. Roberts (Eds), Cold-water Corals and Ecosystems, pp. 895–914. Springer-Verlag, Berlin, Germany.

Bronson, A.W., A.A. Klymiuk, R.A. Stockey, and A.M.F. Tomescu. 2013. A perithecial sordariomycete (Ascomycota, Diaporthales) from the Lower Cretaceous of Vancouver Island, British Columbia, Canada. International Journal of Plant Sciences 174: 278–292.

Brooks, H.K. 1955. Healed wounds and galls on fossil leaves from Wilcox Deposits (Eocene) of Western Tennessee. Psyche 62: 1–9.

Brown, K.B., K.D. Hyde, and D.I. Guest. 1998. Preliminary studies on endophytic fungal communities of *Musa acuminata* species complex in Hong Kong and Australia. Fungal Diversity 1: 27–51.

Brown, R.W. 1936. A fossil shelf-fungus from North Dakota. Journal of the Washington Academy of Sciences 26: 460–462.

Brown, R.W. 1938. Two fossils misidentified as shelf-fungi. Journal of the Washington Academy of Sciences 28: 130–131.

Brown, R.W. 1940. A bracket fungus from the Late Tertiary of southwestern Idaho. Journal of the Washington Academy of Sciences 30: 422–424.

Bruckart, W.L., F.M. Eskandari, D.K. Berner, and M.C. Aime. 2010. Life cycle of *Puccinia acroptili* on *Rhaponticum* (= *Acroptilon*) *repens*. Mycologia 102: 62–68.

Brucker, R.M., and S.R. Bordenstein. 2012. Speciation by symbiosis. Trends in Ecology & Evolution 27: 443–451.

Bruening, F.A., and A.D. Cohen. 2005. Measuring surface properties and oxidation of coal macerals using the atomic force microscope. International Journal of Coal Geology 63: 195–204.

Brundrett, M.C. 2002. Coevolution of roots and mycorrhizas of land plants. New Phytologist 154: 275–304.

Brundrett, M.C. 2004. Diversity and classification of mycorrhizal associations. Biological Reviews 79: 473–495.

Brundrett, M.C. 2009. Mycorrhizal associations and other means of nutrition of vascular plants: Understanding the global diversity of host plants by resolving conflicting information and developing reliable means of diagnosis. Plant and Soil 320: 37–77.

Brundrett, M.C., D.A. Jasper, and N. Ashwath. 1999. Glomalean mycorrhizal fungi from tropical Australia. II. The effect of nutrient levels and host species on the isolation of fungi. Mycorrhiza 8: 315–321.

Bruns, T.D., R. Fogel, T.J. White, and J.D. Palmer. 1989. Accelerated evolution of a false-truffle from a mushroom ancestor. Nature 339: 140–142.

Bruns, T.D., T.M. Szaro, M. Gardes, K.W. Cullings, J.J. Pan, D.L. Taylor, T.R. Horton, A. Kretzer, M. Garbelotto, and Y. Li. 1998. A sequence database for the identification of ectomycorrhizal basidiomycetes by phylogenetic analysis. Molecular Ecology 7: 257–272.

Bruns, T.D., A.E. Arnold, and K.W. Hughes. 2008. Fungal networks made of humans: UNITE, FESIN, and frontiers in fungal ecology. New Phytologist 177: 586–588.

Buchanan, P.K. 2001. A taxonomic overview of the genus *Ganoderma* with special reference to species of medicinal and nutriceutical importance. In: Y.H. Gao, S.F. Zhou, and P.K. Buchanan (eds), *Ganoderma*: Systematics, Biochemistry, Pharmacology and Therapeutics. Proceedings of the International Symposium on Ganoderma Science, pp. 1–8. International Symposium on Ganoderma Science, 27–29 April, 2001, Auckland, New Zealand.

Bucholtz, F. 1912. Beiträge zur Kenntnis der Gattung *Endogone* Link. Beihefte zum Botanischen Centralblatt, Abteilung II 29: 147–224.

Buchwald, N.F. 1970. *Fomes idahoensis* Brown, A fossil polypore fungus from the late Tertiary of Idaho, USA Friesia 9: 339–340.

Budd, G.E., and M.J. Telford. 2009. The origin and evolution of arthropods. Nature 457: 812–817.

Bultman, T.L., and A. Leuchtmann. 2008. Biology of the *Epichloë–Botanophila* interaction: An intriguing association between fungi and insects. Fungal Biology Reviews 22: 131–138.

Bultman, T.L., and P.L. Mathews. 1996. Mycophagy by a millipede and its possible impact on an insect-fungus mutualism. Oikos 75: 67–74.

Bundschuh, M. 2000. Silurische Mikrobohrspuren. Ihre Beschreibung und Verteilung in verschiedenen Faziesräumen (Schweden, Litauen, Großbritannien und USA). Ph.D. Thesis, Fachbereich Geowissenschaften, Johann Wolfgang Goethe-Universität Frankfurt am Main, Germany. 129 pp.

Bureau, E. 1881. Prémices de la flore Éocène du Bois-Gouët (Loire-Inférieure). Bulletin de la Société Géologique de France Nantes 3: 286–293.

Bureau, E., and N. Patouillard. 1893. Additions a la flore Éocène du Bois-Gouët (Loire Inférieure). Bulletin de la Société des Sciences Naturelles de l'Ouest de la France 3: 261–269.

Burford, E.P., M. Kierans, and G.M. Gadd. 2003a. Geomycology: Fungi in mineral substrata. Mycologist 17: 98–107.

Burford, E.P., M. Fomina, and G.M. Gadd. 2003b. Fungal involvement in bioweathering and biotransformation of rocks and minerals. Mineralogical Magazine 67: 1127–1155.

Burford, E.P., S. Hillier, and G.M. Gadd. 2006. Biomineralization of fungal hyphae with calcite ($CaCO_3$) and calcium oxalate mono- and dihydrate in Carboniferous limestone microcosms. Geomicrobiology Journal 23: 599–611.

Burgess, N.D., and D. Edwards. 1988. A new Palaeozoic plant closely allied to *Prototaxites* Dawson. Botanical Journal of the Linnean Society 97: 189–203.

Burgess, N.D., and D. Edwards. 1991. Classification of uppermost Ordovician to Lower Devonian tubular and filamentous macerals from the Anglo-Welsh Basin. Botanical Journal of the Linnean Society 106: 41–66.

Burney, D.A., G.S. Robinson, and L.P. Burney. 2003. *Sporormiella* and the Late Holocene extinctions in Madagascar. Proceedings of the National Academy of Sciences, USA 100: 10800–10805.

Burzin, M.B. 1993. The oldest chytrid (Mycota, Chytridiomycetes *incertae sedis*) from the Upper Vendian of the East European Platform. In B.S. Sokolov (Ed.), Fauna and Ecosystems of the Geological Past, pp. 21–33. Nauka, Moscow. [in Russian].

Butler, E.J. 1939. The occurrence and systematic position of the vesicular-arbuscular type of mycorrhizal fungi. Transactions of the British Mycological Society 22: 274–301.

Butterfield, N.J. 2005a. Probable Proterozoic fungi. Paleobiology 31: 165–182.

Butterfield, N.J. 2005b. Reconstructing a complex early Neoproterozoic eukaryote, Wynniatt Formation, arctic Canada. Lethaia 38: 155–169.

Butterworth, M.A., and R.W. Williams. 1958. The small spore floras of coals in the Limestone Coal Group and Upper Limestone Group of the Lower Carboniferous of Scotland. Transactions of the Royal Society of Edinburgh 63: 353–392.

Cabello, M., L. Gaspar, and R. Pollero. 1994. *Glomus antarcticum* sp. nov., a vesicular-arbuscular mycorrhizal fungus from Antarctica. Mycotaxon 51: 123–128.

Cafaro, M.J. 2005. Eccrinales (Trichomycetes) are not fungi, but a clade of protists at the early divergence of animals and fungi. Molecular Phylogenetics and Evolution 35: 21–34.

Cai, L., T. Giraud, N. Zhang, D. Begerow, G. Cai, and R.G. Shivas. 2011. The evolution of species concepts and species recognition criteria in plant pathogenic fungi. Fungal Diversity 50: 121–133.

Cailleau, G., E.P. Verrecchia, O. Braissant, and L. Emmanuel. 2009. The biogenic origin of needle fibre calcite. Sedimentology 56: 1858–1875.

Cairney, J.W.G. 1992. Translocation of solutes in ectomycorrhizal and saprotrophic rhizomorphs. Mycological Research 96: 135–141.

Cairney, J.W.G. 2000. Evolution of mycorrhiza systems. Naturwissenschaften 87: 467–475.

Cairney, J.W.G., and R.M. Burke. 1996. Physiological heterogeneity within fungal mycelia: An important concept for a functional understanding of the ectomycorrhizal symbiosis. New Phytologist 134: 685–695.

Callow, R.H.T., M.D. Brasier, and D. McIlroy. 2013. Discussion: "Were the Ediacaran siliciclastics of South Australia coastal or deep marine?" by Retallack et al., Sedimentology, 59, 1208–1236. Sedimentology 60: 624–627.

Calo, L., I. García, C. Gotor, and L.C. Romero. 2006. Leaf hairs influence phytopathogenic fungus infection and confer an increased resistance when expressing a *Trichoderma* α-1, 3-glucanase. Journal of Experimental Botany 57: 3911–3920.

Cannon, P.F. 1991. A Revision of *Phyllachora* and Some Similar Genera on the Host Family Leguminosae. International Mycological Institute, Kew, UK, 302 pp.

Cannon, P.F., and P.M. Kirk. 2000. The philosophy and practicalities of amalgamating anamorph and teleomorph concepts. Studies in Mycology 45: 19–25.

Cannon, P.F., and C.M. Simmons. 2002. Diversity and host preference of leaf endophytic fungi in the Iwokrama Forest Reserve, Guyana. Mycologia 94: 210–220.

Cannon, P.F., U. Damm, P.R. Johnston, and B.S. Weir. 2012. *Colletotrichum* – current status and future directions. Studies in Mycology 73: 181–213.

Cano, R.J., and M.K. Borucki. 1995. Revival and identification of bacterial spores in 25- to 40-million-year-old Dominican amber. Science 268: 1060–1064.

Cano, L.M., S. Raffaele, R.H. Haugen, D.G.O. Saunders, L. Leonelli, D. MacLean, S.A. Hogenhout, and S. Kamoun. 2013. Major transcriptome reprogramming underlies floral mimicry induced by the rust fungus *Puccinia monoica* in Boechera stricta. PLoS ONE 8: e75293.

Cantrell, S.A., R.T. Hanlin, and A. Emiliano. 2007. *Periconia variicolor* sp. nov., a new species from Puerto Rico. Mycologia 99: 482–487.

Cantrell, S.A., J.C. Dianese, J. Fell, N. Gunde-Cimerman, and P. Zalar. 2011. Unusual fungal niches. Mycologia 103: 1161–1174.

Cantrill, D.J., and J.G. Douglas. 1988. Mycorrhizal conifer roots from the Lower Cretaceous of the Otway Basin, Victoria. Australian Journal of Botany 36: 257–272.

Cantrill, D.J., and H.J. Falcon-Lang. 2001. Cretaceous (Late Albian) coniferales of Alexander Island, Antarctica. 2. Leaves, reproductive structures and roots. Review of Palaeobotany and Palynology 115: 119–145.

Cantrill, D.J, and I. Poole. 2005. A new Eocene *Araucaria* from Seymour Island, Antarctica: Evidence from growth form and bark morphology. Alcheringa 29: 341–350.

Caporael, L.R. 1976. Ergotism: The Satan loosed in Salem? Science 192: 21–26.

Carroll, G. 1988. Fungal endophytes in stems and leaves: From latent pathogen to mutualistic symbiont. Ecology 69: 2–9.

Carroll, G.C., and F.E. Carroll. 1978. Studies on the incidence of coniferous needle endophytes in the Pacific Northwest. Canadian Journal of Botany 56: 3034–3043.

Carruthers, W. 1873. On *Traquairia*, a Radiolarian Rhizopod from the Coal-Measures. Report of the 42nd Annual Meeting of the British Association for the Advancement of Science (Brighton, August 1872), pp. 126. John Murray, London, UK.

Carruthers, W. 1889. Visit to the Natural History Museum, Cromwell Road, Department of Botany: Saturday Afternoon, March 16th, 1889. Demonstration on Fossil Fungi, Lichens, Mosses, etc. Proceedings of the Geologists' Association 11: xxi–xxiii.

Casadevall, A. 2005. Fungal virulence, vertebrate endothermy, and dinosaur extinction: Is there a connection? Fungal Genetics and Biology 42: 98–106.

Cash, W., and T. Hick. 1879. On fossil fungi from the Lower Coal Measures of Halifax. Proceedings of the Yorkshire Geological and Polytechnic Society 7: 115–121.

Caspary, R., and R. Klebs. 1906. Robert Caspary: Die Flora des Bernsteins und anderer fossiler Harze des ostpreußischen Tertiärs. Nach dem Nachlasse des Verstorbenen bearbeitet von Richard Klebs. Band I. I. Thallophyta. II. Bryophyta. III. Pteridophyta. IV. Gymnospermae. Abhandlungen der Königlich Preußischen Geologischen Landesanstalt. Neue Folge 4: 1–183.

Caspary, R., and R. Klebs. 1907. Atlas von dreissig Tafeln zu der Abhandlung: Robert Caspary: Die Flora des Bernsteins und anderer fossiler Harze des ostpreufsischen Tertiärs. Nach dem Nachlasse des Verstorbenen bearbeitet von Richard Klebs. Abhandlungen der Königlich Preußischen Geologischen Landesanstalt. Neue Folge 4: 1–18.

Castañeda-Posadas, C., L. Calvillo-Canadell, and S.R.S. Cevallos-Ferriz. 2009. Woods from Miocene sediments in Panotla, Tlaxcala, Mexico. Review of Palaeobotany and Palynology 156: 494–506.

Castillo, D.M., and T.E. Pawlowska. 2010. Molecular evolution in bacterial endosymbionts of fungi. Molecular Biology and Evolution 27: 622–636.

Castillo-Guevara, C., J. Sierra, G. Galindo-Flores, M. Cuautle, and C. Lara. 2011. Gut passage of epigeous ectomycorrhizal fungi by two opportunistic mycophagous rodents. Current Zoology 57: 293–299.

Castlebury, L.A., A.Y. Rossman, W.J. Jaklitsch, and L.N. Vasilyeva. 2002. A preliminary overview of the Diaporthales based on large subunit nuclear ribosomal DNA sequences. Mycologia 94: 1017–1031.

Cavalazzi, B., F. Westall, S.L. Cady, R. Barbieri, and F. Foucher. 2011. Potential fossil endoliths in vesicular pillow basalt, Coral Patch Seamount, eastern North Atlantic Ocean. Astrobiology 11: 619–632.

Cavaliere, A.R., and R.S. Alberte. 1970. Fungi in animal shell fragments. Journal of the Elisha Mitchell Society 86: 203–206.

Cázares, E., and J.M. Trappe. 1994. Spore dispersal of ectomycorrhizal fungi on a glacier forefront by mammal mycophagy. Mycologia 86: 507–510.

Cevallos-Ferriz, S.R.S., and R.A. Stockey. 1989. Permineralized fruits and seeds from the Princeton chert (Middle Eocene) of British Columbia: Nymphaeaceae. Botanical Gazette 150: 207–217.

Cevallos-Ferriz, S.R.S., R.A. Stockey, and K.B. Pigg. 1991. The Princeton chert: Evidence for in situ aquatic plants. Review of Palaeobotany and Palynology 70: 173–185.

Chabasse, D. 1998. Origine et interrelation des champignons avec le vivant. Évolution durant les temps géologiques. Journal de Mycologie Médicale 8: 125–138.

Chabaud, M., A. Genre, B.J. Sieberer, A. Faccio, J. Fournier, M. Novero, D.G. Barker, and P. Bonfante. 2011. Arbuscular mycorrhizal hyphopodia and germinated spore exudates trigger Ca^{2+} spiking in the legume and nonlegume root epiderm. New Phytologist 189: 347–355.

Chamier, A.C. 1985. Cell-wall-degrading enzymes of aquatic hyphomycetes: A review. Botanical Journal of the Linnean Society 91: 67–81.

Chandra, A., R.K. Saxena, and M.G.A.P. Setty. 1984. Palynological investigation of the sediment cores from the Arabian Sea. 1. Fungal spores. Biovigyanam 10: 41–58.

Chaney, R.W., and H.L. Mason. 1936. A Pleistocene flora from Fairbanks, Alaska. American Museum Novitates 887: 1–17.

Channing, A. 2003. The Rhynie chert early land plants: palaeo-ecophysiological and taphonomic analogues. Applied Earth Science (Transactions of the Institue of Mining and Metallurgy, section B) 112: B170–B171.

Channing, A., and D. Edwards. 2009a. Yellowstone hot spring environments and the palaeo-ecophysiology of Rhynie chert plants: Towards a synthesis. Plant Ecology and Diversity 2: 111–143.

Channing, A., and D. Edwards. 2009b. Silicification of higher plants in geothermally influenced wetlands: Yellowstone as a Lower Devonian Rhynie analog. PALAIOS 24: 505–521.

Channing, A., and D.E. Wujek. 2010. Preservation of protists within decaying plants from geothermally influenced wetlands of Yellowstone National Park, Wyoming, United States. PALAIOS 25: 347–355.

Channing, A., D. Edwards, and S. Sturtevant. 2004. A geothermally influenced wetland containing unconsolidated geochemical sediments. Canadian Journal of Earth Sciences 41: 809–827.

Channing, A., A. Zamuner, D. Edwards, and D. Guido. 2011. *Equisetum thermale* sp. nov. (Equisetales) from the Jurassic San Agustín hot spring deposit, Patagonia: Anatomy, paleoecology, and inferred paleoecophysiology. American Journal of Botany 98: 680–697.

Chapela, I.H., S.A. Rehner, T.R. Schultz, and U.G. Mueller. 1994. Evolutionary history of the symbiosis between fungus-growing ants and their fungi. Science 266: 1691–1694.

Chapman, F. 1913. Description of new and rare fossils obtained by deep boring in the Mallee. Proceedings of the Royal Society of Victoria 26: 165–191.

Chazottes, V., G. Cabioch, S. Golubic, and G. Radtke. 2009. Bathymetric zonation of modern microborers in dead coral substrates from New Caledonia – Implications for paleodepth reconstructions in Holocene corals. Palaeogeography, Palaeoclimatology, Palaeoecology 280: 456–468.

Cheeke, T.E., T.N. Rosenstiel, and M.B. Cruzan. 2012. Evidence of reduced arbuscular mycorrhizal fungal

colonization in multiple lines of Bt maize. American Journal of Botany 99: 700–707.
Cheewangkoon, R., J.Z. Groenewald, B.A. Summerell, K.D. Hyde, C. To-anun, and P.W. Crous. 2009. Myrtaceae, a cache of fungal biodiversity. Persoonia 23: 55–85.
Chen, J., H.-P. Blume, and L. Beyer. 2000. Weathering of rocks induced by lichen colonization – a review. Catena 39: 121–146.
Chen, S.-F., and C.-Y. Chien. 1995. Some chytrids of Taiwan (I). Botanical Bulletin of Academia Sinica 36: 235–241.
Chen, S.-F., and C.-Y. Chien. 1998. Some chytrids of Taiwan (II). Botanical Bulletin of Academia Sinica 39: 47–56.
Chen, Y, L.D. Caro, M. Mastalerz, A. Schimmelmann, and A. Blandón. 2013. Mapping the chemistry of resinite, funginite and associated vitrinite in coal with micro-FTIR. Journal of Microscopy 249: 69–81.
Cheng, L., F.L. Booker, C. Tu, K.O. Burkey, L. Zhou, H.D. Shew, T.W. Rufty, and S. Hu. 2012. Arbuscular mycorrhizal fungi increase organic carbon decomposition under elevated CO_2. Science 337: 1084–1087.
Chin, K. 2007. The paleobiological implications of herbivorous dinosaur coprolites from the Upper Cretaceous Two Medicine Formation of Montana: Why eat wood? PALAIOS 22: 554–566.
Chin, K., and B.D. Gill. 1996. Dinosaurs, dung beetles, and conifers: Participants in a Cretaceous food web. PALAIOS 11: 280–285.
Chitaley, S.D. 1957. Further report on the fossil microflora from the Mohgaon Kalan Beds of the Madhya Pradesh, India. Proceedings of the National Institute of Science in India 23B: 69–79.
Chitaley, S.D., and G.V. Patil. 1972. An ebenaceous fossil wood infected with deuteromyceteous fungus from the Deccan Intertrappean beds of India. Botanique 3: 99–105.
Chitaley, S.D., and M.T. Sheikh. 1971. An infected grain from the Deccan Intertrappean Cherts of Mohgaonkalan. Journal of Indian Botanical Society 50: 137–142.
Chitaley, S.D., and N.R. Yawale. 1976. Fungal remains from the Deccan intertrappean beds of Mohgaonkalan, India. Proceedings of the Indian Science Congress, Part 3, Section VI: Botany 63: 52.
Chitaley, S.D., and N.R. Yawale. 1978. Fungal remains from the Deccan Intertrappean beds of Mohgaonkalan, India. Botanique (Nagpu) 7: 189–194.
Chlebicki, A., and M.W. Lorenc. 1997. Subfossil *Fomes fomentarius* from a Holocene fluvial deposit in Poland. Holocene 7: 101–103.
Chmura, G.L., P.A. Stone, and M.S. Ross. 2006. Non-pollen microfossils in Everglades sediments. Review of Palaeobotany and Palynology 141: 103–119.
Chomnunti, P., C.L. Schoch, B. Aguirre-Hudson, T.W. Ko-Ko, S. Hongsanan, E.B.G. Jones, R. Kodsueb, R. Phookamsak, E. Chukeatirote, A.H. Bahkali, and K.D. Hyde. 2011. Capnodiaceae. Fungal Diversity 51: 103–134.
Chou, C.K.S. 1978. Penetration of young stems of *Pinus radiata* by *Diplodia pinea*. Physiological Plant Pathology 12: 189–192.
Chouvenc, T., C.A. Efstathion, M.L. Elliott, and N.-Y. Su. 2012. Resource competition between two fungal parasites in subterranean termites. Naturwissenschaften 99: 949–958.
Chrysler, M.A., and C.M. Haenseler. 1936. A Cretaceous fungus: *Xylomites cycadeoideae*. American Journal of Botany 23: 33–36.
Church, A.H. 1919. Thalassiophyta and the subaerial transmigration. Oxford Botanical Memoirs 3: 1–95.
Clark, W.H., and D.C. Prusso. 1986. *Desmidiospora myrmecophila* found infesting the ant *Camponotus semitestaceus*. Mycologia 78: 865–866.
Clarke, R.T. 1965. Fungal spores from Vermejo Formation coal beds (Upper Cretaceous) of central Colorado. Mountain Geologist 2: 85–93.
Clay, K. 1992. Fungal endophytes of plants: biological and chemical diversity. Natural Toxins 1: 147–149.
Clay, K., and C. Schardl. 2002. Evolutionary origins and ecological consequences of endophyte symbiosis with grasses. American Naturalist 160: S99–S127.
Clémençon, H. 2005. Rhizomorph anatomy of *Ossicaulis lignatilis* (Tricholomatales), with special attention to its haustoria-like intrahyphal hyphae. Mycological Progress 4: 167–173.
Cloud, P. 1976. Beginnings of biospheric evolution and their biogeochemical consequences. Paleobiology 2: 351–387.
Cockerell, T.D.A. 1908. Descriptions of Tertiary plants, II. American Journal of Science 26: 537–544.
Cohn, F. 1872. Untersuchungen über Bakterien. Beiträge zur Biologie der Pflanzen 1: 127–244.
Colani, M. 1920. Étude sur les flores tertiaires de quelques gisements de lignite de L'Indochine et du Yunnan. 1. Structure microscopique d'un bois fossile. Bulletin de Service Géologique 8: 397–446.
Colbath, G.K., and H.R. Grenfell. 1995. Review of biological affinities of Paleozoic acid-resistant, organic-walled eukaryotic algal microfossils (including "acritarchs"). Review of Palaeobotany and Palynology 86: 287–314.
Cole, G.T. 1986. Models of cell differentiation in conidial fungi. Microbiological Reviews 50: 95–132.
Collinge, D.B. 2009. Cell wall appositions: The first line of defence. Journal of Experimental Botany 60: 351–352.
Collins, B., and J. Parke. 2008. Spatial and temporal aspects of tylosis formation in tanoak inoculated with *Phytophthora ramorum*. In: S.J. Frankel, J.T. Kliejunas, and K.M. Palmieri (eds), Proceedings of the Sudden Oak Death 3rd Science Symposium, US Department of Agriculture General Technology Report PSW-GTR-214: 335.
Cologgi, D.L., S. Lampa-Pastirk, A.M. Speers, S.D. Kelly, and G. Reguera. 2011. Extracellular reduction of uranium via *Geobacter* conductive pili as a protective cellular mechanism. Proceedings of the National Academy of Sciences, USA 108: 15248–15252.
Comandini, O., and A.C. Rinaldi. 2004. Tracing megafaunal extinctions with dung fungal spores. Mycologist 18: 140–142.
Concha-Lozano, N., P. Gaudon, J. Pages, G. de Billerbeck, D. Lafon, and O. Eterradossi. 2012. Protective effect of

endolithic fungal hyphae on oolitic limestone buildings. Journal of Cultural Heritage 13: 120–127.

Conroy, C.J., and M. van Tuinen. 2003. Extracting time from phylogenies: Positive interplay between fossil and genetic data. Journal of Mammalogy 84: 444–455.

Conti, M.E., and G. Cecchetti. 2001. Biological monitoring: Lichens as bioindicators of air pollution assessment – a review. Environmental Pollution 114: 471–492.

Conwentz, H. 1880. Die fossilen Hölzer von Karlsdorf am Zobten. Ein Beitrag zur Kenntniss der im norddeutschen Diluvium vorkommenden Geschiebehölzer. Schriften der Naturforschenden Gesellschaft in Danzig 4: 1–47.

Conwentz, H. 1890. Monographie der Baltischen Bersteinbäume: Vergleichende Untersuchungen über die Vegetationsorgane und Blüten, sowie über das Harz und die Krankheiten der baltischen Bernsteinbäume. Wilhelm Engelmann, Leipzig. 151 pp.

Conwentz, H. 1896. On English amber and amber generally. Natural Science 9: 99–106.

Cook, E.J., B. van Geel, S. van der Kaars, and J. van Arkel. 2011. A review of the use of non-pollen palynomorphs in palaeoecology with examples from Australia. Palynology 35: 155–178.

Cookson, I.C. 1947. Fossil fungi from Tertiary deposits in the Southern Hemisphere. Part I. Proceedings of the Linnean Society of New South Wales 72: 207–214.

Cookson, I.C., and S.L. Duigan. 1951. Tertiary Araucariaceae from South-Eastern Australia, with notes on living species. Australian Journal of Scientific Research, Series B 4: 415–449.

Corneliussen, E.F., and T.G. Perry. 1970. The ectoproct *Batostoma*? *cornula* (Cumings & Galloway) and its enigmatic intrazooecial spines [Fort Atkinson Limestone (Cincinnatian), Wilmington, Illinois]. Journal of Paleontology 44: 997–1008.

Corner, E.J.H. 1966. A Monograph of Cantharelloid Fungi. Annals of Botany Memoir 2. Oxford University Press, New York, NY. 255 pp.

Corradi, N., and P. Bonfante. 2012. The arbuscular mycorrhizal symbiosis: Origin and evolution of a beneficial plant infection. PLoS Pathogens 8: e1002600.

Couch, J.N. 1939. Technic for collection, isolation and culture of chytrids. Journal of the Elisha Mitchell Scientific Society 55: 208–214.

Couradeau, E., K. Benzerara, D. Moreira, E. Gérard, J. Kaźmierczak, R. Tavera, and P. López-García. 2011. Prokaryotic and eukaryotic community structure in field and cultured microbialites from the alkaline Lake Alchichica (Mexico). PLoS ONE 6: e28767.

Courty, P.-E., F. Walder, T. Boller, K. Ineichen, A. Wiemken, A. Rousteau, and M.-A. Selosse. 2011. Carbon and nitrogen metabolism in mycorrhizal networks and mycoheterotrophic plants in tropical forests: A stable isotope analysis. Plant Physiology 156: 952–961.

Cowley, G.T., and D.J. Colquhoun. 1966. A Pleistocene fungus from South Carolina. Mycologia 58: 483–486.

Cox, G., and F. Sanders. 1974. Ultrastructure of the host-fungus interface in a vesicular-arbuscular mycorrhiza. New Phytologist 73: 901–912.

Cramer, M.D., and H.-J. Hawkins. 2009. A physiological mechanism for the formation of root casts. Palaeogeography, Palaeoclimatology, Palaeoecology 274: 125–133.

Creber, G.T., and S.R. Ash. 1990. Evidence of widespread fungal attack on Upper Triassic trees in the southwestern U.S.A. Review of Palaeobotany and Palynology 63: 189–195.

Crepet, W.L., and G.D. Feldman. 1991. The earliest remains of grasses in the fossil record. American Journal of Botany 78: 1010–1014.

Crespo, A., H.T. Lumbsch, J.-E. Mattsson, O. Blanco, P.K. Divakar, K. Articus, E. Wiklund, P.A. Bawingan, and M. Wedin. 2007. Testing morphology-based hypotheses of phylogenetic relationships in Parmeliaceae (Ascomycota) using three ribosomal markers and the nuclear *RPB1* gene. Molecular Phylogenetics and Evolution 44: 812–824.

Cridland, A.A. 1962. The fungi in cordaitean rootlets. Mycologia 54: 230–234.

Croll, D., M. Giovannetti, A.M. Koch, C. Sbrana, M. Ehinger, P.J. Lammers, and I.R. Sanders. 2009. Nonself vegetative fusion and genetic exchange in the arbuscular mycorrhizal fungus *Glomus intraradices*. New Phytologist 181: 924–937.

Crous, P.W., W. Gams, M.J. Wingfield, and P.S. van Wyk. 1996. *Phaeoacremonium* gen. nov. associated with wilt and decline diseases of woody hosts and human infections. Mycologia 88: 786–796.

Crous, P.W., C.L. Schoch, K.D. Hyde, A.R. Wood, C. Gueidan, G.S. de Hoog, and J.Z. Groenewald. 2009. Phylogenetic lineages in the Capnodiales. Studies in Mycology 64: 17–47.

Crous, P.W., B.A. Summerell, A.C. Alfenas, J. Edwards, I.G. Pascoe, I.J. Porter, and J.Z. Groenewald. 2012. Genera of diaporthalean coelomycetes associated with leaf spots of tree hosts. Persoonia 28: 66–75.

Crowther, T.W., and A.D. A'Bear. 2012. Impacts of grazing soil fauna on decomposer fungi are species-specific and density-dependent. Fungal Ecology 5: 277–281.

Cullings, K.W., T.M. Szaro, and T.D. Bruns. 1996. Evolution of extreme specialization within a lineage of ectomycorrhizal epiparasites. Nature 379: 63–66.

Currah, R.S., and R.A. Stockey. 1991. A fossil smut fungus from the anther of an Eocene angiosperm. Nature 350: 698–699.

Currah, R.S., R.A. Stockey, and B.A. LePage. 1998. An Eocene tar spot on a fossil palm and its fungal hyperparasite. Mycologia 90: 667–673.

Currano, E.D., C.C. Labandeira, and P. Wilf. 2010. Fossil insect folivory tracks paleotemperature for six million years. Ecological Monographs 80: 547–567.

Czeczott, H. 1961. The flora of the Baltic amber and its age. Prace Muzeum Ziemi 4: 139–145.

Czeczuga, B., and E. Muszyńska. 2001. Zoosporic fungi growing on gymnosperm pollen in water of varied trophic state. Polish Journal of Environmental Studies 10: 89–94.

Czeczuga, B., and E. Muszyńska. 2004. Aquatic zoosporic fungi from baited spores of cryptogams. Fungal Diversity 16: 11–22.

Dadachova, E., R.A. Bryan, X. Huang, T. Moadel, A.D. Schweitzer, P. Aisen, J.D. Nosanchuk, and A. Casadevall. 2007. Ionizing radiation changes the electronic properties of melanin and enhances the growth of melanized fungi. PLoS ONE 2: e457.

Daghlian, C.P. 1978. A new melioloid fungus from the Early Eocene of Texas. Palaeontology 21: 171–176.

Dahanayake, K., and W.E. Krumbein. 1985. Ultrastructure of a microbial mat-generated phosphorite. Mineralium Deposita 20: 260–265.

Dahanayake, K., G. Gerdes, and W.E. Krumbein. 1985. Stromatolites, oncolites and oolites biogenically formed in situ. Naturwissenschaften 72: 513–518.

Daniel, J.C., and K. Chin. 2010. The role of bacterially mediated precipitation in the permineralization of bone. PALAIOS 25: 507–516.

Darby, D.G. 1974. Reproductive modes of *Huroniospora microreticulata* from cherts of the Precambrian Gunflint Iron-Formation. Geological Society of America Bulletin 85: 1595–1596.

Daugherty, L.H. 1941. The Upper Triassic flora of Arizona. Carnegie Institution of Washington Publication 526: 1–108.

Davey, M.L., L. Nybakken, H. Kauserud, and M. Ohlson. 2009. Fungal biomass associated with the phyllosphere of bryophytes and vascular plants. Mycological Research 113: 1254–1260.

David, J.C. 2002. A preliminary checklist of the names of fungi above the rank of order. Constancea 83: 30. Published online at http://ucjeps.berkeley.edu/cgi-bin/send_pdf.pl?pdf_file=83.16_T [last accessed August 10, 2013].

Davidson, J.G.N., and G.L. Barron. 1973. Nematophagous fungi: *Haptoglossa*. Canadian Journal of Botany 51: 1317–1323.

Daviero-Gomez, V., H. Kerp, and H. Hass. 2005. *Nothia aphylla*: The issue of clonal development in early land plants. International Journal of Plant Sciences 166: 319–326.

Davis, B., and G.A. Leisman. 1962. Further observations on *Sporocarpon* and allied genera. Bulletin of the Torrey Botanical Club 89: 97–109.

Davis, C.A. 1916. On the fossil algae of the petroleum-yielding shales of the Green River Formation of Colorado and Utah. Proceedings of the National Academy of Sciences, USA 2: 114–119.

Davis, R.H. 2000. *Neurospora*: Contributions of a Model Organism. Oxford University Press, New York, NY. 352 pp.

Davitt, A.J., M. Stansberry, and J.A. Rudgers. 2010. Do the costs and benefits of fungal endophyte symbiosis vary with light availability? New Phytologist 188: 824–834.

Dawson, J.W. 1857. Remarks on a specimen of fossil wood from the Devonian rocks (Gaspé Sandstones) of Gaspé, Canada East. Proceedings of the American Association for the Advancement of Science 2: 174–176.

Dawson, J.W. 1859. On the fossil plants from Devonian rocks of Canada. Quarterly Journal of the Geological Society London 15: 477–488.

De Andrade, M.L. 2003. First descriptions of two new amber species of *Cyphomyrmex* from Mexico and the Dominican Republic (Hymenoptera: Formicidae). Beiträge zur Entomologie 53: 131–139.

De Bary, A. 1879. Die Erscheinung der Symbiose: Vortrag, gehalten auf der Versammlung Deutscher Naturforscher und Aerzte zu Cassel. Verlag von Karl J. Trübner, Strassburg, France. 30 pp.

Debey, M.H., and C.R. von Ettingshausen. 1859. Die urweltlichen Thallophyten des Kreidegebirges von Aachen und Maestricht. Denkschriften der Kaiserlichen Akademie der Wissenschaften, Mathematisch-Naturwissenschaftliche Classe 16: 131–214.

Defossez, E., M.-A. Selosse, M.-P. Dubois, L. Mondolot, A. Faccio, C. Djieto-Lordon, D. McKey, and R. Blatrix. 2009. Ant-plants and fungi: A new threeway symbiosis. New Phytologist 182: 942–949.

Defossez, E., C. Djiéto-Lordon, D. McKey, M.-A. Selosse, and R. Blatrix. 2011. Plant-ants feed their host plant, but above all a fungal symbiont to recycle nitrogen. Proceedings of the Royal Society of London 278B: 1419–1426.

Degawa, Y., and S. Tokumasu. 1997. Zygospore formation in *Mortierella capitata*. Mycoscience 38: 387–394.

de Hoog, G.S., J. Guarro, J. Gené, and M.J. Figueras. 2000. Atlas of Clinical Fungi (Second Edition). Centraalbureau voor Schimmelcultures, Utrecht, The Netherlands, and Universitat Rovira I Virgili, Reus, Spain. 1126 pp.

De la Peña, E., S. Rodríguez-Echeverría, W.H. van der Putten, H. Freitas, and M. Moens. 2006. Mechanism of control of root-feeding nematodes by mycorrhizal fungi in the dune grass *Ammophila arenaria*. New Phytologist 169: 829–840.

De la Providencia, I.E., F.A. de Souza, F. Fernández, N.S. Delmas, and S. Declerck. 2005. Arbuscular mycorrhizal fungi reveal distinct patterns of anastomosis formation and hyphal healing mechanisms between different phylogenic groups. New Phytologist 165: 261–271.

Delaux, P.-M., N. Séjalon-Delmas, G. Bécard, and J.-M. Ané. 2013. Evolution of the plant-microbe symbiotic 'toolkit'. Trends in Plant Science 18: 298–304.

Demchenko, K., T. Winzer, J. Stougaard, M. Parniske, and K. Pawlowski. 2004. Distinct roles of *Lotus japonicus SYMRK* and *SYM15* in root colonization and arbuscule formation. New Phytologist 163: 381–392.

den Bakker, H.C., N.W. VanKuren, J.B. Morton, and T.E. Pawlowska. 2010. Clonality and recombination in the life history of an asexual arbuscular mycorrhizal fungus. Molecular Biology and Evolution 27: 2474–2486.

Dennis, R.L. 1969. Fossil mycelium with clamp connections from the Middle Pennsylvanian. Science 163: 670–671.

Dennis, R.L. 1970. A Middle Pennsylvanian basidiomycete mycelium with clamp connections. Mycologia 62: 578–584.

Dennis, R.L. 1976. *Palaeosclerotium*, a Pennsylvanian age fungus combining features of modern Ascomycetes and Basidiomycetes. Science 192: 66–68.

DePriest, P.T. 2004. Early molecular investigations of lichen-forming symbionts: 1986–2001. Annual Review of Microbiology 58: 273–301.

Descals, E. 2005. Diagnostic characters of propagules of Ingoldian fungi. Mycological Research 109: 545–555.

Desjardin, D.E., M. Capelari, and C. Stevani. 2007. Bioluminescent *Mycena* species from São Paulo, Brazil. Mycologia 99: 317–331.

Desjardin, D.E., A.G. Oliveira, and C.V. Stevani. 2008. Fungi bioluminescence revisited. Photochemical and Photobiological Sciences 7: 170–182.

Desjardin, D.E., B.A. Perry, D.J. Lodge, C.V. Stevani, and E. Nagasawa. 2010. Luminescent *Mycena*: New and noteworthy species. Mycologia 102: 459–477.

Devarajan, P.T., and T.S. Suryanarayanan. 2006. Evidence for the role of phytophagous insects in dispersal of non-grass fungal endophytes. Fungal Diversity 23: 111–119.

Dexheimer, J., S. Gianinazzi, and V. Gianinazzi-Pearson. 1979. Ultrastructural cytochemistry of the host-fungus interfaces in the endomycorrhizal association *Glomus mosseae*/*Allium cepa*. Zeitschrift für Pflanzenphysiologie 92: 191–206.

Dhillion, S.S. 1993. Vesicular-arbuscular mycorrhizas of *Equisetum* species in Norway and the U.S.A.: Occurrence and mycotrophy. Mycological Research 97: 656–660.

Dick, M.W. 1969. Morphology and taxonomy of the Oomycetes, with special reference to Saprolegniaceae, Leptomitaceae and Pythiaceae. New Phytologist 68: 751–775.

Dick, M.W. 1988. Coevolution in the heterokont fungi (with emphasis on the downy mildews and their angiosperm hosts). In K.A. Pirozynski, and D.L. Hawksworth (Eds), Coevolution of Fungi with Plants and Animals, pp. 31–62. Academic Press, London, UK.

Dick, M.W. 2001. Straminipilous fungi: Systematics of the Peronosporomycetes, Including Accounts of the Marine Straminipilous Protists, the Plasmodiophorids, and Similar Organisms. Kluwer Academic Publishers, Dordrecht, The Netherlands. 670 pp.

Dick, M.W. 2002. Towards an understanding of the evolution of the downy mildews. In P.T.N. Spencer-Phillips, U. Gisi, and A. Lebeda (Eds), Advances in Downy Mildew Research (Volume 1, pp. 1–57). Kluwer Academic Publishers, Dordrecht, The Netherlands.

Dickie, I.A., and R.J. Holdaway. 2010. Podocarp roots, mycorrhizas, and nodules. Smithsonian Contributions to Botany 95: 175–187.

Dickson, S. 2004. The *Arum-Paris* continuum of mycorrhizal symbioses. New Phytologist 163: 187–200.

Dierssen, K. 1972. Ein Holzpilz (Polyporaceae s.l.) aus der Unterkreide des Teutoburger Waldes. Osnabrücker Naturwissenschaftliche Mitteilungen 1: 159–164.

Dietrich, D., T. Lampke, and R. Rößler. 2013. A microstructure study on silicified wood from the Permian Petrified Forest of Chemnitz. Paläontologische Zeitschrift 87: 397–407.

Dighton, J. 2003. Fungi in Ecosystem Processes. Marcel Dekker Inc. New York, NY. 424 pp.

Dijkstra, S.J. 1949. Megaspores and some other fossils from the Aachenian (Senonian) in South Limburg, The Netherlands. Mededelingen van de Geologische Stichting N.S. 3: 19–32.

Dilcher, D.L. 1963. Eocene epiphyllous fungi. Science 142: 667–669.

Dilcher, D.L. 1965. Epiphyllous fungi from Eocene deposits in western Tennessee, U.S.A. Palaeontographica 116B: 1–54.

Dilcher, D.L. 1969. *Podocarpus* from the Eocene of North America. Science 164: 299–301.

Ding, S.-T., B.-N. Sun, J.-Y. Wu, and X.-C. Li. 2011. Miocene *Smilax* leaves and associated epiphyllous fungi from Zhejiang, East China and their paleoecological implications. Review of Palaeobotany and Palynology 165: 209–223.

Dingle, J., and P.A. McGee. 2003. Some endophytic fungi reduce the density of pustules of *Puccinia recondita* f. sp. *tritici* in wheat. Mycological Research 107: 310–316.

Döbbeler, P. 1997. Biodiversity of bryophilous ascomycetes. Biodiversity and Conservation 6: 721–738.

Doi, Y. 1983. Neogene epiphyllous fungi on *Fagus* and their living relatives on *Fagus crenata* in northeast Honshu, Japan. Memoirs National Science Museum, Tokyo 16: 53–73.

Doi, Y., and K. Uemura. 1985. Fossil *Microthyrium* on *Buxus* leaf compressions from the Upper Miocene, and its living relative in Japan. Bulletin of the National Science Museum, Tokyo, Series B 11: 127–136.

Dominguez de Toledo, L.S. 1994. Suggestions for describing and illustrating fungal spores. Mycotaxon 52: 259–270.

Domke, W. 1952. Der erste sichere Fund eines Myxomyceten im baltischen Bernstein (*Stemonitis splendens* Rost. fa. *succini* fa. nov. foss.). Mitteilungen aus dem Geologischen Staatsinstitut in Hamburg 21: 154–161.

Donlan, R.M. 2002. Biofilms: Microbial life on surfaces. Emerging Infectious Diseases 8: 881–890.

Dörfelt, H., and U. Schäfer. 1998. Fossile Pilze in Bernstein der alpischen Trias. Zeitschrift für Mykologie 64: 141–151.

Dörfelt, H., and A.R. Schmidt. 2005. A fossil *Aspergillus* from Baltic amber. Mycological Research 109: 956–960.

Dörfelt, H., and A.R. Schmidt. 2006. An Archaic slime mould in Baltic amber. Palaeontology 49: 1013–1017.

Dörfelt, H., and B. Striebich. 2000. *Palaeocybe striata*, ein neuer fossiler Pilz in Bernstein des Tertiär. Zeitschrift für Mykologie 66: 27–34.

Dörfelt, H., A.R. Schmidt, P. Ullmann, and J. Wunderlich. 2003. The oldest fossil myxogastroid slime mould. Mycological Research 107: 123–126.

Dotzler, N., M. Krings, T.N. Taylor, and R. Agerer. 2006. Germination shields in *Scutellospora* (Glomeromycota: Diversisporales, Gigasporaceae) from the 400 million-year-old Rhynie chert. Mycological Progress 5: 178–184.

Dotzler, N., M. Krings, R. Agerer, J. Galtier, and T.N. Taylor. 2008. *Combresomyces cornifer* gen. sp. nov., an endophytic peronosporomycete in *Lepidodendron* from the Carboniferous of central France. Mycological Research 112: 1107–1114.

Dotzler, N., C. Walker, M. Krings, H. Hass, H. Kerp, T.N. Taylor, and R. Agerer. 2009. Acaulosporoid glomeromycotan spores with a germination shield from the 400-million-year-old Rhynie chert. Mycological Progress 8: 9–18.

Dotzler, N., T.N. Taylor, J. Galtier, and M. Krings. 2011. *Sphenophyllum* (Sphenophyllales) leaves colonized by fungi from the Upper Pennsylvanian Grand-Croix cherts of central France. Zitteliana A 51: 3–8.

Doubinger, J., and D. Pons. 1973. Les champignons épiphylles du Tertiaire de Colombie. I. Le gisement de Cerrejón (Paléocène-Éocène). Comptes Rendus du 96e Congrès National des Sociétés Savantes (Toulouse, 1971). Section des Sciences 5: 233–252.

Doubinger, J., and D. Pons. 1975. Les champignons épiphylles de la formation Guaduas (Maestrichtien, Bassin de Boyacá, Colombie). Comptes Rendus du 95e Congrès National des Sociétés Savantes, Reims, Section des Sciences 3: 145–162.

Douzery, E.J.P., E.A. Snell, E. Bapteste, F. Delsuc, and H. Philippe. 2004. The timing of eukaryotic evolution: Does a relaxed molecular clock reconcile proteins and fossils? Proceedings of the National Academy of Sciences, USA 101: 15386–15391.

Doyle, J.A. 2006. Seed ferns and the origin of angiosperms. Journal of the Torrey Botanical Society 133: 169–209.

Drancourt, M., and D. Raoult. 2005. Palaeomicrobiology: Current issues and perspectives. Nature Reviews Microbiology 3: 23–35.

Dreger-Jauffert, F. 1980. Morphological changes of leaves and buds in plum-tree witches' brooms caused by *Taphrina insititiae* (Sad.) Johans. Flora 169: 376–385.

Drigo, B., A.S. Pijl, H. Duyts, A.M. Kielak, H.A. Gamper, M.J. Houtekamer, H.T.S. Boschker, P.L.E. Bodelier, A.S. Whiteley, J.A. van Veen, and G.A. Kowalchuk. 2010. Shifting carbon flow from roots into associated microbial communities in response to elevated atmospheric CO_2. Proceedings of the National Academy of Sciences, USA 107: 10938–10942.

D'Rozario, A., C. Labandeira, W.-Y. Guo, Y.-F. Yao, and C.-S. Li. 2011. Spatiotemporal extension of the Euramerican *Psaronius* component community to the Late Permian of Cathaysia: In situ coprolites in a *P. housuoensis* stem from Yunnan Province, southwest China. Palaeogeography, Palaeoclimatology, Palaeoecology 306: 127–133.

Druzhinina, I.S., V. Seidl-Seiboth, A. Herrera-Estrella, B.A. Horwitz, C.M. Kenerley, E. Monte, P.K. Mukherjee, S. Zeilinger, I.V. Grigoriev, and C.P. Kubicek. 2011. *Trichoderma*: The genomics of opportunistic success. Nature Reviews Microbiology 9: 749–759.

Du, B.-X., D.-F. Yan, B.-N. Sun, X.-C. Li, K.-Q. Dao, and X.-Q. Li. 2012. *Cunninghamia praelanceolata* sp. nov. with associated epiphyllous fungi from the upper Miocene of eastern Zhejiang, S.E. China and their palaeoecological implications. Review of Palaeobotany and Palynology 182: 32–43.

Dubey, R., and N.A. Moonnambeth. 2013. Hyperparasitism of *Isthmospora spinosa* Stevens and *Spiropes melanoplaca* (Berk. & Curtis) Ellis on *Meliola tylophorae – indicae* Hosag. parasitizing *Tylophora indica* (Burm. f.) Merill from India – a new record. Journal on New Biological Reports 2: 64–66.

Duckett, J.G., and R. Ligrone. 2005. A comparative cytological analysis of fungal endophytes in the sporophyte rhizomes and vascularized gametophytes of *Tmesipteris* and *Psilotum*. Canadian Journal of Botany 83: 1443–1456.

Duckett, J.G., K.S. Renzaglia, and K. Pell. 1991. A light and electron microscope study of rhizoid-ascomycete associations and flagelliform axes in British hepatics with observations on the effects of the fungi on host morphology. New Phytologist 118: 233–257.

Dueñas, H. 1979. Estudio palinológico de los 35 mts. superiores de la sección Tarragona, Sabana de Bogotá. Caldasia 12: 539–571.

Dufrêne, Y.F., C.J.P. Boonaert, H.C. van der Mei, H.J. Busscher, and P.G. Rouxhet. 2001. Probing molecular interactions and mechanical properties of microbial cell surfaces by atomic force microscopy. Ultramicroscopy 86: 113–120.

Dugan, F.M. 2008. Fungi in the Ancient World: How Mushrooms, Mildews, Molds, and Yeast Shaped the Early Civilizations of Europe, the Mediterranean, and the Near East. American Phytopathological Society Press, St. Paul, MN. 152 pp.

Duhamel, M., R. Pel, A. Ooms, H. Bücking, J. Jansa, J. Ellers, N.M. van Straalen, T. Wouda, P. Vandenkoornhuyse, and E.T. Kiers. 2013. Do fungivores trigger the transfer of protective metabolites from host plants to arbuscular mycorrhizal hyphae? Ecology 94: 2019–2029.

Duhoux, E., G. Rinaudo, H.G. Diem, F. Auguy, D. Fernandez, D. Bogusz, C. Franche, Y. Dommergues, and B. Huguenin. 2001. Angiosperm *Gymnostoma* trees produce root nodules colonized by arbuscular mycorrhizal fungi related to *Glomus*. New Phytologist 149: 115–125.

Duncan, P.M. 1876. On some unicellular algae parasitic within Silurian and Tertiary corals, with a notice of their presence in *Calceola sandalina* and other fossils. Quarterly Journal of the Geological Society 32: 205–211.

Dunn, M.T., G.W. Rothwell, and G. Mapes. 2003. On Paleozoic plants from marine strata: *Trivena arkansana* (Lyginopteridaceae) gen. et sp. nov., a lyginopterid from the Fayetteville Formation (Middle Chesterian/Upper Mississippian) of Arkansas, USA. American Journal of Botany 90: 1239–1252.

Duplessis, S., C.A. Cuomo, Y.-C. Lin, A. Aerts, E. Tisserant, C. Veneault-Fourrey, D.L. Joly, S. Hacquard, J. Amselem, B.L. Cantarel, R. Chiu, P.M. Coutinho, N. Feau, M. Field, P. Frey, E. Gelhaye, J. Goldberg, M.G. Grabherr, C.D. Kodira, A. Kohler, U. Kües, E.A. Lindquist, S.M. Lucas, R. Mago, E. Mauceli, E. Morin, C. Murat, J.L. Pangilinan, R. Park, M. Pearson, H. Quesneville, N. Rouhier, S. Sakthikumar, A.A. Salamov, J. Schmutz, B. Selles, H. Shapiro, P. Tanguay, G.A. Tuskan, B. Henrissat, Y. Van de Peer, P. Rouzé, J.G. Ellis, P.N. Dodds, J.E. Schein, S. Zhong, R.C. Hamelin, I.V. Grigoriev, L.J. Szabo, and F. Martin. 2011. Obligate biotrophy features unraveled by the genomic analysis of rust fungi. Proceedings of the National Academy of Sciences, USA 108: 9166–9171.

Dupres, V., D. Alsteens, G. Andre, and Y.F. Dufrêne. 2010. Microbial nanoscopy: A closer look at microbial cell surfaces. Trends in Microbiology 18: 397–405.

Durand, J.P. 1962. Rôle et répartition de "*Microcodium*" dans les formations fluvio-lacustres provençales du Crétacé supérieur et de l'Eocène. Compte Rendu Sommaire des Séances de la Société Géologique de France 9: 263–265.

Duringer, P., M. Schuster, J.F. Genise, A. Likius, H.T. Mackaye, P. Vignaud, and M. Brunet. 2006. The first fossil fungus gardens of Isoptera: Oldest evidence of symbiotic termite fungiculture (Miocene, Chad basin). Naturwissenschaften 93: 610–615.

Duringer, P., M. Schuster, J.F. Genise, H.T. Mackaye, P. Vignaud, and M. Brunet. 2007. New termite trace fossils:

Galleries, nests and fungus combs from the Chad basin of Africa (Upper Miocene-Lower Pliocene). Palaeogeography, Palaeoclimatology, Palaeoecology 251: 323–353.

Durugbo, E.U., O.T. Ogundipe, and O.K. Ulu. 2010. Palynological evidence of Pliocene-Pleistocene climatic variations from the Western Niger Delta, Nigeria. International Journal of Botany 6: 351–370.

Eastwood, D.C., D. Floudas, M. Binder, A. Majcherczyk, P. Schneider, A. Aerts, F.O. Asiegbu, S.E. Baker, K. Barry, M. Bendiksby, M. Blumentritt, P.M. Coutinho, D. Cullen, R.P. de Vries, A. Gathman, B. Goodell, B. Henrissat, K. Ihrmark, H. Kauserud, A. Kohler, K. LaButti, A. Lapidus, J.L. Lavin, Y.-H. Lee, E. Lindquist, W. Lilly, S Lucas, E. Morin, C. Murat, J.A. Oguiza, J. Park, A.G. Pisabarro, R. Riley, A. Rosling, A. Salamov, O. Schmidt, J. Schmutz, I. Skrede, J. Stenlid, A. Wiebenga, X. Xie, U. Kües, D.S. Hibbett, D. Hoffmeister, N. Högberg, F. Martin, I.V. Grigoriev, and S.C. Watkinson. 2011. The plant cell wall–decomposing machinery underlies the functional diversity of forest fungi. Science 333: 762–765.

Eckert, A.J., and B.D. Hall. 2006. Phylogeny, historical biogeography, and patterns of diversification for *Pinus* (Pinaceae): Phylogenetic tests of fossil-based hypotheses. Molecular Phylogenetics and Evolution 40: 166–182.

Ediger, V.S. 1981. Fossil fungal and algal bodies from Thrace Basin, Turkey. Palaeontographica 179B: 87–102.

Ediger, V.S., and C. Alişan. 1989. Tertiary fungal and algal palynomorph biostratigraphy of the Northern Thrace Basin, Turkey. Review of Palaeobotany and Palynology 58: 139–161.

Edwards, D. 2003. Xylem in early tracheophytes. Plant, Cell and Environment 26: 57–72.

Edwards, D., and L. Axe. 2012. Evidence for a fungal affinity for *Nematasketum*, a close ally of *Prototaxites*. Botanical Journal of the Linnean Society 168: 1–18.

Edwards, D., and J.B. Richardson. 2004. Silurian and Lower Devonian plant assemblages from the Anglo-Welsh Basin: A palaeobotanical and palynological synthesis. Geological Journal 39: 375–402.

Edwards, D., and V. Rose. 1984. Cuticles of *Nematothallus*: A further enigma. Botanical Journal of the Linnean Society 88: 35–54.

Edwards, D., K.L. Davies, and L. Axe. 1992. A vascular conducting strand in the early land plant *Cooksonia*. Nature 357: 683–685.

Edwards, D., H. Kerp, and H. Hass. 1998. Stomata in early land plants: An anatomical and ecophysiological approach. Journal of Experimental Botany 49: 255–278.

Edwards, D., L. Axe, J. Parkes, and D. Rickard. 2006. Provenance and age of bacteria-like structures on mid-Palaeozoic plant fossils. International Journal of Astrobiology 5: 109–142.

Edwards, D., P.A. Selden, and L. Axe. 2012. Selective feeding in an Early Devonian terrestrial ecosystem. PALAIOS 27: 509–522.

Edwards, D., L. Axe, and R. Honegger. 2013. Contributions to the diversity in cryptogamic covers in the mid-Palaeozoic: *Nematothallus* revisited. Botanical Journal of the Linnean Society 173: 505–534.

Edwards, W.N. 1922. An Eocene microthyriaceous fungus from Mull, Scotland. Transactions of the British Mycological Society 8: 66–72.

Egger, K.N. 2006. The surprising diversity of ascomycetous mycorrhizas. New Phytologist 170: 421–423.

Ehinger, M.O., D. Croll, A.M. Koch, and I.R. Sanders. 2012. Significant genetic and phenotypic changes arising from clonal growth of a single spore of an arbuscular mycorrhizal fungus over multiple generations. New Phytologist 196: 853–861.

Ehlers, E.G., D.V. Stiles, and J.D. Birle. 1965. Fossil bacteria in pyrite. Science 148: 1719–1721.

Ehrlich, H.L., and D.K. Newman. 2008. Geomicrobiology (Fifth Edition). CRC Press, Boca Raton, FL. 628 pp.

Eichwald, K.E. von. 1830. Naturhistorische Skizze von Lithauen, Volynien und Podolien in geognostisch-mineralogischer, botanischer und zoologischer Hinsicht. Joseph Zawadzki, Vilnius. 255 pp.

Ekdale, A.A., R.G. Bromley, and S.G. Pemberton. 1984. Bioerosion. In A.A. Ekdale, R.G. Bromley, and S.G. Pemberton (Eds), Ichnology: Trace Fossils in Sedimentology and Stratigraphy, Short Course No. 15, pp. 108–128. Society of Economic Paleontologists and Mineralogists, Tulsa, OK.

Elad, Y. 1995. Mycoparasitism. In U.S. Singh, K. Kohmoto, and R.P. Singh (Eds), Pathogenesis and Host Specificity in Plant Diseases: Histopathological, Biochemical, Genetic and Molecular Bases, Volume II: Eukaryotes, pp. 289–307. Pergamon Press, London, UK.

Elias, M.K. 1966. Living and fossil algae and fungi, formerly known as structural parts of marine bryozoans. Palaeobotanist 14: 5–18.

Elias, R.J., and D.-J. Lee. 1993. Microborings and growth in Late Ordovician halysitids and other corals. Journal of Paleontology 67: 922–934.

Ellis, D. 1915. Fossil micro-organisms from the Jurassic and Cretaceous rocks of Great Britain. Proceedings of the Royal Society of Edinburgh 35: 110–133.

Ellis, D. 1918. Phycomycetous fungi from the English Lower Coal Measures. Proceedings of the Royal Society of Edinburgh 38: 130–145.

Ellis, J.P. 1977. The genera *Trichothyrina* and *Actinopeltis* in Britain. Transactions of the British Mycological Society 68: 145–155.

Ellis, M.B. 1971. Dematiaceous Hyphomycetes. Commonwealth Mycological Institute, Kew, England. 608 pp.

Ellis, M.B. 1976. More Dematiaceous Hyphomycetes. Commonwealth Mycological Institute, Kew, England. 507 pp.

Ellis, M.B., and J.P. Ellis. 1985. Microfungi on Land Plants: An Identification Handbook. Croom Helm Ltd. London, UK. 818 pp.

Ellis-Evans, J.C. 1985. Fungi from maritime Antarctic freshwater environments. British Antarctic Survey Bulletin 68: 37–45.

El-Saadawy, W.E., and W.S. Lacey. 1979. Observations on *Nothia aphylla* Lyon ex Høeg. Review of Palaeobotany and Palynology 27: 119–147.

Elsik, W.C. 1966a. Degradation of arci in a fossil *Alnus* pollen grain. Nature 209: 825.

Elsik, W.C. 1966b. Biologic degradation of fossil pollen grains and spores. Micropaleontology 12: 515–518.

Elsik, W.C. 1968. Palynology of a Paleocene Rockdale lignite, Milam County, Texas. I. Morphology and taxonomy. Pollen et Spores 10: 263–314.

Elsik, W.C. 1969. Late Neogene palynomorph diagrams, northern Gulf of Mexico. Transactions of the Gulf Coast Association of Geological Societies 19: 509–528.

Elsik, W.C. 1971. Microbiological degradation of sporopollenin. In J. Brooks, P.R. Grant, M. Muir, P. van Gijzel, and G. Shaw (Eds), Sporopollenin, pp. 480–510. Academic Press, London, UK.

Elsik, W.C. 1976a. Fossil fungal spores. In D.J. Weber, and W.M. Hess (Eds), The Fungal Spore: Form and Function, pp. 849–862. John Wiley & Sons, New York, NY.

Elsik, W.C. 1976b. Microscopic fungal remains and Cenozoic palynostratigraphy. Geoscience and Man 15: 115–120.

Elsik, W.C. 1978. Classification and geologic history of the microthyriaceous fungi. In: D.C. Bharadwaj, K.M. Lele, and R.K. Kar (Eds), 4th International Palynological Conference, Lucknow, Proceedings, Volume 1, pp. 331–342. Birbal Sahni Institute of Palaeobotany, Lucknow, India.

Elsik, W.C. 1980. The utility of fungal spores in marginal marine strata of the late Cenozoic, northern Gulf of Mexico. In: D.C. Bharadwaj, H.P. Singh, and R.S. Tiwari (eds), 4th International Palynological Conference, Lucknow, Proceedings, Volume 2, pp. 436–443. Birbal Sahni Institute of Palaeobotany, Lucknow, India.

Elsik, W.C. 1981. Fungal palynomorphs. Louisiana State University Palynology Shortcourse, Baton Rouge, LA. October 4–6. 242 pp.

Elsik, W.C. (Chairman) 1983. Annotated glossary of fungal palynomorphs. American Association of Stratigraphic Palynologists Foundation Contributions Series 11: 1–35.

Elsik, W.C. 1986. Palynology of a Late Pleistocene giant ground sloth locality, southwest Harris County, Texas. Pollen et Spores 28: 77–82.

Elsik, W.C. 1989. The fungal morphotype *Felixites* n. gen. Pollen et Spores 31: 155–159.

Elsik, W.C. 1990. *Hypoxylonites* and *Spirotremesporites* form genera for Eocene to Pleistocene fungal spores bearing a single furrow. Palaeontographica 216B: 137–169.

Elsik, W.C. 1996. Fungi. In J. Jansonius, and D.C. McGregor (Eds), Palynology: Principles and Applications, Volume 1, Principles, pp. 293–305. American Association of Stratigraphic Palynologists Foundation, Dallas, TX.

Elsik, W.C., and D.L. Dilcher. 1974. Palynology and age of clays exposed in Lawrence Clay Pit, Henry County, Tennessee. Palaeontographica 146B: 65–87.

Elsik, W.C., and J. Jansonius. 1974. New genera of Paleogene fungal spores. Canadian Journal of Botany 52: 953–958.

Elsik, W.C., and T.E. Yancey. 2000. Palynomorph biozones in the context of changing paleoclimate, middle Eocene to lower Oligocene of the Northwest Gulf of Mexico. Palynology 24: 177–186.

Elsik, W.C., V.S. Ediger, and Z. Bati. 1990. Fossil fungal spores: *Anatolinites* gen. nov. Palynology 14: 91–103.

Emerson, R. 1941. An experimental study of the life cycles and taxonomy of *Allomyces*. Lloydia 4: 77–144.

Emerson, R., and J.A. Robertson. 1974. Two new members of the Blastocladiaceae. I. Taxonomy, with an evaluation of genera and interrelationships in the family. American Journal of Botany 61: 303–317.

Engelhardt, H. 1881. Ueber die fossilen Pflanzen des Süsswassersandsteins von Grasseth: ein neuer Beitrag zur Kenntniss der fossilen Pflanzen Böhmens. Nova Acta Academiae Caesareae Leopoldino-Carolinae Germanicae Naturae Curiosorum 43: 245–324.

Engelhardt, H. 1887. Ueber *Rossellinia congregata* Beck sp., eine neue Pilzart aus der Braunkohlenformation Sachsens Sitzungsberichte und Abhandlungen der Naturwissenschaftlichen Gesellschaft Isis, Dresden. 4: 33–35.

Engh, I., M. Nowrousian, and U. Kück. 2010. *Sordaria macrospora*, a model organism to study fungal cellular development. European Journal of Cell Biology 89: 864–872.

Epstein, L., and R.L. Nicholson. 2006. Adhesion and adhesives of fungi and oomycetes. In A.M. Smith, and J.A. Callow (Eds), Biological Adhesives, pp. 41–62. Springer-Verlag, Berlin, Germany.

Ercolin, F., and D. Reinhardt. 2011. Successful joint ventures of plants: arbuscular mycorrhiza and beyond. Trends in Plant Science 16: 356–362.

Erdei, B., and M. Lesiak. 1999–2000. A study of dispersed cuticles, fossil seeds and cones from Sarmatian (Upper Miocene) deposits of Sopron-Piusz Puszta (W Hungary). Studia Botanica Hungarica 30/31: 5–26.

Ergolskaya, Z.V. 1936. Petrographic examination of the Barzas coals. Trudy Centralnogo Nauchno-Issledovatelskogo Geologo-Razvedocnogo Instituta (Transactions of the Central Geological and Prospecting Institute) 70: 5–53.

Eriksson, B. 1978. Fossil microthyriaceous fungi from Tervola, northern Finland. Annals of Botany Fennici 15: 122–127.

Eriksson, O. 1966. On *Anthostomella* Sacc., *Entosordaria* (Sacc.) Höhn. and some related genera (Pyrenomycetes). Svensk Botanisk Tidskrift 60: 315–324.

Eriksson, O.E. 1981. The families of bitunicate ascomycetes. Opera Botanica 60: 1–209.

Eriksson, O.E. 2005. Ascomyceternas ursprung och evolution – protolichenes-hypotesen. Svensk Mykologisk Tidskrift 26: 22–33.

Eriksson, O.E. (Ed.), 2006. Outline of Ascomycota – 2006. Myconet 12: 1–82.

Eriksson, O., and D.L. Hawksworth. 1986. An alphabetical list of the generic names of ascomycetes – 1986. Systema Ascomycetum 5(1): 3–111.

Eriksson, O.E., and K. Winka. 1997. Supraordinal taxa of Ascomycota. Myconet 1: 1–16.

Erwin, D.H., M. Laflamme, S.M. Tweedt, E.A. Sperling, D. Pisani, and K.J. Peterson. 2011. The Cambrian conundrum: Early divergence and later ecological success in the early history of animals. Science 334: 1091–1097.

Esser, K. 1982. Cryptogams: Cyanobacteria, Algae, Fungi, Lichens. Cambridge University Press, New York, NY. 610 pp.

Estrada, C., W.T. Wcislo, and S.A. Van Bael. 2013. Symbiotic fungi alter plant chemistry that discourages leaf-cutting ants. New Phytologist 198: 241–251.

Etheridge, R. 1891. On the occurrence of microscopic fungi, allied to the genus *Palaeachlya*, Duncan, in the Permo-Carboniferous rocks of N.S. Wales and Queensland. Records of the Geological Survey of New South Wales 2: 95–99.

Etheridge, R. 1899. On two additional perforating bodies, believed to be thallophytic cryptograms, from the Lower Palaeozoic rocks of New South Wales. Records of the Geological Survey of New South Wales 3: 121–127.

Etheridge, R. 1904. An endophyte (*Stichus mermisoides*) occurring in the test of a Cretaceous bivalve. Records of the Geological Survey of New South Wales 5: 255–257.

Etienne, S., and J. Dupont. 2002. Fungal weathering of basaltic rocks in a cold oceanic environment (Iceland): Comparison between experimental and field observations. Earth Surface Processes and Landforms 27: 737–748.

Ettingshausen, C. von. 1853. Die Tertiaer-floren der Oesterreichischen Monarchie. 2. Die tertiäre Flora von Häring in Tirol. Abhandlungen der kaiserlich-königlichen Geologischen Reichsanstalt Wien 3 (Sect. 3, Nr. 2): 1–118.

Fahey, C., R.A. York, and T.E. Pawlowska. 2012. Arbuscular mycorrhizal colonization of giant sequoia (*Sequoiadendron giganteum*) in response to restoration practices. Mycologia 104: 988–997.

Farr, M.L. 1971. A modified 'Beeli Formula' as identification tool for asterinaceous fungi and their pycnidial stages. Mycopathologia et Mycologia applicata 43: 161–163.

Farrell, M.A., and H.G. Turner. 1932. Bacteria in anthracite coal. Journal of Bacteriology 23: 155–162.

Fassi, B., A. Fontana, and J.M. Trappe. 1969. Ectomycorrhizae formed by *Endogone lactiflua* with species of *Pinus* and *Pseudotsuga*. Mycologia 61: 412–414.

Feddermann, N., R. Finlay, T. Boller, and M. Elfstrand. 2010. Functional divsersity in arbuscular mycorrhiza – the role of gene expresion, phosphorous nutrition and symbiotic efficiency. Fungal Ecology 3: 1–8.

Fedonkin, M.A., and E.L. Yochelson. 2002. Middle Proterozoic (1.5 Ga) *Horodyskia moniliformis* Yochelson and Fedonkin, the oldest known tissue-grade colonial eucaryote. Smithsonian Contributions to Paleobiology 94: 1–29.

Felix, J. 1894. Studien über fossile Pilze. Zeitschrift der Deutschen Geologischen Gesellschaft 46: 269–280.

Feng, Z. 2012. *Ningxiaites specialis*, a new woody gymnosperm from the uppermost Permian of China. Review of Palaeobotany and Palynology 181: 34–46.

Feng, Z., J. Wang, and L.-J. Liu. 2010. First report of oribatid mite (arthropod) borings and coprolites in Permian woods from the Helan Mountains of northern China. Palaeogeography, Palaeoclimatology, Palaeoecology 288: 54–61.

Fernández, N., M.I. Messuti, and S. Fontenla. 2008. Arbuscular mycorrhizas and dark septate fungi in *Lycopodium paniculatum* (Lycopodiaceae) and *Equisetum bogotense* (Equisetaceae) in a Valdivian temperate forest of Patagonia, Argentina. American Fern Journal 98: 117–127.

Fernando, W.G.D., J.X. Zhang, C.Q. Chen, W.R. Remphrey, A. Schurko, and G.R. Klassen. 2005. Molecular and morphological characteristics of *Apiosporina morbosa*, the causal agent of black knot in *Prunus* spp. Canadian Journal of Plant Pathology 27: 364–375.

Ferreira, E.P., M. de Aravjo Carvalho, and M.C. Viviers. 2005. Palinologia (fungos) da Formação Calumbi, Paleoceno da Bacia de Sergipe, Brasil. Arquivos do Museu Nacional, Rio de Janeiro 63: 395–410.

Field, K.J., D.D. Cameron, J.R. Leake, S. Tille, M.I. Bidartondo, and D.J. Beerling. 2012. Contrasting arbuscular mycorrhizal responses of vascular and non-vascular plants to a simulated Palaeozoic CO_2 decline. Nature Communications 3: 835.

Figueiral, I., V. Mosbrugger, N.P. Rowe, A.R. Ashraf, T. Utescher, and T.P. Jones. 1999. The Miocene peat-forming vegetation of northwestern Germany: an analysis of wood remains and comparisons with previous palynological interpretations. Review of Palaeobotany and Palynology 104: 239–266.

Finlay, R., H. Wallander, M. Smits, S. Holmstrom, P. van Hees, B. Lian, and A. Rosling. 2009. The role of fungi in biogenic weathering in boreal forest soils. Fungal Biology Reviews 23: 101–106.

Fiore, M. 1932. Miceti fossili rinvenuti su di una palma (*Latanites* sp.) del Bolca. Bolletino della Società di Naturalista in Napoli 43: 153–156.

Fiore-Donno, A.M., S.I. Nikolaev, M. Nelson, J. Pawlowski, T. Cavalier-Smith, and S.L. Baldauf. 2010. Deep phylogeny and evolution of slime moulds (Mycetozo). Protist 161: 55–70.

Fisher, M.C., T.W.J. Garner, and S.F. Walker. 2009. Global emergence of *Batrachochytrium dendrobatidis* and amphibian chytridiomycosis in space, time, and host. Annual Review of Microbiology 63: 291–310.

Fisher, M.C., D.A. Henk, C.J. Briggs, J.S. Brownstein, L.C. Madoff, S.L. McCraw, and S.J. Gurr. 2012. Emerging fungal threats to animal, plant and ecosystem health. Nature 484: 186–194.

Fitter, A.H. 2005. Darkness visible: Reflections on underground ecology. Journal of Ecology 93: 231–243.

Fitzpatrick, D.A., M.E. Logue, J.E. Stajich, and G. Butler. 2006. A fungal phylogeny based on 42 complete genomes derived from supertree and combined gene analysis. BMC. Evolutionary Biology 6: 99.

Fleischmann, A., M. Krings, H. Mayr, and R. Agerer. 2007. Structurally preserved polypores from the Neogene of North Africa: *Ganodermites libycus* gen. et sp. nov. (Polyporales, Ganodermataceae). Review of Palaeobotany and Palynology 145: 159–172.

Fletcher, B.J., D.J. Beerling, and W.G. Chaloner. 2004. Stable carbon isotopes and the metabolism of the terrestrial Devonian organism *Spongiophyton*. Geobiology 2: 107–119.

Flewelling, A.J., J.A. Johnson, and C.A. Gray. 2013. Isolation and bioassay screening of fungal endophytes from North Atlantic marine macroalgae. Botanica Marina 56: 287–297.

Floudas, D, M. Binder, R. Riley, K. Barry, R.A. Blanchette, B. Henrissat, A.T. Martínez, R. Otillar, J.W. Spatafora, J.S. Yadav, A. Aerts, I. Benoit, A. Boyd, A. Carlson, A. Copeland, P.M. Coutinho, R.P. de Vries, P. Ferreira, K. Findley, B. Foster, J. Gaskell, D. Glotzer, P. Górecki, J. Heitman, C. Hesse, C. Hori, K. Igarashi, J.A. Jurgens, N. Kallen, P. Kersten, A. Kohler, U. Kües, T.K.A. Kumar, A. Kuo, K. LaButti, L.F. Larrondo, E. Lindquist, A. Ling, V.

Lombard, S. Lucas, T. Lundell, R. Martin, D.J. McLaughlin, I. Morgenstern, E. Morin, C. Murat, L.G. Nagy, M. Nolan, R.A. Ohm, A. Patyshakuliyeva, A. Rokas, F.J. Ruiz-Dueñas, G. Sabat, A. Salamov, M. Samejima, J. Schmutz, J.C. Slot, F. John, St., J. Stenlid, H. Sun, S. Sun, K. Syed, A. Tsang, A. Wiebenga, D. Young, A. Pisabarro, D.C. Eastwood, F. Martin, D. Cullen, I.V. Grigoriev, and D.S. Hibbett. 2012. The Paleozoic origin of enzymatic lignin decomposition reconstructed from 31 fungal genomes. Science 336: 1715–1719.

Fogel, M.L., and N. Tuross. 1999. Transformation of plant biochemicals to geological macromolecules during early diagenesis. Oecologia 120: 336–346.

Fohrer, E., and T. Simon. 2002. Baumpilze und Trüffel: Höhere Pilze aus dem Keuper, Teil 1. Fossilien 19: 360–363.

Fohrer, E., and T. Simon. 2003. Baumpilze und Trüffel: Höhere Pilze aus dem Keuper, Teil 2. Fossilien 20: 19–23.

Fomina, M., I.J. Alexander, S. Hillier, and G.M. Gadd. 2004. Zinc phosphate and pyromorphite solubilization by soil plant-symbiotic fungi. Geomicrobiology Journal 21: 351–366.

Foo, E., J.J. Ross, W.T. Jones, and J.B. Reid. 2013. Plant hormones in arbuscular mycorrhizal symbioses: An emerging role for gibberellins. Annals of Botany 111: 769–779.

Försterra, G, L. Beuck, V. Häussermann, and A. Freiwald. 2005. Shallow-water *Desmophyllum dianthus* (Scleractinia) from Chile: Characteristics of the biocoenoses, the bioeroding community, heterotrophic interactions and (paleo)-bathymetric implications. In A. Freiwald, and J.M. Roberts (Eds), Cold-water Corals and Ecosystems, pp. 937–977. Springer-Verlag, Berlin, Germany.

Foster, C.B., M.H. Stephenson, C. Marshall, G.A. Logan, and P.F. Greenwood. 2002. A revision of *Reduviasporonites* Wilson 1962: Description, illustration, comparison and biological affinities. Palynology 26: 35–58.

Fox, G.E., and E. Stackebrandt. 1987. The application of 16S rRNA cataloguing and 5S rRNA sequencing in microbial systematics. Methods in Microbiology 19: 405–458.

Fraaye, R.H.B., and M.W. Fraaye. 1995. Miocene bracket fungi (Basidiomycetes, Aphyllophorales) from The Netherlands. Contributions to Tertiary and Quaternary Geology 32: 27–33.

Frank, A.B. 1877. Ueber die biologischen Verhältnisse des Thallus einiger Krustenflechten. Beiträge zur Biologie der Pflanzen 2: 123–200.

Franken, P., and N. Requena. 2001. Molecular approaches to arbuscular mycorrhiza functioning. In B. Hock (Ed.), The Mycota IX: Fungal Associations, pp. 19–28. Springer-Verlag, Berlin, Germany.

Frantz, U. 1959. Die Pollenflora der Braunkohle von Lohsa/Niederlausitz. PhD thesis Freie Universität Berlin, Germany. 46 pp.

Freckman, D.W. (Ed.). 1982. Nematodes in Soil Ecosystems. University of Texas Press, Austin, TX. 220 pp.

Freeman, K.R., A.P. Martin, D. Karki, R.C. Lynch, M.S. Mitter, A.F. Meyer, J.E. Longcore, and D.R. Simmons. 2009. Evidence that chytrids dominate fungal communities in high-elevation soils. Proceedings of the National Academy of Sciences, USA 106: 18315–18320.

Freire Cruz, A., and T. Ishii. 2011. Arbuscular mycorrhizal fungal spores host bacteria that affect nutrient biodynamics and biocontrol of soilborne plant pathogens. Biology Open 1: 52–57.

Frey-Klett, P., and J. Garbaye. 2005. Mycorrhiza helper bacteria: A promising model for the genomic analysis of fungal–bacterial interactions. New Phytologist 168: 4–8.

Frey-Klett, P., J. Garbaye, and M. Tarkka. 2007. The mycorrhiza helper bacteria revisited. New Phytologist 176: 22–36.

Fry, W.L., and D.J. McLaren. 1959. Fungal filaments in a Devonian limestone from Alberta. Geological Survey of Canada Bulletin 48: 1–9.

Furman, T.E. 1970. The nodular mycorrhizae of *Podocarpus rospigliosii*. American Journal of Botany 57: 910–915.

Furnes, H., N.R. Banerjee, K. Muehlenbachs, H. Staudigel, and M. de Wit. 2004. Early life recorded in Archean pillow lavas. Science 304: 578–581.

Furnes, H., N. McLoughlin, K. Muehlenbachs, N. Banerjee, H. Staudigel, Y. Dilek, M. de Wit, M. Van Kranendonk, and P. Schiffman. 2008. Oceanic pillow lavas and hyaloclastites as habitats for microbial life through time – a review. In Y. Dilek, H. Furnes, and K. Muehlenbachs (Eds), Links Between Geological Processes, Microbial Activities and Evolution of Life: Microbes and Geology, pp. 1–68. Springer, New York, NY.

Gadd, G.M. 2007. Geomycology: Biogeochemical transformations of rocks, minerals, metals and radionuclides by fungi, bioweathering and bioremediation. Mycological Research 111: 3–49.

Gadd, G.M. 2010. Metals, minerals and microbes: Geomicrobiology and bioremediation. Microbiology 156: 609–643.

Gadd, G.M., Y.J. Rhee, K. Stephenson, and Z. Wei. 2012. Geomycology: Metals, actinides and biominerals. Environmental Microbiology Reports 4: 270–296.

Galagan, J.E., M.R. Henn, L.-J. Ma, C.A. Cuomo, and B. Birren. 2005. Genomics of the fungal kingdom: Insights into eukaryotic biology. Genome Research 15: 1620–1631.

Gallaud, I. 1905. Études sur les mycorhizes endotrophes. Revue Générale de Botanique 17: 5–15.

Galleni, L. 1995. How does the Teilhardian vision of evolution compare with contemporary theories? Zygon: Journal of Religion and Science 30: 25–45.

Galtier, J. 1971. Sur les flores du Carbonifère inférieur d'Esnost et du Roannais. Bulletin trimestriel de la Société d'histoire naturelle et des amis du Museum d'Autun 57: 24–28.

Galtier, J., and T.L. Phillips. 1999. The acetate peel technique. In T.P. Jones, and N.P. Rowe (Eds), Fossil Plants and Spores: Modern Techniques, pp. 67–70. Geological Society, London, UK.

Gangwar, P.K., and A.P. Shamshery. 1982. A fossil blastocladialean thallus from the Deccan Intertrappean Series, Mohgaon-Kalan, Madhya Pradesh, India. Bangladesh Journal of Botany 11: 73–76.

Gao, B., and R.S. Gupta. 2012. Phylogenetic framework and molecular signatures for the main clades of the phylum Actinobacteria. Microbiology and Molecular Biology Reviews 76: 66–112.

Garbaye, J. 1994a. Les bactéries auxiliaires de la mycorhization: Une nouvelle dimension de la symbiose mycorhizienne. Acta Botanica Gallica 141: 517–521.

Garbaye, J. 1994b. Helper bacteria: A new dimension to the mycorrhizal symbiosis. New Phytologist 128: 197–210.

García Massini, J.L. 2007a. A possible endoparasitic chytridiomycete fungus from the Permian of Antarctica. Palaeontologica Electronica 10: 16A.

García Massini, J.L. 2007b. A glomalean fungus from the Permian of Antarctica. International Journal of Plant Sciences 168: 673–678.

García-Massini, J.L., M. del, C. Zamaloa, and E.J. Romero. 2004. Fungal fruiting bodies in the Cullen Formation (Miocene) in Tierra del Fuego, Argentina. Ameghiniana 41: 83–90.

García Massini, J., A. Channing, D.M. Guido, and A.B. Zamuner. 2012. First report of fungi and fungus-like organisms from Mesozoic hot springs. PALAIOS 27: 55–62.

Garcia-Pichel, F. 2006. Plausible mechanisms for the boring on carbonates by microbial phototrophs. Sedimentary Geology 185: 205–213.

Gareth Jones, E.B., and K.L. Pang. 2012. Marine Fungi and Fungal-like Organisms. de Gruyter, Boston, MA. 300 pp.

Gargas, A., P.T. DePriest, M. Grube, and A. Tehler. 1995. Multiple origins of lichen symbioses in fungi suggested by SSU rDNA phylogeny. Science 268: 1492–1495.

Garms, R. 1975. Observations on filarial infections and parous rates of anthropophilic blackflies in Guatemala, with reference to the transmission of *Onchocerca volvulus*. Tropenmedizin und Parasitologie 26: 169–182.

Garty, J., C. Giele, and W.E. Krumbein. 1982. On the occurrence of pyrite in a lichen-like inclusion in Eocene amber (Baltic). Paleogeography, Palaeoclimatology, Palaeoecology 39: 139–147.

Garwood, R, A. Ross, D. Sotty, D. Chabard, S. Charbonnier, M. Sutton, and P.J. Withers. 2012. Tomographic reconstruction of neopterous Carboniferous insect nymphs. PLoS ONE 7: e45779.

Gaspard, D. 2011. Endolithic algae, fungi and bacterial activity in Holocene and Cretaceous brachiopod shells – diagenetic consequences. Memoirs of the Association of Australasian Palaeontologists 41: 327–337.

Gatrall, M., and S. Golubic. 1970. Comparative study on some Jurassic and recent endolithic fungi using scanning electron microscope. In: T.P. Crimes, and J.C. Harper (eds), Trace Fossils, Proceedings of the Liverpool Geological Society International Conference, Liverpool University, Jan. 1970, Geological Journal Special Issue No 3, pp. 167–178. Seel House Press, Liverpool, UK.

Gazis, R., and P. Chaverri. 2010. Diversity of fungal endophytes in leaves and stems of wild rubber trees (*Hevea brasiliensis*) in Peru. Fungal Ecology 3: 240–254.

Gazis, R., S. Rehner, and P. Chaverri. 2011. Species delimitation in fungal endophyte diversity studies and its implications in ecological and biogeographic inferences. Molecular Ecology 20: 3001–3013.

Gazzè, S.A., L. Saccone, K.V. Ragnarsdottir, M.M. Smits, A.L. Duran, J.R. Leake, S.A. Banwart, and T.J. McMaster. 2012. Nanoscale channels on ectomycorrhizal-colonized chlorite: Evidence for plant-driven fungal dissolution. Journal of Geophysical Research, Biogeosciences 117: 1–8.

Geiser, D.M., C. Gueidan, J. Miadlikowska, F. Lutzoni, F. Kauff, V. Hofstetter, E. Fraker, C.L. Schoch, L. Tibell, W.A. Untereiner, and A. Aptroot. 2006. Eurotiomycetes: Eurotiomycetidae and Chaetothyriomycetidae. Mycologia 98: 1053–1064.

Gektidis, M. 1999. Development of microbial euendolithic communities: The influence of light and time. Bulletin of the Geological Society of Denmark 45: 147–150.

Gelorini, V., I. Ssemmanda, and D. Verschuren. 2012. Validation of non-pollen palynomorphs as paleoenvironmental indicators in tropical Africa: Contrasting ~200-year paleolimnological records of climate change and human impact. Review of Palaeobotany and Palynology 186: 90–101.

Genise, J.F. 2004. Fungus traces in wood: A rare bioerosional item. In: L.A. Buatois, and M.G. Mángano (eds), Ichnia 2004, 1st International Congress on Ichnology, April 19–23, Abstract Book, pp. 37. Museo Paleontólogico Egidio Feruglio, Trelew, Patagonia, Argentina.

Genise, J.F., A.M. Alonso-Zarza, J.M. Krause, M.V. Sánchez, L. Sarzetti, J.L. Farina, M.G. González, M. Cosarinsky, and E.S. Bellosi. 2010. Rhizolith balls from the Lower Cretaceous of Patagonia: Just roots or the oldest evidence of insect agriculture? Palaeogeography, Palaeoclimatology, Palaeoecology 287: 128–142.

Genise, J.F., E.S. Bellosi, M. Verde, and M.G. González. 2011. Large ferruginized palaeorhizospheres from a Paleogene lateritic profile of Uruguay. Sedimentary Geology 240: 85–96.

Genise, J.F., R. Garrouste, P. Nel, P. Grandcolas, P. Maurizot, D. Cluzel, R. Cornette, A.-C. Fabre, and A. Nel. 2012. *Asthenopodichnium* in fossil wood: Different trace makers as indicators of different terrestrial palaeoenvironments. Palaeogeography, Palaeoclimatology, Palaeoecology 365–366: 184–191.

Genise, J.F., R.N. Melchor, M.V. Sánchez, and M.G. González. 2013. *Attaichnus kuenzelii* revisited: A Miocene record of fungus-growing ants from Argentina. Palaeogeography, Palaeoclimatology, Palaeoecology 386: 349–363.

Gennard, D.E., and C.R. Hackney. 1989. First Irish record of a fossil bracket fungus *Fomes fomentarius* (L. ex Fr.) Kickx. Irish Naturalists' Journal 23: 19–21.

Genre, A., and P. Bonfante. 2005. Building a mycorrhizal cell: How to reach compatibility between plants and arbuscular mycorrhizal fungi. Journal of Plant Interactions 1: 3–13.

Genre, A., M. Chabaud, T. Timmers, P. Bonfante, and D.G. Barker. 2005. Arbuscular mycorrhizal fungi elicit a novel intracellular apparatus in *Medicago truncatula* root epidermal cells before infection. Plant Cell 17: 3489–3499.

Genre, A., M. Chabaud, A. Faccio, D.G. Barker, and P. Bonfante. 2008. Prepenetration apparatus assembly precedes and predicts the colonization patterns of arbuscular mycorrhizal fungi within the root cortex of both *Medicago truncatula* and *Daucus carota*. Plant Cell 20: 1407–1420.

Gensel, P.G. 2008. The earliest land plants. Annual Review of Ecology, Evolution, and Systematics 39: 459–477.

Gerdemann, J.W., and J.M. Trappe. 1974. The Endogonaceae in the Pacific Northwest. Mycologia Memoir 5: 1–76.

German, T.N., and V.N. Podkovyrov. 2011. The role of cyanobacteria in the assemblage of the Lakhanda microbiota. Paleontological Journal 45: 320–332.

Germeraad, J.H. 1979. Fossil remains of fungi, algae and other organisms from Jamaica. Scripta Geologica 52: 1–41.

Gerrienne, P. 1991. Les *Pachytheca* de la Gileppe et de Nonceveux (Dévonien inférieur de Belgique). Annales de la Société Géologique de Belgique 113: 267–285.

Gerrienne, P., M. Fairon-Demaret, and J. Galtier. 1999. A Namurian A (Silesian) permineralized flora from the Carrière du Lion at Engihoul (Belgium). Review of Palaeobotany and Palynology 107: 1–15.

Ghignone, S., A. Salvioli, I. Anca, E. Lumini, G. Ortu, L. Petiti, S. Cruveiller, V. Bianciotto, P. Piffanelli, L. Lanfranco, and Paola Bonfante. 2012. The genome of the obligate endobacterium of an AM fungus reveals an interphylum network of nutritional interactions. ISME Journal 6: 136–145.

Ghosh, P., S.K. Bhattacharya, A. Sahni, R.K. Kar, D.M. Mohabey, and K. Ambwani. 2003. Dinosaur coprolites from the Late Cretaceous (Maastrichtian) Lameta Formation of India: Isotopic and other markers suggesting a C_3 plant diet. Cretaceous Research 24: 743–750.

Gianinazzi, S., A. Gollotte, M.-N. Binet, D. van Tuinen, D. Redecker, and D. Wipf. 2010. Agroecology: The key role of arbuscular mycorrhizas in ecosystem services. Mycorrhiza 20: 519–530.

Gianinazzi-Pearson, V., D. van Tuinen, D. Wipf, E. Dumas-Gaudot, G. Recorbet, Y. Liu, J. Doidy, D. Redecker, and N. Ferrol. 2012. Exploring the genome of glomeromycotan fungi. In B. Hock (Ed.), The Mycota IX, Fungal Associations, pp. 1–21 (Second Edition). Springer Verlag, Berlin, Germany.

Gill, J.L. 2014. Ecological impacts of the late Quaternary megaherbivore extinctions. New Phytologist 201: 1163–1169.

Gill, J.L., K.K. McLauchlan, A.M. Skibbe, S. Goring, C.R. Zirbel, and J.W. Williams. 2013. Linking abundances of the dung fungus *Sporormiella* to the density of bison: Implications for assessing grazing by megaherbivores in palaeorecords. Journal of Ecology 101: 1125–1136.

Giovannetti, M., C. Sbrana, L. Avio, and P. Strani. 2004. Patterns of below-ground plant interconnections established by means of arbuscular mycorrhizal networks. New Phytologist 164: 175–181.

Girard, V., and S.M. Adl. 2011. Amber microfossils: On the validity of species concept. Comptes Rendus Palevol 10: 189–200.

Girard, V., A.R. Schmidt, S. Struwe, V. Perrichot, G. Breton, and D. Néraudeau. 2009. Taphonomy and palaeoecology of mid-Cretaceous amber-preserved microorganisms from southwestern France. Geodiversitas 31: 153–162.

Givulescu, R. 1971. Zwei Microthyriaceen aus dem Neogen Rumäniens. Zeitschrift für Pilzkunde 37: 199–202.

Glasspool, I.J., D. Edwards, and L. Axe. 2006. Charcoal in the Early Devonian: A wildfire-derived Konservat-Lagerstätte. Review of Palaeobotany and Palynology 142: 131–136.

Glaub, I. 1994. Mikrobohrspuren in ausgewählten Ablagerungsräumen des europäischen Jura und der Unterkreide (Klassifikation und Palökologie). Courier Forschungsinstitut Senckenberg 174: 1–324.

Glaub, I. 1999. Paleobathymetric reconstructions and fossil microborings. Bulletin of the Geological Society of Denmark 45: 143–146.

Glaub, I. 2004. Recent and sub-recent microborings from the upwelling area off Mauritania (West Africa) and their implications for palaeoecology. In D. McIlroy (Ed.), The Application of Ichnology to Palaeoenvironmental and Stratigraphic Analysis. Geological Society of London Special Publication 228, pp. 63–76. Geological Society Publishing House, Bath, UK.

Glaub, I., and M. Bundschuh. 1997. Comparative study on Silurian and Jurassic/Lower Cretaceous microborings. Courier Forschungsinstitut Senckenberg 201: 123–135.

Glaub, I., and K. Vogel. 2004. The stratigraphic record of microborings. Fossils and Strata 51: 126–135.

Glaub, I., S. Golubic, M. Gektidis, G. Radtke, and K. Vogel. 2007. Microborings and microbial endoliths: Geological implications. In W. Miller III. (Ed.), Trace Fossils: Concepts, Problems, Prospects, pp. 368–381. Elsevier, Amsterdam.

Glawe, D.A. 2006. Synopsis of Erysiphales (powdery mildew fungi) occurring in the Pacific Northwest. Pacific Northwest Fungi 1: 1–27.

Glazebrook, J. 2005. Contrasting mechanisms of defense against biotrophic and nectrotrophic pathogens. Annual Review of Phytopathology 43: 205–227.

Gleason, F.H., P.M. Letcher, and P.A. McGee. 2004. Some Chytridiomycota in soil recover from drying and high temperatures. Mycological Research 108: 583–589.

Gleason, F.H., P.M. Letcher, and P.A. McGee. 2007. Some aerobic Blastocladiomycota and Chytridiomycota can survive but cannot grow under anaerobic conditions. Australasian Mycologist 26: 57–64.

Gleason, F.H., M. Kagami, E. Lefevre, and T. Sime-Ngando. 2008. The ecology of chytrids in aquatic ecosystems: Roles in food web dynamics. Fungal Biology Reviews 22: 17–25.

Gleason, F.H., C.N. Daynes, and P.A. McGee. 2010a. Some zoosporic fungi can grow and survive within a wide pH range. Fungal Ecology 3: 31–37.

Gleason, F.H., A.V. Marano, P. Johnson, and W.W. Martin. 2010b. Blastocladian parasites of invertebrates. Fungal Biology Reviews 24: 56–67.

Gleason, F.H., F.C. Küpper, J.P. Amon, K. Picard, C.M.M. Gachon, A.V. Marano, T. Sime-Ngando, and O. Lilje. 2011. Zoosporic true fungi in marine ecosystems: A review. Marine and Freshwater Research 62: 383–393.

Glockling, S.L., W.L. Marshall, and F.H. Gleason. 2013. Phylogenetic interpretations and ecological potentials of the Mesomycetozoea (Ichthyosporea). Fungal Ecology 6: 237–247.

Goh, T.K., W.H. Ho, K.D. Hyde, and K.M. Tsui. 1997. Four new species of *Xylomyces* from submerged wood. Mycological Research 101: 1323–1328.

Golubic, S., G. Brent, and T. Le Campion-Alsumard. 1970. Scanning electron microscopy of endolithic algae and fungi using a mulitpurpose casting-embedding technique. Lethaia 3: 203–209.

Golubic, S., R.D. Perkins, and K.J. Lucas. 1975. Boring microorganisms and microborings in carbonate substrates. In R.W. Frey (Ed.), The Study of Trace Fossils, pp. 229–259. Springer-Verlag, New York, NY.

Golubic, S., I. Friedmann, and J. Schneider. 1981. The lithobiontic ecological niche, with special reference to microorganisms. Journal of Sedimentary Petrology 51: 475–478.

Golubic, S., S.E. Campbell, K. Drobne, B. Cameron, W.L. Balsam, F. Cimerman, and L. Dubois. 1984. Microbial endoliths: A benthic overprint in the sedimentary record, and a paleobathymetric cross-reference with foraminifera. Journal of Paleontology 58: 351–361.

Golubic, S., G. Radtke, and T. Le Campion-Alsumard. 2005. Endolithic fungi in marine ecosystems. Trends in Microbiology 13: 229–235.

Golubic, S., G. Radtke, and T. Le Campion-Alsumard. 2007. Endolithic fungi. In B.N. Ganguli, and S.K. Deshmukh (Eds), Fungi: Multifaceted Microbes, pp. 38–48. Anamaya Publishers, New Delhi, India.

Gomes, A.L., and G.W. Fernandes. 1994. Influence of parasitism by *Pilostyles ingae* (Rafflesiaceae) on its host plant, *Mimosa naguirei* (Leguminosae). Annals of Botany 74: 205–208.

Gong, Y.-M., R. Xu, and B. Hu. 2008. Endolithic fungi: A possible killer for the mass extinction of Cretaceous dinosaurs. Science in China Series D: Earth Sciences 51: 801–807.

González García, V., M.A. Portal Onco, and V. Ruben Susan. 2006. Review. Biology and systematics of the form genus *Rhizoctonia*. Spanish Journal of Agricultural Research 4: 55–79.

Goos, R.D., and L. Palm. 1979. Early stages in colony development in the genus *Meliola*. Canadian Journal of Botany 57: 461–464.

Goos, R.D., R.D. Brooks, and B.J. Lamore. 1977. An undescribed hyphomycete from wood submerged in a Rhode Island stream. Mycologia 69: 280–286.

Goodman, R.M., and J.B. Weisz. 2002. Plant-microbe symbioses: An evolutionary survey. In J.T. Staley, and A.L. Reysenbach (Eds), Biodiversity of Microbial Life: Foundation of Earth's Biosphere, pp. 237–287. Wiley-Liss, Inc., New York, NY.

Goodsman, D.W., I. Lusebrink, S.M. Landhäusser, N. Erbilgin, and V.J. Lieffers. 2013. Variation in carbon availability, defense chemistry and susceptibility to fungal invasion along the stems of mature trees. New Phytologist 197: 586–594.

Göppert, H.R. 1836. Die fossilen Farrnkräuter. Verhandlungen der Kaiserlichen Leopoldinisch-Carolinischen Akademie der Naturforscher 17 (Supplement): 1–486.

Göppert, H.R.. 1841–1846. Die Gattungen der fossilen Pflanzen verglichen mit denen der Jetztwelt und durch Abbidungen erläutert (Les genres des plantes fossiles comparés avec ceux du monde moderne expliqués par des figures). Henry and Cohen, Bonn, Germany. 120 pp.

Göppert, H.R. 1855. Die Tertiäre Flora von Schossnitz in Schlesien. Heyn'sche Buchhandlung, Görlitz, Germany. 52 pp.

Göppert, H.R., and G.C. Berendt. 1845. Der Bernstein und die in ihm befindlichen Pflanzenreste der Vorwelt. In G.C. Berendt (Ed.), Die im Bernstein Befindlichen Organischen Reste der Vorwelt (Volume 1, pp. 6–126). Commission der Nicolaischen Buchhandlung, Berlin, Germany.

Gorbushina, A.A. 2007. Life on the rocks. Environmental Microbiology 9: 1613–1631.

Gorbushina, A.A., and W.E. Krumbein. 2000. Rock dwelling fungal communities: Diversity of life styles and colony structure. In J. Seckbach (Ed.), Journey to Diverse Microbial Worlds: Adaptation to Exotic Environments, pp. 317–344. Kluwer, Dordrecht, The Netherlands.

Gorbushina, A.A., and K.A. Palinska. 1999. Biodeteriorative processes on glass: Experimental proof of the role of fungi and cyanobacteria. Aerobiologia 15: 183–191.

Gordon, G.L.R., and M.W. Phillips. 1998. The role of anaerobic gut fungi in ruminants. Nutrition Research Reviews 11: 133–168.

Gostinčar, C., M. Grube, S. de Hoog, P. Zalar, and N. Gunde-Cimerman. 2010. Extremotolerance in fungi: Evolution on the edge. FEMS Microbiology Ecology 71: 2–11.

Govers, F., and M. Gijzen. 2006. *Phytophthora* genomics: The plant destroyers' genome decoded. Molecular Plant Microbe Interactions 19: 1295–1301.

Govindarajulu, M., P.E. Pfeffer, H. Jin, J. Abubaker, D.D. Douds, J.W. Allen, H. Bücking, P.J. Lammers, and Y. Shachar-Hill. 2005. Nitrogen transfer in the arbuscular mycorrhizal symbiosis. Nature 435: 819–827.

Gradstein, F.M., J.G. Ogg, M. Schmitz, and G.M. Ogg (Eds). 2012. The Geologic Time Scale 2012, 2 Volumes. Elsevier, Amsterdam, The Netherlands. 1176 pp.

Graham, A. 1962. The role of fungal spores in palynology. Journal of Paleontology 36: 60–68.

Graham, A. 1965. The Sucker Creek and Trout Creek Miocene floras of southeastern Oregon. Kent State University Bulletin (Research Series IX) 53(12): 9–147.

Graham, A. 1971. The role of myxomyceta spores in palynology (with a brief note on the morphology of certain algal zygospores). Review of Palaeobotany and Palynology 11: 89–99.

Graham, L.E. 1996. Green algae to land plants: An evolutionary transition. Journal of Plant Research 109: 241–251.

Graham, L.E., R.B. Kodner, M.M. Fisher, J.M. Graham, L.W. Wilcox, J.M. Hackney, J. Obst, P.C. Bilkey, D.T. Hanson, and M.E. Cook. 2004. Early land plant adaptations to terrestrial stress: A focus on phenolics. In A.R. Hemsley, and I. Poole (Eds), The Evolution of Plant Physiology: From Whole Plants to Ecosystems, pp. 155–169. Academic Press, New York, NY.

Graham, L.E., M.E. Cook, D.T. Hanson, K.B. Pigg, and J.M. Graham. 2010. Structural, physiological, and stable carbon isotope evidence that the enigmatic Paleozoic fossil *Prototaxites* formed from rolled liverwort mats. American Journal of Botany 97: 268–275.

Graham, L.E., P. Arancibia-Avila, W.A. Taylor, P.K. Strother, and M.E. Cook. 2012. Aeroterrestrial *Coleochaete* (Streptophyta, Coleochaetales) models early plant adaptation to land. American Journal of Botany 99: 130–144.

Graff, P.W. 1932. The morphological and cytological development of *Meliola circinans*. Bulletin of the Torrey Botanical Club 59: 241–266.

Graur, D., and T. Pupko. 2001. The Permian bacterium that isn't. Molecular Biology and Evolution 18: 1143–1146.

Gray, J., and A.J. Boucot. 1994. Early Silurian nonmarine animal remains and the nature of the early continental ecosystem. Acta Palaeontologica Polonica 38: 303–328.

Gray, J., and W.A. Shear. 1992. Early life on land. American Scientist 80: 444–456.

Green, T.G.A., and O.L. Lange. 1994. Photosynthesis in poikilohydric plants: A comparison of lichens and bryophytes. In E.-D. Schulze, and M.M. Caldwell (Eds), Ecophysiology of Photosynthesis, pp. 319–341. Springer-Verlag, Berlin, Germany.

Greenblatt, C.L., J. Baum, B.Y. Klein, S. Nachshon, V. Koltunov, and R.J. Cano. 2004. *Micrococcus luteus* – survival in amber. Microbial Ecology 48: 120–127.

Gregor, H.J. 1980. Die miozänen Frucht- und Samen-floren der Oberpfälzer Braunkohle II. Die Funde aus den Kohlen und tonigen Zwischenmitteln. Palaeontographica 174B: 7–94.

Gregor, H.J. 1994. Neue Pflanzenfossilien aus dem niederrheinischen Tertiär IX. Die niederrheinische Braunkohle – ein literarischer Überblick und neue paläobotanische Befunde aus diversen Tagebauen. Documenta Naturae 89: 20–30.

Gregory, P.H. 1983. Some major epidemics caused by *Phytophthora*. In D.C. Erwin, S. Bartnicki-Garcia, and P.H. Tsao (Eds), *Phytophthora*: Its Biology, Taxonomy, Ecology, and Pathology, pp. 271–278. American Phytopathological Society, St. Paul, MN.

Grenville, D.J., R.L. Peterson, and Y. Piché. 1985a. The development, structure, and histochemistry of sclerotia of ectomycorrhizal fungi. I. *Pisolithus tinctorius*. Canadian Jorunal of Botany 63: 1402–1411.

Grenville, D.J., R.L. Peterson, and Y. Piché. 1985b. The development, structure, and histochemistry of sclerotia of ectomycorrhizal fungi. II. *Paxillus involutus*. Canadian Journal of Botany 63: 1412–1417.

Grenville-Briggs, L., C.M.M. Gachon, M. Strittmatter, L. Sterck, F.C. Küpper, and P. van West. 2011. A molecular insight into algal-oomycete warfare: cDNA analysis of *Ectocarpus siliculosus* infected with the basal oomycete *Eurychasma dicksonii*. PLoS ONE 6: e24500.

Grey, K., E.L. Yochelson, M.A. Fedonkin, and D McB. Martina. 2010. *Horodyskia williamsii* new species, a Mesoproterozoic macrofossil from Western Australia. Precambrian Research 180: 1–17.

Grice, K., H. Lu, P. Atahan, M. Asif, C. Hallmann, P. Greenwood, E. Maslen, S. Tulipani, K. Williford, and J. Dodson. 2009. New insights into the origin of perylene in geological samples. Geochimica et Cosmochimica Acta 73: 6531–6543.

Griffing, D.H., J.S. Bridge, and C.L. Hotton. 2000. Coastal l–fluvial palaeoenvironments and plant palaeoecology of the Lower Devonian (Emsian), Gaspé Bay, Québec, Canada. In P.F. Friend, and B.P.J. Williams (Eds), New Perspectives on the Old Red Sandstone, pp. 61–84. Geological Society, London, UK.

Grimaldi, D.A. 1996. Amber: Window to the Past. Harry N. Abrams, New York, NY. 216 pp.

Grimaldi, D., and M.S. Engel. 2005. Evolution of the Insects. Cambridge University Press, Cambridge. XVI + 755 pp.

Grimaldi, D., C.W. Beck, and J.J. Boon. 1989. Occurrence, chemical characteristics, and paleontology of the fossil resins from New Jersey. American Museum Novitates 2948: 1–27.

Grimaldi, D.A., J.A. Lillegraven, T.W. Wampler, D. Bookwalter, and A. Shedrinsky. 2000a. Amber from Upper Cretaceous through Paleocene strata of the Hanna Basin, Wyoming, with evidence for source and taphonomy of fossil resins. Rocky Mountain Geology 35: 163–204.

Grimaldi, D., A. Shedrinsky, and T.P. Wampler. 2000b. A remarkable deposit of fossiliferous amber from the Upper Cretaceous (Turonian) of New Jersey. In D. Grimaldi (Ed.), Studies on Fossils in Amber, with Particular Reference to the Cretaceous of New Jersey, pp. 1–76. Backhuys Publishers, Kerkwerve, The Netherlands.

Grube, M., and D.L. Hawksworth. 2007. Trouble with lichen: The re-evaluation and re-interpretation of thallus form and fruit body types in the molecular era. Mycological Research 111: 1116–1132.

Grube, M., and K. Winka. 2002. Progress in understanding the evolution and classification of lichenized ascomycetes. Mycologist 16: 67–76.

Grube, M., M. Cardinale, J.V. de Castro, Jr., H. Müller, and G. Berg. 2009. Species-specific structural and functional diversity of bacterial communities in lichen symbioses. ISME Journal 3: 1105–1115.

Grubisha, L.C., J.M. Trappe, R. Molina, and J.W. Spatafora. 2002. Biology of the ectomycorrhizal genus *Rhizopogon*. VI. Re-examination of infrageneric relationships inferred from phylogenetic analyses of ITS sequences. Mycologia 94: 607–619.

Grüss, J. 1931. Die Urform des *Anthomyces reukaufii* und andere Einschlüsse in den Bernstein durch Insekten verschleppt. Wochenschrift für Brauerei 48: 63–68.

Gryganskyi, A.P., R.A. Humber, M.E. Smith, J. Miadlikovska, S. Wu, K. Voigt, G. Walther, I.M. Anishchenko, and R. Vilgalys. 2012. Molecular phylogeny of the Entomophthoromycota. Molecular Phylogenetics and Evolution 65: 682–694.

Güçlü, Ö., H. Bıyık, and A. Şahiner. 2010. Mycoflora identified from loggerhead turtle (*Caretta caretta*) egg shells and nest sand at Fethiye beach, Turkey. African Journal of Microbiology Research 4: 408–413.

Guedegbe, H.J., E. Miambi, A. Pando, J. Roman, P. Houngnandan, and C. Rouland-Lefevre. 2009. Occurrence of fungi in combs of fungus-growing termites (Isoptera: Termitidae, Macrotermitinae). Mycological Research 113: 1039–1045.

Gueidan, C., C. Ruibal, G.S. de Hoog, and H. Schneider. 2011. Rock-inhabiting fungi originated during periods of dry climate in the late Devonian and Middle Triassic. Fungal Biology 115: 987–996.

Guido, D.M., and K.A. Campbell. 2011. Jurassic hot spring deposits of the Deseado Massif (Patagonia, Argentina): Characteristics and controls on regional distribution. Journal of Volcanology and Geothermal Research 203: 35–47.

Guimarães, J.T.F., A.C.R. Nogueira, J.B. Cavalcante Da Silva, Jr., J.L. Soares, and R. Silveira. 2013. Fossil fungi from Miocene sedimentary rocks of the central and coastal Amazon region, North Brazil. Journal of Paleontology 87: 484–492.

Gunderson, J.H., H. Elwood, A. Ingold, K. Kindle, and M.L. Sogin. 1987. Phylogenetic relationships between chlorophytes, chrysophytes and oomycetes. Proceedings of the National Academy of Sciences, USA 84: 5823–5827.

Gupta, A. 1994. Fungal fruiting bodies from Lower Tertiary sediments of Sirmaur district, Himachal Pradesh, India. Botanical Journal of the Linnean Society 115: 247–259.

Gupta, A. 2002. Algal/fungal spores from Early Tertiary sediments of Sirmaur District, Himachal. Pradesh, India. Tertiary Research 21: 123–154.

Habgood, K.S., H. Hass, and H. Kerp. 2004. Evidence for an early terrestrial food web: Coprolites from the Early Devonian Rhynie chert. Transactions of the Royal Society of Edinburgh: Earth Sciences 94: 371–389.

Hacquebard, P.A., and J.R. Donaldson. 1969. Carboniferous coal deposition associated with flood-plain and limnic environments in Nova Scotia. In E.C. Dapples, and M.E. Hopkins (Eds), Environments of Coal Deposition, pp. 143–192. Geological Society of America, Boulder, CO.

Haelewaters, D., P. van Wielink, J.W. van Zuijlen, and A. De Kesel. 2012. New records of Laboulbeniales (Fungi, Ascomycota) for The Netherlands. Entomologische Berichten 72: 175–183.

Hågvar, S., and R. Steen. 2013. Succession of beetles (genus *Cis*) and oribatid mites (genus *Carabodes*) in dead sporocarps of the red-banded polypore fungus *Fomitopsis pinicola*. Scandinavian Journal of Forest Research 28: 436–444.

Halary, S., S.B. Malik, L. Lildhar, C.H. Slamovits, M. Hijri, and N. Corradi. 2011. Conserved meiotic machinery in *Glomus* spp., a putatively ancient asexual fungal lineage. Genome Biology and Evolution 3: 950–958.

Halket, A.C. 1930. The rootlets of '*Amyelon radicans*' Will.; Their anatomy, their apices, and their endophytic fungus. Annals of Botany 44: 865–905.

Hallbauer, D.K., and K.T. van Warmelo. 1974. Fossilized plants in thucholite from Precambrian rocks of the Witwatersrand, South Africa. Precambrian Research 1: 199–212.

Hallbauer, D.K., H.M. Jahns, and H.A. Beltmann. 1977. Morphological and anatomical observations on some Precambrian plants from the Witwatersrand, South Africa. Geologische Rundschau 66: 477–491.

Halling, R.E. 2001. Ectomycorrhizae: Co-evolution, significance, and biogeography. Annals of Missouri Botanical Garden 88: 5–13.

Halme, P., J. Heilmann-Clausen, T. Rama, T. Kosonen, and P. Kunttu. 2012. Monitoring fungal biodiversity – towards an integrated approach. Fungal Ecology 5: 750–758.

Hamamoto, T., and K. Horikoshi. 1994. Characterization of a bacterium isolated from amber. Biodiversity and Conservation 3: 567–572.

Hammond-Kosack, K.E., and J.D.G. Jones. 1996. Resistance gene-dependent plant defense responses. Plant Cell 8: 1773–1791.

Hancock, A., and T. Atthey. 1869. On some curious fossil fungi from the black shale of the Northumberland coal-field. Annals and Magazine of Natural History 4: 221–228.

Hancock, A., and T. Atthey. 1870. On some curious fossil fungi from the black shale of the Northumberland coal-field. Transactions of the Northumberland. Natural History 3: 321–329.

Hannemann, B. 1973. Anhangsorgane der Flechten, ihre Strukturen und ihre systematische. Verteilung. Bibliotheca Lichenologica 1: 1–123.

Hansen, E. 2000. The hidden history of a scented wood. Saudi Aramco World 51: 2–12.

Hansen, J.M. 1980. Morphological characterization of encrusting, palynomorph green algae from the Cretaceous-Tertiary of central West Greenland and Denmark. Grana 19: 67–77.

Hansen, K., T. Læssøe, and D.H. Pfister. 2001. Phylogenetics of the Pezizaceae, with an emphasis on *Peziza*. Mycologia 93: 958–990.

Hansen, K., T. Læssøe, and D.H. Pfister. 2002. Phylogenetic diversity in the core group of *Peziza* inferred from ITS sequences and morphology. Mycological Research 106: 879–902.

Hansford, C.G. 1961. The Meliolineae: A monograph. Beihefte zur Sydowia 2: 1–806.

Hansford, C.G. 1963. Iconographia meliolinearum. Sydowia Annales Mycologici, Ser. II, Beihefte 5: 1–285.

Harikumar, V.S., and V.P. Potty. 2009. Occurrence of *Arum*- and *Paris*-morphological types of arbuscular mycorrhizal colonization in sweet potato. Journal of Root Crops 35: 55–58.

Harman, G.E., C.R. Howell, A. Viterbo, I. Chet, and M. Lorito. 2004. *Trichoderma* species – opportunistic, avirulent plant symbionts. Nature Reviews Microbiology 2: 43–56.

Harper, C.J., B. Bomfleur, A.-L. Decombeix, E.L. Taylor, T.N. Taylor, and M. Krings. 2012. Tylosis formation and fungal interactions in an Early Jurassic conifer from northern Victoria Land, Antarctica. Review of Palaeobotany and Palynology 175: 25–31.

Harper, C.J., T.N. Taylor, M. Krings, and E.L. Taylor. 2013. Mycorrhizal symbiosis in the Paleozoic seed fern *Glossopteris* from Antarctica. Review of Palaeobotany and Palynology 192: 22–31.

Harrington, T.C. 2005. Ecology and evolution of mycophagous bark beetles and their fungal partners. In F.E. Vega, and M. Blackwell (Eds), Ecological and Evolutionary Advances in Insect-Fungal Associations, pp. 257–291. Oxford University Press, New York, NY.

Harrison, M.J. 2005. Signaling in the arbuscular mycorrhizal symbiosis. Annual Review of Microbiology 59: 19–42.

Harrison, M.J., G.R. Dewbre, and J. Liu. 2002. A phosphate transporter from *Medicago truncatula* involved in the acquisition of phosphate released by arbuscular mycorrhizal fungi. Plant Cell 14: 2413–2429.

Hart, J.A. 1988. Rust fungi and host plant coevolution: Do primitive hosts harbor primitive parasites? Cladistics 4: 339–366.

Hartkopf-Fröder, C., J. Rust, T. Wappler, E.M. Friis, and A. Viehofen. 2012. Mid-Cretaceous charred fossil flowers reveal direct observation of arthropod feeding strategies. Biology Letters 8: 295–298.

Harvey, R., A.G. Lyon, and P.N. Lewis. 1969. A fossil fungus from Rhynie chert. Transactions of the British Mycological Society 53: 155–156.

Hass, H., and N.P. Rowe. 1999. Thin sections and wafering. In T.P. Jones, and N.P. Rowe (Eds), Fossil Plants and Spores: Modern Techniques, pp. 76–81. Geological Society, London, UK.

Hass, H., T.N. Taylor, and W. Remy. 1994. Fungi from the Lower Devonian Rhynie chert: Mycoparasitism. American Journal of Botany 81: 29–37.

Hatakeyama, S., K. Tanaka, and Y. Harada. 2005. Bambusicolous fungi in Japan (5): Three species of *Tetraploa*. Mycoscience 46: 196–200.

Hatcher, P.E. 1995. Three-way interactions between plant pathogenic fungi, herbivorous insects and their host plants. Biological Reviews 70: 639–694.

Hause, B., and T. Fester. 2005. Molecular and cell biology of arbuscular mycorrhizal symbiosis. Planta 221: 184–196.

Havinga, A.J. 1971. An experimental investigation into the decay of pollen and spores in various soil types. In J. Brook, P.R. Grant, M.D. Muir, P. Van Gijzel, and G. Shaw (Eds), Sporopollenin, pp. 446–479. Academic Press, London, UK.

Havinga, A.J. 1984. A 20-year experimental investigation into the differential corrosion susceptibility of pollen and spores in various soil types. Pollen et Spores 26: 541–558.

Hawksworth, D.L. 1988. The variety of fungal-algal symbioses, their evolutionary significance, and the nature of lichens. Botanical Journal of the Linnean Society 96: 3–20.

Hawksworth, D.L. 2004. Fungal diversity and its implications for genetic resource collections. Studies in Mycology 50: 9–18.

Hawksworth, D.L. 2011. A new dawn for the naming of fungi: Impacts of decisions made in Melbourne in July 2011 on the future publication and regulation of fungal names. MycoKeys 1: 7–20.

Hawksworth, D.L., and P.E.J. Wiltshire. 2011. Forensic mycology: the use of fungi in criminal investigations. Forensic Science International 206: 1–11.

Hawksworth, D.L., P.M. Kirk, B.C. Sutton, and D.N. Pegler. 1995. Ainsworth and Bisby's Dictionary of The Fungi (Eighth Edition). CAB International, Oxon, UK. 616 pp.

Hawksworth, D.L., T. Iturriaga, and A. Crespo. 2005. Líquenes como bioindicadores inmediatos de contaminación y cambios medio-ambientales en los trópicos. Revista Iberoamericana de Micología 22: 71–82.

Heath, T.A., J.P. Huelsenbeck, and T. Stadler. 2013. The fossilized birth-death process: A coherent model of fossil calibration for divergence time estimation. In: Evolution 2013 Meeting, Abstract, 21–25 June 2013, Snowbird, UT. Retrieved from http://arxiv.org/abs/1310.2968.

Hebsgaard, M.B., M.J. Phillips, and E. Willerslev. 2005. Geologically ancient DNA: Fact or artefact? Trends in Microbiology 13: 212–220.

Heckman, D.S., D.M. Geiser, B.R. Eidell, R.L. Stauffer, N.L. Kardos, and S.B. Hedges. 2001. Molecular evidence for the early colonization of land by fungi and plants. Science 293: 1129–1133.

Hedges, S.B., and S. Kumar. 2004. Precision of molecular time estimates. Trends in Genetics 20: 242–247.

Hedges, S.B., J.E. Blair, M.L. Venturi, and J.L. Shoe. 2004. A molecular timescale of eukaryote evolution and the rise of complex multicellular life. BMC Evolutionary Biology 4: 2–11.

Hedges, S.B., F.U. Battistuzzi, and J.E. Blair. 2006. Molecular timescale of evolution in the Proterozoic. In S. Xiao, and A.J. Kaufman (Eds), Neoproterozoic Geobiology and Paleobiology, pp. 199–229. Springer, Dordrecht, The Netherlands.

Heer, O. 1876, 1877. Beiträge zur fossilen Flora Spitzbergens: gegründet auf die Sammlungen der Schwedischen Expedition vom Jahre 1872 auf 1873. Kongliga Svenska Vetenskaps-Akademiens Handlingar 14: 1–141.

Heimann, M., and M. Reichstein. 2008. Terrestrial ecosystem carbon dynamics and climate feedbacks. Nature 451: 289–292.

Helgason, T., and A.H. Fitter. 2005. The ecology and evolution of the arbuscular mycorrhizal fungi. Mycologist 19: 96–101.

Helgason, T., and A.H. Fitter. 2009. Natural selection and the evolutionary ecology of the arbuscular mycorrhizal fungi (Phylum Glomeromycota). Journal of Experimental Botany 60: 2465–2480.

Henk, D.A., and R. Vilgalys. 2007. Molecular phylogeny suggests a single origin of insect symbiosis in the Pucciniomycetes with support for some relationships within the genus *Septobasidium*. American Journal of Botany 94: 1515–1526.

Hennebert, G.L., and B.C. Sutton. 1994. Unitary parameters in conidiogenesis. In D.L. Hawksworth (Ed.), Ascomycete Systematics: Problems and Perspectives in the Nineties. Nato Life Sciences Series A, Volume 269, pp. 65–76. Plenum Press, New York, NY.

Henskens, F.L., T.G.A. Green, and A. Wilkins. 2012. Cyanolichens can have both cyanobacteria and green algae in a common layer as major contributors to photosynthesis. Annals of Botany 110: 555–563.

Herbst, R., and A. Lutz. 2001. A catalogue of fossil fungi in southern South America. Acta Geológica Lilloana 18: 241–248.

Hermann, T.N. 1979. Nakhodki gribov v Rifee. In B.S. Sokolov (Ed.). Paleontologiia Dokembriia I Rannego Kembriia, pp. 129–136. Akademia Nauka SSSR, Leningrad, Russia.

Hermann, T.N., and V.N. Podkovyrov. 2006. Fungal remains from the Late Riphean. Paleontological Journal 40: 207–214.

Hermann, T.N., and V.N. Podkovyrov. 2008. On the nature of the Precambrian microfossils *Arctacellularia* and *Glomovertella*. Paleontological Journal 42: 655–664.

Hernández, J.R., and J.F. Hennen. 2003. Rust fungi causing galls, witches' brooms, and other abnormal plant growths in northwestern Argentina. Mycologia 95: 728–755.

Hernández Roa, J.J., C.R. Virella, and M.J. Cafaro. 2009. First survey of arthropod gut fungi and associates from Vieques, Puerto Rico. Mycologia 101: 896–903.

Herzer, H. 1893a. A new fungus from the Coal Measures. American Geologist 12: 289–290.

Herzer, H. 1893b. A new fungus from the Coal Measures. American Geologist 11: 365–366.

Herzer, H. 1895. Un nouveau champignon des couches de houille, *Dactyloporus archaeus*. Revue Mycologique 17: 115–117.

Hessburg, P.F., and E.M. Hansen. 1987. Pathological anatomy of black stain root disease of Douglas-fir. Canadian Journal of Botany 65: 962–971.

Hibbett, D.S. 2007. Agaricomycotina. Jelly Fungi, Yeasts, and Mushrooms. Version 20 April 2007. http://tolweb.org/Agaricomycotina/20531/2007.04.20 in The Tree of Life Web Project, http://tolweb.org/.

Hibbett, D.S., and M. Binder. 2002. Evolution of complex fruiting-body morphologies in homobasidiomycetes. Proceedings of the Royal Society of London 269B: 1963–1969.

Hibbett, D.S., and M.J. Donoghue. 1995. Progress toward a phylogenetic classification of the Polyporaceae through parsimony analysis of mitochondrial ribosomal DNA sequences. Canadian Journal of Botany 73 (Suppl.1): s853–s861.

Hibbett, D.S., and M.J. Donoghue. 2001. Analysis of character correlations among wood decay mechanisms, mating systems, and substrate ranges in Homobasidiomycetes. Systematic Biology 50: 215–242.

Hibbett, D.S., and P.B. Matheny. 2009. The relative ages of ectomycorrhizal mushrooms and their plant hosts estimated using Bayesian relaxed molecular clock analyses. BMC Biology 7: 13.

Hibbett, D.S., and R.G. Thorn. 2001. Basidiomycota: Homobasidiomycetes. In D.J. McLaughlin, E.G. McLaughlin, and P.A. Lemke (Eds), The Mycota VIIB: Systematics and Evolution, pp. 121–160. Springer-Verlag, Berlin, Germany.

Hibbett, D.S., A. Tsuneda, and S. Murakami. 1994. The secotioid form of *Lentinus tigrinus*: genetics and development of a fungal morphological innovation. American Journal of Botany 81: 466–478.

Hibbett, D.S., E.M. Pine, E. Langer, G. Langer, and M.J. Donoghue. 1997a. Evolution of gilled mushrooms and puffballs inferred from ribosomal DNA sequences. Proceedings of the National Academy of Sciences, USA 94: 12002–12006.

Hibbett, D.S., D. Grimaldi, and M.J. Donoghue. 1997b. Fossil mushrooms from Miocene and Cretaceous ambers and the evolution of Homobasidiomycetes. American Journal of Botany 84: 981–991.

Hibbett, D.S., M.J. Donoghue, and P.B. Tomlinson. 1997c. Is *Phellinites digiustoi* the oldest homobasidiomycete? American Journal of Botany 84: 1005–1011.

Hibbett, D.S., M. Binder, Z. Wang, and Y. Goldman. 2003. Another fossil agaric from Dominican amber. Mycologia 95: 685–687.

Hibbett, D.S., M. Binder, J.F. Bischoff, M. Blackwell, P.F. Cannon, O.E. Eriksson, S. Huhndorf, T. James, P.M. Kirk, R. Lücking, H.T. Lumbsch, F. Lutzoni, P.B. Matheny, D.J. McLaughlin, M.J. Powell, S. Redhead, C.L. Schoch, J.W. Spatafora, J.A. Stalpers, R. Vilgalys, M.C. Aime, A. Aptroot, R. Bauer, D. Begerow, G.L. Benny, L.A. Castlebury, P.W. Crous, Y.-C. Dai, W. Gams, D.M. Geiser, G.W. Griffith, C. Gueidan, D.L. Hawksworth, G. Hestmark, K. Hosaka, R.A. Humber, K.D. Hyde, J.E. Ironside, U. Kõljalg, C.P. Kurtzman, K.H. Larsson, R. Lichtwardt, J. Longcore, J. Miadlikowska, A. Miller, J.M. Moncalvo, S. Mozley-Standridge, F. Oberwinkler, E. Parmasto, V. Reeb, J.D. Rogers, C. Roux, L. Ryvarden, J.P. Sampaio, A. Schüßler, J. Sugiyama, R.G. Thorn, L. Tibell, W.A. Untereiner, C. Walker, Z. Wang, A. Weir, M. Weiss, M.M. White, K. Winka, Y.-J. Yao, and N. Zhang. 2007. A higher-level phylogenetic classification of the Fungi. Mycological Research 111: 509–547.

Higgins, K.L., P.D. Coley, T.A. Kursar, and A.E. Arnold. 2011. Culturing and direct PCR suggest prevalent host generalism among diverse fungal endophytes of tropical forest grasses. Mycologia 103: 247–260.

Hildebrandt, U., K. Janetta, and H. Bothe. 2002. Towards growth of arbuscular mycorrhizal fungi independent of a plant host. Applied and Environmental Microbiology 68: 1919–1924.

Hildebrandt, U., F. Ouziad, F.-J. Marner, and H. Bothe. 2006. The bacterium *Paenibacillus validus* stimulates growth of the arbuscular mycorrhizal fungus *Glomus intraradices* up to the formation of fertile spores. FEMS Microbiology Letters 254: 258–267.

Hillier, R.D., D. Edwards, and L.B. Morrissey. 2008. Sedimentological evidence for rooting structures in the Early Devonian Anglo-Welsh Basin (UK), with speculation on their producers. Palaeogeography, Palaeoclimatology, Palaeoecology 270: 366–380.

Hilton, J., and R.M. Bateman. 2006. Pteridosperms are the backbone of seed-plant phylogeny. Journal of the Torrey Botanical Society 133: 119–168.

Ho, S. 2008. The molecular clock and estimating species divergence. Nature Education 1: 1–2.

Hobbie, E.A., and C.K. Boyce. 2010. Carbon sources for the Palaeozoic giant fungus *Prototaxites* inferred from modern analogues. Proceedings of the Royal Society of London 277B: 2149–2156.

Hodkinson, I.D., N.R. Webb, and S.J. Coulson. 2002. Primary community assembly on land – the missing stages: Why are the heterotrophic organisms always there first? Journal of Ecology 90: 569–577.

Hodson, E., F. Shahid, J. Basinger, and S. Kaminskyj. 2009. Fungal endorhizal associates of *Equisetum* species from Western and Arctic Canada. Mycological Progress 8: 19–27.

Hoffland, E., T.W. Kuyper, H. Wallander, C. Plassard, A.A. Gorbushina, K. Haselwandter, S. Holmström, R. Landeweert, U.S. Lundström, A. Rosling, R. Sen, M.M.

Smits, P.A.W. van Hees, and N. van Breemen. 2004. The role of fungi in weathering. Frontiers in Ecology and the Environment 2: 258–264.

Hoffman, M.T., and A.E. Arnold. 2010. Diverse bacteria inhabit living hyphae of phylogenetically diverse fungal endophytes. Applied and Environmental Microbiology 76: 4063–4075.

Hofmann, B.A., and J.D. Farmer. 2000. Filamentous fabrics in low-temperature mineral assemblages: Are they fossil biomarkers? Implications for the search for a subsurface fossil record on the early Earth and Mars. Planetary and Space Science 48: 1077–1086.

Hofmann, H.J., and J.P. Grotzinger. 1985. Shelf-facies microbiotas from the Odjick and Rocknest formations (Epworth Group; 1.89 Ga), northwestern Canada. Canadian Journal of Earth Science 22: 1781–1792.

Hofmann, K. 1996. Die mikro-endolithischen Spurenfossilien der borealen Oberkreide Nordwest-Europas und ihre Faziesbeziehungen. Geologisches Jahrbuch, Reihe A, Band A 136: 1–151.

Hoffmann, K., K. Voigt, and P.M. Kirk. 2011. Mortierellomycotina subphyl. nov., based on multi-gene genealogies. Mycotaxon 115: 353–363.

Högberg, P., A.H. Plamboeck, A.F.S. Taylor, and P.M.A. Fransson. 1999. Natural ^{13}C abundance reveals trophic status of fungi and host-origin of carbon in mycorrhizal fungi in mixed forests. Proceedings of the National Academy of Sciences, USA 96: 8534–8539.

Högberg, P., A. Nordgren, N. Buchmann, A.F.S. Taylor, A. Ekblad, M.N. Högberg, G. Nyberg, M. Ottosson-Löfvenius, and D.J. Read. 2001. Large-scale forest girdling shows that current photosynthesis drives soil respiration. Nature 411: 789–792.

Höhnel, F. von. 1917. Fungi imperfecti. Beiträge zur Kenntnis derselben. Hedwigia 59: 236–284.

Höhnel, F. von. 1924. Studien über Hyphomyzeten. Centralblatt für Bakterien- und Parasitenkunde, Abteilung 2, 60: 1–26.

Hollick, A. 1910. A new fossil polypore. Mycologia 2: 93–94.

Holm, L. 1959. Some remarks on Meschinelli's fungus-names. Taxon 8: 66–67.

Honegger, R. 1991. Functional aspects of the lichen symbiosis. Annual Reviews of Plant Physiology and Plant Molecular Biology 42: 553–578.

Honegger, R. 1992. Lichens: Mycobiont-photobiont relationships. In W. Reisser (Ed.). Algae and Symbioses: Plants, Animals, Fungi, Viruses, Interactions Explored, pp. 255–275. Biopress Ltd., Bristol, England.

Honegger, R. 1993. Developmental biology of lichens. New Phytologist 125: 659–677.

Honegger, R. 2000. Simon Schwendener (1829–1919) and the dual hypothesis of lichens. Bryologist 103: 307–313.

Honegger, R., and S. Scherrer. 2008. Sexual reproduction in lichen-forming ascomycetes. In T.H. Nash, III (Ed.). Lichen Biology, pp. 94–103 (Second Edition). Cambridge University Press, Cambridge, UK.

Honegger, R., L. Axe, and D. Edwards. 2013a. Bacterial epibionts and endolichenic actinobacteria and fungi in the Lower Devonian lichen *Chlorolichenomycites salopensis*. Fungal Biology 117: 512–518.

Honegger, R., D. Edwards, and L. Axe. 2013b. The earliest records of internally stratified cyanobacterial and algal lichens from the Lower Devonian of the Welsh Borderland. New Phytologist 197: 264–275.

Hooghiemstra, H. 2012. Non-pollen palynomorphs: From unknown curiosities to informative fossils. Celebrating the scientific career of Bas van Geel. Review of Palaeobotany and Palynology 186: 2–4.

Horn, K., T. Franke, M. Unterseher, M. Schnittler, and L. Beenken. 2013. Morphological and molecular analyses of fungal endophytes of achlorophyllous gametophytes of *Diphasiastrum alpinum* (Lycopodiaceae). American Journal of Botany 100: 2158–2174.

Horodyskyj, L.B., T.S. White, and L.R. Kump. 2012. Substantial biologically mediated phosphorus depletion from the surface of a Middle Cambrian paleosol. Geology 40: 503–506.

Horton, T.R., and T.D. Bruns. 2001. The molecular revolution in ectomycorrhizal ecology: Peeking into the black-box. Molecular Ecology 10: 1855–1871.

Hosagoudar, V.B. 1996. Meliolales of India, Volume I. Botanical Survey of India, Calcutta, India. 363 pp.

Hosagoudar, V.B. 2003a. Digital formula for the identification of Meliolaceae. Sydowia 55: 168–171.

Hosagoudar, V.B. 2003b. Meliolaceous fungi on rare medicinal plants in southern India. Zoos' Print Journal 18: 1147–1154.

Hosagoudar, V.B. 2006. Biogeographical distribution of Meliolaceae members in India. Zoos' Print Journal 21: 2495–2505.

Hosagoudar, V.B. 2008. Meliolales of India. Volume II. Botanical Survey of India, Calcutta. 390 pp.

Hosagoudar, V.B., and D.K. Agarwal. 2008. Taxonomic Studies of Meliolales: Identification Manual. International Book Distributors, Dehradun, India. 263 pp.

Hosagoudar, V.B., and G.R. Archana. 2010. A new species of the genus *Ectendomeliola* (Meliolaceae) from Kerala, India. Journal of Threatened Taxa 2: 1092–1095.

Hosagoudar, V.B., T.K. Abraham, and P. Pushpangadan. 1997. The Meliolineae: A Supplement. Tropical Botanic Garden and Research Institute, Palode, Thiruvananthapuram, Kerala, India. 201 pp.

Hotton, C.L., F.M. Hueber, and C.C. Labandeira. 1996. Plant-arthropod interactions from early terrestrial ecosystems: Two Devonian examples. In: J.E. Repetski (eds), 6th North American Paleontological Convention, Abstracts of Papers, pp. 181. Paleontological Society at the University of Tennessee, Department of Geological Sciences, Knoxville, TN.

Hotton, C.L., F.M. Hueber, D.H. Griffing, and J.S. Bridge. 2001. Early terrestrial plant environments: An example from the Emsian of Gaspé, Canada. In P.G. Gensel, and D. Edwards (Eds), Plants Invade the Land: Evolutionary and Environmental Perspectives, pp. 179–212. Columbia University Press, New York.

Hower, J.C., J.M.K. O'Keefe, T.J. Volk, and M.A. Watt. 2010. Funginite–resinite associations in coal. International Journal of Coal Geology 83: 64–72.

Hower, J.C., J.M.K. O'Keefe, C.F. Eble, T.J. Volk, A.R. Richardson, A.B. Satterwhite, R.S. Hatch, and I.J. Kostova. 2011a. Notes on the origin of inertinite macerals in coals: Funginite associations with cutinite and suberinite. International Journal of Coal Geology 85: 186–190.

Hower, J.C., J.M.K. O'Keefe, C.F. Eble, A. Raymond, B. Valentim, T.J. Volk, A.R. Richardson, A.B. Satterwhite, R.S. Hatch, J.D. Stucker, and M.A. Watt. 2011b. Notes on the origin of inertinite macerals in coal: Evidence for fungal and arthropod transformations of degraded macerals. International Journal of Coal Geology 86: 231–240.

Huang, H.-C., E.G. Kokko, and R.S. Erickson. 1999. Infection of alfalfa pollen by *Botrytis cinerea*. Botanical Bulletin of Academia Sinica 40: 101–106.

Huang, H.-C., E.G. Kokko, and J.-W. Huang. 2003. Infection of afalfa (*Medicago sativa* L.) pollen by mycoparasitic fungi *Coniothyrium minitans* Campbell and *Gliocladium catenulatum* Gilmon and Abbott. Revista Mexicana de Fitopatología 21: 117–122.

Huang, T.-C. 1981. Miocene palynomorphs of Taiwan (VI) – miscellaneous spores and pollen grains. Taiwania 26: 45–57.

Hübers, M., B. Bomfleur, M. Krings, and H. Kerp. 2011. An Early Carboniferous leaf-colonizing fungus. Neues Jahrbuch für Geologie und Paläontologie, Abhandlungen 261: 77–82.

Hudspeth, D.S.S., S.A. Nadler, and M.E.S. Hudspeth. 2000. A *COX2* molecular phylogeny of the Peronosporomycetes. Mycologia 92: 674–684.

Hueber, F.M. 2001. Rotted wood-alga-fungus: The history and life of *Prototaxites* Dawson 1859. Review of Palaeobotany and Palynology 116: 123–158.

Hughes, D.P., S.B. Andersen, N.L. Hywel-Jones, W. Himaman, J. Billen, and J.J. Boomsma. 2011a. Behavioral mechanisms and morphological symptoms of zombie ants dying from fungal infection. BMC Ecology 11: 13.

Hughes, D.P., T. Wappler, and C.C. Labandeira. 2011b. Ancient death-grip leaf scars reveal ant-fungal parasitism. Biology Letters 7: 67–70.

Hughes, S.J. 1966. New Zealand fungi. New Zealand Journal of Botany 4: 333–353.

Hughes, S.J. 1976. Sooty moulds. Mycologia 68: 693–820.

Hughes, S.J. 1979. Relocation of species of *Endophragmia* auct. with notes on relevant generic names. New Zealand Journal of Botany 17: 139–188.

Huhndorf, S.M., A.N. Miller, and F.A. Fernández. 2004. Molecular systematics of the Sordariales: The order and the family Lasiosphaeriaceae redefined. Mycologia 96: 368–387.

Humber, R.A. 2012. Entomophthoromycota: A new phylum and reclassification for entomophthoroid fungi. Mycotaxon 120: 477–492.

Humphreys, C.P., P.J. Franks, M. Rees, M.I. Bidartondo, J.R. Leake, and D.J. Beerling. 2010. Mutualistic mycorrhiza-like symbiosis in the most ancient group of land plants. Nature Communications 1: 103.

Huntley, J.W., S. Xiao, and M. Kowalewski. 2006. 1.3 Billion years of acritarch history: An empirical morphospace approa. Precambrian Research 144: 52–68.

Husband, R., E.A. Herre, S.L. Turner, R. Gallery, and P.W. Young. 2002. Molecular diversity of arbuscular mycorrhizal fungi and patterns of host association over time and space in a tropical forest. Molecular Ecology 11: 2669–2678.

Hustad, V.P., and A.N. Miller. 2011. Phylogenetic placement of four genera within the Leotiomycetes (Ascomycota). North American Fungi 6: 1–13.

Hutchens, E. 2009. Microbial selectivity on mineral surfaces: Possible implications for weathering processes. Fungal Biology Reviews 23: 115–121.

Hutchinson, S.A. 1955. A review of the genus *Sporocarpon* Williamson. Annals of Botany 19: 425–435.

Hutchinson, S.A., and J. Walton. 1953. A presumed ascomycete from the Upper Carboniferous. Nature 172: 36–37.

Hutchison, J.A. 1963. The genus *Entomophthora* in the Western Hemisphere. Transactions of the Kansas Academy of Science 66: 237–254.

Hyakumachi, M., and T. Ui. 1988. Development of the teleomorph of non-self-anastomosing isolates of *Rhizoctonia solani* by a burried-slide method. Plant Pathology 37: 438–440.

Hyde, K.D., and P.F. Cannon. 1999. Fungi Causing Tar Spots on Palms. Mycological Papers, Volume 175. CABI, Wallingford, UK. 114 pp.

Hyde, K.D., and K. Soytong. 2008. The fungal endophyte dilemma. Fungal Diversity 33: 163–173.

Hyde, K.D., E.B.G. Jones, J.-K. Liu, H. Ariyawansa, E. Boehm, S. Boonmee, U. Braun, P. Chomnunti, P.W. Crous, D.-Q. Dai, P. Diederich, A. Dissanayake, M. Doilom, F. Doveri, S. Hongsanan, R. Jayawardena, J.D. Lawrey, Y.-M. Li, Y.-X. Liu, R. Lücking, J. Monkai, L. Muggia, M.P. Nelsen, K.-L. Pang, R. Phookamsak, I.C. Senanayake, C.A. Shearer, S. Suetrong, K. Tanaka, K.M. Thambugala, N.N. Wijayawardene, S. Wikee, H.-X. Wu, Y. Zhang, B. Aguirre-Hudson, S.A. Alias, A. Aptroot, A.H. Bahkali, J.L. Bezerra, D.J. Bhat, E. Camporesi, E. Chukeatirote, C. Gueidan, D.L. Hawksworth, K. Hirayama, S. De Hoog, J.-C. Kang, K. Knudsen, W.-J. Li, X.-H. Li, Z.-Y. Liu, A. Mapook, E.H.C. McKenzie, A.N. Miller, P.E. Mortimer, A.J.L. Phillips, H.A. Raja, C. Scheuer, F. Schumm, J.E. Taylor, Q. Tian, S. Tibpromma, D.N. Wanasinghe, Y. Wang, J.-C. Xu, S. Yacharoen, J.-Y. Yan, and M. Zhang. 2013. Families of Dothideomycetes. Fungal Diversity 63: 1–313.

Hynson, N.A., T.P. Madsen, M.-A. Selosse, I.K.U. Adam, Y. Ogura-Tsujita, M. Roy, and G. Gebauer. 2013. The physiological ecology of mycoheterotrophy. In V.S.F.T. Merckx (Ed.). Mycoheterotrophy: The Biology of Plants Living on Fungi, pp. 297–342. Springer Science + Business Media B.V. New York, NY.

Ibañez, C.G., and A.B. Zamuner. 1996. Hyphomycetes (Deuteromycetes) in cones of *Araucaria mirabilis* (Spegazzini) Windhausen, Middle Jurassic of Patagonia, Argentina. Mycotaxon 59: 137–143.

Iglesias, A., A.B. Zamuner, D.G. Poiré, and F. Larriestra. 2007. Diversity, taphonomy and palaeoecology of an angiosperm flora from the Cretaceous (Cenomanian-Coniacian) in southern Patagonia, Argentina. Palaeontology 50: 445–466.

Ikeda, K., K. Inoue, H. Kitagawa, H. Meguro, S. Shimoi, and P. Park. 2012. The role of the extracellular matrix (ECM) in phytopathogenic fungi: A potential target for disease control.

In C.J. Cumagun (Ed.). Plant Pathology, pp. 131–150. InTech, Rijeka, Croatia.

Illman, W.I. 1984. Zoosporic fungal bodies in the spores of the Devonian fossil vascular plant, *Horneophyton*. Mycologia 76: 545–547.

Imaizumi-Anraku, H., N. Takeda, M. Charpentier, J. Perry, H. Miwa, Y. Umehara, H. Kouchi, Y. Murakami, L. Mulder, K. Vickers, J. Pike, J.A. Downie, T. Wang, S. Sato, E. Asamizu, S. Tabata, M. Yoshikawa, Y. Murooka, G.J. Wu, M. Kawaguchi, S. Kawasaki, M. Parniske, and M. Hayashi. 2005. Plastid proteins crucial for symbiotic fungal and bacterial entry into plant roots. Nature 433: 527–531.

Imhof, S. 2009. Arbuscular, ecto-related, orchid mycorrhizas – three independent structural lineages towards mycoheterotrophy: Implications for classification? Mycorrhiza 19: 357–363.

International Committee for Coal and Organic Petrology (ICCP), 2001. The new inertinite classification (ICCP System 1994). Fuel 80: 459–471.

International Culture Collection of (Vesicular) Arbuscular Mycorrhizal Fungi (INVAM). 2013. Developmental concepts of morphological characters. Retrived from http://invam.caf.wvu.edu/fungi/taxonomy/concepts/dvpterm.htm.

Iqbal, S.H., and Shahbaz. 1990. Influence of aging of *Cycas circinalis* on its vesicular-arbuscular mycorrhiza and endogonaceous spores in the rhizosphere. Transactions of the Mycological Society of Japan 31: 197–206.

Ishchenko, T.A., and A.A. Ishchenko. 1980. Novyy vid roda Orestovia (Thallophyta) iz srednego devona Voronezhskoy anteklizy. Paleontologicheskiy Sbornik (L'vov) 17: 79–83.

Iturralde-Vinent, M.A. 2001. Geology of the amber-bearing deposits of the Greater Antilles. Caribbean Journal of Science 43: 141–167.

Iturralde-Vinent, M.A., and R.D.E. MacPhee. 1996. Age and paleogeographical origin of Dominican amber. Science 273: 1850–1852.

Ivanov, S., E.E. Fedorova, E. Limpens, S. De Mita, A. Genre, P. Bonfante, and T. Bisseling. 2012. *Rhizobium*-legume symbiosis shares an exocytotic pathway required for arbuscule formation. Proceedings of the National Academy of Sciences, USA 109: 8316–8321.

Ivarsson, M., C. Broman, S. Lindblom, and N.G. Holm. 2009. Fluid inclusions as a tool to constrain the preservation conditions of sub-seafloor cryptoendolith. Planetary and Space Science 57: 447–490.

Ivarsson, M., S.P. Kilias, C. Broman, J. Naden, and K. Detsi. 2010. Fossilized microorganisms preserved as fluid inclusions in epithermal veins, Vani Mn-Ba deposit, Milos Island, Greece. Proceedings of the XIX Carpathian Balkan Geological Association (CBGA) Congress 100: 297–307.

Jackson, F.D., X. Jin, and J.G. Schmitt. 2009. Fungi in a Lower Cretaceous turtle egg from China: Evidence of ecological interactions. PALAIOS 24: 840–845.

Jackson, T.A. 1967. Fossil actinomycetes in Middle Precambrian glacial varves. Science 155: 1003–1005.

Jago, J.B., J.G. Gehling, J.R. Paterson, and G.A. Brock. 2012. Comments on Retallack, G.J. 2011: Problematic megafossils in Cambrian palaeosols of South Australia. Palaeontology 55: 913–917.

Jahren, A.H., S. Porter, and J.J. Kuglitsch. 2003. Lichen metabolism identified in Early Devonian terrestrial organisms. Geology 31: 99–102.

Jain, K.P., and R.C. Gupta. 1970. Some fungal remains from the Tertiaries of Kerala Coast. Palaeobotanist 18: 177–182.

Jain, K.P., and R.K. Kar. 1979. Palynology of Neogene sediments around Quilon and Varkala, Kerala Coast, South India 1. Fungal remains. Palaeobotanist 26: 105–118.

Jain, R.K. 1968. Middle Triassic pollen grains and spores from Minas de Petroleo beds of the Cacheuta Formation (Upper Gondwana), Argentina. Palaeontographica 122B: 1–47.

Jakobsen, I. 2004. Hyphal fusion to plant species connections – giant mycelia and community nutrient flow. New Phytologist 164: 4–7.

James, J.F. 1885. Remarks on a supposed fossil fungus from the Coal Measures. Journal of the Cincinnati Society of Natural History 8: 157–158.

James, J.F. 1893a. Fossil fungi. Journal of the Cincinnati Society of Natural History 16: 94–98.

James, J.F. 1893b. Notes on fossil fungi. Journal of Mycology 7: 268–273.

James, T.Y., D. Porter, C.A. Leander, R. Vilgalys, and J.E. Longcore. 2000. Molecular phylogenetics of the Chytridiomycota supports the utility of ultrastructural data in chytrid systematics. Canadian Journal of Botany 78: 336–350.

James, T.Y., P.M. Letcher, J.E. Longcore, S.E. Mozley-Standridge, D. Porter, M.J. Powell, G.W. Griffith, and R. Vilgalys. 2006a. A molecular phylogeny of the flagellated fungi (Chytridiomycota) and description of a new phylum (Blastocladiomycota). Mycologia 98: 860–871.

James, T.Y., F. Kauff, C.L. Schoch, P.B. Matheny, V. Hofstetter, C.J. Cox, G. Celio, C. Gueidan, E. Fraker, J. Miadlikowska, H.T. Lumbsch, A. Rauhut, V. Reeb, A.E. Arnold, A. Amtoft, J.E. Stajich, K. Hosaka, G.-H. Sung, D. Johnson, B. O'Rourke, M. Crockett, M. Binder, J.M. Curtis, J.C. Slot, Z. Wang, A.W. Wilson, A. Schüßler, J.E. Longcore, K. O'Donnell, S. Mozley-Standridge, D. Porter, P.M. Letcher, M.J. Powell, J.W. Taylor, M.M. White, G.W. Griffith, D.R. Davies, R.A. Humber, J.B. Morton, J. Sugiyama, A.Y. Rossman, J.D. Rogers, D.H. Pfister, D. Hewitt, K. Hansen, S. Hambleton, R.A. Shoemaker, J. Kohlmeyer, B. Volkmann-Kohlmeyer, R.A. Spotts, M. Serdani, P.W. Crous, K.W. Hughes, K. Matsuura, E. Langer, G. Langer, W.A. Untereiner, R. Lücking, B. Büdel, D.M. Geiser, A. Aptroot, P. Diederich, I. Schmitt, M. Schultz, R. Yahr, D.S. Hibbett, F. Lutzoni, D.J. McLaughlin, J.W. Spatafora, and R. Vilgalys. 2006b. Reconstructing the early evolution of fungi using a six-gene phylogeny. Nature 443: 818–822.

James, T.Y., A.P. Litvintseva, R. Vilgalys, J.A.T. Morgan, J.W. Taylor, M.C. Fisher, L. Berger, C. Weldon, L. du Preez, and J.E. Longcore. 2009. Rapid global expansion of the fungal disease chytridiomycosis into declining and healthy amphibian populations. PLoS Pathogens 5: e1000458.

Jansen, H., and H.-J. Gregor. 1996. Neufund eines jungtertiären Baumschwammes mit Begleitflora aus der Umgebung von Almelo (Niederlande). Documenta Naturae 107: 1–12.

Jansonius, J. 1976. Palaeogene fungal spores and fruiting bodies of the Canadian Arctic. Geoscience and Man 15: 129–132.

Jansonius, J., and L.V. Hills. 1977. Genera file of fossil spores: Supplement. Special Publication of the Department of Geology, University of Calgary, Canada 1977: 3288–3431.

Jansonius, J., and R.M. Kalgutkar. 2000. Redescription of some fossil fungal spores. Palynology 24: 37–47.

Jansonius, J., S.J. Dijkstra, and W.C. Elsik. 1981. *Geleenites* Dijkstra 1949: A Cretaceous fungal fructification. Pollen et Spores 23: 557–562.

Janssen, A., R. Swennen, N. Podoor, and E. Keppens. 1999. Biological and diagenetic influence in recent and fossil tufa deposits from Belgium. Sedimentary Geology 126: 75–95.

Jansson, H.-B., and G.O. Poinar, Jr. 1986. Some possible fossil nematophagous fungi. Transactions of the British Mycological Society 87: 471–474.

Jany, J.-L., and T.E. Pawlowska. 2010. Multinucleate spores contribute to evolutionary longevity of asexual Glomeromycota. American Naturalist 175: 424–435.

Jargeat, P., C. Cosseau, B. Ola'h, A. Jauneau, P. Bonfante, J. Batut, and G. Bécard. 2004. Isolation, free-living capacities, and genome structure of "*Candidatus* Glomeribacter gigasporarum," the endocellular bacterium of the mycorrhizal fungus *Gigaspora margarita*. Journal of Bacteriology 186: 6876–6884.

Javaux, E.J. 2007. The early eukaryotic fossil record. In G. Jékely (Ed.). Eukaryotic Membranes and Cytoskeleton: Origins and Evolution, pp. 1–19. Springer, Austin, TX.

Javaux, E.J., A.H. Knoll, and M.R. Walter. 2001. Morphological and ecological complexity in early eukaryotic ecosystems. Nature 412: 66–69.

Javaux, E.J., A.H. Knoll, and M.R. Walter. 2004. TEM evidence for eukaryotic diversity in mid-Proterozoic oceans. Geobiology 2: 121–132.

Jayalal, U., P. Wolseley, C. Gueidan, A. Aptroot, S. Wijesundara, and V. Karunaratne. 2012. *Anzia mahaeliyensis* and *Anzia flavotenuis*, two new lichen species from Sri Lanka. Lichenologist 44: 381–389.

Jeffrey, E.C. 1906. The wound reactions of *Brachyphyllum*. Annals of Botany 20: 383–394.

Jeffrey, E.C., and M.A. Chrysler. 1906. The lignites of Brandon. Report of the Vermont State Geologist 5: 1–7.

Jeffries, P., and T.W.K. Young. 1994. Interfungal Parasitic Relationships. CABI, Wallingford, UK. 288 pp.

Jehu, T.J. 1918. Rock-boring organisms as agents in coast erosion. Scottish Geographical Magazine 34: 1–11.

Jen, M.N. 1958. The glaciation of Yulong-shan in China. Erdkunde 12: 308–313.

Jha, N., and N. Aggarwal. 2011. First find of *Trichothyrites*, *Notothyrites* and *Frasnacritetrus* from Permian Gondwana sediments of Godavari Graben, India. Phytomorphology 61: 61–67.

Ji, B., C.A. Gehring, G.W.T. Wilson, R.M. Miller, L. Flores-Rentería, and N. Collins Johnson. 2013. Patterns of diversity and adaptation in Glomeromycota from three prairie grasslands. Molecular Ecology 22: 2573–2587.

Jiang, C., R. Alexander, R.I. Kagi, and A.P Murray. 2000. Origin of perylene in ancient sediments and its geological significance. Organic Geochemistry 31: 1545–1559.

Jiang, R.H.Y., and B.M. Tyler. 2012. Mechanisms and evolution of virulence in oomycetes. Annual Review of Phytopathology 50: 295–318.

Johnson, C.N. 1996. Interactions between mammals and ectomycorrhizal fungi. Trends in Ecology and Evolution 11: 503–507.

Johnson, N.C., and J.H. Graham. 2013. The continuum concept remains a useful framework for studying mycorrhizal functioning. Plant and Soil 363: 411–419.

Johnson, T.W., Jr., and F.K. Sparrow, Jr. 1961. Fungi in Oceans and Estuaries. J. Cramer, Weinheim, Germany. 720 pp.

Johnson, T.W., Jr., R.L. Seymour, and D.E. Padgett. 2002. Biology and systematics of the Saprolegniaceae. Available online at http://bit.ly/1oqeR1C.

Johnston, C.A., P. Groffman, D.D. Breshears, Z.G. Cardon, W. Currie, W. Emanuel, J. Gaudinski, R.B. Jackson, K. Lajtha, K. Nadelhoffer, D. Nelson, Jr., W.M. Post, G. Retallack, and L. Wielopolski. 2004. Carbon cycling in soil. Frontiers in Ecology and the Environment 2: 522–528.

Johnston, P.R. 2000. *Vizella metrosideri* sp. nov. New Zealand Journal of Botany 38: 629–633.

Joly, P. 1964. Le genre *Alternaria*: Recherches physiologiques, biologiques et systématiques. Encyclopédie Mycologique 33: 1–250.

Jones, B., and R.W. Renaut. 2007. Selective mineralization of microbes in Fe-rich precipitates (jarosite, hydrous ferric oxides) from acid hot springs in the Waiotapu geothermal area, North Island, New Zealand. Sedimentary Geology 194: 77–98.

Jones, B., R.W. Renaut, and M.R. Rosen. 1999. Role of fungi in the formation of siliceous coated grains, Waiotapu geothermal area, North Island, New Zealand. PALAIOS 14: 475–492.

Jones, B., R.W. Renaut, and M.R. Rosen. 2000. Stromatolites forming in acidic hot-spring waters, North Island, New Zealand. PALAIOS 15: 450–475.

Jones, B., R.W. Renaut, and M.R. Rosen. 2003. Silicified microbes in a geyser mound: The enigma of low-temperature cyanobacteria in a high-temperature setting. PALAIOS 18: 87–109.

Jones, B., K.O. Konhauser, R.W. Renaut, and R.S. Wheeler. 2004. Microbial silicification in Iodine Pool, Waimangu geothermal area, North Island, New Zealand: Implications for recognition and identification of ancient silicified microbes. Journal of the Geological Society of London 161: 983–993.

Jones, M.D.M., I. Forn, C. Gadelha, M.J. Egan, D. Bass, R. Massana, and T.A. Richards. 2011a. Discovery of novel intermediate forms redefines the fungal tree of life. Nature 474: 200–203.

Jones, M.D.M., T.A. Richards, D.L. Hawksworth, and D. Bass. 2011b. Validation and justification of the phylum name *Cryptomycota* phyl. nov. IMA Fungus 2: 173–175.

Joppa, L.N., D.L. Roberts, and S.L. Pimm. 2011. How many species of flowering plants are there? Proceedings of the Royal Society 278B: 554–559.

Jorgensen, R. 1993. The origin of land plants: A union of alga and fungus advanced by flavonoids? BioSystems 31: 193–207.

Joubert, J.J., and F.H.J. Rijkenberg. 1971. Parasitic green algae. Annual Review of Phytopathology 9: 45–64.

Jovan, S. 2008. Lichen Bioindication of Biodiversity, Air Quality, and Climate: Baseline Results From Monitoring in Washington, Oregon, and California. U.S. Department of Agriculture, Forest Service, Pacific Northwest Research Station. General Technical Report PNW-GTR-737. 115 pp.

Joy, K.W., A.J. Willis, and W.S. Lacey. 1956. A rapid cellulose peel technique in palaeobotany. Annals of Botany 20: 635–637.

Judelson, H.S. 2009. Sexual reproduction in Oomycetes: Biology, diversity, and contributions to fitness. In K. Lamour, and S. Kamoun (Eds), Oomycete Genetics and Genomics: Diversity, Interactions, and Research Tools, pp. 121–138. John Wiley & Sons, Inc., Hoboken, NJ.

Jumpponen, A. 2001. Dark septate endophytes – are they mycorrhizal? Mycorrhiza 11: 207–211.

Jumpponen, A., and J.M. Trappe. 1998a. Dark septate endophytes: A review of facultative biotrophic root-colonizing fungi. New Phytologist 140: 295–310.

Jumpponen, A., and J.M. Trappe. 1998b. Performance of *Pinus contorta* inoculated with two strains of root endophytic fungus, *Phialocephala fortinii*: Effects of synthesis system and glucose concentration. Canadian Journal of Botany 76: 1205–1213.

Jurgens, J.A., R.A. Blanchette, and T.R. Filley. 2009. Fungal diversity and deterioration in mummified woods from the ad Astra Ice Cap region in the Canadian High Arctic. Polar Biology 32: 751–758.

Jurina, A.L., and V.A. Krassilov. 2002. Lichenlike fossils from the Givetian of central Kazakhstan. Paleontological Journal 36: 541–547.

Jurina, A.L., and O.N. Putiatina. 2000. Revision of the genus *Flabellofolium* Stone, 1973 (group of Paleozoic plants with *Ginkgo*-like megaphylls) and the first findings of its Givetian members in central Kazakhstan. Paleontologicheskii Zhurnal 3: 103–110.

Justo, A., and D.S. Hibbett. 2011. Phylogenetic classification of *Trametes* (Basidiomycota, Polyporales) based on a five-marker dataset. Taxon 60: 1567–1583.

Kabanov, P., P. Anadón, and W.E. Krumbein. 2008. *Microcodium*: An extensive review and a proposed non-rhizogenic biologically induced origin for its formation. Sedimentary Geology 205: 79–99.

Kagami, M., A. de Bruin, B.W. Ibelings, and E. van Donk. 2007. Parasitic chytrids: Their effects on phytoplankton communities and food-web dynamics. Hydrobiologia 578: 113–129.

Kaim, A., B.E. Tucholke, and A. Warén. 2012. A new Late Pliocene large provannid gastropod associated with hydrothermal venting at Kane Megamullion, Mid-Atlantic Ridge. Journal of Systematic Palaeontology 10: 423–433.

Kaiser, R. 2006. Flowers and fungi use scents to mimic each other. Science 311: 806–807.

Kalgutkar, R.M. 1993. Paleogene fungal palynomorphs from Bonnet Plume Formation, Yukon Territory. Geological Survey of Canada Bulletin 444: 51–105.

Kalgutkar, R.M. 1997. Fossil fungi from the lower Tertiary Iceberg Bay Formation, Eukeka Sound Group, Axel Heiberg Island, Northwest Territories, Canada. Review of Palaeobotany and Palynology 97: 197–226.

Kalgutkar, R.M., and D.R. Braman. 2008. Santonian to ?earliest Campanian (Late Cretaceous) fungi from the Milk River Formation, southern Alberta, Canada. Palynology 32: 39–61.

Kalgutkar, R.M., and J. Jansonius. 2000. Synopsis of Fossil Fungal Spores, Mycelia and Fructifications. American Association of Stratigraphic Palynologists Foundation, Dallas, TX. 429 pp.

Kalgutkar, R.M., and L. Sigler. 1995. Some fossil fungal form-taxa from the Maastrichtian and Palaeogene ages. Mycological Research 99: 513–522.

Kalgutkar, R.M., and A.R. Sweet. 1988. Morphology, taxonomy and phylogeny of the fossil fungal genus *Pesavis* from northwestern Canada. Contributions to Canadian Paleontology, Geological Survey of Canada Bulletin 379: 117–133.

Kalgutkar, R.M., E.M.V. Nambudiri, and W.D. Tidwell. 1993. *Diplodites sweetii* sp. nov. from the Late Cretaceous (Maastrichtian) Deccan Intertrappean Beds of India. Review of Palaeobotany and Palynology 77: 107–118.

Kamoun, S. 2003. Molecular genetics of pathogenic oomycetes. Eukaryotic Cell 2: 191–199.

Kar, R.K. 1990. Palynology of Miocene and Mio-Pliocene sediments of north-east India. Journal of Palynology 26: 171–217.

Kar, R.K., and R.K. Saxena. 1976. Algal and fungal microfossils from Matanomadh Formation (Palaeocene), Kutch, India. Palaeobotanist 23: 1–15.

Kar, R.K., R.Y. Singh, and S.C.D. Sah. 1972. On some algal and fungal remains from Tura Formation of Garo Hills, Assam. Palaeobotanist 19: 146–154.

Kar, R.K., N. Sharma, A. Agarwal, and R. Kar. 2003. Occurrence of fossil-wood rotters (Polyporales) from the Lameta Formation (Maastrichtian), India. Current Science 85: 37–40.

Kar, R.K., N. Sharma, and U.K. Verma. 2004a. Plant pathogen *Protocolletotrichum* from a Deccan Intertrappean bed (Maastrichtian), India. Cretaceous Research 25: 945–950.

Kar, R.K., N. Sharma, and R. Kar. 2004b. Occurrence of fossil fungi in dinosaur dung and its implication on food habit. Current Science 87: 1053–1056.

Kar, R.K., B.D. Mandaokar, and R. Kar. 2006. Fossil aquatic fungi from the Miocene sediments of Mizoram, northeast India. Current Science 90: 291–292.

Kar, R.K., B.D. Mandaokar, and R.K. Kar. 2010. Fungal taxa from the Miocene sediments of Mizoram, northeast India. Review of Palaeobotany and Palynology 158: 240–249.

Karatygin, I.V., N.S. Snigirevskaya, and K.N. Demchenko. 2006. Species of the genus *Glomites* as plant mycobionts in Early Devonian ecosystems. Paleontological Journal 40: 572–579.

Karatygin, I.V., N.S. Snigirevskaya, and S.V. Vikulin. 2009. The most ancient terrestrial lichen *Winfrenatia reticulata*: A new find and new interpretation. Paleontological Journal 43: 107–114.

Karling, J.S. 1928. Studies in the Chytridiales III. A parasitic chytrid causing cell hypertrophy in *Chara*. American Journal of Botany 15: 485–496.

Karling, J.S. 1977. Chytridiomycetarum Iconographia: An Illustrated and Brief Descriptive Guide to the Chytridiomycetous Genera with a Supplement of the Hyphochytridiomycetes. Lubrecht and Cramer, Monticello, NY. 414 pp.

Karling, J.S. 1981. Predominantly Holocarpic and Eucarpic Simple Biflagellate Phycomycetes. J. Cramer, Vaduz, Liechtenstein. 252 pp.

Karpińska-Kołaczek, M., P. Kołaczek, W. Heise, and G. Worobiec. 2010. *Tetraploa aristata* Berkeley & Broome (Fungi, Pleosporales), a new taxon to Poland. Acta Societatis Botanicorum Poloniae 79: 239–244.

Katinas, V. 1983. Baltijos Gintaras. Mosklas, Vilnius, Lithuania. 111 pp.

Ke, P., and Z.Y. Shi. 1978. Early Tertiary spores and pollen grains from the coastal region of the Bohai. Academy of Petroleum Exploration, Development and Planning Research of the Ministry of Petroleum and Chemical Industries and the Nanjing Institute of Geology, and Paleontology, Chinese Academy of Sciences, Kexue Chubanshe, Peking, China. 177 pp.

Kedves, M. 1959. Palynologische Untersuchungen der Miozänen Braunkohlen der Herend 13 Bohrung. Acta Biologica Academiae Scientiarum Hungaricae 5: 167–179.

Keller, H.W., and S.E. Everhart. 2008. Myxomycete species concepts, monotypic genera, the fossil record, and additional examples of good taxonomic practice. Revista Mexicana de Micología 27: 9–19.

Kellogg, D.W., and E.L. Taylor. 2004. Evidence of oribatid mite detritivory in Antarctica during the Late Paleozoic and Mesozoic. Journal of Paleontology 78: 1146–1153.

Kellogg, E.A. 2001. Evolutionary history of the grasses. Plant Physiology 125: 1198–1205.

Kemler, M., M. Lutz, M. Göker, F. Oberwinkler, and D. Begerow. 2009. Hidden diversity in the non-caryophyllaceous plant-parasitic members of *Microbotryum* (Pucciniomycotina: Microbotryales). Systematics and Biodiversity 7: 297–306.

Kendrick, B. 2003. Analysis of morphogenesis in hyphomycetes: New characters derived from considering some conidiophores and conidia as condensed hyphal systems. Canadian Journal of Botany 81: 75–100.

Kendrick, B., and T.R. Nag Raj. 1979. Morphological terms in fungi imperfecti. In B. Kendrick (Ed.), The Whole Fungus: The Sexual-Asexual Synthesis, Volume 1, pp. 43–55. National Museum of Natural Sciences, National Museums of Canada, Ottawa, and The Kananaskis Foundation, Ottawa, Canada.

Kerp, H., and H. Hass. 2004. De Onder-Devonische Rhynie Chert – het oudste en meest compleet bewaard gebleven terrestrische ecosysteem. Grondboor en Hamer 58: 33–50.

Kerp, H., H. Hass, and V. Mosbrugger. 2001. New data on *Nothia aphylla* Lyon, 1964 ex El-Saadawy et Lacey 1979, a poorly known plant from the Lower Devonian Rhynie chert. In P.G. Gensel, and D. Edwards (Eds), Plants Invade the Land: Evolutionary and Environmental Perspectives, pp. 52–82. Columbia University Press, New York, NY.

Kerp, H., N.H. Trewin, and H. Hass. 2004. New gametophytes from the Early Devonian Rhynie chert. Transactions of the Royal Society of Edinburgh: Earth Sciences 94: 411–428.

Kershaw, K.A., and J.W. Millbank. 1970. Isidia as vegetative propagules in *Peltigera aphthosa* var. *variolosa* (Massal.) Thoms. Lichenologist 4: 214–217.

Kevan, P.G., W.G. Chaloner, and D.B.O. Savile. 1975. Interrelationships of early terrestrial arthropods and plants. Palaeontology 18: 391–417.

Khalvati, M.A., Y. Hu, A. Mozafar, and U. Schmidhalter. 2005. Quantification of water uptake by arbuscular mycorrhizal hyphae and its significance for leaf growth, water relations, and gas exchange of barley subjected to drought stress. Plant Biology 7: 706–712.

Khan, A.G. 1972. Podocarp-type mycorrhizal nodules in *Aesculus indica*. Annals of Botany 36: 229–238.

Khan, A.G., and P.G. Valder. 1972. The occurrence of root nodules in the Ginkgoales, Taxales, and Coniferales. Proceedings of the Linnean Society of New South Wales 97: 35–41.

Kholeif, S.E.A., and P.J. Mudie. 2009. Palynological records of climate and oceanic conditions in the Late Pleistocene and Holocene of the Nile Cone, southeastern Mediterranean, Egypt. Palynology 33: 1–24.

Kiage, L.M., and K.B. Liu. 2009. Palynological evidence of climate change and land degradation in the Lake Baringo area, Kenya, East Africa, since AD 1650. Palaeogeography, Palaeoclimatology, Palaeoecology 279: 60–72.

Kidston, R., and W.H. Lang. 1917. On Old Red Sandstone plants showing structure, from the Rhynie Chert Bed, Aberdeenshire. Part I. *Rhynia Gwynne-Vaughani*, Kidston and Lang. Transactions of the Royal Society of Edinburgh 51: 761–784.

Kidston, R., and W.H. Lang. 1920a. On Old Red Sandstone plants showing structure, from the Rhynie Chert Bed, Aberdeenshire. Part II. Additional notes on *Rhynia Gwynne-Vaughani*, Kidston and Lang; with descriptions of *Rhynia major*, n.sp., and *Hornea lignieri*, n.g. n.sp. Transactions of the Royal Society of Edinburgh 52: 603–627.

Kidston, R., and W.H. Lang. 1920b. On Old Red Sandstone plants showing structure, from the Rhynie Chert Bed, Aberdeenshire. Part III. *Asteroxylon Mackiei*, Kidston and Lang. Transactions of the Royal Society of Edinburgh 52: 643–680.

Kidston, R., and W.H. Lang. 1921a. On Old Red Sandstone plants showing structure, from the Rhynie Chert Bed, Aberdeenshire. Part IV. Restorations of the vascular cryptogams, and discussion of their bearing on the general morphology of the Pteridophyta and the origin of the organisation of land-plants. Transactions of the Royal Society of Edinburgh 52: 831–854.

Kidston, R., and W.H. Lang. 1921b. On Old Red Sandstone plants showing structure, from the Rhynie Chert Bed, Aberdeenshire. Part V. The Thallophyta occurring in the peat-bed; the succession of the plants throughout a vertical section of the bed, and the conditions of accumulation and preservation of the deposit. Transactions of the Royal Society of Edinburgh 52: 855–902.

Kiecksee, A.P., L.J. Seyfullah, H. Dörfelt, J. Heinrichs, H. Süss, and A.R. Schmidt. 2012. Pre-Cretaceous Agaricomycetes yet to be discovered: Reinvestigation of a putative Triassic bracket fungus from southern Germany. Fossil Record 15: 85–89.

Kiers, E.T., M. Duhamel, Y. Beesetty, J.A. Mensah, O. Franken, E. Verbruggen, C.R. Fellbaum, G.A. Kowalchuk, M.M. Hart, A. Bago, T.M. Palmer, S.A. West, P. Vandenkoornhuyse, J. Jansa, and H. Bücking. 2011. Reciprocal rewards stabilize cooperation in the mycorrhizal symbiosis. Science 333: 880–882.

Kirchheimer, F. 1942. *Phycopeltis microthyrioides* n. sp. eine blattbewohnende Alge aus dem Tertiär. Botanisches Archiv 44: 172–204.

Kirk, P.M. 1985. Index of Fungi Supplement: Saccardo's omissio. Commonwealth Mycological Institute, Kew, Surrey, UK: 101.

Kirk, P.M., P.F. Cannon, J.C. David, and J.A. Stalpers (Eds). 2001. Dictionary of The Fungi (Ninth Edition). CAB International Publishing, Oxon, UK. 655 pp.

Kirk, P.M, P.F. Cannon, D.W. Minter, and J.A. Stalpers (Eds). 2008. Dictionary of The Fungi (10th Edition). CAB International Publishing, Wallingford, UK. 771 pp.

Kittelmann, S., G.E. Naylor, J.P. Koolaard, and P.H. Janssen. 2012. A proposed taxonomy of anaerobic fungi (class Neocallimastigomycetes) suitable for large-scale sequence-based community structure analysis. PLoS ONE 7: e36866.

Kizil'ste, L.Y. 2006. Fusinite of fossil coals as an information source about the anatomy of ancient plants. Paleontological Journal 40: 448–452.

Klappa, C.F. 1978. Biolithogenesis of *Microcodium*: Elucidation. Sedimentology 25: 489–522.

Klappa, C.F. 1979a. Lichen stromatolites: Criterion for subaerial exposure and a mechanism for the formation of laminar calcretes (caliche). Journal of Sedimentary Petrology 49: 387–400.

Klappa, C.F. 1979b. Calcified filaments in quaternary calcretes: Organo-mineral interactions in the subaerial vadose environment. Journal of Sedimentary Petrology 49: 955–968.

Klappa, C.F. 1980. Rhizoliths in terrestrial carbonates: Classification, recognition, genesis and significance. Sedimentology 27: 613–629.

Klironomos, J.N. 2000. Host specificity and functional diversity among arbuscular mycorrhizal fungi. In: C.R. Bell, M. Brylinsky, and P. Johnson-Green (eds), Microbial Biosystems: New Frontiers. Proceedings of the 8th International Symposium on Microbial Ecology, pp. 845–851. Atlantic Canada Society for Microbial Ecology, Halifax, Canada.

Kloppholz, S., H. Kuhn, and N. Requena. 2011. A secreted fungal effector of *Glomus intraradices* promotes symbiotic biotrophy. Current Biology 21: 1204–1209.

Klymiuk, A.A., C.J. Harper, D.S. Moore, E.L. Taylor, T.N. Taylor, and M. Krings. 2013a. Reinvestigating Carboniferous "Actinomycetes": Authigenic formation of biomimetic carbonates provides insight into early diagenesis of permineralized plants. PALAIOS 28: 80–92.

Klymiuk, A.A., T.N. Taylor, E.L. Taylor, and M. Krings. 2013b. Paleomycology of the Princeton Chert. I. Fossil hyphomycetes associated with the early Eocene aquatic angiosperm, *Eorhiza arnoldii*. Mycologia 105: 521–529.

Klymiuk, A.A., T.N. Taylor, E.L. Taylor, and M. Krings. 2013c. Paleomycology of the Princeton Chert II. Dark-septate fungi in the aquatic angiosperm *Eorhiza arnoldii* indicate a diverse assemblage of root-colonizing fungi during the Eocene. Mycologia 105: 1100–1109.

Knapp, D.G., A. Pintye, and G.M. Kovács. 2012. The dark side is not fastidious – dark septate endophytic fungi of native and invasive plants of semiarid sandy areas. PLoS ONE 7: e32570.

Knauth, L.P. 2013. Fossils come in to land: Not all at sea. Nature 493: 29.

Knobloch, E., and F. Kotlaba. 1994. *Trametites eocenicus*, a new fossil polypore from the Bohemian Eocene. Czech Mycology 47: 207–213.

Knoll, A.H., S. Golubic, J. Green, and K. Swett. 1986. Organically preserved microbial endoliths from the late Proterozoic of East Greenland. Nature 321: 856–857.

Knoll, A.H., E.J. Javaux, D. Hewitt, and P. Cohen. 2006. Eukaryotic organisms in Proterozoic oceans. Philosophical Transactions of the Royal Society London 361B: 1023–1038.

Knoll, A.H., D.E. Canfield, and K.O. Konhauser (Eds). 2012. Fundamentals of Geobiology. Wiley-Blackwell, Oxford, UK. 456 pp.

Knowlton, F.H. 1923. Revision of the flora of the Green River formation, with descriptions of new species. United States Geological Survey Professional Paper 131F: 133–182.

Kobluk, D.R., and C.F. Kahle. 1978. Geologic significance of boring and cavity-dwelling marine algae. Bulletin of Canadian Petroleum Geology 26: 362–379.

Kobluk, D.R., and M.J. Risk. 1974. Devonian boring algae or fungi associated with micrite tubules. Canadian Journal Earth Sciences 11: 1606–1610.

Koch, E., and A. Slusarenko. 1990. *Arabidopsis* is susceptible to infection by a downy mildew fungus. Plant Cell 2: 437–445.

Köck, C. 1939. Fossile Kryptogamen aus der eozänen Braunkohle des Geiseltales. Nova Acta Leopoldina 6: 333–359.

Koeniguer, J.-C. 1975. Les *Prototaxites* (Nématophytes) Ordoviciens et Dévoniens du Sahara Central. Actes du 99e Congrès National des Sociétés Savantes, Besançon, 1974, Section des Sciences 2: 383–388. Bibliothèque Nationale, Paris, France.

Koeniguer, J.C., and M.V. Locquin. 1979. Un polypore fossile à spores porées du Miocène de Libye: *Archeterobasidium syrtae*, gen. et sp. nov. Comptes Rendus du 104e Congrès National des Sociétés Savantes, Bordeaux, 1979, fasc. I (Paléobotanique), pp. 323–329. Bibliothèque Nationale, Paris, France.

Koestler, R.J., V.H. Koestler, A.E. Charola, and F.E. Nieto-Fernandez (Eds). 2003. Art, Biology, and Conservation:

Biodeterioration of Works of Art, Metropolitan Museum of Art Series. The Metropolitan Museum of Art, New York, NY. 576 pp.

Kohlmeyer, J., and B. Volkmann-Kohlmeyer. 1998. A new marine *Xylomyces* on *Rhizophora* from the Caribbean and Hawaii. Fungal Diversity 1: 159–164.

Koide, R.T. 2000. Functional complementarity in the arbuscular mycorrhizal symbiosis. New Phytologist 147: 233–235.

Koller, R., A. Rodriguez, C. Robin, S. Scheu, and M. Bonkowski. 2013. Protozoa enhance foraging efficiency of arbuscular mycorrhizal fungi for mineral nitrogen from organic matter in soil to the benefit of host plants. New Phytologist 199: 203–211.

Kölliker, A. von. 1860. Über das ausgebreitete Vorkommen von pflanzlichen Parasiten in den Hartgebilden niederer Thiere. Zeitschrift für Wissenschaftliche Zoologie 10: 215–232.

Kolo, K., E. Keppens, A. Préat, and P. Claeys. 2007. Experimental observations on fungal diagenesis of carbonate substrates. Journal of Geophysical Research: Biogeosciences 112(G1): 1–20.

Kolodziej, B., S. Golubic, G. Radtke, and I.I. Bucur. 2008. Fossil record of microendoliths in living coral skeletons. In: A. Uchman (ed.), The 2nd International Congress on Ichnology, 28.07.–28.09.2008, Kraków, Abstract Book and the Intra-Congress Field Trip Guidebook: 64, Kraków. Retrieved from http://www.hlug.de/fileadmin/dokumente/geologie/geologie/poster/Ichnia_poster_Boguslaw_Krakau.pdf

Kolosov, P.N. 2013. Fungiform organisms from the Early Cambrian of the Northern Tien Shan. Paleontological Journal 47: 549–553.

Kopczyński, K. 2006. Zapis kopalny grzybów i organizmów grzybopodobnych. [Fossil record of fungi and pseudofungi]. Przegląd Geologiczny 54: 231–237. [in Polish with an English abstract].

Korf, R.P. 1977. A purported fossil discomycete: *Ascodesmisites*. Mycotaxon 6: 193–194.

Košir, A. 2004. *Microcodium* revisited: Root calcification products of terrestrial plants on carbonate-rich substrates. Journal of Sedimentary Research 74: 845–857.

Koske, R.E. 1985. *Glomus aggregatum* emended: A distinct taxon in the *Glomus fasciculatum* complex. Mycologia 77: 619–630.

Koske, R.E., C.F. Friese, P.D. Olexia, and R.L. Hauke. 1985. Vesicular-arbuscular mycorrhizas in *Equisetum*. Transactions of the British Mycological Society 85: 350–353.

Kothe, E., I. Schlunk, D. Senftleben, and K. Krause. 2013. Ectomycorrhiza-specific gene expression. In F. Kempken (Ed.), The Mycota IX: Agricultural Applications, pp. 295–312 (Second Edition). Springer, Berlin, Germany.

Kottke, I. 2002. Mycorrhizae-rhizosphere determinants of plant communities. In Y. Waisel, A. Eshel, and U. Kafkafi (Eds), Plant Roots: The Hidden Half, pp. 919–932 (Third Edition). Marcel Dekker Inc., New York, NY.

Kottke, I., and M. Nebel. 2005. The evolution of mycorrhiza-like associations in liverworts: An update. New Phytologist 167: 330–334.

Kottke, I., and F. Oberwinkler. 1989. Amplification of root-fungus interface in ectomycorrhizae by Hartig net architecture. Annals of Forest Science 46S: 737s–740s.

Kottke, I., A. Beiter, M. Weiss, I. Haug, F. Oberwinkler, and M. Nebel. 2003. Heterobasidiomycetes form symbiotic associations with hepatics: Jungermanniales have sebacinoid mycobionts while *Aneura pinguis* (Metzgeriales) is associated with a *Tulasnella* species. Mycological Research 107: 957–968.

Kottke, I., J.P. Suárez, P. Herrera, D. Cruz, R. Bauer, I. Haug, and S. Garnica. 2010. Atractiellomycetes belonging to the 'rust' lineage (Pucciniomycotina) form mycorrhizae with terrestrial and epiphytic neotropical orchids. Proceedings of the Royal Society 277B: 1289–1298.

Kough, J.L., R. Molina, and R.G. Linderman. 1985. Mycorrhizal responsiveness of *Thuja, Calocedrus, Sequoia*, and *Sequoiadendron* species of western North America. Canadian Journal of Forest Research 15: 1049–1054.

Kowalchuk, G.A. 2012. Bad news for soil carbon sequestration? Science 337: 1049–1050.

Kowalski, M., G. Hausner, and M.D. Piercey-Normore. 2011. Bioactivity of secondary metabolites and thallus extracts from lichen fungi. Mycoscience 52: 413–418.

Krassilov, V.A. 1967. Early Cretaceous Flora of the Southern Coastal Regions and Its Significance in Stratigraphy. Akademiia Nauk CCCP, Moscow. 264 pp.

Krassilov, V.A. 1981. *Orestovia* and the origin of vascular plants. Lethaia 14: 235–250.

Krassilov, V.A. 2007. Mines and galls on fossil leaves from the Late Cretaceous of southern Negev, Israel. African Invertebrates 48: 13–22.

Krassilov, V.A. 2008. Mine and gall predation as top down regulation in the plant-insect systems from the Cretaceous of Negev, Israel. Palaegeography, Palaeoclimatology, Palaeoecology 261: 261–269.

Krassilov, V.A., and N.M. Makulbekov. 2003. The first finding of Gasteromycetes in the Cretaceous of Mongolia. Paleontological Journal 37: 439–442.

Kraus, M.J., and S.T. Hasiotis. 2006. Significance of different modes of rhizolith preservation to interpreting paleoenvironmental and paleohydrologic settings: examples from Paleogene paleosols, Bighorn Basin, Wyoming, USA. Journal of Sedimentary Research 76: 633–646.

Krause, C., S. Garnica, R. Bauer, and M. Nebel. 2011. Aneuraceae (Metzgeriales) and tulasnelloid fungi (Basidiomycota) – a model for early steps in fungal symbiosis. Fungal Biology 115: 839–851.

Krause, S., V. Liebetrau, S. Gorb, M. Sánchez-Román, J.A. McKenzie, and T. Treude. 2012. Microbial nucleation of Mg-rich dolomite in exopolymeric substances under anoxic modern seawater salinity: New insight into an old enigma. Geology 40: 587–590.

Kräusel, R. 1936. Landbewohnende Algenbäume zur Devonzeit? Berichte der Deutschen Botanischen Gesellschaft 54: 379–385.

Kreisel, H. 1977. *Lenzites warnieri* (Basidiomycetes) im Pleistocän von Thüringen. Feddes Repertorium 88: 365–373.

Kreisel, H., and J. Ansorge. 2009. Subfossile Baumschwämme aus dem Quartär Vorpommerns. Zeitschrift für Mykologie 75: 33–50.

Kretzschmar, M. 1982. Fossil fungi in iron stromatolites from Warstein (Rhenish Massif, Northwestern Germany). Facies 7: 237–260.

Krings, M. 2000. Remains of secretory cavities in pinnules of Stephanian pteridosperms from Blanzy-Montceau (Central France): A comparative study. Botanical Journal of the Linnean Society 132: 369–383.

Krings, M. 2001. Pilzreste auf und in den Fiedern zweier Pteridospermen aus dem Stefan von Blanzy-Montceau (Zentralfrankreich). Abhandlungen des Staatlichen Museums für Mineralogie und Geologie Dresden 46/47: 189–196.

Krings, M., and T.N. Taylor. 2012a. Fungal reproductive units enveloped in a hyphal mantle from the Lower Pennsylvanian of Great Britain, and their relevance to our understanding of Carboniferous fungal "sporocarps". Review of Palaeobotany and Palynology 175: 1–9.

Krings, M., and T.N. Taylor. 2012b. Microfossils with possible affinities to the zygomycetous fungi in a Carboniferous cordaitalean ovule. Zitteliana A 52: 3–7.

Krings, M., and T.N. Taylor. 2013. *Zwergimyces vestitus* (Kidston et W.H. Lang) nov. comb., a fungal reproductive unit enveloped in a hyphal mantle from the Lower Devonian Rhynie chert. Review of Palaeobotany and Palynology 190: 15–19.

Krings, M., N. Dotzler, T.N. Taylor, and J. Galtier. 2007a. A microfungal assemblage in *Lepidodendron* from the Upper Visean (Carboniferous) of central France. Comptes Rendus Palevol 6: 431–436.

Krings, M., T.N. Taylor, H. Hass, H. Kerp, N. Dotzler, and E.J. Hermsen. 2007b. An alternative mode of early land plant colonization by putative endomycorrhizal fungi. Plant Signaling and Behavior 2: 125–126.

Krings, M., T.N. Taylor, H. Hass, H. Kerp, N. Dotzler, and E.J. Hermsen. 2007c. Fungal endophytes in a 400-million-yr-old land plant: Infection pathways, spatial distribution, and host responses. New Phytologist 174: 648–657.

Krings, M., N. Dotzler, and T.N. Taylor. 2009a. *Globicultrix nugax* nov. gen. et nov. spec. (Chytridiomycota), an intrusive microfungus in fungal spores from the Rhynie chert. Zitteliana A 48/49: 165–170.

Krings, M., J. Galtier, T.N. Taylor, and N. Dotzler. 2009b. Chytrid-like microfungi in *Biscalitheca* cf. *musata* (Zygopteridales) from the Upper Pennsylvanian Grand-Croix cherts (Saint-Etienne Basin, France). Review of Palaeobotany and Palynology 157: 309–316.

Krings, M., N. Dotzler, J. Galtier, and T.N. Taylor. 2009c. Microfungi from the upper Visean (Mississippian) of central France: Chytridiomycota and chytrid-like remains of uncertain affinity. Review of Palaeobotany and Palynology 156: 319–328.

Krings, M., N. Dotzler, T.N. Taylor, and J. Galtier. 2009d. A Late Pennsylvanian fungal leaf endophyte from Grand-Croix, France. Review of Palaeobotany and Palynology 156: 449–453.

Krings, M., T.N. Taylor, and J. Galtier. 2009e. An enigmatic microorganism from the Upper Pennsylvanian Grand-Croix cherts (Saint-Etienne Basin, France). Zitteliana A 48/49: 171–173.

Krings, M., N. Dotzler, T.N. Taylor, and J. Galtier. 2010a. A fungal community in plant tissue from the Lower Coal Measures (Langsettian, Lower Pennsylvanian) of Great Britain. Bulletin of Geosciences 85: 679–690.

Krings, M., N. Dotzler, J.E. Longcore, and T.N. Taylor. 2010b. An unusual microfungus in a fungal spore from the Lower Devonian Rhynie chert. Palaeontology 53: 753–759.

Krings, M., N. Dotzler, T.N. Taylor, and J. Galtier. 2010c. Microfungi from the Upper Visean (Mississippian) of central France: Structure and development of the sporocarp *Mycocarpon cinctum* nov. sp. Zitteliana A 50: 127–135.

Krings, M., T.N. Taylor, N. Dotzler, and A.L. Decombeix. 2010d. *Galtierella biscalithecae* nov. gen. et sp., a Late Pennsylvanian endophytic water mold (Peronosporomycetes) from France. Comptes Rendus Palevol 9: 5–11.

Krings, M., T.N. Taylor, J. Galtier, and N. Dotzler. 2010e. A fossil peronosporomycete oogonium with an unusual surface ornament from the Carboniferous of France. Fungal Biology 114: 446–450.

Krings, M., T.N. Taylor, E.L. Taylor, N. Dotzler, and C. Walker. 2011a. Arbuscular mycorrhizal-like fungi in Carboniferous arborescent lycopsids. New Phytologist 191: 311–314.

Krings, M., T.N. Taylor, N. Dotzler, and J. Galtier. 2011b. Fungal remains in cordaite (Cordaitales) leaves from the Upper Pennsylvanian of central France. Bulletin of Geosciences 86: 777–784.

Krings, M., N. Dotzler, and T.N. Taylor. 2011c. Mycoparasitism in *Dubiocarpon*, a fungal sporocarp from the Carboniferous. Neues Jahrbuch für Geologie und Paläontologie, Abhandlungen 262: 241–245.

Krings, M., T.N. Taylor, and N. Dotzler. 2011d. The fossil record of the Peronosporomycetes (Oomycota). Mycologia 103: 445–457.

Krings, M., T.N. Taylor, and J.F. White, Jr. 2011e. Fungal sporocarps from the Carboniferous: An unusual specimen of *Traquairia*. Review of Palaeobotany and Palynology 168: 1–6.

Krings, M., N. Dotzler, J. Galtier, and T.N. Taylor. 2011f. Oldest fossil basidiomycete clamp connections. Mycoscience 52: 18–23.

Krings, M., J. Galtier, T.N. Taylor, and N. Dotzler. 2012a. Microbial endophytes and pollen chamber contents in a fossil seed from the Upper Pennsylvanian Grand-Croix cherts, France. Geologica et Palaeontologica 44: 93–100.

Krings, M., T.N. Taylor, and N. Dotzler. 2012b. Fungal endophytes as a driving force in land plant evolution: Evidence from the fossil record. In D. Southworth (Ed.), Biocomplexity of Plant-Fungal Interactions, pp. 5–27 (First Edition). Wiley-Blackwell, Oxford, UK.

Krings, M., T.N. Taylor, E.L. Taylor, H. Hass, H. Kerp, N. Dotzler, and C.J. Harper. 2012c. Microfossils from the Lower Devonian Rhynie chert with suggested affinities to the Peronosporomycetes. Journal of Paleontology 86: 358–367.

Krings, M., T.N. Taylor, N. Dotzler, and G. Persichini. 2012d. Fossil fungi with suggested affinities to the Endogonaceae from the Middle Triassic of Antarctica. Mycologia 104: 835–844.

Krings, M., J.F. White, Jr., N. Dotzler, and C.J. Harper. 2013a. A putative zygomycetous fungus with mantled zygosporangia and apposed gametangia from the Lower Coal Measures (Carboniferous) of Great Britain. International Journal of Plant Sciences 174: 269–277.

Krings, M., T.N. Taylor, N. Dotzler, and C.J. Harper. 2013b. *Frankbaronia velata* nov. sp., a putative peronosporomycete oogonium containing multiple oospores from the Lower Devonian Rhynie chert. Zitteliana A 53: 23–30.

Krings, M., T.N. Taylor, and N. Dotzler. 2013c. Fossil evidence of the zygomycetous fungi. Persoonia 30: 1–10.

Krings, M., T.N. Taylor, E.L. Taylor, H. Kerp, and N. Dotzler. 2014. First record of a fungal "sporocarp" from the Lower Devonian Rhynie chert. Palaeobiodiversity and Palaeoenvironments. In press. Doi:http://dx.doi.org/10.1007/s12549-013-0135-7

Krishnamurthy, K.V., and D.K. Upreti. 2001. Reproductive biology of lichens. In B.M. Johri, and P.S. Srivastava (Eds), Reproductive Biology of Plants, pp. 127–147. Springer Verlag/Narosa Publishing House, Delhi, India.

Krüger, M., H. Stockinger, C. Krüger, and A. Schüssler. 2009. DNA-based species level detection of Glomeromycota: One PCR primer set for all arbuscular mycorrhizal fungi. New Phytologist 183: 212–223.

Krüger, M., C. Krüger, C. Walker, H. Stockinger, and A. Schüssler. 2012. Phylogenetic reference data for systematics and phylotaxonomy of arbuscular mycorrhizal fungi from phylum to species level. New Phytologist 193: 970–984.

Krumbein, W.E., U. Brehm, G. Gerdes, A.A. Gorbushina, G. Levit, and K.A. Palinska. 2003. Biofilm, biodictyon, biomat microbialites, oolites, stromatolites, geophysiology, global mechanism, parahistology. In W.E. Krumbein, D.M. Paterson, and G.A. Zavarzin (Eds), Fossil and Recent Biofilms: A Natural History of Life on Earth, pp. 1–27. Kluwer Academic Publishers, Dordrecht, The Netherlands.

Ksepka, D.T., M.J. Benton, M.T. Carrano, M.A. Gandolfo, J.J. Head, E.J. Hermsen, W.G. Joyce, K.S. Lamm, J.S.L. Patané, M.J. Phillips, P.D. Polly, M. Van Tuinen, J.L. Ware, R.C.M. Warnock, and J.F. Parham. 2011. Synthesizing and databasing fossil calibrations: Divergence dating and beyond. Biology Letters 7: 801–803.

Kuang, Y.-F., B.K. Kirchoff, and J.-P. Liao. 2012. Presence of the protruding oncus is affected by anther dehiscence and acetolysis technique. Grana 51: 253–262.

Kubota, M., T.P. McGonigle, and M. Hyakumachi. 2005. Co-occurrence of *Arum*- and *Paris*-type morphologies of arbuscular mycorrhizae in cucumber and tomato. Mycorrhiza 15: 73–77.

Kudryavtsev, A.B., J.W. Schopf, D.G. Agresti, and T.J. Wdowiak. 2001. *In situ* laser-Raman imagery of Precambrian microscopic fossils. Proceedings of the National Academy of Sciences, USA 98: 823–826.

Kuldau, G., and C. Bacon. 2008. Clavicipitaceous endophytes: Their ability to enhance resistance of grasses to multiple stresses. Biological Control 46: 57–71.

Kumar, P. 1990. Fungal remains in the Miocene Quilon Beds of Kerala state, South India. Review of Palaeobotany and Palynology 63: 13–28.

Kumar, T.K.A., R. Healy, J.W. Spatafora, M. Blackwell, and D.J. McLaughlin. 2012. *Orbilia* ultrastructure, character evolution and phylogeny of Pezizomycotina. Mycologia 104: 462–476.

Kumaran, K.P.N., M. Shindikar, and R.B. Limaye. 2004. Fossil record of marine manglicolous fungi from Malvan (Konkan) west coast of India. Indian Journal of Marine Sciences 33: 257–261.

Kunzmann, L. 2012. Early Oligocene riparian and swamp forests with a mass occurrence of *Zingiberoideophyllum* (extinct Zingiberales) from Saxony, Central Germany. PALAIOS 27: 765–778.

Küpper, F.C., I. Maier, D.G. Müller, S. Loiseaux-de Goer, and L. Guillou. 2006. Phylogenetic affinities of two eukaryotic pathogens of marine macroalgae, *Eurychasma dicksonii* (Wright) Magnus and *Chytridium polysiphoniae* Cohn. Cryptogamie Algologie 27: 165–184.

Kurtzman, C.P. 2011. Phylogeny of the ascomycetous yeasts and the renaming of *Pichia anomala* to *Wickerhamomyces anomalus*. Antonie van Leeuwenhoek 99: 13–23.

Kurtzman, C.P., J.W. Fell, and T. Boekhout (Eds). 2011. The Yeasts, A Taxonomic Study, Volumes 1–3 (Fifth Edition). Elsevier, San Diego, CA. 2354 pp.

Labandeira, C.C. 1998. Early history of arthropod and vascular plant associations. Annual Review of Earth and Planetary Sciences 26: 329–377.

Labandeira, C.C. 2002. The history of associations between plants and animals. In C.M. Herrera, and O. Pellmyr (Eds), Plant-Animal Interactions: An Evolutionary Approach, pp. 26–74. Blackwell Science, London, UK, pp. 248–261.

Labandeira, C.C. 2005. Invasion of the continents: Cyanobacterial crusts to tree-inhabiting arthropods. Trends in Ecology and Evolution 20: 253–262.

Labandeira, C.C., and E.D. Currano. 2013. The fossil record of plant-insect dynamics. Annual Review of Earth and Planetary Sciences 41: 287–311.

Labandeira, C.C., and T.L. Phillips. 2002. Stem borings and petiole galls from Pennsylvanian tree ferns of Illinois, USA: Implications for the origin of the borer and galler functional-feeding-groups and holometabolous insects. Palaeontographica 264A: 1–84.

Labandeira, C.C., and R. Prevec. 2014. Plant paleopathology and the roles of pathogens and insects. International Journal of Paleopathology 4: 1–16.

Labandeira, C.C., P. Wilf, K.R. Johnson, and F. Marsh. 2007. Guide to Insect (and Other) Damage Types on Compressed Plant Fossils. Version 3.0 Smithsonian Institution, Washington, DC. 25 pp.

Läbe, S, C.T. Gee, C. Ballhaus, and T. Nagel. 2012. Experimental silicification of the tree fern *Dicksonia antarctica* at high temperature with silica-enriched H_2O vapor. PALAIOS 27: 835–841.

Labeyrie, E.S., D.P. Molloy, and R.W. Lichtwardt. 1996. An investigation of Harpellales (Trichomycetes) in New York

State blackflies (Diptera: Simuliidae). Journal of Invertebrate Pathology 68: 293–298.

LaFerrière, J., and R.E. Koske. 1981. Occurrence of VA mycorrhizas in some Rhode Island pteridophytes. Transactions of the British Mycological Society 76: 331–332.

Lak, M., D. Néraudeau, A. Nel, P. Cloetens, V. Perrichot, and P. Tafforeau. 2008. Phase contrast X-ray synchrotron imaging: Opening access to fossil inclusions in opaque amber. Microscopy and Microanalysis 14: 251–259.

Lakova, I. 2000. Dispersed tubular structures and filaments from Late Silurian – Middle Devonian marine deposits of North Bulgaria and Macedonia. Geologica Balcanica 30: 29–42.

Lakova, I., and M.C. Goncuoglu. 2005. Early Ludlovian (early Late Silurian) palynomorphs from the Palaeozoic of Camdag, NW Anatolia, Turkey. Journal of the Earth Sciences Application and Research Centre of Hacettepe University 26: 61–73.

Lambers, H., and F.P. Teste. 2013. Interactions between arbuscular mycorrhizal and non-mycorrhizal plants: Do non-mycorrhizal species at both extremes of nutrient availability play the same game? Plant, Cell and Environment 36: 1911–1915.

Lambers, H., C. Mougel, B. Jaillard, and P. Hinsinger. 2009. Plant-microbe-soil interactions in the rhizosphere: An evolutionary perspective. Plant and Soil 321: 83–115.

Lambert, L.H., T. Cox, K. Mitchell, R.A. Rosselló-Mora, C. Del Cueto, D.E. Dodge, P. Orkand, and R.J. Cano. 1998. *Staphylococcus succinus* sp. nov., isolated from Dominican amber. International Journal of Systematic Bacteriology 48: 511–518.

Landeweert, R., E. Hoffland, R.D. Finlay, T.W. Kuyper, and N. van Breemen. 2001. Linking plants to rocks: Ectomycorrhizal fungi mobilize nutrients from minerals. Trends in Ecology and Evolution 16: 248–254.

Landvik, S., O.E. Eriksson, and M.L. Berbee. 2001. *Neolecta* – A fungal dinosaur? Evidence from β-tubulin amino acid sequences. Mycologia 93: 1151–1163.

Landvik, S., T.K. Schumacher, O.E. Eriksson, and S.T. Moss. 2003. Morphology and ultrastructure of *Neolecta* species. Mycological Research 107: 1021–1031.

Lang, W.H. 1899. The prothallus of *Lycopodium clavatum*, L. Annals of Botany 13: 279–318.

Lang, W.H. 1937. On the plant-remains from the Downtonian of England and Wales. Philosophical Transactions of the Royal Society of London 227B: 245–291.

Lange, R.T. 1969. Recent and fossil epiphyllous fungi of the *Manginula-Shortensis* Group. Australian Journal of Botany 17: 565–574.

Lange, R.T. 1976. Fossil epiphyllous "germlings", their living equivalents and their palaeohabitat indicator value. Neues Jahrbuch für Geologie und Paläontologie, Abhandlungen 151: 142–165.

Lange, R.T. 1978a. Southern Australian Tertiary epiphyllous fungi, modern equivalents in the Australasian region, and habitat indicator value. Canadian Journal of Botany 56: 532–541.

Lange, R.T. 1978b. Correlation of particular southern and northern hemisphere Paleogene floras by the unusual fungal spores *Ctenosporites* and *Pesavis tagluensis*. Pollen et Spores 20: 399–403.

Lange, R.T., and P.H. Smith. 1971. The Maslin Bay flora, South Australia. 3. Dispersed fungal spores. Neues Jahrbuch für Geologie und Paläontologie, Montshefte 11: 663–681.

Lange, R.T., and P.H. Smith. 1975. *Ctenosporites* and other Paleogene fungal spores. Canadian Journal of Botany 53: 1156–1157.

Lapinskas, V. 2007. A brief history of Ergotism: From St. Anthony's fire and St. Vitus' dance until today. Medicinos Teorija ir Praktika 13: 202–206.

LaPolla, J.S., G.M. Dlussky, and V. Perrichot. 2013. Ants and the fossil record. Annual Review of Entomology 58: 609–630.

Lara, E., and L. Belbahri. 2011. SSU rRNA reveals major trends in oomycete evolution. Fungal Diversity 49: 93–100.

Larew, H.G. 1986. The fossil gall record: A brief summary. Proceedings of the Entomological Society of Washington 88: 385–388.

Larsson, K.-H., E. Parmasto, M. Fischer, E. Langer, K.K. Nakasone, and S.A. Redhead. 2006. Hymenochaetales: A molecular phylogeny for the hymenochaetoid clade. Mycologia 98: 926–936.

Larsson, S.G. 1978. Baltic Amber – a Palaeobiological Study. Entomonograph 1. Scandinavian Science Press, Klampenborg, DK. 192 pp.

Latz, W., and B.P. Kremer. 1996. Fossile Schimmelpilze in Gestein. Mikrokosmos 85: 229–232.

Laveine, J.-P. 1986. The size of the frond in the genus *Alethopteris Sternberg* (Pteridospermopsida, Carboniferous). Géobios 19: 49–56.

Laveine, J.-P., and A. Belhis. 2007. Frond architecture of the seed-fern *Macroneuropteris scheuchzeri*, based on Pennsylvanian specimens from the Northern France coal field. Palaeontographica 277B: 1–41.

Lawrey, J.D. 1986. Biological role of lichen substances. Bryologist 89: 111–122.

Lawrey, J.D. 1995. The chemical ecology of lichen mycoparasites: A review. Canadian Journal of Botany 73: 603–608.

Lawrey, J.D., and P. Diederich. 2003. Lichenicolous fungi: Interactions, evolution, and biodiversity. Bryologist 106: 80–120.

Lawrey, J.D., and P. Diederich. 2011. Lichenicolous fungi – worldwide checklist, including isolated cultures and sequences available. Retrieved from http://www.lichenicolous.net.

Lawrey, J.D., M. Binder, P. Diederich, M.C. Molina, M. Sikaroodi, and D. Ertz. 2007. Phylogenetic diversity of lichen-associated homobasidiomycetes. Molecular Phylogenetics and Evolution 44: 778–789.

Leake, J.R. 1994. The biology of myco-heterotrophic ('saprophytic') plants. New Phytologist 127: 171–216.

Leake, J.R., D.P. Donnelly, and L. Boddy. 2002. Interactions between ecto-mycorrhizal and saprotrophic fungi. In M.G.A. van der Heijden, and I. Sanders (Eds), Mycorrhizal Ecology, pp. 345–372. Springer-Verlag, Berlin, Germany.

Leake, J.R., D.D. Cameron, and D.J. Beerling. 2008a. Fungal fidelity in the myco-heterotroph-to-autotroph life cycle of Lycopodiaceae: A case of parental nuture? New Phytologist 177: 572–576.

Leake, J.R., A.L. Duran, K.E. Hardy, I. Johnson, D.J. Beerling, S.A. Banwart, and M.M. Smits. 2008b. Biological weathering in soil: The role of symbiotic root-associated fungi biosensing minerals and directing photosynthate-energy into grain-scale mineral weathering. Mineralogical Magazine 72: 85–89.

Le Calvez, T., G. Burgaud, S. Mahé, G. Barbier, and P. Vandenkoornhuyse. 2009. Fungal diversity in deep-sea hydrothermal ecosystems. Applied and Environmental Microbiology 75: 6415–6421.

Lee, R.E. 1971. The pit connections of some lower red algae: Ultrastructure and phylogenetic significance. British Phycological Journal 6: 29–38.

Lee, Y.S. 2000. Qualitative evaluation of ligninolytic enzymes in xylariaceous fungi. Journal of Microbiology and Biotechnology 10: 462–469.

Legrand, J., D. Pons, H. Nishida, and T. Yamada. 2011. Barremian palynofloras from the Ashikajima and Kimigahama formations (Choshi Group, Outer Zone of south-west Japan). Geodiversitas 33: 87–135.

Lekberg, Y., J. Meadow, J.R. Rohr, D. Redecker, and C.A. Zabinski. 2011. Importance of dispersal and thermal environment for mycorrhizal communities: Lessons from Yellowstone National Park. Ecology 92: 1292–1302.

Lendner, A. 1907. Sur quelques Mucorinées. Bulletin de l'Herbier Boissier, Série II 7: 249–251.

Leonard, G., and T.A. Richards. 2012. Genome-scale comparative analysis of gene fusions, gene fissions, and the fungal tree of life. Proceedings of the National Academy of Sciences, USA 109: 21402–21407.

LePage, B.A., R.S. Currah, and R.A. Stockey. 1994. The fossil fungi of the Princeton chert. International Journal of Plant Sciences 155: 828–836.

LePage, B.A., R.S. Currah, R.A. Stockey, and G.W. Rothwell. 1997. Fossil ectomycorrhizae from the Middle Eocene. American Journal of Botany 84: 410–412.

Lepage, T., D. Bryant, H. Philippe, and N. Lartillot. 2007. A general comparison of relaxed molecular clock models. Molecular Biology and Evolution 24: 2669–2680.

Leppik, E.E. 1955. Evolution of the angiosperms as mirrored in the phylogeny of the rust fungi. Archivum Societatis Zoologicae Botanicae Fennicae 'Vanamo' 9: 149–160.

Leppik, E.E. 1973. Origin and evolution of conifer rusts in the light of continental drift. Mycopathologia et Mycologia applicata 49: 121–136.

Leroux, O., F. Leroux, A. Bagniewska-Zadworna, J.P. Knox, M. Claeys, S. Bals, and R.L.L. Viane. 2011. Ultrastructure and composition of cell wall appositions in the roots of *Asplenium* (Polypodiales). Micron 42: 863–870.

Lesquereux, L. 1877. A species of fungus recently discovered in the shales of the Darlington Coal Bed (Lower Productive Coal Measures, Alleghany River Series) at Cannelton, in Beaver County, Pennsylvania. Proceedings of the American Philosophical Society 17: 173–175.

Letcher, P.M., and M.J. Powell. 2002. Frequency and distribution patterns of zoosporic fungi from moss-covered and exposed forest soils. Mycologia 94: 761–771.

Letcher, P.M., P.A. McGee, and M.J. Powell. 2004. Distribution and diversity of zoosporic fungi from soils of four vegetation types in New South Wales, Australia. Canadian Journal of Botany 82: 1490–1500.

Letcher, P.M., M.J. Powell, J.G. Chambers, J.E. Longcore, P.F. Churchill, and P.M. Harris. 2005. Ultrastructural and molecular delineation of the Chytridiaceae (Chytridiales). Canadian Journal of Botany 83: 1561–1573.

Letcher, P.M., M.J. Powell, P.F. Churchill, and J.G. Chambers. 2006. Ultrastructural and molecular phylogenetic delineation of a new order, the Rhizophydiales (Chytridiomycota). Mycological Research 110: 898–915.

Letcher, P.M., C.G. Vélez, M.E. Barrantes, M.J. Powell, P.F. Churchill, and W.S. Wakefield. 2008a. Ultrastructural and molecular analyses of Rhizophydiales (Chytridiomycota) isolates from North America and Argentina. Mycological Research 112: 759–782.

Letcher, P.M., M.J. Powell, D.J.S. Barr, P.F. Churchill, W.S. Wakefield, and K.T. Picard. 2008b. Rhizophyctidales – a new order in Chytridiomycota. Mycological Research 112: 1031–1048.

Leveau, J.H.J. 2006. Microbial communities in the phyllosphere. In M. Riederer, and C. Mueller (Eds), Biology of the Plant Cuticle, pp. 334–367. Blackwell Publishing, Oxford, UK.

Leveau, J.H.J., and G.M. Preston. 2008. Bacterial mycophagy: definition and diagnosis of a unique bacterial-fungal interaction. New Phytologist 177: 859–876.

Lévesque, C.A. 2011. Fifty years of oomycetes – From consolidation to evolutionary and genomic exploration. Fungal Diversity 50: 35–46.

Lewis, D.H. 1991. Mutualistic symbioses in the origin and evolution of land plants. In L. Margulis, and R. Fester (Eds), Symbiosis as a Source of Evolutionary Innovation, pp. 289–300. MIT Press, Cambridge, MA.

Lewis, L.A., and R.M. McCourt. 2004. Green algae and the origin of land plants. American Journal of Botany 91: 1535–1556.

Li, A.-R., S.E. Smith, F.A. Smith, and K.-Y. Guan. 2012. Inoculation with arbuscular mycorrhizal fungi suppresses initiation of haustoria in the root hemiparasite *Pedicularis tricolor*. Annals of Botany 109: 1075–1080.

Li, Y., K.D. Hyde, R. Jeewon, L. Cai, D. Vijaykrishna, and K. Zhang. 2005. Phylogenetics and evolution of nematode-trapping fungi (Orbiliales) estimated from nuclear and protein coding genes. Mycologia 97: 1034–1046.

Lichtwardt, R.W. 1986. The Trichomycetes: Fungal Associates of Arthropods. Springer-Verlag, New York. 343 pp.

Lichtwardt, R.W. 2012. Evolution of Trichomycetes. In J.K. Misra, J.P. Tiwari, and S.K. Deshmukh (Eds), Systematics and Evolution of Fungi, pp. 107–114. Science Publishers, Enfield, NH.

Lignier, O. 1906. *Radiculites reticulatus*, radicelle fossile de Séquoïnée. Bulletin de la Société Botanique de France, Serie 4, 6: 193–201.

Ligrone, R, A. Carafa, E. Lumini, V. Bianciotto, P. Bonfante, and J.G. Duckett. 2007. Glomeromycotean associations in liverworts: A molecular, cellular, and taxonomic analysis. American Journal of Botany 94: 1756–1777.

Ligrone, R., J.G. Duckett, and K.S. Renzaglia. 2012. Major transitions in the evolution of early land plants: A bryological perspective. Annals of Botany 109: 851–871.

Limaye, R.B., K.P.N. Kumaran, K.M. Nair, and D. Padmalal. 2007. Non-pollen palynomorphs as potential palaeoenvironmental indicators in the Late Quaternary sediments of the west coast of India. Current Science 92: 1370–1382.

Lincoff, G. 1996. Mushroom mania. Natural History 105(4): 32–35.

Lindahl, P.O. 1960. The different types of isidia in the lichen genus *Peltigera*. Svensk Botanisk Tidskrift 54: 565–570.

Lindley, J., and W. Hutton. 1831–33. The Fossil Flora of Great Britain; Or, Figures and Descriptions of the Vegetable Remains Found in a Fossil State in This Country, Volume I. James Ridgeway, London, UK. 218 pp.

Lindley, J., and W. Hutton. 1833–35. The Fossil Flora of Great Britain; Or, Figures and Descriptions of the Vegetable Remains Found in a Fossil State in This Country, Volume II. James Ridgeway and Sons, London, UK. 206 pp.

Lindley, J., and W. Hutton. 1837. The Fossil Flora of Great Britain; Or, Figures and Descriptions of the Vegetable Remains Found in a Fossil State in This Country, Volume III. James Ridgeway and Sons, London, UK. 204 pp.

Lindow, S.E., and M.T. Brandl. 2003. Microbiology of the phyllosphere. Applied and Environmental Microbiology 69: 1875–1883.

Liu, A.G., D. McIlroy, J.J. Matthews, and M.D. Brasier. 2012. A new assemblage of juvenile Ediacaran fronds from the Drook Formation, Newfoundland. Journal of the Geological Society, London 169: 395–403.

Liu, X.-Y., and K. Voigt. 2010. Molecular characters of zygomycetous fungi. In Y. Gherbawy, and K. Voigt (Eds), Molecular Identification of Fungi, pp. 461–488. Springer-Verlag, Berlin, Germany.

Liu, Y., J.W. Leigh, H. Brinkmann, M.T. Cushion, N. Rodriguez-Ezpeleta, H. Philippe, and B.F. Lang. 2006. Phylogenomic analyses support the monophyly of Taphrinomycotina, including *Schizosaccharomyces* fission yeasts. Molecular Biology and Evolution 26: 27–34.

Liu, Y., E.T. Steenkamp, H. Brinkmann, L. Forget, H. Philippe, and B.F. Lang. 2009. Phylogenomic analyses predict sistergroup relationship of nucleariids and Fungi and paraphyly of zygomycetes with significant support. BMC Evolutionary Biology 9: 272.

Liu, Y.J., and B.D. Hall. 2004. Body plan evolution of ascomycetes, as inferred from an RNA polymerase II phylogeny. Proceedings of the National Academy of Sciences, USA 101: 4507–4512.

Locquin, M.V. 1981. Affinites fongiques probables des chitinozoaires devenant chitinomycètes. Cahiers de Micropaléontologie 1: 29–36.

Locquin, M.V. 1982. Les champignons fossiles. 1: Recherches sur quelques organismes fongiques et d'affinités fongiques probables présents au Paléozoïque. Microéditions. Paris, 173 pp.

Locquin, M.V. 1983. Nouvelles recherches sur les champignons fossiles. Comptes rendus du 108e Congrès National des Sociétés Savantes, Grenoble, 1983, fasc. 1 (Sciences de la Terre), tome 2: 179–190. Bibliothèque Nationale, Paris, France.

Locquin, M.V., and J.C. Koeniguer. 1981. Un nouveau polypore fossile du Miocène de Libye: *Fomites libyae* Locquin et Koeniger, gen. et sp. nov. Comptes Rendus du 106e Congrès National des Sociétés Savantes, Perpignan, 1981, fasc. 1 (Paléontologie), pp. 107–117. Bibliothèque Nationale, Paris, France.

Longcore, J.E. 2009. Taxonomic changes since 1960: Chytridiomycota, Blastocladiomycota & Neocallimastigomycota. Retrieved from http://www.umaine.edu/chytrids/Bibliography-Index.html.

Longcore, J.E., A.P. Pessier, and D.K. Nichols. 1999. *Batrachochytrium dendrobatidis* gen. et sp. nov., a chytrid pathogenic to amphibians. Mycologia 91: 219–227.

Loomis, F.B. 1900. Siluric fungi from western New York. Bulletin of the New York State Museum 39: 223–226.

López-Sáez, J.A., and L. López-Merino. 2007. Coprophilous fungi as a source of information of anthropic activities during the prehistory in the Amblés Valley (Àvila, Spain): The archaeopalynological record. Revista Española de Micropaleontología 39: 103–116.

Lücking, R. 1997. The use of foliicolous lichens as bioindicators in the tropics, with special reference to the microclimate. Abstracta Botanica 21: 99–116.

Lücking, R., S. Huhndorf, D.H. Pfister, E.R. Plata, and H.T. Lumbsch. 2009. Fungi evolved right on track. Mycologia 101: 810–822.

Ludwig, R. 1859–1861. Fossile Pflanzen aus der ältesten Abtheilung der Rheinisch-Wetterauer Tertiär-Formation. Palaeontographica 8: 39–154.

Ludwig, R. 1861. Zur Palaeontologie des Ural's: Actinozoen und Bryozoen aus dem Carbon-Kalkstein im Gouvernement Perm. Palaeontographica 10: 17–36.

Lumbsch, H.T., and S.M Huhndorf. 2007. Whatever happened to the pyrenomycetes and loculoascomycetes? Mycological Research 111: 1064–1074.

Lumini, E., V. Bianciotto, P. Jargeat, M. Novero, A. Salvioli, A. Faccio, G. Bécard, and P. Bonfante. 2007. Presymbiotic growth and sporal morphology are affected in the arbuscular mycorrhizal fungus *Gigaspora margarita* cured of its endobacteria. Cellular Microbiology 9: 1716–1729.

Lundgren, J.G. 2009. Nutritional aspects of non-prey foods in the life histories of predaceous Coccinellidae. Biological Control 51: 294–305.

Luttrell, E.S. 1946. The genus *Stomiopeltis* (Hemisphaeriaceae). Mycologia 38: 565–586.

Luttrell, E.S. 1951. Taxonomy of the pyrenomycetes. University of Missouri Studies 24: 1–120.

Luttrell, E.S. 1955. The ascostromatic ascomycetes. Mycologia 47: 511–532.

Luttrell, E.S. 1973. Loculoascomycetes. In G.C. Ainsworth, F.K. Sparrow, and A.S. Sussman (Eds), The Fungi: An Advanced Treatise (Volume IV-A, pp. 134–219). Academic Press, New York, NY.

Lutzoni, F., M. Pagel, and V. Reeb. 2001. Major fungal lineages are derived from lichen symbiotic ancestors. Nature 411: 937–940.

Lutzoni, F., F. Kauff, J.C. Cox, D. McLaughlin, G. Celio, B. Dentinger, M. Padamsee, D. Hibbett, T.Y. James, E. Baloch, M. Grube, V. Reeb, V. Hofstetter, C. Schoch, A.E. Arnold, J. Miadlikowska, J. Spatafora, D. Johnson, S. Hambleton,

M. Crockett, R. Shoemaker, G.H. Sung, R. Lücking, T. Lumbsch, K. O'Donnell, M. Binder, P. Diederich, D. Ertz, C. Gueidan, K. Hansen, R.C. Harris, K. Hosaka, Y.W. Lim, B. Matheny, H. Nishida, D. Pfister, J. Rogers, A. Rossman, I. Schmitt, H. Sipman, J. Stone, J. Sugiyama, R. Yahr, and R. Vilgalys. 2004. Assembling the fungal tree of life: Progress, classification, and evolution of subcellular traits. American Journal of Botany 91: 1446–1480.

Lyck, J.M., and L. Stemmerik. 2000. Palynology and depositional history of the Paleocene? Thyra Ø Formation, Wandel Sea Basin, eastern North Greenland. Geology of Greenland Survey Bulletin 187: 21–49.

Lyon, A.G. 1962. On the fragmentary remains of an organism referable to the Nematophytales, from the Rhynie chert, *Nematoplexus rhyniensis* gen. et sp. nov. Transactions of the Royal Society of Edinburgh 65: 79–87.

Lyons, P.C. 1991. Bacteria-like bodies in coalified Carboniferous xylem – enigmatic microspheroids or possible evidence of microbial saprophytes in a vitrinite precursor? International Journal of Coal Geology 18: 293–303.

Lyons, P.C. 2000. Funginite and secretinite – two new macerals of the inertinite maceral group. International Journal of Coal Geology 44: 95–98.

MacGinitie, H.D. 1937. The flora of the Weaverville beds of Trinity County, California, with descriptions of the plant-bearing beds. Carnegie Institution of Washington Publications 465: 131.

Maciá-Vicente, J.G., H.-B. Jansson, K. Samir, S.K. Abdullah, E. Descals, J. Salinas, and L.V. Lopez-Llorca. 2008. Fungal root endophytes from natural vegetation in Mediterranean environments with special reference to *Fusarium* spp. FEMS Microbiology Ecology 64: 90–105.

Mackie, W. 1913. The rock series of Craigbeg and Ord Hill, Rhynie, Aberdeenshire. Transactions of the Edinburgh Geological Society 10: 205–236.

Magallón, S., and M.J. Sanderson. 2001. Absolute diversification rates in angiosperm clades. Evolution 55: 1762–1780.

Magallón, S., K.W. Hilu, and D. Quandt. 2013. Land plant evolutionary timeline: Gene effects are secondary to fossil constraints in relaxed clock estimation of age and substitution rates. American Journal of Botany 100: 556–573.

Magallon-Puebla, S., and S.R.S. Cevallos-Ferriz. 1993. A fossil earthstar (Geasteraceae; Gasteromycetes) from the Late Cenozoic of Puebla, Mexico. American Journal of Botany 80: 1162–1167.

Mägdefrau, K. 1937. Lebensspuren fossiler "Bohr"-Organismen. Beiträge zur Naturkundlichen Forschung in Südwestdeutschland 2: 54–67.

Mägdefrau, K. 1957. Flechten und Moose im baltischen Bernstein. Berichte der Deutschen Botanischen Gesellschaft 70: 433–435.

Mägdefrau, K. 1966. Die Strukturerhaltung fossiler Pflanzen. Bild der Wissenschaft 12: 988–997.

Mägdefrau, K. 1992. Geschichte der Botanik: Leben und Leistung großer Forscher (Second Edition). Gustav Fischer Verlag, Stuttgart, Germany. 359 pp.

Magnus, P. 1903. Ein von F.W. Oliver nachgewiesener fossiler parasitischer Pilz. Berichte der Deutschen Botanischen Gesellschaft 21: 248–250.

Mahabalé, T.S. 1968. On a fossil species of *Diplodia* from the Deccan Intertrappean series, M.P., India. Palaeobotanist 17: 295–297.

Mahaney, W.C., R.W. Barendregt, C.C.R. Allen, M.W. Milner, and D. Bray. 2013. Coprolites from the Cretaceous Bearpaw Formation of Saskatchewan. Cretaceous Research 41: 31–38.

Maherali, H., and J.N. Klironomos. 2007. Influence of phylogeny on fungal community assembly and ecosystem functioning. Science 316: 1746–1748.

Maheshwari, H.K. 1972. Permian wood from Antarctica and revision of some lower Gondwana wood taxa. Palaeontographica 138B: 1–43.

Maia, L.C., and J.W. Kimbrough. 1994. Ultrastructural studies on spores of *Glomus intraradices*. International Journal of Plant Science 155: 689–698.

Maillet, F., V. Poinsot, O. Andre, V. Puech-Pagès, A. Haouy, M. Gueunier, L. Cromer, D. Giraudet, D. Formey, A. Niebel, E. Andres Martinez, H. Driguez, G. Bécard, and J. Dénarié. 2011. Fungal lipochitooligosaccharide symbiotic signals in arbuscular mycorrhiza. Nature 469: 58–63.

Maithy, P.K. 1973. Micro-organisms from the Bushimay System (Late Pre-Cambrian) of Kanshi, Zaire. Palaeobotanist 22: 133–149.

Makonde, H.M., H.I. Boga, Z. Osiemo, R. Mwirichia, J.B. Stielow, M. Göker, and H.-P. Klenk. 2013. Diversity of Termitomyces associated with fungus-farming termites assessed by cultural and culture-independent methods. PLoS ONE 8: e56464.

Malloch, D.W., K.A. Pirozynski, and P.H. Raven. 1980. Ecological and evolutionary significance of mycorrhizal symbioses in vascular plants (A Review). Proceedings of the National Academy of Sciences, USA 77: 2113–2118.

M'Alpi, D. 1903. On the so-called petrified mushroom. Victorian Naturalist 20: 14–16.

Mamet, B.L., A. Roux, and W.W. Nassichuk. 1987. Algues carbonifères et permiennes de l'Arctique canadien. Geological Survey of Canada Bulletin 342(1–22): 92–99.

Mandal, A, N. Samajpati, and S. Bera. 2011. A new species of *Meliolinites* (fossil Meliolales) from the Neogene sediments of sub-Himalayan West Bengal, India. Nova Hedwigia 92: 435–440.

Mandyam, K., and A. Jumpponen. 2005. Seeking the elusive function of the root-colonising dark septate endophytic Fungi. Studies in Mycology 53: 173–189.

Mani, M.S. 1964. Ecology of Plant Galls. Dr. W. Junk Publishers, The Hague, The Netherlands. 434 pp.

Marano, A.V., M.M. Steciow, M.L. Arellano, A.M. Arambarri, and M.V. Sierra. 2007. El género *Nowakowskiella* (Cladochytriaceae, Chytridiomycota) en ambientes de la Pcia, de Buenos Aires (Argentina): Taxonomía, frequencia y abundancia de las especies encontradas. Boletín de la Sociedad Argentina de Botánica 42: 13–24.

Marano, A.V., C.L.A. Pires-Zottarelli, M.D. Barrera, M.M. Steciow, and F.H. Gleason. 2011. Diversity, role in decomposition, and succession of zoosporic fungi and straminipiles on submerged decaying leaves in a woodland stream. Hydrobiologia 659: 93–109.

Marano, A.V., F.H. Gleason, F. Bärlocher, C.L.A. Pires-Zottarelli, O. Lilje, S.K. Schmidt, S. Rasconi, M. Kagami, M.D. Barrera, T. Sime-Ngando, S. Boussiba, J.I. de Souza, and J.E. Edwards. 2012. Quantitative methods for the analysis of zoosporic fungi. Journal of Microbiological Methods 89: 22–32.

Marcinkiewicz, T. 1979. Fungi-like forms on Jurassic megaspores (Utwory grzybopodobne na megasporach jurajskich). Acta Palaeobotanica 20: 123–128.

Margulis, L. 1993. Symbiosis in Cell Evolution. W.H. Freeman, New York. 452 pp.

Margulis, L., and K.V. Schwartz. 1998. Five kingdoms: An Illustrated Guide to the Phyla of Life on Earth (Third Edition). W.H. Freeman & Company, San Francisco, CA. 520 pp.

Mark, D.F., C.M. Rice, A.E. Fallick, N.H. Trewin, M.R. Lee, A. Boyce, and J.K.W. Lee. 2011. $^{40}Ar/^{39}Ar$ dating of hydrothermal activity, biota and gold mineralization in the Rhynie hot-spring system, Aberdeenshire, Scotland. Geochimica et Cosmochimica Acta 75: 555–569.

Mark, D.F., C.M. Rice, and N.H. Trewin. 2013. Discussion on 'A high-precision U–Pb age constraint on the Rhynie Chert Konservat-Lagerstätte: Time scale and other implications'. Journal of the Geological Society, London 170: 701–703.

Markham, P., and A.J. Collinge. 1987. Woronin bodies of filamentous fungi. FEMS Microbiology Letters 46: 1–11.

Marleau, J., J. Dalpé, M. St-Arnaud, and M. Hijri. 2011. Spore development and nuclear inheritance in arbuscular mycorrhizal fungi. BMC. Evolutionary Biology 11: 51.

Márquez, L.M., R.S. Redman, R.J. Rodriguez, and M.J. Roossinck. 2007. A virus in a fungus in a plant: Three-way symbiosis required for thermal tolerance. Science 315: 513–515.

Marshall, W.A. 1996. Aerial dispersal of lichen soredia in the maritime Antarctic. New Phytologist 134: 523–530.

Martel, A., A. Spitzen-van der Sluijs, M. Blooi, W. Bert, R. Ducatelle, M.C. Fisher, A. Woeltjes, W. Bosman, K. Chiers, F. Bossuyt, and F. Pasmans. 2013. *Batrachochytrium salamandrivorans* sp. nov. causes lethal chytridiomycosis in amphibians. Proceedings of the National Academy of Sciences, USA 110: 15325–15329.

Martill, D.M. 1989. Fungal borings in neoselachian teeth from the Lower Oxford Clay of Peterborough. Mercian Geologists 12: 1–4.

Martin, L.L., C.M.R. Friedman, and L.A. Phillips. 2012. Fungal endophytes of the obligate parasitic dwarf mistletoe *Arceuthobium americanum* (Santalaceae) act antagonistically in vitro against the native fungal pathogen *Cladosporium* (Davidiellaceae) of their host. American Journal of Botany 99: 2027–2034.

Martín-González, A., J. Wierzchos, J.C. Gutiérrez, J. Alonso, and C. Ascaso. 2009. Microbial Cretaceous park: Biodiversity of microbial fossils entrapped in amber. Naturwissenschaften 96: 551–564.

Martín-Rodrigues, N., S. Espinel, J. Sanchez-Zabala, A. Ortíz, C. González-Murua, and M.K. Duñabeitia. 2013. Spatial and temporal dynamics of the colonization of *Pinus radiata* by *Fusarium circinatum*, of conidiophora development in the pith and of traumatic resin duct formation. New Phytologist 198: 1215–1227.

Martin-Sanchez, P.M., A. Nováková, F. Bastian, C. Alabouvette, and C. Saiz-Jimenes. 2012. Two new species of the genus *Ochroconis, O. lascauxensis* and *O. anomala* isolated from black stains in Lascaux Cave, France. Fungal Biology 116: 574–589.

Martínez, A. 1968. Microthyriales (Fungi, Ascomycetes) fósiles del Cretácico inferior de la provincia de Santa Cruz, Argentina. Ameghiniana 5: 257–263.

Martínez-Hernández, E., and A.C. Tomasini-Ortiz. 1989. Esporas, hifas y otros restos de hongos fósiles de la cuenca Carbonífera de Fuentes-Río Escondido (Campaniano-Maastrichtiano), Estado de Coahuila. Universidad Nacional Autónoma de México, Instituto de Geología, Revista 8: 235–242.

Marx, D.H. 1972. Ectomycorrhizae as biological deterrents to pathogenic root infections. Annual Review of Phytopathology 10: 429–454.

Marynowski, L., E. Szełęg, M.O. Jędrysek, and B.R.T. Simoneit. 2011. Effects of weathering on organic matter: Part II: Fossil wood weathering and implications for organic geochemical and petrographic studies. Organic Geochemistry 42: 1076–1088.

Marynowski, L., J. Smolarek, A. Bechtel, M. Philippe, S. Kurkiewicz, and B.R.T. Simoneit. 2013. Perylene as an indicator of conifer fossil wood degradation by wood-degrading fungi. Organic Geochemistry 59: 143–151.

Mason, H.L. 1934. Pleistocene flora of the Tomales Formation. Studies of the Pleistocene paleobotany of California, Contributions Paleontology IV, Carnegie Institution of Washington Publication 415: 81–179.

Massicotte, H.B., L.H. Melville, L.E. Tackaberry, and R.L. Peterson. 2008. A comparative study of mycorrhizas in several genera of Pyroleae (Ericaceae) from western Canada. Botany 86: 610–622.

Mata, J.L., K.W. Hughes, and R.H. Petersen. 2004. Phylogenetic placement of *Marasmiellus juniperinus*. Mycoscience 45: 214–221.

Matekwor Ahulu, E., M. Nakata, and M. Nonaka. 2005. *Arum*- and *Paris*-type arbuscular mycorrhizas in a mixed pine forest on sand dune soil in Niigata Prefecture, central Honshu, Japan. Mycorrhiza 15: 129–136.

Matheny, P.B., J.A. Gossmann, P. Zalar, T.K.A. Kumar, and D.S. Hibbett. 2006. Resolving the phylogenetic position of the Wallemiomycetes: An enigmatic major lineage of Basidiomycota. Canadian Journal of Botany 84: 1794–1805.

Matsunaga, K.K.S., R.A. Stockey, and A.M.F. Tomescu. 2013. *Honeggeriella complexa* gen. et sp. nov., a heteromerous lichen from the Lower Cretaceous of Vancouver Island (British Columbia, Canada). American Journal of Botany 100: 450–459.

Matsuura, K, and T. Yashiro. 2010. Parallel evolution of termite-egg mimicry by sclerotium-forming fungi in distant termite groups. Biological Journal of the Linnean Society 100: 531–537.

Matsuura, K., T. Yashiro, K. Shimizu, S. Tatsumi, and T. Tamura. 2009. Cuckoo fungus mimics termite eggs by producing the cellulose-digesting enzyme beta-glucosidase. Current Biology 19: 30–36.

Mayerhofer, M.S., G. Kernaghan, and K.A. Harper. 2013. The effects of fungal root endophytes on plant growth: A meta-analysis. Mycorrhiza 23: 119–128.

McAndrews, J.H., and C.L. Turton. 2010. Fungal spores record Iroquoian and Canadian agriculture in 2nd millennium A.D. sediment of Crawford Lake, Ontario, Canada. Vegetation History and Archaeobotany 19: 495–501.

McCreadie, J.W., C.E. Beard, and P.H. Adler. 2005. Context-dependent symbiosis between black flies (Diptera: Simuliidae) and trichomycete fungi (Harpellales: Legeriomycetaceae). Oikos 108: 362–370.

McGee, P.A., S. Bullock, and B.A. Summerell. 1999. Structure of mycorrhizae of the Wollemi pine (*Wollemia nobilis*) and related Araucariaceae. Australian Journal of Botany 47: 85–95.

McGonigle, T.P. 1995. The significance of grazing on fungi in nutrient cycling. Canadian Journal of Botany 73(Suppl): S1370–S1376.

McIlroy de la Rosa, J.P., P.A. Warke, and B.J. Smith. 2012. Lichen-induced biomodification of calcareous surfaces: Bioprotection versus biodeterioration. Progress in Physical Geography 37: 325–351.

McIlroy de la Rosa, J.P., M. Casares Porcel, and P.A. Warke. 2013. Mapping stone surface temperature fluctuation: Implications for lichen distribution and biomodification on historic stone surfaces. Journal of Cultural Heritage 14: 346–353.

McKellar, R.C., A.P. Wolfe, K. Muehlenbachs, R. Tappert, M.S. Engel, T. Cheng, and G.A. Sánchez-Azofeifa. 2011. Insect outbreaks produce distinctive carbon isotope signatures in defensive resins and fossiliferous ambers. Proceedings of the Royal Society 278B: 3219–3224.

McLaughlin, D.J. 1976. On *Palaeosclerotium* as a link between ascomycetes and basidiomycetes. Science 193: 602.

McLaughlin, D.J., D.S. Hibbett, F. Lutzoni, J.W. Spatafora, and R. Vilgalys. 2009. The search for the fungal tree of life. Trends in Microbiology 17: 488–497.

McLean, R.C. 1912. A group of rhizopods from the Carboniferous Period. Proceedings of the Cambridge Philosophical Society 16: 493–513.

McLean, R.C. 1922. On the fossil genus *Sporocarpon*. Annals of Botany 36: 71–90.

McLoughlin, N., M.D. Brasier, D. Wacey, O.R. Green, and R.S. Perry. 2007. On biogenicity criteria for endolithic microborings on Early Earth and beyond. Astrobiology 7: 10–26.

McLoughlin, N., H. Staudigel, H. Furnes, B. Eickmann, and M. Ivarsson. 2010. Mechanisms of microtunneling in rock substrates: Distinguishing endolithic biosignatures from abiotic microtunnels. Geobiology 8: 245–255.

McMahon, T.A., L.A. Brannelly, M.W.H. Chatfield, P.T.J. Johnson, M.B. Joseph, V.J. McKenzie, C.L. Richards-Zawacki, M.D. Venesky, and J.R. Rohr. 2013. Chytrid fungus *Batrachochytrium dendrobatidis* has nonamphibian hosts and releases chemicals that cause pathology in the absence of infection. Proceedings of the National Academy of Sciences, USA 110: 210–215.

McMaster, T.J. 2012. Atomic force microscopy of the fungi-mineral interface: Applications in mineral dissolution, weathering and biogeochemistry. Current Opinion in Biotechnology 23: 562–569.

McNamara, C.J., and R. Mitchell. 2005. Microbial deterioration of historic stone. Frontiers in Ecology and the Environment 3: 445–451.

McNamara, M.E., D.E.G. Briggs, and P.J. Orr. 2012. The controls on the preservation of structural color in fossil insects. PALAIOS 27: 443–454.

McNeill, J., F.R. Barrie, W.R. Buck, V. Demoulin, W. Greauter, D.L. Hawksworth, P.S. Herendeen, S. Knapp, K. Marhold, J. Prado, W.F. Prud'homme van Reine, G.F. Smith, J.H. Wiersema and N.J. Turland (eds). 2012. International Code of Nomenclature for Algae, Fungi, and Plants (Melbourne Code) adopted by the 18th International Botanical Congress Melbourne, Australia, July 2011. Retrieved from http://www.iapt-taxon.org/nomen/main.php.

Mechaber, W.L, D.B. Marshall, R.A. Mechaber, R.T. Jobe, and F.S. Chew. 1996. Mapping leaf surface landscapes. Proceedings of the National Academy of Sciences, USA 93: 4600–4603.

Medeanic, S., and M.B. Silva. 2010. Indicative value of non-pollen palynomorphs (NPPs) and palynofacies for palaeoreconstructions: Holocene peat, Brazil. International Journal of Coal Geology 84: 248–257.

Mehdiabadi, N.J., and T.R. Schultz. 2010. Natural history and phylogeny of the fungus-farming ants (Hymenoptera: Formicidae: Myrmicinae: Attini). Myrmecological News 13: 37–55.

Mehdiabadi, N.J., U.G. Mueller, S.G. Brady, A.G. Himler, and T.R. Schultz. 2012. Symbiont fidelity and the origin of species in fungus-growing ants. Nature Communications 3: 1–7.

Mendelson, C.V., and J.W. Schopf. 1992a. Proterozoic and selected Early Cambrian acritarchs. In J.W. Schopf, and C. Klein (Eds), The Proterozoic Biosphere, pp. 219–232. Cambridge University Press, Cambridge, NY.

Mendelson, C.V., and J.W. Schopf. 1992b. Proterozoic and selected Early Cambrian microfossils and microfossil-like objects. In J.W. Schopf, and C. Klein (Eds), The Proterozoic Biosphere, pp. 865–951. Cambridge University Press, Cambridge, NY.

Mendoza, L., J.W. Taylor, and L. Ajello. 2002. The class Mesomycetozoea: A heterogeneous group of microorganisms at the animal-fungal boundary. Annual Review of Microbiology 56: 315–344.

Meng, S., T. Torto-Alalibo, M.C. Chibucos, B.M. Tyler, and R.A. Dean. 2009. Common processes in pathogenesis by fungal and oomycete plant pathogens, described with Gene Ontology terms. BMC Microbiology 9: S7.

Menge, A. 1858. Beitrag zur Bernsteinflora. Neueste Schriften der Naturforschenden Gesellschaft in Danzig 6: 1–18.

Menkis, A., J. Allmer, R. Vasiliauskas, V. Lygis, J. Stenlid, and R. Finlay. 2004. Ecology and molecular characterization of dark septate fungi from roots, living stems, coarse and fine woody debris. Mycological Research 108: 965–973.

Merckx, V., M.I. Bidartondo, and N.A. Hynson. 2009. Mycoheterotrophy: When fungi host plants. Annals of Botany 104: 1255–1261.

Merckx, V.S.F.T., S.B. Janssens, N.A. Hynson, C.D. Specht, T.D. Bruns, and E.F. Smets. 2012. Mycoheterotrophic interactions are not limited to a narrow phylogenetic range of arbuscular mycorrhizal fungi. Molecular Ecology 21: 1524–1532.

Merckx, V.S.F.T., J.V. Freudenstein, J. Kissling, M.J.M. Christenhusz, R.E. Stotler, B. Crandall-Stotler, N. Wickett, P.J. Rudall, H. Maas-van de Kamer, and P.J.M. Maas. 2013. Taxonomy and classification. In V.S.F.T. Merckx (Ed.), Mycoheterotrophy: The Biology of Plants Living on Fungi, pp. 19–101. Springer, New York, NY.

Meschinelli, A. 1892. Fungi fossiles. In P.A. Saccardo (Ed.), Sylloge Fungorum Omnium Hucusque Cognitorum, Supplementum Universale, Volume 10, Part 2, pp. 741–808. R. Friedländer and Sohn, Berlin, Germany.

Meschinelli, A. 1898. Fungorum Fossilium Omnium Hucusque Cognitorum Iconographia 31 Tabulis Exornata, Volumen Unicum. Typis Aloysii Fabris, Venice, Italy. 144 pp.

Messuti, M.I., R. Vidal-Russell, G.C. Amico, and L.E. Lorenzo. 2012. *Chaenothecopsis quintralis*, a new species of calicioid fungus. Mycologia 104: 1222–1228.

Meyer, J. 1987. Plant Galls and Gall Inducers. Gebrüder Borntraeger, Berlin, Germany. 291 pp.

Meyer, K.M., and J.H.J. Leveau. 2012. Microbiology of the phyllosphere: A playground for testing ecological concepts. Oecologia 168: 621–629.

Meyer-Berthaud, B., S.E. Scheckler, and J. Wendt. 1999. *Archaeopteris* is the earliest known modern tree. Nature 398: 700–701.

Meyer-Berthaud, B., A. Soria, and A.L. Decombeix. 2010. The land plant cover in the Devonian: A reassessment of the evolution of the tree habit. In M. Vecoli, G. Clément, and B. Meyer-Berthaud (Eds), The Terrestrialization Process: Modelling Complex Interactions at the Biosphere-Geosphere Interface, pp. 59–70. The Geological Society, London, UK.

Michaelides, J., and B. Kendrick. 1982. The bubble-trap propagules of *Beverwykella*, *Helicoon* and other aero-aquatic fungi. Mycotaxon 14: 247–260.

Mikheyev, A.S., U.G. Mueller, and P. Abbot. 2006. Cryptic sex and many-to-one coevolution in the fungus-growing ant symbiosis. Proceedings of the National Academy of Sciences, USA 103: 10702–10706.

Mikola, P. 1988. Ectendomycorrhiza of conifers. Silva Fennica 22: 19–27.

Milanello do Amaral, M., and G. Ceccantini. 2011. The endoparasite *Pilostyles ulei* (Apodanthaceae-Cucurbitales) influences wood structure in three host species of *Mimosa*. IAWA Journal 32: 1–13.

Millay, M.A., and T.N. Taylor. 1978. Chytrid-like fossils of Pennsylvanian age. Science 200: 1147–1149.

Miller, A.S., and P. Jeffries. 1994. Ultrastructural observations and a computer model of the helicoidal appearance of the spore wall of *Glomus geosporum*. Mycological Research 98: 307–321.

Miller, M.B., and B.L. Bassler. 2001. Quorum sensing in bacteria. Annual Review of Microbiology 55: 165–199.

Mindell, R.A., R.A. Stockey, G. Beard, and R.S. Currah. 2007. *Margaretbarromyces dictyosporus* gen. sp. nov.: A permineralized corticolous ascomycete from the Eocene of Vancouver Island. British Columbia. British Mycological Society 111: 680–684.

Minerdi, D., V. Bianciotto, and P. Bonfante. 2002. Endosymbiotic bacteria in mycorrhizal fungi: From their morphology to genomic sequences. Plant and Soil 244: 211–219.

Misra, J.K. 2012. Systematics of *Stachylina* and *Smittium* – the two largest genera of Harpellales, Zygomycota. In J.K. Misra, J.P. Tiwari, and S.K. Deshmukh (Eds), Systematics and Evolution of Fungi, pp. 115–158. Science Publishers, Enfield, NH.

Misra, J.K., and R.W. Lichtwardt. 2000. Illustrated Genera of Trichomycetes: Fungal Symbionts of Insects and Other Arthropods. Science Publishers, Enfield, NH. 165 pp.

Mitchell, R.and C.J. McNamara (Eds). 2010. Cultural Heritage Microbiology. Fundamental Studies in Conservation Science. ASM Press, Washington, DC. 326 pp.

Mitton, J.B., and S.M. Ferrenberg. 2012. Mountain pine beetle develops an unprecedented summer generation in response to climate warming. American Naturalist 179: E163–E171.

Mizubuti, E.S.G., and W.E. Fry. 2006. Potato late blight. In B.M. Cooke, D.G. Jones, and B. Kaye (Eds), The Epidemiology of Plant Diseases, pp. 445–471 (Second Edition). Springer, Dordrecht, The Netherlands.

Moczydłowska, M., E. Landing, W. Zang, and T. Palacios. 2011. Proterozoic phytoplankton and timing of chlorophyte algae origins. Palaeontology 54: 721–733.

Molina, R., and J.M. Trappe. 1994. Biology of the ectomycorrhizal genus, *Rhizopogon*. I. Host associations, host-specificity and pure culture syntheses. New Phytologist 126: 653–675.

Molnár, K., and E. Farkas. 2010. Current results on biological activities of lichen secondary metabolites: A review. Zeitschrift für Naturforschung 65C: 157–173.

Moncalvo, J.-M. 2000. Systematics of *Ganoderma*. In J. Flood, P.D. Bridge, and M. Holderness (Eds), *Ganoderma* Diseases of Perennial Crops, pp. 23–45. CABI Publishing, Oxfordshire, UK.

Moncalvo, J.-M., and L. Ryvarden. 1997. A Nomenclatural Study of the Ganodermataceae Donk (Synopsis Fungorum 11). Benjamin Sats og Trykk, Oslo, Norway. 114 pp.

Mondo, S.J., K.H. Toomer, J.B. Morton, Y. Lekberg, and T.E. Pawlowska. 2012. Evolutionary stability in a 400-million-year-old heritable facultative mutualism. Evolution 66: 2564–2576.

Money, N.P. 1998. More g's than the space shuttle: Ballistospore discharge. Mycologia 90: 547–558.

Money, N.P. 2007. The Triumph of the Fungi: A Rotten History. Oxford University Press, New York. 216 pp.

Mongkolsamrit, S., N. Kobmoo, K. Tasanathai, A. Khonsanit, W. Noisripoom, P. Srikitikulchai, R. Somnuk, and J.J.

Luangsa-ard. 2012. Life cycle, host range and temporal variation of *Ophiocordyceps unilateralis/Hirsutella formicarum* on Formicine ants. Journal of Invertebrate Pathology 111: 217–224.

Monthoux, O., and K. Lundstrom-Baudais. 1979. Polyporacées des sites néolithiques de Clairvaux et Charavines (France). Candollea 34: 153–166.

Moore, D. 2013. Fungal Biology in the Origin and Emergence of Life. Cambridge University Press, New York, NY, 236 pp.

Moore, J. 2002. Parasites and the Behavior of Animals. Oxford University Press, New York, NY, 338 pp.

Moore, L.R. 1963a. Microbiological colonization and attack on some Carboniferous miospores. Palaeontology 6: 349–372.

Moore, L.R. 1963b. On some micro-organisms associated with the scorpion *Gigantoscorpio willsi* Størmer. Skrifter utgitt av det Norske Videnskaps-Akademi i Oslo, Matematisk-Naturvidenskapelig Klasse 9: 4–14.

Moore, L.R. 1964. The microbiology, mineralogy and genesis of a tonstein. Proceedings of the Yorkshire Geological Society 34: 235–292.

Moore, P.D. 1996. Invertebrates and mycophagy. Nature 381: 372–373.

Moore, T.A., and J.C. Shearer. 2003. Peat/coal type and depositional environment–are they related? International Journal of Coal Geology 56: 233–252.

Moore, T.A., J.C. Shearer, and S.L. Miller. 1996. Fungal origin of oxidised plant material in the Palangkaraya peat deposit, Kalimantan Tengah, Indonesia: Implications for 'inertinite' formation in coal. International Journal of Coal Geology 30: 1–23.

Moore-Landecker, E. 1982. Fundamentals of The Fungi (Second Edition). Prentice Hall, Upper Saddle River, New Jersey, 550 pp.

Moran, K., H.L. Hilbert-Wolf, K. Golder, H.F. Malenda, C.J. Smith, L.P. Storm, E.L. Simpson, M.C. Wizevich, and S.E. Tindall. 2010. Attributes of the wood-boring trace fossil *Asthenopodichnium* in the Late Cretaceous Wahweap Formation, Utah, USA. Palaeogeography, Palaeoclimatology, Palaeoecology 297: 662–669.

Morgenstern, I., S. Klopman, and D.S. Hibbett. 2008. Molecular evolution and diversity of lignin degrading heme peroxidases in the Agaricomycetes. Journal of Molecular Evolution 66: 243–257.

Morris, J.L., V.P. Wright, and D. Edwards. 2011. Siluro-Devonian landscapes of southern Britain: The stability and nature of early vascular plant habitats. Journal of the Geological Society, London 169: 173–190.

Morrison, B.T., and D.A. English. 1967. The significance of mycorrhizal nodules of *Agathis australis*. New Phytologist 66: 245–250.

Morton, J.B. 1985. Variation in mycorrhizal and spore morphology of *Glomus occultum* and *Glomus diaphanum* as influenced by plant host and soil environment. Mycologia 77: 192–204.

Morton, J.B. 1988. Taxonomy of VA mycorrhizal fungi: Classification, nomenclature, and identification. Mycotaxon 32: 267–324.

Morton, J.B. 1990a. Evolutionary relationships among arbuscular mycorrhizal fungi in the Endogonaceae. Mycologia 82: 192–207.

Morton, J.B. 1990b. Species and clones of arbuscular mycorrhizal fungi (Glomales, Zygomycetes): Their role in macro- and microevolutionary processes. Mycotaxon 37: 493–515.

Morton, J.B., and G.L. Benny. 1990. Revised classification of arbuscular mycorrhizal fungi (Zygomycetes): A new order, Glomales, two new suborders, Glomineae and Gigasporineae, and two new families, Acaulosporaceae and Gigasporaceae, with an emendation of Glomaceae. Mycotaxon 37: 471–491.

Morton, J.B., and S.P. Bentivenga. 1994. Levels of diversity in endomycorrhizal fungi (Glomales, Zygomycetes) and their role in defining taxonomic and non-taxonomic groups. Plant and Soil 159: 47–59.

Moskal-del Hoyo, M., M. Wachowiak, and R.A. Blanchette. 2010. Preservation of fungi in archaeological charcoal. Journal of Archaeological Science 37: 2106–2116.

Mosse, B., and G.D. Bowen. 1968. A key to the recognition of some *Endogone* spore types. Transactions of the British Mycological Society 51: 469–483.

Mossman, D.J., W.E.L. Minter, A. Dutkiewicz, D.K. Hallbauer, S.C. George, Q. Hennigh, T.O. Reimer, and F.D. Horscroft. 2008. The indigenous origin of Witwatersrand "carbon". Precambrian Research 164: 173–186.

Motomura, H., M.-A. Selosse, F. Martos, A. Kagawa, and T. Yukawa. 2010. Mycoheterotrophy evolved from mixotrophic ancestors: Evidence in *Cymbidium* (Orchidaceae). Annals of Botany 106: 573–581.

Mountfort, D.O., and C.G. Orpin (Eds). 1994. Anaerobic Fungi: Biology, Ecology, and Function. Marcel Dekker Inc., New York, NY. 304 pp.

Mudie, P.J., A. Rochon, A.E. Aksu, and H. Gillespie. 2002. Dinoflagellate cysts, freshwater algae and fungal spores as salinity indicators in Late Quaternary cores from Marmara and Black seas. Marine Geology 190: 203–231.

Mueller, G.M., G.F. Bills, and M.S. Foster (Eds). 2004. Biodiversity of Fungi: Inventory and Monitoring Methods. Elsevier Academic Press, Burlington, MA. 777 pp.

Mueller, U.G., and N. Gerardo. 2002. Fungus-farming insects: Multiple origins and diverse evolutionary histories. Proceedings of the National Academy of Sciences, USA 99: 15247–15249.

Mueller, U.G., N.M. Gerardo, D.K. Aanen, D.L. Six, and T.R. Schultz. 2005. The evolution of agriculture in insects. Annual Review of Ecology, Evolution, and Systematics 36: 563–595.

Mugambi, G.K., and S.M. Huhndorf. 2009. Molecular phylogenetics of Pleosporales: Melanommataceae and Lophiostomataceae re-circumscribed (Pleosporomycetidae, Dothideomycetes, Ascomycota). Studies in Mycology 64: 103–121.

Mukherjee, D. 2012. Facultative fungal remains from Miocene lignite coal of Neyveli Tamilnadu India. International. Journal of Geology, Earth and Environmental Sciences 2: 1–15.

Müller, E., and W. Löffler. 1992. Mykologie. Thieme, Stuttgart, Germany. 367 pp.

Müller, M.M., M. Varama, J. Heinonen, and A.M. Hallaksela. 2002. Influence of insects on the diversity of fungi in decaying spruce wood in managed and natural forests. Forest Ecology and Management 166: 165–181.

Muñoz, Z., A. Moret, and S. Garcés. 2008. The use of *Verticillium dahliae* and *Diplodia scrobiculata* to induce resistance in *Pinus halepensis* against *Diplodia pinea* infection. European Journal of Plant Pathology 120: 331–337.

Murray, G. 1877. A fossil *Peronospora* of the Palaeozoic Age. The Academy, Nov. 17 1877: 475.

Murray, P. 1831. Account of the *Arbusculites argentea*, from the Carboniferous limestone of Innerteil, near to Kirkcaldy, in Fifeshire. Edinburgh new Philosophical Journal 11: 147–149.

Musotto, L.L., M.V. Bianchinotti, and A.M. Borromei. 2012. Pollen and fungal remains as environmental indicators in surface sediments of Isla Grande de Tierra del Fuego, southernmost Patagonia. Palynology 36: 162–179.

Muszyńska, B., K. Sulkowska-Ziaja, and A. Wójcik. 2013. Levels of physiologically active indole derivatives in the fruiting bodies of some edible mushrooms (Basidiomycota) before and after thermal processing. Mycoscience 54: 312–326.

Muthukumar, T., and K. Prabha. 2012. Fungal asociations in gametophytes and young sporophytic roots of the fern *Nephrolepis exaltata*. Acta Botanica Croatica 71: 139–146.

Muthukumar, T., and V. Tamilselvi. 2010. Occurrence and morphology of endorhizal fungi in crop species. Tropical and Subtropical Agroecosystems 12: 593–604.

Muthukumar, T., and K. Udaiyan. 2002. Arbuscular mycorrhizas in cycads of southern India. Mycorrhiza 12: 213–217.

Nadon, G.C. 1998. Magnitude and timing of peat-to-coal compaction. Geology 26: 727–730.

Nagahama, T., H. Sato, M. Shimazu, and J. Sugiyama. 1995. Phylogenetic divergence of the entomophthoralean fungi: Evidence from nuclear 18S ribosomal RNA gene sequences. Mycologia 87: 203–209.

Nagano, Y., and T. Nagahama. 2012. Fungal diversity in deep-sea extreme environments. Fungal Ecology 5: 463–471.

Nagovitsin, K. 2009. *Tappania*-bearing association of the Siberian platform: Biodiversity, stratigraphic position and geochronological constraints. Precambrian Research 173: 137–145.

Nagpal, R., A.K. Puniya, G.W. Griffith, G. Goel, M. Puniya, J.P. Seghal, and K. Singh. 2009. Anaerobic rumen fungi: Potential and applications. In G.G. Khachatourians, D.K. Arora, T.P. Rajendran, and A.K. Srivastava (Eds), Agriculturally Important Microorganisms, Volume II, pp. 375–393. Academic World International, Dubai, UAE.

Nandi, B., S. Banerjee, and A. Sinha. 2003. Fossil Xylariaceae spores from the Cretaceous and Tertiary sediments of northeastern India. Acta Palaeontologica Sinica 42: 56–67.

Nara, K. 2006. Ectomycorrhizal networks and seedling establishment during early primary succession. New Phytologist 169: 169–178.

Nascimbene, J., P.L. Nimis, and H. Thüs. 2013. Lichens as bioindicators in freshwater ecosystems – challenges and perspectives. Annali di Botanica 3: 45–50.

Nasdala, L., O. Beyssac, J.W. Schopf, and B. Bleisteiner. 2012. Application of Raman-based images in the Earth sciences. In A. Zoubir (Ed.), Raman Imaging: Techniques and Applications. Springer Series in Optical Sciences, 168 pp. 145–187. Springer-Verlag, Berlin, Germany.

Nathorst, A.G. 1879. Om floran i Skånes kolförande bildningar. I. Floran vid Bjuf (andra häftet). Sveriges Geologiska Undersökning C33: 55–82.

Naumann, M., A. Schüßler, and P. Bonfante. 2010. The obligate endobacteria of arbuscular mycorrhizal fungi are ancient heritable components related to the *Mollicutes*. ISME Journal 4: 862–871.

Near, T.J., and M.J. Sanderson. 2004. Assessing the quality of molecular divergence time estimates by fossil calibrations and fossil-based model selection. Philosophical Transactions of the Royal Society 359B: 1477–1483.

Near, T.J., D.I. Bolnick, and P.C. Wainwright. 2005. Fossil calibrations and molecular divergence time estimates in centrarchid fishes (Teleostei: Centrarchidae). Evolution 59: 1768–1782.

Nease, F.R., and F.A. Wolf. 1975. Fossilized fungi as evidence of continental drift. Mycopathologia 57: 15–18.

Nel, A., and J. Prokop. 2006. New fossil gall midges from the earliest Eocene French amber (Insecta, Diptera, Cecidomyiidae). Geodiversitas 28: 37–54.

Nel, P., A.R. Schmidt, C. Bässler, and A. Nel. 2013. Fossil thrips of the family Uzelothripidae suggest 53 million years of morphological and ecological stability. Acta Palaeontologica Polonica 58: 609–614.

Nelder, M.P., C.E. Beard, P.H. Adler, S.-K. Kim, and J.W. McCreadie. 2006. Harpellales (Zygomycota: Trichomycetes) associated with black flies (Diptera: Simuliidae): World review and synthesis of their ecology and taxonomy. Fungal Diversity 22: 121–169.

Němejc, F. 1959. Paleobotanika I. Nakladatelství Československé akademie věd Praha, Czech Republic. 402 pp.

Neuendorf, K.K.E., J.P. Mehl, Jr., and J.A. Jackson. 2005. Glossary of Geology (Fifth Edition). American Geological Institute, Alexandria, VA. 779 pp.

Neuy-Stolz, G. 1958. Zur Flora der Niederrheinischen Bucht während der Hauptflözbildung unter besonderer Berücksichtigung der Pollen und Pilzreste in den hellen Schichten. Fortschritte in der Geologie von Rheinland und Westfalen 2: 503–525.

Newsham, K.K., A.H. Fitter, and A.R. Watkinson. 1995. Arbuscular mycorrhiza protect an annual grass from root pathogenic fungi in the field. Journal of Ecology 83: 991–1000.

Nguyen, Tu, T.T., S. Derenne, C. Largeau, A. Mariotti, H. Bocherens, and D. Pons. 2000. Effects of fungal infection on lipid extract composition of higher plant remains: Comparison of shoots of a Cenomanian conifer, uninfected and infected by extinct fungi. Organic Geochemistry 31: 1743–1754.

Nichols, D.J., and K.R. Johnson. 2008. Plants at the K-T Boundary. Cambridge University Press, Cambridge, MA. 292 pp.

Nicolson, T.H. 1967. Vesicular-arbuscular mycorrhiza – A universal plant symbiosis. Science Progress, Oxford 55: 561–581.

Nielsen, K.S.S., J.K. Nielsen, and R.G. Bromley. 2003. Palaeoecological and ichnological significance of microborings in quaternary foraminifera. Palaeontologia Electronica 6: 1–13. http://palaeo-electronica.org/paleo/2003_1/ichno/issue1_03.htm.

Niinomi, S., S. Takamatsu, and M. Havrylenko. 2008. Molecular data do not support a southern hemisphere base of *Nothofagus* powdery mildews. Mycologia 100: 716–726.

Nikel, S. 2011. Baumpilze in Kieselhölzern aus dem basalen Schilfsandstein des württembergischen Keupers (kmSt), Trias. Jahreshefte der Gesellschaft für Naturkunde in Württemberg 167: 111–162.

Niklas, K.J. 1976. Chemotaxonomy of *Prototaxites* and evidence for possible terrestrial adaptation. Review of Palaeobotany and Palynology 22: 1–17.

Niklas, K.J., and U. Kutschera. 2010. The evolution of the land plant life cycle. New Phytologist 185: 27–41.

Niklas, K.J., and V. Smocovitis. 1983. Evidence for a conducting strand in early Silurian (Llandoverian) plants: implications for the evolution of the land plants. Paleobiology 9: 126–137.

Nishida, H., and J. Sugiyama. 1993. Phylogenetic relationships among *Taphrina*, *Saitoella* and other higher fungi. Molecular Biology and Evolution 10: 431–436.

Nishida, H., and J. Sugiyama. 1994. Archiascomycetes: Detection of a major new lineage within the Ascomycota. Mycoscience 35: 361–366.

Nishida, M., T. Ohsawa, and H. Nishida. 1990. Anatomy and affinities of the petrified plants from the Tertiary of Chile (VI). Botanical Magazine 103: 255–268.

Nissenbaum, A., and A. Horowitz. 1992. The Levantine amber belt. Journal of African Earth Sciences 14: 295–300.

Nobre, T., N.A. Koné, S. Konaté, K.E. Linsenmair, and D.K. Aanen. 2011. Dating the fungus-farming termites' mutualism shows a mixture between ancient codiversification and recent symbiont dispersal across divergent hosts. Molecular Ecology 20: 2619–2627.

Norris, G. 1986. Systematic and stratigraphic palynology of Eocene to Pliocene strata in the Imperial Nuktak C-22 well, Mackenzie Delta Region, District of Mackenzie. NWT Bulletin of the Geological Survey of Canada 340: 1–89.

Norris, G. 1997. Paleocene-Pliocene deltaic to inner shelf palynostratigraphic zonation, depositional environments and paleoclimates in the Imperial Adgo F-28 well, Beaufort-Mackenzie Basin. Bulletin of the Geological Survey of Canada 523: 1–71.

Nurfadilah, S., N.D. Swarts, K.W. Dixon, H. Lambers, and D.J. Merritt. 2013. Variation in nutrient-acquisition patterns by mycorrhizal fungi of rare and common orchids explains diversification in a global biodiversity hotspot. Annals of Botany 111: 1233–1241.

Oberwinkler, F. 1970. Die Gattungen der Basidiolichenen. Vorträge aus dem Gesamtgebiet der Botanik, Neue Folge 4: 139–169.

Oberwinkler, F. 2012. Evolutionary trends in Basidiomycota. Stapfia 96: 45–104.

O'Dell, T.E., H.B. Massicotte, and J.M. Trappe. 1993. Root Colonization of *Lupinus latifolius* Agardh. and *Pinus contorta* Dougl. by *Phialocephala fortinii* Wang & Wilcox. New Phytologist 124: 93–100.

O'Donnell, K.L., J.J. Ellis, C.W. Hesseltine, and G.R. Hooper. 1977. Azygosporogenesis in *Mucor azygosporus*. Canadian Journal of Botany 55: 2712–2720.

Oehl, F., G.A. da Silva, B.T. Goto, and E. Sieverding. 2011a. Glomeromycota: Three new genera and glomoid species reorganized. Mycotaxon 116: 75–120.

Oehl, F., E. Sieverding, J. Palenzuela, K. Ineichen, and G. Alves da Silva. 2011b. Advances in Glomeromycota taxonomy and classification. IMA Fungus 2: 191–199.

Oehl, F., C. Castillo, D. Schneider, V. Säle, and E. Sieverding. 2012. *Ambispora reticulata*, a new species in the Glomeromycota from mountainous areas in Switzerland and Chile. Journal of Applied Botany and Food Quality 85: 129–133.

Ohm, R.A., N. Feau, B. Henrissat, C.L. Schoch, B.A. Horwitz, K.W. Barry, B.J. Condon, A.C. Copeland, B. Dhillon, F. Glaser, C.N. Hesse, I. Kosti, K. LaButti, E.A. Lindquist, S. Lucas, A.A. Salamov, R.E. Bradshaw, L. Ciuffetti, R.C. Hamelin, G.H.J. Kema, C. Lawrence, J.A. Scott, J.W. Spatafora, B.G. Turgeon, P.J.G.M. de Wit, S. Zhong, S.B. Goodwin, and I.V. Grigoriev. 2012. Diverse lifestyles and strategies of plant pathogenesis encoded in the genomes of eighteen Dothideomycetes fungi. PLoS Pathogens 8: e1003037.

O'Keefe, J., J.C. Hower, R. Hatch, R.H. Bartley, and S.E. Bartley. 2011a. Fungal forms in Miocene Eel River coals: Correlating between reflected light petrography and palynology. Geological Society of America Abstracts with Programs 43: 501.

O'Keefe, J.M.K., J.C. Hower, R.F. Finkelman, J.W. Drew, and J.D. Stucker. 2011b. Petrographic, geochemical, and mycological aspects of Miocene coals from the Nováky and Handlová mining districts, Slovakia. International Journal of Coal Geology 87: 268–281.

Olempska, E. 1986. Endolithic microorganisms in Ordovician ostracod valves. Acta Palaeontologica Polonica 31: 229–236.

Oliver, F.W. 1903. Notes on fossil fungi. New Phytologist 2: 49–53.

Olivier, D.L. 1978. *Retiarius* gen. nov.: *Phyllosphere* fungi which capture wind-borne pollen grains. Transactions of the British Mycological Society 71: 193–201.

Olsen, P.E., and D. Baird. 1990. The ichnogenus *Atreipus* and its significance for Triassic biostratigraphy. In K. Padian (Ed.), The Beginning of the Age of Dinosaurs. Faunal change across the Triassic-Jurassic Boundary, pp. 61–88. Cambridge University Press, New York, NY.

Omacini, M., E.J. Chaneton, C.M. Ghersa, and C.B. Müller. 2001. Symbiotic fungal endophytes control insect host-parasite interaction webs. Nature 409: 78–81.

Omacini, M., T. Eggers, M. Bonkowski, A.C. Gange, and T.H. Jones. 2006. Leaf endophytes affect mycorrhizal status and growth of co-infected and neighbouring plants. Functional Ecology 20: 226–232.

Ono, Y., and J.F. Hennen. 1983. Taxonomy of the Chaconiaceous genera (Uredinales). Transactions of the Mycological Society of Japan 24: 369–402.

Onofri, S., L. Zucconi, L. Selbmann, G.S. de Hoog, A. de los Rios, R. Ruisi, and M. Grube. 2007. Fungal associations at the cold edge of life. In J. Seckbach (Ed.), Algae and Cyanobacteria in Extreme Environments. Cellular Origin, Life in Extreme Habitats and Astrobiology Series, Volume 11, pp. 739–757. Springer, Dordrecht, The Netherlands.

Öpik, M., M. Moora, M. Zobel, Ü. Saks, R. Wheatley, F. Wright, and T. Daniell. 2008. High diversity of arbuscular mycorrhizal fungi in a boreal herb-rich coniferous forest. New Phytologist 179: 867–876.

Orellana, R., J.J. Leavitt, L.R. Comolli, R. Csencsits, N. Janot, K.A. Flanagan, A.S. Gray, C. Leang, M. Izallalen, T. Mester, and D.R. Lovley. 2013. U(VI) reduction by diverse outer surface c-type cytochromes of *Geobacter sulfurreducens*. Applied and Environmental Microbiology 79: 6369–6374.

Osborn, J.M., T.N. Taylor, and J.F. White, Jr. 1989. *Palaeofibulus* gen. nov., a clamp-bearing fungus from the Triassic of Antarctica. Mycologia 81: 622–626.

Osborn, T.G.B. 1909. The lateral roots of *Amyelon radicans*, Will., and their mycorhiza. Annals of Botany 23: 603–611.

Osono, T. 2006. Role of phyllosphere fungi of forest trees in the development of decomposer fungal communities and decomposition processes of leaf litter. Canadian Journal of Microbiology 52: 701–716.

Ostonen, I., and K. Lõhmus. 2003. Proportion of fungal mantle, cortex and stele of ectomycorrhizas in *Picea abies* (L.) Karst. in different soils and site conditions. Plant and Soil 257: 435–442.

Otjen, L., and R.A. Blanchette. 1984. *Xylobolus frustulatus* decay of oak: Patterns of selective delignification and subsequent cellulose removal. Applied and Environmental Microbiology 47: 670–676.

Otrosina, W.J., and M. Garbelotto. 2010. *Heterobasidion occidentale* sp. nov. and *Heterobasidion irregulare* nom. nov.: A disposition of North American Heterobasidion biological species. Fungal Biology 114: 16–25.

Padamsee, M., T.K.A. Kumar, R. Riley, M. Binder, A. Boyd, A.M. Calvo, K. Furukawa, C. Hesse, S. Hohmann, T.Y. James, K. LaButti, A. Lapidus, E. Lindquist, S. Lucas, K. Miller, S. Shantappa, I.V. Grigoriev, D.S. Hibbett, D.J. McLaughlin, J.W. Spatafora, and M.C. Aime. 2012. The genome of the xerotolerant mold *Wallemia sebi* reveals adaptations to osmotic stress and suggests cryptic sexual reproduction. Fungal Genetics and Biology 49: 217–226.

Padovan, A.C.B., G.F.O. Sanson, A. Brunstein, and M.R.S. Briones. 2005. Fungi evolution revisited: Application of the penalized likelihood method to a Bayesian fungal phylogeny provides a new perspective on phylogenetic relationships and divergence dates of Ascomycota groups. Journal of Molecular Evolution 60: 726–735.

Pain, A., and C. Hertz-Fowler. 2008. Genomic adaptation: A fungal perspective. Nature Reviews Microbiology 6: 572–573.

Palacios-Chávez, R. 1983. The fungal spores of Lower and Middle Miocene of north of Chiapas (Mexico). Boletín de la Sociedad Mexicana de Micología 17: 33–42.

Pampaloni, L. 1902. I resti organici nel Disodile di Melilli in Sicilia. Palaeontographia Italica 8: 121–130.

Paracer, S., and V. Ahmadjian. 2000. Symbiosis: An Introduction to Biological Associations (Second Edition). Oxford University Press, New York, NY, 304 pp.

Paradkar, S.A. 1974. Pollen and fungal spore association on a fossil leaf from the Deccan Intertrappean beds of India. Journal of Palynology 10: 119–125.

Parbery, D.G., and R.F.N. Langdon. 1963. Studies on graminicolous species of *Phyllachora* Fckl. III. The relationship of certain scolecospores to species of *Phyllachora*. Australian Journal of Botany 11: 141–151.

Parbery, I.H., and J.F. Brown. 1986. Sooty moulds and black mildews in extra-tropical rainforests. In N.J. Fokkema, and J. van der Heuvel (Eds), Microbiology of the Phyllosphere, pp. 101–120. Cambridge University Press, New York, NY.

Pareek, H.S. 1966. Fusinized resins in Gondwana (Permian) coals of India. Economic Geology 61: 137–146.

Parfrey, L.W., D.J.G. Lahr, A.H. Knoll, and L.A. Katz. 2011. Estimating the timing of early eukaryotic diversification with multigene molecular clocks. Proceedings of the National Academy of Sciences, USA 108: 13624–13629.

Parham, J.F., P.C.J. Donoghue, C.J. Bell, T.D. Calway, J.J. Head, P.A. Holroyd, J.G. Inoue, R.B. Irmis, W.G. Joyce, D.T. Ksepka, J.S.L. Patané, N.D. Smith, J.E. Tarver, M. van Tuinen, Z. Yang, K.D. Angielczyk, J.M. Greenwood, C.A. Hipsley, L. Jacobs, P.J. Makovicky, J. Müller, K.T. Smith, J.M. Theodor, R.C.M. Warnock, and M.J. Benton. 2012. Best practices for justifying fossil calibrations. Systematic Biology 61: 346–359.

Paris, F., and J. Verniers. 2005. Chitinozoa. In R.C. Selley, L.R.M. Cocks, and I.R. Plimer (Eds), Encyclopedia of Geology, Volume 3, pp. 428–440. Elsevier Academic Press, Amsterdam, The Netherlands.

Parmasto, E., I. Parmasto, and T. Möls. 1987. Variation of Basidiospores in the Hymenomycetes and Its Significance to Their Taxonomy. Bibliotheca Mycologia Series, Volume 115 Gebrüder Borntraeger Verlag, Stuttgart, Germany. 168 pp.

Parmeter, J.R. Jr. 1970. *Rhizoctonia solani*, Biology and Pathology. University of California Press, Berkeley, CA. 255 pp.

Parniske, M. 2008. Arbuscular mycorrhiza: The mother of plant root endosymbioses. Nature Reviews Microbiology 6: 763–775.

Parry, S.F., S.R. Noble, Q.G. Crowley, and C.H. Wellman. 2011. A high-precision U-Pb age constraint on the Rhynie chert Konservat-Lagerstätte: Time scale and other implications. Journal of the Geological Society of London 168: 863–872.

Parry, S.F., S.R. Noble, Q.G. Crowley, and C.H. Wellman. 2013. Reply to Discussion on 'A high-precision U–Pb age constraint on the Rhynie Chert Konservat-Lagerstätte: Time scale and other implications' *Journal*, Vol. 168, 863–872. Journal of the Geological Society of London 170: 703–706.

Parsons, M.G., and G. Norris. 1999. Paleogene fungi from the Caribou Hills, Mackenzie Delta, northern Canada. Palaeontographica 250B: 77–167.

Patil, G.V., and R.B. Singh. 1974. An infected stem from the Deccan Intertrappean beds of Mohgaonkalan. Botanique 5: 141–145.

Patil, R.S., and C.G.K. Ramanujam. 1988. Fungal flora of the carbonaceous clays from Tonakkal area, Kerala. Geological Survey of India Special Publication 11: 261–270.

Paul, B., and M.M. Steciow. 2004. *Saprolegnia multispora*, a new oomycete isolated from water samples taken in a river in the Burgundian region of France. FEMS Microbiology Letters 237: 393–398.

Pawlowska, T.E. 2005. Genetic processes in arbuscular mycorrhizal fungi. FEMS Microbiology Letters 251: 185–192.

Pawlowska, T.E. 2007. How the genome is organized in the Glomeromycota. In J. Heitman, J.W. Kronstad, J.W. Taylor, and L.A. Casselton (Eds), Sex in Fungi: Molecular Determination and Evolutionary Implications, pp. 419–430. ASM Press, Washington, DC.

Pawlowska, T.E., and J.W. Taylor. 2004. Organization of genetic variation in individuals of arbuscular mycorrhizal fungi. Nature 427: 733–737.

Peacock, K.A. 2011. Symbiosis in ecology and evolution. In K. deLaplante, B. Brown, and K.A. Peacock (Eds), Handbook of the Philosophy of Science, Volume 11: Philosophy of Ecology, pp. 219–250. Elsevier, Oxford, England.

Pearce, C.A., P. Reddell, and K.D. Hyde. 1999. A revision of *Phyllachora* (Ascomycotina) on hosts in the angiosperm family Asclepiadaceae, including *P. gloriana* sp. nov. on *Tylophora benthamii* from Australia. Fungal Diversity 3: 123–138.

Peckmann, J., W. Bach, K. Behrens, and J. Reitner. 2008. Putative cryptoendolithic life in Devonian pillow basalt, Rheinisches Schiefergebirge, Germany. Geobiology 6: 125–135.

Pedrão Ferreira, E., M. De Araujo Carvalho, and M.C. Viviers. 2005. Palinologia (fungos) da formação Calumbi, Paleoceno da Bacia de Sergipe, Brasil. Arquivos do Museu National, Rio de Janeiro 63: 395–410.

Pegler, D.N. 1996. Hyphal analysis of basidiomata. Mycological Research 100: 129–142.

Pegler, D.N., T. Laessoe, and B.M. Spooner. 1995. British Puffballs, Earthstars and Stinkhorns. Royal Botanic Gardens, Kew, UK, 255 pp.

Peintner, U., R. Pöder, and T. Pümpel. 1998. The iceman's fungi. Mycological Research 102: 1153–1162.

Peláez, F., M.J. Martínez, and A.T. Martínez. 1995. Screening of 68 species of basidiomycetes for enzymes involved in lignin degradation. Mycological Research 99: 37–42.

Peñalver, E., X. Delclòs, and C. Soriano. 2007. A new rich amber outcrop with palaeobiological inclusions in the Lower Cretaceous of Spain. Cretaceous Research 28: 791–802.

Pennisi, E. 2004. The secret life of fungi. Science 304: 1620–1622.

Penny, J.S. 1947. Studies on the conifers of the Magothy flora. American Journal of Botany 34: 281–296.

Pereira-Carvalho, R.C., G. Sepúlveda-Chavera, E.A.S. Armando, C.A. Inácio, and J.C. Dianese. 2009. An overlooked source of fungal diversity: Novel hyphomycete genera on trichomes of cerrado plants. Mycological Research 113: 261–274.

Pérez, F., C. Castillo-Guevara, G. Galindo-Flores, M. Cuautle, and A. Estrada-Torres. 2012. Effect of gut passage by two highland rodents on spore activity and mycorrhiza formation of two species of ectomycorrhizal fungi (*Laccaria trichodermophora* and *Suillus tomentosus*). Botany 90: 1084–1092.

Pérez, L.I., P.E. Gundel, C.M. Ghersa, and M. Omacini. 2013. Family issues: Fungal endophyte protects host grass from the closely related pathogen *Claviceps purpurea*. Fungal Ecology 6: 379–386.

Perkins, R.D., and S.D. Halsey. 1971. Geologic significance of microboring fungi and algae in Carolina shelf sediments. Journal of Sedimentary Research 41: 843–853.

Perrichot, V. 2004. Early Cretaceous amber from south-western France: Insight into the Mesozoic litter fauna. Geologica Acta 2: 9–22.

Perry, R.S., N. Mcloughlin, B.Y. Lynne, M.A. Sephton, J.D. Oliver, C.C. Perry, K. Campbell, M.H. Engel, J.D. Farmer, M.D. Brasier, and J.T. Staley. 2007. Defining biominerals and organominerals: Direct and indirect indicators of life. Sedimentary Geology 201: 157–179.

Persiel, I. 1960. Beschreibung neuer Arten der Gattung *Chytriomyces* und einiger seltener niederer Phycomyceten. Archiv für Mikrobiologie 36: 283–305.

Petersen, A.B., and S. Rosendahl. 2000. Phylogeny of the Peronosporomycetes (Oomycota) based on partial sequences of the large ribosomal subunit (LSU rDNA). Mycological Research 104: 1295–1303.

Petersen, R.H. 1974. The rust fungus life cycle. Botanical Review 40: 453–513.

Peterson, E.B. 2000. An overlooked fossil lichen (Lobariaceae). Lichenologist 32: 298–300.

Peterson, H.E. 1910. An account of Danish freshwater Phycomycetes, with biological and systematical remarks. Annals of Mycology 8: 494–560.

Peterson, K.J., B. Waggoner, and J.W. Hagadorn. 2003. A fungal analog for Newfoundland Ediacaran fossils? Integrative and Comparative Biology 43: 127–136.

Peterson, R.L., and M.L. Farquhar. 1994. Mycorrhizas – integrated development between roots and fungi. Mycologia 86: 311–326.

Peterson, R.L., H.B. Massicotte, and L.H. Melville. 2004. Mycorrhizas: Anatomy and Cell Biology. National Research Council (NRC) Research Press, Ottawa, Ontario, Canada. 173 pp.

Peterson, R.L., C. Wagg, and M. Pautler. 2008. Associations between microfungal endophytes and roots: Do structural features indicate function? Botany 86: 445–456.

Petkovits, T., L.G. Nagy, K. Hoffmann, L. Wagner, I. Nyilasi, T. Griebel, D. Schnabelrauch, H. Vogel, K. Voigt, C. Vágvölgyi, and T. Papp. 2011. Data Partitions, bayesian analysis and phylogeny of the zygomycetous fungal family Mortierellaceae, inferred from nuclear ribosomal DNA sequences. PLoS ONE 6: e27507.

Petrak, F. 1950. Über *Loranthomyces* v. Höhn. und einige andere Gattungen der Trichothyriaceen. Sydowia 4: 163–174.

Phadtare, N.R. 1989. Palaeoecologic significance of some fungi from the Miocene of Tanakpur (U.P.) India. Review of Palaeobotany and Palynology 59: 127–131.

Phillips, S.E., and P.G. Self. 1987. Morphology, crystallography and origin of needle-fibre calcite in quaternary pedogenic calcretes of South Australia. Australian Journal of Soil Research 25: 429–444.

Phillips, T.L., and J. Galtier. 2005. Evolutionary and ecological perspectives of Late Paleozoic ferns Part I. Zygopteridales. Review of Palaeobotany and Palynology 135: 165–203.

Phillips, T.L., and J. Galtier. 2011. Evolutionary and ecological perspectives of late Paleozoic ferns Part II. The genus *Ankyropteris* and the Tedeleaceae. Review of Palaeobotany and Palynology 164: 1–29.

Phillott, A.D., C.J. Parmenter, and C.J. Limpus. 2004. Occurrences of mycobiota in eastern Australian sea turtle nests. Memoirs of the Queensland Museum 49: 701–703.

Phipps, C.J. 2001. The Evolution of Epiphyllous Fungal Communities with an Emphasis on the Miocene of Idaho, Ph.D. Dissertation. University of Kansas, Lawrence, KS. 151 pp.

Phipps, C.J. 2007. *Entopeltacites remberi* sp. nov. from the Miocene of Clarkia, Idaho, USA. Review of Palaeobotany and Palynology 145: 193–200.

Phipps, C.J., and W.C. Rember. 2004. Epiphyllous fungi from the Miocene of Clarkia, Idaho: Reproductive structures. Review of Palaeobotany and Palynology 129: 67–79.

Phipps, C.J., and T.N. Taylor. 1996. Mixed arbuscular mycorrhizae from the Triassic of Antarctica. Mycologia 88: 707–714.

Phoenix, V.R., R.W. Renaut, B. Jones, and F.G. Ferris. 2005. Bacterial S-layer preservation and rare arsenic-antimony-sulphide bioimmobilization in siliceous sediments from Champagne Pool hot spring, Waiotapu, New Zealand. Journal of the Geological Society of London 162: 323–332.

Pia, J. 1927. Thallophyta. In M. Hirmer (Ed.). Handbuch der Paläobotanik: Band I – Thallophyta, Bryophyta, Pteridophyta, pp. 31–136. R. Oldenbourg, München, Germany.

Piepenbring, M., R. Bauer, and F. Oberwinkler. 1998a. Teliospores of smut fungi: General aspects of teliospore walls and sporogenesis. Protoplasma 204: 155–169.

Piepenbring, M., R. Bauer, and F. Oberwinkler. 1998b. Teliospores of smut fungi: Teliospore connections, appendages, and germ pores studied by electron microscopy; phylogenetic discussion of characteristics of teliospores. Protoplasma 204: 202–218.

Piepenbring, M., R. Bauer, and F. Oberwinkler. 1998c. Teliospores of smut fungi: Teliospore walls and the development of ornamentation studied by electron microscopy. Protoplasma 204: 170–201.

Piercey-Normore, M.D., and P.T. DePriest. 2001. Algal switching among lichen symbioses. American Journal of Botany 88: 1490–1498.

Pilger, T.J., and J.R. Thomasson. 2005. Scanning electron microscopy of fungi specimens from the Elam Bartholomew Herbarium. Transactions of the Kansas Academy of Science 108: 172.

Pinho, D.B., O.L. Pereira, A.L. Firmino, M. da Silva, W.G. Ferreira-Junior, and R.W. Barreto. 2012. New Meliolaceae from the Brazilian Atlantic forest 1. Species on hosts in the families Asteraceae, Burseraceae, Euphorbiaceae, Fabaceae and Sapindaceae. Mycologia 104: 121–137.

Piperno, D.R., and H.-D. Sues. 2005. Dinosaurs dined on grass. Science 310: 1126–1128.

Pires-Zottarelli, C.L.A., and A.L. Gomes. 2007. Contribuição para o conhecimento de Chytridiomycota da "Reserva Biológica de Paranapiacaba", Santo André, SP, Brasil (Contribution to the knowledge of the Chytridiomycota from the "Reserva Biológica de Paranapiacaba", Santo André, State of São Paulo, Brazil). Biota Neotropica 7: 309–329.

Pirozynski, K.A. 1976a. Fossil fungi. Annual Review of Phytopathology 14: 237–246.

Pirozynski, K.A. 1976b. Fungal spores in fossil record. Biological Memoirs 1: 104–120.

Pirozynski, K.A. 1978. Fungal spores through the ages – a mycologist's view. In: 4th International Palynological Conference, Lucknow, Proceedings (Trudy Mezdunarodnoy Palinologicheskoy Konferentsiy), Izvestija, pp. 327–330. Akademiia Nauk CCCR, Moscow, USSR.

Pirozynski, K.A. 1981. Interactions between fungi and plants through the ages. Canadian Journal of Botany 59: 1824–1827.

Pirozynski, K.A., and Y. Dalpé. 1989. Geological history of the Glomaceae with particular reference to mycorrhizal symbiosis. Symbiosis 7: 1–36.

Pirozynski, K.A., and D.W. Malloch. 1975. The origin of land plants: A matter of mycotrophism. BioSystems 6: 153–164.

Pirozynski, K.A., and R.A. Schoemaker. 1970. Some Asterinaceae and Meliolaceae on conifers in Canada. Canadian Journal of Botany 48: 1321–1328.

Pirozynski, K.A., and L.K. Weresub. 1979. The classification and nomenclature of fossil fungi. In B. Kendrick (Ed.), The Whole Fungus: The Sexual-Asexual Synthesis, Volume 2, pp. 653–688. National Museum of Natural Sciences, National Museums of Canada, Ottawa, and The Kananaskis Foundation, Ottawa, Canada.

Pirozynski, K.A., A. Carter, and R.G. Day. 1984. Fungal remains in Pleistocene ground squirrel dung from Yukon Territory, Canada. Quaternary Research 22: 375–382.

Pirozynski, K.A., D.M. Jarzen, A. Carter, and R.G. Day. 1988. Palynology and mycology of organic clay balls accompanying mastodon bones – New Brunswick, Canada. Grana 27: 123–139.

Pivato, B., P. Offre, S. Marchelli, B. Barbonaglia, C. Mougel, P. Lemanceau, and G. Berta. 2009. Bacterial effects on arbuscular mycorrhizal fungi and mycorrhiza development as influenced by the bacteria, fungi, and host plant. Mycorrhiza 19: 81–90.

Playford, G, J.F. Rigby, and D.C. Archibald. 1982. A Middle Triassic flora from the Moolayember Formation, Bowen Basin, Queensland. Geological Survey of Queensland Publication 380: 1–52.

Plett, J.M., M. Kemppainen, S.D. Kale, A. Kohler, V. Legué, A. Brun, B.M. Tyler, A.G. Pardo, and F. Martin. 2011. A secreted effector protein of *Laccaria bicolor* is required for symbiosis development. Current Biology 21: 1197–1203.

Podhalańska, T., and J. Nõlvak. 1995. Endolithic trace-fossil assemblage in Lower Ordovician limestones from northern Estonia. GFF (Geologiska Föreningens i Stockholm Förhandlingar) 117: 225–231.

Pöggeler, S., M. Nowrousian, and U. Kück. 2006. Fruiting-body development in Ascomycetes. In U. Kües, and R. Fischer (Eds), The Mycota II: Growth, Differentiation and Sexuality, pp. 325–355. Springer-Verlag, Berlin, Germany.

Poinar, G.O., Jr. 1992. Life in Amber. Stanford University Press, Stanford, CA. 350 pp.

Poinar, G.O., Jr. 2001. Fossil puffballs (Gasteromycetes: Lycoperdales) in Mexican amber. Historical Biology 15: 219–222.

Poinar, G.O., Jr. 2003. Coelomycetes in Dominican and Mexican amber. Mycological Research 107: 117–122.

Poinar, G.O., Jr. 2010. Palaeoecological perspectives in Dominican amber. Annales de la Société Entomologique de France 46: 23–52.

Poinar, G.O., Jr. 2011. The Evolutionary History of Nematodes – As Revealed in Stone, Amber and Mummies. (Nematology Monographs and Perspectives, Volume 9). Koninklijke Brill NV, Leiden, The Netherlands. 429 pp.

Poinar, G., Jr. 2014. Bird's nest fungi (Nidulariales: Nidulariaceae) in Baltic and Dominican amber. Fungal Biology 118: 325–329.

Poinar, G.O., Jr., and R. Buckley. 2007. Evidence of mycoparasitism and hypermycoparasitism in Early Cretaceous amber. Mycological Research 111: 503–506.

Poinar, G.O., Jr., and R. Poinar. 1992. The Amber Forest: A Reconstruction of a Vanished World. Princeton University Press, Princeton, NJ. 239 pp.

Poinar, G.O., Jr., and R. Poinar. 2005. Fossil evidence of insect pathogens. Journal of Invertebrate Pathology 89: 243–250.

Poinar, G.O., Jr., and R. Singer. 1990. Upper Eocene gilled mushroom from the Dominican Republic. Science 248: 1099–1101.

Poinar, G.O., Jr., and G.M. Thomas. 1982. An entomophthoralean fungus from Dominican amber. Mycologia 74: 332–334.

Poinar, G.O., Jr., and G.M. Thomas. 1984. A fossil entomogenous fungus from Dominican amber. Experientia 40: 578–579.

Poinar, G.O., Jr., B.M. Waggoner, and U.-C. Bauer. 1993. Terrestrial soft-bodied protists and other microorganisms in Triassic amber. Science 259: 222–224.

Poinar, G.O., Jr., E.B. Peterson, and J.L. Platt. 2000. Fossil *Parmelia* in New World amber. Lichenologist 32: 263–269.

Poinar, G.O., Jr., H. Kerp, and H. Hass. 2008. *Palaeonema phyticum* gen. n., sp. n. (Nematoda: Palaeonematidae fam. n.), a Devonian nematode associated with early land plants. Nematology 10: 9–14.

Pons, D., and E. Boureau. 1977. Les champignons épiphylles d'un *Frenelopsis* du Cenomanian moyen de l'Anjou (France). Revue de Mycologie 41: 349–358.

Pontoppidan, M.-B., W. Himaman, N.L. Hywel-Jones, J.J. Boomsma, and D.P. Hughes. 2009. Graveyards on the move: The spatio-temporal distribution of dead *Ophiocordyceps*-infected ants. PLoS ONE 4: e4835.

Porter, C.L., and G. Zebrowski. 1937. Lime-loving molds from Australian sands 29: 252–257.

Porter, S.M. 2006. The Proterozoic fossil record of heterotrophic eukaryotes. In S. Xiao, and A.J. Kaufman (Eds), Neoproterozoic Geobiology and Paleobiology, pp. 1–21. Springer, Berlin, Heidelberg, Germany.

Porter, T.M., W. Martin, T.Y. James, J.E. Longcore, F.H. Gleason, P.H. Adler, P.M. Letcher, and R. Vilgalys. 2011. Molecular phylogeny of the Blastocladiomycota (Fungi) based on nuclear ribosomal DNA. Fungal Biology 115: 381–392.

Potonié, R., and H. Venitz. 1934. Zur Mikrobotanik des miocänen Humodils der niederrheinischen Bucht. Arbeiten aus dem Institut für Paläobotanik und Petrographie der Brennsteine 5: 5–54.

Poulsen, M., A.E.F. Little, and C.R. Currie. 2009. Fungus-growing ant-microbe symbiosis: Using microbes to defend beneficial associations within symbiotic communities. In J.F. White, Jr., and M.S. Torres (Eds), Defensive Mutualism in Microbial Symbiosis, pp. 149–166. CRC Press, Boca Raton, FL.

Powell, C.L., N.H. Trewin, and D. Edwards. 2000. Palaeoecology and plant succession in a borehole through the Rhynie cherts, Lower Old Red Sandstone, Scotland. In P.F. Friend and B.P.J. Williams (Eds), New Perspectives on the Old Red Sandstone. Geological Society of London Special Publication 180, pp. 439–457. Geological Society Publishing House, Bath, UK.

Powell, M.J. 1976. Development of the discharge apparatus in the fungus *Entophlyctis*. Archives of Microbiology 111: 59–71.

Powell, M.J. 1978. Phylogenetic implications of the microbody-lipid globule complex in zoosporic fungi. BioSystems 10: 167–180.

Powell, M.J. 1993. Looking at mycology with a Janus face: A glimpse at Chytridiomycetes active in the environment. Mycologia 85: 1–20.

Powell, M.J., and L. Gillette. 1987. Septal Structure of the chytrid *Rhizophlyctis harderi*. Mycologia 79: 635–639.

Powell, M.J., and S. Roychoudhury. 1992. Ultrastructural organization of *Rhizophlyctis harderi* zoospores and redifinition of the type 1 microbody-lipid globule complex. Canadian Journal of Botany 70: 750–761.

Pozo, M.J., C. Cordier, E. Dumas-Gaudot, S. Gianinazzi, J.M. Barea, and C. Azcón-Aguilar. 2002. Localized versus systemic effect of arbuscular mycorrhizal fungi on defence responses to *Phytophthora* infection in tomato plants. Journal of Experimental Botany 53: 525–534.

Prabhuji, S.K., A. Tripathi, A. Singh, and B. Chaturvedi. 2012. Sexual reproduction in water moulds. II. Genus *Allomyces* (family Blastocladiaceae). Vegetos 25: 292–297.

Prager, A., A. Barthelmes, M. Theuerkauf, and H. Joosten. 2006. Non-pollen palynomorphs from modern Alder carrs and their potential for interpreting microfossil data from peat. Review of Palaeobotany and Palynology 141: 7–31.

Prasad, B., and R. Asher. 2001. Acritarch biostratigraphy and lithostratigraphic classification of Proterozoic and Lower Paleozoic sediments (Pre-unconformity sequence) of Ganga Basin, India. Paleontographica Indica 5: 1–151.

Prasad, V., C.A.E. Strömberg, H. Alimohammadian, and A. Sahni. 2005. Dinosaur coprolites and the early evolution of grasses and grazers. Science 310: 1177–1180.

Pratt, L.M., T.L. Phillips, and J.M. Dennison. 1978. Evidence of non-vascular land plants from the Early Silurian (Llandoverian) of Virginia, U.S.A. Review of Palaeobotany and Palynology 25: 121–149.

Préat, A., B. Mamet, C. De Ridder, F. Boulvain, and D. Gillan. 2000. Iron bacterial and fungal mats, Bajocian stratotype (Mid-Jurassic, northern Normandy, France). Sedimentary Geology 137: 107–126.

Preece, T.F., and A.J. Hick. 1990. An Introductory Scanning Electron Microscope Atlas of Rust Fungi. Farrand Press, London, UK, 240 pp.

Preece, T.F., and A.J. Hick. 1994. British gall-causing rust fungi. In M.A.J. Williams (Ed.), Plant Galls: Organisms, Interactions, Populations. Systematics, pp. 57–66. Clarendon Press, Oxford, UK.

Pressel, S., R. Ligrone, and J.G. Duckett. 2008a. The ascomycete *Rhizoscyphus ericae* elicits a range of host responses in the rhizoids of leafy liverworts: An experimental and cytological analysis. Fieldiana: Botany, n.s. 47: 59–72.

Pressel, S., R. Ligrone, J.G. Duckett, and E.C. Davis. 2008b. A novel ascomycetous endophytic association in the rhizoids of the leafy liverwort family, Schistochilaceae (Jungermanniidae, Hepaticopsida). American Journal of Botany 95: 531–541.

Pressel, S., M.I. Bidartondo, R. Ligrone, and J.G. Duckett. 2010. Fungal symbioses in bryophytes: New insights in the Twenty First Century. Phytotaxa 9: 238–253.

Preston, L.J., and M.J. Genge. 2010. The Rhynie Chert, Scotland, and the search for life on Mars. Astrobiology 10: 549–560.

Price, R.J., and B.V.A. Thorne. 1977. The micropaleontology, palynology and stratigraphy of the IOE Ikattok J-17 well. Robertson Research (North America) Limited. Exploration Report 181: 1–20.

Priess, K., T. Le Campion-Alsumard, S. Golubic, F. Gadel, and B.A. Thomassin. 2000. Fungi in corals: Black bands and density-banding of *Porites lutea* and *P. lobata* skeleton. Marine Biology 136: 19–27.

Prieto, M., and M. Wedin. 2013. Dating the diversification of the major lineages of Ascomycota (Fungi). PLoS ONE 8: e65576.

Prillinger, H., K. Lopandic, W. Schweigkofler, R. Deak, H.J.M. Aarts, R. Bauer, K. Sterflinger, G.F. Kraus, and A. Maraz. 2002. Phylogeny and systematics of the fungi with special reference to the Ascomycota and Basidiomycota. In: M. Breitenbach, R. Crameri, and S.B. Lehrer (eds), Fungal Allergy and Pathogenicity. Chemical Immunology 81: 207–295.

Pringle, A, S.N. Patek, M. Fischer, J. Stolze, and N.P. Money. 2005. The captured launch of a ballistospore. Mycologia 97: 866–871.

Printzen, C., and H.T. Lumbsch. 2000. Molecular evidence for the diversification of extant lichens in the Late Cretaceous and Tertiary. Molecular Phylogenetics and Evolution 17: 379–387.

Privé-Gill, C., H. Thomas, and P. Lebret. 1999. Fossil wood of *Sindora* (Leguminosae, Caesalpiniaceae) from the Oligo-Miocene of Saudi Arabia: Paleobiogeographical considerations. Review of Palaeobotany and Palynology 107: 191–199.

Prokhorov, V.P., and V.V. Bodyagin. 2007. The ecology of aero-aquatic hyphomycetes. Moscow University Biological Sciences Bulletin 62: 15–20.

Pujana, R.R., J.L.G. Massini, R.R. Brizuela, and H.P. Burrieza. 2009. Evidence of fungal activity in silicified gymnosperm wood from the Eocene of southern Patagonia (Argentina). Géobios 42: 639–647.

Pumplin, N., and M.J. Harrison. 2009. Live-cell imaging reveals periarbuscular membrane domains and organelle location in *Medicago truncatula* roots during arbuscular mycorrhizal symbiosis. Plant Physiology 151: 809–819.

Punt, W., S. Blackmore, S. Nilsson, and A. Le Thomas. 1994. Glossary of pollen and spore terminology. LPP Contributions Series 1: 1–71.

Purdy, L.H., and B.A. Purdy. 1982. Ancient polypores from an archaeological wet site in Florida. Botanical Gazette 143: 551–553.

Purin, S., and M.C. Rillig. 2008. Parasitism of arbuscular mycorrhizal fungi: Reviewing the evidence. FEMS Microbiology Letters 279: 8–14.

Purvis, O.W. 2000. Lichens. The Natural History Museum, London, UK, 112 pp.

Quirk, J., D.J. Beerling, S.A. Banwart, G. Kakonyi, M.E. Romero-Gonzalez, and J.R. Leake. 2012. Evolution of trees and mycorrhizal fungi intensifies silicate mineral weathering. Biology Letters 8: 1006–1011.

Rabatin, S.C., and L.H. Rhodes. 1982. *Acaulospora bireticulata* inside oribatid mites. Mycologia 74: 859–861.

Radforth, N.W. 1958. Palaeobotanical evaluation of fossil wood in Onakawana lignites. Transactions of the Royal Society of Canada, Third Series 52: 41–53.

Radhika, K.P., and B.F. Rodrigues. 2007. Arbuscular mycorrhizas in association with aquatic and marshy plant species in Goa, India. Aquatic Botany 86: 291–294.

Radtke, G. 1991. Die mikroendolithischen Spurenfossilien im Alt-Tertiär West-Europas und ihre palökologische Bedeutung. Courier Forschungsinstitut Senckenberg 138: 1–185.

Radtke, G., I. Glaub, K. Vogel, and S. Golubic. 2010. A new dichotomous microboring: *Abeliella bellafurca* isp. nov., distribution, variability and biological origin. Ichnos 17: 25–33.

Raghavendra, A.K.H., and G. Newcombe. 2013. The contribution of foliar endophytes to quantitative resistance to *Melampsora* rust. New Phytologist 197: 909–918.

Raghukumar, C., and S. Raghukumar. 1991. Fungal invasion of massive corals. Marine Ecology 12: 251–260.

Raghukumar, C., and S. Raghukumar. 1998. Barotolerance of fungi isolated from deep-sea sediments of the Indian Ocean. Aquatic Microbial Ecology 15: 153–163.

Ragozina, A.L. 1993. Cyanobacterial mats from Lower Cambrian (Drevnikh?) phosphorites of Mongolia. In

B.S. Sokolov (Ed.), Fauna I Ekosistemy Geologicheskogo Proshlogo (Fauna and Ecosystems of the Geological Past), pp. 33–36. Nauka, Moscow, Russia. Pl. 3–5.

Raguso, R.A., and B.A. Roy. 1998. 'Floral' scent production by *Puccinia* rust fungi that mimic flowers. Molecular Ecology 7: 1127–1136.

Ramanujam, C.G.K. 1982. Recent advances in the study of fossil fungi. In D.C. Bharadwaj (Ed.), Recent advances in Cryptogamic Botany. Part II: Fossil Cryptogams, pp. 287–301. Palaeobotanical Society, Lucknow, India.

Ramanujam, C.G.K., and P. Ramachar. 1963. Sporae dispersae of the rust fungi (Uredinales) from the Miocene lignite of South India. Current Science 32: 271–272.

Ramanujam, C.G.K., and P. Ramachar. 1980. Recognizable spores of rust fungi (Uredinales) from Neyveli Lignite, Tamil Nadu. Records of the Geological Survey of India 113: 80–85.

Ramanujam, C.G.K., and K.P. Rao. 1973. On some microthyriaceous fungi from a Tertiary lignite of South India. Palaeobotanist 20: 203–209.

Ramanujam, C.G.K., and K.P. Rao. 1978. Fungal spores from the Neogene strata of Kerala in South India. In: D.C. Bharadwaj, et al. (Eds), Proceedings of the 4th International Palynology Conference, 1976–1977, pp. 291–304. Birbal Sahni Institute of Palaeobotany, Lucknow, India.

Ramanujam, C.G.K., and K. Srisailam. 1978. Fossil fungal spores from the Neogene beds around Cannanore in Kerala state. Botanique (Nagpur) 9: 119–138.

Rao, A.R. 1958. Fungal remains from some Tertiary deposits of India. Palaeobotanist 7: 43–46.

Rao, K.P., and C.G.K. Ramanujam. 1976. A further record of microthyriaceous fungi from the Neogene deposits of Kerala in South India. Geophytology 6: 98–104.

Rao, M.R. 2003. *Kalviwadithyrites*, a new fungal fruiting body from Sindhudurg Formation (Miocene) of Maharashtra, India. Palaeobotanist 52: 117–119.

Rao, S., Y. Chan, D.C. Lacap, K.D. Hyde, S.B. Pointing, and R.L. Farrell. 2012. Low-diversity fungal assemblage in an Antarctic Dry Valleys soil. Polar Biology 35: 567–574.

Rasmussen, H.N., and F.N. Rasmussen. 2009. Orchid mycorrhiza: Implications of a mycophagous life style. Oikos 118: 334–345.

Raven, J.A. 2000. Land plant biochemistry. Philosophical Transactions of the Royal Society 355B: 833–846.

Raven, J.A., and D. Edwards. 2001. Roots: Evolutionary origins and biogeochemical significance. Journal of Experimental Botany 52: 381–401.

Raven, P.H., R.F. Evert, and S.E. Eichhorn. 1999. Biology of Plants (Sixth Edition). W.H. Freeman and Company, New York, NY, 944 pp.

Rayner, A.D.M., and L. Boddy. 1988. Fungal Decomposition of Wood: Its Biology and Ecology. John Wiley & Sons, Chichester, UK, 602 pp.

Read, D.J. 1983. The biology of mycorrhiza in the Ericales. Canadian Journal of Botany 61: 985–1004.

Read, D.J., J.G. Duckett, R. Francis, R. Ligrone, and A. Russell. 2000. Symbiotic fungal associations in 'lower' land plants. Philosophical Transactions of the Royal Society 355B: 815–831.

Read, N.D., and K.M. Lord. 1991. Examination of living fungal spores by scanning electron microscopy. Experimental Mycology 15: 132–139.

Réblová, M., and K.A. Seifert. 2004. *Cryptadelphia* (Trichosphaeriales), a new genus for holomorphs with *Brachysporium* anamorphs and clarification of the taxonomic status of *Wallrothiella*. Mycologia 96: 343–367.

Reddy, P.R., C.G.K. Ramanujam, and K. Srisailam. 1982. Fungal fructifications from Neyveli lignite, Tamil Nadu – their stratigraphic and palaeoclimatic significance. Records of the Geological Survey of India 114: 112–122.

Redecker, D. 2002. New views on fungal evolution based on DNA markers and the fossil record. Research in Microbiology 153: 125–130.

Redecker, D., and P. Raab. 2006. Phylogeny of the Glomeromycota (arbuscular mycorrhizal fungi): Recent developments and new gene markers. Mycologia 98: 885–895.

Redecker, D., R. Kodner, and L.E. Graham. 2000a. Glomalean fungi from the Ordovician. Science 289: 1920–1921.

Redecker, D., J.B. Morton, and T.D. Bruns. 2000b. Molecular phylogeny of the arbuscular mycorrhizal fungi *Glomus sinuosum* and *Sclerocystis coremioides*. Mycologia 92: 282–285.

Redecker, D., R. Kodner, and L.E. Graham. 2002. *Palaeoglomus grayi* from the Ordovician. Mycotaxon 84: 33–37.

Redecker, D., A. Schüßler, H. Stockinger, S.L. Stürmer, J.B. Morton, and C. Walker. 2013. An evidence-based consensus for the classification of arbuscular mycorrhizal fungi (Glomeromycota). Mycorrhiza 23: 515–531.

Redhead, S.A. 1979. Mycological observations: 1, on *Cristulariella*; 2, on *Valdensinia*; 3, on *Neolecta*. Mycologia 71: 1248–1253.

Rehner, S.A., A.M. Minnis, G.H. Sung, J.J. Luangsaard, L. Devotto, and R.A. Humber. 2011. Phylogeny and systematics of the anamorphic, entomopathogenic genus *Beauveria*. Mycologia 103: 1055–1073.

Reihman, M.A., and J.T. Schabilion. 1976a. Cuticles of two species of *Alethopteris*. American Journal of Botany 63: 1039–1046.

Reihman, M.A., and J.T. Schabilion. 1976b. Two species of *Alethopteris* from Iowa coal balls. Proceedings of the Iowa Academy of Science 83: 10–19.

Reinhard, K.J., and V.M. Bryant, Jr. 1992. Coprolite analysis: A biological perspective on archaeology. Papers in Natural Resources 46: 245–288.

Reininger, V., and T.N. Sieber. 2012. Mycorrhiza reduces adverse effects of dark septate endophytes (DSE) on growth of conifers. PLoS ONE 7: e42865.

Relman, D.A. 2008. 'Til death do us part': Coming to terms with symbiotic relationships. Nature Reviews Microbiology 6: 721–724.

Remy, W., and H. Hass. 1996. New information on gametophytes and sporophytes of *Aglaophyton major* and inferences about possible environmental adaptations. Review of Palaeobotany and Palynology 90: 175–193.

Remy, W., and R. Remy. 1980a. Devonian gametophytes with anatomically preserved gametangia. Science 208: 295–296.

Remy, W., and R. Remy. 1980b. *Lyonophyton rhyniensis* nov. gen. et nov. spec., ein Gametophyt aus dem Chert von Rhynie (Unterdevon, Schottland). Argumenta Palaeobotanica 6: 37–72.
Remy, W., T.N. Taylor, H. Hass, and H. Kerp. 1994a. Four hundred-million-year-old vesicular arbuscular mycorrhizae. Proceedings of the National Academy of Sciences, USA 91: 11841–11843.
Remy, W., T.N. Taylor, and H. Hass. 1994b. Early Devonian fungi: A blastocladalean fungus with sexual reproduction. American Journal of Botany 81: 690–702.
Renault, B. 1883. Cours de Botanique fossile fait au Muséum d'histoire naturelle, Vol. 3. Masson, Paris. 328 pp.
Renault, B. 1894a. Sur quelques nouveaux parasites des *Lépidodendrons*. Société d'Histoire Naturelle d'Autun. Procès-verbaux des Séances de 1893: 168–178.
Renault, B. 1894b. Bactéries des temps primaries. Bulletin de la Société d'Histoire Naturelle d'Autun 7: 433–468.
Renault, B. 1895a. Chytridinées fossiles du Dinantien (culm). Revue de Mycologique 17: 158–161.
Renault, B. 1895b. Parasites des écorces de *Lépidodendrons*. Le Naturaliste 2: 77–78.
Renault, B. 1895c. Sur quelques *Micrococcus* du Stéphanien, terrain houiller supérieur. Comptes Rendus Hebdomadaires des Séances de l'Académie des Sciences, Paris 120: 217–220.
Renault, B. 1896a. Recherches sur les Bactériacées fossiles. Annales des Sciences Naturelles, Botanique, Série 8(2): 275–349.
Renault, B. 1896b. Bassin Houiller et Permien d'Autun et d'Épinac. Fascicule IV: Flore Fossile, Deuxième Partie. Etudes des Gîtes Minéraux de la France. Imprimerie Nationale, Paris, France. 578 pp.
Renault, B. 1896c. Notice sur les Travaux Scientifiques de M. Bernard Renault. Imprimerie Dejussieu Père et fils, Autun, France. 162 pp.
Renault, B. 1896d. Les Bactériacées de la houille. Comptes Rendus Hebdomadaires des Séances de l'Académie des Sciences 123: 953–955.
Renault, B. 1897. Les Bactériacées des bogheads. Comptes Rendus Hebdomadaires des Séances de l'Académie des Sciences 124: 1315–1318.
Renault, B. 1898. Les microrganismes des lignites. Comptes Rendus Hebdomadaires des Séances de l'Académie des Sciences 126: 1828–1831.
Renault, B. 1899. Sur quelques microorganisms des combustibles fossils. Atlas. Imprimerie Théolier. J. Thomas et Cie, Saint-Étienne, France.
Renault, B. 1900. Sur quelques microorganisms des combustibles fossiles. Text. Imprimerie Théolier. J. Thomas et Cie, Saint-Étienne, France. 460 pp.
Renault, B. 1901. Du rôle de quelques bactériacées fossiles au point de vue géologique. Congrès Géologique International, Comptes Rendus de la VIIIe Session, en France, 1ère fasc: 646–663.
Renault, B. 1903. Sur quelques nouveaux champignons et algues fossiles, de l'époque houillère. Comptes rendus hebdomadaires des séances de l'Académie des Sciences 136: 904–907.
Renault, B., and C.E. Bertrand. 1885. *Grilletia spherospermii*, Chytridiacée fossile du terrain houiller supérieur. Comptes Rendus Hebdomadaires des Séances de l'Académie des Sciences, Paris 100: 1306–1308.
Renker, C., P. Otto, K. Schneider, B. Zimdars, M. Maraun, and F. Buscot. 2005. Oribatid mites as potential vectors for soil microfungi: Study of mite-associated fungal species. Microbial Ecology 50: 518–528.
Reolid, M., and M.I. Benito. 2012. Belemnite taphonomy (Upper Jurassic, western Tethys) Part 1: Biostratinomy. Palaeogeography, Palaeoclimatology, Palaeoecology 358–360: 72–88.
Retallack, G.J. 1994. Were the Ediacaran fossils lichens? Paleobiology 20: 523–544.
Retallack, G.J. 2001. Soils of the past: An introduction to paleopedology (Second Edition). Blackwell, Science Ltd., Oxford, UK, 520 pp.
Retallack, G.J. 2009. Cambrian-Ordovician non-marine fossils from South Australia. Alcheringia 33: 355–391.
Retallack, G.J. 2011. Problematic megafossils in Cambrian palaeosols of South Australia. Palaeontology 54: 1223–1242.
Retallack, G.J. 2012. Were Ediacaran siliciclastics of South Australia coastal or deep marine? Sedimentology 59: 1208–1236.
Retallack, G.J. 2013a. Ediacaran life on land. Nature 493: 89–92.
Retallack, G.J. 2013b. Comment on "Trace fossil evidence for Ediacaran bilaterian animals with complex behaviors" by Chen et al. [Precambrian Res. 224 (2013) 690–701]. Precambrian Research 231: 383–385.
Retallack, G.J., K.L. Dunn, and J. Saxby. 2013a. Problematic Mesoproterozoic fossil *Horodyskia* from Glacier National Park, Montana, USA. Precambrian Research 226: 125–142.
Retallack, G.J., E.S. Krull, G.D. Thackray, and D. Parkinson. 2013b. Problematic urn-shaped fossils from a Paleoproterozoic (2.2 Ga) paleosol in South Africa. Precambrian Research 235: 71–87.
Retallick, R.W.R., and V. Miera. 2007. Strain differences in the amphibian chytrid *Batrachochytrium dendrobatidis* and non-permanent, sub-lethal effects of infection. Diseases of Aquatic Organisms 75: 201–207.
Rex, G.M. 1986. The preservation and palaeoecology of the Lower Carboniferous silicified plant deposits at Esnost, near Autun, France. Geobios 19: 773–800.
Reynolds, D.R., and D.L. Dilcher. 1984. A folicolous alga of Eocene age. Review of Palaeobotany and Palynology 43: 397–403.
Rice, A.V., M.N. Thormann, and D.W. Langor. 2007. Virulence of, and interactions among, mountain pine beetle associated blue-stain fungi on two pine species and their hybrids in Alberta. Canadian Journal of Botany 85: 316–323.
Rice, A.V., M.N. Thormann, and D.W. Langor. 2008. Mountain pine beetle-associated blue-stain fungi are differentially adapted to boreal temperatures. Forest Pathology 38: 113–123.
Rice, C.M., W.A. Ashcroft, D.J. Batten, A.J. Boyce, J.B.D. Caulfield, A.E. Fallick, M.J. Hole, E. Jones, M.J. Pearson, G. Rogers, J.M. Saxton, F.M. Stuart, N.H. Trewin, and G.

Turner. 1995. A Devonian auriferous hot spring system, Rhynie, Scotland. Journal of the Geological Society, London 152: 229–250.

Rice, C.M., N.H. Trewin, and L.I. Anderson. 2002. Geological setting of the Early Devonian Rhynie cherts, Aberdeenshire, Scotland: An early terrestrial hot spring system. Journal of the Geological Society of London 159: 203–214.

Rice, D.W., A.J. Alverson, A.O. Richardson, G.J. Young, M.V. Sanchez-Puerta, J. Munzinger, K. Barry, J.L. Boore, Y. Zhang, C.W. dePamphilis, E.B. Knox, and J.D. Palmer. 2013. Horizontal transfer of entire genomes viamitochondrial fusion in the angiosperm Amborella. Science 342: 1468–1473.

Richard, F. 1967. Découverte d'un horizon â *Microcodium* dans la série carbonatée Crétaceo-Tertiaire de Göcek (province de Muğla, Turquie). Comptes rendus de l'Académie des Sciences, Paris 264: 1133–1136.

Richards, T.A., D.M. Soanes, M.D.M. Jones, O. Vasieva, G. Leonard, K. Paszkiewicz, P.G. Foster, N. Hall, and N.J. Talbot. 2011. Horizontal gene transfer facilitated the evolution of plant parasitic mechanisms in the oomycetes. Proceedings of the National Academy of Sciences, USA 108: 15258–15263.

Richards, T.A., M.D.M. Jones, G. Leonard, and D. Bass. 2012. Marine fungi: Their ecology and molecular diversity. Annual Review of Marine Science 4: 495–522.

Richardson, A.R., C.F. Eble, J.C. Hower, and J.M.K. O'Keefe. 2012. A critical re-examination of the petrology of the No. 5 Block coal in eastern Kentucky with special attention to the origin of inertinite macerals in the splint lithotypes. International Journal of Coal Geology 98: 41–49.

Richardson, M. 2009. The ecology of the Zygomycetes and its impact on environmental exposure. Clinical Microbiology and Infection 15(Supplement 5): 2–9.

Richter, G. 1994. Bacteria and bacteria-like structures from the oil-shale of Messel. Kaupia 4: 21–28.

Riederer, M., and L. Schreiber. 1995. Waxes: the transport barriers of plant cuticles. In R.J. Hamilton (Ed.), Waxes: Chemistry, Molecular Biology and Functions, pp. 130–156. The Oily Press, West Ferry, Dundee, Scotland.

Rikkinen, J. 1999. Two new species of resinicolous *Chaenothecopsis* (Mycocaliciaceae) from western North America. Bryologist 102: 366–369.

Rikkinen, J. 2003. Calicioid lichens from European Tertiary amber. Mycologia 95: 1032–1036.

Rikkinen, J., and G.O. Poinar., Jr. 2000. A new species of resinicolous *Chaenothecoptis* (Mycocaliciaceae, Ascomycota) from 20 million year old Bitterfeld amber, with remarks on the biology of resinicolous fungi. Mycological Research 104: 7–15.

Rikkinen, J., and G.O. Poinar., Jr. 2001. Fossilised fungal mycelium from Tertiary Dominican amber. Mycological Research 105: 890–896.

Rikkinen, J., and G.O. Poinar., Jr. 2002a. Fossilised *Anzia* (Lecanorales, lichen-forming Ascomycota) from European Tertiary amber. Mycological Research 106: 984–990.

Rikkinen, J., and G.O. Poinar., Jr. 2002b. Yeast-like fungi in Dominican amber. Karstenia 42: 29–32.

Rikkinen, J., and G.O. Poinar., Jr. 2008. A new species of *Phyllopsora* (Lecanorales, lichen-forming Ascomycota) from Dominican amber, with remarks on the fossil history of lichens. Journal of Experimental Botany 59: 1007–1011.

Rikkinen, J., H. Dörfelt, A.R. Schmidt, and J. Wunderlich. 2003. Sooty moulds from European Tertiary amber, with notes on the systematic position of *Rosaria* ('Cyanobacteria'). Mycological Research 107: 251–256.

Riley, R., and N. Corradi. 2013. Searching for clues of sexual reproduction in the genomes of arbuscular mycorrhizal fungi. Fungal Ecology 6: 44–49.

Rinaldi, A.C., O. Comandini, and T.W. Kuyper. 2008. Ectomycorrhizal fungal diversity: Separating the wheat from the chaff. Fungal Diversity 33: 1–45.

Rioux, D., and A.R. Biggs. 1994. Cell wall changes in host and nonhost systems: Microscopic aspects. In O. Petrini, and G.B. Ouellette (Eds), Host Wall Alterations by Parasitic Fungi, pp. 31–44. APS Press, St. Paul, MN.

Ritchie, D. 1947. The formation and structure of the zoospores in *Allomyces*. Journal of the Elisha Mitchell Scientific Society 63: 168–205.

Roberts, J.A., P.C. Bennett, L.A. González, G.L. Macpherson, and K.L. Milliken. 2004. Microbial precipitation of dolomite in methanogenic groundwater. Geology 32: 277–280.

Roberts, R.G., J.A. Robertson, and R.T. Hanlin. 1984. *Ascotricha xylina*: Its occurrence, morphology, and typification. Mycologia 76: 963–968.

Robinson, J.M. 1990. Lignin, land plants, and fungi: Biological evolution affecting Phanerozoic oxygen balance. Geology 18: 607–610.

Robison, C.R., and C.P. Person. 1973. A silicified semiaquatic dicotyledon from the Eocene Allenby Formation of British Columbia. Canadian Journal of Botany 51: 1373–1377.

Rodriguez, R.J., J.F. White, Jr., A.E. Arnold, and R.S. Redman. 2009. Fungal endophytes: Diversity and functional roles. New Phytologist 182: 314–330.

Rodríguez de Sarmiento, M.N., J. Durango de Cabrera, and N. Vásquez de Ramallo. 1998. Hojas fósiles de Angiospermas con hongos, Terciario del Chubut, Argentina. Acta Geológica Lilloana 18: 131–133.

Roger, A.J., and L.A. Hug. 2006. The origin and diversification of eukaryotes: Problems with molecular phylogenetics and molecular clock estimation. Philosophical Transactions of the Royal Society 361B: 1039–1054.

Rogers, T.R. 2008. Treatment of zygomycosis: Current and new options. Journal of Antimicrobial Chemotherapy 61(Supplement 1): i35–i39.

Rohfritsch, O. 2008. Plants, gall midges, and fungi: A three-component system. Entomologia Experimentalis et Applicata 128: 208–216.

Rohr, D.M., A.J. Boucot, J. Miller, and M. Abbott. 1986. Oldest termite nest from the Upper Cretaceous of west Texas. Geology 14: 87–88.

Rosendahl, C.O. 1943. Some fossil fungi from Minnesota. Bulletin of the Torrey Botanical Club 70: 126–138.

Rosenstock, N.P. 2009. Can ectomycorrhizal weathering activity respond to host nutrient demands? Fungal Biology Reviews 23: 107–114.

Rosling, A., and N. Rosenstock. 2008. Ectomycorrhizal fungi in mineral soil. Mineralogical Magazine 72: 127–130.

Rosling, A., R.D. Finlay, and G.M. Gadd. 2009a. Editorial: Geomycology. Fungal Biology Reviews 23: 91–93.

Rosling, A., T. Roose, A.M. Herrmann, F.A. Davidson, R.D. Finlay, and G.M. Gadd. 2009b. Approaches to modelling mineral weathering by fungi. Fungal Biology Reviews 23: 138–144.

Rosling, A., F. Cox, K. Cruz-Martinez, K. Ihrmark, G.-A. Grelet, B.D. Lindahl, A. Menkis, and T.Y. James. 2011. Archaeorhizomycetes: Unearthing an ancient class of ubiquitous soil fungi. Science 333: 876–879.

Rossi, W., and A. Weir. 2007. New species of *Stigmatomyces* from various continents. Mycologia 99: 139–143.

Rossi, W., M. Kotrba, and D. Triebel. 2005. A new species of *Stigmatomyces* from Baltic amber, the first fossil record of Laboulbeniomycetes. Mycological Research 109: 271–274.

Rossman, A.Y., and M.E. Palm. 2006. Why are *Phytophthora* and other Oomycota not true Fungi? Outlooks on Pest Management 17: 217–219.

Rossman, A.Y., D.F. Farr, and L.A. Castlebury. 2007. A review of the phylogeny and biology of the Diaporthales. Mycoscience 48: 135–144.

Rothwell, G.W. 1972. *Palaeosclerotium pusillum* gen. et sp. nov., a fossil eumycete from the Pennsylvanian of Illinois. Canadian Journal of Botany 50: 2353–2356.

Rouse, G.E. 1962. Plant microfossils from the Burrard Formation of western British Columbia. Micropaleontology 8: 187–218.

Rouse, G.E., and P.S. Mustard. 1997. Nomenclatural note and corrections. Palynology 21: 207–208.

Roux, W. 1887. Ueber eine im Knochen lebende Gruppe von Fadenpilzen (*Mycelites ossifragus*). Zeitschrift für Wissenschaftliche Zoologie 45: 769–802.

Rowley, J.R., J.S. Rowley, and J.J. Skvarla. 1990. Corroded exines from Havinga's leaf mold experiment. Palynology 14: 53–79.

Roy, B.A. 1993. Floral mimicry by a plant pathogen. Nature 362: 56–58.

Roy, B.A. 1994. The use and abuse of pollinators by fungi. Trends in Ecology and Evolution 9: 335–339.

Rozema, J., B. van Geel, L.O. Björn, J. Lean, and S. Madronich. 2002. Toward solving the UV puzzle. Science 296: 1621–1622.

Rubinstein, C.V., P. Gerrienne, G.S. de la Puente, R.A. Astini, and P. Steemans. 2010. Early Middle Ordovician evidence for land plants in Argentina (eastern Gondwana). New Phytologist 188: 365–369.

Ruibal, C., C. Gueidan, L. Selbmann, A.A. Gorbushina, P.W. Crous, J.Z. Groenewald, L. Muggia, M. Grube, D. Isola, C.L. Schoch, J.T. Staley, F. Lutzoni, and G.S. de Hoog. 2009. Phylogeny of rock-inhabiting fungi related to Dothideomycetes. Studies in Mycology 64: 123–133.

Ruiz-Lozano, J.M., R. Azcon, and M. Gomez. 1995. Effects of arbuscular-mycorrhizal *Glomus* species on drought tolerance: Physiological and nutritional plant responses. Applied and Environmental Microbiology 61: 456–460.

Rungjindamai, N., J. Sakayaroj, N. Plaingam, S. Somrithipol, and E.B.G. Jones. 2008. Putative basidiomycete teleomorphs and phylogenetic placement of the coelomycete genera: *Chaetospermum*, *Giulia* and *Mycotribulus* based on nu-rDNA sequences. Mycological Research 112: 802–810.

Russel, J., and S. Bulman. 2005. The liverwort *Marchantia foliacea* forms a specialized symbiosis with arbuscular mycorrhizal fungi in the genus *Glomus*. New Phytologist 165: 567–579.

Russell, A.J., M.I. Bidartondo, and B.G. Butterfield. 2002. The root nodules of the Podocarpaceae harbour arbuscular mycorrhizal fungi. New Phytologist 156: 283–295.

Ruyter-Spira, C., S. Al-Babili, S. van der Krol, and H. Bouwmeester. 2013. The biology of strigolactones. Trends in Plant Science 18: 72–83.

Ryan, B.D., F. Bungartz, and T.H. Nash., III. 2002. Morphology and anatomy of the lichen thallus. In T.H. Nash III, B.D. Ryan, C. Gries, and F. Bungartz (Eds), Lichen flora of the Greater Sonoran Desert Region, Volume 1. Lichens unlimited, pp. 8–34. Arizona State University, Tempe, AZ.

Ryberg, M., and P.B. Matheny. 2012. Asynchronous origins of ectomycorrhizal clades of Agaricales. Proceedings of the Royal Society of Biological Sciences 279B: 2003–2011.

Ryvarden, L. 1991. Genera of Polypores. Nomenclature and Taxonomy (Synopsis Fungorum 5, Fungiflora). Grønlands Grafiske A/S, Oslo, Norway. 363 pp.

Ryvarden, L., and R.L. Gilbertson. 1993. European Polypores, Part 1 (Synopsis Fungorum 6, Fungiflora). Grønlands Grafiske A/S, Oslo, Norway. 387 pp.

Sabah, A., I. Dakua, P. Kumar, W.S. Mohammed, and J. Dutta. 2012. Growth of templated gold microwires by self organization of colloids on *Aspergillus niger*. Digest Journal of Nanomaterials and Biostructures 7: 583–591.

Saccardo, P.A., G.B. Traverso, and A. Trotter. 1882–1931. Sylloge Fungorum Omnium Hucusque Cognitorum. Digessit P.A. Saccordo, Vols 1–25; Imprint varies.

Sadowski, E.-M., C. Beimforde, M. Gube, J. Rikkinen, H. Singh, L.J. Seyfullah, J. Heinrichs, P.C. Nascimbene, J. Reitner, and A.R. Schmidt. 2012. The anamorphic genus *Monotosporella* (Ascomycota) from Eocene amber and from modern *Agathis* resin. Fungal Biology 116: 1099–1110.

Saenz, G.S., and J.W. Taylor. 1999. Phylogenetic relationships of *Meliola* and *Meliolina* inferred from nuclear small subunit rRNA sequences. Mycological Research 103: 1049–1056.

Sahni, B., and H.S. Rao. 1943. A silicified flora from the Intertrappean cherts around Sausar in the Deccan. Proceedings of the National Academy of Sciences, India 13: 36–75.

Saint Martin, S., J.-P. Saint Martin, V. Girard, D. Grosheny, and D. Néraudeau. 2012. Filamentous micro-organisms in Upper Cretaceous amber (Martigues, France). Cretaceous Research 35: 217–229.

Saint Martin, S., J.-P. Saint Martin, V. Girard, and D. Néraudeau. 2013. Organismes filamenteux de l'ambre du Santonien de Belcodène (Bouches-du-Rhône, France). Annales de Paléontologie 99: 339–360.

Salard-Cheboldaeff, M., and M.V. Locquin. 1980. Champignons présents au Tertiaire le long du littoral de l'Afrique équatoriale. Comptes Rendus du 105e Congrès National des Sociétés Savantes, Caen, 1980, Section des Sciences 1: 183–195.

Saldajeno, M.G.B., W.A. Chandanie, M. Kubota, and M. Hyakumachi. 2008. Effects of interactions of arbuscular mycorrhizal fungi and beneficial saprophytic mycoflora on plant growth and disease protection. In Z.A. Siddiqui, M.S. Akhtar, and K. Futai (Eds), Mycorrhizae: Sustainable Agriculture and Forestry, pp. 211–226. Springer Science + Business Media B.V., Dordrecht, The Netherlands.

Saldajeno, M.G.B., M. Ito, and M. Hyakumachi. 2012. Interaction between the plant growth-promoting fungus *Phoma* sp. GS8–2 and the arbuscular mycorrhizal fungus *Glomus mosseae*: Impact on biocontrol of soil-borne diseases, microbial population, and plant growth. Australasian Plant Pathology 41: 271–281.

Salmon, E.S. 1903. *Cercosporites* sp., a new fossil fungus. Journal of Botany, London 41: 127–130.

Samant, B. 2000. Fungal remains form the Bhavnagar lignite, Gujarat, India. Geophytology 28: 11–18.

Samant, B., and D.M. Mohabey. 2009. Palynoflora from Deccan volcano-sedimentary sequence (Cretaceous-Palaeogene transition) of central India: Implications for spatio-temporal correlation. Journal of Biosciences 34: 811–823.

Samant, B., and P.M. Tapaswi. 2000. Fungal remains from the Surat lignite deposits (Early Eocene) of Gujarat, India. Gondwana. Geological Magazine 15: 25–30.

Samorini, G. 1992. The oldest representation of hallucinogenic mushrooms in the world – Sahara desert, 9000–7000 BP. Integration 2/3: 69–78.

Samuels, G.J., and M. Blackwell. 2001. Pyrenomycetes – fungi with perithecia. In D.J. McLaughlin, and E.G McLaughlin (Eds), The Mycota, VII. Part A. Systematics and evolution, pp. 221–255. Springer Verlag, Berlin.

Sander, P.M., C.T. Gee, J. Hummel, and M. Clauss. 2010. Mesozoic plants and dinosaur herbivory. In C.T. Gee (Ed.), Plants in Mesozoic Time: Morphological Innovations, Phylogeny, Ecosystems, pp. 331–359. University of Indiana Press, Bloomington, IN.

Sanders, I.R. 2011. Fungal sex: Meiosis machinery in ancient symbiotic fungi. Current Biology 21: R896–R897.

Sanders, I.R., and D. Croll. 2010. Arbuscular mycorrhiza: The challenge to understand the genetics of the fungal partner. Annual Review of Genetics 44: 271–292.

Sanders, I.R., and A.H. Fitter. 1992. Evidence for differential responses between host-fungus combinations of vesicular–arbuscular mycorrhizas from a grassland. Mycological Research 96: 415–419.

Sanders, W.B. 1994. Role of lichen rhizomorphs in thallus propagation and substrate colonization. Cryptogamic Botany 4: 283–289.

Sankaranarayanan, K., M.N. Timofeeff, R. Spathis, T.K. Lowenstein, and J.K. Lum. 2011. Ancient microbes from halite fluid inclusions: Optimized surface sterilization and DNA extraction. PLoS ONE 6: e20683.

Santamaria, S. 2001. Los Laboulbeniales, un grupo enigmático de hongos parásitos de insectos. Lazaroa 22: 3–19.

Santiago-Rodriguez, T.M., A.R. Patrício, J.I. Rivera, M. Coradin, A. Gonzalez, G. Tirado, R.J. Cano, and G.A. Toranzos. 2013. *luxS* in bacteria isolated from 25- to 40-million-year-old amber. FEMS Microbiology Letters 350: 117–124.

Sapp, J. 2004. The dynamics of symbiosis: An historical overview. Canadian Journal of Botany 82: 1046–1056.

Sappenfield, A., M.L. Droser, and J.G. Gehling. 2011. Problematica, trace fossils, and tubes within the Ediacara Member (South Australia): Redefining the Ediacaran trace fossil record one tube at a time. Journal of Paleontology 85: 256–265.

Sasaki, T., M. Kawamura, and H. Ishikawa. 1996. Nitrogen recycling in the brown planthopper, *Nilaparvata lugens*: Involvement of yeast-like endosymbionts in uric acid metabolism. Journal of Insect Physiology 42: 125–129.

Saunders, M., A.E. Glenn, and L.M. Kohn. 2010. Exploring the evolutionary ecology of fungal endophytes in agricultural systems: Using functional traits to reveal mechanisms in community processes. Evolutionary Applications 3: 525–537.

Savile, D.B.O. 1978. Paleoecology and convergent evolution in rust fungi (Uredinales). BioSystems 10: 31–36.

Savile, D.B.O. 1979. Fungi as aids in higher plant classification. Botanical Review 45: 377–503.

Saxena, R.K. 2006. A Catalogue of Tertiary Fungi From India (1989–2005), Diamond Jubilee Special Publication. Birbal Sahni Institute of Palaeobotany, Lucknow, India. 37 pp.

Saxena, R.K. 2009. Substitute names for later homonyms of five species and validation of the names of eight species of fossil fungi from Indian Tertiary sediments. Mycotaxon 110: 47–51.

Saxena, R.K. 2012. Validation of names of fossil fungi from Tertiary sediments of India. Novon 22: 223–226.

Saxena, R.K., and A.P. Bhattacharyya. 1990. Palynological investigation of the Dharmsala sediments in Dharmsala area, Kangra District, Himachal Pradesh. Geophytology 19: 109–116.

Saxena, R.K., and S. Khare. 1991. Fungal remains from the Neyveli Formation of Tiruchirapalli District, Tamil Nadu, India. Geophytology 21: 37–43.

Saxena, R.K., and N.K. Misra. 1990. Palynological investigation of the Ratnagiri Beds of Sindhu Durg District, Maharashtra. Palaeobotanist 38: 263–276.

Saxena, R.K., and P.S. Ranhotra. 2009. Palynofloral study of the intertrappean bed exposed at a new locality in Kutch district, Gujarat and its implications on palaeoenvironment and age. Journal of the Geological Society of India 74: 690–696.

Saxena, R.K., and S. Sarkar. 1986. Morphological study of *Frasnacritetrus* Taugourdeau emend. from Tertiary sediments of Himachal Pradesh, India. Review of Palaeobotany and Palynology 46: 209–225.

Saxena, R.K., and S.K.M. Tripathi. 2005. Fossil fungal spores. In: Challenges in Indian Palaeobiology: Current Status, Recent Developments and Future Directions; Diamond Jubilee National Conference (15–16 Nov.), Abstract Volume, pp. 124–125. Birbal Sahni Institute of Palaeobotany, Lucknow, India.

Saxena, R.K., and S.K.M. Tripathi. 2011. Indian fossil fungi. Palaeobotanist 60: 1–208.

Saxena, R.K., and G.K. Trivedi. 2009. Palynological investigation of the Kopili Formation (Late Eocene) in

North Cachar Hills, Assam, India. Acta Palaeobotanica 49: 253–277.

Schaarschmidt, F. 1966. Die Keuperflora von Neuewelt bei Basel. V. Ein Ascomycet in *Pterophyllum*. Schweizerische Paläontologische Abhandlungen (Mémoires Suisses de Paléontologie) 84: 67–79.

Scheckler, S.E., and J. Galtier. 2003. Tyloses and ecophysiology of the Early Carboniferous progymnosperm tree *Protopitys buchiana*. Annals of Botany 91: 739–747.

Scheublin, T.R., and M.G.A. van der Heijden. 2006. Arbuscular mycorrhizal fungi colonize nonfixing root nodules of several legume species. New Phytologist 172: 732–738.

Scheublin, T.R., K.P. Ridgway, J.P.W. Young, and M.G. van der Heijden. 2004. Nonlegumes, legumes, and root nodules harbor different arbuscular mycorrhizal fungal communities. Applied and Environmental Microbiology 70: 6240–6246.

Schickmann, S., A. Urban, K. Kräutler, U. Nopp-Mayr, and K. Hackländer. 2012. The interrelationship of mycophagous small mammals and ectomycorrhizal fungi in primeval, disturbed and managed Central European mountainous forests. Oecologia 170: 395–409.

Schigel, D.S. 2008. Collecting and rearing fungivorous Coleoptera. Revue d'écologie (La Terre et la Vie) 63: 7–12.

Schigel, D.S. 2012. Fungivory and host associations of Coleoptera: A bibliography and review of research approaches. Mycology 3: 258–272.

Schmalhausen, J. 1883. Beiträge zur Tertiär-Flora Süd-West-Russlands. In W. Dames, and E. Kayser (Eds), Palaeontologische Abhandlungen, pp. 284–336. Druck und Verlag von G. Reimer, Berlin, Germany.

Schmid, R. 1976. Septal pores in *Prototaxites*, an enigmatic Devonian plant. Science 191: 287–288.

Schmidt, A.R., and D.L. Dilcher. 2007. Aquatic organisms as amber inclusions and examples from a modern swamp forest. Proceedings of the National Academy of Sciences, USA 104: 16581–16585.

Schmidt, A.R., and H. Dörfelt. 2007. Evidence of Cenozoic Matoniaceae from Baltic and Bitterfeld amber. Review of Palaeobotany and Palynology 144: 145–156.

Schmidt, A.R., and U. Schäfer. 2005. *Leptotrichites resinatus* new genus and species: A fossil sheathed bacterium in alpine Cretaceous amber. Journal of Paleontology 79: 175–184.

Schmidt, A.R., H. von Eynatten, and M. Wagreich. 2001. The Mesozoic amber of Schliersee (southern Germany) is Cretaceous in age. Cretaceous Research 22: 423–428.

Schmidt, A.R., E. Ragazzi, O. Coppellotti, and G. Roghi. 2006. A microworld in Triassic amber. Nature 444: 835.

Schmidt, A.R., H. Dörfelt, and V. Perrichot. 2007. Carnivorous fungi from Cretaceous amber. Science 318: 1743.

Schmidt, A.R., H. Dörfelt, and V. Perrichot. 2008. *Palaeoanellus dimorphus* gen. et sp. nov. (Deuteromycotyna): A Cretaceous predatory fungus. American Journal of Botany 95: 1328–1334.

Schmidt, A.R., V. Perrichot, M. Svojtka, K.B. Anderson, K.H. Belete, R. Bussert, H. Dörfelt, S. Jancke, B. Mohr, E. Mohrmann, P.C. Nascimbene, A. Nel, P. Nel, E. Ragazzi, G. Roghi, E.E. Saupe, K. Schmidt, H. Schneider, P.A. Selden, and N. Vávra. 2010a. Cretaceous African life captured in amber. Proceedings of the National Academy of Sciences, USA 107: 7329–7334.

Schmidt, A.R., H. Dörfelt, S. Struwe, and V. Perrichot. 2010b. Evidence for fungivory in Cretaceous amber forests from Gondwana and Laurasia. Palaeontographica 283B: 157–173.

Schmidt, A.R., S. Jancke, E.E. Lindquist, E. Ragazzi, G. Roghi, P.C. Nascimbene, K. Schmidt, T. Wappler, and D.A. Grimaldi. 2012. Arthropods in amber from the Triassic Period. Proceedings of the National Academy of Sciences, USA 109: 14796–14801.

Schmidt, A.R., C. Beimforde, L.J. Seyfullah, S.-E. Wege, H. Dörfelt, V. Girard, H. Grabenhorst, M. Gube, J. Heinrichs, A. Nel, P. Nel, V. Perrichot, J. Reitner, and J. Rikkinen. 2014. Amber fossils of sooty moulds. Review of Palaeobotany and Palynology 200: 53–64.

Schmidt, H. 1992. Mikrobohrspuren ausgewählter Faziesbereiche der tethyalen und germanischen Trias (Beschreibung, Vergleich und bathymetrische Interpretation). Frankfurter Geowisenschaftliche Arbeiten, Serie A 12: 1–228.

Schmidt, S., J.A. Raven, and C. Paungfoo-Lonhienne. 2013. The mixotrophic nature of photosynthetic plants. Functional Plant Biology 40: 425–438.

Schmidt, W.J. 1964. Kanälchen von *Mycelites ossifragus* in einer fossilen Schwanen-Eieschale und die Abhängigkeit ihres Verlaufs von der Calcitstruktur. Zeitschrift für Zellforschung 62: 635–638.

Schmitt, I., G. Mueller, and H.T. Lumbsch. 2005. Ascoma morphology is homoplaseous and phylogenetically misleading in some pyrenocarpous lichens. Mycologia 97: 362–374.

Schmitz-Münker, M., and J.L. Franzen. 1988. Die Rolle von Bakterien im Verdauungstrakt mitteleozäner Vertebraten aus der Grube Messel bei Darmstadt und ihr Beitrag zur Fossildiagenese und Evolution. Courier Forschungsinstitut Senckenberg 107: 129–146.

Schneider, J. 1976. Biological and Inorganic Factors in the Destruction of limestone Coasts. Contributions to Sedimentology 6. Schweizerbart, Stuttgart, 112 pp.

Schneider, K., C. Renker, and M. Maraun. 2005. Oribatid mite (Acari, Oribatida) feeding on ectomycorrhizal fungi. Mycorrhiza 16: 67–72.

Schoch, C.L., R.A. Shoemaker, K.A. Seifert, S. Hambleton, J.W. Spatafora, and P.W. Crous. 2006. A multigene phylogeny of the Dothideomycetes using four nuclear loci. Mycologia 98: 1041–1052.

Schoch, C.L., G.-H. Sung, F. López-Giráldez, J.P. Townsend, J. Miadlikowska, V. Hofstetter, B. Robbertse, P.B. Matheny, F. Kauff, Z. Wang, C. Gueidan, R.M. Andrie, K. Trippe, L.M. Ciufetti, A. Wynns, E. Fraker, B.P. Hodkinson, G. Bonito, J.Z. Groenewald, M. Arzanlou, G.S. de Hoog, P.W. Crous, D. Hewitt, D.H. Pfister, K. Peterson, M. Gryzenhout, M.J. Wingfield, A. Aptroot, S.-O. Suh, M. Blackwell, D.M. Hillis, G.W. Griffith, L.A. Castlebury, A.Y. Rossman, H.T. Lumbsch, R. Lücking, B. Büdel, A. Rauhut, P. Diederich, D. Ertz, D.M. Geiser, K. Hosaka, P. Inderbitzin, J. Kohlmeyer, B. Volkmann-Kohlmeyer, L. Mostert, K. O'Donnell, H.

Sipman, J.D. Rogers, R.A. Shoemaker, J. Sugiyama, R.C. Summerbell, W. Untereiner, P.R. Johnston, S. Stenroos, A. Zuccaro, P.S. Dyer, P.D. Crittenden, M.S. Cole, K. Hansen, J.M. Trappe, R. Yahr, F. Lutzoni, and J.W. Spatafora. 2009a. The Ascomycota tree of life: A phylum-wide phylogeny clarifies the origin and evolution of fundamental reproductive and ecological traits. Systematic Biology 58: 224–239.

Schoch, C.L., P.W. Crous, J.Z. Groenewald, E.W.A. Boehm, T.I. Burgess, J. de Gruyter, G.S. de Hoog, L.J. Dixon, M. Grube, C. Gueidan, Y. Harada, S. Hatakeyama, K. Hirayama, T. Hosoya, S.M. Huhndorf, K.D. Hyde, E.B.G. Jones, J. Kohlmeyer, A. Kruys, Y.M. Li, R. Lücking, H.T. Lumbsch, L. Marvanová, J.S. Mbatchou, A.H. McVay, A.N. Miller, G.K. Mugambi, L. Muggia, M.P. Nelsen, P. Nelson, C.A. Owensby, A.J.L. Phillips, S. Phongpaichit, S.B. Pointing, V. Pujade-Renaud, H.A. Raja, E. Rivas Plata, B. Robbertse, C. Ruibal, J. Sakayaroj, T. Sano, L. Selbmann, C.A. Shearer, T. Shirouzu, B. Slippers, S. Suetrong, K. Tanaka, B. Volkmann-Kohlmeyer, M.J. Wingfield, A.R. Wood, J.H.C. Woudenberg, H. Yonezawa, Y. Zhang, and J.W. Spatafora. 2009b. A class-wide phylogenetic assessment of Dothideomycetes. Studies in Mycology 64: 1–15.

Schoknecht, J.D., and J.L. Crane. 1977. Revision of *Torula* and *Hormiscium* species, *Torula occulta, T. diversa, T. elasticae, T. bigemina* and *Hormiscium condensatum* reexamined. Mycologia 69: 533–546.

Schönborn, W., H. Dörfelt, W. Foissner, L. Krienitz, and U. Schäfer. 1999. A fossilized microcenosis in Triassic amber. Journal of Eukaryotic Microbiology 46: 571–584.

Schopf, J.W. 1968. Microflora of the Bitter Springs Formation, Late Precambrian, Central Australia. Journal of Paleontology 42: 651–688.

Schopf, J.W., and C. Klein (Eds). 1992. The Proterozoic Biosphere: A Multidisciplinary Study. Cambridge University Press, Cambridge, MA. 1348 pp.

Schopf, J.W., A.B. Kudryavtsev, D.G. Agresti, A.D. Czaja, and T.J. Wdowiak. 2005. Raman imagery: A new approach to assess the geochemical maturity and biogenicity of permineralized Precambrian fossils. Astrobiology 5: 333–371.

Schopf, J.W., A.B. Tripathi, and A.B. Kudryavtsev. 2006. Three-dimensional confocal optical imagery of Precambrian microscopic organisms. Astrobiology 6: 1–16.

Schopf, J.W., A.B. Kudryavtsev, and V.N. Sergeev. 2010. Confocal laser scanning microscopy and Raman imagery of the Late Neoproterozoic Chichkan microbiota of South Kazakhstan. Journal of Paleontology 84: 402–416.

Schopf, J.W., A.B. Kudryavtsev, A.B. Tripathi, and A.D. Czaja. 2011. Three-dimensional morphological (CLSM) and chemical (Raman) imagery of cellularly mineralized fossils. In P.A. Allison and D.J. Bottjer (Eds), Taphonomy: Process and Bias Through Time, Topics in Geobiology, 32, pp. 457–486. Springer Science + Business Media B.V., Dordrecht, The Netherlands.

Schrank, E. 1999. Palynology of the dinosaur beds of Tendaguru (Tanzania) – preliminary results. Fossil Record (Mitteilungen aus dem Museum für Naturkunde in Berlin, Geowissenschaftliche Reihe) 2: 171–183.

Schubert, T.S. 1981. *Strigula* Fries, the plant parasitic lichen. Plant Pathology Circular No. 227 (Florida Department of Agriculture and Consumer Services Contribution No. 511): 2 pp.

Schultz, T.R. 2007. The fungus-growing ant genus *Apterostigma* in Dominican amber. In R.R. Snelling, B.L. Fisher, and P.S. Ward (Eds), Advances in Ant Systematics (Hymenoptera: Formicidae): Homage to E.O. Wilson – 50 Years of Contributions, Issue 80, pp. 425–426. Memoirs of the American Entomological Institute, Gainesville, FL.

Schultz, T.R., and S.G. Brady. 2008. Major evolutionary transitions in ant agriculture. Proceedings of the National Academy of Sciences, USA 105: 5435–5440.

Schulz, B., and C. Boyle. 2005. The endophytic continuum. Mycological Research 109: 661–686.

Schulz, B., and C. Boyle. 2006. What are endophytes?. In B. Schulz, C. Boyle, and T. Sieber (Eds), Microbial Root Endophytes, pp. 1–13. Springer-Verlag, Berlin, Germany.

Schulz, B., C. Boyle, and T. Sieber (Eds). 2006. Microbial Root Endophytes. Springer-Verlag, Berlin, Germany. 367 pp.

Schulz, M.S., D. Vivit, C. Schulz, J. Fitzpatrick, and A. White. 2010. Biologic origin of iron nodules in a marine terrace chronosequence, Santa Cruz, California. Soil Science Society of America Journal 74: 550–564.

Schumann, G., W. Manz, J. Reitner, and M. Lustrino. 2004. Ancient fungal life in North Pacific Eocene oceanic crust. Geomicrobiology Journal 21: 241–246.

Schüßler, A. 2000. *Glomus claroideum* forms an arbuscular mycorrhiza-like symbiosis with the hornwort *Anthoceros punctatus*. Mycorrhiza 10: 15–21.

Schüßler, A., and M. Kluge. 2001. *Geosiphon pyriforme*, an endocytosymbiosis between fungus and cyanobacteria, and its meaning as a model system for arbuscular mycorrhizal research. In B. Hock (Ed.), The Mycota IX: Fungal Associations, pp. 151–161. Springer Verlag, Berlin, Germany.

Schüßler, A., and C. Walker (Eds). 2010. The Glomeromycota. A species list with new families and new genera. The Royal Botanic Garden Edinburgh, The Royal Botanic Garden Kew, Botanische Staatssammlung Munich, and Oregon State University. Electronic edition updated 2011, retrieved from www.amf-phylogeny.com; species list updated 2013, retrieved from http://schuessler.userweb.mwn.de/amphylo/.

Schüßler, A., and C. Walker. 2011. Evolution of the 'plant-symbiotic' fungal phylum, Glomeromycota. In S. Pöggeler, and J. Wöstemeyer (Eds), The Mycota XIV: Evolution of Fungi and Fungal-like Organisms, pp. 163–185. Springer-Verlag, Berlin, Germany.

Schüßler, A., E. Schnepf, D. Mollenhauer, and M. Kluge. 1995. The fungal bladders of the endocyanosis *Geosiphon pyriforme*, a *Glomus*-related fungus: Cell wall permeability indicates a limiting pore radius of only 0.5 nm. Protoplasma 185: 131–139.

Schüßler, A., P. Bonfante, E. Schnepf, D. Mollenhauer, and M. Kluge. 1996. Characterization of the *Geosiphon pyriforme* symbiosome by affinity techniques: Confocal laser scanning microscopy (CLSM) and electron microscopy. Protoplasma 190: 53–67.

Schüßler, A., D. Schwarzott, and C. Walker. 2001. A new fungal phylum, the Glomeromycota: Phylogeny and evolution. Mycological Research 105: 1413–1421.

Schüßler, A., M. Krüger, and C. Walker. 2011. Revealing natural relationships among arbuscular mycorrhizal fungi: Culture line BEG47 represents *Diversispora epigaea*, not *Glomus versiforme*. PLoS ONE 6: e23333.

Schwartzman, D.W. 2010. Was the origin of the lichen symbiosis triggered by declining atmospheric carbon dioxide levels? Bibliotheca Lichenologica 105: 191–196.

Schwarze, F.W.M.R. 2007. Wood decay under the microscope. Fungal Biology Reviews 21: 133–170.

Schwarze, F.W.M.R., C. Mattheck, and J. Engels. 2000. Fungal Strategies of Woody Decay in Trees. Springer, Heidelberg, Germany. 185 pp.

Schweiger, P.F., H. Rouhier, and B. Söderström. 2002. Visualisation of ectomycorrhizal rhizomorph structure using laser scanning confocal microscopy. Mycological Research 106: 349–354.

Schweinfurth, S.P. 2009. An introduction to coal quality. In: B.S. Pierce and K.O. Dennen (eds), The National Coal Resource Assessment Overview, pp. 16. U.S. Geological Survey Professional Paper 1625-F.

Schweitzer, H.-J. 2000. Neue Pflanzenfunde im Unterdevon der Eifel (Deutschland). Senckenbergiana Lethaea 80: 371–395.

Schweitzer, J. 2003. Die Landnahme der Pflanzen. Decheniana 156: 177–216.

Schwendemann, A.B., T.N. Taylor, E.L. Taylor, M. Krings, and N. Dotzler. 2009. *Combresomyces cornifer* from the Triassic of Antarctica: Evolutionary stasis in the Peronosporomycetes. Review of Palaeobotany and Palynology 154: 1–5.

Schwendemann, A.B., T.N. Taylor, E.L. Taylor, and M. Krings. 2010. Organization, anatomy, and fungal endophytes of a Triassic conifer embryo. American Journal of Botany 97: 1873–1883.

Schwendemann, A.B., A.-L. Decombeix, T.N. Taylor, E.L. Taylor, and M. Krings. 2011. Morphological and functional stasis in mycorrhizal root nodules as exhibited by a Triassic conifer. Proceedings of the National Academy of Sciences, USA 108: 13630–13634.

Scott, A.C. (Ed.) 1987. Coal and Coal-bearing Strata: Recent Advances. (Geological Society of London Special Publication 32. Proceedings of a conference held at the University of London, 8–10 April 1986). Blackwell, Oxford, London, Edinburgh, Boston, Palo Alto, Melbourne. viii + 332 pp.

Scott, A.C. 2000. The pre-Quaternary history of fire. Palaeogeography, Palaeoclimatology, Palaeoecology 164: 297–345.

Scott, A.C. 2002. Coal petrology and the origin of coal macerals: A way ahead? International Journal of Coal Geology 50: 119–134.

Scott, A.C., J. Galtier, and G. Clayton. 1984. Distribution of anatomically-preserved floras in the Lower Carboniferous in western Europe. Transactions of the Royal Society of Edinburgh 75: 311–340.

Scott, A.C., J. Stephenson, and M.E. Collinson. 1994. The fossil record of leaves with galls. In M.A.J. Williams (Ed.), Plant Galls: Systematics Association Special, Volume 49, pp. 447–470. Clarendon Press, Oxford, UK.

Scott, A.C., N. Pinter, M.E. Collinson, M. Hardiman, R.S. Anderson, A.P.R. Brain, S.Y. Smith, F. Marone, and M. Stampanoni. 2010. Fungus, not comet or catastrophe, accounts for carbonaceous spherules in the Younger Dryas "impact layer". Geophysical Research Letters 37: L14302.

Scott, R. 1911. On *Traquairia*. Annals of Botany 25: 459–467.

Scott, R.A. 1956. *Cryptocolax* – a new genus of fungus (Aspergillaceae) from the Eocene of Oregon. American Journal of Botany 43: 589–593.

Seaver, F.J. 1916. North American species of *Ascodesmis*. Mycologia 8: 1–4.

Seidl, M.F., G. Van der Ackerveken, F. Govers, and B. Snel. 2012. Reconstruction of oomycete genome evolution identifies differences in evolutionary trajectories leading to present-day large gene families. Genome Biology and Evolution 4: 199–211.

Seifert, K.A., and W. Gams. 2011. The genera of Hyphomycetes – 2011 update. Persoonia 27: 119–129.

Seifert, K.A., and G.J. Samuels. 2000. How should we look at anamorphs? Studies in Mycology 45: 5–18.

Seifert, K., G. Morgan-Jones, W. Gams, and B. Kendrick. 2011. The Genera of Hyphomycetes. CBS Biodiversity Series, Volume 9. Centraalbureau voor Schimmelcultures (CBS), Fungal Biodiversity Centre (KNAW), Royal Netherlands Academy of Arts and Sciences, Utrecht, The Netherlands. 997 pp.

Seilacher, A., L.A. Buatois, and M.G. Mángano. 2005. Trace fossils in the Ediacaran-Cambrian transition: Behavioral diversification, ecological turnover and environmental shift. Palaeogeography, Palaeoclimatology, Palaeoecology 227: 323–356.

Sekimoto, S., D. Rochon, J.E. Long, J.M. Dee, and M.L. Berbee. 2011. A multigene phylogeny of *Olpidium* and its implications for early fungal evolution. BMC Evolutionary Biology 11: 331.

Selbmann, L., G.S. de Hoog, A. Mazzaglia, E.L. Friedmann, and S. Onofri. 2005. Fungi at the edge of life: Cryptoendolithic black fungi from Antarctic desert. Studies in Mycology 51: 1–32.

Selkirk, D.R. 1972. Fossil *Manginula*-like fungi and their classification. Proceedings of the Linnean Society of New South Wales 97: 141–149.

Selkirk, D.R. 1975. Tertiary fossil fungi from Kiandra, New South Wales. Proceedings of the Linnean Society of New South Wales 100: 70–94.

Sellstedt, A., K. Huss-Danell, and A.-S. Ahlqvist. 1986. Nitrogen fixation and biomass production in symbioses between *Alnus incana* and *Frankia* strains with different hydrogen metabolism. Physiologia Plantarum 66: 99–107.

Selosse, M.-A. 2002. *Prototaxites*: A 400 myr old giant fossil, a saprophytic holobasidiomycete, or a lichen? Mycological Research News 106: 642–644.

Selosse, M.-A., and F. Le Tacon. 1998. The land flora: A phototroph-fungus partnership? Trends in Ecology and Evolution 13: 15–20.

Selosse, M.-A., and M. Roy. 2009. Green plants that feed on fungi: Facts and questions about mixotrophy. Trends in Plant Science 14: 64–70.

Selosse, M.-A., F. Richard, X. He, and S.W. Simard. 2006. Mycorrhizal networks: *Des liaisons dangereuses*? Trends in Ecology and Evolution 21: 621–628.

Sen, R. 2005. Towards a multifunctional rhizosphere concept: Back to the future? New Phytologist 168: 266–268.

Seo, G.S., and P.M. Kirk. 2000. Ganodermataceae: Nomenclature and classification. In J. Flood, P.D. Bridge, and M. Holderness (Eds), *Ganoderma* Diseases of Perennial Crops, pp. 3–22. CABI Publishing, Wallingford, UK.

Sérusiaux, E. 1998. Further observations on the lichen genus *Strigula* in New Zealand. Bryologist 101: 147–152.

Seward, A.C. 1898. Fossil Plants for Students of Botany and Geology, Volume I. Cambridge University Press, Cambridge, UK. 452 pp.

Shade, A., P.S. McManus, and J. Handelsman. 2013. Unexpected diversity during community succession in the apple flower microbiome. mBio 4: e00602–12.

Shah, M.A., Z.A. Reshi, and N. Rasool. 2010. Plant invasions induce a shift in Glomalean spore diversity. Tropical Ecology 51: 317–323.

Sharma, B.D., and R.P. Tripathi. 1999. *Sclerocystis* type fossil fungal sporocarp from the Early Devonian of Rhynie chert. Indian Fern Journal 16: 19–23.

Sharma, B.D., D.R. Bohra, and R. Harsh. 1993. Vesicular arbuscular mycorrhizae association in Lower Devonian plants of the Rhynie chert. Phytomorphology 43: 105–110.

Sharma, N., R.K. Kar, A. Agarwal, and R. Kar. 2005. Fungi in dinosaurian (*Isisaurus*) coprolites from the Lameta Formation (Maastrichtian) and its reflection on food habit and environment. Micropaleontology 51: 73–82.

Shaw, P.J.A. 1992. Fungi, fungivores, and fungal food webs. In G.C. Carroll, and D.T. Wicklow (Eds), The Fungal Community: Its Organization and Role in the Ecosystem, pp. 295–310. Marcel Dekker, New York.

Shear, W.A. 1991. The early development of terrestrial ecosystems. Nature 351: 283–289.

Shear, W.A., and P.A. Selden. 2001. Rustling in the undergrowth: Animals in early terrestrial ecosystems. In P.G. Gensel, and D. Edwards (Eds), Plants Invade the Land: Evolutionary and Environmental Perspectives, pp. 29–51. Columbia University Press, New York, NY.

Shearer, C.A., D.H. Langsam, and J.E. Longcore. 2004. Fungi in freshwater habitats. In G.M. Mueller, G.F. Bills, and M.S. Foster (Eds), Biodiversity of Fungi: Inventory and Monitoring Methods, pp. 513–532. Elsevier Academic Press, Burlington, MA.

Shearer, C.A., E. Descals, B. Kohlmeyer, J. Kohlmeyer, L. Marvanová, D. Padgett, D. Porter, H.A. Raja, J.P. Schmit, H.A. Thorton, and H. Voglymayr. 2007. Fungal biodiversity in aquatic habitats. Biodiversity and Conservation 16: 49–67.

Sheffy, M.V., and D.L. Dilcher. 1971. Morphology and taxonomy of fungal spores. Palaeontographica 133B: 34–51.

Shepherd, R.W., and G.J. Wagner. 2007. Phylloplane proteins: Emerging defenses at the aerial frontline? Trends in Plant Science 12: 51–56.

Shepherd, R.W., and G.J. Wagner. 2012. Fungi and leaf surfaces. In D. Southworth (Ed.). Biocomplexity of Plant–Fungal Interactions, pp. 131–154. Wiley-Blackwell, Oxford, UK.

Sherwood, M.A. 1985. Preliminary checklist of the fungi of the Clarkia flora, Unit 2, P-33. In C.J. Smiley (Ed.). Late Cenozoic History of the Pacific Northwest: Interdisciplinary Studies on the Clarkia Fossil Beds of Northern Idaho, pp. 241–244. American Association for the Advancement of Science (AAAS), Pacific Division, San Francisco, CA.

Sherwood-Pike, M.A. 1985. *Pelicothallos* Dilcher, an overlooked fossil lichen. Lichenologist 17: 114–115.

Sherwood-Pike, M.A. 1988. Freshwater fungi: Fossil record and paleoecological potential. Palaeogeography, Palaeoclimatology, Palaeoecology 62: 271–285.

Sherwood-Pike, M.A. 1990. Fossil evidence for fungus-plant interactions. In A.J. Boucot (Ed.). Evolutionary Paleobiology of Behavior and Coevolution, pp. 118–123. Elsevier, Amsterdam, The Netherlands.

Sherwood-Pike, M.A. 1991. Fossils as keys to evolution in fungi. Biosystems 25: 121–129.

Sherwood-Pike, M.A., and J. Gray. 1985. Silurian fungal remains: Probable records of the class Ascomycetes. Lethaia 18: 1–20.

Sherwood-Pike, M.A., and J. Gray. 1988. Fossil leaf-inhabiting fungi from northern Idaho and their ecological significance. Mycologia 80: 14–22.

Shete, R.H., and A.R. Kulkarni. 1977. Deuteromycetous fungal spore balls from the Deccan Intertrappean beds of Mahurzari, Maharashtra. Science and Culture 43: 458–460.

Shi, G.L., Z.Y. Zhou, and Z.M. Xie. 2010. A new *Cephalotaxus* and associated epiphyllous fungi from the Oligocene of Guangxi, South China. Review of Palaeobotany and Palynology 161: 179–195.

Shih, P.M., and N.J. Matzke. 2013. Primary endosymbiosis events date to the later Proterozoic with cross-calibrated phylogenetic dating of duplicated ATPase proteins. Proceedings of the National Academy of Sciences, USA 110: 12355–12360.

Shih, P.M., D. Wu, A. Latifi, S.D. Axen, D.P. Fewer, E. Talla, A. Calteau, F. Cai, N. Tandeau de Marsac, R. Rippka, M. Herdman, K. Sivonen, T. Coursin, T. Laurent, L. Goodwin, M. Nolan, K.W. Davenport, C.S. Han, E.M. Rubin, J.A. Eisen, T. Woyke, M. Gugger, and C.A. Kerfeld. 2013. Improving the coverage of the cyanobacterial phylum using diversity-driven genome sequencing. Proceedings of the National Academy of Sciences, USA 110: 1053–1058.

Sieber, T.N. 2007. Endophytic fungi in forest trees: Are they mutualists? Fungal Biology Reviews 21: 75–89.

Silliman, B.R., and S.Y. Newell. 2003. Fungal farming in a snail. Proceedings of the National Academy of Sciences, USA 100: 15643–15648.

Silva-Hanlin, D.M.W., and R.T. Hanlin. 1998. The order Phyllachorales: Taxonomic review. Mycoscience 39: 97–104.

Simon, L., J. Bousquet, R.C. Lévesque, and M. Lalonde. 1993. Origin and diversification of endomycorrhizal fungi and coincidence with vascular pland plants. Nature 363: 67–69.

Simoncsics, P. 1959. Palynologische Untersuchungen an der Miozänen Braunkohle des Salgótarjáner Kohlenreviers. Acta Biologica Academiae Scientiarum Hungaricae 5: 181–199.

Singer, R. 1977. An interpretation of *Palaeosclerotium*. Mycologia 69: 850–854.

Singer, R., and S. Archangelsky. 1957. Un nuevo hongo fosíl de los bosques petrificados de Santa Cruz, prov, Patagonia. Ameghiniana 1: 40–41.

Singer, R., and S. Archangelsky. 1958. A petrified Basidiomycete from Patagonia. American Journal of Botany 45: 194–198.

Singh, H.P., R.K. Saxena, and M.R. Rao. 1986. Palynology of the Barail (Oligocene) and Surma (Lower Miocene) sediments exposed along Sonapur-Badarpur Road Section, Jaintia Hills (Meghalaya) and Cachar (Assam). Part II. Fungal remains. Palaeobotanist 35: 93–105.

Singh, L.P., S.S. Gill, and N. Tuteja. 2011. Unraveling the role of fungal symbionts in plant abiotic stress tolerance. Plant Signalling and Behavior 6: 175–191.

Singh, P., C. Raghukumar, R.M. Meena, P. Verma, and Y. Shouche. 2012. Fungal diversity in deep-sea sediments revealed by culture-dependent and culture-independent approaches. Fungal Ecology 5: 543–553.

Singh, S.K., and M.S. Chauhan. 2008. Fungal remains from the Neogene sediments of Mahuadanr Valley, Latehar district, Jharkhand, India and their palaeoclimatic significance. Journal of the Palaeontological Society of India 53: 73–81.

Singhai, L.C. 1972. *Deccanosporium eocenum* gen. et. sp. nov.: A petrified fungus from the Deccan Intertrappean series. Journal of the Ravishankar University 1: 24–28.

Singhai, L.C. 1974. Fossil fungi from the Deccan Intertrappean beds of Madhya Pradesh, India. Journal of Biological Sciences 17: 92–102.

Singhai, L.C. 1975. On a petrified fungus *Palaeophthora mohgaonensis* gen. et sp. nov. from the Deccan Intertrappean beds of Mohgaonkalan. Proceedings of the 62nd Indian Science Congress, New Delhi 3: 41.

Singhai, L.C. 1978. *Palaeophthora mohgaonensis* Singhai – A fossil fungus from the Deccan Intertrappean beds of Mohgaonkalan, Chhindwara District, M.P., India. Palaeobotanist 25: 481–485.

Sjamsuridzal, W., H. Nishida, H. Ogawa, M. Kakishima, and J. Sugiyama. 1999. Phylogenetic positions of rust fungi parasitic on ferns: Evidence from 18S rDNA sequence analysis. Mycoscience 40: 21–27.

Skarby, A., and J.R. Rowley. 1993. Decay in Upper Cretaceous Normapolles exines. Grana Supplement 2: 66–69.

Skirgiełło, O.A. 1961. Flora kopalna Turowa kolo Bogatynia II. [Fossil flora of Turow near Bogatyni II]. (Ascomycetes – Roselliniaceae, Amphisphaeriaceae, ?Meliolaceae), Basidiomycetes (Polyporaceae) Prace Muzeum Ziemi 4: 5–12 [in Polish and English].

Slater, B.J., S. McLoughlin, and J. Hilton. 2012. Animal-plant interactions in a Middle Permian permineralised peat of the Bainmedart Coal Measures, Prince Charles Mountains, Antarctica. Palaeogeography, Palaeoclimatology, Palaeoecology 363–364: 109–126.

Slater, B.J., S. McLoughlin, and J. Hilton. 2013. Peronosporomycetes (Oomycota) from a Middle Permian permineralised peat within the Bainmedart Coal Measures, Prince Charles Mountains, Antarctica. PLoS ONE 8: e70707.

Smith, B.J., and K. Sivasithamparam. 2003. Morphological studies of *Ganoderma* (Ganodermataceae) from the Australasian and Pacific regions. Australian Systematic Botany 16: 487–503.

Smith, F.A., and S.E. Smith. 1997. Structural diversity in (vesicular)-arbuscular mycorrhizal symbioses. New Phytologist 137: 373–388.

Smith, F.A., and S.E. Smith. 2011. What is the significance of the arbuscular mycorrhizal colonisation of many economically important crop plants? Plant and Soil 348: 63–79.

Smith, F.A., and S.E. Smith. 2013. How useful is the mutualism-parasitism continuum of arbuscular mycorrhizal functioning? Plant and Soil 363: 7–18.

Smith, J. 1896. On the discovery of fossil microscopic plants in the fossil amber of the Ayrshire Coal-Field. Transactions of the Geological Society of Glasgow 10: 318–323.

Smith, M.R., and N.J. Butterfield. 2013. A new view on *Nematothallus*: Coralline red algae from the Silurian of Gotland. Palaeontology 56: 345–357.

Smith, P.H. 1978. Fungal spores of the genus *Ctenosporites* from the early Tertiary of southern England. Palaeontology 21: 717–722.

Smith, P.H. 1980. Trichothyriaceous fungi from the Early Tertiary of southern England. Palaeontology 23: 205–212.

Smith, P.H., and P.R. Crane. 1979. Fungal spores of the genus *Pesavis* from the Lower Tertiary of Britain. Botanical Journal of the Linnean Society 79: 243–248.

Smith, S.E., and D.J. Read. 2008. Mycorrhizal Symbiosis (Third Edition). Academic Press, London, UK. 800 pp.

Smith, S.E., and F.A. Smith. 2011. Roles of arbuscular mycorrhizas in plant nutrition and growth: New paradigms from cellular to ecosystem scales. Annual Review of Plant Biology 62: 227–250.

Smith, S.Y., and R.A. Stockey. 2007. Establishing a fossil record for the perianthless Piperales: *Saururus tuckerae* sp. nov. (Saururaceae) from the Middle Eocene Princeton Chert. American Journal of Botany 94: 1642–1657.

Smith, S.Y., R.S. Currah, and R.A. Stockey. 2004. Cretaceous and Eocene poroid hymenophores from Vancouver Island, British Columbia. Mycologia 96: 180–186.

Smith, W.G. 1877. A fossil *Peronospora*. Gardener's Chronicle 8(Oct.): 499–500.

Smith, W.G. 1878. A fossil fungus. Journal of Horticulture, Cottage Gardener and Home Farmer 34: 84–86.

Smith, W.G. 1884. Diseases of Field and Garden Crops Chiefly Such as are Caused by Fungi. Macmillan and Co., London, UK, 353 pp.

Smits, M.M. 2009. Scale matters? Exploring the effect of scale on fungal–mineral interactions. Fungal Biology Reviews 23: 132–137.

Smits, M.M., S. Bonneville, S. Haward, and J.R. Leake. 2008. Ectomycorrhizal weathering, a matter of scale? Mineralogical Magazine 72: 131–134.

Smits, M.M., A.M. Herrmann, M. Duane, O.W. Duckworth, S. Bonneville, L.G. Benning, and U. Lundström. 2009. The fungal–mineral interface: Challenges and considerations of micro-analytical developments. Fungal Biology Reviews 23: 122–131.

Smits, M.M., S. Bonneville, L.G. Benning, S.A. Banwart, and J.R. Leake. 2012. Plant-driven weathering of apatite – the role of an ectomycorrhizal fungus. Geobiology 10: 445–456.

Smoot, E.L., and T.N. Taylor. 1983. Filamentous microorganisms from the Carboniferous of North America. Canadian Journal of Botany 61: 2251–2256.

Soanes, D.M., T.A. Richards, and N.J. Talbot. 2007. Insights from sequencing fungal and oomycete genomes: What can we learn about plant disease and the evolution of pathogenicity? The Plant Cell 19: 3318–3326.

Solms-Laubach, H.G. 1891. Fossil Botany: Being an Introduction to Palaeophytology from the Standpoint of the Botanist. Clarendon Press, Oxford, UK. 401 pp.

Somrithipol, S., J. Sakayaroj, N. Rungjindamai, N. Plaingam, and E.B.G. Jones. 2008. Phylogenetic relationship of the coelomycete genus *Infundibulomyces* based on nuclear rDNA data. Mycologia 100: 735–741.

Søndergaard, M., and S. Laegaard. 1977. Vesicular-arbuscular mycorrhiza in some aquatic vascular plants. Nature 268: 232–233.

Song, B., T.-H. Li, and Y.-H. Shen. 2003. Two new *Meliola* species from China. Fungal Diversity 12: 173–177.

Song, Z.-C., and L. Cao. 1994. Late Cretaceous fungal spores from King George Island, Antarctica. In Y. Shen (Ed.). Stratigraphy and Palaeontology of Fildes Peninsula, King George Island, Antarctica, pp. 37–49. Science Press, Beijing, China.

Song, Z.-C., and F. Huang. 2002. On the classification of fossil fungi. Acta Palaeontologica Sinica 41: 471–477.

Song, Z.-C., Y. Zheng, M. Li, Y. Zhang, W. Wang, D. Wang, C. Zhao, S. Zhou, Z. Zhu, and Y. Zhao. 1999. Fossil spores and pollen of China, Volume 1. Late Cretaceous and Tertiary spores and pollen. [In Chinese with English Summary.] Science Press, Beijing, China. 910 pp.

Soomro, S., S.M. Leghari, R. Lashari, A.W. Rajar, and Q.D. Abbasi. 2010. Fossil fungal spores from Brown Coal of Sonda, District Thatta, Sindh, Pakistan. Sindh University Research Journal, Science Series 42: 73–84.

Spanos, N.P., and J. Gottlieb. 1976. Ergotism and the Salem Village witch trials. Science 194: 1390–1394.

Spanu, P.D., J.C. Abbott, J. Amselem, T.A. Burgis, D.M. Soanes, K. Stüber, E. Ver Loren van Themaat, J.K.M. Brown, S.A. Butcher, S.J. Gurr, M.-H. Lebrun, C.J. Ridout, P. Schulze-Lefert, N.J. Talbot, N. Ahmadinejad, C. Ametz, G.R. Barton, M. Benjdia, P. Bidzinski, L.V. Bindschedler, M. Both, M.T. Brewer, L. Cadle-Davidson, M.M. Cadle-Davidson, J. Collemare, R. Cramer, O. Frenkel, D Godfrey, J. Harriman, C. Hoede, B.C. King, S. Klages, J. Kleemann, D. Knoll, P.S. Koti, J. Kreplak, F.J. López-Ruiz, X. Lu, T. Maekawa, S. Mahanil, C. Micali, M.G. Milgroom, G. Montana, S. Noir, R.J. O'Connell, S. Oberhaensli, F. Parlange, C. Pedersen, H. Quesneville, R. Reinhardt, M. Rott, S. Sacristán, S.M. Schmidt, M. Schön, P. Skamnioti, H. Sommer, A. Stephens, H. Takahara, H. Thordal-Christensen, M. Vigouroux, R. Weßling, T. Wicker, and R. Panstruga. 2010. Genome expansion and gene loss in powdery mildew fungi reveal tradeoffs in extreme parasitism. Science 330: 1543–1546.

Sparrow, F.K., Jr. 1960. Aquatic Phycomycetes (Second Edition). University of Michigan Press, Ann Arbor, MI. 1185 pp.

Spatafora, J.W., G.-H. Sung, D. Johnson, C. Hesse, B. O'Rourke, M. Serdani, R. Spotss, F. Lutznoi, V. Hofstetter, J. Miadlikowska, V. Reeb, C. Gueidan, E. Fraker, T. Lumbsch, R. Lücking, I. Schmitt, K. Hosaka, A. Aptroot, C. Roux, A.N. Miller, D.M. Geiser, J. Hafellner, G. Hestmark, A.E. Arnold, B. Büdel, A. Rauhut, D. Hewitt, W.A. Untereiner, M.S. Cole, C. Scheidegger, M. Shultz, H. Sipman, and C.L. Schoch. 2006. A five-gene phylogeny of Pezizomycotina. Mycologia 98: 1018–1028.

Spatafora, J.W., C.A. Owensby, G.W. Douhan, E.W.A. Boehm, and C.L. Schoch. 2012. Phylogenetic placement of the ectomycorrhizal genus *Cenococcum* in Gloniaceae (Dothideomycetes). Mycologia 104: 758–765.

Spears, D.A. 2012. The origin of tonsteins, an overview, and links with seatearths, fireclays and fragmental clay rocks. International Journal of Coal Geology 94: 22–31.

Spegazzini, C.L. 1889. Fungi Puiggariani. Pugillus 1. Boletin de la Academia Nacional de Ciencias en Córdoba 11: 381–622.

Spencer, J. 1893. Recreations in fossil botany. (Sporocarpons and Zygosporites.). Hardwicke's Science-Gossip: An Illustrated Medium of Interchange and Gossip for Students and Lovers of Nature 19: 155–158.

Spencer, M.A., M.C. Vick, and M.W. Dick. 2002. Revision of *Aplanopsis*, *Pythiopsis*, and 'subcentric' *Achlya* species (Saprolegniaceae) using 18S rDNA and morphological data. Mycological Research 106: 549–560.

Speranza, M., J. Wierzchos, J. Alonso, L. Bettuchi, A. Martín-González, and C. Ascaso. 2010. Traditional and new microscopy techniques applied to the study of microscopic fungi included in amber. In A. Méndez-Vilas and J. Díaz (eds), Microscopy: Science, Technology, Application and Education, Volume 2, pp. 1135–1145. Formatex, Badajoz, Spain.

Spring, S. 2002. The genera *Leptothrix* and *Sphaerotilus*. In M. Dworkin (ed.). The Prokaryotes: An evolving Electronic Resource for the Microbiological Community. Release 3.9. Springer-Verlag, New York, NY.

Srivastava, A.K. 1993. Evidence of fungal parasitism in the *Glossopteris* flora of India. 12 Congrès International de la Stratigraphie Géologie du Carbonifère et Permien, Comptes Rendus, Volume 2: 141–146.

Srivastava, A.K. 2007. Fossil evidence of gall-inducing arthropod-plant interactions in the Indian subcontinent. Oriental Insects 41: 213–222.

Srivastava, A.K., and R. Tewari. 1994. Possible evidence of bacterial degradation in *Glossopteris* flora of India. Palaeobotanist 42: 174–177.

Srivastava, R., and R.K. Kar. 2004. Record of new petrified epiphyllous fungi (*Polyhyphaethyrites*) from the Deccan Intertrappean beds of Mohgaon Kalan, Madhya Pradesh, India. Current Science 87: 866–867.

Srivastava, R., D.K. Kapgate, and S. Chatterjee. 2009. Permineralized fungal remains in the fossil wood of *Barringtonia* from the Deccan Intertrappean sediments of Yavatmal District, Maharashtra, India. Palaeobotanist 58: 11–19.

Srivastava, S.C., A.K. Srivastava, A.P. Bhattacharyya, and R. Tewari. 1999. Degraded Permian palynomorphs from North-East Himalaya, India. Permophiles 33: 32–36.

Srivastava, S.K. 1968. Fungal elements from the Edmonton Formation (Maestrichtian), Alberta, Canada. Canadian Journal of Botany 46: 1115–1118.

Srivastava, S.K. 1976. Biogenic infection in Jurassic spores and pollen. Geoscience and Man 15: 95–100.

Stach, E., and W. Pickhardt. 1957. Pilzreste (Sklerotinit) in paläozoischen Steinkohlen. Paläontologische Zeitschrift 31: 139–162.

Stach, E., and W. Pickhardt. 1964. Tertiäre und karbonische Pilzreste (Sklerotinit). Fortschritte in der Geologie von Rheinland und Westfalen 12: 377–392.

Stackebrandt, E., C. Koch, O. Gvozdiak, and P. Schumann. 1995. Taxonomic dissection of the genus *Micrococcus*: *Kocuria* gen. nov., *Nesterenkonia* gen. nov., *Kytococcus* gen. nov., *Dermacoccus* gen. nov., and *Micrococcus* Cohn 1872 gen. emend. International Journal of Systematic Bacteriology 45: 682–692.

Stanevich, A.M., E.N. Maksimova, T.A. Kornilova, A.M. Mazukabzov, and D.P. Gladkochub. 2007. Microfossils of the Late Proterozoic Debengdinskaya formation of the Olenekskiy uplift. Bulletin of the Tomsk Polytechnic University 311: 9–14.

Stanevich, A.M., A.A. Postnikov, T.A. Kornilova, A.A. Terleev, and N.V. Popov. 2013. Bacterial, fungal, and algal microfossils in the Lower Proterozoic Baikal region of Siberia (Udokan and Sayany Mountains). Paleontological Journal 47: 977–983.

Stanosz, G.R., D.R. Smith, M.A. Guthmiller, and J.C. Stanosz. 1997. Persistence of *Sphaeropsis sapinea* on or in asymptomatic shoots of red and jack pines. Mycologia 89: 525–530.

Steiger, M., A.E. Charola, and K. Sterflinger. 2011. Weathering and deterioration. In S. Siegesmund, and R. Snethlage (Eds), Stone in Architecture: Properties, Durability, pp. 227–316 (Fourth Edition). Springer-Verlag, Berlin, Germany.

Stein, W.E., G.D. Harmon, and F.M. Hueber. 1993. *Spongiophyton* from the Lower Devonian of North America reinterpreted as a lichen. Abstract. American Journal of Botany 80: 93.

Steiner, M.B., Y. Eshet, M.R. Rampino, and D.M. Schwindt. 2003. Fungal abundance spike and the Permian-Triassic boundary in the Karoo Supergroup (South Africa). Palaeogeography, Palaeoclimatology, Palaeoecology 194: 405–414.

Steinhauer, H. 1818. On fossil reliquia of unknown vegetables in the coal strata. Transactions of the American Philosophical Society, New Series 1: 265–297.

Stephens, W.E. 2012. Whewellite and its key role in living systems. Geology Today 28: 180–185.

Stephenson, S.L., and H. Stempen. 1994. Myxomycetes: A Handbook of Slime Molds. Timber Press, Portland, OR. 200 pp.

Stephenson, S.L., M. Schnittler, and Y.K. Novozhilov. 2008. Myxomycete diversity and distribution from the fossil record to the present. Biodiversity and Conservation 17: 285–301.

Sterflinger, K. 2000. Fungi as geologic agents. Geomicrobiology Journal 17: 97–124.

Sternberg, K.M. Graf von. 1820. Versuch einer geognostischen botanischen Darstellung der Flora der Vorwelt. Vol. 1(1). F. Fleischer, Leipzig. 24 pp.

Sternberg, S. 1994. The emerging fungal threat. Science 266: 1632–1634.

Stidd, B.M., and K. Cosentino. 1975. *Albugo*-like oogonia from the American Carboniferous. Science 190: 1092–1093.

Stireman, J.O., III, H. Devlin, T.G. Carr, and P. Abbot. 2010. Evolutionary diversification of the gall midge genus *Asteromyia* (Cecidomyiidae) in a multitrophic ecological context. Molecular Phylogenetics and Evolution 54: 194–210.

Stocker-Wörgötter, E., and J.A. Elix. 2004. Experimental studies of lichenized fungi: Formation of rare depsides and dibenzofuranes by the culture mycobiont of *Bunodophoron patagonicum* (Sphaerophoraceae, lichenized Ascomycota). Bibliotheca Lichenologica 88: 659–669.

Stockey, R.A., G.W. Rothwell, H.D. Addy, and R.S. Currah. 2001. Mycorrhizal association of the extinct conifer *Metasequoia milleri*. Mycological Research 105: 202–205.

Stokland, J.N., J. Siitonen, and B.G. Jonsson. 2012. Biodiversity in Dead Wood. Cambridge University Press, Cambridge, UK. 509 pp.

Stone, G.N., R.W.J.M. van der Ham, and J.G. Brewer. 2008. Fossil oak galls preserve ancient multitrophic interactions. Proceedings of the Royal Society 275B: 2213–2219.

Stone, J., and G. Carroll. 1985. Observations on the development of ascocarps in *Phaeocryptopus gaeumannii* and on the possible existence of an anamorphic state. Sydowia, Annales Mycologici Ser. II 38: 317–323.

Stone, J.K., C.W. Bacon, and J.F. White., Jr. 2000. An overview of endophytic microbes: endophytism defined. In C.W. Bacon, and J.F. White (Eds), Microbial Endophytes, pp. 3–29. Marcel Dekker, New York, NY.

Stone, J.K., J.D. Polishook, and J.R.J. White. 2004. Endophytic fungi. In G. Mueller, G.F. Bills, and M.S. Foster (Eds), Biodiversity of Fungi: Inventory and Monitoring Methods, pp. 241–270. Elsevier Academic Press, Burlington, MA.

Stone, J.K., B.R. Capitano, and J.L. Kerrigan. 2008. The histopathology of *Phaeocryptopus gaeumannii* on Douglas-fir needles. Mycologia 100: 431–444.

Stopes, M.C. 1935. On the petrology of banded bituminous coals. Fuel 14: 4–13.

Stopes, M.C., and K. Fujii. 1911. Studies on the structure and affinities of Cretaceous plants. Philosophical Transactions of the Royal Society of London 201B: 1–90.

Størmer, L. 1963. *Gigantoscorpio willsi*: A new scorpion from the Lower Carboniferous of Scotland and its associated preying microorganisms. Skrifter utgitt av det Norske Videnskaps-Akademi i Oslo, Matematisk-Naturvidenskapelig Klasse 8: 1–171.

Straus, A. 1950. *Trametites undulatus* n. sp., ein Baumschwamm aus der rheinischen Braunkohle. Braunkohle Wärme und Energie 2: 342.

Straus, A. 1952a. Beiträge zur Pliocänflora von Willershausen III. Die niederen Pflanzengruppen bis zu den Gymnospermen. Palaeontographica 93B: 1–44.

Straus, A. 1952b. Pilze aus dem Pliozän von Willershausen (Kr. Osterode, Harz). Zeitschrift für Pilzkunde 21: 11–14.

Straus, A. 1956. Pilze und Coniferen aus dem Pliozän von Willershausen. Mitteilungen der Deutschen Dendrologischen Gesellschaft 59: 34–37.

Straus, A. 1961. Neue Pyrenomyceten aus dem Pliocän von Willershausen. Zeitschrift für Pilzkunde 27: 88–90.

Stretch, R.C., and H.A. Viles. 2002. The nature and rate of weathering by lichens on lava flows on Lanzarote. Geomorphology 47: 87–94.

Strittmatter, M., C.M.M. Gachon, and F.C. Küpper. 2009. Ecology of lower oomycetes. In K. Lamour, and S. Kamoun (Eds), Oomycete Genetics and Genomics: Diversity, Interactions, and Research Tools, pp. 25–46. John Wiley & Sons, Inc., Hoboken, NJ.

Strother, P.K. 1988. New species of *Nematothallus* from the Silurian Bloomsburg Formation of Pennsylvania. Journal of Paleontology 62: 967–982.

Strother, P.K. 1993. Clarification of the genus *Nematothallus* Lang. Journal of Paleontology 67: 1090–1094.

Strother, P.K. 2000. The cryptospore record indicates a Cambrian origin for land plants. Annual meeting of the Botanical Society of America, Portland, Oregon, 6–10 August [Abstract].

Strother, P.K., and A. Traverse. 1979. Plant microfossils from Llandoverian and Wenlockian rocks of Pennsylvania. Palynology 3: 1–21.

Strother, P.K., G. Wood, W. Taylor, and J. Beck. 2004. Middle Cambrian cryptospores and the origin of land plants. Memoir of the Association of Australasian Palaeontologists 29: 99–113.

Strother, P.K., L. Battison, M.D. Brasier, and C.H. Wellman. 2011. Earth's earliest non-marine eukaryotes. Nature 473: 505–509.

Strullu-Derrien, C., and D.-G. Strullu. 2007. Mycorrhization of fossil and living plants. Comptes Rendus Palevol 6: 483–494.

Strullu-Derrien, C., J.-P. Rioult, and D.-G. Strullu. 2009. Mycorrhizas in Upper Carboniferous *Radiculites*-type cordaitalean rootlets. New Phytologist 182: 561–564.

Strullu-Derrien, C., P. Kenrick, J.P. Rioult, and D.G. Strullu. 2011. Evidence of parasitic Oomycetes (Peronosporomycetes) infecting the stem cortex of the Carboniferous seed fern *Lyginopteris oldhamia*. Proceedings of the Royal Society 278B: 675–680.

Strullu-Derrien, C., S. McLoughlin, M. Philippe, A. Mørk, and D.G. Strullu. 2012. Arthropod interactions with bennettitalean roots in a Triassic permineralized peat from Hopen, Svalbard Archipelago (Arctic). Palaeogeography, Palaeoclimatology, Palaeoecology 348–349: 45–58.

Strullu-Derrien, C., P. Kenrick, S. Pressel, J.G. Duckett, J.P. Rioult, and D.G. Strullu. 2014. Fungal associations in *Horneophyton ligneri* from the Rhynie Chert (c. 407 million year old) closely resemble those in extant lower land plants: novel insights into ancestral plant–fungus symbioses. New Phytologist, Doi:http://dx.doi.org/10.1111/nph.12805.

Stubblefield, S.P., and H.P. Banks. 1983. Fungal remains in the Devonian trimerophyte *Psilophyton dawsonii*. American Journal of Botany 70: 1258–1261.

Stubblefield, S.P., and T.N. Taylor. 1983. Studies of Paleozoic fungi. I. The structure and organization of *Traquairia* (Ascomycota). American Journal of Botany 70: 387–399.

Stubblefield, S.P., and T.N. Taylor. 1984. Fungal remains in the lycopod megaspore *Triletes rugosus* (Loose) Schopf. Review of Palaeobotany and Palynology 41: 199–204.

Stubblefield, S.P., and T.N. Taylor. 1986. Wood decay in silicified gymnosperms from Antarctica. Botanical Gazette 147: 116–125.

Stubblefield, S.P., and T.N. Taylor. 1988. Recent advances in palaeomycology. New Phytologist 108: 3–25.

Stubblefield, S.P., T.N. Taylor, C.E. Miller, and G.T. Cole. 1983. Studies of Carboniferous fungi. II. The structure and organization of *Mycocarpon, Sporocarpon, Dubiocarpon* and *Coleocarpon* (Ascomycotina). American Journal of Botany 70: 1482–1498.

Stubblefield, S.P., T.N. Taylor, C.E. Miller, and G.T. Cole. 1984. Studies of Paleozoic fungi. III. Fungal parasitism in a Pennsylvanian gymnosperm. American Journal of Botany 71: 1275–1282.

Stubblefield, S.P., T.N. Taylor, and C.B. Beck. 1985a. Studies of Paleozoic fungi. IV. Wood-decaying fungi in *Callixylon newberryi* from the Upper Devonian. American Journal of Botany 72: 1765–1774.

Stubblefield, S.P., T.N. Taylor, and C.E. Miller. 1985b. Studies of Paleozoic fungi IV: Wall ultrastructure of fossil endogonaceous chlamydospores. Mycologia 77: 83–96.

Stubblefield, S.P., C.E. Miller, T.N. Taylor, and G.T. Cole. 1985c. *Geotrichites glaesarius*, a conidial fungus from Tertiary Dominican amber. Mycologia 77: 11–16.

Stubblefield, S.P., T.N. Taylor, and J.M. Trappe. 1987a. Vesicular-arbuscular mycorrhizae from the Triassic of Antarctica. American Journal of Botany 74: 1904–1911.

Stubblefield, S.P., T.N. Taylor, and R.L. Seymour. 1987b. A possible endogonaceous fungus from the Triassic of Antarctica. Mycologia 79: 905–906.

Stürmer, S.L. 2012. A history of the taxonomy and systematics of arbuscular mycorrhizal fungi belonging to the phylum Glomeromycota. Mycorrhiza 22: 247–258.

Suetrong, S., C.L. Schoch, J.W. Spatafora, J. Kohlmeyer, B. Volkmann-Kohlmeyer, J. Sakayaroj, S. Phongpaichit, K. Tanaka, K. Hirayama, and E.B.G. Jones. 2009. Molecular systematics of the marine Dothideomycetes. Studies in Mycology 64: 155–173.

Sugiyama, J. 1987. Pleomorphic Fungi: The Diversity and Its Taxonomic Implications. Elsevier, Amsterdam, Oxford, New York, Tokyo. 342 pp.

Sugiyama, J., K. Hosaka, and S.-O. Suh. 2006. Early diverging Ascomycota: Phylogenetic divergence and related evolutionary enigmas. Mycologia 98: 996–1005.

Suh, S.-O., and M. Blackwell. 1999. Molecular phylogeny of the cleistothecial fungi placed in Cephalothecaceae and Pseudeurotiaceae. Mycologia 91: 836–848.

Suh, S.-O., M. Blackwell, C.P. Kurtzman, and M.-A. Lachance. 2006. Phylogenetics of Saccharomycetales, the ascomycete yeasts. Mycologia 98: 1006–1017.

Sundram, S., S. Meon, I.A. Seman, and R. Othman. 2011. Symbiotic interaction of endophytic bacteria with arbuscular mycorrhizal fungi and its antagonistic effect on *Ganoderma boninense*. Journal of Microbiology 49: 551–557.

Sung, G.-H., N.L. Hywel-Jones, J.-M. Sung, J.J. Luangsa-ard, B. Shrestha, and J.W. Spatafora. 2007. Phylogenetic classification of *Cordyceps* and the clavicipitaceous fungi. Studies in Mycology 57: 5–59.

Sung, G.-H., G.O. Poinar, Jr., and J.W. Spatafora. 2008. The oldest fossil evidence of animal parasitism by fungi supports a Cretaceous diversification of fungal-arthropod symbioses. Molecular Phylogenetics and Evolution 49: 495–502.

Sutton, B.C. 1980. The Coelomycetes: Fungi Imperfecti with Pycnidia, Acervuli and Stromata. Commonwealth Mycological Institute, Kew, UK. 696 pp.

Sutton, B.C. 1996. Conidiogenesis, classification, and correlation. In B.C. Sutton (Ed.). A Century of Mycology, pp. 135–160. British Mycological Society, Cambridge University Press, Cambridge, UK.

Suzuki, Y. 1910. On the structure and affinities of two new conifers and a new fungus from the Upper Cretaceaous of Hokkaidō (Yezo). Botanical Magazine 24: 181–196.

Sverdrup, H. 2009. Chemical weathering of soil minerals and the role of biological processes. Fungal Biology Reviews 23: 94–100.

Swann, E.C., E.M. Frieders, and D.J. McLaughlin. 2001. Urediniomycetes. In D.J. McLaughlin, E.G. McLaughlin, and P.E. Lemke (Eds), The Mycota VIIB: Systematics and Evolution, pp. 37–56. Springer-Verlag, Berlin, Germany.

Swanson-Hysell, N.L., A.C. Maloof, J.L. Kirschvink, D.A.D. Evans, G.P. Halverson, and M.T. Hurtgen. 2012. Constraints on Neoproterozoic paleogeography and Paleozoic orogenesis from paleomagnetic records of the Bitter Springs Formation, Amadeus Basin, Central Australia. American Journal of Science 312: 817–884.

Swart, H.J. 1975a. Callosities in fungi. Transactions of the British Mycological Society 64: 511–515.

Swart, H.J. 1975b. Australian leaf-inhabiting fungi: VII. Further studies in *Vizella*. Transactions of the British Mycological Society 64: 301–306.

Swatzell, L.J., M.J. Powell, and J.Z. Kiss. 1996. The relationship of endophytic fungi to the gametophyte of the fern *Schizaea pusilla*. International Journal of Plant Sciences 157: 53–62.

Swe, A., J. Li, K.Q. Zhang, S.B. Pointing, R. Jeewon, and K.D. Hyde. 2011. Nematode-trapping fungi. Current Research in Environmental & Applied Mycology 1: 1–26.

Tadych, M., M.S. Torres, and J.F. White., Jr. 2009. Diversity and ecological roles of clavicipitaceous endophytes of grasses. In J.F. White Jr., and M.S. Torres (Eds), Defensive Mutualism in Microbial Symbioses, pp. 247–256. CRC Press, Boca Raton, FL.

Tafforeau, P., R. Boistel, E. Boller, A. Bravin, M. Brunet, Y. Chaimanee, P. Cloetens, M. Feist, J. Hoszowska, J.-J. Jaeger, R.F. Kay, V. Lazzari, L. Marivaux, A. Nel, C. Nemoz, X. Thibault, P. Vignaud, and S. Zabler. 2006. Applications of X-ray synchrotron microtomography for non-destructive 3D studies of paleontological specimens. Applied Physics A 83: 195–202.

Takamatsu, S. 2013. Origin and evolution of the powdery mildews (Ascomycota, Erysiphales). Mycoscience 54: 75–86.

Takamatsu, S., S. Niinomi, M. Harada, and M. Havrylenko. 2010. Molecular phylogenetic analyses reveal a close evolutionary relationship between *Podosphaera* (Erysiphales: Erysiphaceae) and its rosaceous hosts. Persoonia 24: 38–48.

Tanabe, Y., M.M. Watanabe, and J. Sugiyama. 2005. Evolutionary relationships among basal fungi (Chytridiomycota and Zygomycota): Insights from molecular phylogenetics. Journal of General and Applied Microbiology 51: 267–276.

Tanai, T. 1987. A bracket fungus from the Miocene, west of Kobe City, western Japan. Journal of Japanese Botany 62: 1–6.

Tanaka, K., K. Hirayama, H. Yonezawa, S. Hatakeyama, Y. Harada, T. Sano, T. Shirouzu, and T. Hosoya. 2009. Molecular taxonomy of bambusicolous fungi: Tetraplosphaeriaceae, a new pleosporalean family with *Tetraploa*-like anamorphs. Studies in Mycology 64: 175–209.

Tanesaka, E., H. Masuda, and K. Kinugawa. 1993. Wood degrading ability of basidiomycetes that are wood descomposers, litter descomposers, or mycorrhizal symbionts. Mycologia 85: 347–354.

Tang, Y., and B. Lian. 2012. Diversity of endolithic fungal communities in dolomite and limestone rocks from Nanjiang Canyon in Guizhou karst area, China. Canadian Journal of Microbiology 58: 685–693.

Tanner, L.H., and S.G. Lucas. 2013. Degraded wood in the Upper Triassic Petrified Forest Formation (Chinle Group), northern Arizona: Differentiating fungal rot from arthropod boring. New Mexico Museum of Natural History and Science Bulletin 61: 582–588.

Taprab, Y., T. Johjima, Y. Maeda, S. Moriya, S. Trakulnaleamsai, N. Naparatnaraporn, M. Ohkuma, and T. Kudo. 2005. Symbiotic fungi produce laccases potentially involved in phenol degradation in fungus combs of fungus-growing termites in Thailand. Applied and Environmental Microbiology 71: 7696–7704.

Tarkka, M.T., and P. Frey-Klett. 2008. Mycorrhiza helper bacteria. In A. Varma (ed.). Mycorrhiza: Genetics and Molecular Biology, Eco-Function, Biotechnology, Eco-Physiology, Structure and Systematics, pp. 113–132 (Third Edition). Springer Verlag, Berlin, Germany.

Taugourdeau, P. 1968. Sur un curieux microfossile incertae sedis du Frasnien du Boulonnais *Frasnacritetrus* nov. gen. (Acritarche). Cahiers de Micropaléontologie, Série I. 10: 1–4.

Taylor, A.F.S. 2006. Common mycelial networks: Life-lines and radical addictions. New Phytologist 169: 6–8.

Taylor, A.F.S., and I. Alexander. 2005. The ectomycorrhizal symbiosis: Life in the real world. Mycologist 19: 102–112.

Taylor, J.W. 1995. Making the Deuteromycota redundant: A practical integration of mitosporic and meiosporic fungi. Canadian Journal of Botany 73: 754–759.

Taylor, J.W., and M.L. Berbee. 2006. Dating divergences in the Fungal Tree of Life: Review and new analyses. Mycologia 98: 838–849.

Taylor, L.L., J.R. Leake, J. Quirk, K. Hardy, S.A. Banwart, and D.J. Beerling. 2009. Biological weathering and the long-term carbon cycle: Integrating mycorrhizal evolution and function into the current paradigm. Geobiology 7: 171–191.

Taylor, L.L., S. Banwart, J. Leake, and D.J. Beerling. 2011. Modeling the evolutionary rise of ectomycorrhiza on subsurface weathering environments and the geochemical carbon cycle. American Journal of Science 311: 369–403.

Taylor, L.L., S.A. Banwart, P.J. Valdes, J.R. Leake, and D.J. Beerling. 2012. Evaluating the effects of terrestrial ecosystems, climate and carbon dioxide on weathering over geological time: A global-scale process-based approach. Philosophical Transactions of the Royal Society 367B: 565–582.

Taylor, P.D., and M.A. Wilson. 2003. Palaeoecology and evolution of marine hard substrate communities. Earth-Science Reviews 62: 1–103.

Taylor, T.N. 1982. The origin of land plants: A paleobotanical perspective. Taxon 31: 155–177.

Taylor, T.N. 1988. The origin of land plants: Some answers, more questions. Taxon 37: 805–833.

Taylor, T.N. 1993a. Fungi. In M.J. Benton (Ed.), The Fossil Record, Volume 2, pp. 9–13. Chapman and Hall, London, UK.

Taylor, T.N. 1993b. The role of Late Paleozoic fungi in understanding the terrestrial paleoecosystem. In S. Archangelsky (Ed.), Comptes Rendu 12e Congrès International de Géologie du Carbonifère-Permian, Vol. 2, pp. 147–154. Buenos Aires, Argentina. 22–27 September, 1991.

Taylor, T.N. 1994. The fossil history of ascomycetes. In D.L. Hawksworth (Ed.). Ascomycete Systematics: Problems and Perspectives in the Nineties, pp. 167–174. Plenum Press, New York, NY.

Taylor, T.N., and J.M. Osborn. 1996. The importance of fungi in shaping the paleoecosystem. Review of Palaeobotany and Palynology 90: 249–262.

Taylor, T.N., and A.C. Scott. 1983. Interactions of plants and animals during the Carboniferous. BioScience 33: 488–493.

Taylor, T.N., and E.L. Taylor. 1993. The Biology and Evolution of Fossil Plants. Prentice Hall, Englewood Cliffs, NJ. 982 pp.

Taylor, T.N., and E.L. Taylor. 1997. The distribution and interactions of some Paleozoic fungi. Review of Palaeobotany and Palynology 95: 83–94.

Taylor, T.N., and J.F. White., Jr. 1989. Fossil fungi (Endogonaceae) from the Triassic of Antarctica. American Journal of Botany 76: 389–396.

Taylor, T.N., W. Remy, and H. Hass. 1992a. Fungi from the Lower Devonian Rhynie chert: Chytridiomycetes. American Journal of Botany 79: 1233–1241.

Taylor, T.N., H. Hass, and W. Remy. 1992b. Devonian fungi: Interactions with the green alga *Palaeonitella*. Mycologia 84: 901–910.

Taylor, T.N., W. Remy, and H. Hass. 1992c. Parasitism in a 400-million-year-old green alga. Nature 357: 493–494.

Taylor, T.N., W. Remy, and H. Hass. 1994a. *Allomyces* in the Devonian. Nature 367: 601.

Taylor, T.N., J. Galtier, and B.J. Axsmith. 1994b. Fungi from the Lower Carboniferous of central France. Review of Palaeobotany and Palynology 83: 253–260.

Taylor, T.N., W. Remy, H. Hass, and H. Kerp. 1995a. Fossil arbuscular mycorrhizae from the Early Devonian. Mycologia 87: 560–573.

Taylor, T.N., H. Hass, W. Remy, and H. Kerp. 1995b. The oldest fossil lichen. Nature 378: 244.

Taylor, T.N., H. Hass, and H. Kerp. 1997. A cyanolichen from the Lower Devonian Rhynie chert. American Journal of Botany 84: 992–1004.

Taylor, T.N., H. Hass, and H. Kerp. 1999. The oldest fossil ascomycetes. Nature 399: 648.

Taylor, T.N., S.D. Klavins, M. Krings, E.L. Taylor, H. Kerp, and H. Hass. 2004. Fungi from the Rhynie chert: A view from the dark side. Transactions of the Royal Society of Edinburgh: Earth Sciences 94: 457–473.

Taylor, T.N., H. Hass, H. Kerp, M. Krings, and R.T. Hanlin. 2005a. Perithecial ascomycetes from the 400 million year old Rhynie chert: An example of ancestral polymorphism. Mycologia 97: 269–285.

Taylor, T.N., M. Krings, S.D. Klavins, and E.L. Taylor. 2005b. *Protoascon missouriensis*, a complex fossil microfungus revisited. Mycologia 97: 725–729.

Taylor, T.N., H. Kerp, and H. Hass. 2005c. Life history biology of early land plants: Deciphering the gametophyte phase. Proceedings of the National Academy of Sciences, USA 102: 5892–5897.

Taylor, T.N., M. Krings, and H. Kerp. 2006. *Hassiella monospora* nov. gen. et sp., a microfungus from the 400 million year old Rhynie chert. Mycological Research 110: 628–632.

Taylor, T.N., E.L. Taylor, and M. Krings. 2009. Paleobotany: The Biology and Evolution of Fossil Plants. Academic Press, New York, NY. 1230 pp.

Taylor, T.N., E.L. Taylor, A.-L. Decombeix, A. Schwendemann, R. Serbet, I. Escapa, and M. Krings. 2010. The enigmatic Devonian fossil *Prototaxites* is not a rolled-up liverwort mat: Comment on the paper by Graham et al. (AJB 97: 268–275). American Journal of Botany 97: 1075–1078.

Taylor, T.N., M. Krings, N. Dotzler, and J. Galtier. 2011. The advantage of thin section preparations over acetate peels in the study of late Paleozoic fungi and other microorganisms. PALAIOS 26: 239–244.

Taylor, T.N., M. Krings, J. Galtier, and N. Dotzler. 2012. Fungal endophytes in *Astromyelon*-type (Sphenophyta, Equisetales, Calamitaceae) roots from the Upper Pennsylvanian of France. Review of Palaeobotany and Palynology 171: 9–18.

Taylor, T.N., M. Krings, and E.L. Taylor. 2014. Fungal Diversity in the Fossil Record. In D.J. McLaughlin, and J.W. Spatafora (Eds), The Mycota VIIB: Systematics and Evolution. Springer, In press.

Taylor, W.A., and P.K. Strother. 2008. Ultrastructure of some Cambrian palynomorphs form the Bright Angel Shale, Arizona, USA. Review of Palaeobotany and Palynology 151: 41–50.

Taylor, W.A., and P.K. Strother. 2009. Ultrastructure, morphology, and topology of Cambrian palynomorphs from the Lone Rock Formation, Wisconsin, USA. Review of Palaeobotany and Palynology 153: 296–309.

Taylor, W.A., and C.H. Wellman. 2009. Ultrastructure of enigmatic phytoclasts (banded tubes) from the Silurian-Lower Devonian: Evidence for affinities and role in early terrestrial ecosystems. PALAIOS 24: 167–180.

Taylor, W.A., C. Free, C. Boyce, R. Helgemo, and J. Ochoada. 2004. SEM analysis of *Spongiophyton* interpreted as a fossil lichen. International Journal of Plant Sciences 165: 875–881.

Tedersoo, L., K. Hansen, B.A. Perry, and R. Kjøller. 2006. Molecular and morphological diversity of pezizalean ectomycorrhiza. New Phytologist 170: 581–596.

Tedersoo, L., T.W. May, and M.E. Smith. 2010. Ectomycorrhizal lifestyle in fungi: Global diversity, distribution, and evolution of phylogenetic lineages. Mycorrhiza 20: 217–263.

Tehon, L.R. 1937. Preservation of fungi in ancient wood. Transactions of the Illinois State Academy of Sciences 30: 147–149.

Tellenbach, C., and T.N. Sieber. 2012. Do colonization by dark septate endophytes and elevated temperature affect pathogenicity of oomycetes? FEMS Microbiology Ecology 82: 157–168.

Teramoto, M., B. Wu, and T. Hogetsu. 2012. Transfer of ^{14}C-photosynthate to the sporocarp of an ectomycorrhizal fungus *Laccaria amethystina*. Mycorrhiza 22: 219–225.

Tesei, D., G. Marzban, K. Zakharova, D. Isola, L. Selbmann, and K. Sterflinger. 2012. Alteration of protein patterns in black rock inhabiting fungi as a response to different temperatures. Fungal Biology 116: 932–940.

Teterevnikova-Babayan, D.N., and M.G. Taslakhchian. 1973. New species of fossil fungi occurring in Armenia. Mycology and Phytology 7: 180–188. [in Russian].

Thaxter, R. 1888. The Entomophthoreae of the United States. Memoirs of the Boston Society of Natural History 4: 133–201.

Thaxter, R. 1891. On certain new or peculiar North American Hyphomycetes. II. *Helicocephalum, Gonatorrhodiella, Desmidiospora* nov. genera and *Everhartia lignatilis* n. sp. Botanical Gazette 16: 201–205.

Thaxter, R. 1922. A revision of the Endogoneae. Proceedings of the American Academy of Arts and Sciences 57: 291–351.

Thines, M., and S. Kamoun. 2010. Oomycete-plant coevolution: recent advances and future prospects. Current Opinion in Plant Biology 13: 427–433.

Thomas, G.M., and G.O. Poinar., Jr. 1988. A fossil *Aspergillus* from Eocene Dominican amber. Journal of Paleontology 62: 141–143.

Thomas, J. 2009. Taxonomy (broad classification). In: J. Thomas, Meliolaceous Fungi in Peppara and Neyyar Wildlife Sanctuaries in Kerala State, Ph.D. Thesis, pp. 36–88. Kannur University, Kerala, India.

Thomas, K. 1848. On the amber beds of East Prussia. Annals and Magazine of Natural History, Series 2, Volume 2(12): 369–380.

Thompson, J.N. 2006. Mutualistic webs of species. Science 312: 372–373.

Thompson, R.H., and D.E. Wujek. 1997. Trentepohliales: *Cephaleuros, Phycopeltis*, and *Stomatochroon*. Morphology, Taxonomy, and Ecology. Science Publishers Inc., Enfield, NH. 149 pp.

Thorn, R.G., M. Scholler, and W. Gams. 2008. Have carnivorous fungi been found in Cretaceous amber? Mycological Research 112: 611–612.

Tibbett, M., and F.E. Sanders. 2002. Ectomycorrhizal symbiosis can enhance plant nutrition through improved access to discrete organic nutrient patches of high resource quality. Annals of Botany 89: 783–789.

Tibell, L. 1992. Crustose lichens as indicators of forest continuity in boreal coniferous forests. Nordic Journal of Botany 12: 427–450.

Tibell, L. 2001. Photobiont association and molecular phylogeny of the lichen genus *Chaenotheca*. Bryologist 104: 191–198.

Tibell, L., and A. Koffman. 2002. *Chaenotheca nitidula*, a new species of calicioid lichen from northeastern North America. Bryologist 105: 353–357.

Tibell, L., and M. Wedin. 2000. Mycocaliciales, a new order for nonlichenized calicioid fungi. Mycologia 92: 577–581.

Tiffney, B.H. 1981. Re-evaluation of *Geaster florissantensis* (Oligocene, North America). Transactions of the British Mycological Society 76: 493–495.

Tiffney, B.H., and E.S. Barghoorn. 1974. The fossil record of the fungi. Occasional Papers of the Farlow Herbarium of Cryptogamic Botany 7: 1–42.

Timdal, E. 2008. Studies on *Phyllopsora* (Ramalinaceae) in Peru. Lichenologist 40: 337–362.

Timoféev, B.V. 1970. Une découverte de Phycomycetes dans le Précambrien. Review of Palaeobotany and Palynology 10: 79–81.

Ting, W.S., and A. Nissenbaum. 1986. Fungi in the Lower Cretaceous Amber from Israel. Special Publication by the Exploration and Development Research Center, Chinese Petroleum Corporation, Miaoli, Taiwan. 27 pp.

Tipping, R. 1987. The origins of corroded pollen grains at five early postglacial pollen sites in western Scotland. Review of Palaeobotany and Palynology 53: 151–161.

Tisserant, E., M. Malbreil, A. Kuo, A. Kohler, A. Symeonidi, R. Balestrini, P. Charron, N. Duensing, N. Frei dit Frey, V. Gianinazzi-Pearson, L.B. Gilbert, Y. Handa, J.R. Herr, M. Hijri, R. Koul, M. Kawaguchi, F. Krajinski, P.J. Lammers, F.G. Masclaux, C. Murat, E. Morin, S. Ndikumana, M. Pagni, D. Petitpierre, N. Requena, P. Rosikiewicz, R. Riley, K. Saito, H. San Clemente, H. Shapiro, D. van Tuinen, G. Bécard, P. Bonfante, U. Paszkowski, Y.Y. Shachar-Hill, G.A. Tuskan, P.W. Young, I.R. Sanders, B. Henrissat, S.A. Rensing, I.V. Grigoriev, N. Corradi, C. Roux, and F. Martin. 2013. Genome of an arbuscular mycorrhizal fungus provides insight into the oldest plant symbiosis. Proceedings of the National Academy of Sciences, USA 110: 20117–20122.

Titov, A., and L. Tibell. 1993. *Chaenothecopsis* in the Russian Far East. Nordic Journal of Botany 13: 313–329.

Tobler, D.J., A. Stefánsson, and L.G. Benning. 2008. In-situ grown silica sinters in Icelandic geothermal areas. Geobiology 6: 481–502.

Toju, H., S. Yamamoto, H. Sato, A.S. Tanabe, G.S. Gilbert, and K. Kadowaki. 2013. Community composition of root-associated fungi in a *Quercus*-dominated temperate forest: "Codominance" of mycorrhizal and root-endophytic fungi. Ecology and Evolution 3: 1281–1293.

Tomescu, A.M.F., and G.W. Rothwell. 2006. Wetlands before tracheophytes: Thalloid terrestrial communities of the Early

Silurian Passage Creek biota (Virginia). In S.F. Greb, and W.A. DiMichele (Eds), Wetlands through Time, Geological Society of America Special Paper 399 pp. 41–56. Geological Society of America, Boulder, CO.

Tomescu, A.M.F., R. Honegger, and G.W. Rothwell. 2008. Earliest fossil record of bacterial-cyanobacterial mat consortia: The Early Silurian Passage Creek biota (440 Ma, Virginia, USA). Geobiology 6: 120–124.

Tomescu, A.M.F., L.M. Pratt, G.W. Rothwell, P.K. Strother, and G.C. Nadon. 2009. Carbon isotopes support the presence of extensive land floras pre-dating the origin of vascular plants. Palaeogeography, Palaeoclimatology, Palaeoecology 283: 46–59.

Tomescu, A.M.F., R.W. Tate, N.G. Mack, and V.J. Calder. 2010. Simulating fossilization to resolve the taxonomic affinities of thalloid fossils in Early Silurian (ca. 425 Ma) terrestrial assemblages. Bibliotheca Lichenologica 105: 183–189.

Toome, M., and M.C. Aime. 2012. Pucciniomycetes. Version 30 January 2012 (under construction). http://tolweb.org/Pucciniomycetes/51246/2012.01.30 in The Tree of Life Web Project, http://tolweb.org/.

Torrecillas, E., M.M. Alguacil, and A. Roldán. 2012. Host preferences of arbuscular mycorrhizal fungi colonizing annual herbaceous plant species in semiarid Mediterranean prairies. Applied and Environmental Microbiology 78: 6180–6186.

Traverse, A. 2007. Paleopalynology (Second Edition). Springer, Dordrecht, The Netherlands. 813 pp.

Traverse, A., and S.R. Ash. 1994. Well-preserved fungal spores from Jurassic rocks of Hells Canyon on the Idaho-Oregon border. Journal of Paleontology 68: 664–668.

Trewin, N.H. 1996. The Rhynie cherts: An early Devonian ecosystem preserved by hydrothermal activity. In G.R. Bock and J.A. Goode (Eds), Evolution of Hydrothermal Ecosystems on Earth (and Mars?) – Ciba Foundation Symposium, Volume 202, pp. 131–149. John Wiley and Sons, Chichester, UK.

Trewin, N.H., and A.H. Knoll. 1999. Preservation of Devonian chemotrophic filamentous bacteria in calcite veins. PALAIOS 14: 288–294.

Tribollet, A. 2007. The boring microflora in modern coral reef ecosystems: A review of its roles. In: M. Wisshak, and L. Tapanila (eds), Current Developments in Bioerosion, pp. 67–94. Erlangen Conference Series. Springer-Verlag, Berlin, Germany.

Tribollet, A., S. Golubic, G. Radtke, and J. Reitner. 2011. On microbiocorrosion. In J. Reitner, N.-V. Quéric, and G. Arp (Eds), Advances in Stromatolite Geobiology, Lecture Notes in Earth Sciences 131 pp. 265–276. Springer-Verlag, Berlin, Germany.

Trinci, A.P.J., and A.J. Collinge. 1974. Occlusion of the septal pores of damaged hyphae of *Neurospora crassa* by hexagonal crystals. Protoplasma 80: 57–67.

Tripathi, A. 2001. Fungal remains from Early Cretaceous Intertrappean Beds of Rajmahal Formation in Rajmahal Basin, India. Cretaceous Research 22: 565–574.

Tripathi, S.K.M. 1988. Algal and fungal remains from Jowai-Sonapur Road Section (Palaeocene-Eocene), Meghalaya. Palaeobotanist 37: 63–76.

Tripathi, S.K.M. 2012. The systematic and evolutionary perspectives of fossil fungi. In J.K. Misra, J.P. Tewari, and S.K. Deshmukh (Eds), Systematics and Evolution of Fungi, pp. 15–27. Science Publishers, Enfield, NH.

Tripathi, S.K.M., and Saxena, R.K. 2005. Fossil microthyriaceous fungi. In: Challenges in Indian Palaeobiology: Current Status, Recent Developments and Future Directions, Diamond Jubilee National Conference (15–16 Nov.) Abstract Volume, 182 pp. Birbal Sahni Institute of Palaeobotany, Lucknow, India.

Trivedi, B.S., and C.L. Verma. 1970. Fungal remains from Tertiary coal bed of Malaya. Journal of Palynology 5: 68–73.

Trivedi, B.S., S.K. Chaturvedi, and C.L. Verma. 1973. A new fossil fungus *Ascodesmisites malayensis* gen. et sp. nov. from Tertiary coals of Malaya. Geophytology 3: 126–129.

Trueman, C.N., and N. Tuross. 2002. Trace elements in recent and fossil bone apatite. In M.J. Kohn, J. Rakovan, and J.M. Hughes (Eds), Phosphates: Geochemical, Geobiological, and Materials Importance, Reviews in Mineralogy and Geochemistry, Volume 48, pp. 489–521. Mineralogical Society of America, Washington, DC.

Tsui, C.K.M., S. Sivichai, and M.L. Berbee. 2006. Molecular systematics of *Helicoma*, *Helicomyces* and *Helicosporium* and their teleomorphs inferred from rDNA sequences. Mycologia 98: 94–104.

Tsuneda, A. 1983. Fungal Morphology and Ecology – Mostly Scanning Electron Microscopy. Tottori Mycological Institute, Tokyo, Japan. 320 pp.

Tucker, S.L., and N.J. Talbot. 2001. Surface attachment and pre-penetration stage development by plant pathogenic fungi. Annual Review of Phytopathology 39: 385–417.

Tuovila, H., J.R. Cobbinah, and J. Rikkinen. 2011. *Chaenothecopsis khayensis*, a new resinicolous calicioid fungus on African mahogany. Mycologia 103: 610–615.

Tuovila, H., A.R. Schmidt, C. Beimforde, H. Dörfelt, H. Grabenhorst, and J. Rikkinen. 2013. Stuck in time – a new *Chaenothecopsis* species with proliferating ascomata from *Cunninghamia* resin and its fossil ancestors in European amber. Fungal Diversity 58: 199–215.

Turnau, K., T. Anielska, and A. Jurkiewicz. 2005. Mycothallic/mycorrhizal symbiosis of chlorophyllous gametophytes and sporophytes of a fern, *Pellaea viridis* (Forsk.) Prantl (Pellaeaceae, Pteridales). Mycorrhiza 15: 121–128.

Tynni, R. 1977. Microthyriaceae-sieni, Tertiäärinen mikrofossilityyppi lapista. Geological Survey of Finland, Report of Investigation 17: 31–36 [in Finnish].

Uchman, A., A. Gaigalas, M. Melešytė, and V. Kazakauskas. 2007. The trace fossil *Asthenopodichnium lithuanicum* isp. nov. from Late Neogene brown-coal deposits, Lithuania. Geological Quarterly 51: 329–336.

Uhl, D., A. Abu Hamad, H. Kerp, and K. Bandel. 2007. Evidence for palaeo-wildfire in the Late Permian palaeoptropics – charcoalified wood from the Um Irna Formation of Jordan. Review of Palaeobotany and Palynology 144: 221–230.

Ulloa, M., and R.T. Hanlin. 2012. Illustrated Dictionary of Mycology (Second Edition). APS (American Phytopathological Society) Press, St. Paul, MN. 784 pp.

Unger, F. von. 1841–47. Chloris Protogaea: Beiträge zur Flora der Vorwelt. Engelmann, Leipzig, Germany. 150 pp.

Unterseher, M. 2011. Diversity of fungal endophytes in temperate forest trees. In A.M. Pirttilä, and A.C. Frank (Eds), Endophytes of Forest Trees: Biology and Applications, Forestry Sciences, Volume 80, pp. 31–46. Springer Science + Business Media B.V., Dordrecht, The Netherlands.

Upchurch, G.R., Jr., and R.A. Askin. 1989. Latest Cretaceous and earliest Tertiary dispersed plant cuticles from Seymour Island. Antarctic Journal of the United States 24: 7–10.

Upreti, D.K., and S. Chatterjee. 2007. Significance of lichens and their secondary metabolites: A review. In B.N. Ganguli, and S.K. Deshmukh (Eds), Fungi: Multifaceted Microbes, pp. 169–188. Anamaya Publishers, New Delhi, India.

U'Ren, J.M., F. Lutzoni, J. Miadlikowska, A.D. Laetsch, and A.E. Arnold. 2012. Host and geographic structure of endophytic and endolichenic fungi at a continental scale. American Journal of Botany 99: 898–914.

Usher, K.M., B. Bergman, and J.A. Raven. 2007. Exploring cyanobacterial mutualisms. Annual Review of Ecology, Evolution, and Systematics 38: 255–273.

Usuki, F., and K. Narisawa. 2007. A mutualistic symbiosis between a dark septate endophytic fungus, *Heteroconium chaetospira*, and a nonmycorrhizal plant, Chinese cabbage. Mycologia 99: 175–184.

Uzuhashi, S., M. Tojo, and M. Kakishima. 2010. Phylogeny of the genus *Pythium* and description of new genera. Mycoscience 51: 337–365.

Vajda, V., and S. McLoughlin. 2004. Fungal proliferation at the Cretaceous-Tertiary boundary. Science 303: 1489.

van Aarle, I.M., T.R. Cavagnaro, S.E. Smith, F.A. Smith, and S. Dickson. 2005. Metabolic activity of *Glomus intraradices* in Arum- and Paris-type arbuscular mycorrhizal colonization. New Phytologist 166: 611–618.

Van der Ham, R.W.J.M., and R.W. Dortangs. 2005. Structurally preserved ascomycetous fungi from the Maastrichtian type area (NE Belgium). Review of Palaeobotany and Palynology 136: 48–62.

Van der Hammen, T. 1954a. The development of Colombian flora throughout geologic periods: I, Maestrichtian to Lower Tertiary. Boletín Geológico 2: 49–106.

Van der Hammen, T. 1954b. Principios para la nomenclatura palinológica sistemática. Boletín Geológico 2: 1–21.

van der Heijden, M.G.A., R.D. Bardgett, and N.M. van Straalen. 2008. The unseen majority: Soil microbes as drivers of plant diversity and productivity in terrestrial ecosystems. Ecology Letters 11: 296–310.

Van der Merwe, M.M., J. Walker, L. Ericson, and J.J. Burdon. 2008. Coevolution with higher taxonomic host groups within the *Puccinia*/*Uromyces* rust lineage obscured by host jumps. Mycological Research 112: 1387–1408.

Van Driel, K.G.A. 2007. Septal Pore Caps in Basidiomycetes, Composition and Ultrastructure. Doctoral Thesis, Utrecht University, Utrecht, The Netherlands. 133 pp.

van Geel, B. 1978. A palaeoecological study of Holocene peat bog sections in Germany and The Netherlands, based on the analysis of pollen, spores and macro- and microscopic remains of fungi, algae, cormophytes and animals. Review of Palaeobotany and Palynology 25: 1–120.

van Geel, B. 2001. Non-pollen palynomorphs. In J.P. Smol, H.J.B. Birks, and W.M. Last (Eds), Tracking Environmental Change Using Lake Sediments. Terrestrial, Algal and Siliceous Indicators, Volume 3, pp. 99–119. Kluwer Academic Publishers, Dordrecht, The Netherlands.

van Geel, B., and A. Aptroot. 2006. Fossil ascomycetes in Quaternary deposits. Nova Hedwigia 82: 313–329.

van Geel, B., A.G. Klink, J.P. Pals, and J. Wiegers. 1986. An Upper Eemian lake deposit from Twente, eastern Netherlands. Review of Palaeobotany and Palynology 47: 31–61.

van Geel, B., A. Aptroot, and D. Mauquoy. 2006. Sub-fossil evidence for fungal hyperparasitism (*Isthmospora spinosa* on *Meliola ellisii*, on *Calluna vulgaris*) in a Holocene intermediate ombrotrophic bog in northern-England. Review of Palaeobotany and Palynology 141: 121–126.

Vann, S.R., and R.A. Taber. 1985. *Annellophora* leaf spot of date palm in Texas. Plant Disease 69: 903–904.

Vannette, R.L., and M.D. Hunter. 2009. Mycorrhizal fungi as mediators of defence against insect pests in agricultural systems. Agricultural and Forest Entomology 11: 351–358.

Van Tieghem, P. 1879. Sur le ferment butyrique (*Bacillus Amylobacter*) à l'époque de la houille. Comptes rendus hebdomadaires des séances de l'Académie des Sciences 89: 1102–1104.

Varadpande, D.G., and V. Sampath. 1981. A new fossil sphaeropsidalean fungus from the Deccan Intertrappean beds of Madhya Pradesh, India. Botanical journal of Maharashtra Vidnyan Mandir, India 16: 15–22.

Varma, C.P., and M.S. Rawat. 1963. A note on some diporate grains recovered from Tertiary horizons of India and their potential marker value. Grana Palynologica 4: 130–139.

Varma, Y.N.R., and R.S. Patil. 1985. Fungal remains from the Tertiary carbonaceous clays of Tonakkal area, Kerala. Geophytology 15: 151–158.

Vasconcelos, C., J.A. McKenzie, S. Bernasconi, D. Grujic, and A.J. Tiens. 1995. Microbial mediation as a possible mechanism for natural dolomite formation at low temperatures. Nature 377: 220–222.

Vasiliauskas, R., A. Menkis, R.D. Finlay, and J. Stenlid. 2007. Wood-decay fungi in fine living roots of conifer seedlings. New Phytologist 174: 441–446.

Veiga, R.S.L., A. Faccio, A. Genre, C.M.J. Pieterse, P. Bonfante, and M.G.A. van der Heijden. 2013. Arbuscular mycorrhizal fungi reduce growth and infect roots of the non-host plant *Arabidopsis thaliana*. Plant, Cell & Environment 36: 1926–1937.

Vélez, C.G., P.M. Letcher, S. Schultz, M.J. Powell, and P.F. Churchill. 2011. Molecular phylogenetic and zoospore ultrastructural analyses of *Chytridium olla* establish the limits of a monophyletic Chytridiales. Mycologia 103: 118–130.

Venkatachala, B.S., and R.K. Kar. 1969. Palynology of the Tertiary sediments of Kutch. 1. Spores and pollen from bore-hole no. 14. Palaeobotanist 17: 157–178.

Verbruggen, E., S.D. Veresoglou, I.C. Anderson, T. Caruso, E.C. Hammer, J. Kohler, and M.C. Rillig. 2013. Arbuscular mycorrhizal fungi – short-term liability but long-term benefits for soil carbon storage? New Phytologist 197: 366–368.

Veresoglou, S.D., and M.C. Rillig. 2012. Suppression of fungal and nematode plant pathogens through arbuscular mycorrhizal fungi. Biology Letters 8: 214–217.

Veresoglou, S.D., G. Menexes, and M.C. Rillig. 2012. Do arbuscular mycorrhizal fungi affect the allometric partition of host plant biomass to shoots and roots? A meta-analysis of studies from 1990 to 2010. Mycorrhiza 22: 227–235.

Veronese, P., M.T. Ruiz, M.A. Coca, A. Hernandez-Lopez, H. Lee, J.I. Ibeas, B. Darnsz, J.M. Pardo, P.M. Hasegawa, R.A. Bressan, and M.L. Narasimhan. 2003. In defense against pathogens. Both plant sentinels and foot soldiers need to know the enemy. Plant Physiology 131: 1580–1590.

Verrecchia, E.P. 2000. Fungi and sediments. In R.E. Riding, and S.M. Awramik (Eds), Microbial Sediments, pp. 68–75. Springer-Verlag, Berlin, Germany.

Verrecchia, E.P., and K.E. Verrecchia. 1994. Needle-fiber calcite: A critical review and a proposed classification. Journal of Sedimentary Research A64: 650–664.

Vijaya, and K.L. Meena. 1996. Corpuscules in the Permian pollen from India. Journal of the Palaeontological Society of India 41: 57–61.

Vijaya, and S. Murthy. 2012. Palynomorphs and oribatid mites–from the Denwa Formation, Satpura Basin, Madhya Pradesh, India. International Journal of Geosciences 3: 195–205.

Vishnu-Mittre, 1973. Studies of fungal remains from the Flandrian deposits in the Whittlesey Mere region, Hunts., England. Palaeobotanist 22: 89–109.

Visscher, H., H. Brinkhuis, D.L. Dilcher, W.C. Elsik, Y. Eshet, C.V. Looy, M.R. Rampino, and A. Traverse. 1996. The terminal Paleozoic fungal event: Evidence of terrestrial ecosystem, destabilization and collapse. Proceedings of the National Academy of Sciences, USA 93: 2155–2158.

Visscher, H., M.A. Sephton, and C.V. Looy. 2011. Fungal virulence at the time of the end-Permian biosphere crisis? Geology 39: 883–886.

Voets, L., I.E. de la Providencia, and S. Declerck. 2006. Glomeraceae and Gigasporaceae differ in their ability to form hyphal networks. New Phytologist 172: 185–188.

Vogel, K., and C.E. Brett. 2009. Record of microendoliths in different facies of the Upper Ordovician in the Cincinnati Arch region USA: The early history of light-related microendolithic zonation. Palaeogeography, Palaeoclimatology, Palaeoecology 281: 1–24.

Vogel, K., S. Golubic, and C.E. Brett. 1987. Endolith associations and their relation to facies distribution in the Middle Devonian of New York State, U.S.A. Lethaia 20: 263–290.

Vogel, K., M. Bundschuh, I. Glaub, K. Hofmann, G. Radtke, and H. Schmidt. 1995. Hard substrate ichnocoenoses and their relations to light intensity and marine bathymetry. Neues Jahrbuch für Geologie und Paläontologie Abhandlungen 195: 49–61.

Vogel, K., M. Gektidis, S. Golubic, W.E. Kiene, and G. Radtke. 2000. Experimental studies on microbial bioerosion at Lee Stocking Island, Bahamas and One Tree Island, Great Barrier Reef. Australia: Implications for paleoecological reconstructions. Lethaia 33: 190–204.

Vohník, M., J.J. Sadowsky, P. Kohout, Z. Lhotáková, F. Nestby, and M. Kolarík. 2012. Novel root-fungus symbiosis in Ericaceae: Sheathed ericoid mycorrhiza formed by a hitherto undescribed basidiomycete with affinities to Trechisporales. PLoS ONE 7: e39524.

Vorholt, J.A. 2012. Microbial life in the phyllosphere. Nature Reviews Microbiology 10: 828–840.

Vreeland, R.H., W.D. Rosenzweig, and D.W. Powers. 2000. Isolation of a 250 million-year-old halotolerant bacterium from a primary salt crystal. Nature 407: 897–900.

Vreeland, R.H., W.D. Rosenzweig, T. Lowenstein, C. Satterfield, and A. Ventosa. 2006. Fatty acid and DNA analyses of Permian bacteria isolated from ancient salt crystals reveal differences with their modern relatives. Extremophiles 10: 71–78.

Wacey, D., N. McLoughlin, M.R. Kilburn, M. Saunders, J.B. Cliff, C. Kong, M.E. Barley, and M.D. Brasier. 2013. Nanoscale analysis of pyritized microfossils reveals differential heterotrophic consumption in the ~1.9-Ga Gunflint chert. Proceedings of the National Academy of Sciences, USA 110: 8020–8024.

Waggoner, B. 1993. Fossil actinomycetes and other bacteria in Eocene amber from Washington State, USA. Tertiary Research 14: 155–160.

Waggoner, B.M. 1994. Fossil actinomycete in Eocene-Oligocene Dominican amber. Journal of Paleontology 68: 398–401.

Waggoner, B.M. 1995. Ediacaran lichens: A critique. Paleobiology 21: 393–397.

Waggoner, B.M. 1996. Bacteria and protists from Middle Cretaceous amber of Ellsworth County, Kansas. PaleoBios 17: 20–26.

Waggoner, B.M., and G.O. Poinar., Jr. 1992. A fossil myxomycete plasmodium from Eocene-Oligocene amber of the Dominican Republic. Journal of Protozoology 39: 639–642.

Waggoner, B.M., and M.F. Poteet. 1996. Unusual oak leaf galls from the Middle Miocene of northwestern Nevada. Journal of Paleontology 70: 1080–1084.

Wagner, C.A., and T.N. Taylor. 1981. Evidence for endomycorrhizae in Pennsylvanian age plant fossils. Science 212: 562–563.

Wagner, C.A., and T.N. Taylor. 1982. Fungal chlamydospores from the Pennsylvanian of North America. Review of Palaeobotany and Palynology 37: 317–328.

Wagner, R.H., and M.P. Castro. 1998. *Neuropteris obtusa*, a rare but widespread Late Carboniferous pteridosperm. Palaeontology 41: 1–22.

Wainwright, M., S. Alharbi, and N.C. Wickramasinghe. 2006. How do microorganisms reach the stratosphere? International Journal of Astrobiology 5: 13–15.

Wainwright, M., A. Laswd, and F. Alshammari. 2009. Bacteria in amber coal and clay in relation to lithopanspermia. International Journal of Astrobiology 8: 141–143.

Walker, C. 1983. Taxonomic concepts in the Endogonaceae: Spore wall characteristics in species descriptions. Mycotaxon 18: 443–455.

Walker, C. 1985. *Endogone lactiflua* forming ectomycorrhizas with *Pinus contorta*. Transactions of the British Mycological Society 84: 353–355.

Walker, C. 1987. Formation and dispersal of propagules of endogonaceous fungi. In G. Pegg, and P. Ayers (Eds), Fungal Infection in Plants, pp. 269–284. Cambridge University Press, Cambridge, MA.

Walker, C., and A. Schüßler. 2004. Nomenclatural clarifications and new taxa in the Glomeromycota. Mycological Research 108: 981–982.

Walker, C., M. Vestberg, F. Demircik, H. Stockinger, M. Saito, H. Sawaki, I. Nishmura, and A. Schüssler. 2007. Molecular phylogeny and new taxa in the Archaeosporales (Glomeromycota): *Ambispora fennica* gen. sp. nov., Ambisporaceae fam. nov., and emendation of *Archaeospora* and Archaeosporaceae. Mycological Research 111: 137–153.

Walker, S.E., and W. Miller., III. 1992. Organism-substrate relations: Toward a logical terminology. PALAIOS 7: 236–238.

Walton, J. 1928. A method for preparing fossil plants. Nature 122: 571.

Wanderlei-Silva, D., E.R. Neto, and R. Hanlin. 2003. Molecular systematics of the Phyllachorales (Ascomycota, fungi) based on 18S ribosomal DNA sequences. Brazilian Archives of Biology and Technology 46: 315–322.

Wang, B., and Y.-L. Qiu. 2006. Phylogenetic distribution and evolution of mycorrhizas in land plants. Mycorrhiza 16: 299–363.

Wang, B., L.H. Yeun, J.-Y. Xue, Y. Liu, J.-M. Ané, and Y.-L. Qiu. 2010. Presence of three mycorrhizal genes in the common ancestor of land plants suggests a key role of mycorrhizas in the colonization of land by plants. New Phytologist 186: 514–525.

Wang, C.J.K. and R.A. Zabel (Eds). 1990. Identification Manual for Fungi from Utility Poles in the Eastern United States. American Type Culture Collection, Rockville, MD. 356 pp.

Wang, H.K., A. Aptroot, P.W. Crous, K.D. Hyde, and R. Jeewon. 2007. The polyphyletic nature of Pleosporales: An example from *Massariosphaeria* based on rDNA and RBP2 gene phylogenies. Mycological Research 111: 1268–1276.

Wang, J., H.W. Pfefferkorn, Y. Zhang, and Z. Feng. 2012. Permian vegetational Pompeii from Inner Mongolia and its implications for landscape paleoecology and paleobiogeography of Cathaysia. Proceedings of the National Academy of Sciences, USA 109: 4927–4932.

Wang, X., M. Krings, and T.N. Taylor. 2010. A thalloid organism with possible lichen affinity from the Jurassic of northeastern China. Review of Palaeobotany and Palynology 162: 591–598.

Wang, Y., E.D. Tretter, R.W. Lichtwardt, and M.M. White. 2013. Overview of 75 years of *Smittium* research, establishing a new genus for *Smittium culisetae*, and prospects for future revisions of the 'Smittium' clade. Mycologia 105: 90–111.

Wang, Z., P.R. Johnston, S. Takamatsu, J.W. Spatafora, and D.S. Hibbett. 2006. Toward a phylogenetic classification of the Leotiomycetes based on rDNA data. Mycologia 98: 1065–1075.

Wang, Z., P.R. Johnston, Z.L. Yang, and J.P. Townsend. 2009. Evolution of reproductive morphology in leaf endophytes. PLoS ONE 4: e4246.

Wang, Z.-Q. 1991. Advances on the Permo-Triassic lycopods in north China. I. An *Isoetes* from the Mid-Triassic in northern Shaanxi Province. Palaeontographica 222B: 1–30.

Wappler, T., and T. Denk. 2011. Herbivory in early Tertiary Arctic forests. Palaeogeography, Palaeoclimatology, Palaeoecology 310: 283–295.

Ward, H.M. 1883. XVI. On the morphology and the development of the perithecium of *Meliola*, a genus of tropical epiphyllous fungi. Philosophical Transactions of the Royal Society of London 174: 583–599.

Warnock, R.C.M., Z.-H. Yang, and P.C.J. Donoghue. 2012. Exploring uncertainty in the calibration of the molecular clock. Biology Letters 8: 156–159.

Watanabe, K., H. Nishida, and T. Kobayashi. 1999. Cretaceous Deuteromycetes on a cycadeoidalean bisexual cone. International Journal of Plant Sciences 160: 435–443.

Waterhouse, G.M. 1975. Key to the species of *Entomophthora* Fres. Bulletin of the British Mycological Society 9: 15–41.

Waters, C.M., and B.L. Bassler. 2005. Quorum sensing: Cell-to-cell communication in bacteria. Annual Review of Cell and Developmental Biology 21: 319–346.

Webb, J.A., and W.B.K. Holmes. 1982. Three new thalloid fossils from the Middle Triassic of eastern Australia. Proceedings of the Royal Society of Queensland 93: 83–88.

Wedin, M., H. Döring, and G. Gilenstam. 2004. Saprotrophy and lichenization as option for the same fungal species on different substrata: Environmental plasticity and fungal lifestyles in the *Stictis-Conotrema* complex. New Phytologist 164: 459–465.

Wehner, J., P.M. Antunes, J.R. Powell, T. Caruso, and M.C. Rillig. 2011. Indigenous arbuscular mycorrhizal fungal assemblages protect grassland host plants from pathogens. PLoS ONE 6: e27381.

Wei, J., D. Peršoh, and R. Agerer. 2010. Four ectomycorrhizae of Pyronemataceae (Pezizomycetes) on Chinese Pine (*Pinus tabulaeformis*): Morpho-anatomical and molecular-phylogenetic analyses. Mycological Progress 9: 267–280.

Weinstein, R.N., D.H. Pfister, and T. Iturriaga. 2002. A phylogenetic study of the genus *Cookeina*. Mycologia 94: 673–682.

Weir, A., and M. Blackwell. 2001. Molecular data support the Laboulbeniales as a separate class of Ascomycota, Laboulbeniomycetes. Mycological Research 105: 1182–1190.

Weir, A., and M. Blackwell. 2005. Fungal biotrophic parasites of insects and other arthropods. In F.E. Vega, and M. Blackwell (Eds), Insect-Fungal Associations: Ecology and Evolution, pp. 119–148. Oxford University Press Inc., New York, NY.

Weir, J.T., and D. Schluter. 2008. Calibrating the avian molecular clock. Molecular Ecology 17: 2321–2328.

Weiss, A.F., and P.Y. Petrov. 1994. Main patterns in the facies-ecological distribution of microfossils in the Riphean basins of Siberia. Stratigrafiya Geologicheskaya korrelyatsiya 2: 97–129.

Weiss, F.E. 1904a. A probable parasite of Stigmarian rootlets. New Phytologist 3: 63–68.

Weiss, F.E. 1904b. A mycorhiza from the Lower Coal-Measures. Annals of Botany 18: 255–264.

Weiss, F.E. 1906. On the tyloses of *Rachiopteris corrugata*. New Phytologist 5: 82–85.

Weiss, M., Z. Sýkorová, S. Garnica, K. Riess, F. Martos, C. Krause, F. Oberwinkler, R. Bauer, and D. Redecker. 2011. Sebacinales everywhere: Previously overlooked ubiquitous fungal endophytes. PLoS ONE 6: e16793.

Welch, J.J., and L. Bromham. 2005. Molecular dating when rates vary. Trends in Ecology and Evolution 20: 320–327.

Weldon, C., L.H. du Preez, A.D. Hyatt, R. Muller, and R. Speare. 2004. Origin of the amphibian chytrid fungus. Emerging Infectious Diseases 10: 2100–2105.

Wellman, C.H. 1995. "Phytodebris" from Scottish Silurian and Lower Devonian continental deposits. Review of Palaeobotany and Palynology 84: 255–279.

Wellman, C.H. 2004. Palaeoecology and palaeophytogeography of the Rhynie chert plants: Evidence from integrated analysis of *in situ* and dispersed spores. Proceedings of the Royal Society of London 271B: 985–992.

Wellman, C.H. 2006. Spore assemblages from the Lower Devonian 'Lower Old Red Sandstone' deposits of the Rhynie outlier, Scotland. Transactions of the Royal Society of Edinburgh: Earth Sciences 97: 167–211.

Wellman, C.H., and J. Gray. 2000. The microfossil record of early land plants. Philosophical Transactions of the Royal Society 355B: 717–732.

Wellman, C.H., P.L. Osterloff, and U. Mohiuddin. 2003. Fragments of the earliest land plants. Nature 425: 282–285.

Wellman, C.H., H. Kerp, and H. Hass. 2006. Spores of the Rhynie chert plant *Aglaophyton (Rhynia) major* (Kidston and Lang) D.S. Edwards, 1986. Review of Palaeobotany and Palynology 142: 229–250.

Weresub, L.K., and K.A. Pirozynski. 1979. Pleomorphism of fungi as treated in the history of mycology and nomenclature. In B. Kendrick (Ed.), The Whole Fungus: The Sexual-Asexual Synthesis, Volume 1, pp. 17–25. National Museum of Natural Sciences, National Museums of Canada, Ottawa, and The Kananaskis Foundation, Ottawa, Canada.

Westall, F., S.T. de Vries, W. Nijman, V. Rouchon, B. Orberger, V. Pearson, J. Watson, A. Verchovsky, I. Wright, J.-N. Rouzaud, D. Marchesini, and A. Severine. 2006. The 3.466 Ga "Kitty's Gap Chert," an early Archean microbial ecosystem. In W.U. Reimold, and R.L. Gibson (Eds), Processes on the Early Earth, Geological Society of America Special Paper 405, pp. 105–131. Geological Society of America, Boulder, CO.

Wheeler, E.A., and C.G. Arnette., Jr. 1994. Identification of Neogene woods from Alaska-Yukon. Quaternary International 22/23: 91–102.

Whipps, J.M., P. Hand, D. Pink, and G.D. Bending. 2008. Phyllosphere microbiology with special reference to diversity and plant genotype. Journal of Applied Microbiology 105: 1744–1755.

White, J.F., Jr., and T.N. Taylor. 1988. Triassic fungus from Antarctica with possible ascomycetous affinities. American Journal of Botany 75: 1495–1500.

White, J.F., Jr., and T.N. Taylor. 1989a. An evaluation of sporocarp structure in the Triassic fungus *Endochaetophora*. Review of Palaeobotany and Palynology 61: 341–345.

White, J.F., Jr., and T.N. Taylor. 1989b. Triassic fungi with suggested affinities to the Endogonales (Zygomycotina). Review of Palaeobotany and Palynology 61: 53–61.

White, J.F., Jr., and T.N. Taylor. 1989c. A Trichomycete-like fossil from the Triassic of Antarctica. Mycologia 81: 643–646.

White, J.F., Jr., and T.N. Taylor. 1991. Fungal sporocarps from Triassic peat deposits in Antarctica. Review of Palaeobotany and Palynology 67: 229–236.

White, M.M., and D.B. Strongman. 2012. New species of *Spartiella* and *Legeriosimilis* from mayflies and other arthropod-associated trichomycetes from Nova Scotia, Canada. Botany 90: 1195–1203.

White, M.M., T.Y. James, K. O'Donnell, M.J. Cafaro, Y. Tanabe, and J. Sugiyama. 2006. Phylogeny of the Zygomycota based on nuclear ribosomal sequence data. Mycologia 98: 872–884.

Whitford, A.C. 1914. IV. On a new fossil fungus from the Nebraska Pliocene. Paper from the University Studies series (The University of Nebraska) 11: 181–183.

Whitford, A.C. 1916. A description of two new fossil fungi. Nebraska Geological Survey Publication 7: 85–92.

Wieland, G.R. 1934. A silicified shelf fungus from the Lower Cretaceous of Montana. American Museum Novitates 725: 1–13.

Wilding, N., N.M. Collins, P.M. Hammond, and J.F. Webber. 1989. Insect-Fungus Interactions. Academic Press, London, UK. 344 pp.

Wilf, P., C.C. Labandeira, K.R. Johnson, and N.R. Cúneo. 2005. Richness of plant-insect associations in Eocene Patagonia: A legacy for South American biodiversity. Proceedings of the Academy of Sciences, USA 102: 8944–8948.

Wilkinson, H.P. 2003. Fossil actinomycete filaments and fungal hyphae in dicotyledonous wood from the Eocene London Clay, Isle-of-Sheppey, Kent, England. Botanical Journal of the Linnean Society 142: 383–394.

Williams, J.L. 1985. Miocene epiphyllous fungi from northern Idaho. In C.J. Smiley, A.E. Leviton, and M. Berson (Eds), Late Cenozoic History of the Pacific Northwest: Interdisciplinary Studies on the Clarkia Fossil Beds of Northern Idaho, pp. 139–142. Pacific Division of the American Association for the Advancement of Science, San Francisco, CA.

Williamson, W.C. 1878. On the organization of the fossil plants of the coal-measures. Part IX. Philosophical Transactions of the Royal Society of London 169: 319–364.

Williamson, W.C. 1880. On the organization of the fossil plants of the coal-measures. Part X. Including an examination of the supposed radiolarians of the Carboniferous rocks. Philosophical Transactions of the Royal Society of London 171B: 493–539.

Williamson, W.C. 1881. On the organization of the fossil plants of the coal-measures: Part XI. Philosophical Transactions of the Royal Society London 172: 283–305.

Williamson, W.C. 1883. On the organization of the fossil plants of the coal-measures: Part XII. Philosophical Transactions of the Royal Society of London 174: 459–475.
Williamson, W.C. 1888. On some anomalous cells developed within the interior of the vascular and cellular tissues of the fossil plants of the coal-measures. Annals of Botany 2: 315–323.
Wilson, C.G., and P.W. Sherman. 2010. Anciently asexual bdelloid rotifers escape lethal fungal parasites by drying up and blowing away. Science 327: 574–576.
Wilson, C.G., and P.W. Sherman. 2013. Spatial and temporal escape from fungal parasitism in natural communities of anciently asexual bdelloid rotifers. Proceedings of the Royal Society 280B: 20131255.
Wilson, D. 1995. Fungal endophytes which invade insect galls: Insect pathogens, benign saprophytes, or fungal inquilines? Oecologia 103: 255–260.
Wilson, L.R. 1962. A Permian fungus spore type from the Flowerpot Formation of Oklahoma. Oklahoma Geology Notes 22: 91–96.
Wilson, L.R. 1965. *Rhizophagites*, a fossil fungus from the Pleistocene of Oklahoma. Oklahoma Geology Notes 25: 257–260.
Winther, J.L., and W.E. Friedman. 2007. Arbuscular mycorrhizal symbionts in *Botrychium* (Ophioglossaceae). American Journal of Botany 94: 1248–1255.
Winther, J.L., and W.E. Friedman. 2008. Arbuscular mycorrhizal associations in Lycopodiaceae. New Phytologist 177: 790–801.
Winther, J.L., and W.E. Friedman. 2009. Phylogenetic affinity of arbuscular mycorrhizal symbionts in *Psilotum nudum*. Journal of Plant Research 122: 485–496.
Wisshak, M. 2006a. Introduction. In M. Wisshak (Ed.), High-Latitude Bioerosion: The Kosterfjord Experiment, pp. 1–36. Springer-Verlag, Berlin, Germany.
Wisshak, M. 2006b. Bioerosion patterns. In M. Wisshak (Ed.), High-Latitude Bioerosion: The Kosterfjord Experiment, pp. 49–114. Springer-Verlag, Berlin, Germany.
Wisshak, M., A. Freiwald, T. Lundälv, and M. Gektidis. 2005. The physical niche of the bathyal *Lophelia pertusa* in a non-bathyal setting: Environmental controls and palaeoecological implications. In A. Freiwald, and J.M. Roberts (Eds), Cold-Water Corals and Ecosystems, pp. 979–1001. Springer, Berlin, Germany.
Wisshak, M., B. Seuß, and A. Nützel. 2008. Evolutionary implications of an exceptionally preserved Carboniferous microboring assemblage in the Buckhorn Asphalt lagerstätte (Oklahoma, USA). In M. Wisshak, and L. Tapanila (Eds), Current Developments in Bioerosion, pp. 21–54. Springer, Berlin, Germany.
Wisshak, M., A. Tribollet, S. Golubic, J. Jakobsen, and A. Freiwald. 2011. Temperate bioerosion: ichnodiversity and biodiversity from intertidal to bathyal depths (Azores). Geobiology 9: 492–520.
Wittlake, E.B. 1969. Fossil phylloxerid plant galls. Arkansas Academy of Science Proceedings 23: 164–167.
Wolf, F.A. 1967. Fungus spores in East African lake sediments. V. Mycologia 59: 397–404.
Wolf, F.A. 1969a. Naming fossil fungi. Association of Southeastern Biologists Bulletin 16: 106–108.
Wolf, F.A. 1969b. Nonpetrified fungi in Late Pleistocene sediment from eastern North Carolina. Journal of the Elisha Mitchell Scientific Society 85: 41–44.
Wolf, F.A. 1969c. A rust and an alga in Eocene sediment from western Kentucky. Journal of Elisha Mitchell Science Society 85: 57–58.
Wolf, F.A., and S.S. Cavaliere. 1966. Fungus spores in East African lake sediments. III. Journal of the Elisha Mitchell Scientific Society 82: 149–154.
Wolf, F.A., and F.R. Nease. 1970. Nonpetrified fossils from the site of a phosphate mine in Eastern North Carolina. Journal of the Elisha Mitchell Scientific Society 86: 44–48.
Wolf, F.A., and F.T. Wolf. 1947. The Fungi: In Two Volumes, Volume II. John Wiley & Sons Inc., New York, NY. Chapman & Hall Limited, London, UK, 538 pp.
Wolfe, B.E., R.E. Tulloss, and A. Pringle. 2012. The irreversible loss of a decomposition pathway marks the single origin of an ectomycorrhizal symbiosis. PLoS ONE 7: e39597.
Wolinski, H. 2003. Confocal imaging reveals structural detail of the cell nucleus and ascospore formation in lichenized fungi. Mycological Research 107: 989–995.
Wood, G.D., and W.C. Elsik. 1999. Paleoecologic and stratigraphic importance of the fungus *Reduviasporonites stoschianus* from the 'Early-Middle' Pennsylvanian of the Copacabana Formation, Peru. Palynology 23: 43–53.
Wood, G.D., A.M. Gabriel, and J.C. Lawson. 1996. Palynological techniques – processing and microscopy. In J. Jansonius and D.C. McGregor (Eds), Palynology: Principles and Applications, Volume 1, pp. 29–50. American Association of Stratigraphic Palynologists Foundation.
Wood, J.M., and P.W. Basson. 1972. Specimens resembling *Microcodium elegans* Glück from Paleozoic shales of Missouri. American Midland Naturalist 87: 207–214.
Wood, T.G., and R.J. Thomas. 1989. The mutualistic association between Macrotermitinae and *Termitomyces*. In N. Wilding, N.M. Collins, P.M. Hammond, and J.F. Webber (Eds), Insects-fungus Interactions, pp. 69–92. Academic Press, New York, NY.
Worobiec, E., G. Worobiec, and P. Gedl. 2009. Occurrence of fossil bamboo pollen and a fungal conidium of *Tetraploa* cf. *aristata* in Upper Miocene deposits of Józefina (Poland). Review of Palaeobotany and Palynology 157: 211–217.
Worobiec, G., and E. Worobiec. 2013. Epiphyllous fungi from the Oligocene shallow-marine deposits of the Krabbedalen Formation, Kap Brewster, central East Greenland. Acta Palaeobotanica 53: 165–179.
Wright, V.P. 1984. The significance of needle-fibre calcite in a Lower Carboniferous palaeosol. Geological Journal 19: 23–32.
Wright, V.P. 1986. The role of fungal biomineralization in the formation of Early Carboniferous soil fabrics. Sedimentology 33: 831–838.
Wu, B., H. Maruyama, M. Teramoto, and T. Hogetsu. 2012. Structural and functional interactions between extraradical mycelia of ectomycorrhizal *Pisolithus* isolates. New Phytologist 194: 1070–1078.

Wu, C.-G. 1993. Glomales of Taiwan: III. A comparative study of spore ontogeny in *Sclerocystis* (Glomaceae, Glomales). Mycotaxon 47: 25–39.

Wu, H.X., C.L. Schoch, S. Boonmee, A.H. Bahkali, P. Chomnunti, and K.D. Hyde. 2011. A reappraisal of Microthyriaceae. Fungal Diversity 51: 189–248.

Wu, T. 2011. Can ectomycorrhizal fungi circumvent the nitrogen mineralization for plant nutrition in temperate forest ecosystems? Soil Biology and Biochemistry 43: 1109–1117.

Wu, W.Y., C.L. Hotton, and C.C Labandeira. 2007. Fungal fossils and plant-fungi interactions from a 300 million-year-old coal-ball deposit. In: The 2007 Annual Meeting of Mycological Society of America, pp. 110–111. Mycological Society of America, Baton Rogue, LA.

Xiao, G., S.-H. Ying, P. Zheng, Z.-L. Wang, S. Zhang, X.-Q. Xie, Y. Shang, R.J. St. Leger, G.-P. Zhao, C. Wang, and M.-G. Feng. 2012. Genomic perspectives on the evolution of fungal entomopathogenicity in Beauveria bassiana. Scientific Reports 2: 483.

Xiao, S. 2013. Fossils come in to land: Muddying the waters. Nature 493: 28–29.

Xiao, S., A.H. Knoll, A.J. Kaufman, L. Yin, and Y. Zhang. 1997. Neoproterozoic fossils in Mesoproterozoic rocks? Chemostratigraphic resolution of a biostratigraphic conundrum from the North China Platform. Precambrian Research 84: 197–220.

Xiao, S., X. Yuan, M. Steiner, and A.H. Knoll. 2002. Macroscopic carbonaceous compressions in a terminal Proterozoic shale: A systematic reassessment of the Miaohe biota, South China. Journal of Paleontology 76: 347–376.

Yan, Y.-Z., and S.-X. Zhu. 1992. Discovery of acanthomorphic acritarchs from the Baicaoping Formation in Yongji, Shanxi and its geological significance. Acta Micropalaeontologica Sinica 9: 267–282.

Yang, E., L. Xu, Y. Yang, X. Zhang, M. Xiang, C. Wang, Z. An, and X. Liu. 2012. Origin and evolution of carnivorism in the Ascomycota (fungi). Proceedings of the National Academy of Sciences, USA 109: 10960–10965.

Yang, H., Y. Zang, Y. Yuan, J. Tang, and X. Chen. 2012. Selectivity by host plants affects the distribution of arbuscular mycorrhizal fungi: Evidence from ITS rDNA sequence metadata. BMC Evolutionary Biology 12: 50.

Yang, Y., E. Yang, Z. An, and X. Liu. 2007. Evolution of nematode-trapping cells of predatory fungi of the Orbiliaceae based on evidence from rRNA-encoding DNA and multiprotein sequences. Proceedings of the National Academy of Sciences, USA 104: 8379–8384.

Yang, Z.L. 2011. Molecular techniques revolutionize knowledge of basidiomycete evolution. Fungal Diversity 50: 47–58.

Yao, Y.J., D.N. Pegler, and T.W.K. Young. 1996. Genera of Endogonales. Royal Botanic Gardens, Surrey, UK. 233 pp.

Yarwood, C.E. 1973. Pyrenomycetes: Erysiphales. In G.C. Ainsworth, F.K. Sparrow, and A.S. Sussman (Eds), The Fungi: An Advanced Treatise, Volume IV-A, pp. 71–86. Academic Press, New York, NY.

Yeloff, D., D. Charman, B. van Geel, and D. Mauquoy. 2007. Reconstruction of hydrology, vegetation and past climate change in bogs using fungal microfossils. Review of Palaeobotany and Palynology 146: 102–145.

Yin, L.-M. 1997. Acanthomorphic acritarchs from Meso-Neoproterozoic shales of the Ruyang Group, Shanxi, China. Review of Palaeobotany and Palynology 98: 15–25.

Yin, L.M., Y.L. Zhao, L.Z. Bian, and J. Peng. 2013. Comparison between cryptospores from the Cambrian Log Cabin Member, Pioche Shale, Nevada, USA and similar specimens from the Cambrian Kaili Formation, Guizhou, China. Science China, Earth Sciences 56: 703–709.

Young, J.P.W. 2009. Kissing cousins: Mycorrhizal fungi get together. New Phytologist 181: 751–753.

Young, P.A. 1926. Penetration phenomena and facultative parasitism in *Alternaria*, *Diplodia*, and other fungi. Botanical Gazette 81: 258–279.

Young, T.W.K. 1969. Electron and phase-contrast microscopy of spores in two species of the genus *Mycotypha* (Mucorales). Journal of General Microbiology 55: 243–249.

Yousten, A.A., and K.E. Rippere. 1997. DNA similarity analysis of a putative ancient bacterial isolate obtained from amber. FEMS Microbiology Letters 152: 345–347.

Yu, T.E.J.-C., K.N. Egger, and R.L. Peterson. 2001. Ectendomycorrhizal associations – characteristics and functions. Mycorrhiza 11: 167–177.

Yuan, X., S. Xiao, and T.N. Taylor. 2005. Lichen-like symbiosis 600 million years ago. Science 308: 1017–1020.

Yuan, Z.-L., C.-L. Zhang, F.-C. Lin, and C.P. Kubicek. 2010. Identity, diversity, and molecular phylogeny of the endophytic mycobiota in the roots of rare wild rice (*Oryza granulate*) from a nature reserve in Yunnan, China. Applied and Environmental Microbiology 76: 1642–1652.

Zak, B. 1973. Classification of ectomycorrhizae. In G.C. Marks, and T.T. Kozlowski (Eds), Ectomycorrhizae: Their Ecology and Physiology, pp. 43–76. Academic Press Inc., New York, NY.

Zalessky, M.D. 1915. Histoire naturelle d'un charbon. Trudy Geologicheskago Komiteta (Mémoires du Comité Géologique), Nouvelle série 139: 1–74.

Zambell, C.B., J.M. Adams, M.L. Gorring, and D.W. Schwartzman. 2012. Effect of lichen colonization on chemical weathering of hornblende granite as estimated by aqueous elemental flux. Chemical Geology 291: 166–174.

Zang, W.-L., and M.R. Walter. 1992. Late Proterozoic and Cambrian Microfossils and Biostratigraphy, Amadeus Basin, Central Australia. Memoir of the Association of Australasian Palaeontologists, Volume 12. Association of Australasian Palaeontologists, Brisbane, Australia. 132 pp.

Zanin, Y.N., V.A. Luchinina, and E.A. Zhegallo. 2002. Traces of copropel-loving organisms in phosphatized coprolite. Russian Geology and Geophysics 43: 400–403.

Zebrowski, G. 1936. New genera of Cladochytriaceae. Annals of the Missouri Botanical Garden 23: 553–564.

Zeff, M.L., and R.D. Perkins. 1979. Microbial alteration of Bahamian deep-sea carbonates. Sedimentology 26: 175–201.

Zhang, N., L.A. Castlebury, A.N. Miller, S.M. Huhndorf, C.L. Schoch, K.A. Seifert, A.Y. Rossman, J.D. Rogers, J. Kohlmeyer, B. Volkmann-Kohlmeyer, and G.-H. Sung. 2006. An overview of the systematics of the Sordariomycetes based on a four-gene phylogeny. Mycologia 98: 1076–1087.

Zhang, Q., R. Yang, J. Tang, H. Yang, S. Hu, and X. Chen. 2010. Positive feedback between mycorrhizal fungi and plants influences plant invasion success and resistance to invasion. PLoS ONE 5: e12380.

Zhang, T., Y.-Q. Zhang, H.-Y. Liu, Y.-Z. Wei, H.-L. Li, J. Su, L.-X. Zhao, and L.-Y. Yu. 2013. Diversity and cold adaptation of culturable endophytic fungi from bryophytes in the Fildes Region, King George Island, maritime Antarctica. FEMS Microbiology Letters 341: 52–61.

Zhang, X.G., and B.R. Pratt. 2008. Microborings in Early Cambrian phosphatic and phosphatized fossils. Palaeogeography, Palaeoclimatology, Palaeoecology 267: 185–195.

Zhang, Y., and S. Golubic. 1987. Endolithic microfossils (cyanophyta) from early Proterozoic stromatolites, Hebei, China, Acta Micropaleontologica Sinica 4: 1–12.

Zhang, Y., C.L. Schoch, J. Fournier, P.W. Crous, J. de Gruyter, J.H.C. Woudenberg, K. Hirayama, K. Tanaka, S.B. Pointing, J.W. Spatafora, and K.D. Hyde. 2009. Multi-locus phylogeny of Pleosporales: a taxonomic, ecological and evolutionary re-evaluation. Studies in Mycology 64: 85–102.

Zhang, Z.-Y. 1980. Lower Tertiary fungal spores from Lunpola Basin of Xizang, China. Acta Palaeontologica Sinica 19: 296–301.

Zhao, G.Z., X.Z. Liu, and W.P. Wu. 2007. Helicosporous hyphomycetes from China. Fungal Diversity 26: 313–524.

Zheng, S.-L., and W. Zhang. 1986. The cuticles of two fossil cycads and epiphytic fungi. Acta Botanica Sinica 28: 427–436. [in Chinese].

Zhi, X.-Y., W.-J. Li, and E. Stackebrandt. 2009. An update of the structure and 16S rRNA gene sequence-based definition of higher ranks of the class Actinobacteria, with the proposal of two new suborders and four new families and emended descriptions of the existing higher taxa. International Journal of Systematic and Evolutionary Microbiology 59: 589–608.

Zhou, D., and K.D. Hyde. 2001. Host-specificity, host-exclusivity, and host-recurrence in saprobic fungi. Mycological Research 105: 1449–1457.

Zhou, L.-W., and Y.-C. Dai. 2012. Recognizing ecological patterns of wood-decaying polypores on gymnosperm and angiosperm trees in northeast China. Fungal Ecology 5: 230–235.

Zhu, Z.H., L.Y. Wu, P. Xi., Z.C. Song, and Y.Y. Zhang. 1985. A research on Tertiary palynology from the Qaidam Basin. Qinghai Province. In Research Institute of Exploration and Development, Qinghai Petroleum Administration, and Nanjing Institute of Geology and Palaeontology (Eds), pp. 297. Petroleum Industry Press, Beijing, China. [in Chinese].

Zhuang, W.-Y., and C.-Y. Liu. 2012. What an rRNA secondary structure tells about phylogeny of fungi in Ascomycota with emphasis on evolution of major types of ascus. PLoS ONE 7: e47546.

Zhuravlev, A.Y., J.A.G. Vintaned, and A.Y. Ivantsov. 2009. First finds of problematic Ediacaran fossil *Gaojiashania* in Siberia and its origin. Geological Magazine 146: 775–780.

Ziegler, R. 1992. Komplex-thallöse, fossile Organismen mit blattflechtenartigem Bau aus dem mittleren Keuper (Trias, Karn) Unterfrankens. In: J. Kovar-Eder (Ed.), Palaeovegetational Development in Europe and Regions Relevant to its Palaeofloristic Evolution, Proceedings of the Pan-Europe Palaeobotany Conference, Vienna, 19–23 September 1991, pp. 341–349. Museum of Natural History Vienna, Vienna, Austria.

Zijlstra, J.D., P. Van't Hof, J. Baar, G.J.M. Verkley, R.C. Summerbell, I. Paradi, W.G. Braakhekke, and F. Berendse. 2005. Diversity of symbiotic root endophytes of the *Helotiales* in ericaceous plants and the grass, *Deschampsia flexuosa*. Studies in Mycology 53: 147–162.

Zimmerman, N.B., and P.M. Vitousek. 2012. Fungal endophyte communities reflect environmental structuring across a Hawaiian landscape. Proceedings of the National Academy of Sciences, USA 109: 13022–13027.

Zjawiony, J.K. 2004. Biologically active compounds from Aphyllophorales (Polypore) fungi. Journal of Natural Products 67: 300–310.

Zoubir, A. (Ed.), 2012. Raman Imaging: Techniques and Applications, Springer Series in Optical Sciences 168. Springer-Verlag, Berlin, Germany. 383 pp.

Zuckerkandl, E., and L. Pauling. 1962. Molecular disease, evolution and genetic heterozygosity. In M. Kasha, and B. Pullman (Eds), Horizons in Biochemistry, pp. 189–225. Academic Press, New York, NY.

Zuffardi-Comerci, R. 1934. *Fomes (Polyporus) mattirolii* n. sp. nel Miocene della Libia. Missione Scientifica della Reale Accademia d'Italia a Cufra (1931-IX) 3: 58–60.

INDEX

Note: Page numbers followed by "*f*" and "*t*" refer to figures and tables, respectively.

A
Absidia spinosa, 84–85, 85*f*
Absidia spinosa var. *azygospora*, 84–85, 85*f*
Acaulospora, 240
Acaulosporoid spore types, 115–116
Acer pseudoplatanus, 3
Achlyites, 270–271
Actinomycetes, 265–267
  fossil, 265–267
Actinomycodium, 266
Aecidites, 194
Agaricales, 178–181
Agaricomycotina, 174–175, 177–193
  Agaricales, 178–181
  Boletales, 181–182
  Hymenochaetales, 182–183
  wood rot, 187–193
Agathis, 153
Aglaophyton major, 34–35, 35*f*, 48–49, 62–63, 70*f*, 72, 115, 120–121, 122*f*
Agonomycetes, 237
Ahmadjian, Vernon, 202*f*
Aimia, 31–32
Alectoria succini, 212
Alethopteris, 169
Alexopoulos, Constantine J., 10*f*
Algacites intertextus, 5
Allomyces, 72
Allomyces arbuscula, 72
  life cycle of, 70*f*
Allomyces macrogynus, 72
Alternaria, 229–233
Alternation of generations, 69, 72
Amber, 24, 139–142, 156–157, 161, 163, 211–213, 263, 266–268, 270
  fossils, 99–100
Amerospores, 224–227
Amphisphaerella, 227
Amphisphaerella dispersella, 223
Amyelon, 60, 119–120
Anamorph, 11–12
Anatolinites, 229–230
Andrews, Henry N., Jr., 120*f*
Animal-fungus interactions, 239–243
  coprolites, 241–243, 241*f*
  Glomeromycota, 240–241
Annella, 263
Annella capitata, 263*f*
Annellophora, 230–231
Annosus, 187
Annosus syrtae, 187
Antarcticycas, 125
Antarcticycas schopfii, 126*f*
Ants–fungus interactions, 245–246
Anzia, 212
Anzia electra, 211
Apiosporina, 227–228
Appianoporites vancouverensis, 182–183, 182*f*, 183*f*
Appositions, 59–60
Aquatic ecosystems, 41–43
Aquilaria, 3
Araucaria, 140, 153
Araucaria marenssii, 140
Araucarioxylon, 189–191, 189*f*, 190*f*, 191*f*
Arbuscular mycorrhizae, 119–126
  evolution of, 126–128
Arbuscules, 106–107
Arbusculites argentea, 232
Arbusculites dicotylophylli, 232
Archaeoglomus, 242
Archaeomarasmius, 180–181
Archaeomarasmius leggetti, 180–181, 180*f*
Archagaricon, 5–6
Archechytridium operculatum, 111
Archephoma cycadeoidellae, 237–238
Archeterobasidium syrtae, 185, 187
Arctacellularia, 229
Arcyria sulcata, 267–268
Armatella caulicola, 143*f*
Arthroon rochei, 194–195, 195*f*
Arthropod
  coprolites, 241–242
  interaction with fungi, 243–250
Arum-type glomeromycota, 105–107, 105*f*, 125
  biodiversity of, 107–108
  ecology of, 107
  systematics of, 107–108
Asarum canadense, 107*f*
Ascocarp, 130–132, 141, 143, 146–147, 152–153, 162–163
Ascodesmis, 137
Ascodesmisites malayensis, 137
Ascomata, 130–131, 135, 138–142, 145–147, 149, 151, 153–159, 161
Ascomycetes
  Cenozoic, 137–163
  Mesozoic, 137–163
  Paleozoic, 133–136
Ascomycota, 11, 129
  Cenozoic ascomycetes, 137–163
  endophytes, 163–170
  epiphylls, 170–171
  geologic history of, 131–133
  life cycle of, 131*f*
  Mesozoic ascomycetes, 137–163
  microthyriaceous fungi, 148–163
  Paleozoic ascomycetes, 133–136
  Pezizomycotina. *See* Pezizomycotina
  phylogenetic relationships within, 130*f*
  sporocarps, 136
Ascotricha xylina, 138
Asexual spores, 112–116
Ash, Sidney R., 191*f*
Aspergillites, 139

Aspergillites torulosus, 139
Aspergillus, 24, 138–139
Aspergillus collembolorum, 139, 139*f*
Aspergillus janus, 139
Aspidella, 32
Asterina kosciuskensis, 159–160
Asterinites colombiensis, 150–151
Asteromites mexicanus, 237, 238*f*
Asterothyrites, 153, 153*f*, 155, 159
Asterothyrites canadensis, 153
Asterothyrites minutus, 153
Asterothyrites ostiolatus, 153
Asteroxylon mackiei, 114–115, 133, 133*f*
Asthenopodichnium lignorum, 251, 251*f*
Asthenopodichnium lithuanicum, 251
Asthenopodichnium xylobiontum, 251
Astraeus, 178–179
Astromyelon, 167
Aulographum, 155
Aureofungus yaniguaensis, 180
Axe, Lindsey, 218*f*

B

Bacillus agilis, 262
Bacillus permians, 262
Bacillus sphaericus, 263
Bacteria, 261–265
  fossil, 262–265
Bacterial mycophagy, 262
Barghoorn, Elso S., 8–9, 9*f*
Barr, Donald J. S., 44*f*
Barringtonioxylon, 138
Basidiocarp, 173–176, 178–181, 183–187, 197
Basidiomycota, 11, 173
  Agaricomycotina, 177–193
  ectomycorrhizae, 195–199
  fossil, 175–177
  life cycle of, 175*f*
  phylogenetic relationships within, 174*f*
  Pucciniomycotina, 193–194
  Ustilaginomycotina, 194–195
Basidiosporites, 225
Basidiosporites fournieri, 225
Basidiosporites ovalis, 225
Basidiosporites sadasivanii, 225
Basinger, James S., 198*f*
Batrachochytrium dendrobatidis, 41–43
Baxter, Robert W., 89*f*
Beauveria, 244, 244*f*
Berbee, Mary L., 29*f*
Berry, Edward W., 273*f*
Bertrand, Paul, 50*f*
Betula, 150–151, 156–157
Biodiversity, 1, 9–10
Bioerosion and rock weathering, 254–255
Biostratigraphy, 223
Birsiomyces pterophylli, 138
Biscalitheca, 51–52, 51*f*, 52*f*
Blanchette, Robert A., 188*f*
Blastocladiomycota, 41, 44–45, 69
Boletales, 181–182
Bonfante, Paula, 111*f*
Botryodiplodia mohgaoensis, 137
Botryopteris, 54, 55*f*, 265–266
Botryopteris antiqua, 176–177
Botryopteris tridentata, 263*f*, 266*f*
Boullard, Bernard, 47*f*
Bovista plumbea, 179
Brachycladium, 5–6
Brachyphyllum patens, 147
Brachysporisporites, 229–230, 230*f*
Brachysporisporites antarcticus, 229–230
Brachysporisporites inuvikensis, 229–230
Brachysporisporites magnus, 229–230
Brachysporium, 229–230
Bracket fungi, 174–175, 177, 183, 185
Braselton, James P., 43*f*
Brefeldiellites, 153–154
Brefeldiellites argentina, 153–154
Brefeldiellites fructiflabella, 153–154
Bretonia hardingheni, 267
Brown, Ronald W., 184*f*
Brundrett, Mark C., 125*f*
Bruns, Thomas D., 195*f*
Butterfield, Nicholas J., 78*f*
Buxus pliocenica, 159
Byttnertiopsis daphnogenes, 245

C

Cafaro, Matías J., 249*f*
Calicium, 212, 213*f*
Callimothallus, 154–155
Callimothallus australis, 154
Callimothallus corralesensis, 154
Callimothallus dilcheri, 154
Callimothallus pertusus, 154–155, 154*f*
Callixylon, 188–189
Callixylon newberryi, 188–189, 189*f*
Calluna vulgaris, 145
Camarosporium, 237
Cannanorosporonites, 230
Capnodiales, 161–162
Capnophialophora, 161
Carbon spherules, 25–26
Carboniferous chytrids, 49–58
  paleozoic pollen grains, 57–58
  plant spores, 55–56
  problems in naming, 51–52
  seeds, 58
  vegetative tissues, 52–55
Carboniferous coals, 17, 21
Carboniferous spores, 81–85, 116–117
Carboniferous–Permian complexes, 276–281
Cardonia stellata, 266, 266*f*
Carnivorous fungi, 247–248, 247*f*
Carpolithes umbonatus, 5
Cash, William, 271*f*
Cashhickia acuminata, 128, 167, 168*f*
Castanopsis, 139–140
Casts, 18
Catenochytridium, 59
Caudosphaera, 31–32
Cenococcum, 197–198
Cenozoic ascomycetes, 137–163
Cenozoic chytrids, 65–66
Cenozoic coals, 17
Cenozoic lichens, 211–214
Cenozoic spores, 117–118
Cephaleuros villosus, 213–214
Cephaleuros, 213–214
Cephalotaxus, 155–157
Cephalotheca, 139–140
Ceratosporella, 147–148
Cercosporites, 139
Cevallos-Ferriz, Sergio R.S., 179*f*
Chaenotheca, 212, 213*f*
Chaenothecopsis, 140, 140*f*
Chaenothecopsis bitterfeldensis, 140
Chaenothecopsis parvula, 139–140
Chaetosphaerites, 230, 230*f*
Chaetosphaerites bilychnis, 230
Chaetosphaerites pollenisimilis, 230
Chaetothyriales, 31
Chaetothyriomycetidae, 130–131, 135
Charcoal, 17–18
Charnia, 32
Charniodiscus, 32
Chemical weathering, 201–202, 206–207
Chitaley, Shya D., 66*f*
Chlorolichenomycites salopensis, 207
Chroococcidiopsis, 209
Chroococcus, 209
Chrysobalanus, 153–154, 213–214
Chytridiomycosis, 1–2
Chytridiomycota, 41, 69
  Carboniferous chytrids, 49–58
  Cenozoic chytrids, 65–66

life history, stages in, 45*f*
Mesozoic chytrids, 65–66
parasitic chytrids, 58–65
Permian chytrids, 49–58
phylogenetic relationships within, 45*f*
Precambrian chytrids, 45–46
Rhynie chert chytrids, 46–49
Chytridium, 65
Chytrids, 44–66
Carboniferous, 49–58
Cenozoic, 65–66
life history, stages in, 45*f*
Mesozoic, 65–66
parasitic, 58–65
Permian, 49–58
phylogenetic relationships within, 45*f*
Precambrian, 45–46
Rhynie chert, 46–49
Chytriomyces mammilifer, 53–54
Chytriomyces reticulatus, 53–54
Cladistics, 27
Cladochytrium, 49–51, 55–56, 58–59
Cladophlebis, 158–159
Cladosporites bipartitus, 161–162
Cladosporites fasciculatus, 161–162, 237
Cladosporites ligni-perditor, 161–162
Cladosporites oligocaenicum, 161–162
Cladosporium, 161–162
Clamp connection, 177*f*
Clasterosporium, 238
Clavascina orlovensis, 17
Claviceps purpurea, 2–3, 37
Clay, Keith, 164*f*
Coal, 15–17
Coal balls, 15, 21–23, 21*f*
Coelomycetes, 237–238
Coleochaete, 35
Colletotrichum, 146
Collybia, 181
Combresomyces, 97–98, 280–281
Combresomyces caespitosus, 280–281
Combresomyces cornifer, 278–280, 279*f*, 280*f*
Combresomyces rarus, 280–281, 282*f*
Combresomyces williamsonii, 279
Compressions, 14–15
Conidia, 129–130, 133–134, 137–142, 146–148, 161–162, 169
Conostoma, 58
Cookeina tricholoma, 229
Cookeina, 229
Cookson, Isabel C., 152*f*
Cooksonia, 34*f*
Coprinites dominicana, 179–180, 179*f*
Coprinus, 179–180
Coprolites, 26, 26*f*, 241–243, 241*f*
arthropod, 241–242
large animal, 242–243
Cordaites, 25
Cordana musae, 147–148
Cordana pauciseptata, 147–148
Creber, Geoffrey T., 191*f*
Cribrites aureus, 137
Cridland, Arthur A., 120*f*
Crustose lichens, 203
Cryphonectria parasitica, 146
Cryptocolax clarnensis, 139–140
Cryptodidymosphaerites princetonensis, 145–146, 146*f*
Cryptomeriopsis mesozoica, 147
Cryptomycota, 4
Cryptospore, 33–35, 33*f*
Crystals, 26
Ctenosporites, 236–237
Ctenosporites eskerensis, 236–237
Ctenosporites sherwoodiae, 236–237
Culcitalna achraspora, 236
Cunninghamia, 140, 155, 211
Curvularia, 242
Cyanobacteria, 39
Cyanolichenomycites devonicus, 207
Cycadeoidella, 137
Cycadeoidella japonica, 237–238

D

Dactyloporus archaeus, 6, 7*f*, 183–184
Daedaleites volhynicus, 5
Daghlian, Charles P., 145*f*
Dalpé, Yolande, 112*f*
Daohugouthallus ciliiferus, 210, 210*f*
Dark septate endophytes (DSEs), 166–168
Darwin, Charles, 5
Davidiella, 161–162
Dawson, John William, 216*f*
de Bary, Heinrich Anton, 36
Deccanosporium eocenum, 237
Decomposition, 1, 3–5
Dendromyceliates rajmahalensis, 237
Dendromyceliates splendus, 237
Dennis, Robert L., 176*f*
Desmidiospora, 137
Deuteromycota, 129–130
Diaporthales, 146
Dicellaesporites, 227–228, 228*f*
Dicellaeporisporites poratus, 23*f*
Dicellaesporites aculeolatus, 227–228
Dicellaesporites antarcticus, 227–228
Dicellaesporites himachalensis, 227–228
Dicellaesporites popovii, 227–228
Dick, Michael W., 269*f*
Dictyospores, 232–233
Dictyosporites, 232, 233*f*
Dictyosporites dicotylophylli, 232
Dictyosporites dictyosus, 232
Dictyosporites globimuriformis, 232
Dictyosporium, 232, 236–237
Dictyozamites, 153–156
Didymoporisporonites, 228, 228*f*
Didymoporisporonites conicus, 228
Didymoporisporonites discors, 228
Didymoporisporonites psilatus, 228
Didymosphaeria, 145–146
Didymospores, 227–232, 227*f*
Diet, 240–243, 245–246
Dilcher, David L., 144*f*
Dinosaur–fungus interactions, 242–243, 250
Diospyros, 155
Diplodia, 137
Diplodia intertrappea, 137
Diplodites, 137
Diplodites mohgaoensis, 137
Diplodites rodei, 137
Diplodites sahnii, 137
Diplodites sweetii, 137
Diploneurospora, 228, 228*f*
Diploneurospora tewarii, 228
Diplophlyctis, 51–52
Diporicellaesporites, 230–231, 230*f*
Diporicellaesporites aequabilis, 230–231
Diporicellaesporites dilcheri, 230–231
Diporicellaesporites jansonii, 230–231
Diporicellaesporites pluricellus, 230–231
Diporisporites, 225
Diporisporites bhavnagarensis, 225
Diporisporites communis, 225
Diporisporites giganticus, 225
Discoascina perforata, 17
Diskagma, 205–206
Diskagma buttonii, 205–206
Divergence times, 28*f*, 31–34
Dodgella priscus, 258–259
Dörfelt, Heinrich, 268*f*
Dothideomycetes, 31, 130–131, 146–162
Dothideomycetidae, 135, 146–147
Dotzler, Nora, 115*f*
Dubiocarpon, 89–90, 90*f*, 91*f*, 96
Dyadospora, 228

Dyadosporites, 228, 229*f*
Dyadosporites cannanorensis, 228
Dyadosporites ellipsus, 228
Dyadosporites grandiporus, 228
Dyadosporites megaporus, 228
Dyadosporites sahnii, 228

E

Ectomycorrhizae, 195–199
  fossil, 197–199
Ectomycorrhizal (ECM) symbioses, 30–31
Edwards, Dianne, 217*f*
Eggs, fungal interactions in, 250
Elfvingia, 187
Elsikisporonites, 234
Elsikisporonites tubulatus, 234
Endobiotic mycoparasites, 58–59
Endochaetophora, 96
Endochaetophora antarctica, 94–96, 95*f*
Endogonales, 88
Endogone, 75*f*, 82–84, 96–97, 97*f*, 128
Endogone lactiflua, 83*f*, 86
Endolithic fungi, 250, 254–259
Endophytes, 163–170
  dark septate, 166
  fossil fungal, 165–170
  leaf, 168–170
  root, 166–168
Endosymbionts, fossil bacteria as, 264
Enigmatic fossils, 97–99
Enterobryus, 249–250
Entomophthora, 97*f*, 100, 245*f*
Entopeltacites, 149
Entopeltacites attenuatus, 149
Entopeltacites cooksoniae, 149
Entopeltacites irregularis, 149
Entopeltacites maegdefravii, 149
Entopeltacites remberi, 149, 150*f*
Entophlyctis, 51–52, 64–66
Entropezites patricii, 181, 181*f*, 251
Eomelanomyces cenococcoides, 197–198, 197*f*
Eomycetopsis, 31–32, 45–46
Eomycetopsis robusta, 46*f*
Eorhiza arnoldii, 168, 169*f*, 231, 236
Eosaccharomyces, 31–32
Epibiontic chytrid zoosporangium, 57*f*
Epibiotic chytrid zoosporangia, 50*f*
Epibiotic mycoparasites, 58–59
Epicoccum deccanensis, 138
Epiphylls, 170–171
Equisetum, 20–21, 103–104, 167
Erysiphales, 162–163
  fossil, 162–163
Erysiphe, 162
Erysiphites melilli, 162
Erysiphites protogaeus, 162
Eurotiomycetes, 138–141
  fossil, 139–140
Eurychasma, 269
Euthythyrites, 155
Euthythyrites keralensis, 155
Evolutionary rate, 28
Excipulites, 14–15
Exesipollenites, 263*f*
Exesisporites, 225
Exesisporites annulatus, 225
Exesisporites neogenicus, 225
Exesisporites verrucatus, 225

F

Fagus, 170
Farghera robusta, 206–207
Felixites, 230
Felixites playfordii, 230
Ficus kiewiensis, 162
First land plants, 33–36
Flabellitha elinae, 210
Flakea papillata, 203*f*
Flechtenstoffe, 202
Foliose lichens, 203
Fomes, 185, 187
Fomes idahoensis, 185
Fomes mattirolii, 184–185
Fomes pinicola, 185
Fossil
  actinomycetes, 265–267
  bacteria, 262–265
  ectomycorrhizae, 197–199
  Erysiphales, 162–163
  Eurotiomycetes, 139–140
  evidence, early, 31–33
  fungal endophytes, 165–170
  fungi, naming, 11–12
  leaf endophytes, 169
  lichens, 214–220
  Meliolales, 143–145
  Microthyriaceae, 151–152
  Mycetozoa, 267–268
  peronosporomycetes, 269–282
  root endophytes, 166–168
  rusts, 194
  smuts, 194–195
  Sordariomycetes, 141–142
  taxa, 272*t*
  trace (ichnofossils), 251
  wood rot, 188–193
Fossil glomeromycota, 111–128
  arbuscular mycorrhizae, 119–126
  Carboniferous spores, 116–117
  Cenozoic spores, 117–118
  Mesozoic spores, 117–118
  Permian spores, 116–117
  Rhynie chert asexual spores, 112–116
  root nodules, 118–119
  sporocarps, 109*f*, 118
  subfossil spores, 117–118
Foveoletisporonites, 231, 231*f*
Foveoletisporonites indicus, 231
Foveoletisporonites miocenicus, 231
Fractisporonites, 229
Frank, Albert Bernhard, 36
Frankbaronia, 274–275
Frankbaronia polyspora, 274–275, 274*f*, 276*f*
Frankbaronia velata, 274–276, 275*f*
Frankia, 265
Frasnacritetrus, 147–148, 235
Frasnacritetrus conatus, 235, 235*f*
Frenelopsis, 150–151
Fruticose, 203
Fungal interactions, 239
  with animals, 239–243
  with arthropods, 243–250
  in eggs, 250
  with geosphere, 254–259
  with plants, 251–254
  trace fossils (ichnofossils), 251
Fungal spores, 221
  Agonomycetes, 237
  Amerospores, 224–227
  Coelomycetes, 237–238
  Dictyospores, 232–233
  Didymospores, 227–232, 227*f*
  Helicospores, 233–235, 234*f*
  Hyphomycetes, 236–237
  naming, 222–223
  in paleoecology, 223–224
  Scolecospores, 233
  Staurospores, 235
  in stratigraphy, 223
  taxa, 224–235
Fungi, 1–2, 26
  Basidiomycota. *See* Basidiomycota
  carnivorous, 247–248, 247*f*
  and first land plants, 33–36
  gasteroid, 178–179
  microthyriaceous, 148–163

timetree of, 28*f*
Funginite, 16, 16*f*
Fungus-animal interactions, 239–243
coprolites, 241–243, 241*f*
Glomeromycota, 240–241
Fungus gardens, 245–246
Fusiformisporites, 229, 229*f*
Fusiformisporites keralensis, 229

G

Galls, 251–254
Galtier, Jean, 193*f*
Galtierella biscalithecae, 277, 278*f*
Gametothallus, 69, 71
Ganoderma, 185–186
Ganoderma adspersum, 185
Ganoderma applanatum, 185
Ganoderma lucidum, 185
Ganodermataceae, 185, 187
Ganodermites libycus, 185–186, 186*f*, 187*f*
García-Massini, Juan L., 118*f*
Gasteroid fungi, 178–179
Gasteromyces farinosus, 178
Geaster florissantensis, 178
Geasterites, 178
Geastroidea lobata, 178–179
Geastrum, 178–179
Geastrum tepexensis, 178–179, 179*f*
Geleenites, 138
Geleenites fascinus, 138
Geobiology, 4–5
Geochemical evidence, 25
Geochemistry, 25
Geologic time, 4–5
Geomicrobiology, 4–5
Geomycology, 4–5
Geosiphon, 32, 127
Geosiphon pyriformis, 205–206
Geosphere–fungus interactions, 254–259
bioerosion and rock weathering, 254–255
rhizosphere, 255–256
substrate boring, 256–259
Geotrichites glaesarius, 243
Geotrichum, 243
Gerdemann, James W., 104*f*
German, Tamara, 77*f*
Gigantoscorpio willsi, 239–240
Gigaspora gigantea, 109*f*
Gigaspora margarita, 264–265
Gigaspora ramisporophora, 110*f*
Gigasporites myriamyces, 125, 126*f*
Gigasporoid spore types, 114–115
Gilled mushrooms, 179–181
Ginkgo, 210–211
Global Boundary Stratotype Sections and Points (GSSPs), 12
Globicultrix nugax, 58–59, 59*f*
Gloeocapsa, 209
Glomeribacter, 264
Glomeromycota, 60–61, 103
arbuscular mycorrhizae, 119–128
biology of, 104–108
Carboniferous spores, 116–117
Cenozoic spores, 117–118
character of, 108–110
fossil, 111–128
interaction with animals, 240–241
Mesozoic spores, 117–118
Permian spores, 116–117
phylogenetic relationships within, 104*f*
reproduction of, 110–111
Rhynie chert asexual spores, 112–116
root nodules, 118–119
sporocarps, 118
subfossil spores, 117–118
Glomites, 123
Glomites cycestris, 125
Glomites rhyniensis, 79, 120–123
Glomites sporocarpoides, 123
Glomites vertebrariae, 125
Glomoid spore, 110*f*
Glomorphites intercalaris, 116–117, 117*f*
Glomus, 110–112, 116–118
Glomus fasciculatum, 117–118
Glomus intraradices, 110–111, 264
Glomus vesicles, 109*f*
Glossopteris, 170–171, 189–191, 189*f*, 190*f*, 191*f*, 241–242
Gonapodya, 72
Göppert, Heinrich Robert, 6*f*
Granodiporites, 228
Grilletia sphaerospermii, 49–51, 58
Grosmannia clavigera, 246
Gymnosporangium, 194
Gymnostoma, 118
Gyrinops, 3

H

Halifaxia taylorii, 82–84, 83*f*
Haptoglossa, 269
Harper, Carla J., 192*f*
Hartig net, 198–199, 198*f*
Hass, Hagen, 121*f*
Hassiella monospora, 273–274, 274*f*
Hawksworth, David L., 11*f*
Heer, Oswald, 151*f*
Helicoma, 234–235
Helicomyces, 234–235
Helicoonites goosii, 234, 235*f*
Helicospores, 233–235, 234*f*
Helicosporium, 234–235
Heterobasidiomycetes, 174–175
Heterobasidion, 191
Heterobasidion annosum, 187
Heterotrophs, 1–2
Hibbett, David S., 181*f*
Hick, Thomas, 271*f*
Hirsutella, 141–142
Hollick, Arthur, 183*f*
Holoarthric conidia, 243, 243*f*
Holocarpic chytrid zoosporangium, 48*f*, 49*f*
Holomorph, 11
Honegger, Rosemarie, 208*f*
Honeggeriella complexa, 210–211, 211*f*
Horodyskia, 32
Hosagoudar, Virupakshagouda B., 143*f*
Host responses, to parasitic chytrids, 59–61
Hueber, Francis M., 214*f*
Huroniospora, 31–32
Hymenochaetales, 182–183
Hymenostilbe, 141–142
Hyperparasite, 251
Hypertrophy, 64–65
Hyphal mantle, 80–84, 86*f*, 87–88
Hyphochytridiomycetes, 44–45
Hyphochytrium catenoides, 57–58
Hyphomycetes, 236–237
Hypocreomycetidae, 134–135
Hypoxylonites, 225–226, 226*f*
Hypoxylonites brazosensis, 225–226
Hypoxylonites chaiffetzii, 225–226
Hypoxylonites oblongus, 225–226
Hysterites, 14–15

I

Ichnofossils, 251
Ichnotaxa, 101
Impressions, 13–14
Inapertisporites, 194–195, 225–226, 226*f*, 230–231
Inapertisporites argentinus, 226
Inapertisporites circularis, 226
Inapertisporites granulatus, 226
Inapertisporites trivedii, 226
Inapertisporites variabilis, 226
Incolaria securiformis, 6, 183–184

Index Fungorum, 5
International Chronostratigraphic Chart, 12
International Commission on Stratigraphy (IUGS), 12
Involutisporonites, 234–235, 234*f*
Involutisporonites chowdhryi, 234
Involutisporonites trapezoides, 234
Isoetes, 264
Isthmospora spinosa, 145

J
Jansonius, Jan, 11*f*
Jimwhitea circumtecta, 85–88, 86*f*, 87*f*
Jumpponen, Ari, 166*f*

K
Kalgutkar, Ramakant M., 10*f*
Kalviwadithyrites saxenae, 156–157
Karling, John S., 43*f*
Karlingomyces, 51–52
Kendrick, W. Bryce, 222*f*
Kidston, Robert, 46*f*
Krispiromyces discoides, 64–65, 65*f*
Kryphiomyces catenulatus, 72, 73*f*

L
Laboulbeniomycetes, 163
Laccaria trichodermophora, 37
Lacrimasporonites, 226, 227*f*
Lacrimasporonites sondensis, 226
Lang, William H., 46*f*
Large animal coprolites, 242–243
Larix, 38
Lasiostrobus polysacci, 251–252
Latanites, 145–146
Laurinoxylon, 161–162
Lawrey, James D., 203*f*
Leaf endophytes, 168–170
  fossil, 169
Leguminosites, 253*f*
Leisman, Gilbert A., 89*f*
Lembrosiopsis, 155
Lemonniera, 235
Lentinus tigrinus, 177
Lenzites warnieri, 185
Leotiomycetes, 162–163
LePage, Ben A., 198*f*
Lepidodendron, 5–6, 49–53, 65–66, 262*f*
Leptonema, 7
Leptostromites ellipticus, 237, 238*f*
Leptothrix, 267
Leptothyrites dominicanus, 237, 238*f*
Leptotrichites resinatus, 267
Lesquereux, Leo, 5, 5*f*
Lichens, 32–33, 36, 201
  Cenozoic, 211–214
  crustose, 203
  evolution of, 205
  foliose, 203
  fossils, 214–220
  Mesozoic, 210–211
  Paleozoic, 206–210
  Precambrian evidence of, 205–206
  reproduction of, 204–205
Lichtwardt, Robert W., 248*f*
Lignite, 16
Liriodendron, 211
Lithomucorites miocenicus, 99
Lithopolyporales, 185
Lithopolyporales zeerabadensis, 185
Lithouncinula, 162
Lobaria, 213–214
Lobaria pulmonaria, 203*f*
Lolium perenne, 37
Longcore, Joyce E., 43*f*
Lophosphaeridium, 77–78
Lücking, Robert, 202*f*
Lycoperdites tertiarius, 178, 178*f*
Lycophytes, 116, 123–124
Lyonomyces pyriformis, 64–65
Lyonophyton rhyniensis, 106*f*
Lyons, Paul C., 16*f*

M
Macerals, 15–16
Mackie, William, 19, 19*f*
Magnolia, 211
Maiasaura, 242–243
Majasphaeridium, 31–32
Manginula, 149
Manginula memorabilis, 149
Marasmiellus, 180–181
Marasmius, 180–181
Margaretbarromyces dictyosporus, 147, 148*f*
Mariopteris nervosa, 169
Mariusia andegavensis, 150
Martha J. Powell, 44*f*
McLaughlin, David J., 136*f*
Medullosa, 169
Meliola, 142–143
Meliola anfracta, 160*f*
Meliola cyclobalanopsina, 142*f*
Meliola ellisii, 145
Meliolales, 142–145
  fossil, 143–145
Meliolinites anfracta, 144, 144*f*
Meliolinites dilcheri, 144–145, 145*f*
Meliolinites siwalika, 145
Meliolinites spinksii, 144
Meniscoideisporites cretacea, 237–238
Mesozoic ascomycetes, 137–163
Mesozoic chytrids, 65–66
Mesozoic lichens, 210–211
Mesozoic spores, 117–118
Metacapnodium succinum, 161, 161*f*
Metarhizium, 37
Metarhizium robertsii, 37
Metasequoia, 211
Metasequoia glyptostroboides, 125–126, 127*f*
Metasequoia milleri, 125–126, 126*f*, 198–199
Microbial mats, 1
Microborings, 46, 257–259
Micrococcus, 262–263
Micrococcus paludis, 262
Microcodium, 256, 256*f*
Microfavichnus alveolatus, 173–174
Microfossils. *See also* Fossil
  Precambrian, 76–79
Microthallites, 155
Microthallites cooksoniae, 155
Microthallites lutosus, 155
Microthallites spinulatus, 155
Microthyriaceae, 151–161
  classification of, 152
  fossil, 151–152
Microthyriaceous fungi, 148–163
Microthyriacites, 155–157
Microthyriacites baqueroensis, 155–156
Microthyriacites cooksoniae, 156–157, 158*f*
Millay, Michael A., 57*f*
Milleromyces rhyniensis, 64–65, 64*f*
Molecular clocks, 28–31
Molinaea asterinoides, 228
Money, Nicholas P., 176*f*
Monoporisporites, 226–227, 227*f*
Monoporisporites basidii, 227
Monoporisporites minutus, 227
Monosporiosporites ovalis, 225
Monotosporella, 141, 229–230
Monotosporella doerfeltii, 141
Mortierella capitata, 82–84
Mucor combrensis, 98
Mucor tenuis, 77–78
Mucorites, 31–32

Mucorites combrensis, 98, 99*f*
Mucorites ripheicus, 77–78, 77*f*, 78*f*
Mucorodium, 266
Mucorodium palaeomycoides, 98, 99*f*
Multicellaesporites, 231, 231*f*
Multicellaesporites bilobus, 231
Multicellaesporites dilcheri, 231
Mummified wood, 24–25
Mushrooms, gilled, 179–181
Mutualists, 8–9
Muyocopron, 150–151
Mycelia Sterilia, 222, 237
Mycena, 180–181
Mycetophagites atrebora, 181, 181*f*, 251
Mycetozoa, 267–268
  fossil, 267–268
Mycobiont, 201–205, 208–211
Mycocarpon, 89–90, 90*f*, 96–97
Mycocarpon asterineum, 93–94, 94*f*, 96–97
Mycocarpon cinctum, 92–94, 93*f*
Mycocarpon rhyniense, 92–93
Mycoparasite
  endobiotic, 58–59
  epibiotic, 58–59
Mycorrhiza helper bacteria (MHB), 264
Mycorrhizae, 37–39
Mycosphaeroides, 31–32
Mycostratigraphy, 223
Mycozygosporites laevigatus, 99
Myriostroma, 178–179
Myxomycites mangini, 267

N

Nathorst, Alfred G., 152*f*
Nematasketum, 217–218, 217*f*
Nematode-fungus interactions, 247–248
Nematophytes, 214–215, 217–220
Nematoplexus rhyniensis, 218, 219*f*
Nematothallopsis, 218
Nematothallopsis gotlandii, 218
Nematothallus, 216–218
Nematothallus williamii, 218*f*
Neocallimastigomycota, 41–43, 43*f*
Neolecta, 130, 135
Neolecta vitellina, 135
Neotyphodium, 37, 166
Neotyphodium occultans, 37
Neurospora, 228
Nicholsonella, 259
Nilssoniopteris, 158–159
Nocardioformis dominicanus, 266
Nodules, 25
Non-pollen palynomorph (NPP), 221, 223
Nostoc, 32, 201–202
Nothia aphylla, 61–64, 61*f*, 62*f*, 63*f*, 122–123
Notophytum krauselii, 118, 119*f*, 120*f*
Notothyrites cephalotaxi, 155–156
Nowakowskiella, 55–56, 58–59, 65

O

Oleinites, 155
Oliver, Francis W., 53*f*
Olpidium, 49–51, 65
Onakawananus varitas, 137
Oochytrium, 65–66
Oochytrium lepidodendri, 7*f*, 49–53, 51*f*
Oogonium–antheridium complexes, 273–276
Oomycetes, 268–269
Ophiocordyceps, 141–142
Ophiocordyceps unilateralis, 245
Ordovicimyces, 259, 270–271
Orestovia, 132
Ornasporonites, 231, 231*f*
Ornasporonites inaequalis, 231
Ornatifilum, 132
Ornatifilum granulatum, 132
Ornatifilum lornensis, 132
Orthogonum lineare, 258–259
Osborn, Jeffrey M., 177*f*
Ossicaulis lignatilis, 216
Ovularia, 162
Ovularites barbouri, 162

P

Pachytheca, 218–219
Paenibacillus validus, 264
Palaeachlya, 270–271
Palaeachlya perforane, 259
Palaeancistrus martinii, 176–177
Palaeoagaricites antiquus, 181, 181*f*, 251
Palaeoamphisphaerella, 227
Palaeoamphisphaerella keralensis, 227
Palaeoamphisphaerella pirozynskii, 227
Palaeoanellus dimorphus, 247–248, 247*f*, 248*f*
Palaeo-Aspergillus multiseriate, 139
Palaeoblastocladia milleri, 69–73, 70*f*, 71*f*, 72*f*
  life cycle of, 72*f*
Palaeocurvularia variabilis, 242
Palaeocybe striata, 181
Palaeodikaryomyces baueri, 100, 100*f*
Palaeodiplodites yezoensis, 137
Palaeofibulus, 177
Palaeofibulus antarctica, 177, 177*f*
Palaeoglomus grayi, 111–112
Palaeomyces, 113*f*
Palaeomyces gracilis, 98, 99*f*
Palaeomyces majus, 55–56, 98
Palaeomyces vestitus, 80
Palaeomycites, 98, 112
Palaeonitella, 64–65
Palaeonitella cranii, 64–65, 64*f*, 65*f*
Palaeopericonia, 238
Palaeopericonia fritzschei, 238
Palaeoperone, 270–271
Palaeophthora mohgaonensis, 270–271
Palaeorhiphidium, 32
Palaeosclerotium pusillum, 135–136, 135*f*, 136*f*
Palaeosordaria lagena, 138
Paleobasidiospora taugourdeauii, 111
Paleocatenaria disjoncta, 111
Paleoecology, 259
  fungal spores in, 223–224
Paleomicrobiology, 4
Paleomycology
  defined, 4–5
  history of, 5–10
Paleoophiocordyceps coccophagus, 141–142, 244, 244*f*
Paleopathology, 4
Paleopyrenomycites devonicus, 30, 133–135, 133*f*, 134*f*
Paleoserenomyces allenbyensis, 145–146, 146*f*
Paleozoic, 33–34
Paleozoic ascomycetes, 133–136
Paleozoic lichens, 206–210
Paleozoic pollen grains, chytrids in, 57–58
Palynology, 221, 224, 228
Panus tigrinus, 177
Paramicrothallites, 155
Paramicrothallites irregularis, 154
Parapolyporites japonica, 184–185
Parasitic chytrids, 58–65
  endobiotic mycoparasites, 58–59
  epibiotic mycoparasites, 58–59
  host responses to, 59–61
  hypertrophy, 64–65
  Nothia aphylla, 61–64
Paris-type glomeromycota, 105–107, 106*f*, 119–120, 123, 125–126
Parmathyrites, 156
Parmathyrites indicus, 156

Parmathyrites ramanujamii, 156
Parmathyrites tonakkalensis, 156
Parmathyrites turaensis, 156, 156*f*
Parmelia, 212
Parmelia ambra, 212
Parmelia isidiiveteris, 212, 212*f*
Pecopteris, 165*f*, 169
Pelicothallos, 213–214
Pelicothallus villosus, 213–214
Penicillium, 5–6, 129, 229, 250–251
Periconia, 238
Perisporiacites, 138
Perisporiacites larundae, 138
Perisporiacites shaheziensis, 138
Perisporiacites varians, 138
Perisporiacites zamiophylli, 138
Permian chytrids, 49–58
Permian spores, 116–117
Permineralizations, 18–23
Peronosporoides, 7, 270–271
Peronosporites antiquarius, 270–271, 271*f*
Peronosporites palmae, 271–273, 273*f*
Peronosporoides palmi, 271–273
Peronosporomycetes, 7–8, 32, 44–45, 268–282
  fossil, 269–282
Persea pseudocarolinensis, 149
Pesavis tagluensis, 235*f*, 236–237
Peterson, R. Larry, 38*f*
Petrifactions, 18–23
Petrosphaeria japonica, 137
Petsamomyces polymorphus, 76–77, 270
Petsamomyces, 45–46, 76–77
Pezizasporites taiwanensis, 137
Pezizites, 137
Pezizomycotina, 138–163
  Dothideomycetes, 146–162
  Eurotiomycetes, 138–141
  Meliolales, 142–145
  Phyllachorales, 145–146
  Sordariomycetes, 141–142
Phaeoacremonium parasiticum, 3
Phaeocryptopus, 147
Phaeolus, 182–183
Phialophora parasitica, 3
Phlyctochytrium, 64–65
Phoma, 237–238
Photobiont, 201–212, 216–217
Phragmothyrites, 155–157
Phragmothyrites concentricus, 156–157, 157*f*
Phragmothyrites kangukensis, 156*f*
Phragmothyrites ramanujamii, 156–157
Phycomyces nitens, 97–98
Phycopeltis, 158–159
Phyllachora, 233
Phyllachorales, 145–146
Phyllopsora, 212
Phyllopsora dominicanus, 212
Phylogenetic systematic, 27–28
Phytophthora, 277
Phytophthora infestans, 268–269
Pia, Julius von, 8, 9*f*
Picea abies, 3, 166
Pilobolus, 3
Pilostyles, 251–252
Pinus, 38, 251–252
Pinus taeda, 168–169
Piromyces communis, 43*f*
Pirozynski, Kris A., 5, 6*f*
Pityophyllum, 156–157
Plant fossils, formation and preservation of, 13–26
  carbon spherules, 25–26
  casts, 18
  charcoal, 17–18
  coal, 15–17
  compressions, 14–15
  coprolites, 26
  crystals, 26
  fungi, 26
  impressions, 13–14
  nodules, 25
  permineralizations, 18–23
  petrifactions, 18–23
  rock surfaces, 25
  unaltered material, 23–25
Plant spores, chytrids in, 55–56
Plant–fungus interactions, 251–254
Pleistocene, 23
Pleosporales, 147–148
Pleosporites shiraianus, 137, 147
Pleosporomycetidae, 146–147
Pleotrachelus, 65–66
Plochmopeltinites, 157–158
Plochmopeltinites cooksoniae, 157–158
Plochmopeltinites masonii, 157–158
Pluricellaesporite, 231–232, 231*f*
Pluricellaesporite eocenicus, 231–232
Pluricellaesporite globatus, 231–232
Pluricellaesporite typicus, 231–232
Pluricellaesporites psilatus, 231–232
Podkovyrov, Victor N., 77*f*
Podocarpus, 153–154
Podozamites, 14*f*
Poinar, George O., Jr., 10*f*
Polycellaesporonites, 232, 232*f*
Polycellaesporonites acuminatus, 232
Polycellaesporonites alternariatus, 232
Polycellaesporonites bellus, 232
Polymorphyces, 262–263
Polyporales, 183–187
Polyporites bowmanii, 183–184, 184*f*
Polyporites bowmanni, 5
Polyporites brownii, 185
Polyporites stevensonii, 185
Populus, 223
Precambrian microfossils, 31–34, 45–46, 76–79
Preservation, of fossil plants, 13–26
  carbon spherules, 25–26
  casts, 18
  charcoal, 17–18
  coal, 15–17
  compressions, 14–15
  coprolites, 26
  crystals, 26
  fungi, 26
  impressions, 13–14
  nodules, 25
  permineralizations, 18–23
  petrifactions, 18–23
  rock surfaces, 25
  unaltered material, 23–25
Proplebeia dominicana, 263
Propythium, 270–271
Prosphyracephala succini, 163, 163*f*, 245
Proterozoic, 30
Protoascon missouriensis, 84–85, 84*f*, 85*f*, 162
Protocolletotrichum, 146
Protocolletotrichum deccanensis, 146
Protoerysiphe, 162
Protomycena electra, 180
Protomycena, 180–181
Protophysarum balticum, 267–268, 268*f*
Protophysarum phloiogenum, 267–268
Prototaxites, 17–18, 175–176, 214–218, 214*f*, 239–240
Prototaxites southworthii, 215*f*
Psaronius, 54, 54*f*, 167, 167*f*, 176–177, 177*f*, 241
Pseudopolyporus carbonicus, 183–184
Pseudotsuga menziesii, 168–169
Psilophyton dawsonii, 114*f*
Pterophyllum longifolium, 138
Pteropus brachyphylli, 147, 147*f*
Puccinia, 194, 229–230

Puccinia monoica, 193
Pucciniales, 193–194
Pucciniasporonites, 194
Pucciniasporonites arcotensis, 194
Pucciniomycotina, 174–175, 193–194
Puccinites, 194
Pyrenula papilligera, 204*f*
Pythites, 270–271

Q
Quatsinoporites cranhamii, 182–183, 182*f*, 183*f*
Quercus, 156–157
Quorum sensing, 39

R
Radiculites, 123
Ratnagiriathyrites, 158
Ratnagiriathyrites hexagonalis, 158
Redecker, Dirk, 113*f*
Reduviasporonites, 229
Rember, William C., 150*f*
Remy, Renate, 121*f*
Remy, Winfreid, 121*f*
Renault, Bernard, 7–8, 7*f*
Resinite, 17
Retallack, Gregory J., 207*f*
Reticulatasporites, 226
Reymanella globosa, 263
Rhizoclosmatium aurantiacum, 42*f*
Rhizoctonia, 229
Rhizoctonia nandorii, 237
Rhizoctonia solani, 237
Rhizomorpha sigillaria, 6
Rhizomorpha sigillariae, 183–184
Rhizophagites, 112
Rhizophagites acinus, 112
Rhizophagus irregularis, 110–111
Rhizophidites triassicus, 66
Rhizophydium, 56, 65
Rhizophydium pollinis-pini, 66
Rhizopogon, 198–199
Rhizopus, 79–80, 79*f*
Rhizopus nigricans, 101
Rhizosphere, 254–259
Rhynia gwynne-vaughanii, 123
Rhynie chert, 19–21, 19*f*, 34–35, 79–80
  asexual spores, 112–116
  chytrids, 46–49
  ecosystem, 20*f*
  fossil bacteria in, 264–265
Rikkinen, Jouko, 141*f*
Roannaisia bivitilis, 92–93, 93*f*
Rock surfaces, 25
Rock weathering, bioerosion and, 254–255
Root endophytes, 166–168
  fossil, 166–168
Root nodules, 118–119
Rossellinia congregata, 137
Ruflorinia, 153–154
Rusts, 194

S
Saccharomyces cerevisiae, 110–111
Saccharomycotina, 130
Saccomorpha clava, 258–259, 259*f*, 270*f*
Saccomorpha terminalis, 270
Sapindus, 150–151
Saprolegnia, 275*f*
Saururus tuckerae, 194–195
Saxena, Rajendra K., 153*f*
Schedonorus phoenix, 166
Schmidt, Alexander R., 139*f*
Schüßler, Arthur, 105*f*
Sclerocystis, 109*f*, 118, 119*f*
Scleroderma, 181–182
Scleroderma echinosporites, 181–182
Scolecospores, 233
Scolecosporites, 233, 234*f*
Scolecosporites longus, 233
Scolecosporites maslinensis, 233
Scutellospora, 114–115
Scutellospora cerradensis, 110*f*
Scutellosporites devonicus, 114–115, 114*f*
Seeds, chytrids in, 58
Selosse, Marc-André, 108*f*
Septochytrium, 51–52
Septoglomus, 116
Septosporium, 232
Sequoiadendron giganteum, 104–105
Serenomyces, 145–146
Seward, Albert C., 8, 8*f*
Shortensis, 149
Shuiyousphaeridium, 31–32
Sigillaria, 6
Sirophoma, 137–138
Slime molds, 267–268
Smilax, 150–151, 155–157
Smith, Selena Y., 182*f*
Smuts, 194–195
Sooty molds, 161
Sordariomycetes, 134–135, 141–142
  fossil, 141–142
Sordariomycetidae, 134–135
Spartiella aurensis, 248
Spataporthe taylorii, 141, 142*f*
Sphaerites, 14–15, 151–152
Sphaerites suessii, 14*f*
Sphaeropsis, 137
Sphaerospermum, 58
Sphaerotilus, 267
Sphenophyllum, 54–55
Spinosporonites, 232
Spinosporonites indicus, 232
Spirotremesporites, 226
Spizellomyces, 56
Spizellomyces pseudodichotomus, 43*f*
Spongiophyton, 218–219
Spongiophyton nanum, 219*f*
Spores, 16*f*
Sporidesmium, 143–144
Sporocarpon, 89–90, 95
Sporocarpon cellulosum, 90*f*
Sporocarps, 80–99, 109*f*, 118, 136
Sporothallus, 69–71
Staphlosporonites, 232–233
Staphlosporonites allomorphus, 232–233
Staphlosporonites billelsikii, 232–233
Staphlosporonites delumbus, 232–233
Staphlosporonites dichotomus, 232–233
Staphylococcus, 263
Staurospores, 235
Stemonitis splendens, 267–268
Stemphylium, 232
Stichus, 258–259
Stigmaria, 65, 116, 123–124
Stigmaria ficoides, 124*f*, 125*f*
Stigmatomyces, 163, 245
Stigmatomyces succini, 163, 163*f*
Stockey, Ruth A., 23*f*
Stolophorites lineatus, 101, 101*f*
Stomiopeltis, 150–151
Stomiopeltis aspersa, 150–151
Stomiopeltis plectilis, 150–151
Stomiopeltites, 150–151, 151*f*
Stomiopeltites amorphos, 150–151
Stomiopeltites cretacea, 150–151
Stopes, Marie C., 16*f*
Stratigraphy, fungal spores in, 223
Streptomyces, 267
Streptothrix, 5–6
Striadiporites, 229
Striadyadosporites, 229
Striasporonites, 229
Strigula, 213–214
Stromatolites, 32
Strullu-Derrien, Christine, 281*f*
Stubblefield, Sara P., 89*f*

Subfossil spores, 117–118
Sublagenicula, 56
Substrate boring, 256–259
Suillus, 198–199
Suillus tomentosus, 37
Symbiosis, 36–39, 201–202, 204–205, 207–209
Synanamorph, 11–12
Synchytrium, 41–43, 55, 65
Synchytrium endobiolicum, life cycle of, 42*f*
Synchytrium permicus, 55

T

Taenioxylon, 138
Taphrinomycotina, 130
Tappania, 31–32, 78–79, 78*f*
Tappania plana, 78
Taxa
  fossil, 272*t*
  incertae sedis, 137–138
  interaction with fungus, 258–259
Taxus, 215–216
Taylor, John W., 29*f*
Teleomorph, 11–12
Tephromeia atra, 204*f*
Termitomyces, 246
Terrestrialization, 34
Tetrachaetum, 235
Tetraploa, 147–148, 235
Tetraploa aristata, 147–148
Thallites, 210
Thallus morphology and structure, 202–204
Thanatephorus cucumeris, 237
Thielaviopsis, 231
Thuchomyces, 205–206
Thuchomyces lichenoides, 205–206
Tiffney, Bruce H., 8–9, 9*f*
Torula, 230
Trace fossils, 251
Trametites, 185
Trametites eocenicus, 185
Trappe, James M., 104*f*
Traquairia, 90–92, 91*f*, 92*f*, 95–96, 95*f*, 162
Traquairia carruthersii, 92, 95
Traquairia ornatus, 92
Traquairia spenceri, 92
Traquairia williamsonii, 91*f*, 92
Traverse, Alfred, 222*f*
Trebouxia, 201–202, 207
Trentepohlia, 201–202
Trewin, Nigel H., 20*f*
Triassic, 85–88
Tribolites, 235
Tribolites tetrastonyx, 235
Tricellaesporonites, 233
Tricellaesporonites granulatus, 233
Tricellaesporonites semicircularis, 233
Trichoderma, 251
Trichomycetes, 248–250
Trichopeltinites, 158–159
Trichopeltinites fusilis, 158–159, 159*f*
Trichopeltinites kiandrensis, 158–159
Trichopeltinites nilssonioptericola, 158–159
Trichopeltinites pulcher, 158–159
Trichothyrina, 159
Trichothyrites, 159
Trichothyrites airensis, 159
Trichothyrites hordlensis, 159
Trichothyrites ostiolatus, 159
Trichothyrites padappakarensis, 159*f*
Trichothyrites pleistocaenicus, 159
Triletes rugosus, 56, 56*f*
Tripathi, Suryakant M., 153*f*
Trivena arkansana, 252–254
Tubefia, 234–235

U

Uhlia allenbyensis, 145–146
Ulvella, 155
Unaltered material, 23–25
Uncinulites baccarinii, 162
Uromyces, 194
Urophlyctis, 65
Urophlyctites oliveranus, 169
Usnea, 202–203
Usnea sanguinea, 204*f*
Ustilaginomycotina, 174–175, 194–195
Ustilago deccani, 194–195
Uzelothrips eocenicus, 242

V

Vegetative tissues, chytrids in, 52–55
Vendomyces, 45–46
Verrusporonites, 229
Verschlussband, 174
Vertebraria, 125, 126*f*
Vesicles, 109*f*
Viracarpon, 137
Viscospora, 116
Vizella, 149–150
Vizella memorabilis, 149*f*
Vizella metrosideri, 149–150

W

Walker, Christopher, 108*f*
Wallemiomycetes, 174–175
Weiss, Frederick E., 117*f*
White, James F., Jr., 95*f*
Wieland, George R., 184*f*
Williamson, William C., 90*f*
Winfrenatia reticulata, 79, 208–209, 208*f*, 209*f*
Wollemia nobilis, 38
Wood rot, 187–193
  fossil, 188–193
Woronium, 49–51

X

Xylariomycetidae, 134–135
Xylomites, 151–152
Xylomites asteriformis, 14*f*
Xylomites cycadeoideae, 138
Xylomites intermedius, 151–152
Xylomyces, 231–232, 236
Xylomyces chlamydosporis, 231–232

Z

Zizyphoides, 150–151, 156–157
Zoosporangium, 42*f*, 47*f*, 48*f*, 69–72, 71*f*
Zwergimyces, 80
Zwergimyces vestitus, 80–81, 80*f*, 93–94
Zygomycetes, 75
  amber fossils, 99–100
  enigmatic fossils, 97–99
  iochnotaxa, 101
  Precambrian microfossils, 76–79
  Rhynie chert, 79–80
  sporocarps, 80–99
  zygosporangium-gametangia complexes, 80–99
Zygopteris, 5–6
Zygopteris illinoiensis, 176–177
Zygosporangium-gametangia complexes, 80–99
Zygosporites, 97–98, 98*f*, 279–280, 281*f*

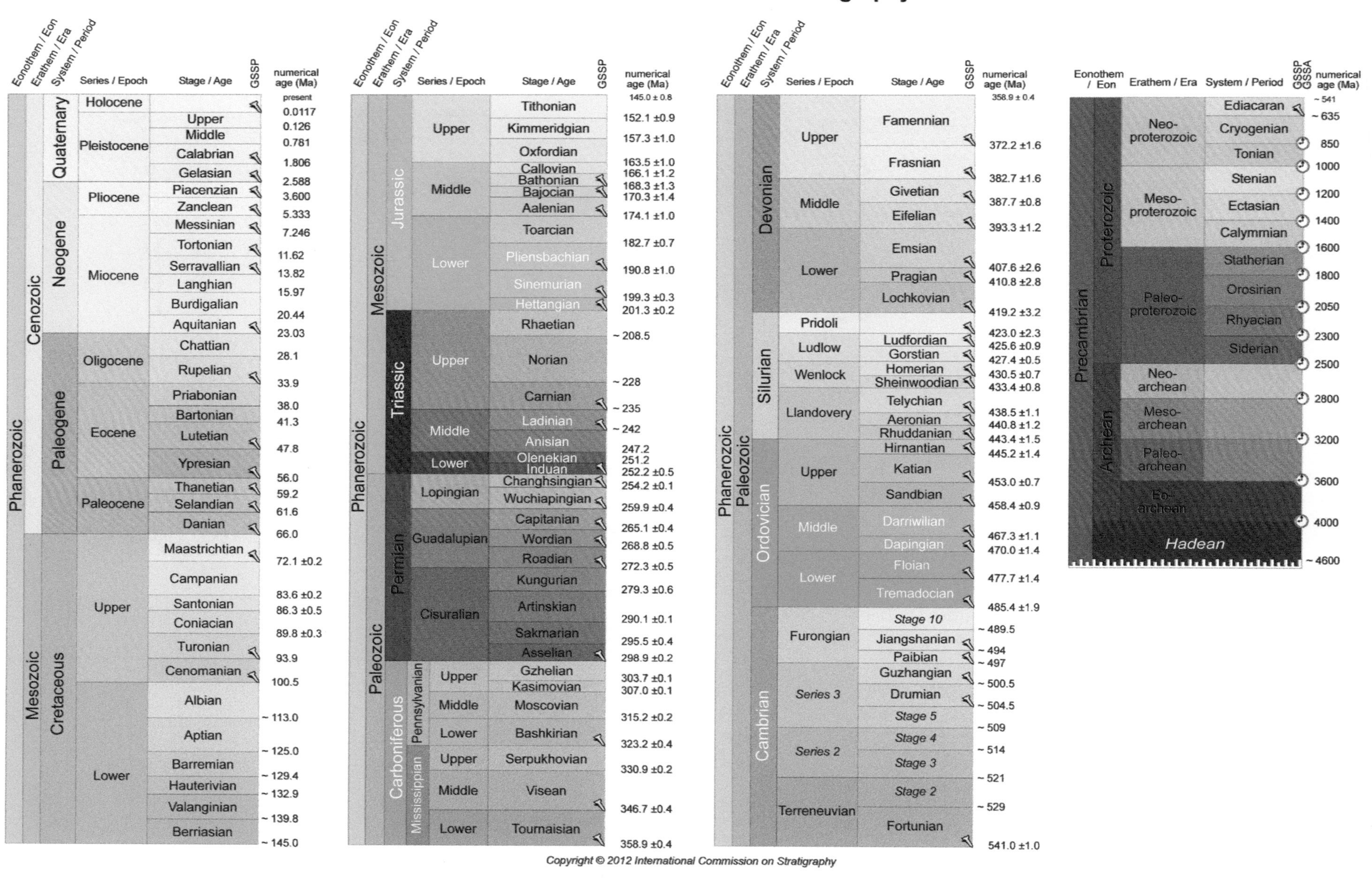
INTERNATIONAL CHRONOSTRATIGRAPHIC CHART
International Commission on Stratigraphy
Eonothem / Eon
Erathem / Era
System / Period
Series / Epoch
Stage / Age
GSSP
numerical age (Ma)
Phanerozoic
Cenozoic
Quaternary
Holocene
present
0.0117
Pleistocene
Upper
0.126
Middle
0.781
Calabrian
1.806
Gelasian
2.588
Neogene
Pliocene
Piacenzian
3.600
Zanclean
5.333
Miocene
Messinian
7.246
Tortonian
11.62
Serravallian
13.82
Langhian
15.97
Burdigalian
20.44
Aquitanian
23.03
Paleogene
Oligocene
Chattian
28.1
Rupelian
33.9
Eocene
Priabonian
38.0
Bartonian
41.3
Lutetian
47.8
Ypresian
56.0
Paleocene
Thanetian
59.2
Selandian
61.6
Danian
66.0
Mesozoic
Cretaceous
Upper
Maastrichtian
72.1 ±0.2
Campanian
83.6 ±0.2
Santonian
86.3 ±0.5
Coniacian
89.8 ±0.3
Turonian
93.9
Cenomanian
100.5
Lower
Albian
~ 113.0
Aptian
~ 125.0
Barremian
~ 129.4
Hauterivian
~ 132.9
Valanginian
~ 139.8
Berriasian
~ 145.0
145.0 ± 0.8
Jurassic
Upper
Tithonian
152.1 ±0.9
Kimmeridgian
157.3 ±1.0
Oxfordian
163.5 ±1.0
Middle
Callovian
166.1 ±1.2
Bathonian
168.3 ±1.3
Bajocian
170.3 ±1.4
Aalenian
174.1 ±1.0
Lower
Toarcian
182.7 ±0.7
Pliensbachian
190.8 ±1.0
Sinemurian
199.3 ±0.3
Hettangian
201.3 ±0.2
Triassic
Upper
Rhaetian
~ 208.5
Norian
~ 228
Carnian
~ 235
Middle
Ladinian
~ 242
Anisian
247.2
Lower
Olenekian
251.2
Induan
252.2 ±0.5
Paleozoic
Permian
Lopingian
Changhsingian
254.2 ±0.1
Wuchiapingian
259.9 ±0.4
Guadalupian
Capitanian
265.1 ±0.4
Wordian
268.8 ±0.5
Roadian
272.3 ±0.5
Cisuralian
Kungurian
279.3 ±0.6
Artinskian
290.1 ±0.1
Sakmarian
295.5 ±0.4
Asselian
298.9 ±0.2
Carboniferous
Pennsylvanian
Upper
Gzhelian
303.7 ±0.1
Kasimovian
307.0 ±0.1
Middle
Moscovian
315.2 ±0.2
Lower
Bashkirian
323.2 ±0.4
Mississippian
Upper
Serpukhovian
330.9 ±0.2
Middle
Visean
346.7 ±0.4
Lower
Tournaisian
358.9 ±0.4
358.9 ± 0.4
Devonian
Upper
Famennian
372.2 ±1.6
Frasnian
382.7 ±1.6
Middle
Givetian
387.7 ±0.8
Eifelian
393.3 ±1.2
Lower
Emsian
407.6 ±2.6
Pragian
410.8 ±2.8
Lochkovian
419.2 ±3.2
Silurian
Pridoli
423.0 ±2.3
Ludlow
Ludfordian
425.6 ±0.9
Gorstian
427.4 ±0.5
Wenlock
Homerian
430.5 ±0.7
Sheinwoodian
433.4 ±0.8
Llandovery
Telychian
438.5 ±1.1
Aeronian
440.8 ±1.2
Rhuddanian
443.4 ±1.5
Ordovician
Upper
Hirnantian
445.2 ±1.4
Katian
453.0 ±0.7
Sandbian
458.4 ±0.9
Middle
Darriwilian
467.3 ±1.1
Dapingian
470.0 ±1.4
Lower
Floian
477.7 ±1.4
Tremadocian
485.4 ±1.9
Cambrian
Furongian
Stage 10
~ 489.5
Jiangshanian
~ 494
Paibian
~ 497
Series 3
Guzhangian
~ 500.5
Drumian
~ 504.5
Stage 5
~ 509
Series 2
Stage 4
~ 514
Stage 3
~ 521
Terreneuvian
Stage 2
~ 529
Fortunian
541.0 ±1.0
Eonothem / Eon
Erathem / Era
System / Period
GSSP GSSA
numerical age (Ma)
Precambrian
Proterozoic
Neo-proterozoic
Ediacaran
~ 541
Cryogenian
~ 635
Tonian
850
1000
Meso-proterozoic
Stenian
1200
Ectasian
1400
Calymmian
1600
Paleo-proterozoic
Statherian
1800
Orosirian
2050
Rhyacian
2300
Siderian
2500
Archean
Neo-archean
2800
Meso-archean
3200
Paleo-archean
3600
Eo-archean
4000
Hadean
~ 4600
Copyright © 2012 International Commission on Stratigraphy

INTERNATIONAL CHRONOSTRATIGRAPHIC CHART

International Commission on Stratigraphy

Eonothem / Eon	Erathem / Era	System / Period	Series / Epoch	Stage / Age	GSSP	numerical age (Ma)
						present
Phanerozoic	Cenozoic	Quaternary	Holocene		x	0.0117
			Pleistocene	Upper		0.126
				Middle		0.781
				Calabrian	x	1.806
				Gelasian	x	2.588
		Neogene	Pliocene	Piacenzian	x	3.600
				Zanclean	x	5.333
			Miocene	Messinian	x	7.246
				Tortonian	x	11.62
				Serravallian	x	13.82
				Langhian		15.97
				Burdigalian		20.44
				Aquitanian	x	23.03
		Paleogene	Oligocene	Chattian		28.1
				Rupelian	x	33.9
			Eocene	Priabonian		38.0
				Bartonian		41.3
				Lutetian	x	47.8
				Ypresian	x	56.0
			Paleocene	Thanetian	x	59.2
				Selandian	x	61.6
				Danian	x	66.0
	Mesozoic	Cretaceous	Upper	Maastrichtian	x	72.1 ±0.2
				Campanian		83.6 ±0.2
				Santonian		86.3 ±0.5
				Coniacian		89.8 ±0.3
				Turonian	x	93.9
				Cenomanian	x	100.5
			Lower	Albian		~ 113.0
				Aptian		~ 125.0
				Barremian		~ 129.4
				Hauterivian		~ 132.9
				Valanginian		~ 139.8
				Berriasian		~ 145.0

Eonothem / Eon	Erathem / Era	System / Period	Series / Epoch	Stage / Age	GSSP	numerical age (Ma)
						145.0 ± 0.8
Phanerozoic	Mesozoic	Jurassic	Upper	Tithonian		152.1 ±0.9
				Kimmeridgian		157.3 ±1.0
				Oxfordian		163.5 ±1.0
			Middle	Callovian		166.1 ±1.2
				Bathonian	x	168.3 ±1.3
				Bajocian	x	170.3 ±1.4
				Aalenian	x	174.1 ±1.0
			Lower	Toarcian		182.7 ±0.7
				Pliensbachian	x	190.8 ±1.0
				Sinemurian	x	199.3 ±0.3
				Hettangian	x	201.3 ±0.2
		Triassic	Upper	Rhaetian		~ 208.5
				Norian		~ 228
				Carnian	x	~ 235
			Middle	Ladinian	x	~ 242
				Anisian		247.2
			Lower	Olenekian		251.2
				Induan	x	252.2 ±0.5
	Paleozoic	Permian	Lopingian	Changhsingian	x	254.2 ±0.1
				Wuchiapingian	x	259.9 ±0.4
			Guadalupian	Capitanian	x	265.1 ±0.4
				Wordian	x	268.8 ±0.5
				Roadian	x	272.3 ±0.5
			Cisuralian	Kungurian		279.3 ±0.6
				Artinskian		290.1 ±0.1
				Sakmarian		295.5 ±0.4
				Asselian	x	298.9 ±0.2
		Carboniferous – Pennsylvanian	Upper	Gzhelian		303.7 ±0.1
				Kasimovian		307.0 ±0.1
			Middle	Moscovian		315.2 ±0.2
			Lower	Bashkirian	x	323.2 ±0.4
		Carboniferous – Mississippian	Upper	Serpukhovian		330.9 ±0.2
			Middle	Visean	x	346.7 ±0.4
			Lower	Tournaisian	x	358.9 ±0.4

Eonothem / Eon	Erathem / Era	System / Period	Series / Epoch	Stage / Age	GSSP	numerical age (Ma)
						358.9 ± 0.4
Phanerozoic	Paleozoic	Devonian	Upper	Famennian	x	372.2 ±1.6
				Frasnian	x	382.7 ±1.6
			Middle	Givetian	x	387.7 ±0.8
				Eifelian	x	393.3 ±1.2
			Lower	Emsian	x	407.6 ±2.6
				Pragian	x	410.8 ±2.8
				Lochkovian	x	419.2 ±3.2
		Silurian	Pridoli		x	423.0 ±2.3
			Ludlow	Ludfordian	x	425.6 ±0.9
				Gorstian	x	427.4 ±0.5
			Wenlock	Homerian	x	430.5 ±0.7
				Sheinwoodian	x	433.4 ±0.8
			Llandovery	Telychian	x	438.5 ±1.1
				Aeronian	x	440.8 ±1.2
				Rhuddanian	x	443.4 ±1.5
		Ordovician	Upper	Hirnantian	x	445.2 ±1.4
				Katian	x	453.0 ±0.7
				Sandbian	x	458.4 ±0.9
			Middle	Darriwilian	x	467.3 ±1.1
				Dapingian	x	470.0 ±1.4
			Lower	Floian	x	477.7 ±1.4
				Tremadocian	x	485.4 ±1.9
		Cambrian	Furongian	*Stage 10*		~ 489.5
				Jiangshanian	x	~ 494
				Paibian	x	~ 497
			Series 3	Guzhangian	x	~ 500.5
				Drumian	x	~ 504.5
				Stage 5		~ 509
			Series 2	*Stage 4*		~ 514
				Stage 3		~ 521
			Terreneuvian	*Stage 2*		~ 529
				Fortunian	x	541.0 ±1.0

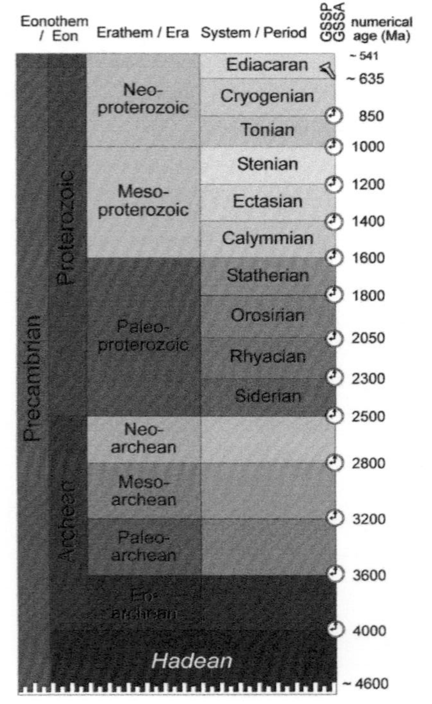

Eonothem / Eon
Erathem / Era
System / Period
GSSP GSSA
numerical age (Ma)
Precambrian
Proterozoic
Archean
Neo-proterozoic
Meso-proterozoic
Paleo-proterozoic
Neo-archean
Meso-archean
Paleo-archean
Hadean
Ediacaran
Cryogenian
Tonian
Stenian
Ectasian
Calymmian
Statherian
Orosirian
Rhyacian
Siderian
~ 541
~ 635
850
1000
1200
1400
1600
1800
2050
2300
2500
2800
3200
3600
4000
~ 4600

Printed and bound by CPI Group (UK) Ltd, Croydon, CR0 4YY
08/05/2025
01864920-0001